Advanced Structured Materials

Volume 221

Common engineering materials are reaching their limits in many applications, and new developments are required to meet the increasing demands on engineering materials. The performance of materials can be improved by combining different materials to achieve better properties than with a single constituent, or by shaping the material or constituents into a specific structure. The interaction between material and structure can occur at different length scales, such as the micro, meso, or macro scale, and offers potential applications in very different fields.

This book series addresses the fundamental relationships between materials and their structure on overall properties (e.g., mechanical, thermal, chemical, electrical, or magnetic properties, etc.). Experimental data and procedures are presented, as well as methods for modeling structures and materials using numerical and analytical approaches. In addition, the series shows how these materials engineering and design processes are implemented and how new technologies can be used to optimize materials and processes.

Advanced Structured Materials is indexed in Google Scholar and Scopus.

Holm Altenbach · Victor A. Eremeyev
Editors

Dynamics of Discrete and Continuum Structures and Media

Springer

Editors
Holm Altenbach
Faculty of Mechanical Engineering
Otto-von-Guericke-Universität Magdeburg
Magdeburg, Germany

Victor A. Eremeyev
DICAAR
Università degli Studi di Cagliari
Cagliari, Italy

ISSN 1869-8433 ISSN 1869-8441 (electronic)
Advanced Structured Materials
ISBN 978-3-031-75625-2 ISBN 978-3-031-75626-9 (eBook)
https://doi.org/10.1007/978-3-031-75626-9

This Springer imprint is published by the registered company Springer Nature Switzerland AG
The registered company address is: Gewerbestrasse 11, 6330 Cham, Switzerland

Preface

Fig. 1 Prof. Alexey Porubov, Bologna, Italy, Euromech solid mechanics conference, 2018

Alexey Victorovich Porubov was born on 16th July 1964, in Leninigrad, now Saint Petersburg, Russia, sometimes called Northern Venice, one of the largest cities in Russia. He graduated in 1987 from the Leningrad Polytechnical Institute, now known as Peter the Great St. Petersburg Polytechnic University. The master's entitled "*Nonlinear Boussinesq equation wave solutions on inviscid liquid layer surface*" was written under the supervision of Prof. A. M. Samsonov. In 1996 he defended his doctoral thesis (candidate of physical mathematical sciences) "*Nonlinear waves* *on thin viscous inhomogeneous liquid layer*" in the same university. Then in 2007 he defended his habilitation (doctoral dissertation) entitled "*Generation of strain solitary waves in nonlinear solids*" at the Institute for Problems in Mechanical Engineering of the Russian Academy of Sciences, St. Petersburg, Russia.

Alexey Porubov is a known scientist in the area of nonlinear waves in solids and fluids. He contributed in the problems of convecting fluids, localized nonlinear strain waves in solids, including experimental observation of strain solitons. He also tries to establish a connection between nonlinear mechanics and mathematics solving mechanical problems and finding solutions to partial differential nonlinear non-integrable equations, developing combined analyticao-numerical methods and applications of the control methods.

He is on the list of the most frequently cited Russian scientists.

Alexey Porubov began his scientific career at the A. F. Ioffe Physical Technical Institute of the Russian Academy of Sciences 1987 and worked there for 20 years in the positions of young research fellow, research fellow, senior research fellow. Then since 2007 he works in the Institute of Problems of Mechanical Engineering of the Russian Academy of Sciences of the Russian Academy of Sciences, where he became the head of the Laboratory of Micromechanics of Materials in 2016. At the same time he worked as a part-time professor at Peter the Great St. Petersburg Polytechnic University and St. Petersburg State University.

Fig. 2 Alexey Porubov at home with his cat Tisha

Alexey Porubov is the member of the Euromech, the member of the Russian National Committee on Theoretical and Applied Mechanics, and the member of the Russian Acoustical Society.

Alexey Porubov has many international contacts, having worked as a visiting researcher and later as a visiting professor in the following institutions

- Department of Mathematics and Statistics, Edinburgh University (UK), 1995, 1999;
- Instituto Pluridisciplinar, Universidad Complutense de Madrid (Spain), 1997–1998 (postdoctoral position), 1999;
- University of Torino (Italy), short-term visits during 2000–2004;
- University of Kyoto (Japan), 2003;
- CNRS Visiting Research Associateship at the Laboratoire de Modelisation en Mecanique, Universite Pierre et Marie Curie, Paris (France), 2003;
- Kyushu Unversity (Japan), 2004;
- CMLA Ecole Normal Superior de Cachan, Cachan (France), 2005–2014 (short-term visits);
- Institute Jean le Rond d'Alembert, Universite Pierre et Marie Curie, Paris (France), 2002–2014;
- Institute of General Mechanics of the RWTH Aachen, (Germany), 2011–2014;
- Centre de mathematiques et de Leurs Applications, Ecole Normale Superieure de Cachan, Cachan (France), 2014.

His scientific interests relate to the following topics:

- Analytical description of nonlinear waves in fluids and solids including soft biological tissues;
- Derivation of exact, asymptotic and numerical solutions to nonlinear nonintegrable equations;
- Continuum limits of discrete systems;
- Dynamics of metamaterials;
- Feedback control for nonlinear partial differential equations.

He has published more than a hundred papers in peer-reviewed journals and several monographs, see the references below. During his scientific career, he has received grants from INTAS, Russian Foundation for Basic Research, Russian Science Foundation, Russian Ministry of Science and Education, DAAD, German Research Foundation, Paris City Council, Japanese Society for Promotion of Science, International Science Foundation and some others.

This volume of the Advanced Structured Materials Series is dedicated to the 60th birthday of Prof. Alexey Porubov and contains a selection of scientific papers prepared by papers by his friends and colleagues from different countries. It is devoted to actual research in dynamics considering discrete and continuum models of continuum and structures.

Magdeburg, Germany
Cagliari, Italy
July 2024

Holm Altenbach
Victor A. Eremeyev

References

1. Altenbach H, Berezovski A, dell'Isola F, Porubov A (eds) (2023 a) Sixty Shades of Generalized Continua: Dedicated to the 60th Birthday of Prof. Victor A. Eremeyev, Advanced Structured Materials, vol 170. Springer, Cham
2. Altenbach H, Eremeyev VA, Pavlov IS, Porubov AV (eds) (2020) Nonlinear Wave Dynamics of Materials and Structures, Advanced Structured Materials, vol 122. Springer, Cham
3. Altenbach H, Irschik H, Porubov AV (eds) (2023 b) Progress in Continuum Mechanics, Advanced Structured Materials, vol 196. Springer, Cham
4. Altenbach H, Belyaev A, Eremeyev VA, Krivtsov A, Porubov AV (eds) (2019) Dynamical Processes in Generalized Continua and Structures, Advanced Structured Materials, vol 103. Springer, Cham
5. Christov CI, Maugin GA, Porubov AV (2007) On Boussinesq's paradigm in nonlinear wave propagation. Comptes Rendus—Mecanique 335(9–10):521–535
6. dell'Isola F, Eremeyev VA, Porubov A (eds) (2018) Advances in Mechanics of Microstructured Media and Structures, Advanced Structured Materials, vol 87. Springer, Cham
7. Porubov AV (2003) Amplification of nonlinear strain waves in solids. World Scientific, New Jersey
8. Porubov AV (2009) Localization of nonlinear strain waves: asymptotic and numerical methods (in Russian). Fizmatlit, Moscow
9. Porubov AV (2021) Wave modulation in a nonlinear acoustic metamaterial. International Journal of Non-Linear Mechanics 137
10. Porubov AV (2023a) Dynamics and control of band gaps in a mass-in mass metamaterial model with an extra attached mass. Continuum Mechanics and Thermodynamics 35(6):2325–2336
11. Porubov AV (2023b) Nonlinear wave localization in an acoustic metamaterial. Communications in Nonlinear Science and Numerical Simulation 119
12. Porubov AV (2024) Nonlinear wave localization in an acoustic metamaterial with attached masses through one element of main chain. Mechanics Research Communications 139
13. Porubov AV, Aero EL, Maugin GA (2009) Two approaches to study essentially nonlinear and dispersive properties of the internal structure of materials. Physical Review E - Statistical, Nonlinear, and Soft Matter Physics 79(4)
14. Porubov AV, Andrianov IV, Danishevs'kyy VV (2012) Nonlinear strain wave localization in periodic composites. International Journal of Solids and Structures 49(23–24):3381–3387
15. Porubov AV, Antonov ID, Fradkov AL (2019) Boundary control of nonlinear strain waves in di-atomic crystal layer. Wave Motion 91
16. Porubov AV, Antonov ID, Indeitsev DI, Fradkov AL (2018) Mechanical system allowing distributive control with feedback. Mechanics Research Communications 93:124–127
17. Porubov AV, Belyaev AK, Polyanskiy VA (2021) Nonlinear hybrid continuum–discrete dynamic model of influence of hydrogen concentration on strength of materials. Continuum Mechanics and Thermodynamics 33(4):933–941

18. Porubov AV, Bouche D, Bonnaud G (2013) Analytical solutions to detect the scheme dispersion for the coupled nonlinear equations. Communications in Nonlinear Science and Numerical Simulation 18(10):2679–2688
19. Porubov AV, Krivtsov AM (2022a) Dispersive propagation of localized waves in a mass-in-mass metamaterial lattice. Continuum Mechanics and Thermodynamics 34(6):1475–1483
20. Porubov AV, Krivtsov AM (2022b) Dispersive propagation of localized waves in a mass-in-mass metamaterial lattice. Continuum Mechanics and Thermodynamics 34(6):1475–1483
21. Porubov AV, Maugin GA (2005) Longitudinal strain solitary waves in presence of cubic non-linearity. International Journal of Non-Linear Mechanics 40(7):1041–1048
22. Porubov AV, Maugin GA (2006) Propagation of localized longitudinal strain waves in a plate in the presence of cubic nonlinearity. Physical Review E—Statistical, Nonlinear, and Soft Matter Physics 74(4)
23. Porubov AV, Maugin GA (2011) Application of non-linear strain waves to the study of the growth of long bones. International Journal of Non-Linear Mechanics 46(2):387–394
24. Porubov AV, Parker DF (1999) Some general periodic solutions to coupled nonlinear schrödinger equations. Wave Motion 29(2):97–109
24. Porubov AV, Pastrone F (2004) Non-linear bell-shaped and kink-shaped strain waves in microstructured solids. International Journal of Non-Linear Mechanics 39(8):1289–1299
25. Porubov AV, Pastrone F, Maugin GA (2004) Selection of two-dimensional nonlinear strain waves in micro-structured media. Comptes Rendus—Mecanique 332(7):513–518
26. Porubov AV, Samsonov AM, Velarde MG, Bukhanovsky AV (1998) Strain solitary waves in an elastic rod embedded in another elastic external medium with sliding. Physical Review E 58(3):3854
27. Porubov AV, Tsuji H, Lavrenov IV, Oikawa M (2005) Formation of the rogue wave due to non-linear two-dimensional waves interaction. Wave Motion 42(3):202–210
28. Porubov AV, Velarde MG (2000) Dispersive-dissipative solitons in nonlinear solids. Wave Motion 31(3):197–207
29. Samsonov AM, Dreiden GV, Porubov AV, Semenova IV (1998) Longitudinal-strain soliton focusing in a narrowing nonlinearly elastic rod. Physical Review B 57(10):5778–5787
30. Sokolov AA, Müller WH, Porubov AV, Gavrilov SN (2021) Heat conduction in 1d harmonic crystal: Discrete and continuum approaches. International Journal of Heat and Mass Transfer 176

Contents

Chapter 1
Simulation of Ice Floe and Surface Fluid Wave Effect on the Dynamics of a Floater

Andrei K. Abramian, Nikolay M. Bessonov, and Anastasia A. Chevrichkina

Abstract The paper studies an approach to numerical simulation of a semi-infinite ice floe effect on a floater dynamics. The authors have solved the problem numerically in a 2D formulation. The ice floe is considered as a semi-infinite beam with a bending stiffness equal to the cylindrical bending stiffness of a plate. Behavior of a floater under the actions of concentrated non-stationary and stationary harmonic forces is studied. Water is simulated as an ideal, incompressible fluid, and its flow is irrotational. The authors use nonlinear boundary conditions on the fluid free surface. The values of displacements and angles of rotation of the floater are found. It is found that at non-stationary regime maximum displacements and rotation angles of a floater can exceed their values at stationary regimes. The authors have revealed some cases when at nonlinear conditions on a fluid free surface floater's displacements and angles of rotation exceed the displacements and angles of rotation of the floater at linear conditions. The calculations proved that it is not necessary to take into account the floater bending rigidity for parameters corresponding to real offshore structures.

Keywords Ice floe · Surface waves · Oscillation · Floater

1.1 Introduction

The operation of floating-type structures in the presence of ice floe in various loads acting on structures has been studied in many papers. Historically, the first studies were devoted to ships' behavior near ice floe to ensure safe navigations [1–4]. Currently, floating oil platforms and nuclear power stations become the subject of research [5–8]. For the problem statement, a floating structure is simulated as a rigid body [4–7] or as a plate (beam) [1–3, 8]. Harmonic force [2–7], external force

A. K. Abramian (✉) · N. M. Bessonov · A. A. Chevrichkina
Institute for Problems in Mechanical Engineering, Saint Petersburg, Russia
e-mail: andabr33@yahoo.co.uk

H. Altenbach and V. A. Eremeyev (eds.), *Dynamics of Discrete and Continuum Structures and Media*, Advanced Structured Materials 221,
https://doi.org/10.1007/978-3-031-75626-9_1

moment [8], and progressive waves [3–6] are considered as external loads on a floating object. Water is considered as an ideal incompressible heavy fluid, and the ice floe as a plate floating on its surface. In papers [1–4, 6, 7], the authors used linear boundary conditions on the fluid surface. In [1–4, 6, 7], various analytical methods were applied to obtain solutions and in [8–10] numerical methods were used. The result of solving the problems was. In this paper, we consider issues that have not been considered previously in other papers. One of these questions is, for example, the question of the behavior of a floating body and an ice cover in a non-stationary regime. In the aforementioned papers, the displacement dependences (including angle of rotation) of both the floating body and the ice floe on the frequencies of the external forces and on the spatial coordinate were found. The present paper considers questions that have not been considered before for the best knowledge of the authors. One of the questions is a floating body and an ice floe behavior in a non-stationary regime. The results obtained in this work reveal that the maximum displacements of a floating body and an ice floe in the non-stationary regime can be significantly greater than in the stationary regime. The other question is the floating body elasticity influence on the maximum displacements at external forces applied to it. The authors found parameters of the considered system when nonlinear boundary conditions on the fluid surface are significant. In the present paper, the authors are solving the problem using the numerical method. The reason for not applying the analytical method is that a mathematically rigorous solution of the problem requires the construction of a solution that not only takes into account the singularity that occurs at the edge of the ice plate, but also obtains a solution that correctly describes waves that go to infinity. As far as the authors of this article know, researchers have not yet been able to overcome both difficulties and obtain a satisfactory analytical solution.

1.2 The Problem Statement

Consider the problem of the ice-floater interaction through surface waves (see Fig. 1.1).

The author simulates an ice floe as an elastic plate flowing an incompressible fluid [1]. The floater is considered as an elastic plate with cylindrical rigidity much greater than that of ice. Elastic plates on the water surface are separated by regions with free surfaces. The problem is studied in 2D formulation. The model is assumed to be 2D, which implies that the out-of-plane dimension of the floater is much bigger than the distance l between the ice floe and the floater. In this regard, instead of the equations of the plates, the equations of the beam type are used. The cylindrical stiffness of the plate is taken as the bending stiffness of the beam. The Cartesian coordinate system in the (x, y) plane is used. The water is considered as an ideal, incompressible fluid, and its flow as irrotational [1]. The potential of fluid velocities Φ satisfies the Laplace equation in the region occupied by the fluid:

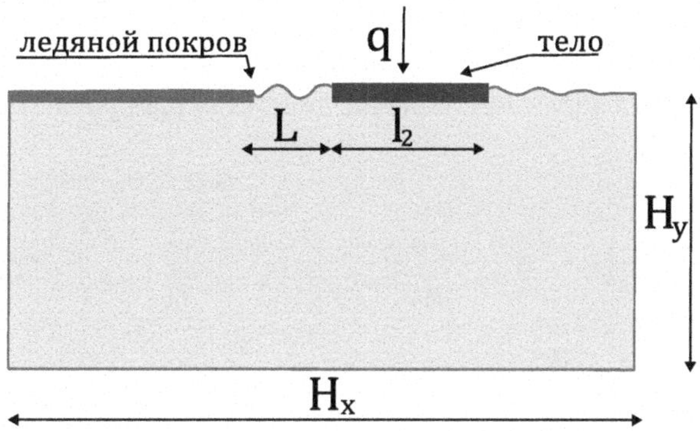

Fig. 1.1 Layout of the ice floe and the body simulating the floating structure in the calculated area

$$\Delta\Phi = 0, \tag{1.1}$$

$$\frac{\partial\Phi}{\partial t} + \frac{|\nabla\Phi|^2}{2} = -\frac{1}{\rho}p - gy, \tag{1.2}$$

where $p(x, y, t)$ is the fluid pressure, ρ is the fluid density, and g is acceleration of gravity.

Assume that here and below, the pressure p is counted from the atmospheric pressure. Below, we confine ourselves to small transversal oscillations of the ice floe; thus the motion equation of the ice beam is

$$\rho_1 h \frac{\partial^2 w}{\partial t^2} = -D\frac{\partial^4 w}{\partial x^4} + p + q, \tag{1.3}$$

where $w(x, t)$ is vertical displacement of the surface from the equilibrium horizontal position, ρ_1, h are ice density and thickness, $p(x, t)$ is the fluid pressure, $q(x, t)$ is the external force applied to the body, $D = Eh^3/(12(1-\nu^2))$ is the cylindrical rigidity of the plate, and E, ν are Young's modulus and Poisson's ratio respectively. Note that the cylindrical stiffness of the plate is given as a variable along the water surface $D = D(x)$, and it equals either the cylindrical stiffness of the ice or the floating body. It can be taken on the fluid free surface, and it is small compared to the cylindrical stiffness of ice. Similarly, the density $\rho(x)$ and the thickness of the beam $h(x)$ are given as variables and equal to either the density and thickness of the ice or the floater. On the free surface it is much less than the ice parameters. Test calculations have revealed that the ratio of the indicated quantities corresponding to the surface of the water and ice should be on the order of 10^4 or less. As a test problem, the problem of the dynamics of a single finite beam in contact with a basin of finite depth and infinite length along the x coordinate was taken. The harmonic force is applied at the beam's center. This approach allows describing the ice beam, the free surface, and the floating body by the general Eq. (1.3) with boundary conditions on the surface, bottom, and side surfaces of the calculated region.

1.3 Numerical Method

The equations describing the motions of the system have the form:

$$\Delta\Phi = 0, \tag{1.4}$$

$$\frac{\partial\Phi}{\partial x}|_{x=0} = 0, \quad \frac{\partial\Phi}{\partial x}|_{x=H_x} = 0, \quad \frac{\partial\Phi}{\partial y}|_{y=-H_y} = 0, \quad \frac{\partial\Phi}{\partial y}|_{y=0} = \frac{dw}{dt}, \tag{1.5}$$

$$\rho_1 h\frac{\partial^2 w}{\partial t^2} = -D\frac{\partial^4 w}{\partial x^4} + q - \rho\left(\frac{\partial\Phi}{\partial t} + \alpha\frac{|\nabla\Phi|^2}{2}\right) - \rho g w - \mu\frac{\partial w}{\partial t}, \tag{1.6}$$

$$\frac{\partial w}{\partial x}|_{x=0} = 0, \quad w|_{x=0} = 0, \quad \frac{\partial w}{\partial x}|_{x=H_x} = 0, \quad w|_{x=H_x} = 0, \tag{1.7}$$

where H_x, H_y are width and depth of the calculated region.

The initial conditions are as follows:

$$\Phi|_{t=0} = 0, \quad w|_{t=0} = 0, \quad \frac{dw}{dt}|_{t=0} = 0.$$

In the literature, the case of small surface displacement is usually considered, and it is assumed that

$$\frac{d\Phi}{dt} \approx \frac{\partial\Phi}{\partial t}. \tag{1.8}$$

This study considers a more general, nonlinear case when

$$\frac{d\Phi}{dt} = \frac{\partial\Phi}{\partial t} + \alpha\frac{|\nabla\Phi|^2}{2}. \tag{1.9}$$

Here $\alpha = 0$ for the linear case, and $\alpha = 1$ for the nonlinear case. The system of Eqs. (1.4)–(1.7) is solved numerically. A 2D orthogonal non-uniform grid $I \times J$ is introduced with steps Δx_i and Δy_i in x and y, respectively, where $i = 1, \dots I$, $j = 1, \dots J$. To improve the accuracy of calculations, the grid thickens along the x axis in the vicinity of the body ends and the edges of the ice floe, and it expands near the left and right sides of the computational domain. Vertically, the grid thickens near the water surface. In real life, the size of the x region extends to "infinity". However, in numerical simulation, one has to set these dimensions limited although they are as far apart as possible. To move away the boundaries of the computational domain without increasing the number of grid nodes, a non-uniform grid is used with a gradually increasing step toward the boundaries of the domain. However, starting from the moment equal to the time of the surface waves passage to the boundaries of the computational domain and back, the solution begins to be affected by the waves reflected from the boundaries of the domain. To reduce the influence of the reflection effect, damping boundary conditions were introduced at the boundaries of the region, which damped the incoming waves. Such a numerical technique has proven itself in practice and is often used to reduce the influence of reflection from boundaries [11, 12]. To introduce damping boundary conditions, the term $-\mu\partial w/\partial t$ is added to the right side of Eq. (1.6), where μ is the coefficient of external damping. The experience of test calculations in study the μ effect on the degree of the wave reflection from the boundaries shows that the optimal solution reducing the effect of reflected waves by 85–95%, is to set μ on an area with a length of about 5% of the size of the entire area. In the rest (central) part of the computational domain, the value of μ is set equal to zero. Test calculations reveals that a very sharp change in the value of ţ does not give the desired result, but, on the contrary, leads to the formation of additional reflected waves at the boundary. As test calculations show, a smooth change of μ from zero to $10^3 \ldots 10^4$ kg/(m^{2}s) near the calculation domain boundary is optimum. When calculating numerically, the system (1.4)–(1.7) is divided into two subsystems (1.4), (1.5), and (1.6), (1.7), which were solved sequentially in two stages at each time step. At each time step, the nodal values for the potential $\Phi^n_{i,j}$ are calculated, where n is the number of the time step. At the grid nodes corresponding to the fluid surface, the deviations of the surface from the equilibrium horizontal position w^n_i are determined. The sequence of calculations is as follows. Let the values potential $\Phi^n_{i,1}$ at all nodes and vertical deviations at near-surface nodes w^n_i and w^{n-1}_i be known from the previous time step Δt or taken from the initial conditions.

- Using the values $w^n_i - w^{n-1}_i$, we find the velocities $(dw/dt)^n \approx (w^n_i - w^{n-1}_i)/\Delta t, i = 1 \ldots I$ at all surface nodes.
- The found values $(dw/dt)^n$ are substituted into the boundary condition (1.5), then system (1.4), (1.5) is solved numerically and new values $\Phi^{n+1}_{i,j}$ are found at all grid nodes.
- Using the obtained values of $\Phi^{n+1}_{i,j}$ and the values of $\Phi^n_{i,j}$, j taken from the previous step time, the values $(d\Phi/dt)^{n+1}$ are calculated at all near-surface grid nodes using difference approximations of relations (1.8) or (1.9).
- The obtained values $(d\Phi/dt)^{n+1}$ are substituted into Eq. (1.6), which is solved numerically, and the values w^{n+1}_i are found at all near-surface nodes. Then the process is repeated.

The iterative method is used for solving the system (1.4) and (1.5) numerically. To solve Eq. (1.6) the substitution $M = \partial^2 w/\partial x^2$ x2 is used. This allows replacing Eq. (1.6) of the fourth order in x, with the system of two second-order equations. This system is approximated by a three-point implicit difference scheme and solved using the matrix sweep method.

1.4 Results

As the ice presents on one side of the floater, the problem is not symmetric with respect to the floater. The reflected waves only come from the left side of the floater, which leads to the floater displacement and its rotation because of an asymmetric

Table 1.1 Values of physical parameters, geometry parameters of the basin, and the floater. l_2 is the length of the floater, and $h_2\rho_2$ is the product of the thickness and density of the floater

H_y, m	H_x, m	h, m	ρ_1, kg/m^3	ν_{ice}	E_{ice}, GPa	$h_2\rho_2$, kg/m^2	l_2, m
100	200	1	925	0.3	5	3218	30

pressure. The tangent of the floater rotation angle is calculated as the ratio of the difference in the displacements of its ends to its length. This is true in the case of a large bending rigidity of the floater. The displacement and rotation angle of the floater and ice are calculated for the case when a concentrated non-stationary harmonic force is applied in the center of the floater. It is given in the following form:

$$q = A\sin(\omega t)\delta(x - H_x/2)H(t),$$

where A is the force amplitude, ω is the circular frequency of the force, $\delta(x)$ is the delta function, floater is located in the center of basin with the width H_x, and $H(t)$ is the Heaviside function:

$$H(t) = \begin{cases} 0\ , t < 0 \\ 1\ , t \geqslant 0 \end{cases}.$$

The force values are taken as in [8], which makes it possible to compare the results obtained in [8] with the results of our paper. The values of the main parameters are given in Table 1.1.

Figures 1.2 and 1.3 show typical dependences of the maximum displacement of a floater on the force frequency at different ice thicknesses in unsteady and stationary regimes, respectively. In these calculations, the non-stationary regime is the regime in which oscillations are established from the beginning of the countdown to the moment of their stabilization. Stationary regime is a regime when the maximum oscillation amplitude does not change for at least 10 periods. All figures show the maximum displacements or angles of rotation in stationary or non-stationary regimes.

The dependencies are obtained for the distance from the ice edge to the floater with $L = 15m$. The dependence of the displacement of the left edge of the floater on the frequency in Fig. 1.3 shows that there are frequency ranges in which the effect of reflected waves on these movements is visible. The dependencies are obtained for the distance from the ice edge to the floater with $L = 15$ m. The dependence of the displacement of the left edge of the floater on the frequency in Fig. 1.3 shows that there are frequency ranges in which the effect of reflected waves on these movements is visible. When the floater oscillates, it loses energy in the form of radiated waves and some of this energy is trapped in the gap between the floater and ice in the form of a standing wave [8]. Within these frequency bands, the length of a surface wave is approximately the same as gap length. Simultaneously the amount of energy transferring to the wave significantly increases the floater displacement. It should be noted then when the reflected waves are in-phase with the vertical motions of the

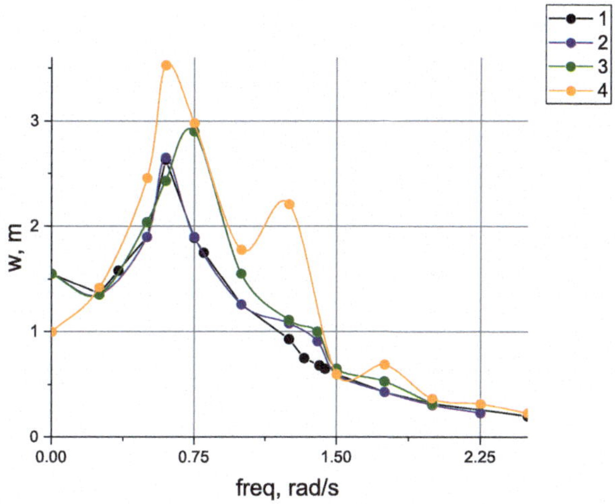

Fig. 1.2 Displacement of the left edge of the floater via the frequency of the force with an amplitude of 294300 N/m in non-stationary regime for different ice thicknesses h. Line 1—$h = 0$— no ice; line 2—$h = 0.1$ m; line 3—$h = 1$ m; line 4—$h = 10$ m

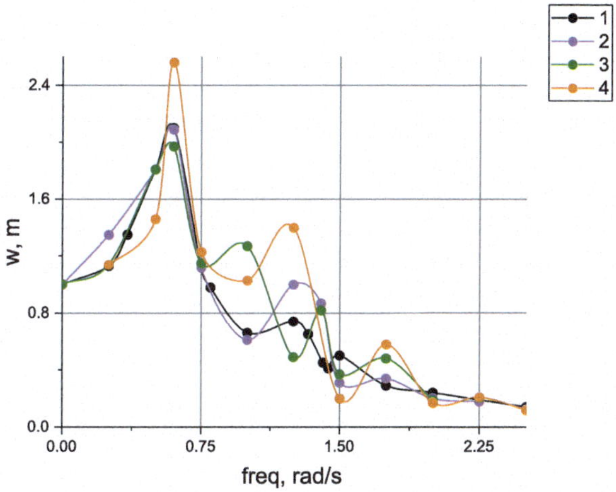

Fig. 1.3 3 Displacement of the left edge of the floater via the frequency of the force with an amplitude of 294300 N/m in stationary regime for different ice thicknesses h. Line 1—$h = 0$— no ice; line 2—$h = 0.1$ m; line 3—$h = 1$ m; line 4—$h = 10$ m

floater, resonance occurs and the floater's response increases. If the reflected waves are in antiphase with the floater, anti-resonance occurs and its response lowers. The graphs in Figs. 1.2 and 1.3 reveal that it is important to take into account the non-stationary regime, since the maximum displacements in it exceed the maximum displacements in the stationary regime. It should be noted with an increase in the ice thickness the displacement of the floater also increases. At the same time, calculations show that the displacements of the ice edge decrease with the ice thickness increase due to the plate cylindrical rigidity increase. As the curves in Fig. 1.4 reveal, the number of resonant peaks increases with decreasing of cylindrical stiffness. The obtained results qualitatively coincides with the results obtained in [4] for the case of a free-floating plate.

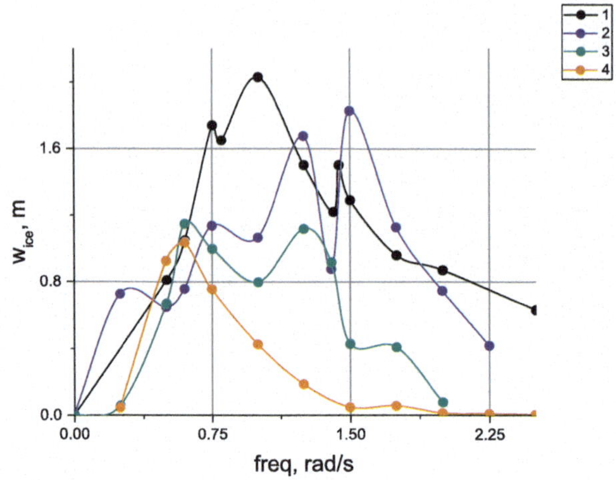

Fig. 1.4 Ice edge displacement via force frequency with amplitude 294300 N/m in stationary regime for different ice thicknesses h. Line 1—h—no ice; line 2—$h = 0.1$ m; line 3—$h = 1$ m; line 4—$h = 10$ m

Most of the calculations performed by the authors of this paper confirm the conclusion made in [8] that nonlinear conditions' effect on the fluid surface is insignificant. The results of the linear problem performed by the authors are in good agreement with the results obtained in [8] with the help of the analytical method of expansion in terms of vertical eigenmodes in the steady state for a harmonic external excitation. Due to the good agreement between the values of the dependences obtained by numerical and analytical methods, the solution obtained by analytical formulas is not shown on the graphs. However, for some values of the parameters of the considered system, it occurs that the nonlinear boundary conditions lead to a difference from results obtained in the linear condition case. The displacements of the floater edge at linear and nonlinear conditions on the fluid surface are shown in Fig. 1.5 for $L = 10$ m and $A = 250000$ N/m. The graphs in Fig. 1.5 show that at $\omega = 0.5$ rad/s, the floater displacement amplitude in the nonlinear case is greater than in the linear case.

The authors found the differences in the values of the floater maximum angle of rotation in the linear and nonlinear cases in the stationary regime of frequencies in the range of 0.6–1.3 rad/s, as the curves in Fig. 1.6 present.

However, in the wide range of system parameter values the difference between the floater displacements and rotation angles for linear and nonlinear boundary conditions is not observed. Figure 1.7 shows a typical dependence of the angles of rotation on the acting force frequency in a stationary linear regime for various distances from the ice edge to the floater. To interpret the results presented in Fig. 1.7, it is necessary to note that due to the symmetry of the floater itself, and the absence of ice, the angle of the floater rotation under force loading is zero. In addition, the graphs in Fig. 1.7 show that the ice has little effect on the magnitude of the floater rotation angle up to the frequency $\omega = 0.55$ rad/s. At frequencies greater than 0.55 rad/s, the effect of the ice is significant and manifests itself, among other things, by resonant peaks. The ice effect on the rotation angle is absent because of the reflected/incident wave

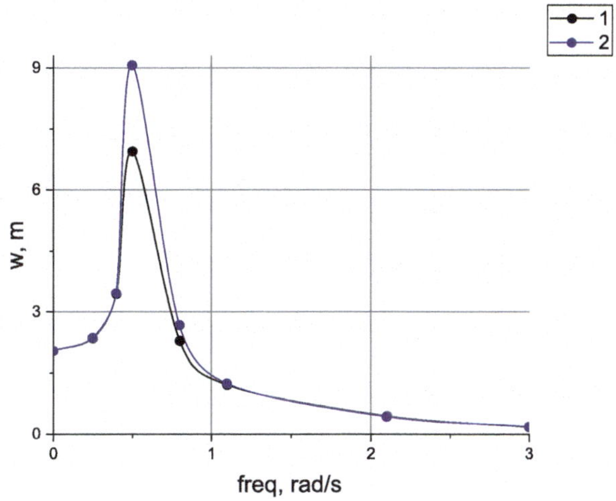

Fig. 1.5 Dependence of the left edge floater displacement in the linear and nonlinear cases via the frequency of the force with A = 250000 N/m in stationary regime for $L = 10$. Line 1—$\alpha = 0$; line 2—$\alpha = 1$

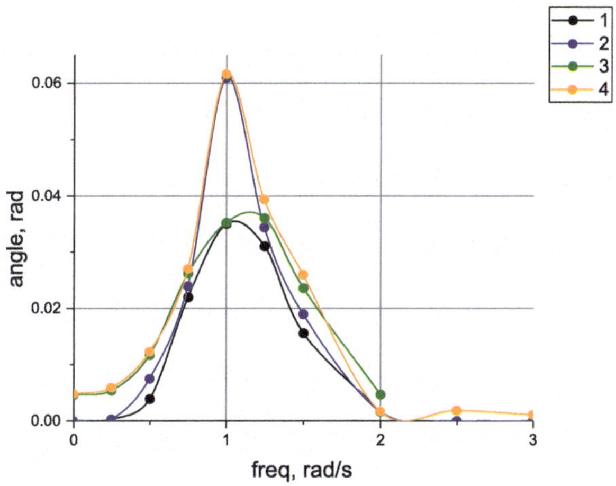

Fig. 1.6 Dependence of the floater angle of rotation on the frequency of the force with A = 294300 N/m for $L = 5$ m. Line 1—$\alpha = 0$ stationary regime; line 2—$\alpha = 1$ stationary regime ; line 3—$\alpha = 0$ nonstationary regime; line 4—$\alpha = 1$ nonstationary regime

interaction on the edge of the ice. The work [10] shows that when a low-frequency wave is incident normal to the edge its energy is transferred to a floating plate, while in case a high-frequency wave is incident and its energy is completely reflected into the fluid. Figure 1.7 presents a similar behavior at frequencies below 0.55 rad/s; the waves are almost not reflected in this case. Non-propagating waves forming in the space between the ice and the floater decay quickly along the coordinate and do not affect it. Therefore, the propagating reflected surface waves are the main in the process of a wave/floater interaction.

The paper also considers the floater displacement dependence on the frequencies at different distances between it and the ice edge. Figure 1.8 presents dependences of the maximum displacement of the left edge of the floater on the harmonic force frequency with A = 200000 N/m for various distances to the ice edge.

Fig. 1.7 Dependence of the floater rotation angle on the frequency of the force with A = 200000 N/m in stationary regime. Distance effect for nonlinear case. Line 1—L = 15 m, α= 0; line 2—L = 15 m, α = 1; line 3—L = 1 m, α = 0; line 4—L = 1 m, α = 1 = 1

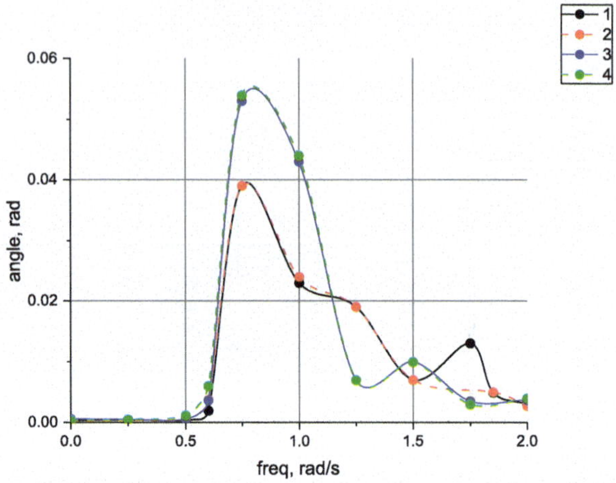

Fig. 1.8 Dependence of the ice edge displacement on the of the force with A = 200000 N/m in stationary regime. Distance effect for nonlinear case. Line 1—L = 15 m, α = 0; line 2—L = 15 m, α = 1; line 3—L = 1 m, α = 0; line 4—L = 1 m, α = 1

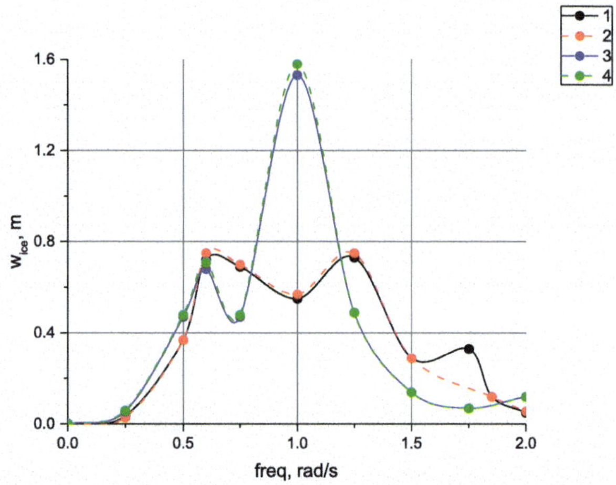

Figure 1.8 shows that decrease of the distance to the ice edge increases maximum displacements of the floater at frequencies in the range from 0.75 to 1.25 rad/s. However, calculations using other parameters of the system do not indicate clear dependence of the displacement amplitude on the distance to the ice edge. The result obtained can be caused by the complex pattern of waves reflected from the ice edge. The wave pattern depends on length of the interacting waves. The result interpretation is difficult because the problem has no symmetry. The authors consider another factor presumably affecting displacement of the ice and the floater, namely the floater's cylindrical rigidity. The calculations revealed that the floater cylindrical rigidity starts affecting the maximum displacements of both the ice and the floater only if its value significantly decreases, by 1000 times compared with the parameters of offshore structures selected in [8]. Therefore, we can consider the behavior of a real offshore structure in the problem under consideration as the behavior of a rigid body.

Regarding the influence of the floater mass, an increase in the mass of the floater leads to a decrease in the magnitude of the resonant peaks, but the resonances themselves occurred at the same frequencies as at other masses. The work [8] obtained the same result.

1.5 Conclusion and Discussion

In the present paper, the authors propose an approach to the numerical solution of the problem studying effect of an ice cover and fluid surface waves on a floater dynamics. Based on the numerical solution obtained, an analysis of behavior of the floater displacements and floater rotation angles under the action of a harmonic concentrated force on the latter is carried out. It is shown that in the non-stationary loading regime, the values of the maximum displacements and rotation angles of the floater can differ significantly from the values determined in the stationary regime. It is also shown that the assumption made earlier in [8] on the possibility of using linear boundary conditions on the liquid surface can be used for most of the possible practical system parameters. At the same time, for a number of parameters and at certain frequencies, the nonlinear boundary conditions lead to greater values of the maximum displacements and rotation angles of the floater than in the linear formulation. The flexural rigidity of a floater begins to affect it's displacements and angles of rotation on a fluid surface at its values far from those that floating offshore structures can have. The influence of the distance between the ice edge and the floater requires additional analysis. It should be noted that such important factors as 3D formulation of the problem, ice cover inhomogeneities (old ice inclusions, cavities filled with brine, etc.), slush, moments acting on a floater, and speed of its approaching the ice have not been considered yet. The authors intend to take into account all these factors in their future works.

Acknowledgements This research is supported by the Russian Science Foundation via grant No. 23-29-00459 (https://rscf.ru/project/23-29-00459).

References

1. Kheisin DE (1967) Ice cover dynamics. Leningrad. Gidrometeorologicheskoye izd-vo (in Russian)
2. Li ZF, Wu GX (2021) Hydrodynamic force on a ship floating on the water surface near a semi-infinite ice sheet. Phys Fluids 33(12):127101. https://doi.org/10.1063/5.0071972
3. Li ZF, Wu GX, Ren K (2021) Interactions of waves with a body floating in an open water channel confined by two semi-infinite ice sheets. J Fluid Mech 917:A19. https://doi.org/10.1017/jfm.2021.274

4. Nelli F, Bennetts LG, Skene DM, Monty JP, Lee JH, Meylan MH, Toffoli A (2017) Reflection and transmission of regular water waves by a thin, floating plate. Wave Motion 70:209–221. https://doi.org/10.1016/j.wavemoti.2016.09.003
5. Taylor RE, Ohkusu M (2000) Green functions for hydroelastic analysis of vibrating free-free beams and plates. Appl Ocean Res 22(5):295–314. https://doi.org/10.1016/S0141-1187(00)00018-3
6. Sturova IV (2002) The action of periodic surface pressures on a floating elastic platform. J Appl Math Mech 66(1):71–81. https://doi.org/10.1016/S0021-8928(02)00010-2
7. Tkacheva LA (2005) Impact of a periodic load on a floating elastic plate. Izvestiya Rossiyskoy academii nauk. Mekhanika Zhidkosti i Gaza (2):132–146 (in Russian)
8. Keijdener C (2019) The effect of hydrodynamics on the interaction between floating structures and flexible ice floes. Thesis for: PhD, Delft University of Technology. https://doi.org/10.4233/uuid:a66b84b9-6c66-4c9c-a2e1-87163244ae07
9. Keijdener C, Barbosaa JMO, Metrikine AV (2017) The influence of level ice on the frequency domain response of floaters. Cold Reg Sci Technol 143(1):112–125. https://doi.org/10.1016/j.coldregions.2017.09.004
10. Fox C, Squire VA (1990) Reflection and transmission characteristics at the edge of shore fast sea ice. J Geophys Res 95(C7):11629–11639. https://doi.org/10.1029/JC095iC07p11629
11. Samarskiy AA, Nikolayev ES (1978) Methods for solving grid equations. Nauka, Moscow (in Russian)
12. Potter D (1973) Computational physics. John Wiley

Chapter 2
Biocompatible Structured Materials for Tissue Engineering Based on Ethylene–Vinyl Acetate Copolymers: From Controllable Waviness and Microroughness to Spatially Modulated Electrostatics and Diffusion Programmable by Electron Beam Modifications

Pavel L. Aleksandrov, Oleg V. Gradov, Margarita A. Gradova, Irina A. Maklakova, Anatoly A. Popov, Anna V. Ratnovskaya, and Popov G. Shelenkov

Abstract The article begins with a comprehensive review of the topic of EVA-containing polymer composites for tissue engineering on ethylene–vinyl acetate copolymer-based systems. The article focuses on the problem of textural and (bio)chemical biocompatibility of the material. As a solution to the first problem, it is proposed to use electron microscopy and correlation-spectral image analysis, which emulates laser diffraction (2D FFT implemented with the FFTW library in QAVIS software). The article also describes the modernization and digitization of an electron microscope, thanks to which such analysis becomes possible in a video stream and in real time or post factum. Also eleven tables of analysis results are given so that each specialist in the field of cell biology and regenerative medicine can independently assess the suitability of certain polymer compositions with different ratios of components for his tasks.

P. L. Aleksandrov
Shemyakin-Ovchinnikov Institute of Bioorganic Chemistry of the Russian Academy of Sciences (IBCh RAS), Moscow, Russian Federation

O. V. Gradov (✉) · M. A. Gradova · I. A. Maklakova
N.N. Semenov Federal Research Center for Chemical Physics, Russian Academy of Sciences (FRCCP RAS), Moscow, Russian Federation
e-mail: o.v.gradov@gmail.com

A. A. Popov · A. V. Ratnovskaya · P. G. Shelenkov
Institute Institute of Biochemical Physics Named After N.M. Emanuel of the Russian Academy of Sciences (IBCP RAS), Moscow, Russian Federation

H. Altenbach and V. A. Eremeyev (eds.), *Dynamics of Discrete and Continuum Structures and Media*, Advanced Structured Materials 221,
https://doi.org/10.1007/978-3-031-75626-9_2

Keywords Ethylene–vinyl acetate · EVA and PE compositions · EVA implants · Tissue engineering · Correlation spectral analysis KSA · FFTW library · QAVIS software · 2D FFT

2.1 Introduction

Ethylene–vinyl acetate (EVA) copolymer-based systems are of significant importance in regenerative medicine, prosthetics, tissue engineering, and design of biocompatible implants. Modern biomedicine also applies hydroxyapatite/EVA composites for production of composite scaffolds for simultaneous substitution or repairing of the articular cartilage and underlying bone tissue [1–5]. Most of the authors consider such materials as non-biodegradable implantable structures [6]. However, attempts are regularly made to enhance biodegradability of EVA-containing blend composites [7]. Sometimes there are even works where EVA-containing biocompatible systems are represented as the novel trend in polymer engineering, while this is not correct, since all such works are usually based on the conventional achievements of the late 1980th/early 1990th, which are usually forgotten during the search of the current literature. For this reason, we should start our study of EVA-containing polymer composites with a brief review of the classical works providing the necessary argumentation and containing the main prerequisites of the novel approaches which have been recently proposed.

First of all the works of C. W. Daniel and coauthors should be mentioned, focused on histogenesis or morphogenesis mediated by the EVA-based implants, from vascularization to neuromorphogenesis:

1. Neural morphogenesis control, including controlled release of the nerve growth factor (NGF) from a biocompatible ethylene–vinyl acetate copolymer implant [8, 9];
2. Retinal regeneration [10] including growth factor-induced regeneration after retinectomy using EVAc implants containing bFGF (basic fibroblast growth factor);
3. Ductal growth control using transforming growth factors (such as TGF-β1 or TGF-13 from cytokine protein family) [11–13] based on EVA-containing implants with TGF (a special role in this process can be played by either E-cadherins or P-cadherins [14] and also a competitive agent EGF—epidermal growth factor, which can inhibit ductal morphogenesis [15] after EGF release from EVA);
4. Mast cells (a kind of stem cells) induction [16] (compared to the controls, suggesting that dexamethasone (DMS) is released from the EVA/DMS composite), which can play an important role in the inflammatory response to the intraperitoneal implants in experimental rats and other laboratory animals.

Of significant interest is the release of neurochemical agents from the EVA implants, for example, development of the dopamine-containing polymer systems based on the EVA matrixes [17, 18] and neuropharmacological investigations such as drug release from the polymers in brain tissue [19] (for example adenosine-releasing EVA implants in lateral ventricles of rats and phenytoin controlled release for reducing of the experimental seizures in rats [20], controlled release of dexamethasone from the intracranial implants of pure EVA [21], EVA-mediated suramin chemotherapy in the treatment of experimental brain tumours [22], etc.) or protection from excitotoxicity [23]. The works considering histoendocrinological applications of EVA and its composites are less relevant nowadays (for example—luteinizing hormone—releasing hormone analogue (LHRHa) and testosterone release from the EVA implants [24] or controlled release of insulin from the EVA copolymer coatings [25]), controlled release of vitamins and other bioactive low-molecular compounds/ small molecules (this trend is actual until now [26]).

The requirements of the biological optimality for the materials containing EVA include: electrobiophysical optimality of the interfaces (this is especially necessary for neuromorphogenesis, since the neurons are electrically active cells); membrane-mimetic properties such as diffusion parameters (which is especially important for the implementation of morphogenesis as a reaction–diffusion process) and permeability for gaseous and liquid media; rheological and micromechanical optimality (soft matter, porous structure, etc.) In this aspect, EVA-containing materials are rather suitable research objects, since the properties of their prototypes have already been studied previously in details. Gas permeability has been studied not only for EVA itself, but also for a number of its nanocomposites and mixed matrix membranes [27–35]. Liquid permeability has also been studied for both low molecular agents and biological regulators such as hormones (including neurohormones and steroids [36, 37]). The dependence of permeability on the methods of synthesis, molding, and sample preparation has also been addressed. Ionic permeability has also been investigated [38] and was found to satisfy the passive requirements of membrane mimetic engineering. Oxygen permeability studies and measurements [39, 40] are of particular interest according to the criteria of redox biology in implantology.

The main applied and fundamental studies of the diffusion processes in EVA-containing systems were carried out in the 1970–1980s [41–46], until the first half of the 1990s, when the trend of EVA application in implantology with the controlled drug release has emerged. Therefore, it would be reasonable to believe that many biomedical advances in this area are based on the previous achievements in physical chemistry or polymer sciences. In those times, diffusion of different molecular agents through EVA copolymers [47]; diffusion through the different blend membranes and membraneous structures containing EVA [48–52]; diffusion through the EVA-containing microcomposites and nanocomposites [30, 53, 54]; and diffusion in solution [55] have been studied. Diffusion was studied using various agents: dyes (in dye-labeled poly (vinyl acetate-co-ethylene) latex and latex films [56]); isotopes [57]; classical pharmaceutical and toxicological agents [58, 59]; mineral oils [60, 61]. Not diffusion itself, but also thermal diffusion by the thermal field-flow fractionation [62] has been considered. Therefore, in the field of diffusion membrane-mimetic properties and controlled release of different agents in the EVA-containing polymer structures comprehensive information is now accessible.

In biological systems ion exchange and diffusion of ions through the membrane are associated with bioelectrogenesis and electrical properties of biointerfaces which determine both bioenergetics and reactivity. Therefore, for the development of biocompatible implants, in addition to permeability, the data on the intrinsic electrophysical properties of the implant material and its interface with a biological tissue are also rather useful. The corresponding electrophysical properties of the EVA-containing composites are also well studied to date.

Since the early 1960s dielectric properties of EVAc copolymers and copolymer blends [63–67] have been repeatedly investigated, as well as the influence of the blend compositions on the dielectric properties [68]. In the current works, EVA-containing composites are predominantly considered [69–73] and, partially, nanocomposites, especially nanocarbon-based/nanotube-filled ones [74–82]. The effect of different types of carbon fillers on dielectric properties and electromagnetic efficiency of (nano)carbon-filled materials [83] were studied separately, multifractal analysis of the fracture morphology of poly(ethylene-co-vinyl acetate)/carbon black conductive composites [84] was also performed. Current methodological opportunities allow to measure a variety of the properties and electrophysical response of the polymer systems containing EVA, as well as to find similarities in their behavior with the other electrophysically active materials for the design of implants and scaffolds, such as PVDF, PHB, etc., with their electrophysical optimality resulting not from the electroneutrality, but from the piezoelectric properties. EVA is not an exception: for example, hydroxyapatite embedded blends of EVA and millable polyurethane can be used as piezoelectric or pyroelectric materials [85], a porous poly (ethylene-co-vinyl acetate) copolymer films demonstrate pseudo-piezoelectric effect [86]. Thus, it is possible to speak about the control of a number of properties of such implantable materials using an electric field. It is also possible to create controlled release systems such as artificial muscles and other piezoelectric actuators using polymer carriers based on EVA. It is possible to implement with the high accuracy on such systems both signal transduction via an electric channel (up to the breakdown threshold) and macrokinetically complex electrochemical actuation using different liquids or soft mater as the signal transmission/transduction medium. This is possible, since the interface polarization and dielectric properties, including dielectric relaxation [87–91], loss curve broadening and correlation parameters [92] of EVA required for the operation of actuators are now well understood.

The early works in the field of the direct current conductivity and electrode polarization in dielectric study of poly (ethylene-co-vinyl acetate)-poly (vinyl chloride) blends [93] have provided prerequisites for the transition from the additive dielectric spectroscopy technologies to the local dielectric spectroscopy which has been widely applied in the last decade for characterization of the EVA-based polymer systems [94, 95]. Since actuation for the fine (bio-)mechanics and organ-specific target delivery is based on the local tissue excitation (which is similar to the "local reaction of protoplasm" in biological systems), local measurements of the dielectric properties, in our opinion, provide a direct way to the local control of such systems. Moreover, if the subject of excitation is EVA itself, the rest part of the polymer composite can be inert, elastic, but not signal-conducting (although the blends composed of EVA and millable polyurethane can be used as piezoelectric or pyroelectric materials as a whole). There are also several examples of the complex studies addressing dielectric, thermal, and mechanical properties of the EVA-containing systems based on rubbers/elastomeric substances simultaneously [96–98]. For this reason, electromechanical mobility of such systems should be studied in correlation with the local charge state. The simplest way to supply the local charge to the polymer surface includes exposure to the focused electron beam, similar to that used in the scanning (raster/sweep) scheme of a scanning electron microscope.

The study of the electron beam effects on EVA, including the electron beam modification of EVA is already a classical area of work [99–103], in contrast to the more recent research area of the microwave effect on EVA [104]. As in the previously mentioned research trends, EVA copolymers [105–109] and blends were initially investigated (and, accordingly, electron beam-induced cross linking [110–112], interest in which has been renewed in the twenty-first century [113–116]), then the research focus was moved to EVA composites and, in particular, nanocomposites (including clay filled or mineral fiber reinforced ones) [117–122]. Interest in EVA-containing polymer blends remains due to the technological optimality of the electron beam irradiation induced compatibilization [123] and the multifactor nature of the irradiation effect observed [124–126]. The aforementioned EVA-based elastomeric composites can be also obtained using electron-beam technology [127]. Using the electron beam exposure methods it is possible to change simultaneously qualitatively different properties of the material from the microscopic phase characteristics to the additive mechanical properties. An example of the exotic effect of the electron beam treatment is the enhancement of physical properties of the flame retarded EVA copolymer/composites after the electron beam irradiation and intumescent flame retardant formation techniques [12–131]. Thus, it can be assumed that knowing the criteria for varying the required properties, it is possible to adapt the material to be applied for the production of an implant, prosthesis, scaffold, etc., thereby realizing the optimal ratios of a number of electrophysical and structural properties of the polymer material necessary for a good interface contact with the biological tissues.

Repeated application of the scanning electron microscopy techniques for the study of the EVA-containing systems since the XX century [132–135] and until the second decade of the XXI century [136–139] together with the appropriate approaches to the chemical characterization of the polymer structures obtained (for example, wave-dispersive or energy-dispersive spectroscopy for characterizing the local chemical composition of the samples) makes it possible to speak about the electron beam treatment of such systems. This also applies to the structures such as low density polyethylene/ethylene vinyl acetate filled composites, which are widely used for the wire or cable applications [140].

Generation of the local charge at an arbitrary point of the sample allows observing the charge gradients and breakdown effects. In this regard, the installations which attempt to integrate wide-range dielectric spectroscopy with position-sensitive techniques, in particular, electrostatic force microscopy (which has been previously performed in spatial and temporal separation [141]) are considered rather promising.

In fact, charge formation/space charge formation [142, 143] (especially in multilayer systems [144]) as the processes associated with a double electric layer do not make sense without the charge (including space charge) accumulation [145] at the surface in the electrical double layer. This means that direct observation of the breakdown phenomena associated with the space charge formation in EVA and its copolymers under the electron beam is possible [146]. Analysis of the literature shows that, indeed, the breakdown effect is observed in EVA copolymers both in the room-temperature region as a thermal breakdown [147] (the theory of which has been developed by N. N. Semenov) and in the low-temperature conditions (as a non-thermal breakdown), in accordance with the work of the Japanese researchers published in the 1970s [148–150], including the phenomena widely reproduced under the liquid nitrogen temperature [151]. There is a known relationship between the electric breakdown and the space charges in EVA-incorporated media (including polyethylene [152]), regardless of the nature of the electric discharge (either an air dielectric barrier discharge or corona discharge or low-pressure plasma [153, 154]. Therefore, observing the charge wandering or discharges on the surface of the EVA-incorporated media is a useful tool for characterization and determination of the charge properties (at least, the charge sign, the presence of a region of capacitances/sources and drains, localized threshold voltages, etc.) of the polymer material. This is a good alternative to the aforementioned approach integrating the data of dynamic force microscopy and the measurements in the frequency domain of the dielectric parameters.

The features of electrophysical phenomena, including charging/discharging or breakdown, are usually determined by the morphological features of the heterogeneous sample or its surface. Typically, in the polymer systems containing EVA, morphology, rheology or mechanics, thermal and other properties are interrelated [155–166]. In this regard, it is possible to correlate microelectromechanical/electrorheological phenomena with the surface morphology in situ—during electron microscopy. In this case, the breakdown will only visualize the morphological features of the sample surface invisible to the eye. Then it would be possible to obtain colocalization maps of the electrophysical parameters and morphology of the sample surface. In the case of simultaneous application of the microprobe with chemical composition mapping, one can also find out which chemical constituents are responsible for the electrophysical/breakdown-inducing defect. At the same time, it should be mentioned that secondary emission X-ray radiation (EDS) can also operate as the polymer modification factor (at least for a sufficiently short-wavelength gamma spectral region [167]). Therefore, the clidonographic [168, 169] approach to characterizing samples containing EVA can be not only useful in assessing their applicability in creating scaffolds and other biomedical structures (which is known to be determined by the VA content—for medical applications, its content should be 5–20%), but also in predicting electrophysical properties and charging/discharging behavior according to the chemical composition [170–172].

2.2 Materials and Methods

Reprogramming of the SEM Registration System Based on DIGIC III Processor. For SEM image and video registration a DIY system was adapted with a Canon A590IS CCD sensor (maximum resolution 3264 × 2448 pixels in static mode; maximum video recording resolution 640 × 480 with a maximum video frame rate of 30 frames per second when using the MJPEG video codec) and a Canon DIGIC III image processor, synchronized directly from the control system of the electron microscope through a special cable based on a universal bus cable. The registration system was mounted on a 3D-printed module, compatible in size with the Canon A590IS mount. The key requirement for the image registration system implemented on the DIGIC processor was supporting of CHDK—a resident program that expands the control functions of Canon digital cameras. CHDK program was installed to the SD card as follows:

1. On a PC with pre-installed JAVA, the installer of CHDK ("STICK") was unpacked and one of its files—either STICK.BAT or STICK.CMD was started (one of them may not provide a successful CHDK installation).
2. Testing was performed using a random SEM image.
3. After completion, the flag on the memory card was transferred to the "Lock" position.
4. Using the camera buttons <Print> and <Menu> CHDK menu was entered.
5. Pressing <Set> allowed to enter the <Enhanced photo operations> section.
6. Using the <Set> , <Up> and <Down> buttons, the following camera parameters were set:

    a. Disable overrides—No
    b. Override TV type—long exp
    c. Long exp value—the exposure required by the SEM (in the case of Jeol JSM-T330A it is 38 s for scanning)
    d. Override AV, Override ISO, Subj dist—the following values for aperture, ISO and focal length were applied: 20, 12 and 250. After setting the numerical values of the above parameters, checkboxes should appear to the left of the values, confirming the values entered
    e. Disable overrides on start—the corresponding checkmark should be removed, since this option makes CHDK save the values entered after the camera is turned off.

7. Double clicking <Menu> returns to the main menu, and then using <Set> , <Up>, <Down> find "CHDK settings" section, "Remote parameters" subsection.
8. "Enable remote", "Switch type—one push", "Control mode—normal" parameters were set.

Synchronization Line Upgrade

A new synchronization line was developed to control the photoregistration procedure. To synchronize with SEM, pulses with a voltage of 3–5 V and a current of up to 10 milliamperes were used, supplied through the power lines of the USB input of the camera with the standard polarity. The appearance of the signal is equivalent to a partial pressing the photographing button, and the disappearance of the signal is equivalent to pressing the button completely (i.e., starting the photographing). When creating a synchronization line between Jeol JSM-T330A and Canon A590IS using a USB-A—mini-USB-B cable, the following operations were performed:

1. USB-A connector was cut off, after which it was required to free, strip and tin the black and red wires (color marking is standard for the universal serial bus), the white and green wires can be cut off.
2. The cap on the right side of the front panel of the electron microscope was removed and the screws for fixing the strip with the <Shutter> button were unscrewed.
3. The ground and power lines were found on the board nearest to the <Shutter> button using easily detectable designations on the pinout of any microcircuit: ground is designated as Gnd, power as Vcc. The <Shutter> button has two switch contact groups, one of which is not used. To the latter unused normally open pair of contacts the red wire of the USB cable and the power line found in accordance with the above instructions were soldered.
4. The black wire was soldered to the ground line.
5. The criterion for correctness during testing of the system was that when the cable was connected to the camera and the <Shutter> button was pressed, shooting automatically starts.

The advantages of Canon A590IS include the ability to simultaneously connect a USB cable, a power supply and an analog video signal transduction cable. Therefore, we also implemented an image output to the analog monitor from the analog signal transmission cable. It is also possible to use a camera that does not have an additional power input and video output, but it is less convenient.

Creating an Adapter for Optical Registration Tract from the CRT

A special adapter was created for optical connection of the camera to the microscope screen. Its parameters depend on the system used. For the case of the Jeol JSM-T330A—Canon A590IS system, the following model was used: https://www.tinkercad.com/things/2i7MgsxqgKt-sem (Fig. 2.1). The model was printed on a standard type 3D printer with black PLA filament FDplast. There are no specific requirements for the surface treatment of such products. However, when preparing a 3D model for printing, it is necessary to make sure that no supports are added inside the side tube. The rest of the settings are standard.

The adapter is attached to the film camera mounting system of the microscope, and the Canon A590IS is attached using a cable tie that goes through the side tube (Fig. 2.2).

Power Supply Optimization. A power supply unit (3 V, plus in the center) was connected to the registration system, allowing it to operate without using batteries or accumulators. In this case, automatic shutdown was disabled in the camera settings, and the screen was turned off in the CHDK settings (<Print> - > <Menu> - > “CHDK

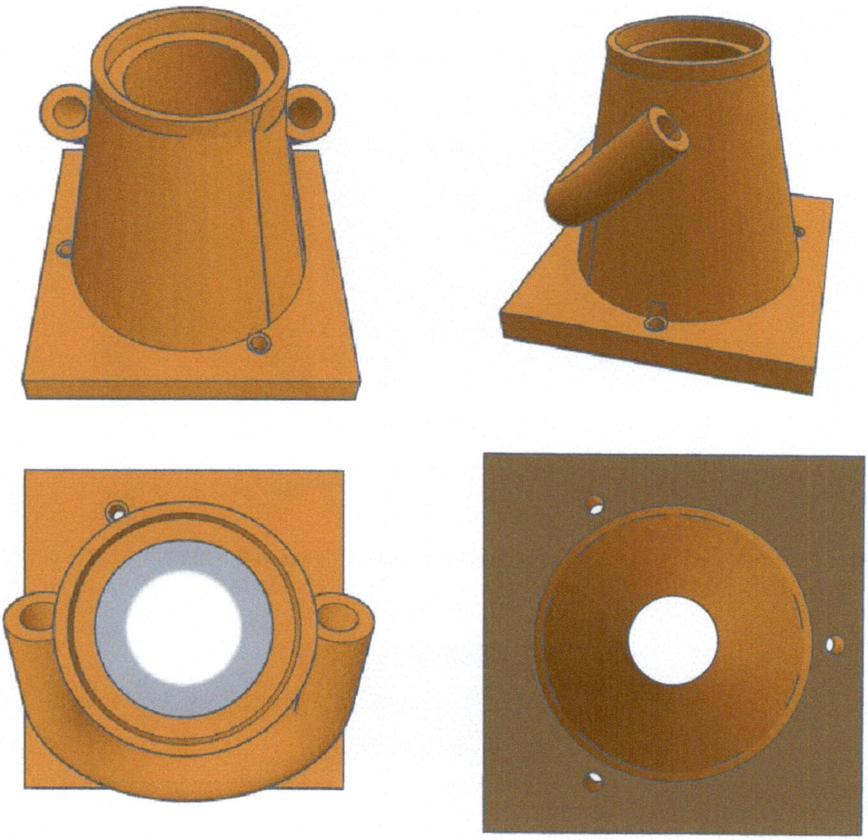

Fig. 2.1. 3D printable adapter for optical registration tract from the CRT of the JEOL SEM (developed by Pavel L. Aleksandrov in 2018–2019)

settings", then "Disable LCD off" is selected and using the <Set> button is switched to the "always on" mode).

Output To An External Liquid Crystal Or Plasma Monitor. It is also possible to improve the view of the small screen of the microscope by connecting to the camera a TV with analog video input via AVC-BC300 or STV-250N cable or a standard camcorder cable (see the scheme below) or the Scart input (via the same cable and Video-Scart adapter). The above scheme is shown in details in Fig. 2.3, and the resulting view of the system after upgrade is shown in Fig. 2.4.

Photography Process After Upgrade. The following procedures are required to obtain digital SEM images from JEOL JSM-T330A:

1. Adjustment of the image parameters is carried out as usual using the knobs and buttons on the electron microscope control panel.
2. The minimum Spot size is set, brightness and contrast are matched.

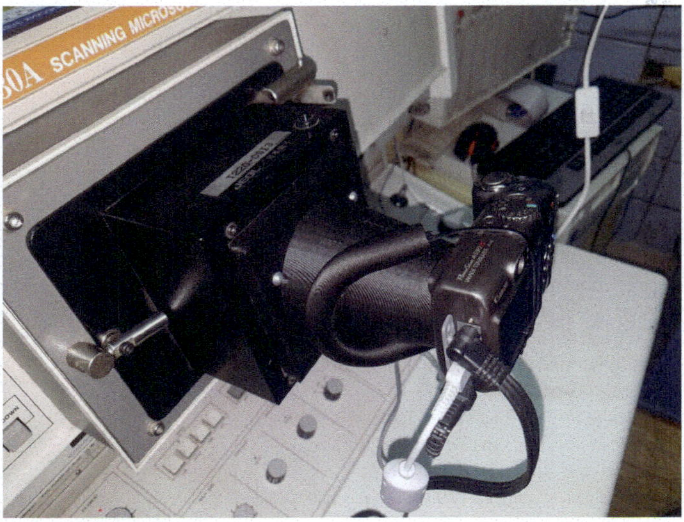

Fig. 2.2 Canon A590IS fixation on the 3D printable adapter for optical registration tract from the CRT

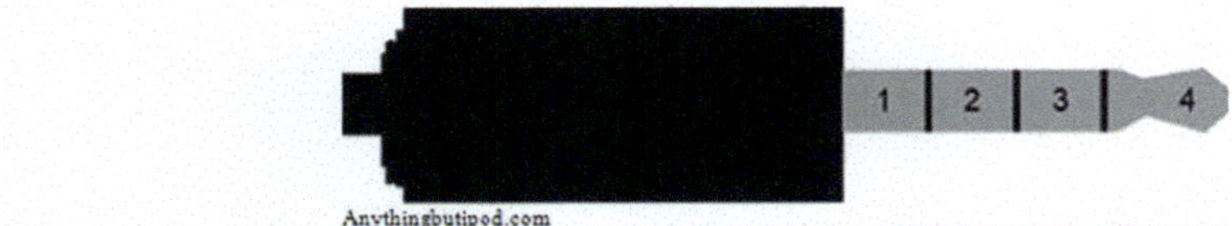

	1	2	3	4
Archos, Gigabeat, Creative Zen Vision Series, Cown iAudio, Apple iBook	Ground	Video	Right	Left
Standard Camcorder Cable	Right	Ground	Video	Left
iPod Video	Video	Ground	Left	Right
Zune	Video	Ground	Right	Left

Fig. 2.3 Contact scheme for the Canon A590IS

3. The exposure time of frames is adjusted (in the case of Jeol JSM-T330A press the “Qucik” key located to the left of the “Shutter” button).
4. The photocamera cover is closed.
5. The camera turns on. In the case of Canon A590IS, the camera should be in “M” (manual) mode and manual focus should be selected (approximately 20–25).
6. For Jeol JSM-T330A, it is necessary to achieve blanking of the screen persistence. For this, the “Slow 1” key is dropped first; without releasing it, one needs to wait 3–4 s; after that the “Shutter” key is lowered, the “Slow 1” key is released, then the “Shutter” key is also released.
7. If everything has been done correctly, the camera takes a frame from the microscope screen during the exposure time, then during the same time it takes a dark frame, subtracts the latter fro the former, and saves it at the SD card.

Fig. 2.4 General view of the installation after upgrade

8. The images obtained can be conveniently processed by IrfanView softwave in the following algorithm: "File → Batch conversion / rename → Advanced". In particular, monochromatization, cropping, contrasting (if necessary) can be carried out through this algorithm in batch processing of the entire image series stereotypically.

Correlation Spectral Analysis. Correlation-spectral analysis was obtained by the QAVIS software, developed by our Far Eastern colleagues (see below) in real time mode or after image acquisitions [https://oias.poi.dvo.ru/qavis/]. Most important details about this technique are provided in the articles of the software authors [173–179]. Most important microscale applications (excluding geophysica and marine sciences) developed by our colleagues are provided in the articles [180–190].

2.3 Results

Results (any multi-scale electron micrographs of the composites and 2D FFT patterns/laser diffraction emulation patterns generated using QAVIS software) for different concentrations of components (EVA and PE) are provided in the tables (Fig. 2.5).

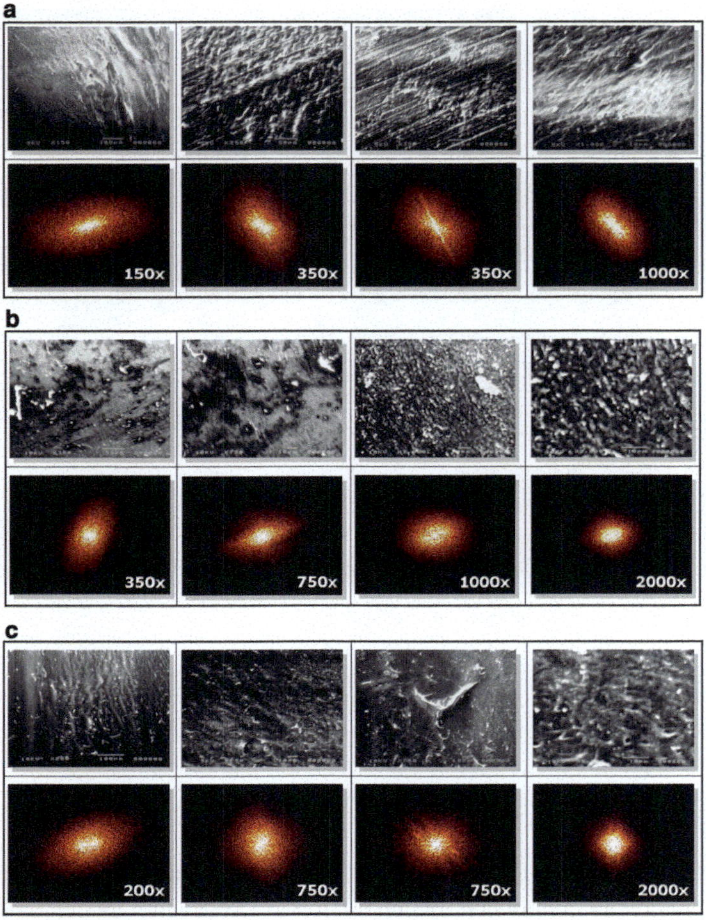

Fig. 2.5 **a** EVA 20% + PE 80% (multi-scale electron micrographs of the composites and 2D FFT patterns / laser diffraction emulation patterns). **b** EVA 50% + PE 50% (multi-scale electron micrographs of the composites and 2D FFT patterns / laser diffraction emulation patterns). **c** EVA 90% + PE 10% (multi-scale electron micrographs of the composites and 2D FFT patterns/laser diffraction emulation patterns). **d** "Composite 20–80" Shelenkov's [189–191] know-how (multi-scale electron micrographs of the composites and 2D FFT patterns/laser diffraction emulation patterns). **e** "Composite 30–70" Shelenkov's [189–191] know-how (multi-scale electron micrographs of the composites and 2D FFT patterns/laser diffraction emulation patterns). **f** "Composite 40–60 (Shelenkov's [189–191] know-how) (multi-scale electron micrographs of the composites and 2D FFT patterns/laser diffraction emulation patterns). **g** "Composite 50–50" Shelenkov's [189–191] know-how (multi-scale electron micrographs of the composites and 2D FFT patterns/laser diffraction emulation patterns). **h** "Composite 60–40" Shelenkov's [189–191] know-how (multi-scale electron micrographs of the composites and 2D FFT patterns/laser diffraction emulation paterns). **i** "Composite 80–20" Shelenkov's [189–191] know-how (multi-scale electron micrographs of the composites and 2D FFT patterns/laser diffraction emulation patterns). **j** "Composite 90–10" Shelenkov's [189–191] know-how (multi-scale electron micrographs of the composites and 2D FFT patterns/laser diffraction emulation patterns). **k** Standard extruded polyethylene (PE) with voids/pores (multi-scale electron micrographs of the composites and 2D FFT patterns/laser diffraction emulation patterns)

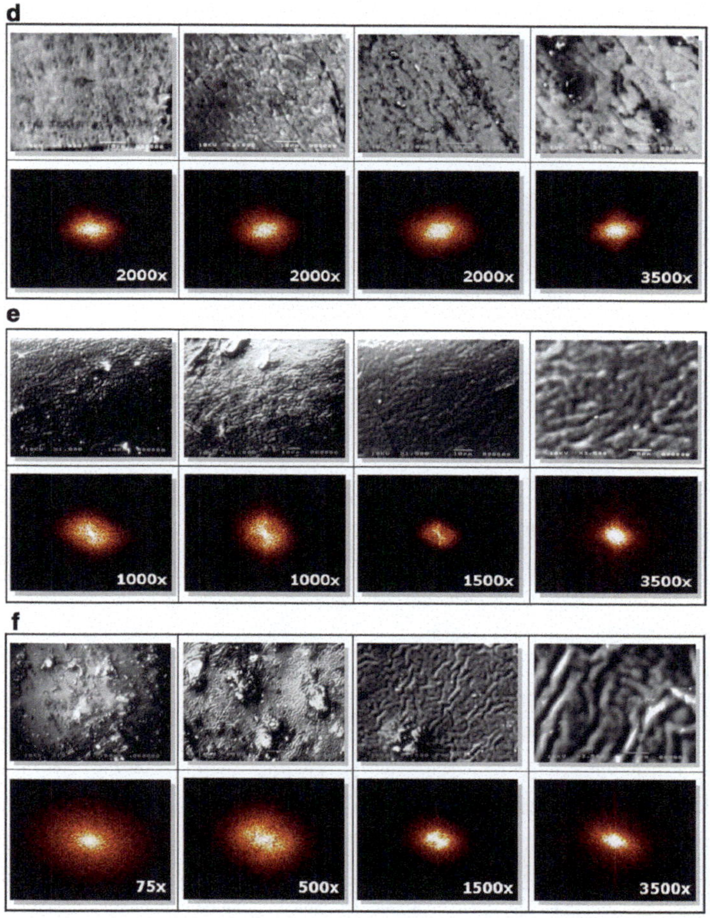

Fig. 2.5 (continued)

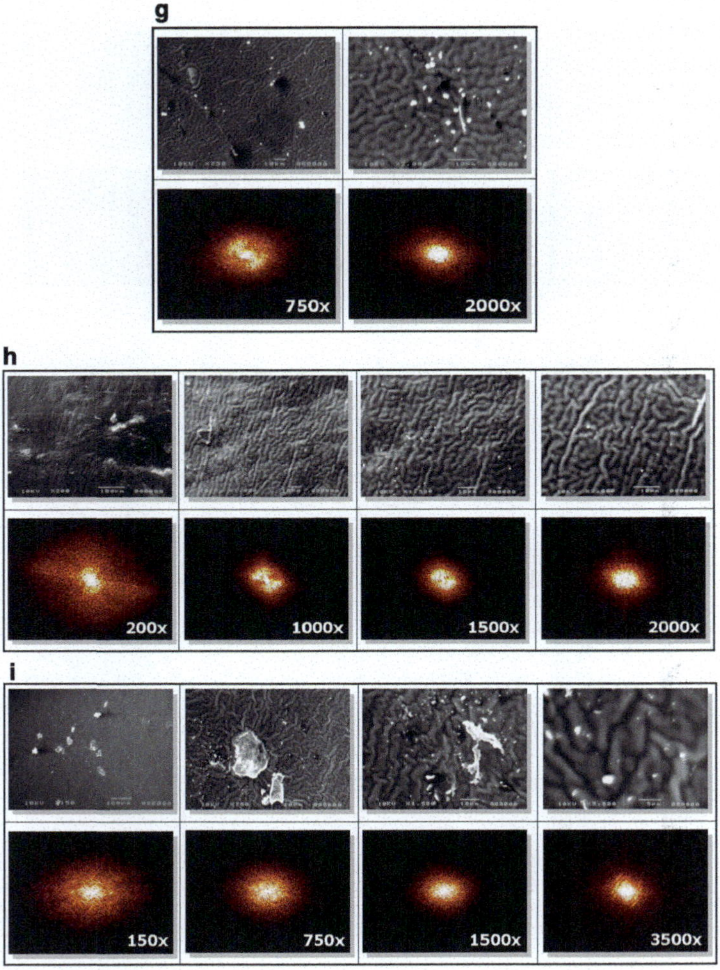

Fig. 2.5 (continued)

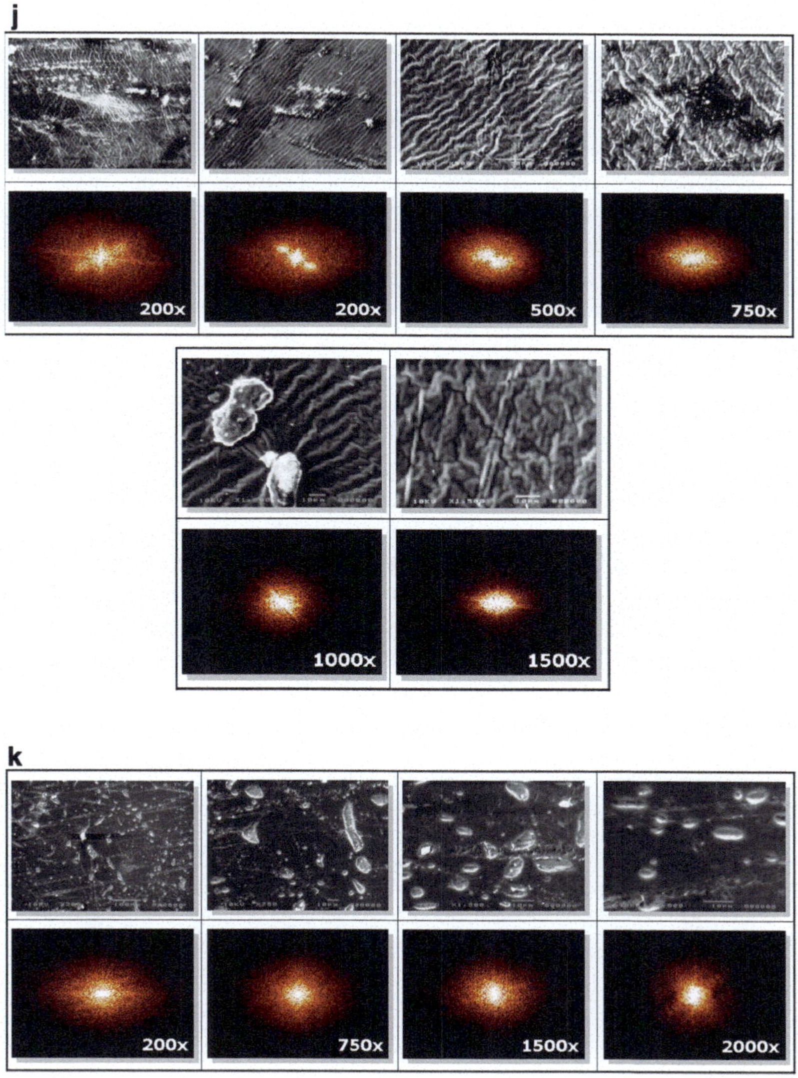

Fig. 2.5 (continued)

2.4 Summary

According to the data of electron microscopy and correlation spectral analysis, it can be seen that the samples differ quite strongly, depending on the content of the components. Differences in microroughness and sometimes in texture porosity indicate the possibility of cell localization on them. It is well known that the development and spread of cells depends on the texture of the surface on which they are located. Correlation-spectral analysis for these samples shows in some cases completely diffuse scattering, and in others the presence of an ordered texture. As you can see from the tables, this phenomenon can be observed at different magnification scales. When the magnification scales are optimal for the cellular level, and a texture [192, 193] good enough for cell proliferation is observed, this composite can be considered biocompatible in terms of textural criteria. In some cases, you can observe the regularity and periodicity of the texture, which is inherent in scaffolds. In some other cases, one can see grooves similar to those arising in open topologies of laboratories on a chip and microfluidic and nanofluidic biological devices [194–196]. Thus, focusing on the data we obtained and tables for correlation spectral analysis, any specialist in the field of regenerative medicine and cell biology can determine which of these structures of the composite and which of the ratios of the concentrations of components can be optimal for cell biology and medicine. In principle, it does not yet follow that only these concentrations and textures will be biologically effective and optimal, but this gives certainty in the sample statistics or the nomenclature of cell carriers. These studies and research investigations require continuation. You can create a machine learning database to automatically identify suitable samples for cell biology. It is possible to create a system for analog pattern recognition by diffractograms and two-dimensional Fourier spectra for the same purposes. The technical possibilities for this have existed for a long time, since such studies have been conducted since the 1970s, but they still remain actual/relevant [197–223]. Therefore, in our opinion, the further development of the topic is not frightening time (or resource) consuming.

Acknowledgements The authors express their gratitude to:

• Mikhail K. Filippov for reconstructing the old SEM device for its use in electrophysical and electrophysiological experiments (since 2018);

• Valentina N. Seliverstova, Alexander V. Kholyavko, Vladimir S. Kalinin, and Evgeniy N. Kulpin (Department of Experimental Technology of the Federal Research Center for Chemical Physics of the Russian Academy of Sciences) for the production of unique components/details for electron microscopes according to the drawings of Mikhail K. Filippov and Oleg V. Gradov.

References

1. Lim JI, Lee YK (2009) EVA-enhanced embedding medium for histological analysis of 3D porous scaffold material. Micron 40(7):756–760. https://doi.org/10.1016/j.micron.2009.04.002
2. Zhou S, Zhang L, Wang YY, Zuo Y, Gao SB, Li YB (2011) Fabrication of hydroxyapatite/ethylene-vinyl acetate/polyamide 66 composite scaffolds by the injection-molding method. Polym-Plast Technol Eng 50(10):1047–1054. https://doi.org/10.1080/03602559.2011.557824
3. Luo X, Zhang L, Morsi Y, Zou Q, Wang Y, Gao S, Li Y (2011) Hydroxyapatite/polyamide 66 porous scaffold with an ethylene vinyl acetate surface layer used for simultaneous substitute and repair of articular cartilage and underlying bone. Appl Surf Sci 257(23):9888–9894

4. Luo X, Tan S, Zhou X, Dong H (2014) Preparation and characterization of ethylene vinyl acetate/ hydroxyapatite/polyamide 66 composite scaffold for simultaneous substitute and repair for articular cartilage and underlying bone. Mater Rev 02 (Chinese abstract paper)
5. Parikh A, Anand U, Ugwu MC, Feridooni T, Massoud E, Agu RU (2014) Drug-eluting nasal implants: formulation, characterization, clinical applications and challenges. Pharmaceutics 6(2):249–267. https://doi.org/10.3390/pharmaceutics6020249
6. Chew SA, Danti S (2017) Biomaterial-based implantable devices for cancer therapy. Adv Healthcare Mater 6(2):1600766. https://doi.org/10.1002/adhm.201600766
7. Chaiwutthinan P, Chauyjuljit S, Thipkham N, Kowalski CP, Boonmahitthisud A (2019) Poly (lactic acid)/ethylene vinyl acetate copolymer blend composites with wood flour and wollastonite: physical properties, morphology, and biodegradability. J Vinyl Add Tech 25(4):313–327. https://doi.org/10.1002/vnl.21697
8. Powell EM, Sobarzo MR, Saltzman WM (1990) Controlled release of nerve growth factor from a polymeric implant. Brain Res 515(1–2):309–311. https://doi.org/10.1016/0006-8993(90)90612-F
9. Hoffman D, Wahlberg L, Aebischer P (1990) NGF released from a polymer matrix prevents loss of ChAT expression in basal forebrain neurons following a fimbria-fornix lesion. Exp Neurol 110(1):39–44. https://doi.org/10.1016/0014-4886(90)90049-X
10. Park CM, Hollenberg MJ (1993) Growth factor-induced retinal regeneration in vivo. Int Rev Cytol 146:49–74. https://doi.org/10.1016/S0074-7696(08)60379-4
11. Silberstein GB, Daniel CW (1987) Reversible inhibition of mammary gland growth by transforming growth factor-β. Science 237(4812):291–293. https://doi.org/10.1126/science.3474783
12. Daniel CW, Silberstein GB, Van Horn K, Strickland P, Robinson S (1989) TGF-β1-induced inhibition of mouse mammary ductal growth: developmental specificity and characterization. Dev Biol 135(1):20–30. https://doi.org/10.1016/0012-1606(89)90154-1
13. Daniel CW, Robinson S, Silberstein GB (1996) The role of TGF-β in patterning and growth of the mammary ductal tree. J Mammary Gland Biol Neoplasia 1:331–341. https://doi.org/10.1007/BF02017389
14. Daniel CW, Strickland P, Friedmann Y (1995) Expression and functional role of E-and P-cadherins in mouse mammary ductal morphogenesis and growth. Dev Biol 169(2):511–519. https://doi.org/10.1006/dbio.1995.1165
15. Coleman S, Daniel CW (1990) Inhibition of mouse mammary ductal morphogenesis and down-regulation of the EGF receptor by epidermal growth factor. Dev Biol 137(2):425–433. https://doi.org/10.1016/0012-1606(90)90267-M
16. Christenson L, Wahlberg L, Aebischer P (1991) Mast cells and tissue reaction to intraperitoneally implanted polymer capsules. J Biomed Mater Res 25(9):1119–1131. https://doi.org/10.1002/jbm.820250906
17. During MJ, Freese A, Sabel BA, Saltzman WM, Deutch A, Roth RH, Langer R (1989) Controlled release of dopamine from a polymeric brain implant: in vivo characterization. Ann Neurol Official J Am Neurol Assoc Child Neurol Soc 25(4):351–356. https://doi.org/10.1002/ana.410250406
18. Fung LK, Saltzman WM (1997) Polymeric implants for cancer chemotherapy. Adv Drug Deliv Rev 26(2–3):209–230. https://doi.org/10.1016/S0169-409X(97)00036-7
19. Saltzman WM, Radomsky ML (1991) Drugs released from polymers: diffusion and elimination in brain tissue. Chem Eng Sci 46(10):2429–2444. https://doi.org/10.1016/0009-2509(91)80036-X
20. Tamargo RJ, Rossell LA, Kossoff EH, Tyler BM, Ewend MG, Aryanpur JJ (2002) The intracerebral administration of phenytoin using controlled-release polymers reduces experimental seizures in rats. Epilepsy Res 48(3):145–155. https://doi.org/10.1016/S0920-1211(01)00330-8
21. Reinhard CS, Radomsky ML, Saltzman WM, Hilton J, Brem H (1991) Polymeric controlled release of dexamethasone in normal rat brain. J Control Release 16(3):331–339. https://doi.org/10.1016/0168-3659(91)90010-B

22. Bellinzona M, Roser F, Matthies C, Samii M, Saini M (2004) Biopolymer-mediated suramin chemotherapy in the treatment of experimental brain tumours. Acta Oncol 43(3):259–263. https://doi.org/10.1080/02841860310023129
23. Sanberg PR, Emerich DF, Aebischer P, Amisetti SM, Ouellette W, Koutouzis TK, Cahill DW, Norman AB (1993) Substance P containing polymer implants protect against striatal excitotoxicity. Brain Res 628(1–2):327–329. https://doi.org/10.1016/0006-8993(93)90974-R
24. Henry JC, McLean E, Mayer I, Donaldson EM (1998) Induction of precocious maturation in masculinized Atlantic salmon by treatment with sustained-release LHRHa and testosterone. Aquacult Int 6:261–268. https://doi.org/10.1023/A:1009255815481
25. Brown L, Munoz C, Siemer L, Edelman E, Langer R (1986) Controlled release of insulin from polymer matrices: control of diabetes in rats. Diabetes 35(6):692–697. https://doi.org/10.2337/diab.35.6.692
26. Zhang C, Easteal AJ, Edmonds NR, Liang G, Razzak M, Leech W (2010) In vitro and mechanism study of poly (ethylene-co-vinyl acetate)-based implant for sustained release of vitamin B12. Macromol Res 18:653–659. https://doi.org/10.1007/s13233-010-0711-x
27. Senuma A (2000) Gas permeability coefficients of ethylene-vinyl acetate copolymer-modified poly (dimethylsiloxane) membranes. Double-column approach for two-phase materials. Macromol Chem Phys 201(5):568–576. https://doi.org/10.1002/(SICI)1521-3935(20000301)201:5%3C568::AID-MACP568%3E3.0.CO;2-J
28. Selvin TP, Seno J, Murukan B, Santhosh AA, Sabu T, Weimin Y, Sri B (2010) Poly (ethylene-co-vinyl acetate)/calcium phosphate nanocomposites: thermo mechanical and gas permeability measurements. Polym Compos 31(6):1011–1019. https://doi.org/10.1002/pc.20887
29. Mousavi SA, Gholizadeh M, Sedghi S, Pourafshari-Chenar M, Barmala M, Soltani A (2010) Effects of preparation conditions on the morphology and gas permeation properties of polyethylene (PE) and ethylene vinyl acetate (EVA) films. Chem Eng Res Des 88(12):1593–1598. https://doi.org/10.1016/j.cherd.2010.03.013
30. Palakattukunnel ST, Thomas S, Sreekumar PA, Bandyopadhyay S (2011) Poly (ethylene-co-vinyl acetate)/calcium phosphate nanocomposites: contact angle, diffusion and gas permeability studies. J Polym Res 18:1277–1285. https://doi.org/10.1007/s10965-010-9530-1
31. Thomas PS, Sreekumar PA, Aprem AS, Thomas S (2011) Poly (ethylene-co-vinyl acetate)/calcium phosphate nanocomposites: mechanical, gas permeability, and molecular transport properties. J Appl Polym Sci 120(4):1974–1983. https://doi.org/10.1002/app.33238
32. Goodarzi V, Jafari SH, Khonakdar HA, Ghalei B, Mortazavi M (2013) Assessment of role of morphology in gas permselectivity of membranes based on polypropylene/ethylene vinyl acetate/clay nanocomposite. J Membr Sci 445:76–87. https://doi.org/10.1016/j.memsci.2013.04.073
33. Adelnia H, Bidsorkhi HC, Ismail AF, Matsuura T (2015) Gas permeability and permselectivity properties of ethylene vinyl acetate/sepiolite mixed matrix membranes. Sep Purif Technol 146:351–357. https://doi.org/10.1016/j.seppur.2015.03.060
34. Abdollahi M, Khoshbin M, Biazar H, Khanbabaei G (2017) Preparation, morphology and gas permeation properties of carbon dioxide-selective vinyl acetate-based Polymer/Poly (ethylene oxide-b-amide 6) blend membranes. Polymer 121:274–285. https://doi.org/10.1016/j.polymer.2017.06.033
35. Shirinia M, Abdollahi M, Omidkhah M (2020) Simultaneous enhancement of CO_2 permeability and CO_2/CH_4 and CO_2/N_2 selectivity via incorporating dense, rubbery and CO_2-philic vinyl acetate-based copolymers into poly (ethylene oxide-b-amide 6) membranes. React Funct Polym 154:104673. https://doi.org/10.1016/j.reactfunctpolym.2020.104673
36. Chowdary KPR, Babu JS (2003) Permeability of ethylene vinyl acetate copolymer microcapsules: effect of solvents. Indian J Pharm Sci 65(1):62–66
37. Zhen LWF, Zhao-Yin JIN (1991) Modification of ethylene-vinyl acetate copolymer and its effect on permeability of some steroids. Chin J Pharm 6 (bibliographically unidentified reprint)
38. Han JH, Suh KS, Kim SJ (1996) Ion permeability and physical properties of ethylene vinyl acetate copolymer/ethylene acrylic acid copolymer blends. Trans-Korean Inst Electr Eng 45:1282–1287

39. Urresti O, González A, Fernández-Berridi MJ, Iruin JJ, Irusta L (2011) Oxygen permeability through poly (ethylene-co-vinyl acetate)/clay nanocomposites prepared by microwave irradiation. J Membr Sci 373(1–2):173–177. https://doi.org/10.1016/j.memsci.2011.03.003
40. Hosseinkhanli H, Sharif A, Aalaie J, Khalkhali T, Akhlaghi S (2013) Oxygen permeability and the mechanical and thermal properties of (low-density polyethylene)/poly (ethylene-co-vinyl acetate)/organoclay blown film nanocomposites. J Vinyl Add Tech 19(2):132–139. https://doi.org/10.1002/vnl.20329
41. Nakagawa T (1974) Permeation, diffusion and solution of gases in films of ethylene-vinyl acetate copolymers and their hydrolyzed films. Nippon Kagaku Kaishi 1:147–154. https://doi.org/10.1246/nikkashi.1974.147
42. Suzuki F, Noguchi M, Onozato K, Takahashi N (1982) On the similarity of alkali saponification of ethylene-vinyl acetate copolymer in heterogeneous system with case ii type diffusion. Sen'i Gakkaishi 38(1):T28–T34. https://doi.org/10.2115/fiber.38.T28
43. Kagayama A, Mustafa R, Akaho E, Khawam N, Truelove J, Hussain A (1984) Mechanism of diffusion of compounds through ethylene vinyl acetate copolymers I. Kinetics of diffusion of l-chloro-4-nitrobenzene, 3, 4-dimethylphenol and 4-hexylresorcinol. Int J Pharm 18(3):247–258. https://doi.org/10.1016/0378-5173(84)90140-6
44. Maurin M, Hussain A, Dittert L (1987) Mechanism of diffusion through ethylene-vinyl acetate copolymers. 4. Effect of vinyl-acetate content of copolymer on diffusion and partition-coefficients. J Pharm Sci 76(11):S282–S282. PMID: 2726908
45. Maurin MB, Dittert LW, Hussain AA (1989) Mechanism of diffusion of monosubstituted benzoic acids through ethylene-vinyl acetate copolymers. Prog Clin Biol Res 292:279–282
46. Skirda L (1989) Study of the structure–morphology and diffusion properties of ethylene vinyl acetate copolymer and low density polyethylene. Vysokomol. Soedin., Kratk. Soobshcheniya 31(4):300–305. (in Russian)
47. Maurin MB, Dittert LW, Hussain AA (1992) Mechanism of diffusion of monosubstituted benzoic acids through ethylene-vinyl acetate copolymers. J Pharm Sci 81(1):79–84. https://doi.org/10.1002/jps.2600810116
48. Chen SX, Lostritto RT (1996) Diffusion of benzocaine in poly (ethylene-vinyl acetate) membranes: effects of vehicle ethanol concentration and membrane vinyl acetate content. J Control Release 38(2–3):185–191. https://doi.org/10.1016/0168-3659(95)00119-0
49. Joseph A, Mathai AE, Thomas S (2003) Sorption and diffusion of methyl substituted benzenes through cross-linked nitrile rubber/poly (ethylene co-vinyl acetate) blend membranes. J Membr Sci 220(1–2):13–30. https://doi.org/10.1016/S0376-7388(03)00175-3
50. Kumar SA, Kumaran MG, Thomas S (2006) Sorption and diffusion of arenes through poly (ethylene-co-vinyl acetate) membranes. J Mater Sci 41:4892–4900. https://doi.org/10.1007/s10853-006-0315-1
51. Moly KA, Bhagawan SS, George SC, Thomas S (2007) Sorption and diffusion of aromatic solvents through linear low density polyethylene–ethylene vinyl acetate blend membranes. J Mater Sci 42:4552–4561. https://doi.org/10.1002/polb.21262
52. Kumar SA, Kumaran MG, Thomas S (2007) Sorption and diffusion of normal alkanes through poly (ethylene-co-vinyl acetate) membranes. J Polym Sci, Part B: Polym Phys 45(17):2470–2480. https://doi.org/10.1002/polb.21262
53. Muralidharan MN, Kumar SA, Thomas S (2008) Morphology and transport characteristics of poly (ethylene-co-vinyl acetate)/clay nanocomposites. J Membr Sci 315(1–2):147–154. https://doi.org/10.1016/j.memsci.2008.02.013
54. Ramesan MT (2015) Effects of magnetite nanoparticles on morphology, processability, diffusion and transport behavior of ethylene vinyl acetate nanocomposites. Int J Plast Technol 19:368–380. https://doi.org/10.1007/s12588-016-9130-y
55. Ahn S, Kim EH, Lee C (2005) Diffusion-Ordered NMR Spectroscopy of Poly ([ethylene-co-vinyl acetate]-graft-vinyl chloride) in Solution. Bull Korean Chem Soc 26(2):331–333. https://doi.org/10.5012/bkcs.2005.26.2.331
56. Wu J, Tomba JP, Oh JK, Winnik MA, Farwaha R, Rademacher J (2005) Synthesis of dye-labeled poly (vinyl acetate-co-ethylene)(EVA) latex and polymer diffusion in their latex films. J Polym Sci, Part A: Polym Chem 43(22):5581–5596. https://doi.org/10.1002/pola.21029

57. Animitsa IE, Volosentseva LI, Zhukovsky VM (1999) Diffusion behaviour of 22Na in the poly (ethylene oxide)–poly (vinyl acetate) system. Solid State Ionics 119(1–4):245–249. https://doi.org/10.1016/S0167-2738(98)00510-4
58. Yang YS, Peng HY, Qi GR, Yang SL (2002) Diffusion mechanism of drug through semi-crystalline ethylene-vinyl acetate copolymer: Free volume theory. Acta Polymerica Sinica 2:242–244
59. Bobiak JP, Koenig JL (2005) Regions of interest in FTIR imaging applications: diffusion of nicotine into ethylene-co-vinyl acetate films. J Control Release 106(3):329–338. https://doi.org/10.1016/j.jconrel.2005.05.011
60. Bellili A, David N, Vandame B, Wang Q, Goutille Y, Richaud E (2012) Diffusion and solubility of mineral oils through ethylene-vinyl acetate copolymer. Polym Testing 31(2):236–247. https://doi.org/10.1016/j.polymertesting.2011.11.003
61. Richaud E, Bellili A, Goutille Y (2012) Diffusion of mineral oils in ethylene-vinyl acetate copolymer. AIP Conf Proc 1459:42–44. https://doi.org/10.1063/1.4738392
62. Jeon SJ, Lee DW (1995) Thermal diffusion and molecular weight calibration of poly (ethylene-co-vinyl acetate) s by thermal field-flow fractionation. J Polym Sci, Part B: Polym Phys 33(3):411–416. https://doi.org/10.1002/polb.1995.090330309
63. Haddock T, Fischer J (1968) Dielectric sealing of ethylene vinyl acetate copolymers. SPE J 24(4):41
64. Macknigh WJ, Tetrault R (1971) Dielectric relaxation in partially hydrolyzed ethylene-vinyl acetate copolymers. In: Abstracts of papers of the American chemical society, p 105
65. Baird ME, Houston E (1975) Dielectric behaviour of ethylene-vinyl acetate copolymers. Polymer 16(4):308–310. https://doi.org/10.1016/0032-3861(75)90177-9
66. Shokr FS, Al-Gahtany SA (2014) Dielectric properties of conductive polypyrrole/ethylene vinyl acetate copolymer blends. J Elastomers Plast 46(5):470–480. https://doi.org/10.1177/0095244312469960
67. Kaviarasi A, Kumari MLV, Anandhavadivel A (2018) Thermal mechanical and dielectric studies on thin films of ethylene vinyl acetate copolymer with varying concentration of vinyl acetate. Int J Chem Sci 16(4):296
68. Kundu PP, Choudhury RNP, Tripathy DK (1999) Influence of blend composition on the physical, flame retardancy, dielectric, aging, and solvent resistance properties of poly [ethylene (vinyl acetate)] and polychloroprene. J Appl Polym Sci 71(4):551–556. https://doi.org/10.1002/(SICI)1097-4628(19990124)71:4%3C551::AID-APP5%3E3.0.CO;2-7
69. Agoudjil B, Ibos L, Candau Y, Majesté JC (2008) A comparative analysis of dielectric, rheological and thermophysical behaviour of ethylene vinyl acetate/BaTiO3 composites. J Phys D Appl Phys 41(5):055407. https://doi.org/10.1088/0022-3727/41/5/055407
70. Rahaman M, Chaki TK, Khastgir D (2012) Consideration of interface polarization in the modeling of dielectric property for ethylene vinyl acetate (EVA)/polyaniline conductive composites prepared through in-situ polymerization of aniline in EVA matrix. Eur Polymer J 48(7):1241–1248. https://doi.org/10.1016/j.eurpolymj.2012.04.016
71. Rahaman M, Chaki TK, Khastgir D (2014) Polyaniline/ethylene vinyl acetate composites as dielectric sensor. Polym Eng Sci 54(7):1632–1639. https://doi.org/10.1002/pen.23714
72. Bertasius P, Macutkevic J, Banys J, Gaidukovs S, Barkane A, Vaivodiss R (2020) Synergy effects in dielectric and thermal properties of layered ethylene vinyl acetate composites with carbon and Fe3O4 nanoparticles. J Appl Polym Sci 137(24):48814. https://doi.org/10.1002/app.48814
73. Padhi S, Jena DP, Nayak NC (2020) Dielectric behaviour of ethylene vinyl acetate/cenosphere composites. Mater Today Proc 30:355–359. https://doi.org/10.1016/j.matpr.2020.02.693
74. George JJ, Bhadra S, Bhowmick AK (2010) Influence of carbon-based nanofillers on the electrical and dielectric properties of ethylene vinyl acetate nanocomposites. Polym Compos 31(2):218–225. https://doi.org/10.1002/pc.20788
75. Zheng LN, Hao CC (2012) Effect of carbon nanofibers incorporation concentration and orientation on dielectric properties in carbon nanofibers/ethylene-vinyl acetate copolymer composites. Qingdao Keji Daxue Xuebao(Ziran Kexue Ban)/J Qingdao Univ Sci Technol (Natural Science Edition) 33(5):468–473

76. Valentová H, Ilčíková M, Czaniková K, Špitalský Z, Šlouf M, Nedbal J, Omastová M (2014) Dynamic mechanical and dielectric properties of ethylene vinyl acetate/carbon nanotube composites. J Macromol Sci Part B 53(3):496–512. https://doi.org/10.1080/00222348.2013.846814
77. Pal K, Kim JK (2015) Enhanced percolation threshold and dielectric of multi-walled carbon nanotube reinforced ethylene-vinyl acetate: the role of fillers based capacitor applications. Adv Sci Eng Med 7(1):26–32. https://doi.org/10.1166/asem.2015.1653
78. Padhi S, Achary P, Nayak NC (2016) Dielectric behaviour of halloysite nanotubes filled ethylene-vinyl acetate copolymer vulcanizates. Adv Sci Lett 22(2):533–537. https://doi.org/10.1166/asl.2016.6848
79. Zhang T, Yang J, Zhang N, Huang T, Wang Y (2017) Achieving large dielectric property improvement in poly (ethylene vinyl acetate)/thermoplastic polyurethane/multiwall carbon nanotube nanocomposites by tailoring phase morphology. Ind Eng Chem Res 56(13):3607–3617. https://doi.org/10.1021/acs.iecr.6b04763
80. Bhuyan B, Roy A, Srivastava SK, Mittal V (2018) Multiwalled carbon nanotube/montmorillonite hybrid filled ethylene-co-vinyl acetate nanocomposites with enhanced mechanical properties, thermal stability, and dielectric response. Polym Eng Sci 58(7):1155–1165. https://doi.org/10.1002/pen.24677
81. Gaidukovs S, Zukulis E, Bochkov I, Vaivodiss R, Gaidukova G (2018) Enhanced mechanical, conductivity, and dielectric characteristics of ethylene vinyl acetate copolymer composite filled with carbon nanotubes. J Thermoplast Compos Mater 31(9):1161–1180. https://doi.org/10.1177/0892705717734603
82. Barkāne A, Gaidukovs S, Kajaks J, Platnieks O (2019) Dielectric properties of ethylene vinyl acetate copolymer composites filled with carbon nanotubes, graphene nanoplatelets and iron oxide nanoparticles. Key Eng Mater 800:195–199. https://doi.org/10.4028/www.scientific.net/KEM.800.195
83. Sohi NJS, Rahaman M, Khastgir D (2011) Dielectric property and electromagnetic interference shielding effectiveness of ethylene vinyl acetate-based conductive composites: effect of different type of carbon fillers. Polym Compos 32(7):1148–1154. https://doi.org/10.1002/pc.21133
84. Zhang YH, Bai BF, Chen JB, Shen CY, Li JQ (2010) Multifractal analysis of fracture morphology of poly (ethylene-co-vinyl acetate)/carbon black conductive composite. Appl Surf Sci 256(23):7151–7155. https://doi.org/10.1016/j.apsusc.2010.05.042
85. Shafeeq VH, Subash CK, Varghese S, Unnikrishnan GP (2020) Nanohydroxyapatite embedded blends of ethylene-co-vinyl acetate and millable polyurethane as piezoelectric materials: dielectric, viscoelastic and mechanical features. Polym Int 69(12):1256–1266. https://doi.org/10.1002/pi.6070
86. Ennawaoui C, Lifi H, Hajjaji A, Samuel C, Rguiti M, Touhtouh S, Azim C, Courtois C (2019) Dielectric and mechanical optimization properties of porous poly (ethylene-co-vinyl acetate) copolymer films for pseudo-piezoelectric effect. Polym Eng Sci 59(7):1455–1461. https://doi.org/10.1002/pen.25132
87. Buerger DE, Boyd RH (1989) Subglass relaxation processes: dielectric relaxation in vinyl acetate/ethylene copolymers. Macromolecules 22(6):2699–2705. https://doi.org/10.1021/ma00196a029
88. Smith GD, Liu F, Devereaux RW, Boyd RH (1992) Subglass relaxations: mechanical vs. dielectric relaxation in poly (vinyl acetate) and its ethylene copolymers. Macromolecules 25(2):703–708. https://doi.org/10.1021/ma00028a032
89. Safari Ardi M, McQueen DH (1996) Time-domain dielectric relaxation in polyamide 6, poly (vinyl chloride), ethylene-vinyl acetate copolymer and poly (vinylidene fluoride). J Mater Sci 31:357–362. https://doi.org/10.1007/BF01139152
90. Dash BK, Achary PGR, Nayak NC (2015) Dielectric relaxation behaviour of ethylene-vinyl acetate–exfoliated graphene nanoplatelets (xGnP) composites. J Mater Sci: Mater Electron 26:7244–7254. https://doi.org/10.1007/s10854-015-3351-2

91. Nayak NC, Dash BK, Parida BN, Padhee R (2018) Dielectric relaxation behavior of exfoliated graphite nanoplatelets filled ethylene vinyl acetate copolymer and ethylene propylene diene terpolymer blend. J Mater Sci: Mater Electron 29:1955–1963. https://doi.org/10.1007/s10854-017-8106-9
92. Rellick GS, Runt J (1986) A dielectric study of poly (ethylene-co-vinyl acetate)–poly (vinyl chloride) blends. II. Loss curve broadening and correlation parameters. J Polym Sci Part B: Polym Phys 24(2):313–324. https://doi.org/10.1002/polb.1986.090240208
93. Rellick GS, Runt J (1988) A dielectric study of poly (ethylene-co-vinyl acetate)-poly (vinyl chloride) blends. III. Direct current conductivity and electrode polarization. J Polym Sci Part B: Polym Phys 26(7):1425–1438. https://doi.org/10.1002/polb.1988.090260707
94. Leyva ME, Garcia FG, Soares BG (2012) Study of the blend of diglycidyl ether of bisphenol-A with copolymer of poly (ethylene-co-vinyl acetate)-g-poly (methyl methacrylate) by dielectric spectroscopy and differential scanning calorimeter. Matéria (Rio de Janeiro) 17:955–960. https://doi.org/10.1590/S1517-70762012000200002
95. Corrales TP, Laroze D, Zardalidis G, Floudas G, Butt HJ, Kappl M (2013) Dynamic heterogeneity and phase separation kinetics in miscible poly (vinyl acetate)/poly (ethylene oxide) blends by local dielectric spectroscopy. Macromolecules 46(18):7458–7464. https://doi.org/10.1021/ma4007158
96. Ma Z, Jiang P, Wang L, Yang J (2010) Effect of styrene–ethylene–butadiene–styrene and its synergetic effect with ethylene vinyl acetate on the mechanical, thermal, dielectric, and water-treeing behaviors of crosslinked polyethylene. J Appl Polym Sci 118(4):2350–2357. https://doi.org/10.1002/app.32068
97. Mujal-Rosas R, Orrit-Prat J, Ramis-Juan X, Marin-Genesca M, Rahhali A (2011) Study on dielectric, thermal, and mechanical properties of the ethylene vinyl acetate reinforced with ground tire rubber. J Reinf Plast Compos 30(7):581–592. https://doi.org/10.1177/0731684411399135
98. Lu Y, Lin Q, Ren W, Zhang Y (2015) Investigation on the preparation and properties of low-dielectric ethylene-vinyl acetate rubber/mesoporous silica composites. J Polym Res 22:1–10. https://doi.org/10.1007/s10965-015-0694-6
99. Datta SK, Bhowmick AK, Chaki TK, Majali AB, Despande RS (1996) Electron beam initiated modification of ethylene vinyl acetate using trimethylolpropane trimethacrylate. Polymer 37(1):45–55. https://doi.org/10.1016/0032-3861(96)81598-9
100. Datta SK, Chaki TK, Khastgir D (1996) Effect of electron beam radiation on mechanical and electrical properties of poly (ethylene-co-vinyl acetate). Die Angewandte Makromolekulare Chemie: Appl Macromol Chem Phys 238(1):105–117. https://doi.org/10.1002/apmc.1996.052380110
101. Sabet M, Hassan A, Wahit MU, Ratnam CT (2010) Mechanical, thermal and electrical properties of ethylene vinyl acetate irradiated by an electron-beam. Polym-Plast Technol Eng 49(6):589–594. https://doi.org/10.1080/03602551003652755
102. Harish Kumar HG, Mathad RD, Sarma KSS, Sanjeev G, Haramagatti CR (2011) Electron-beam irradiation effects on poly (ethylene-co-vinyl acetate) polymer. Radiat Eff Defects Solids 166(3):198–207. https://doi.org/10.1080/10420150.2010.538927
103. Sabet M, Soleimani H (2017) The impact of electron beam irradiation on low density polyethylene and ethylene vinyl acetate. IOP Conf Ser Mater Sci Eng 204:012005. https://doi.org/10.1088/1757-899X/204/1/012005
104. Craciun G, Ighigeanu D (2013) The mechanical properties of ethylene-vinyl-acetate copolymer/triallylcyanurate mixture cross-linked by electron beam and microwave processing. Revista de Pielarie Incaltaminte 13(1):49–60
105. Matsui T, Shimoda M, Osajima Y (1992) Mechanical changes of electron beam irradiated ethylene–vinyl acetate copolymer (EVA) film (I). Polym Int 29(2):85–90. https://doi.org/10.1002/pi.4990290204
106. Matsui T, Shimoda M, Osajima Y (1992) Structural analysis of electron beam irradiated ethylene-vinyl acetate copolymer (EVA) film (II). Polym Int 29(2):91–95. https://doi.org/10.1002/pi.4990290205

107. Dutta SK, Bhowmick AK, Mukunda PG, Chaki TK (1995) Thermal degradation studies of electron beam cured ethylene-vinyl acetate copolymer. Polym Degrad Stab 50(1):75–82. https://doi.org/10.1016/0141-3910(95)00125-6
108. Dutta SK, Bhowmick AK, Chaki TK (1996) Structure property relationship of ethylene vinyl acetate copolymer grafted with triallyl cyanurate by electron beam radiation. Radiat Phys Chem 47(6):913–926. https://doi.org/10.1016/0969-806X(95)00160-Y
109. Shaltout NA (2009) Effect of electron beam irradiation and ethylene vinyl acetate copolymer on the properties of NBR/HDPE blends. J Macromol Sci Part A 47(2):160–165. https://doi.org/10.1080/10601320903458671
110. Chattopadhyay S, Chaki TK, Bhowmick AK (2001) Structural characterization of electron-beam crosslinked thermoplastic elastomeric films from blends of polyethylene and ethylene-vinyl acetate copolymers. J Appl Polym Sci 81(8):1936–1950. https://doi.org/10.1002/app.1626
111. Datta SK, Chaki TK, Bhowmick AK (1996) Electron beam initiated grafting and crosslinking of ethylene vinyl acetate copolymer. Part-I: structural characterization. Rubber Chem Technol 69(1):120–129. https://doi.org/10.5254/1.3538352
112. Datta SK, Chaki TK, Khastgir D, Bhowmick AK (1996) Electron beam initiated grafting and crosslinking of ethylene vinyl acetate copolymer. Part II: Mechanical and electrical properties. Polym Polym Compos 4(6):419–428. https://doi.org/10.1177/096739119600400605
113. Wang B, Tai Q, Nie S, Zhou K, Tang Q, Hu Y, Song L (2011) Electron beam irradiation cross linking of halogen-free flame-retardant ethylene vinyl acetate (EVA) copolymer by silica gel microencapsulated ammonium polyphosphate and char-forming agent. Ind Eng Chem Res 50(9):5596–5605. https://doi.org/10.1021/ie102394q
114. Wang S, Zhang Y, Zhang Y, Zhang C, Li E (2004) Crosslinking of polyvinyl chloride by electron beam irradiation in the presence of ethylene–vinyl acetate copolymer. J Appl Polym Sci 91(3):1571–1575. https://doi.org/10.1002/app.13265
115. Sharif J, Dahlan KZM, Yunus WMZW (2007) Electron beam crosslinking of poly (ethylene-co-vinyl acetate)/clay nanocomposites. Radiat Phys Chem 76(11–12):1698–1702. https://doi.org/10.1016/j.radphyschem.2007.02.091
116. Manaila E, Stelescu MD, Niculescu-Aron GI, Gurau D (2013) Proving reliability and reproducibility of the technology for grafting and crosslinking blends based on ethylene vinyl acetate copolymer and triallyl cyanurate by electron beam irradiation, using statistical analysis of experimental data obtained. *Revista de* Pielarie Incaltaminte (Leather and Footwear Journal) 13(4):265–282. https://www.revistapielarieincaltaminte.ro/vol13_4.html
117. Gad YH (2009) Improving the properties of poly (ethylene-co-vinyl acetate)/clay composite by using electron beam irradiation. Nucl Instrum Methods Phys Res, Sect B 267(21–22):3528–3534. https://doi.org/10.1016/j.nimb.2009.08.010
118. Bee ST, Sin LT, Hoe TT, Ratnam CT, Bee SL, Rahmat AR (2018) Study of montmorillonite nanoparticles and electron beam irradiation interaction of ethylene vinyl acetate (EVA)/de-vulcanized waste rubber thermoplastic composites. Nucl Instrum Methods Phys Res, Sect B 423:97–110. https://doi.org/10.1016/j.nimb.2018.03.013
119. Munusamy Y, Ismail H, Mariatti M, Ratnam CT (2009) Effect of electron beam irradiation on the properties of ethylene-(vinyl acetate) copolymer/natural rubber/organoclay nanocomposites. J Vinyl Add Tech 15(1):39–46. https://doi.org/10.1002/vnl.20174
120. Hwang TY, Lee S, Kang PH, Park KH, Ahn Y, Lee JW (2011) The effect of electron beam irradiation on the dispersion and properties of poly (ethylene-co-vinyl acetate)/clay nanocomposites. Macromol Res 19:1151–1156. https://doi.org/10.1007/s13233-011-1104-5
121. Hamid F, Akhbar S, Ku Hamid KH, Faizal ARM (2014) Effect of electron beam irradiation on mechanical and thermal properties of ethylene vinyl acetate/polyamide 6/high density polyethylene nanocomposite. Appl Mech Mater 493:703–708. https://doi.org/10.4028/www.scientific.net/AMM.493.703
122. Mousavi SM, Aghili A, Hashemi SA, Goudarzian N, Bakhoda Z, Baseri S (2016) Improved morphology and properties of nanocomposites, linear low density polyethylene, ethylene-co-vinyl acetate and nano clay particles by electron beam. Polym Renew Resour 7(4):135–153. https://doi.org/10.1177/204124791600700402

123. Entezam M, Aghjeh MKR, Ghaffari M (2017) Electron beam irradiation induced compatibilization of immiscible polyethylene/ethylene vinyl acetate (PE/EVA) blends: mechanical properties and morphology stability. Radiat Phys Chem 131:22–27. https://doi.org/10.1016/j.radphyschem.2016.10.016
124. Sabet M, Hassan A, Ratnam CT (2013) Electron-beam irradiation of low density polyethylene/ethylene vinyl acetate blends. J Polym Eng 33(2):149–161. https://doi.org/10.1515/polyeng-2012-0132
125. Bee ST, Hassan A, Ratnam CT, Tee TT, Lee TS (2014) Interactions of montmorillonite and electron beam irradiation in enhancing the properties of alumina trihydrate–added polyethylene and ethylene vinyl acetate blends. J Compos Mater 48(10):1155–1171. https://doi.org/10.1177/0021998313484073
126. Ahmed J, Zhang Y (2021) Effect of electron beam irradiation on the thermal and mechanical properties of ethylene-vinyl acetate copolymer/polyamide blends. Polym Polym Compos 29(6):714–723. https://doi.org/10.1177/0967391120930108
127. Chattopadhyay S, Chaki TK, Bhowmick AK (2001) Development of new thermoplastic elastomers from blends of polyethylene and ethylene–vinyl acetate copolymer by electron-beam technology. J Appl Polym Sci 79(10):1877–1889. https://doi.org/10.1002/1097-4628(20010307)79:10%3C1877::AID-APP170%3E3.0.CO;2-B
128. Wang B, Song L, Hong N, Tai Q, Lu H, Hu Y (2011) Effect of electron beam irradiation on the mechanical and thermal properties of intumescent flame retarded ethylene-vinyl acetate copolymer/organically modified montmorillonite nanocomposites. Radiat Phys Chem 80(11):1275–1281. https://doi.org/10.1016/j.radphyschem.2011.06.008
129. Wang B, Zhou K, Wang L, Song L, Hu Y, Hu S (2012) Enhancement on physical properties of flame retarded ethylene-vinyl acetate copolymer/ferric pyrophosphate composites through electron beam irradiation. Compos Part B Eng 43(2):641–646. https://doi.org/10.1016/j.compositesb.2011.08.027
130. Wang B, Wang X, Shi Y, Tang G, Tang Q, Song L, Hu Y (2012) Effect of vinyl acetate content and electron beam irradiation on the flame retardancy, mechanical and thermal properties of intumescent flame retardant ethylene-vinyl acetate copolymer. Radiat Phys Chem 81(3):308–315. https://doi.org/10.1016/j.radphyschem.2011.10.021
131. Bee ST, Hassan A, Ratnam CT, Tee TT, Sin LT (2013) Investigation of nano-size montmorillonite on electron beam irradiated flame retardant polyethylene and ethylene vinyl acetate blends. Nucl Instrum Methods Phys Res, Sect B 299:42–50. https://doi.org/10.1016/j.nimb.2013.01.040
132. Jyo Y, Nozaki C, Matsuo M (1971) Electron microscopy of poly (vinyl chlorides) and ethylene-vinyl acetate rubber systems. Macromolecules 4(4):517–518. https://doi.org/10.1021/ma60022a033
133. Shim JS, Yeo JK (1977) Styrene-grafted ethylene-vinyl acetate copolymer ii. glass transition behavior via thermal and dynamic mechanical analysis, and morphology via electron microscopy. Polymer (Korea) 1(3):159. https://www.dbpia.co.kr/Journal/articleDetail?nodeId=NODE06122586
134. Hans PJ (1986) Electron Microscopy of Blends of Poly (p-vinyl Phenol) with Poly (vinyl Acetate) and Various Ethylene-co-vinyl Acetate Polymers (Doctoral dissertation, Pennsylvania State University)
135. Mitsuishi K (1997) Acoustic microscopy evaluation of polypropylene/ethylene vinyl acetate filled with calcium carbonate. Seikei-Kakou 9(6):434–441. https://doi.org/10.4325/seikeikakou.9.434
136. Mihaylova MD, Nedkov TE, Krestev VP, Kresteva MN (2001) Supermolecular structure of poly (propylene-co-ethylene)/poly (ethylene-co-vinyl acetate) blends irradiated with fast electrons. SEM, SAXS and DSC studies. Eur Polym J 37(11):2177–2186. https://doi.org/10.1016/S0014-3057(01)00118-5
137. De Souza CM, Tavares MIB (2002) High-resolution solid-state NMR and SEM study of the interaction behavior of poly (ethylene-co-vinyl acetate)/poly (vinyl acetate) blends. J Appl Polym Sci 86(1):116–124. https://doi.org/10.1002/app.10925

138. Sadeghi M, Mousavi SA, Hashemi MM, Chenar MP, Azad RR (2007) Effects of the preparation conditions on ethylene/vinyl acetate membrane morphology with the use of scanning electron microscopy. J Appl Polym Sci 105(5):2683–2688. https://doi.org/10.1002/app.26353
139. Czaniková K, Krupa I, Račko D, Šmatko V, Campo EM, Pavlova E, Omastová M (2015) In situ electron microscopy of Braille microsystems: photo-actuation of ethylene vinyl acetate/carbon nanotube composites. Mater Res Express 2(2):025601. https://doi.org/10.1088/2053-1591/2/2/025601
140. Sabet M, Hassan A, Ratnam CT (2012) Electron beam irradiation of low density polyethylene/ethylene vinyl acetate filled with metal hydroxides for wire and cable applications. Polym Degrad Stab 97(8):1432–1437. https://doi.org/10.1016/j.polymdegradstab.2012.05.010
141. Kummali MM, Alegria A, Miccio LA, Colmenero J (2013) Study of the dynamic heterogeneity in poly (ethylene-ran-vinyl acetate) copolymer by using broadband dielectric spectroscopy and electrostatic force microscopy. Macromolecules 46(18):7502–7512. https://doi.org/10.1021/ma4012522
142. Suzuoki Y, Muto H, Mizutani T, Ieda M (1987) The role of space charge in the electrical conduction of ethylene-vinyl acetate copolymers. J Phys D Appl Phys 20(8):1053–1058. https://doi.org/10.1088/0022-3727/20/8/011
143. Lee SH, Park JK, Han JH, Suh KS (1997) Space charge behaviour in maleic anhydride grafted polyethylene/ethylene-vinyl-acetate copolymer laminates. J Phys D Appl Phys 30(1):1–4. https://doi.org/10.1088/0022-3727/30/1/001
144. Suh KS, Kim JY, Kang CG (1995) Charge formation in polyethylene/ethylene vinyl acetate laminates. Polymer Korea 19:495–495
145. Li J, Xu H, Du B, Fu M, Hou S (2016) Space charge accumulation characteristics in ethylene-vinyl acetate copolymer filled with ZnO nanoparticles. In: 2016 IEEE international conference on dielectrics (ICD-2016), pp 180–183. https://doi.org/10.1109/ICD.2016.7547574
146. Kawamura H, Nawata M, Ieda M (1984) Treeing breakdown phenomena associated with the space charge formation in ethylene-vinyl acetate copolymers. Trans Inst Electr Eng Jpn A 104(2):81–88. https://doi.org/10.1541/ieejfms1972.104.81
147. Murakami Y, Hozuml N, Nagao M (2004) Electrical properties of condensed matter-surface temperature measurement and analysis of thermal breakdown with ethylene-vinyl acetate copolymer in room-temperature region. Japan J Appl Phys Part I: Regular Pap Short Notes 43(9):6184–6187
148. Nagao M, Sawa G, Ieda M (1976) Electrical breakdown of polyethylene, ethylene-vinyl acetate copolymer and their blend in low-temperature region. Jpn J Appl Phys 15(3):547–548. https://doi.org/10.1143/JJAP.15.547
149. Nagao M, Toyoshima S, Sawa G, Ieda M (1977) Dielectric breakdown of ethylene-vinyl acetate composit polymers in low-temperature region. Trans Inst Electr Eng Jpn A 97(12):617–622. https://doi.org/10.1541/ieejfms1972.97.617
150. Ieda M, Nagao M, Sawa G (1979) Dielectric breakdown of polyethylene and ethylene-vinyl acetate composite polymers. In: International conference on dielectric materials, measurements and applications, vol 3, pp 165–188
151. Cai G, Xu J (1991) Breakdown properties of polyethylene-ethylene vinyl acetate copolymer composite under liquid nitrogen temperature. In: Proceedings of the 3rd international conference on properties and applications of dielectric materials, pp 419–422. https://doi.org/10.1109/ICPADM.1991.172086
152. Tu D, Kan L, Kao KC (1988) Electric breakdown and space charges in ethylene-vinyl-acetate (EVA) incorporated polyethylene. In: Proceedings of the second international conference on properties and applications of dielectric materials, pp 598–601. https://doi.org/10.1109/ICPADM.1988.38468
153. Martínez-García A, Sánchez-Reche A, Gisbert-Soler S, Landete-Ruiz MD, Torregrosa-Maciá R, Martín-Martínez JM (2008) Durability of corona discharge and low-pressure air plasma-treated ethylene–vinyl acetate copolymers (EVAs). Int J Adhes Adhes 28(1–2):38–46. https://doi.org/10.1016/j.ijadhadh.2007.03.001

154. Dixon D, Morrison R, Lemoine P, Meenan BJ (2008) Long term effects of air dielectric barrier discharge treatment on the surface properties of ethylene vinyl acetate (EVA). J Adhes Sci Technol 22(7):717–731. https://doi.org/10.1163/156856108X305877
155. Varghese H, Bhagawan SS, Rao SS, Thomas S (1995) Morphology, mechanical and viscoelastic behaviour of blends of nitrile rubber and ethylene-vinyl acetate copolymer. Eur Polym J 31(10):957–967. https://doi.org/10.1016/0014-3057(95)00056-9
156. Peon J, Vega JF, Del Amo B, Martınez-Salazar J (2003) Phase morphology and melt viscoelastic properties in blends of ethylene/vinyl acetate copolymer and metallocene-catalysed linear polyethylene. Polymer 44(10):2911–2918. https://doi.org/10.1016/S0032-3861(03)00176-9
157. Khonakdar HA, Wagenknecht U, Jafari SH, Hässler R, Eslami H (2004) Dynamic mechanical properties and morphology of polyethylene/ethylene vinyl acetate copolymer blends. Adv Polym Technol J Polym Process Inst 23(4):307–315. https://doi.org/10.1002/adv.20019
158. Peeterbroeck S, Alexandre M, Jérôme R, Dubois P (2005) Poly (ethylene-co-vinyl acetate)/clay nanocomposites: effect of clay nature and organic modifiers on morphology, mechanical and thermal properties. Polym Degrad Stab 90(2):288–294. https://doi.org/10.1016/j.polymdegradstab.2005.03.023
159. Bhattacharyya AR, Ghosh AK, Misra A, Eichhorn KJ (2005) Reactively compatibilised polyamide6/ethylene-co-vinyl acetate blends: mechanical properties and morphology. Polymer 46(5):1661–1674. https://doi.org/10.1016/j.polymer.2004.12.012
160. Dikobe DG, Luyt AS (2007) Effect of poly (ethylene-co-glycidyl methacrylate) compatibilizer content on the morphology and physical properties of ethylene vinyl acetate–wood fiber composites. J Appl Polym Sci 104(5):3206–3213. https://doi.org/10.1002/app.26080
161. Cui L, Ma X, Paul DR (2007) Morphology and properties of nanocomposites formed from ethylene-vinyl acetate copolymers and organoclays. Polymer 48(21):6325–6339. https://doi.org/10.1016/j.polymer.2007.08.040
162. Ardhyananta H, Ismail H, Takeichi T (2007) Effects of organoclay loading and ethylene glycol on mechanical, morphology and thermal properties of ethylene vinyl acetate/organoclay nanocomposites. J Reinf Plast Compos 26(8):789–806. https://doi.org/10.1177/0731684407076722
163. Faker M, Aghjeh MR, Ghaffari M, Seyyedi SA (2008) Rheology, morphology and mechanical properties of polyethylene/ethylene vinyl acetate copolymer (PE/EVA) blends. Eur Polymer J 44(6):1834–1842. https://doi.org/10.1016/j.eurpolymj.2008.04.002
164. Dikobe DG, Luyt AS (2009) Morphology and properties of polypropylene/ethylene vinyl acetate copolymer/wood powder blend composites. Express Polym Lett 3(3):190–199. https://doi.org/10.3144/expresspolymlett.2009.24
165. Marchante V, Benavente V, Marcilla A, Martínez-Verdú FM, Beltrán MI (2013) Ethylene vinyl acetate/nanoclay-based pigment composites: Morphology, rheology, and mechanical, thermal, and colorimetric properties. J Appl Polym Sci 130(4):2987–2994. https://doi.org/10.1002/app.39422
166. Dutta J, Naskar K (2014) Investigation of morphology, mechanical, dynamic mechanical and thermal behaviour of blends based on ethylene vinyl acetate (EVA) and thermoplastic polyurethane (TPU). RSC Adv 4(105):60831–60841. https://doi.org/10.1039/C4RA07823C
167. Sabu T, Gupta BR, De SK (1987) Mechanical properties, surface morphology and failure mode of γ-ray irradiated blends of polypropylene and ethylene-vinyl acetate rubber. Polym Degrad Stab 18(3):189–212. https://doi.org/10.1016/0141-3910(87)90001-2
168. Valdman A (1998) HV-insulation of simple forms. IEEE Trans Power Deliv 13(3):849–854. https://doi.org/10.1109/61.686983
169. Giao TN, Jordan JB (1968) Modes of corona discharges in air. IEEE Trans Power Appar Syst 5:1207–1215. https://doi.org/10.1109/TPAS.1968.292211
170. Balagurov BY (2015) Electrophysical properties of composites. URSS, Moscow
171. Gradov OV, Gradova MA, Olkhov AA, Iordanskiy AL (2020) Charge propagation along the polymer fiber of polyhydroxybutyrate: is it possible to apply the cable model? Key Eng Mater 869:246–258. https://doi.org/10.4028/www.scientific.net/KEM.869.246

172. Orehov TC, Gradov OV (2016) Hybridization of COBAC, QSPR/QSAR and SBGN technologies: the unity of theory and practice for biomedical technique design and biochemical diagnostic information analysis. J Med Bioeng 5(2):128–132. https://doi.org/10.18178/jomb.5.2.128-132
173. Dubina VA, Fischenko VK, Konstantinov OG, Mitnik LM (2011) Integration of satellite data and ground based video data in the monitoring systems. Curr Probl Remote Sense Earth Space (Sovremennye problemy distantsionnogo zondirovaniya Zemli iz kosmosa) 8(3):214–222 [in Russian]
174. Goncharova AA, Fischenko VK (2011) QAVIS—program for quick image and video analysis. In: Proceedings of 8th open German-Russian workshop on pattern recognition and image understanding (OGRW-8 2011), p 85
175. Goncharova AA, Fischenko VK, Dubina VA (2012) [Use the express-analysis program QAVIS for the satellite monitoring] (Primenenie programmy jekspressanaliza izobrazhenij i video qavis v zadachah sputnikovogo monitoringa). Curr Probl Remote Sens Earth Space (Sovremennye problemy distantsionnogo zondirovaniya Zemli iz kosmosa) 9(3):293–298 [in Russian]
176. Fischenko VK, Goncharova AA, Dolgikh GI, Zimin PS, Subote AE, Klescheva NA, Golik AV (2021) Express image and video analysis technology QAVIS: application in system for video monitoring of Peter the Great Bay (Sea of Japan/East Sea). J Mar Sci Eng 9(10):1073. https://doi.org/10.3390/jmse9101073
177. Fischenko V, Mitnik L, Dolgikh G, Goncharova A, Zimin P, Klescheva N, Golik A (2022) QAVIS technology: measuring wave processes in coastal zones based on the analysis of internet video broadcast. In: IGARSS 2022–2022 IEEE international geoscience and remote sensing symposium. IEEE, pp 6801–6804. https://doi.org/10.1109/IGARSS46834.2022.9883487
178. Fishchenko VK, Goncharova AA, Zimin PS, Golik AV (2023) Application of the QAVIS program for remote monitoring of marine processes in coastal ocean zones based on the analysis of video broadcasts on the internet. Phys Geospheres XVIII:265–265. https://doi.org/10.34906/9785604968383.127
179. Gradov OV, Nasirov PA, Goncharova AA, Fischenko VK, Jablokov AG (2018) On-chip lensless holographic trichoscopy/trichometry technologies: microinterferometric, Fourier spectral and correlographic techniques for clinical trichology. Morphologia 12(2):7–21. https://doi.org/10.26641/1997-9665.2018.2.7-21
180. Gradov OV, Nasirov PA, Jablokov AG (2018) Lensless on-chip-hemocytometry with secondary processing of cell images in the framework of an unconventional photometric model. Photonics Russia 12(7):716–729. https://doi.org/10.22184/1993-7296.2018.12.7.716.729
181. Adamovich ED, Gradov OV, Nasirov PA (2018) Liquid metal microfluidics with morphometric monioring: from liquid metal antenn towards the radiofrequency polarography. Fund Probl Radioeng Dev Constr 18(3):650–653 (in Russian). https://doi.org/10.13140/RG.2.2.33474.63682
182. Gradov OV (2018) Towards the correlation-spectral multi-channel imaging. Central Eur J Zool 4(1):13–21
183. Gradov OV, Aleksandrov PL, Gradova MA (2019) — Study of mineral samples relevant for desert locations using software correlation spectral analysis of scanning electron microscopy registers: from 2D Fourier spectra to online analysis of statistics of integral spatial characteristics. Softw Syst Comput Methods 4:125–171. https://doi.org/10.7256/2454-0714.2019.4.31379
184. Gorshenev V, Gradov O, Gradova M (2019) Differential estimation of structural properties of biomimetic materials for tissue engineering using real time correlation spectral analysis and structure-preserving maps (morphisms) in category theory framework. Genes Cells 14(Supplement 1):68–69. https://doi.org/10.23868/gc122415
185. Gradov OV, Nasirov FA, Orekhov FK (2024) Towards robotized multi-axis lens-less holographic trichoscopy and trichometry: from five-axis arm-manipulator systems based on robotic

feodorov stages to quaternion metrological setups. J Robot Auto Res 5(1):01–10 (Reprint/Translation)
186. Orekhov FK, Rubtsova NN (2024) Application of digital correlation-spectral analysis for identification, classifiaction and separation of plant roots and plant root tissues: simple low-cost alternatives for laser diffractometer based 2D FFT pattern recognition techniques biogeosystem technique. 11(1):27–81. https://doi.org/10.13187/bgt.2024.1.27
187. Gradov O (2024) Spectrozonal and multispectral raster lens-less microscopy of aquatic organisms: towards new contact optical methods of environmental biotesting using standard model organisms in variable environments. Proc SPIE 13279:132791Y. https://doi.org/10.1117/12.3041302
188. Biryukov V, Prince A, Maklakova I, Gradov O (2024) Metallographic dpssl-assisted lensless microscopy of si-al systems using a multiangle planar analyzer based on a polarizing microscope rotating table and a digital image correlator/2d fft spectrometer. Proc SPIE 13279:1327927. https://doi.org/10.1117/12.3041314
189. Shelenkov PG, Pantyukhov PV, Popov AA (2020) Mechanical properties of superconcentrates based on ethylene-vinyl acetate copolymer and microcrystalline cellulose. Mater Sci Forum 992:306–310. https://doi.org/10.4028/www.scientific.net/MSF.992.306
190. Shelenkov P, Pantyukhov P, Popov A (2018) Highly filled biocomposites based on ethylene-vinyl acetate copolymer and wood flour. IOP Conf Ser Mater Sci Eng 369:012043. https://doi.org/10.1088/1757-899X/369/1/012043
191. Shelenkov PG, Pantyukhov PV, Popov AA (2021) Mechanical properties of bio-composites based on polymer blends of ethylene-vinyl acetate copolymer and polyethylene with natural fillers. Solid State Phenom 316:159–163. https://doi.org/10.4028/www.scientific.net/SSP.316.159
192. Gorchenev VN, Olkhov AA, Gradov OV, Gradova MA, Aleksandrov PL (2019) Emergent topological approach for integration of physical properties of solid and soft matter media in porous scaffolds and tissue-engineered constructs: case of PHB and HAp with multi-angle CLSEM visualization. Genes Cells 14(Supplement 1):69. https://doi.org/10.23868/gc122418
193. Gradov OV (2019) Multiangle laser porosimetry and electron beam porosimetry for scaffolds, decellularized extracellular matrix samples and tissue-like models, including ESEM-based and CLEM-based realizations. Genes Cells 14(S1):69–70. https://doi.org/10.23868/gc122421
194. Gradov OV, Gradova MA (2015) Reaction-diffusion optoelectronics based on dispersed semiconductors. J Phys Conf Ser 643:012072. https://doi.org/10.1088/1742-6596/643/1/012072
195. Gradov OV, Gradova MA (2018) Additive manufacturing of self-organized crystal structures as novel materials for sensor mesofluidic chips based on layer-by-layer growth and their microcrystallomorphological analysis. Comput Nanotechnol (1):172–174. https://doi.org/10.24410/enpg-y079
196. Gradov OV, Gradova MA (2018) Soft matter reaction-diffusion and ferrofluid patterns as dynamic microchannels for optoelectronic lab-on-a-chip with the field-controlled geometry/topology. Comput Nanotechnol (4):75–77. https://doi.org/10.24410/2brk-0y77
197. Vasilenko GI (1977) Holographic pattern recognition. Izd. Sovetskoe Radio, Moscow
198. Girnyk V, Kurashov V, Nakhodkin N(1982). Application of digital holographic filters for optimization of the coherent-optical pattern-recognition methods. 2. Experimental results. Opt Spek 52(6):1034–1039
199. Casasent D (1983) New research in holographic pattern recognition. Proc SPIE 353:6–11. https://doi.org/10.1117/12.933942
200. Korolev AE, Nazarov VN, Staselko DI (1986) High speed holographic pattern recognition in resonant atomic media. Tech Phys Lett 12(6):302
201. Glaser I, Cohen M, Weiss V, Bernstein H, Friesem AA (1987) Pattern recognition with incoherent holographic correlators. Proc SPIE 813:481–482. https://doi.org/10.1117/12.967358
202. Caulfield HJ (1987) Page oriented holographic memories and optical pattern recognition. Proc SPIE 754:74–77. https://doi.org/10.1117/12.939969

203. Kutzner J (1988) Holographic pattern recognition for interpretation of ultrasonic indications. Materialpruefung 30(11/12):379–381
204. Rabal HJ, Furlan WD, Garavaglia M (1988) Astigmatic holographic processor for local pattern recognition. Opt Commun 65(5):343–346. https://doi.org/10.1016/0030-4018(88)90099-5
205. Valera MSI, Kvasnik F (1988) The use of bleached holographic filters in optical pattern recognition. J Phys D Appl Phys 21(10S):S187–S190. https://doi.org/10.1088/0022-3727/21/10S/054
206. Thalmann R (1989) Book review: optical pattern recognition using holographic techniques (Electronic Systems Engineering Series). Opt Lasers Eng 11(3):217–219. https://doi.org/10.1016/0143-8166(89)90032-8
207. Ho HS, Lee SS (1990) Rotation invariant pattern recognition based on holographic associative memory. Proc SPIE 1319:424. https://doi.org/10.1117/12.34786
208. Mok FH, Stoll HM (1992) Holographic inner-product processor for pattern recognition. Proc SPIE 1701:312–322. https://doi.org/10.1117/12.138341
209. Hampp NA, Thoma R, Braeuchle CR, Oesterhelt D (1993) Real-time holographic pattern recognition with bacteriorhodopsin films. Proc SPIE 1732:260–270. https://doi.org/10.1117/12.140405
210. Von Bally G, Vukicevic D, Demoli N, Bjelkhagen H, Wernicke G, Dahms U, Gruber W, Sommerfeld W (1994) Holography and holographic pattern recognition for preservation and evaluation of cultural-historic sources. Naturwissenschaften 81(12):563–565. https://doi.org/10.1007/BF01140009
211. Mikaelian AL, Kiselyov BS, Novoselov BA, Okonov DE (1995) Photothermoplastic-based holographic memory for pattern recognition. Proc SPIE 2490:350–359. https://doi.org/10.1117/12.205793
212. Das PK, DeCusatis CM, Shoop BL, Litynski DM (1998) Acousto-photorefractive holographic interferometric correlator for progressive pattern recognition using wavelet transforms. Proc SPIE 3470:214–225. https://doi.org/10.1117/12.326864
213. Lu TT, Mintzer DT, Kostrzewski AA, Lin FSH (1996) Compact holographic optical neural network system for real-time pattern recognition. Opt Eng 35(8):2122–2131. https://doi.org/10.1117/1.600792
214. Chao TH, Zhou H, Reyes G (1999) High-density holographic memory and its applications to optical pattern recognition. Proc SPIE 3864:181–183. https://doi.org/10.1117/12.997668
215. Cai L, Liu HK (2000) Massive parallel optical pattern recognition and retrieval via a two-stage high-capacity multichannel holographic random access memory system. Opt Eng 39(2):401–408. https://doi.org/10.1117/1.602378
216. Gu C (2002) Holographic memory for high-density data storage and high-speed pattern recognition. Proc SPIE 4929:198–207. https://doi.org/10.1117/12.483214
217. Zhou Y, Tao SQ, Wang DY (2002) Pattern recognition algorithms based on volume holographic image-databases. Chin J Lasers 29(4):359–362
218. Xue Q, Tan W, Yan Y, He Q (2002) Wavelet invariant pattern recognition system based on the volume holographic correlator. Proc SPIE 4459:197–202. https://doi.org/10.1117/12.454021
219. Dubois F, Minetti C, Monnom O, Yourassowsky C, Legros JC, Kischel P (2002) Pattern recognition with a digital holographic microscope working in partially coherent illumination. Appl Opt 41(20):4108–4119. https://doi.org/10.1364/AO.41.004108
220. Guerrero RA (2004) Power law response of volume holographic pattern recognition to partial images. Opt Commun 239(4–6):303–310. https://doi.org/10.1016/j.optcom.2004.06.008
221. Sarkisov SS, Kukhtareva T, Kukhtarev NV, Curley MJ, Edwards V, Creer M (2013) Rapid detection of malignant bio-species using digital holographic pattern recognition and nanophotonics. Proc SPIE 8570:85700M. https://doi.org/10.1117/12.1000111
222. Skrynnik AA, Oganessian VA, Jablokov AG, Gradov OV (2018) System for semiautomatic tissue classification based on optical diffractometer for biopolymer structure analysis. Morphologia 12(3):164–171. https://doi.org/10.26641/1997-9665.2018.3.164-171
223. Kozhevnikov A, Gertz F, Dudko G, Filimonov Y, Khitun A (2015) Pattern recognition with magnonic holographic memory device. Appl Phys Lett 106(14):142409. https://doi.org/10.1063/1.4917507

Chapter 3
60 Years of Asymptotology—Preliminary Results. Or "in Memoriam?"

Igor V. Andrianov and Lelya A. Khajiyeva

Abstract In 1962, Martin Kruskal coined the term "asymptotology". As it is known, asymptotic methods, the essence of which is the study of phenomena in limiting cases, are at the heart of natural sciences of nature. It is now a well-developed part of pure mathematics, closely related to applied mathematics. Kruskal, in his lecture, drew the attention of the audience to the fact that, for all the diversity of asymptotic methods, it is possible to identify some common features in them and study them on their own, regardless of the objects to which they are applied. Such meta-science should constitute the subject of asymptotology. It is necessary to say directly that over the past 60 years such ideology, although it attracted (and still attracts) certain attention, has not become mainstream. Does this mean that Kruskal's idea is currently of purely historical interest? It should be emphasized that we are talking about asymptotology as a kind of meta-science. Asymptotic methods as the most important tool of scientific analysis are not threatened by Big Data and AI. In

---

Moliere has one of his characters observe that for more than forty years he has been talking prose without knowing it. It is doubtful that he benefited from the discovery, but I hope that you will be more fortunate and not disappointed in having by now discovered that asymptotology is what you have been practicing all along!
Kruskal [32]

It is appropriate to recall the opinion of Kolmogorov, who insisted that a full-fledged scientific work can consist of introduced concepts alone—with the presumption of their value.
Uspensky [51, *p. 464*]

In memory of R. G. Barantsev and L. I. Manevitch, adepts of asymptotology.

I. V. Andrianov (✉)
Chair and Institute of General Mechanics, RWTH Aachen University, Eilfschornsteinstraße 18, D-52062 Aachen, Germany
e-mail: igor.andrianov@gmail.com

L. A. Khajiyeva
Department of Mathematical and Computer Modelling, Al-Farabi Kazakh National University, Almaty 050040, Kazakhstan
e-mail: khadle@mail.ru

H. Altenbach and V. A. Eremeyev (eds.), *Dynamics of Discrete and Continuum Structures and Media*, Advanced Structured Materials 221,
https://doi.org/10.1007/978-3-031-75626-9_3

this paper we tried to analyze the opinions of prominent users of asymptotic methods about the future of asymptotology.

Keywords Asymptotology · Asymptotic method · Applied mathematics · Mathematical modeling · Analytical approach · Numerical approach

3.1 Introduction

60 years ago, Kruskal coined the term "Asymptotology" [30]. "*By asymptotology I mean something much broader than asymptotics, but including it; pending further elaboration, I would briefly define asymptotology as the art of dealing with applied mathematical systems in limiting cases* "[30]. (Here and below italicized are quotes from Kruskal's article). (Cf: "Approximations is an art, and famous names are usually associated with successful approximations" [50, p. 2]).

For Kruskal, the appeal to such generalizations was not accidental. "It is in the nature of science that most of its participants tend to specialize to such a degree that they forever sit in their own valleys and pan for specks of gold in the local river. Many never even raise their heads to observe the hills that flank their particular valley. Martin was one of the few who climbed those hills, explored other valleys and realized that the scientific disciplines form an inter-locking jigsaw whose picture tells a much bigger story. Not only did he make major contributions to the areas sketched in this memoir, he also had a lifelong interest in what he called 'asymptotology'" [11].

Over the years, the term "Asymptotology" has not appeared very often in the literature [4–14, 19, 23, 33, 34, 42, 55]. This gives us reason to wonder how viable this very concept is. Nevertheless, it is interesting and useful to reread and think over the article of a renowned scientist once again. "A classic conference paper by the great applied mathematician Martin Kruskal, a pioneer of solitons and much else. His topic in the paper is „Asymptotology " and his insights are wise and funny " (Strogatz St., from Internet). We have given ourselves that pleasure and here are the main points of the article with our comments.

> The first point to note here is that asymptotology is an art, at best a quasi-science, but not a science.

That is, it is not a description of specific techniques of asymptotic analysis, but a system of views on the analysis of problems of applied mathematics.

> First of all, Kruskal defines what he means by "applied mathematical systems".

The aspect of the definition of asymptotology just given which is most in need of explanation is the concept of applied mathematical system. An applied mathematical system is merely the mathematical description of a physical (or occasionally biological or other) system in which the variables expressing the state of the system are complete. The importance of formulating problems in terms of complete state variables constitutes a preliminary principle, not particularly of asymptotology but of applied mathematics in general, the Principle of Classification (or, perhaps better, of Determinism) .... This Principle is obviously closely related to the notion of a well posed problem emphasized by Hadamard. Its particular relevance to asymptotology comes about because only after one has singled out ("determined") an individual solution (or completely "classified" the family of solutions) can one reasonably inquire into its asymptotic behavior.

To our mind, applied mathematical systems have an important distinguishing feature. Since they are related to the description of the external world, and this description is always incomplete, such systems are usually asymptotic in nature. In other words, their formulation includes not only correctly defined boundary value problems, but also a set of asymptotic constraints and estimates. Whenever such system is changed, one should keep track of whether we remain within the original asymptotic assumptions. This is the fundamental difference between applied mathematical systems and pure mathematical systems [3].

Kruskal formulated 7 principles of asymptotology. We present them with our comments.

1. *The Principle of Simplification.*

A real simplification of the system should be pursued, allowing for an analytical solution or drastically reduced computing time.

2. *The Principle of Recursion.*

Having obtained a simplified solution, one has to consider the possibilities of its refinement by constructing higher approximations or by means of iterative methods.

3. *The Principle of Interpretation.*

It is a matter of introducing asymptotic integration parameter(s) into the system, according to which the analysis will be carried out. This is more art than science. M. Fedoryuk: "Let us try to find the asymptotic form of some (non-trivial) solution. It is necessary first of all to guess (and no other word will do) in what form to search for the asymptotics form. Of course, this stage—guessing the form of the asymptotic form is not subject to any formalization. Analogy, experiment, numerical simulations, physical considerations, intuition, "random" guesswork; these are the arsenal of means used by any research worker" [20, p. 143].

Nowadays, it can also be advised to use methods of computer algebra involving neural networks and AI [1], but it is unlikely that it will ever be possible to fully formalize the search for the form of asymptotics.

It is one of the most troublesome difficulties of asymptotological practice to find an appropriate asymptotic form. It is comforting that "If no parameters in the world were very large or very small, science would reduce to an exhaustive list of everything" [48]. So, seek and ye shall find. Remembering that "Unobvious choices of an expansion parameter can greatly extend the range and usefulness of perturbation series" [16]. (For examples of such choices, see [4, 5, 34]).

4. *The Principle of Wild Behaviour.*

Unusual behavior of the limit system may be a consequence of poor choice of parameters of asymptotic expansion or asymptotic procedure.

5. *The Principle of Annihilation.*

Which instructs us to find a complete set of annihilators of the terms which persist in the limit, apply them to the original system, and then go to the limit after multiplying by an appropriate function of EPS so that the now dominant terms persist in the limit.

It means that the limit system may be incorrect or have another degeneracy. In this case it is necessary to find a way to regularize it.

6. *The Principle of Maximal Balance.*

Partly as a consequence of Professor Friedrichs ' comment at the conclusion of my lecture, I now feel that "Minimal Simplification" is more appropriate here. (Footnote in [30], p. 18).

The system may include parameters whose orders are unknown in advance. The principle of minimal simplification suggests that they should be of the same order. In other words, one should not neglect any terms of the equations without sufficient reason. Another possibility is to investigate all possible relations of parameter orders and their corresponding simplified systems, i.e., to construct an "asymptotic portrait" of the system [5]. For this purpose, Kruskal proposed to use Newton's polygon (after Kruskal's paper, the term "Newton-Kruskal diagram" even appeared in some papers and book, i.e. [53]). In modern asymptotology, the method of Newton's polyhedron has become widespread due to the possibility of using computer mathematics for its construction [31, 46].

7. *The Principle of Mathematical Nonsense.*

If, in the course of an asymptotological analysis, a mathematically nonsensical expression appears, this indicates that the asymptotology has not been done correctly or at least not carried out fully (although even incomplete it may be satisfactory for one's purposes).

For instance, one should take into account the possibility of appearance of fictitious asymptotic solutions, which are not approximations of the exact ones [25].

3.2 From the Idealization Principle to Asymptotology

A key concept of Western science is *the principle of idealization.* Any real system or phenomenon consists of a large (often infinite) number of subsystems or less significant phenomena. An attempt to describe them in their entirety is hopeless если не отбросить несущественные при данном анализе детали, т.е. если не идеализировать систему. The principle of idealization was apparently first explicitly formulated by Galileo. Speaking about the motion of bodies, he wrote:

> Hence, in order to handle this matter in a scientific way, it is necessary to cut loose from these difficulties [air resistance, friction, etc.] and having discovered and demonstrated the theorems, in the case of no resistance, to use them and apply them with such limitations as experience will teach. [quoted in acc. with 28, p. 102].
>
> White or red, bitter or sweet, sounding or silent, sweet-smelling or evil-smelling are names for certain effects upon the sense-organs; they can no more be ascribed to the external objects than can the tickling or the pain caused sometimes by touching these objects. If ears, tongues, and noses are removed, I am of the opinion that shape, quantity, and motion would remain, but there would be an end of smells, tastes, and sounds, which abstractedly from the living creature, I take to be mere words [quoted in acc. with 28, pp. 101, 102].

The non-triviality of such an approach lies in the fact that, as mentioned by Kline [28, p.102],"The mathematical method of abstraction is indeed a step away from reality, but paradoxically, it leads back to reality with greater power than if all the factors actually present are taken into account at once" (see also [27]).

Apparently, the seeds of the idealization principle were laid down long before Galileo. "The ancient Greek philosophers displayed a remarkable, wholly original ability to see the world divided into objects, and objects into qualities. The non-triviality of this vision is difficult for us to appreciate now that our civilization's upbringing has largely accustomed us to it. ... Theoretical thinking came, of course, not from those who were able to see the connections (religio) in the world of matters and the interpenetration in the world of concepts, but from those who were able to gut, anatomize the world to find out what it had inside, and to investigate current concepts as unchangeable things. The brilliant Greeks invented length without width, space without time, atoms without qualities" [52].

In my opinion, Galileo's inestimable merit lies in the conscious application of the principle of idealization to the quantitative description of the real world.

The principle of idealization lies at the heart of the natural sciences, which is not always properly reflected in teaching. "According to Andronov, Mandelstam believed that issues of idealization should occupy a fundamental place in all teaching of physics, both in school and university education. Even a school student should realize that in any physical theory we work with ideal models of real things and processes" [43].

Ideal models provide a rough description of reality. To refine it, a *perturbation method* was developed. P. S. Laplace explained the idea of the method as follows: "The simplest way to analyze the various perturbations (*in the motion of the planets*—I.A.) is to imagine a planet moving along an ellipse, the elements of which change smoothly, and at the same time to imagine that the real planet oscillates around this imaginary line along a very small trajectory, the properties of which depend on its periodic perturbations" [32, p.141].

The idealization principle has become a key concept of *asymptotic mathematics.* Asymptotic formulas describe this phenomenon the more accurately the smaller (or larger) some system-describing parameters become. The method of idealization gives a solution in limiting cases when they are exactly equal to zero (or infinity). The search for and use of such parameters is one of the main problems of science.

H. Poincaré wrote about the search for parameters of asymptotic integration: "Scientists have tried to find it in the two extremes, in the infinitely great and in the infinitely small. The astronomer has found it because the distances of the stars are immense, so great that each of them appears only as a point and qualitative differences disappear, and because a point is simpler than a body which has shape and qualities. The physicist, on the other hand, has sought the elementary phenomenon in an imaginary division of bodies into infinitely small atoms, because the conditions of the problem, which undergo slow and continuous variations as we pass from one point of the body to another, may be regarded as constant within each of these little atoms." [38, p.19].

Let me remind that H. Poincaré and T. J. Stieltjes gave a mathematical definition of asymptotical series and thus justified many results of physicists and mechanical engineers, which had been considered as non-strict before that.

The development of asymptotic mathematics led to the realization that "*Asymptotic description* is not only a convenient tool in the mathematical analysis of nature, *it has a more fundamental significance* " [21]. Kruskal generalized this point of view and proposed the concept of *asymptotology* [30] as a set of general principles for asymptotic analysis of mathematical models of natural sciences.

Besides the principle of idealization, the foundation of Western science is based on the principle of hierarchy. This is not surprising, since "In accordance with the monotheistic principle, Christianity (regardless of the desire of the nurtured) establishes in the consciousness the principle of hierarchical construction of nature, which requires continuous completion of the system of interrelated parts of knowledge around some semantic center" [52].

The combination of the principles of idealization and hierarchy leads to reductionism. «The reductionist hypothesis may still be a topic for controversy among philosophers, but among the great majority of active scientists I think it is accepted without question» [2]. At least this is true in physics, although even here there are some indications of the limitations of such an approach [2]. This is especially valid for biology and social sciences. "Classical art, like photography, insisted on the principle of a detailed image, while modern art tends, abstracting from details, to operate with symbols, thus emphasizing the most essential thing in the subject. Both of these principles are represented in science. Modern fashion undoubtedly prefers to delve deep into the subject, increasing the degree of accuracy of the tools used. This method is extremely effective, but in the unrestrained pursuit of details, you can lose sight of the entirety." [45].

Perhaps arguments about reductionism are best left to biologists and philosophers, but the principles of idealization and hierarchy seem to me to be integral to asymptotology.

3.3 Manevitch's Asymptotic Principles [4, 5, 7, 8]

Based on engineering and research experience and relying on Kruskal's principles, L. I. Manevitch formulated the following heuristic recommendations. In our opinion, they successfully reflect the essence of the asymptotic paradigm.

1. In any system there is a parameter (or it can be "entered by hand"), by which it is expedient to carry out the decomposition. Preferably, it should have a clear physical meaning. (Cf: "The applied mathematician has the technical ability to use asymptotic methods when the occasion demands. More important, he possesses the power of formulating problems so that asymptotic methods can be used when the occasion *permits*" [44]).
2. Asymptotic mathematics includes both asymptotic analysis (i.e., decomposition by certain parameters) and synthesis (use of summation and interpolation methods [5–8, 47].
3. Every effective model of applied mathematics is asymptotic in nature. "The Book of Nature is written asymptotically." Without considering the asymptotic constraints behind the model, any modification of the model may lead to erroneous conclusions ("surprises" [35–37] or paradoxes [14, 15]). To evaluate the accuracy of an approximate model, it is important to understand its place in the hierarchy of asymptotics.
4. The principle of asymptotic additivity is fulfilled: if the system has a meaningful asymptotics at small values of some parameter, then a reasonable asymptotics can be constructed at large values of this parameter (and vice versa).

3.4 Experts' Views on the Present and Future of Asymptotology

The term "soliton" coined by Kruskal (together with N. Zabuski) [54] defined a whole epoch in nonlinear physics. The fate of "asymptotology" was less successful.

We decided to discuss asymptotology with experts working in the fields of pure and applied mathematics, mechanics, biology, physics, chemistry, as well as those who utilize asymptotic mathematics in their research. Andrianov sent out them the following letter.

Dear Colleague,

I would like to discuss the following question with you.

This year marks the 60th anniversary of Kruskal's report, at which he introduced the term "asymptotology" [30]. Consistent supporter of asymptotology was Leonid Manevitch, which I noted in a paper [4]. He made me an enthusiast of A. too, so we even wrote a book on this topic [7, 8]. In the article dedicated to the 40th anniversary of the concept [19], it was expressed the opinion that all this is of purely historical interest. Like, computer revolution, AI, big data technologies reduce the value of A. to zero.

I certainly do not agree with this point of view. But let's look at the situation realistically. The concept of A. is used explicitly very rare. Maybe A.'s place really is in books on the history of physics and mechanics and in Kruskal's biography? What is your opinion on this matter?

The received insightful answers provide serious food for thought.

Professor G. Rega

Thank you for your message and for kindly involving me in a scientific discussion. And, of course, many thanks also for providing me with the main related material, mostly your joint book with L. Manevitch on Asymptotology [8], which I was only aware of from cited references. I have just browsed through it and I will certainly look at it with pleasure, which of course will need some time.

As you know, I have worked quite a lot with asymptotic methods (in particular with multiple time scales), however essentially as an aware (though possibly clever) user of them for solving mechanical (and also coupled) problems.

As far as I understand, in Kruskal 1962 paper [30], asymptotology is proposed as a "special branch of knowledge, intermediate, in some sense, between science and art" (as referred to in the Introduction of your book), via the formulation of a number of 'scientific' principles, thereby not only encompassing asymptotic methods but acquiring the status of a higher and well-identified disciplinary framework.

I am not able (at least presently) to say whether this is indeed a fully sustainable 'scientific' position (nourished by a certainly intriguing and fashionable "vision") or something a bit too ambitious, even though all the way recognizing that the rationale and originality of asymptotology stands just in its being somehow at the crossroad between 'art'—which is certainly present in asymptotics—and 'science'.

In this respect, it seems to me that one should distinguish between asymptotics (as a rich and widely used collection of methods) and asymptotology (as the above-mentioned more general framework). There might be some doubts about the full 'scientific' status of the latter, but there is definitely no doubt about the scientific status of the former, even though various aspects of its implementation (likely connected with the 'artistic' side of their nature) can be considered questionable and have been actually the object of discussion and possible criticisms (among others, I remember some general pitfalls noticed by Ferdinand Verhulst [53,A] in the formulation and implementation of the multiple scale method versus the averaging one). But this does not mean that multiple scale (and averaging too) is not—and will indeed continue to be—extremely useful and successful in an enormous variety of dynamical situations—if not even always—for getting an overall understanding and reliable description of the basic aspects of some main nonlinear phenomena, to be of course complemented with/verified through computational and/or experimental investigations.

The situation could be partially different as regards asymptotology, considered as a more general disciplinary framework.

I don't know if you agree with the above distinction, at least in part, and, in such a case, it's not clear to me whether your concerns refer to asymptotics or to asymptology. In any case, unless I missed it, I did not find in the Dewar 2002 paper [19] the (neither explicit nor, I would say, implicit) opinion that "all this is of purely historical interest. Like, modern computers, artificial intelligence, big data reduce the value of A. to zero", as you are mentioning. It seems to me that he basically pointed out how asymptotology "can lead to erroneous conclusions if misapplied", this being possibly due to some 'art' content involved in its conceptual status and/or to the improper or incorrect satisfaction of some of its underlying principles.

This is all I can say on the matter, for the moment, hoping to have somehow contributed (at least a little bit) to the discussion.

I do agree with you that meaningful physical theories and correct engineering schemes are asymptotic in nature, as is the process of our thinking, even though in this last respect we might be somehow worthily conditioned by our scientific education.

Professor A. V. Shatrov

Perhaps, the complexity of the definition and formation of asymptotology is due to its interdisciplinarity. A. M. Il'in written [24] "The idea that the asymptotic analysis includes two basic steps has gaine wide acceptance. The first is the actual construction of the asymptotics. One has to choose the form in which the formal asymptotic solution (f.a.s) is to be sought, and specify the way of constructing this f.a.s.

The second step includes the justification of the constructed asymptotics, i.e., a proof that the f.a.s. obtained is indeed an asymptotic expansion of the solution of the problem. …

The first part, i.e., the construction of the asymptotics, is certainly of interest to experts in many different areas, e.g., physisits, engineers or anyone who has to deal with large or small parameters in his or her problems, while the second part is mainly of interest to a much narrow community of pure mathematicians ".

Professor F. Verhulst

I have to confess that I do not like much the term Asymptotology, it reminds me of astrology and other fake-sciences. I prefer Asymptotics or Asymptotic Theory. I know this trend to do away with analysis and use numerics only. That is the way for lazy and ignorant people. I am preparing a paper for Nonlinear Dynamics where I address this matter in the last section, Conclusions [53, B]. Of course, more politely than formulated here.

And yes, you can understand Asymptotology as a way of understanding reality. But then your interpretation of nature has a flavor of synthesis. First Cartesian observation from the ego, then modeling what we observe and see in experiments and then the analysis of the models. At least 3 layers. Asymptotology would integrate this?

About the attitude of engineers. I called it a trend to restrict to numerics. Of course the engineers who want to understand will look deeper at problems. In fact, my conclusions were written out of maybe too much irritation.

Professor V. Mityushev

I do not agree with the enforced point of view that A.'s conception has to be placed in the history of physics and mechanics. I have the arguments which can be shortly outlines in the following way:

1. Dull computations may be based and properly performed in the framework of the asymptotic theory only. This concerns the theory of homogenization which allows to properly define the effective constants. Other pure dull numerical methods can lead to a methodological mistake.
2. The possibilities of computers are sometimes exaggerated. It is not based on real computational experiments. One can see publications where a general computational scheme is written, a simple example is given and it is declared that everything can be calculated. Such a "modern scientific opinion" looks reasonable for non-specialists. After the movie Matrix one can think that the World can be placed in a computer and calculated https://www.youtube.com/watch?v=JOI opqMhhzc.

   But nobody thinks that the ratio of the sizes of the World to the size of such a computer is reduced to zero.
3. Dull computations are just an element of the fashion trends. If one calculates 2 $\times$ 2 by means of a computer and prepares a nice picture, it looks modern and cool.

Everybody knows (at least who can read and understand a popular medical text) that kids and others shouldn't drink Coca Cola for their health. But advertising everyday tells us to drink it. I don't drink Coca Cola at all, but I am forced to live in the Coca Cola world. So, let them compute everything and drink their Coca Cola. I shall follow asymptotic and other mathematical methods using computational tools at their proper place and usage.

Professor A. N. Gorban

About the terms.

(1) The new interdisciplinary term aims to build a bridge between research activities in different fields with intrinsic logical/methodological affinities.
(2) It must resist the natural centrifugal tendency for researchers to encapsulate themselves in "communities of mutual citation" in order to ensure their sustainability. Such communities often arise "naturally": the world is divided into "own-relevant", "foreign-relevant" and "irrelevant". A new term should either break down some of these barriers or, on the contrary, serve as a germ for the formation of a new community.
3) After the introduction, a game of "snowball" begins—someone sticks, someone falls off. If there is enough sticking, the term takes root.

(4) Some snowballs first proliferate, then dissolve (synergy—has come full circle before our eyes, another example is self-organized criticality).
(5) Some reach a plateau—and a new scientific discipline is institutionalized. Interestingly, this requires first of all a clearly articulated MAIN PROBLEM, which the new discipline is associated with. Then comes the unity of methods (not strictly necessary), object (also not strictly necessary), etc.
(6) Asymptotology was not given a problem to solve. And a stable community rarely arises without a specific purpose, and the main problem itself is not clearly defined.
(7) Maybe you should try to become a classic by setting the MAIN PROBLEM of asymptotology. With examples, etc. it would be a good way to make your mark.

Professor I. Elishakoff

Asymptotology is here to satay so far as there are small or big parameters. You ought to continue effort on its propagation and dissemination, as one of the "last" Mohicans.

Professor B. Dewar

I sometimes fear asymptotology is a dying art; but I hope your book will help keep the flame alive.

Professor V. Pilipchuk

I think Kruskal's asymtotology is not associated with any asymptotics in the formal sense at all. Rather, he meant the idea of additionality—look for extremes, they are always simpler.

Professor S. Gluzman

I suggest paying great attention to the Manevitch 's programme. Experience shows that it works, at least as a general methodology to follow. However, even telling what to look for is not only enlightening but "profitable"'. It appears that following such a methodology is very useful in practice [22, 24].

One can argue that there are some other valuable ideas advanced by V. I. Yukalov, such as self-similarity and conditions of convergence. They do help to implement the programme of Manevitch in concrete terms. When imposed on asymptotic expansions we arrive at accurate approximate formulas. And one should not forget about the old workhorse of Pade approximations, especially to look for the ways to employ them in the context of long truncations in combination with other approximants.

Conclusion

The bottom line is that asymptotic methods are essential in physics, mechanics, and engineering, and the development of computer mathematics does not pose a threat to them.

Moreover, computer algebra opens up new possibilities for asymptotic mathematics. But what, in fact, is the matter with asymptotology? As Prof. Gorban noted, it is appropriate to compare asymptotology with cybernetics.

In 1958, A. N. Kolmogorov asserted in an article in the Great Soviet Encyclopedia [28]: " The question of whether cybernetics has a right to exist as an independent scientific discipline is reduced to the question of whether the general properties of these processes in machines, living organisms, and their associations can be the subject of a sufficiently meaningful unified theory. This question should be answered with full certainty in the affirmative, although only the first steps have been taken in the direction of the systematic construction of cybernetics."

In the case of cybernetics, the great mathematician seems to have been wrong. But can the general properties of asymptotic methods be the subject of a sufficiently meaningful unified theory? In 1976, R. G. Barantsev answered in the affirmative [11]: "Despite the diversity of asymptotic methods, some common features can be identified in them and studied independently of the objects to which they are applied. The science of asymptotic methods itself should be the subject of asymptotology. It does not exist yet." He also proposed to define asymptotic ideology through the triad of *accuracy-locality-simplicity* [7, 8, 13]. In other words, the antagonistic conflict between "accuracy-simplicity" is resolved through "locality"—simple asymptotic models are only valid in a narrow range of changing the parameters. I think this way of posing the question has a deep meaning. However, the idea has not yet received serious development.

Asymptotology as a science at the moment does not exist. So far (?) it remains at the level of quasi-science (see Kruskal's definition).

Sometimes the question of the existence and importance of asymptotology is replaced by the question of the importance of the asymptotic methods themselves. Let us emphasize once again that when we talk about asymptotology, we mean a kind of metascience. No one questions the importance of asymptotic methods in science and engineering [13, 18, 29]. Asymptotic methods are the main tools of analysis in the natural and technical sciences. Development of IT do not devalue them, but lead to new opportunities for synergy of asymptotic mathematics with ANN [1], AI and Big Data. Asymptotic mathematics, among others features, is a set of heuristic guidelines, which are very useful to the applied mathematician and indispensable when teaching a physicist, mechanic, or engineer [7, 8, 18, 21, 29, 42–44]. But can *asymptotology* exist as an *independent scientific discipline*?

The interesting and fruitful idea connection between asymptotology and Pólya's heuristics [39–41] was pointed out by Babich [9]. In our opinion, this idea is worthy of detailed thought.

The rules of asymptotology do not give a recipe for solving a particular problem. But they remind that the obtained asymptotic representation is not unique and they initiate the search for an asymptotically equivalent function with a wider applicability area. Or they will prompt to reveal on what asymptotic simplifications (possibly implicit) the used model is based. This, in turn, will help to avoid any surprises when changing it [3].

A few words about the word “asymptotology”, which looks slightly overloaded. G. G. Malinetsky noted that the term “asymptology” sounds more natural [33]. This is certainly true, but the term was initiated by the original Kruskal’s paper [30], now and it is difficult to change it. It remains to be consoled by the fact that there are many examples of unfortunate scientific terms.

The problem of precise definition of general concepts is typical not only for asymptotology. For example, what is the “theory of nonlinear oscillations”? M. Cartwright wrote on this subject [17]: “Since about 1939 my researches have been, in collaboration with Professor J. E. Littlewood, mainly in a field of mathematics sometimes described as non-linear mechanics, but as most of the equations in mechanics are non-linear, it is better described as non-linear vibrations or oscillations. It is rather a curious branch of mathematics developed by different people from different standpoints, straight mechanics, radio oscillations, pure mathematics and servomechanisms of automatic control theory. ...The problems vary from practical problems which can be solved by fairly straightforward approximation by linear equations to highbrow problems in topology and the theory of sets of points. *There is probably no one competent to treat the subject adequately in all these aspects, but we try to exchange problems, ideas and method”* (italics are ours—authors).

Maybe the true value of asymptotology as some general concept can be described by the last phrase?

Acknowledgements We are deeply grateful to Professors B. Dewar, I. Elishakoff, S. Gluzman, A.N. Gorban, V. Mityushev, V. Pilipchuk, G. Rega, A.V. Shatrov, F. Verhulst, who provided written answers to our questions.

We also thank Professors O.V. Gendelman, J. Kaplunov, Yu.V. Mikhlin, A.I. Zemlynukhin for valuable discussions. These discussions were all the more valuable to us because very often opinions of our colleagues were directly opposite to ours. Discussio mater veritas est.

Author Contributions Conceptualization, I. A.; methodology, I. A. and L. K.; investigation, I. A. and L. K.; writing and original draft preparation, I. A. and L. K.; writing, review, and editing, I. A. and L. K.; formal analysis, I. A. and L. K. All authors have read and agreed to the published version of the manuscript.

Funding This research is funded by the Science Commitee of the Ministry of Science and Higher Education of the Republic of Kazakhstan (Grant No. AP23490543).

Institutional Review Board Statement: Not applicable.

Informed Consent Statement: Not applicable.

Data Availability Statement: Data are contained within the article.

Conflicts of Interest: The authors declare no conflict of interest.

References

1. Abdusalamov R, Kaplunov J, Itskov M (2023) Discovering asymptotic expansions for problems in mechanics using symbolic regression. Mech Res Comm 133(1):104197
2. Anderson PW (1972) More is different. Broken symmetry and the nature of the hierarchical structure of science. Science 177(4047):393–396

3. Andrianov IV (2021) Mathematical models in pure and applied mathematics. In: Abramyan A, Andrianov I, Gaiko V (eds) Nonlinear dynamics of discrete and continuous systems. Springer Nature, Cham, pp 15–30
4. Andrianov IV (2022) The pursuit of simplicity: the scientific heritage of Professor Leonid I. Manevitch Int J Non-Lin Mech 142:103998
5. Andrianov IV, Awrejcewicz J (2024) Asymptotic methods for engineers. CRC Press, Taylor & Francis, Boca Raton
6. Andrianov IV, Awrejcewicz J, Danishevs'kyy VV, Ivankov AO (2014) Asymptotic methods in the theory of plates with mixed boundary conditions. Wiley, Chichester
7. Andrianov IV, Barantsev RG, Manevitch LI (2004) Asymptotic mathematics and synergetics. URSS, Moscow (in Russian)
8. Andrianov IV, Manevitch LI (2002) Asymptotology: ideas, methods, and applications. Kluwer, Dordrecht, Boston, London
9. Babich VM (2005) A unique and interesting book. St. Petersburg University 2(3691) (in Russian).
10. Babich VM, Buldirev VS (1982) The art of asymptotics. Vestnik Leningrad Univ Math 10:227–235
11. Barantsev RG (1976) On asymptotology. Bull LGU 1:69–77 (in Russian)
12. Barantsev RG (2004) Once again on asymptotology. Bull St. Petersburg State Univ Ser 1(2):79–85 (in Russian)
13. Barantsev RG (1989) Asymptotic versus classical mathematics. In: Topics in mathematics analysis. World Scientific, Singapore, pp 49–64
14. Betyaev SK (1995) Hydrodynamics: problems and paradoxes. Phys Usp 38:287–316
15. Birkhoff G (1960) Hydrodynamics: a study in logic, fact, and similitude, 2nd edn. Princeton Univ. Press, Princeton, NJ
16. Boyd J (2019) Strongly nonlinear perturbation theory for solitary waves and bions. Evol Equ Control Theory 8:1–29
17. Cartwright ML (1952) Non-linear vibrations: a chapter in mathematical history. Math Gazette 36(316):80–88
18. Crighton RG (1994) Asymptotics—an indispensable complement to thought, computation and experiment in applied mathematical modeling. In: Fasano A, Primicerio M (eds) Proceedings of the 7th European conference on mathematics in industry. B.G. Teubner, Stuttgart, pp 3–19
19. Dewar RL (2002) Asymptotology—a cautionary tale. ANZIAM J. 44:33–40
20. Fedoryuk MV (1989) Asymptotic methods in analysis. In: Gamkrelidze RV (ed) Analysis I. integral representations and asymptotic methods. Springer, Berlin, pp 83–192
21. Friedrichs KO (1955) Asymptotic phenomena in mathematical physics. Bull Am Math Soc 61:485–504
22. Gluzman S (2020) Nonlinear approximations to critical and relaxation processes. Axioms 9:126
23. Gluzman S (2022) Asymptotics and summation of the effective properties of suspensions, simple liquids and composites. Symmetry 14:1912
24. Gorban AN, Radulesku O, Zinovyev A (2010) Asymptotology of chemical reaction networks. Chem Eng Sc 65(7):2310–2324
25. Il'in AM (1992) Matching of asymptotic expansions of solution of boundary value problem. AMS, Providence
26. Kalyakin LA (2014) Phantom asymptotic solutions. Ufa Math J 6(2):44–65
27. Kline M (1985) Mathematics and the search for knowledge. Oxford University Press, New York, Oxford
28. Kline M (1982) Mathematics: the loss of certainty. Oxford University Press, New York, Oxford
29. Kolmogorov AN (1958) The "Cybernetics". Article for the 51st additional volume of the Great Soviet Encyclopedia, submitted for printing on April 28 (in Russian)
30. Krauss LM (2007) Fear of physics: a guide for the perplexed. Basic Books, London
31. Kruskal M (1963) Asymptotology. In: Drobot S, Viebrock PA (eds) Mathematical models in physical sciences (Proc. Conf. of that name at University of Notre Dame, Apr. 15–17, 1962). Prentice-Hall, Englewood Cliffs, NJ, pp 17–47, with discussion, pp 47–48

32. Kudryashov NA, Kutukov AA (2020) Application of a computer algebra system for constructing Newton polygons for ordinary differential equations. In: Misyurin SY, Arakelian V, Avetisyan AI (eds) Advanced technologies in robotics and intelligent systems. Proceedings of ITR 2019. mechanisms and machine science 80. Springer Nature, Cham, pp 365–373
33. Laplace S (1982) The system of the world. Nauka, LO, Leningrad (in Russian)
34. Malinetsky GG, Georgievich Barantsev R (2020) Accuracy—simplicity—locality of application. Comput Res Model 12(5):955–959 (in Russian)
35. Manevitch LI, Gendelman OV (2011) Tractable models of solid mechanics: formulation, analysis and interpretation. Springer, Berlin, Heidelberg, London, New York
36. Peierls RE (1979) Surprises in theoretical physics. Princeton Univ. Press, Princeton, NJ
37. Peierls RE (1980) Model-making in physics. Contemp Phys 21:3–17
38. Peierls R (1991) More surprises in theoretical physics. Princeton Univ. Press, Princeton, NJ
39. Poincaré JH (1946) The foundations of science. The Science Press, Lancaster
40. Pólya G (1981) Mathematical discovery: on understanding, learning and teaching problem solving. Wiley, New York
41. Pólya G (1954) Mathematics and plausible reasoning. In 2 volumes. Vol. 1: Induction and analogy in mathematics. Vol. 2: Patterns of Plausible Inference. Princeton Univ. Press, Princeton, NJ
42. Pólya G (2004) How to solve it. Princeton Univ. Press, Princeton, NJ
43. Rega G (2023) Memories of Professor Leonid Manevitch. Russ J Nonliner Dyn 19(3):439–443
44. Rytov SM (1990) In the laboratory of oscillations. In: Memories of Academician M.A. Leontovich. Nauka, Moscow, pp 35–48 (in Russian)
45. Segel LA (1966) The importance of asymptotic analysis in applied mathematics. Am Math Monthly 73:7–14
46. Selye H (1964) From dream to discovery. McGraw-Hill, On Being a Scientist
47. Shamrovskii AD (1979) Asymptotic integration of static equation of the theory of elasticity in Cartesian coordinates with automated search of integration parameters. J Appl Math Mech 43(5):925–934
48. Slepyan LI, Yakovlev YuS (1980) Integral transformations in nonstationary problems of mechanics. Sudostroyenie, Leningrad (in Russian)
49. Trefethen LN (1998) Maxims about numerical mathematics, computers, science, and life. SIAM News 4
50. Uspensky VA (2020) Works on Nonmathematics, 2nd edn, revised and expanded. In 5 volumes. Vol. 1: In Memory of the Departed. Publication Apparatus. OGI: Fund Mathematical Etudes. Moscow (in Russian)
51. Van Dyke M (1975) Perturbation methods in fluid mechanics. The Parabolic Press, Stanford, Ca
52. Verhulst F (2015) A. Profits and pitfalls of timescales in asymptotics. SIAM Rev 57(2):255–274. B. Verhulst F (2023) Multiple timing and spatial scaling for bifurcations. Nonlin Dyn 111(2):1–15
53. Voronel A (2003) Theological roots of scientific search. In: Voronel A (ed) Together and apart. MET, Minsk, pp 538–553 (in Russian)
54. White RB (2010) Asymptotic analysis of differential equations. Imperial College Press
55. Zabusky NJ, Kruskal MD (1965) Interaction of solitons in a collisionless plasma and the recurrence of initial states. Phys Rev Lett 15:240–243
56. Zinovyev A (2015) Overcoming complexity of biological systems: from data analysis to mathematical modeling. Math Model Nat Phenom 10(3):186–205

Chapter 4
On the Velocity of Twin Boundary

Arkadi Berezovski

Abstract The velocity of twin boundaries is estimated in the framework of continuum mechanics using the internal variables approach. It is shown that the slow motion of twin boundaries can be reproduced using a single internal variable. To comprehend the fast dynamics of twin boundaries, the concept of dual internal variables is employed.

4.1 Introduction

Twin boundary velocity is a macroscopic characteristic of the twinning process. The velocity is measurable, but experimental results range from near-zero up to sound velocity of a material.

Despite differences in experimental settings, there is evidence of rapid twin tip propagation [17, 21, 28, 36, 39]. At the same time, various researchers have reported moderate twin boundary velocities ranging from 30 to 300 m/s [9, 31, 32, 34]. Even smaller values (from 0 to 14 m/s) for the velocity of the twin boundary were experimentally recorded recently [25, 33]. It is desirable that the scatter of the macroscopic quantity be described in terms of continuum mechanics.

There have been numerous theoretical and computational studies conducted on twin boundary motion. In phase field models [3, 14, 16, 19, c.f.], crystal plasticity models [23, e.g.], or combinations thereof [20, 29], individual twin boundaries are not taken into account. The standard approach to the continuum description of twin (or phase) boundaries is to use a multi-well energy landscape. Each minimum on the strain energy density represents an individual equilibrium phase or variant. It is necessary, however, to assume that the energy landscape is piecewise quadratic with a discontinuity in second derivatives [2], or to deal with higher-order polynomials

A. Berezovski (✉)
Department of Cybernetics, School of Science, Tallinn University of Technology, Ehitajate tee 5, 19086 Tallinn, Estonia
e-mail: arkadi.berezovski@taltech.ee

H. Altenbach and V. A. Eremeyev (eds.), *Dynamics of Discrete and Continuum Structures and Media*, Advanced Structured Materials 221,
https://doi.org/10.1007/978-3-031-75626-9_4

for elastic energy (the Landau-energy landscape framework [11, 15, 27, 40, c.f.]). Phase boundary velocity cannot be predicted uniquely in either case.

With the help of internal variables, this study clarifies how to represent twin boundary propagation at both slow and rapid speeds in the continuum mechanics framework. To keep things simple, we concentrate on a uniaxial scenario with a scalar transverse displacement field $u(x)$.

4.2 Uniaxial Motion

In uniaxial setting, all fields are presumably dependent on a single coordinate and time. Transverse and longitudinal motions are uncoupled. We ignore longitudinal motion and focus entirely on shear strains. Restricting by the transverse displacement

$$u = u(x, t), \tag{4.1}$$

we represent the balance of linear momentum in the isothermal case (without body forces) as

$$\rho v_t = \sigma_x. \tag{4.2}$$

Here σ is the shear stress, $v = u_t$ is the transverse particle velocity, ρ is the constant matter density, and subscripts denote derivatives. The kinematic compatibility condition reads

$$\varepsilon_t = v_x, \tag{4.3}$$

where $\varepsilon = u_x$ is the shear strain.

4.2.1 Jump Conditions

Assume that a certain location x_0 corresponds to a potentially moving boundary $\mathcal{S}$ separating the variants. Within the context of continuum description, the following jump relation is satisfied at this boundary [2, e.g.]:

$$V_N [\![\rho v]\!] + [\![\sigma]\!] = 0. \tag{4.4}$$

Here V_N is the velocity of the moving twin boundary, and $[\![B]\!] = B^+ - B^-$, $B^\pm$ are uniform limits of a field B in approaching the boundary from its positive and negative sides, respectively.

The latter relation is a consequence of the balance of linear momentum (4.2). A more complicated relationship is valid for the driving force f_S acting on the boundary [2]

$$f_S = [\![W]\!] - \langle\sigma\rangle [\![\varepsilon]\!], \tag{4.5}$$

where W is the free energy per unit volume and $\langle ... \rangle$ denotes the arithmetical mean.

4.2.2 Velocity of the Boundary

The velocity of the boundary is given by the relationship

$$\rho V_N^2 [\![\varepsilon]\!] = [\![\sigma]\!], \tag{4.6}$$

which follows from jump relation (4.4) and that for the jump of the kinematic compatibility condition (4.3)

$$[\![v]\!] = -[\![\varepsilon]\!] V_N. \tag{4.7}$$

The direction of the boundary's velocity is determined by the entropy inequality on the boundary [2]

$$f_S V_N \geq 0. \tag{4.8}$$

It should be emphasized that the initial-boundary value problem lacks uniqueness because the stress jump in Eq. (4.6) is unknown in advance [1]. The difficulty can be overcome by introducing a new constitutive concept, namely a kinetic relation, which relates the driving force with the velocity of the boundary [2]. However, things become even more complicated when twins are involved.

4.3 Isothermal Twins Equilibrium

We examine a material that could have two distinct twin variants. We consider both variants to be energetically equivalent, meaning that the energy value of the material points in both variants is the same in equilibrium. It is also assumed that the elastic parameters for both variants are the same. The elastic free energy per unit volume in a uniaxial condition is the quadratic function of shear strain

$$W = \frac{G}{2}\varepsilon^2. \tag{4.9}$$

Here G is the shear modulus of the material and subscript denotes the space derivative. It is evident that for both $+\varepsilon$ and $-\varepsilon$, the energy value is the same.

Assume that the variants occupy adjacent parts of a body separated by a twin boundary. We will look into the conditions needed for the variants to coexist in an equilibrium state. While there is no motion implied by the equilibrium state ($u = 0$), shear strains can be non-zero in both variants. Positive strains are related with variant 1, and negative strains with variant 2. It is commonly known that the strains in distinct twins differ from each other by the so-called transformation strain ε^{tr}. Due to the symmetry, the shear strain values can be given as follows:

$$\varepsilon_1 = \frac{1}{2}\varepsilon^{tr}, \quad \varepsilon_2 = -\frac{1}{2}\varepsilon^{tr}. \tag{4.10}$$

This ensures that the free energy remains equal in both variants. Here lower indices indicate twin variants. It is reasonable to suppose that in equilibrium, the stress value is zero. However, the stress is linear in strain

$$\sigma = \frac{\partial W}{\partial \varepsilon} = G\varepsilon, \tag{4.11}$$

which results in the following values of the stress in distinct variants:

$$\sigma_1 = G\varepsilon_1 = \frac{G}{2}\varepsilon^{tr}, \quad \sigma_2 = G\varepsilon_2 = -\frac{G}{2}\varepsilon^{tr}. \tag{4.12}$$

Consequently, stress equality cannot be realized in the equilibrium under macroscopic representation. Hence, twin equilibria and twin boundary motion become problematic for the macroscopic exposition. It is evident that the continuum description of twin boundary dynamics must be extended.

4.3.1 Internal Variables

The idea of working with internal variables is not new. It has been used extensively in the modeling of shape memory alloys [4, 10, 37, 38]. However, individual phase boundaries were generally avoided because the internal variable was the martensitic volume fraction with a specified transformation kinetics [24, 30, 35]. Different versions of transformation kinetics have been reviewed in [18, 26].

Another internal variable methodology is used in this work [6–8]. Here, the distinction between the twins is characterized by internal variables. They can be used to estimate the velocities of twin boundaries.

We begin with a single internal variable case.

4.4 Single Internal Variable

In the single internal variable case, the free energy W per unit volume is specified as a quadratic function of the shear strain and the internal variable, γ [22, e.g.]

$$W = \frac{G}{2}\varepsilon^2 + \frac{1}{2}A\varepsilon\gamma + \frac{1}{2}C\gamma^2, \tag{4.13}$$

with material parameters $A = G\varepsilon^{tr}$ and C.

We assume that internal variable in twin variants has alternate signs in order to distinguish them. If variant 1 has the negative sign for the internal variable, then in variant 2 it should be positive. The shear stress is, by definition, determined as

$$\sigma = \frac{\partial W}{\partial \varepsilon} = G\left(\varepsilon + \frac{1}{2}\varepsilon^{tr}\gamma\right). \tag{4.14}$$

If $\gamma = -1$ then the value of the shear stress

$$\sigma_1 = G\left(\varepsilon_1 - \frac{1}{2}\varepsilon^{tr}\right) \tag{4.15}$$

corresponds to variant 1. For variant 2 we have, accordingly, $\gamma = 1$ and

$$\sigma_2 = G\left(\varepsilon_2 + \frac{1}{2}\varepsilon^{tr}\right). \tag{4.16}$$

4.4.1 Equilibrium Conditions

Each twin variant's state is determined by the values of shear strains ε_1 and ε_2, as well as internal variable values γ_1 and γ_2. It is reasonable to suppose that both twins are stress-free in equilibrium. The following shear strain values correspond to zero stress levels:

$$\varepsilon_1 = \frac{1}{2}\varepsilon^{tr}, \quad \varepsilon_2 = -\frac{1}{2}\varepsilon^{tr}. \tag{4.17}$$

The driving force acting at the twin boundary reduces to the jump of the free energy in the stress-free equilibrium

$$f_S = [\![W]\!] - \langle\sigma\rangle [\![\varepsilon]\!] = [\![W]\!]. \tag{4.18}$$

Calculating the free energy jump

$$
\begin{aligned}
(W_2 - W_1) &= \left(\frac{G}{2}\varepsilon_2^2 + \frac{1}{2}A\varepsilon_2\gamma_2 + \frac{1}{2}C\gamma_2^2\right) - \\
&- \left(\frac{G}{2}\varepsilon_1^2 + \frac{1}{2}A\varepsilon_1\gamma_1 + \frac{1}{2}C\gamma_1^2\right) = \\
&= \frac{G}{2}\left(\varepsilon_2^2 - \varepsilon_1^2\right) + \frac{1}{2}G\varepsilon^{tr}\left(\varepsilon_2\gamma_2 - \varepsilon_1\gamma_1\right) + \frac{1}{2}C\left(\gamma_2^2 - \gamma_1^2\right) + \sigma_0\varepsilon^{tr} = \\
&= \frac{G}{2}(\varepsilon_2 - \varepsilon_1)(\varepsilon_2 + \varepsilon_1) + \frac{1}{2}G\varepsilon^{tr}\left(\varepsilon_2\gamma_2 - \varepsilon_1\gamma_1\right) + \\
&+ \frac{C}{2}(\gamma_2 - \gamma_1)(\gamma_2 + \gamma_1) = 0,
\end{aligned}
\tag{4.19}
$$

we obtain zero value due to conditions $\varepsilon_1 + \varepsilon_2 = 0$ and $\gamma_2 + \gamma_1 = 0$. Therefore, there is no driving force at the twin boundary in equilibrium, as expected. The introduction of the internal variable allows the twins to achieve a stress-free equilibrium.

4.5 Slow Motion

In order to ascertain the possible motion of the twin boundary, we must compute the driving force in a non-equilibrium situation

$$
\begin{aligned}
f_S = [\![W]\!] - \langle\sigma\rangle [\![\varepsilon]\!] &= \left(\frac{G}{2}\varepsilon_2^2 + \frac{1}{2}A\varepsilon_2\gamma_2 + \frac{1}{2}C\gamma_2^2\right) - \\
&- \left(\frac{G}{2}\varepsilon_1^2 + \frac{1}{2}A\varepsilon_1\gamma_1 - \frac{1}{2}C\gamma_1^2\right) - \frac{1}{2}(\sigma_2 + \sigma_1)(\varepsilon_2 - \varepsilon_1) = \\
&= \frac{G}{2}\left(\varepsilon_2^2 - \varepsilon_1^2\right) + \frac{1}{2}G\varepsilon^{tr}\left(\varepsilon_2\gamma_2 - \varepsilon_1\gamma_1\right) + \frac{1}{2}C\left(\gamma_2^2 - \gamma_1^2\right) + \\
&+ \frac{1}{2}\left(G(\varepsilon_2 + \frac{1}{2}\varepsilon^{tr}\gamma_2) + G(\varepsilon_1 + \frac{1}{2}\varepsilon^{tr}\gamma_1)\right)(\varepsilon_2 - \varepsilon_1) = \\
&= \frac{1}{2}G\varepsilon^{tr}\left(\varepsilon_2\gamma_2 - \varepsilon_1\gamma_1\right) + \frac{1}{2}C\left(\gamma_2^2 - \gamma_1^2\right) + \frac{1}{2}G\varepsilon^{tr}\langle\gamma\rangle(\varepsilon_2 - \varepsilon_1),
\end{aligned}
\tag{4.20}
$$

which is no longer zero. The twin boundary velocity is then computed as

$$
V_N^2 = \frac{[\![\sigma]\!]}{\rho\,[\![\varepsilon]\!]} = \frac{G(\varepsilon_2 + \frac{1}{2}\varepsilon^{tr}\gamma_2) - G(\varepsilon_1 + \frac{1}{2}\varepsilon^{tr}\gamma_1)}{\rho\,[\![\varepsilon]\!]} = \frac{G}{\rho}\left(1 + \frac{\varepsilon^{tr}}{[\![\varepsilon]\!]}\frac{[\![\gamma]\!]}{2}\right).
\tag{4.21}
$$

4.5.1 Asymptotics of the Strain Jump

For a small deviation of the actual shear strain jump from the value of the transformation strain, the shear strain jump can be represented in the form of an asymptotic expansion

$$[\![\varepsilon]\!] = \varepsilon^{tr}(1 + \delta\epsilon^1 + \cdots), \tag{4.22}$$

with a small parameter δ. In the first approximation, the twin boundary's velocity in this case is

$$V_N^2 = \frac{G}{\rho}\left(1 + \frac{\varepsilon^{tr}}{[\![\varepsilon]\!]}\frac{[\![\gamma]\!]}{2}\right) \approx \frac{G}{\rho}\delta\epsilon^1, \tag{4.23}$$

because the value of $[\![\gamma]\!]$ is close to 2 in the vicinity of twin boundary. It is evident that in a particular situation, the twin boundary velocity can be as small as that dictated by the value of δ. The slow motion of the twin boundary, corresponding to the single internal variable model, correlates with the results of quasi-static experiments [12, 13, 25, 33]. However, the single internal variable model is unsuitable for fast dynamics of twin boundaries since it uses a parabolic evolution equation for the internal variable [6, 7].

4.6 Dual Internal Variables and Fast Dynamics

We extend the diffusional description by using the dual internal variable technique [6, 8]. The complete theory of dual internal variables is presented in [8]. In a simplified form of the theory, the free energy density is determined by internal variables γ and ψ using a quadratic function

$$W = \frac{G}{2}\varepsilon^2 + \frac{1}{2}A\varepsilon\gamma - \frac{1}{2}C\gamma^2 - \frac{1}{2}D\psi^2, \tag{4.24}$$

where material parameters $A = G\varepsilon^{tr}$, C, and D are constant. Negative signs for coefficients C and D are applied in the correspondence to [5].

Shear stress values are calculated as before

$$\sigma = \frac{\partial W}{\partial \varepsilon} = G\left(\varepsilon + \frac{1}{2}\varepsilon^{tr}\gamma\right). \tag{4.25}$$

4.6.1 Driving Force and Twin Boundary Velocity

For dual internal variables, the driving force is expressed as

$$
\begin{aligned}
f_S = & [\![W]\!] - \langle\sigma\rangle [\![\varepsilon]\!] = \left(\frac{G}{2}\varepsilon_2^2 + \frac{1}{2}A\varepsilon_2\gamma_2 - \frac{1}{2}C\gamma_2^2 - \frac{1}{2}D\psi_2^2\right) - \\
& - \left(\frac{G}{2}\varepsilon_1^2 + \frac{1}{2}A\varepsilon_1\gamma_1 - \frac{1}{2}C\gamma_1^2 - \frac{1}{2}D\psi_1^2\right) - \frac{1}{2}(\sigma_2+\sigma_1)(\varepsilon_2-\varepsilon_1) = \\
& = \frac{G}{2}\left(\varepsilon_2^2 - \varepsilon_1^2\right) + \frac{1}{2}G\varepsilon^{tr}\left(\varepsilon_2\gamma_2 - \varepsilon_1\gamma_1\right) - \frac{1}{2}C\left(\gamma_2^2 - \gamma_1^2\right) - \\
& - \frac{1}{2}D\left(\psi_2^2 - \psi_1^2\right) - \frac{1}{2}\left(G(\varepsilon_2 + \frac{1}{2}\varepsilon^{tr}\gamma_2) + G(\varepsilon_1 + \frac{1}{2}\varepsilon^{tr}\gamma_1)\right)(\varepsilon_2 - \varepsilon_1) = \\
& = \frac{1}{2}G\varepsilon^{tr}\left(\varepsilon_2\gamma_2 - \varepsilon_1\gamma_1\right) - C\,[\![\gamma]\!]\,\langle\gamma\rangle - D\,[\![\psi]\!]\,\langle\psi\rangle - \frac{1}{2}G\varepsilon^{tr}\langle\gamma\rangle\,[\![\varepsilon]\!].
\end{aligned}
$$

The value of the twin boundary velocity is determined accordingly

$$
V_N^2 = \frac{[\![\sigma]\!]}{\rho\,[\![\varepsilon]\!]} = \frac{G(\varepsilon_2 + \frac{1}{2}\varepsilon^{tr}\gamma_2) - G(\varepsilon_1 + \frac{1}{2}\varepsilon^{tr}\gamma_1)}{\rho\,[\![\varepsilon]\!]} = \frac{G}{\rho}\left(1 + \frac{\varepsilon^{tr}\,[\![\gamma]\!]}{2\,[\![\varepsilon]\!]}\right). \tag{4.26}
$$

4.6.2 Governing Equations

The main distinction between single and dual internal variables approaches is that the evolution equation for the internal variable γ in the dual internal variable theory is hyperbolic [6, 7]

$$
\frac{1}{D}\gamma_{tt} = C\gamma_{xx} - \frac{1}{2}A\varepsilon_x. \tag{4.27}
$$

It is convenient to introduce the time rate of the internal variable γ

$$
\gamma_t = w. \tag{4.28}
$$

In terms of the time rate, w, and the internal variable γ, we have the system of first-order equations

$$
I w_t = C\gamma_x - \frac{1}{2}A\varepsilon_x, \tag{4.29}
$$

$$
\gamma_t = w_x, \tag{4.30}
$$

with $I = 1/D$.

Jump relations corresponding to Eqs. (4.29)–(4.30)

$$V_N\,[\![Iw]\!] + [\![C\gamma]\!] = \frac{1}{2}A\,[\![\varepsilon]\!], \tag{4.31}$$

$$V_N\,[\![\gamma]\!] + [\![w]\!] = 0, \tag{4.32}$$

can be combined to obtain the expression for the twin boundary velocity

$$V_N^2\,[\![I\gamma]\!] = [\![C\gamma]\!] - \frac{1}{2}A\,[\![\varepsilon]\!]. \tag{4.33}$$

Rearranging the latter relationship

$$(IV_N^2 - C)\,[\![\gamma]\!] = -\frac{1}{2}A\,[\![\varepsilon]\!], \tag{4.34}$$

we can express the ratio between jumps of internal variable and shear strain

$$\frac{[\![\gamma]\!]}{[\![\varepsilon]\!]} = -\frac{1}{2}\frac{A}{(IV_N^2 - C)}. \tag{4.35}$$

Remembering the expression for the twin boundary velocity (Eq. 4.26)

$$V_N^2 = \frac{G}{\rho}\left(1 + \frac{\varepsilon^{tr}\,[\![\gamma]\!]}{2\,[\![\varepsilon]\!]}\right), \tag{4.36}$$

we can eliminate jumps to obtain the relation

$$\frac{\rho}{G}V_N^2 - 1 = -\frac{\varepsilon^{tr}}{4}\frac{A}{(IV_N^2 - C)}, \tag{4.37}$$

which leads to the quadratic equation for the square of the twin boundary velocity

$$V_N^4 - \left(\frac{C}{I} + \frac{G}{\rho}\right)V_N^2 + \frac{G}{I\rho}\left(C + \frac{A\varepsilon^{tr}}{4}\right) = 0. \tag{4.38}$$

The solution of the quadratic equation

$$V_N^2 = \frac{1}{2}\left(\frac{C}{I} + \frac{G}{\rho}\right) \pm \sqrt{\frac{1}{4}\left(\frac{C}{I} - \frac{G}{\rho}\right)^2 - \frac{G}{I\rho}\frac{A\varepsilon^{tr}}{4}} \tag{4.39}$$

depends only on material parameters.

If the transformation strain is sufficiently small then the velocity of the twin boundary is close to the shear wave speed in the material. This type of twin boundary dynamics is correlated with the propagation of twin tips [17, 36, 39].

4.7 Conclusions

Twin boundary motion and equilibrium of twins are not well described by classical continuum mechanics. The internal variable models utilized in the paper give further insight on the motion of twin boundaries. The simplest possible scenario (uniaxial motion with only two twin variants) was chosen to explain the experimentally observed discrepancies in twin boundary velocities. By incorporating internal variables into the continuum description of twin boundary motion, twin slow and fast dynamics can be theoretically distinguished. The velocity of a slow diffusional motion of a twin boundary is reproduced in the case of a single internal variable. In contrast, the dual internal variable approach provides fast dynamics of a twin boundary.

Acknowledgements This research was supported by the Estonian Research Council under Research Project RPG1227.

References

1. Abeyaratne R, Knowles JK (1990) On the driving traction acting on a surface of strain discontinuity in a continuum. J Mech Phys Solids 38(3):345–360. https://doi.org/10.1016/0022-5096(90)90003-M
2. Abeyaratne R, Knowles JK (2006) Evolution of phase transitions: a continuum theory. Cambridge University Press
3. Amirian B, Jafarzadeh H, Abali BE, Reali A, Hogan JD (2022) Phase-field approach to evolution and interaction of twins in single crystal magnesium. Comput Mech 70(4):803–818. https://doi.org/10.1007/s00466-022-02209-3
4. Auricchio F, Lubliner J (1997) A uniaxial model for shape-memory alloys. Int J Solids Struct 34(27):3601–3618. https://doi.org/10.1016/S0020-7683(96)00232-6
5. Berezovski A (2016) On the Mindlin microelasticity in one dimension. Mech Res Commun 77:60–64. https://doi.org/10.1016/j.mechrescom.2016.09.005
6. Berezovski A (2018) Internal variables associated with microstructures in solids. Mech Res Commun 93:30–34. https://doi.org/10.1016/j.mechrescom.2017.07.011
7. Berezovski A (2023) Constitutive modeling with single and dual internal variables. Entropy 25(5):721. https://doi.org/10.3390/e25050721
8. Berezovski A, Ván P (2017) Internal variables in thermoelasticity. Springer
9. Bowden F, Cooper R (1962) Velocity of twin propagation in crystals. Nature 195(4846):1091–1092. https://doi.org/10.1038/1951091a0
10. Brinson L, Huang M (1996) Simplifications and comparisons of shape memory alloy constitutive models. J Intell Mater Syst Struct 7(1):108–114. https://doi.org/10.1177/1045389X9600700112

11. Falk F (1980) Model free energy, mechanics, and thermodynamics of shape memory alloys. Acta Metallurgica 28(12):1773–1780. https://doi.org/10.1016/0001-6160(80)90030-9
12. Faran E, Shilo D (2011) The kinetic relation for twin wall motion in NiMnGa. J Mech Phys Solids 59(5):975–987. https://doi.org/10.1016/j.jmps.2011.02.009
13. Faran E, Shilo D (2016) Ferromagnetic shape memory alloys-challenges, applications, and experimental characterization. Exp Tech 40:1005–1031. https://doi.org/10.1007/s40799-016-0098-5
14. Grilli N, Cocks AC, Tarleton E (2020) A phase field model for the growth and characteristic thickness of deformation-induced twins. J Mech Phys Solids 143:104061. https://doi.org/10.1016/j.jmps.2020.104061
15. Gu Y, Chen LQ, Heo TW, Sandoval L, Belak J (2013) Phase field model of deformation twinning in tantalum: parameterization via molecular dynamics. Scripta Materialia 68(7):451–454. https://doi.org/10.1016/j.scriptamat.2012.11.022
16. Hu S, Henager CH, Chen L (2010) Simulations of stress-induced twinning and de-twinning: a phase field model. Acta Materialia 58(19):6554–6564. https://doi.org/10.1016/j.actamat.2010.08.020
17. Kannan V, Hazeli K, Ramesh K (2018) The mechanics of dynamic twinning in single crystal magnesium. J Mech Phys Solids 120:154–178. https://doi.org/10.1016/j.jmps.2018.03.010
18. Khandelwal A, Buravalla V (2009) Models for shape memory alloy behavior: an overview of modeling approaches. Int J Struct Chang Solids 1(1):111–148
19. Levitas VI, Roy AM, Preston DL (2013) Multiple twinning and variant-variant transformations in martensite: Phase-field approach. Phys Rev B 88(5). https://doi.org/10.1103/physrevb.88.054113
20. Liu C, Shanthraj P, Diehl M, Roters F, Dong S, Dong J, Ding W, Raabe D (2018) An integrated crystal plasticity-phase field model for spatially resolved twin nucleation, propagation, and growth in hexagonal materials. Int J Plast 106:203–227. https://doi.org/10.1016/j.ijplas.2018.03.009
21. Mahajan S, Williams D (1973) Deformation twinning in metals and alloys. Int Metall Rev 18(2):43–61. https://doi.org/10.1179/imtlr.1973.18.2.43
22. Maugin GA, Muschik W (1994) Thermodynamics with internal variables. Part I. General concepts. J Non Equilib Thermodyn 19:217–249. https://doi.org/10.1515/jnet.1994.19.3.217
23. Mirkhani H, Joshi SP (2014) Mechanism-based crystal plasticity modeling of twin boundary migration in nanotwinned face-centered-cubic metals. J Mech Phys Solids 68:107–133. https://doi.org/10.1016/j.jmps.2014.03.006
24. Mirzaeifar R, DesRoches R, Yavari A (2011) Analysis of the rate-dependent coupled thermo-mechanical response of shape memory alloy bars and wires in tension. Contin Mech Thermodyn 23(4):363–385. https://doi.org/10.1007/s00161-011-0187-8
25. Mizrahi A, Shilo D, Faran E (2020) Variability of twin boundary velocities in 10M Ni–Mn–Ga measured under μs-scale force pulses. Shape Mem Superelasticity 6:45–53. https://doi.org/10.1007/s40830-020-00264-4
26. Paiva A, Savi MA (2006) An overview of constitutive models for shape memory alloys. Math Probl Eng 2006. https://doi.org/10.1155/MPE/2006/56876
27. Rasmussen K, Lookman T, Saxena A, Bishop A, Albers R, Shenoy S (2001) Three-dimensional elastic compatibility and varieties of twins in martensites. Phys Rev Lett 87(5):055704. https://doi.org/10.1103/PhysRevLett.87.055704
28. Reed-Hill RE, Abbaschian R, Abbaschian L (1973) Physical metallurgy principles. Van Nostrand, New York
29. Rezaee-Hajidehi M, Sadowski P, Stupkiewicz S (2022) Deformation twinning as a displacive transformation: Finite-strain phase-field model of coupled twinning and crystal plasticity. J Mech Phys Solids 163:104855. https://doi.org/10.1016/j.jmps.2022.104855
30. Sadjadpour A, Bhattacharya K (2007) A micromechanics-inspired constitutive model for shape-memory alloys. Smart Mater Struct 16(5):1751. https://doi.org/10.1088/0964-1726/16/5/030
31. Saren A, Nicholls T, Tellinen J, Ullakko K (2016) Direct observation of fast-moving twin boundaries in magnetic shape memory alloy Ni-Mn-Ga 5 M martensite. Scripta Materialia 123:9–12. https://doi.org/10.1016/j.scriptamat.2016.04.004

32. Saren A, Laitinen V, Vinogradova M, Ullakko K (2023) Twin boundary mobility in additive manufactured magnetic shape memory alloy 10m ni-mn-ga. Acta Materialia 246:118666. https://doi.org/10.1016/j.actamat.2022.118666
33. Shilo D, Faran E, Karki B, Müllner P (2021) Twin boundary structure and mobility. Acta Materialia 220:117316. https://doi.org/10.1016/j.actamat.2021.117316
34. Smith AR, Tellinen J, Ullakko K (2014) Rapid actuation and response of Ni-Mn-Ga to magnetic-field-induced stress. Acta Materialia 80:373–379. https://doi.org/10.1016/j.actamat.2014.06.054
35. Song Z (2020) Analytical study on phase transition of shape memory alloy wire under uniaxial tension. Int J Eng Sci 152:103295. https://doi.org/10.1016/j.ijengsci.2020.103295
36. Takeuchi T (1966) Dynamic propagation of deformation twins in iron single crystals. J Phys Soc Jpn 21(12):2616–2622. https://doi.org/10.1143/JPSJ.21.2616
37. Tanaka K, Nagaki S (1982) A thermomechanical description of materials with internal variables in the process of phase transitions. Ingenieur-Archiv 51(5):287–299. https://doi.org/10.1007/BF00536655
38. Tanaka K, Kobayashi S, Sato Y (1986) Thermomechanics of transformation pseudoelasticity and shape memory effect in alloys. Int J Plast 2(1):59–72. https://doi.org/10.1016/0749-6419(86)90016-1
39. Williams D, Reid G (1971) A dynamic study of twin-induced brittle fracture. Acta Metallurgica 19(9):931–937. https://doi.org/10.1016/0001-6160(71)90086-1
40. Zoubková K, Seiner H, Sedlák P, Villa E, Tahara M, Hosoda H, Chernenko V (2022) Nonlinear elastic behavior of Ni-Fe-Ga (Co) shape memory alloy and Landau-energy landscape reconstruction. Acta Materialia 224:117530. https://doi.org/10.1016/j.actamat.2021.117530

Chapter 5
Understanding the Artefacts Induced by Numerical Methods for Step Initial Conditions

Daniel Bouche and Guy Bonnaud

Abstract Numerical results provided by simulating a fluid-flow equation with a discontinuous initial condition are reported within 1D geometry. Three different equations have been used: the linear advection equation, the nonlinear Burgers equation and the adiabatic Navier-Stokes 3 equation system. The numerical results are compared to analytical results got from the equivalent partial differential equation associated with the numerical method used. By doing so, the shape of the numerical result, diffusive or oscillating, the oscillation wavelength, the over/undershoot amplitude and damping of the parasitic oscillations are explained qualitatively and quantitatively. This allows to highlight the differences between the various numerical methods.

5.1 Introduction

To predict how evolves over time *t,* a physical quantity *u*, defined at initial time, over a specific domain in space, for example, a pressure field, we have to solve one or a set of Partial Differential Equations (PDE). To this end, it is often necessary to resort to numerical simulation on a computer. As even the most powerful computers can only process a large, but finite, set of numbers, the PDE must be converted to a discrete problem: the values of its solution are sought in a discrete set of points in the domain, and at discrete values of time. This operation can only give an approximate solution. The goal of numerical analysis is to design discretization methods, called numerical schemes, and to ensure that they provide a solution close to the PDE solution, when the discretization Δx in space and Δt in time is fine enough. For regular enough initial conditions u_0, the Lax–Richtmyer theorem allows to estimate the consistency error of the scheme. For a stable and consistent scheme of order p in space and time, it is

D. Bouche (✉) · G. Bonnaud
ENS Paris-Saclay, Université Paris-Saclay, Gif-sur-Yvette, France
e-mail: Daniel.BOUCHE@cea.fr

H. Altenbach and V. A. Eremeyev (eds.), *Dynamics of Discrete and Continuum Structures and Media*, Advanced Structured Materials 221,
https://doi.org/10.1007/978-3-031-75626-9_5

bounded by $C(t)\left[(\Delta x)^{\mathrm{P}} + (\Delta t)^{\mathrm{P}}\right]$ times the norm of the initial condition in Sobolev space $\mathrm{S}^{\mathrm{P}+1}$. However, for non-regular enough initial conditions, such a step $H(x)$, where H is the Heaviside function, this theorem does not apply, and the error can be much larger. Moreover, we know from practice that for such type of conditions numerical schemes generate artefacts, such as smearing, for so-called dissipative schemes or parasitic oscillations for so-called dispersive schemes. In this paper, we explain how it is possible to predict and quantify these artefacts, which are the main contribution to the error of the schemes. Let us begin with the simplest case, namely that of the linear advection in Sect. 5.2, before dealing with nonlinear advection Sect. 5.3, and with Navier–Stokes equation system in Sect. 5.4. A short Sect. 5.5 will summarize other works for control of numerical oscillations and the last section will gather our concluding remarks.

5.2 Linear Advection Equation

We define a regular mesh along the x-axis, characterized by a node index i and a cell size Δx. We study various numerical schemes within finite-difference discretization framework for a linear partial derivative equation (PDE) describing the time-evolution of a quantity $u(x, t)$, at regular time intervals (time step Δt). The partial derivatives in space and time are replaced by differential quotients, i.e. an algebraic sum of the values u_i of u at various mesh nodes, weighted by a set of coefficients. The number of points within the sum and the coefficients are specific to each numerical scheme.

We focus in this section on the 1D advection equation associated to a constant real positive velocity V:

$$\frac{\partial u}{\partial t} + L(u) = 0 \quad \text{with}\, L(u) = V\,\frac{\partial u}{\partial x}. \tag{5.1}$$

5.2.1 *Method of Differential Approximation*

To study the numerical scheme, we use the method of differential approximation, described in the book of Shokin [1]. For practical purpose, we keep only one or two terms for the differential approximation [2]. The scheme is approximated by the so-called equivalent PDE

$$\frac{\partial u}{\partial t} + V\frac{\partial u}{\partial x} = \frac{c}{n}\frac{\partial^n u}{\partial x^n} + \frac{d}{n+1}\frac{\partial^{n+1} u}{\partial x^{n+1}} \equiv -\varepsilon, \tag{5.2}$$

where $n - 1$ denotes the order of precision of the method and the right-hand side is, with the opposite sign, the consistency error ε of the scheme. An even-order space derivative, multiplied by a positive term, brings dissipation, while an odd-order derivative brings dispersion. When the first term of the above development is of odd order, the method is essentially dissipative. When this order is even, the method is essentially dispersive.

5.2.2 Some Representative Schemes and Their Equivalent Equations

We compare different schemes with stencils involving 2M+1 points in Table 5.1: the left column gives the scheme, the right column the consistency error ε of the scheme divided by the velocity V. We present the results for schemes of order 1 (upwind), 2 (LW, BW and LF), 3 (LW3), 4 (Balakin: we only provide the coefficients for this scheme).

Henceforward, we shall denote by C_{n+1} the coefficient of the term $\Delta x^n \partial^{n+1} u/\partial x^{n+1}$. The lower the coefficients, the better the scheme. The dissipative character of a scheme stems from the terms in Δx, Δx^3 and Δx^5, which must be negative, positive and negative, respectively. Note that, for $\Delta x = 1$, all the terms are null: all schemes become simple translations and then solve exactly the advection equation (Fig. 5.1).

For the first-order method, dissipation (C_2 term) varies linearly with Δx from a maximum at $\Delta x = 0$.

For second-order schemes, LW and BW have the same damping coefficient C_4 (= 3/64) at $\Delta x = 0.5$. The BW method can be highly dissipative at low Δx. Comparing the $|C_3|$ coefficients shows that BW is the most dispersive method.

5.2.3 Solution of the Equivalent PDE

To analyse the various schemes, we focus on the two-term equivalent PDE recalled below:

$$\frac{\partial u}{\partial t} + V\frac{\partial u}{\partial x} = \frac{c}{n}\frac{\partial^n u}{\partial x^n} + \frac{d}{n+1}\frac{\partial^{n+1} u}{\partial x^{n+1}}. \tag{5.3}$$

By using the Fourier transform, we get an integral representation of the solution, which reads for a scheme of even order $n-1$ (mainly dispersive by $\partial^n/\partial x^n$ derivative)

$$u(x,t) = u(x,0) * \int_{-\infty}^{\infty} \mathrm{dk}\; \mathrm{e}^{\mathrm{ik}(x-Vt)} e^{\pm \mathrm{ictk}^n/n} e^{-\mathrm{dtk}^{n+1}/(n+1)} \tag{5.4}$$

Table 5.1 (Left) discretization methods and number of points, (right) consistency error

Upwind: M = 1	
$\frac{u_i^{n+1}-u_i^n}{\Delta t}+V\frac{u_i^n-u_{i-1}^n}{\Delta x}=0$	$\frac{\varepsilon}{V}=\frac{\Delta x}{2}(\eta-1)\frac{\partial^2 u}{\partial x^2}+\frac{\Delta x^2}{6}(2\eta^2-3\eta+1)\frac{\partial^3 u}{\partial x^3}+O(\Delta x^3)$
Lax–Wendroff (LW): M = 1	
$\frac{u_i^{n+1}-u_i^n}{\Delta t}+\frac{V}{2\Delta x}\left(u_{i+1}^n 1-u_{i-1}^n\right) - \frac{V^2\Delta t}{2\Delta x^2}\left(u_{i+1}^n-2u_i^n+u_{i-1}^n\right)=0$	$\frac{\varepsilon}{V}=\frac{\Delta x^2}{6}(-\eta^2+1)\frac{\partial^3 u}{\partial x^3}+\frac{\Delta x^3}{8}\eta(-\eta^2+1)\frac{\partial^4 u}{\partial x^4}+O(\Delta x^4)$
Beam–Warming (BW): M = 2	
$\frac{u_i^{n+1}-u_i^n}{\Delta t}+\frac{V}{2\Delta x}(1-\eta)u_{i-2}^n + \frac{V}{\Delta x}(-2+\eta)u_{i-1}^n-\frac{V}{2\Delta x}(\eta-3)u_i^n=0$	$\frac{\varepsilon}{V}=\frac{\Delta x^2}{6}(-\eta^2+3\eta-2)\frac{\partial^3 u}{\partial x^3}+\frac{\Delta x^3}{8}(-\eta^3+4\eta^2-5\eta+2)\frac{\partial^4 u}{\partial x^4}+O(\Delta x^4)$
Leapfrog (LF): M = 1	
$\frac{u_i^{n+1}-u_i^{n-1}}{2\Delta t}+v\frac{u_{i+1}^n-u_{i-1}^n}{2\Delta x}=0$	$\frac{\varepsilon}{V}=\frac{\Delta x^2}{6}(-\eta^2+1)\frac{\partial^3 u}{\partial x^3}+\frac{\Delta x^4}{120}(9\eta^4-10\eta^2+1)\frac{\partial^5 u}{\partial x^5}+O(\Delta x^6)$
Corrected Lax–Wendroff: M = 2	
Addition of a discretized third-order derivative to the Lax–Wendroff method in order to eliminate the first term of the consistency error	$\frac{\varepsilon}{V}=\frac{\Delta x^3}{24}(\eta^3-2\eta^2-\eta+2)\frac{\partial^4 u}{\partial x^4}+\frac{\Delta x^4}{60}(2\eta^4-5\eta^3+5\eta-2)\frac{\partial^5 u}{\partial x^5}+O(\Delta x^5)$
Balakin: M = 2	
$(a_{-2},a_{-1},a_0,a_1,a_2)=\left[\frac{-1}{24}\left(-1+\eta^2\right)(2+\eta), \frac{-1}{6}(1+\eta)\left(4-\eta^2\right), \frac{5\eta}{4}-\frac{\eta^3}{4}, \frac{1}{6}(1-\eta)\left(4-\eta^2\right), -\frac{1}{24}\left(-1+\eta^2\right)(-2+\eta)\right]$	$\frac{\varepsilon}{V}=\frac{\Delta x^4}{120}(-\eta^4+5\eta^2-4)\frac{\partial^5 u}{\partial x^5}+\frac{\Delta x^5}{144}\eta(-\eta^4+5\eta^2-4)\frac{\partial^6 u}{\partial x^6}+O(\Delta x^6)$

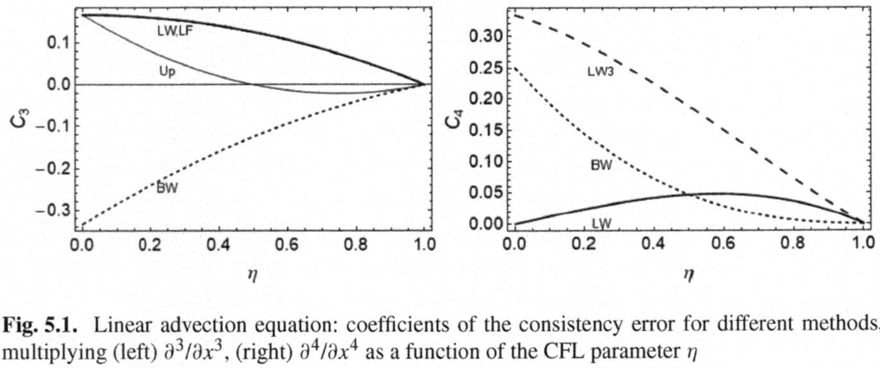

Fig. 5.1. Linear advection equation: coefficients of the consistency error for different methods, multiplying (left) $\partial^3/\partial x^3$, (right) $\partial^4/\partial x^4$ as a function of the CFL parameter η

and for a scheme of odd-order $n-1$ (mainly dissipative)

$$u(x,t) = u(x,0) * \int_{-\infty}^{\infty} \mathrm{dk}\, \mathrm{e}^{\mathrm{ik}(x-Vt)} e^{-\mathrm{ctk}^n/n} e^{\pm \mathrm{idtk}^{n+1}/(n+1)}, \tag{5.5}$$

where * denotes the convolution product. The integrals define generalized Airy functions of order n:

$$\text{odd } n \;:\; \mathrm{Ai}^{n,\,b}(y) = \int_{-\infty}^{\infty} \mathrm{dk}\, \mathrm{e}^{\mathrm{iky}+\mathrm{ik}^n/n - bk^{n+1}/(n+1)}, \tag{5.6}$$

$$\text{even } n \;:\; \mathrm{Ai}^{n,b}(y) = \int_{-\infty}^{\infty} \mathrm{dk}\, \mathrm{e}^{\mathrm{iky}-k^n/n+\mathrm{i}b'k^{n+1}/(n+1)}. \tag{5.7}$$

The number b (resp. b') measures the dissipation (resp. dispersion) ratio of a dissipative (resp. dispersive) Airy function. Chin and Hedstrom, whose papers are referenced in Ref. [3], have tabulated these special functions, along with more general functions with orders of dispersion p and dissipation q differing by more than one unit. Moreover, these functions have asymptotic expansions for large values of their arguments in terms of exponential and trigonometric functions, which make it possible to predict the amplitude and damping of oscillations, as detailed in Ref. [2]. The usual Airy function is obtained for $n = 3$ and $b = 0$.

As an example, for $n = 3$ and for a descending step, we get as the solution:

$$1 - \mathrm{PAi}^{3,\mathrm{b}}(y) \;\text{ with }\; \mathrm{PAi}^{3,\mathrm{b}}(y) = \int_{-\infty}^{\infty} \mathrm{dk} \frac{1}{i2\pi k} e^{\mathrm{iky}+\mathrm{ik}^3/3-\mathrm{bk}^4/4} \tag{5.8}$$

with $\mathrm{PA_i}$ displayed below for a range of b-values. We observe the damping of the oscillations by increasing b and then the modification of the function profile when b approaches 1, featuring an undershoot in addition to an overshoot (Fig. 5.2).

5.2.4 Comparison of Theoretical and Numerical Results

Figure 5.3 compares the theoretical results to numerical simulations for a descending step initial condition. The agreement is good, except for the leapfrog scheme. It has been shown in Ref. [3] that the equivalent equation provides a good approximation of the solution of a scheme only if a dissipation term is included. As the leapfrog scheme does not feature any dissipation, the good agreement for all the studied schemes, except for the leapfrog, is well understood. The disagreements between

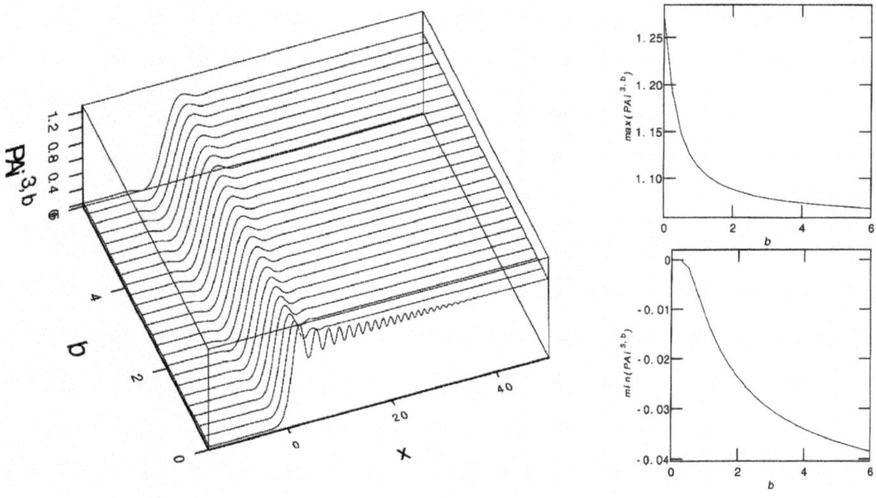

Fig. 5.2. Influence of dissipation in the PAi^{3,b} function: (left) x-profiles, (right) maximum and minimum values

the exact solution and the equivalent equation approximation are analysed in Ref. [4]. The main reasons are the inability of the equivalent equation to deal with high frequencies and the presence of a parasitic mode.

More precise comparisons of the solutions of a large set of numerical schemes with their equivalent equation approximations is provided in Ref. [3], but this book mainly focuses on Green's functions, i.e. solutions of a scheme with a Dirac-like initial condition $u_i(0) = 1$ for $i = 0$, $u_i(0) = 0$ otherwise. Results presented in the book and in the present article are therefore complementary.

The upwind scheme smears the step; its solution is well described by an error function. Primitive of generalized Airy function of order 3 are good approximations for second-order schemes. The artefact is an oscillatory bump and the sign of the dispersion term determines its position at the top (LW) or bottom (BW) of the step. The fourth-order dissipation governs the damping of these oscillations. The profile is monotonous on the bottom (LW) or on the top (BW); this result can be exploited, by using the min-mod limiter, to combine BW and LW in a composite scheme with monotonous profiles (3). The third-order scheme (LW3) generates mirror symmetric bumps. Rising to the fourth order, the main artefact is a bump at the top, but a small undershoot appears, as can be observed on the results for the Balakin scheme (Fig. 5.3).

Equivalent equation with two terms is somehow complicated, because the solution is expressed by using non-standard functions. It is tempting to use the simplest approximation, keeping only the dominant term in the equivalent equation. For odd-order methods, the dissipation term is dominant, and the solution looks like:

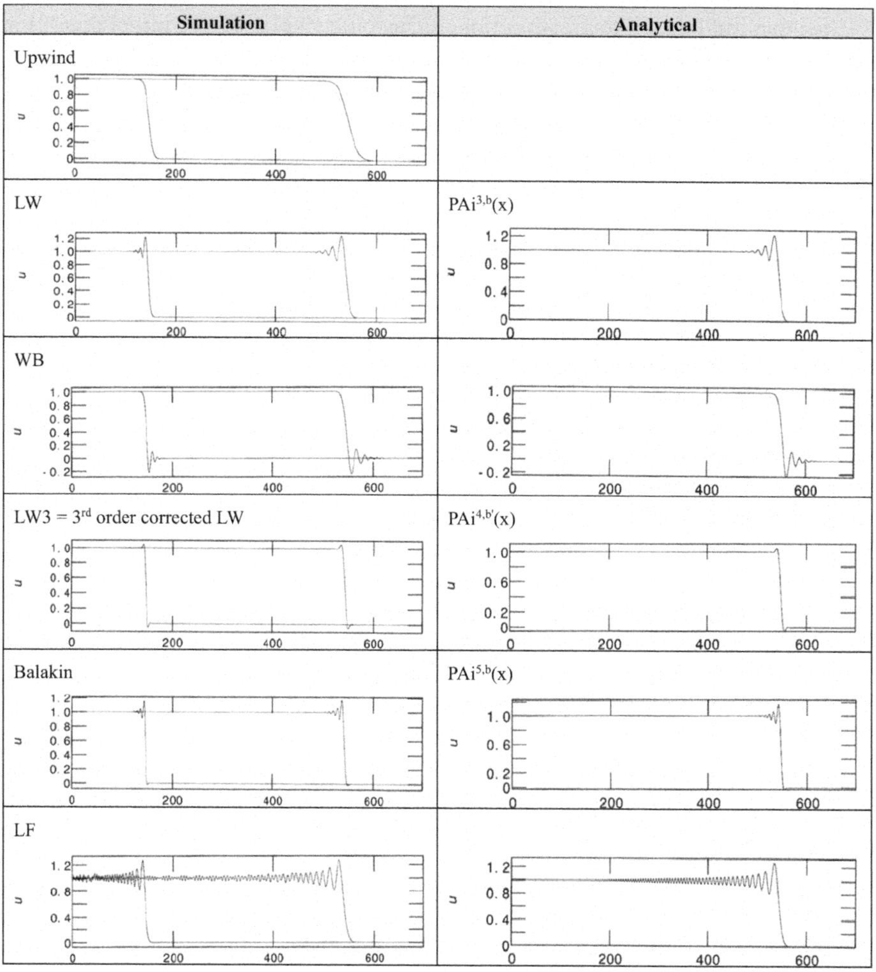

Fig. 5.3. Linear advection equation: (left) snapshot of the profiles after propagation at times 150 and 55 for various numerical methods. (right) Analytical profile using the coefficients of the consistency error by means of generalized Airy function. $u(x,0)$ is a step function. $\Delta x = 1,\ V = 1,\ \eta = 0.49$

$$u(x,t) = u(x,0) * \int_{-\infty}^{\infty} \mathrm{dk}\ \mathrm{e}^{\mathrm{ik}(x-Vt)} e^{-\mathrm{dtk}^n/n} = u(x,0) * \mathrm{Ai}^{n,0}\left[\frac{(x-Vt)}{(\mathrm{d}t)^{1/n}}\right]. \quad (5.9)$$

For the first-order method: $\left(\frac{\partial}{\partial t} + V\frac{\partial}{\partial x}\right)u = \frac{b}{2}\frac{\partial^2 u}{\partial x^2}$ with $b = -V\,\Delta x(\eta - 1)$, the solution reads

$$u(x,t) = u(x,0) * \exp\left[-\frac{(x-Vt)^2}{bt}\right]. \quad (5.10)$$

The equation looks like an advection–diffusion equation. If the initial condition is a step function, the method gives rise to a smooth profile described by an error function with argument $O\left(\sqrt{bt}\right)$. This method is diffusive but stays monotonous.

For weakly dissipative even-order methods, the behaviour of the solution is approximately described by the integral:

$$u(x,t) = u(x,0) * \int_{-\infty}^{\infty} \mathrm{dk} e^{ik(x-Vt)} e^{\pm \mathrm{ictk}^n/n} = u(x,0) * \mathrm{Ai}^{n,0}\left[\frac{\pm(x-Vt)}{(ct)^{1/n}}\right]. \quad (5.11)$$

As both the scheme and the one-term equivalent equation include dissipation, this approximation is correct, but less accurate than that given by the two-term equivalent equation. Note that, according to Godunov theorem, the schemes of order 3 or higher are not monotonous. Oscillations occur for third-order schemes. However, they are exponentially damped, as can be seen on Fig. 5.3. They generate mirror symmetric bumps of amplitude close to 0.05 on the top and the bottom of the step.

For the second-order method, the one-term equivalent equation reads $\left(\frac{\partial}{\partial t} + V\frac{\partial}{\partial x}\right)u = -\frac{c}{3}\frac{\partial^3 u}{\partial x^3}$

$$u(x,t) = u(x,0) * \frac{1}{2\pi}\int_{-\infty}^{\infty} \mathrm{dk} e^{ik(x-Vt)+\mathrm{ick}^3 t/3}. \quad (5.12)$$

The Airy function $A_i\left[(x-Vt)/(ct)^{1/3}\right]$ can be recognized. By contrast with the solution of the upwind scheme, it oscillates. For $c > 0$ (resp. < 0), the oscillations appear on $x < 0$ (resp. > 0). If the initial condition is a descending step function, the solution reads

$$u(x,t) = \frac{1}{(ct)^{1/3}}\left\{1 - \mathrm{PAi}^{3,0}\left[\frac{(x-Vt)}{(ct)^{1/3}}\right]\right\}. \quad (5.13)$$

The smearing of the step grows with time as $(ct)^{1/3}$ for second-order schemes [4]. Note however that, as no dissipative term is included in the equivalent equation, this approximation is not accurate [3]. Comparing the simulation result to the analytical result shows that the oscillations are underestimated.

5.2.5 *Conclusion for Linear PDE*

In the present section, we have shown how to predict the artefacts introduced by numerical schemes solving the advection equation, as a function of their order, through the knowledge of the dissipation and dispersion coefficients. The prediction is accurate, and the method gives good results for other linear PDE, such as the wave equation (5.3). For the nonlinear case, the situation is less simple. Hopefully, the high skills of Alexey Porubov in nonlinear PDE was instrumental in shedding some light on this difficult subject. In the section below, we deal with the nonlinear advection equation, and with Burgers equation by adding to it a diffusion term.

5.3 Burgers Equation

We consider the Burgers equation

$$\frac{\partial u}{\partial t} + u\frac{\partial u}{\partial x} = b\frac{\partial^2 u}{\partial x^2}. \tag{5.14}$$

This equation, that we shall now write $Lu = 0$ has an exact shock solution (or kink) of the form:

$$u_0(\theta) = V + 2\text{bp th}[p(x - Vt)], \tag{5.15}$$

where p and V are the parameters defined by the limits of u at infinity:

$$V = \frac{u_\infty + u_{-\infty}}{2}, p = \frac{u_\infty - u_{-\infty}}{4b}. \tag{5.16}$$

When the Burgers equation is discretized by using a second-order scheme such as Lax–Wendroff (LW) or Beam–Warming (BW), the equivalent equation contains different types of nonlinear terms in his right-hand side, and is not exactly solvable. Relying on the results obtained for the advection equation, we can however get some insight of the effect of these additional terms. It turns out that the main effect of these schemes is to introduce a small dispersion term. Let us study its influence on the shock profile.

5.3.1 Influence of a Small Dispersion Term

We expect that a small dispersion term proportional to $\partial^3 u/\partial x^3$ produces bumps on the top or the bottom of the shock according to the sign of dispersion. To quantify this effect, we used a perturbation method. To this end, a travelling wave solution of $L(u) = -s\delta\ \partial^3 u/\partial x^3$, where δ is a small positive parameter, is sought in the form of increasingly higher order correction terms [5]:

$$u(\theta) = u_0(\theta) + \delta u_1(\theta) + \cdots, \quad \text{where } \theta = x - Vt. \tag{5.17}$$

We substitute this series into $L(u)$, sorted by increasing powers p of δ, and obtain a first-order ordinary differential equation for each δ^p. For the leading order, the equation is homogeneous and its solution is $u_0(\theta)$ previously given. For the next order, the equation becomes inhomogeneous, namely u_1 is the solution of

$$\frac{\partial}{\partial\theta}\left(-Vu_1 + u_0 u_1 - b\frac{\partial u_1}{\partial\theta}\right) = -s\delta\frac{\partial^3 u_0}{\partial\theta^3}, \tag{5.18}$$

which reads

$$u_1 = 4p^2 s\frac{\log[\cosh(p\theta)]}{\cosh^2(p\theta)}. \tag{5.19}$$

According to the sign of s, the dispersion term produces, as in the linear case, for the LW and BW schemes, bumps on top or bottom of the shock, as shown in Fig. 2 of Ref. [5]. More details are gathered in the next subsection.

5.3.2 Comparison with Numerical Simulations for LW and BW Scheme

The first-order differential approximation for LW scheme turns out to be [5]

$$Lu = -\varepsilon = \frac{\Delta t^2}{24}\frac{\partial^3(u^4)}{\partial x^3} - \frac{\Delta x^2}{12}\frac{\partial^3(u^2)}{\partial x^3}. \tag{5.20}$$

By linearizing this equation around the value u_{top} at the top of the shock, we get a positive dispersion term (because of the stability condition for the scheme imposing $|\eta_{\text{top}}| < 1$), as no surprise by coherence with Table 5.1

$$\varepsilon_d = -\frac{u_{\text{top}}}{6}\Delta x^2\left(\eta_{\text{top}}^2 - 1\right) \quad \text{with} \quad \eta_{\text{top}} = \frac{u_{\text{top}}\,\Delta t}{\Delta x}. \tag{5.21}$$

The results are presented below and in Fig. 5 of Ref. [5], for a descending step initial condition. The exact solution is a shock with front propagating at velocity $V = 0.5$. As expected, the LW scheme generates a spike on the top of the shock which grows by reducing the CFL factor η_{top}. The oscillations of the spike are damped by the dissipation associated to the fourth-order derivative; the stronger oscillations for $\eta = 0.2$ compared to $\eta = 0.5$ are due to the smaller C_4 coefficient as shown in Fig. 5.4.

The first-order differential approximation for BW scheme turns out to be [5]

$$Lu = -\varepsilon = \frac{\Delta t^2}{24}\frac{\partial^3(u^4)}{\partial x^3} - \frac{\Delta t\Delta x}{6}\frac{\partial^3(u^3)}{\partial x^3} + \frac{\Delta x^2}{6}\frac{\partial^3(u^2)}{\partial x^3}. \tag{5.22}$$

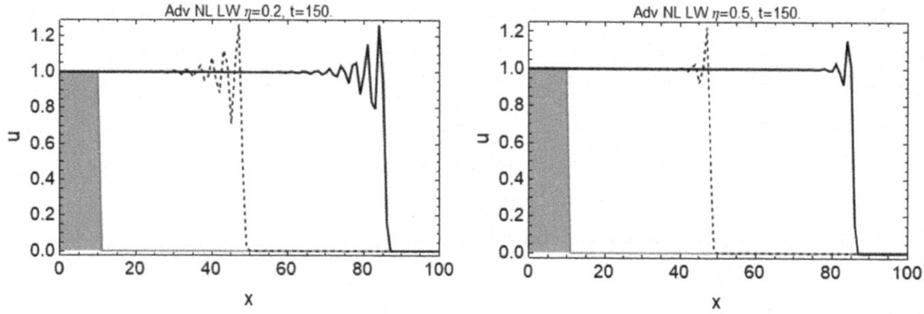

Fig. 5.4. Nonlinear advection equation: snapshot of the profile after propagation at times 0 (grey area), 75 (dotted) and 150 (full line). Lax–Wendroff scheme is used for two different CFL numbers $\eta = \max(u_0)\Delta t/\Delta x$. u(x,0) is a step function. $\Delta x = 1$

By linearizing this equation around the value u_{top} at the top of the step, we get a negative dispersion term:

$$\varepsilon_d = -\frac{u_{\text{top}}}{6}\Delta x^2\left(\eta_{\text{top}}^2 - 3\eta_{\text{top}} + 2\right). \tag{5.23}$$

The results are given below and in Fig. 8 of [5]. As expected, the BW scheme generates a spike on the bottom of the shock. The high fourth-order dissipation of this scheme for low values of η is responsible for damping of the oscillations (Fig. 5.5)

From a qualitative point of view, the results for Burgers equation for these dissipative second-order schemes are close to those for advection equation. An important difference, however, is the width of smearing of the shock: it remains approximately constant in time for Burgers, whereas it grows as $t^{1/3}$ in the linear case (see Sect. 5.2).

The method is in qualitative agreement with the numerical results: it correctly predicts the position of the artefact and its height for LW and BW scheme but disagrees about the spatial profile: the asymptotic solution predicts a bump, whereas

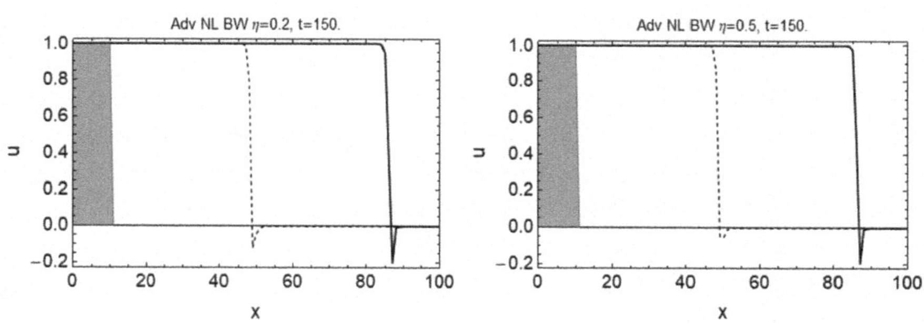

Fig. 5.5. Nonlinear advection equation: snapshot of the profile after propagation at times 0 (grey area), 75 (dotted) and 150 (full line). Beam–Warming scheme is used for two different CFL numbers. $u(x, 0)$ is a step function. $\Delta x = 1$

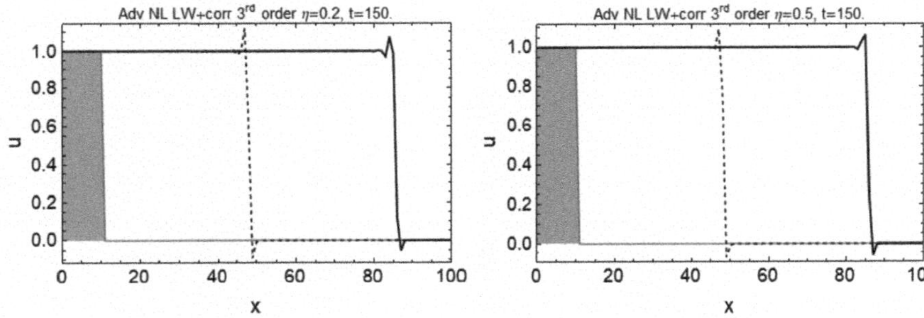

Fig. 5.6. Nonlinear advection equation: snapshot of the profile after propagation at times 0 (grey area), 75 (dotted) and 150 (full line). The third-order corrected Lax–Wendroff scheme is used for two different CFL numbers. $u(x,0)$ is a step function. $\Delta x = 1$

the numerical overshoots or undershoots show oscillations. We have not yet found how to correct this disagreement. This question deserves further work.

5.3.3 Scheme of Order 3

The same method was also applied to the LW3 scheme. The equivalent equation for this scheme and its approximate solution are given in Ref. [5]. The numerical shock profiles of this scheme, shown below and in Fig. 12 of Ref. [5] exhibit mirror symmetric bumps both on the upper and lower parts of the shock. The scheme generates the artefacts similar to that obtained in the linear case. The agreement between the theoretical results and the numerical simulation are good, as detailed in Ref. [5] (Fig. 5.6).

5.3.4 Leapfrog Scheme

The results for the leapfrog scheme are affected by strong oscillations; the front speed appears moreover higher than the correct one. Adding a linear viscosity, by adding the operator $\beta \ \partial^2 u/\partial x^2$,, regularizes the solution, by damping the oscillations. Nevertheless, the leapfrog scheme remains improper to catch a shock front (Fig. 5.7).

In the nonlinear case, the absence of dissipation generates artefacts that completely spoil the results. Further evidence is given in Subsect. 5.3.6, for triangular-shape initial conditions.

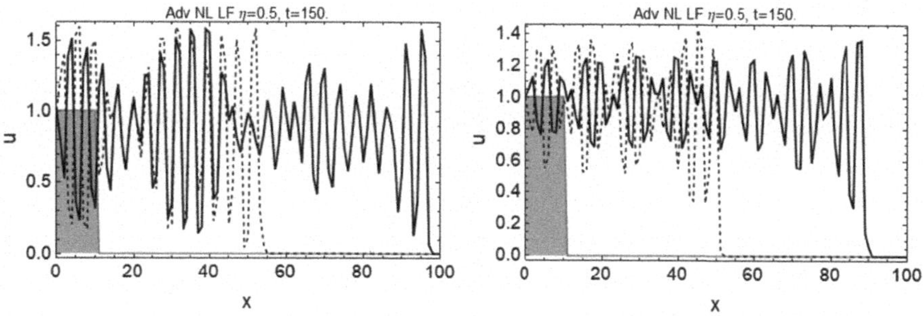

Fig. 5.7. Nonlinear advection equation: snapshot of the profile after propagation at times 0 (grey area), 75 (dotted) and 150 (full line). The leapfrog scheme is used. (right) a linear diffusion term is added, with $\beta = -0.005$. $u(x, 0)$ is a step function

5.3.5 *Influence of a Small Nonlinear Cubic Term, Correction of LW Scheme*

By the same method, the solution for a nonlinear cubic term was obtained in Ref. [6]. More precisely, the equation

$$Lu = -\delta\alpha \frac{\partial^2 (u^3)}{\partial x^2} \tag{5.24}$$

was solved. It turns out that, for positive values of α, the cubic term generates a hat-shape profile on the top of the shock. Adding to LW scheme a difference term of the form $\beta\left[\left(u_{i+1}^n\right)^3 - 2\left(u_i^n\right)^3 + \left(u_{i-1}^n\right)^3\right]$ which is a centred discretisation of the above cubic term, it was possible, with a well-chosen value of ε, to compensate the effect of the dispersion term. As a result, the overshoot of the corrected LW scheme is very small, as can be seen in Fig. 15 of Ref. [6]. This method of compensation was further developed in Ref. [6], where it is shown that it is also possible to compensate the dissipation.

5.3.6 *Comparison of Schemes for Triangular Initial Conditions*

We compare in this subsection the numerical results to the exact solutions for a triangular-shape initial condition to illustrate the crucial role of dissipation. Whenever the solution is single-valued, the nonlinear advection equation can be solved by the method of characteristics. For the specific choice of an initial triangular hat profile with δ full-length basis, centred on x = 0:

$$u(x, 0) = V\left(1 - \frac{|x|}{\delta/2}\right) \tag{5.25}$$

the strong solution of the PDE reads, by using dimensionless distance ξ and time τ:

$$u(x, t) = \frac{1 - \xi}{1 - \tau} \quad \text{with } \xi = \frac{x}{\delta/2}, \tau = \frac{t}{\delta/2V} \tag{5.26}$$

from $t = 0$ till the breaking time associated to first crossing of characteristics, namely $t = -1/\partial v[u_0(x)]/\partial x$ that is $\tau = 1$. Later on, a shock propagates, the features of which can be determined via a weak solution provided by the integrated PDE around the shock leading to the shock position for $\tau \geq 1$:

$$\xi_s = \sqrt{2(1 + \tau)} - 1. \tag{5.27}$$

As the mass $\int_{-\infty}^{\infty} \mathrm{d}x\, u(x, t)$ is kept constant equal to $\int_{-\infty}^{\infty} \mathrm{d}x\, u(x, 0)$ and as the profile of $u(x,t)$ is a square triangle, the peak amplitude reads

$$A = \sqrt{\frac{2}{1 + \tau}}. \tag{5.28}$$

Figure 5.8 compares the exact solution to the numerical simulations for upwind, LW, BW and LW3 schemes. As the exact solution develops a shock, we find as expected, the smearing of the shock by the upwind scheme, the spike on the top for LW scheme, and at the bottom for BW scheme and on both sides for the LW3 scheme. Because of its higher fourth-order dissipation, the undershoot for the BW scheme is visible only close to the shock front (Fig. 5.8).

Once more, the predicted artefacts are visible, but do not change the nature of the solution. Although this paragraph is not strictly in the realm of this paper, because non-dissipative schemes are outside the scope of equivalent equation method, we present some results, as a kind of warning: they can give disastrous results! For the non-dissipative leapfrog scheme, the numerical solution looks like a train of solitons, reminding, though they are not exactly bell-shaped, the soliton solution of Korteweg–de Vries equation. Note that the fastest soliton overtakes the shock. The figure corresponds to a shorter simulation time limited to $t = 200$, since an overflow appears beyond that time. In the figure below, the profile at time 200 is invisible, since all the solitons have left on the frame on the right side! (Fig. 5.9).

Adding sufficient viscosity, by discretizing the operator $\beta\ \partial^2/\partial x^2$ transforming therefore the advection equation into a Burgers equation, rules out the problem: the resulting solution approximates much better the shock solution of the advection equation, with the usual parasitic oscillations on the top of the shock induced by dispersion of the method. It is worthwhile to notice that discretizing $\partial^2/\partial x^2$ by the three-level Dufort–Frankel method $\left[u_{i-1}^n - \left(u_i^{n+1} + u_i^{n-1}\right) + u_{i+1}^n\right]/\Delta x^2$ is much more efficient than using the usual explicit method $\left[u_{i-1}^n - 2u_i^n + u_{i+1}^n\right]/\Delta x^2$ because of the gain of stability (Fig. 5.10).

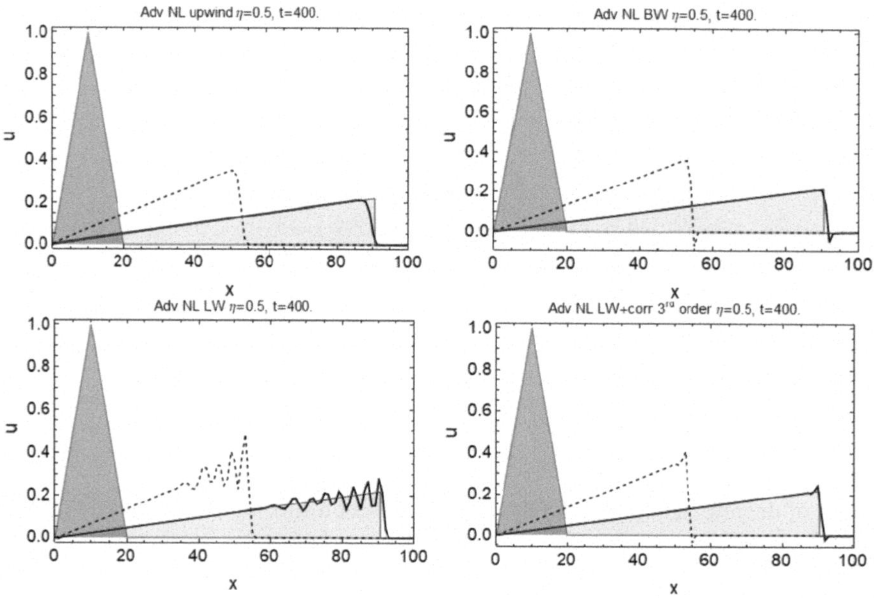

Fig. 5.8. Nonlinear advection equation: snapshot of the profile after propagation at times 0 (dark grey area), 133 (dotted) and 400 (black line). $u(x,0)$ is a tent function. The profile of the exact solution is shown with the light-grey-filled square triangle. (top-left) Upwind scheme, (top-right) Beam–Warming, (bottom-left) Lax–Wendroff scheme, (bottom-right) Lax–Wendroff with third-order correction

Fig. 5.9. Nonlinear advection equation: snapshot of the profile after propagation at times 0 (dark grey triangle), 67 (dotted). $u(x,\ 0)$ is a tent function. The profile of the exact solution is shown with the light-grey-filled square triangle at time $t = 200$. A Leapfrog scheme is used

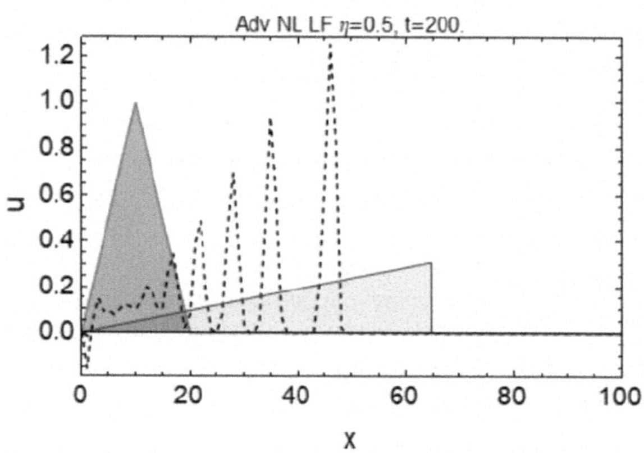

5.3.7 *Conclusion on Burgers Equation*

Though the equivalent equation to schemes of orders 2 and 3 are not exactly solvable, it is possible to predict the artefacts generated by these schemes by solving it approximately by a perturbation method. By using this approximate solution, the

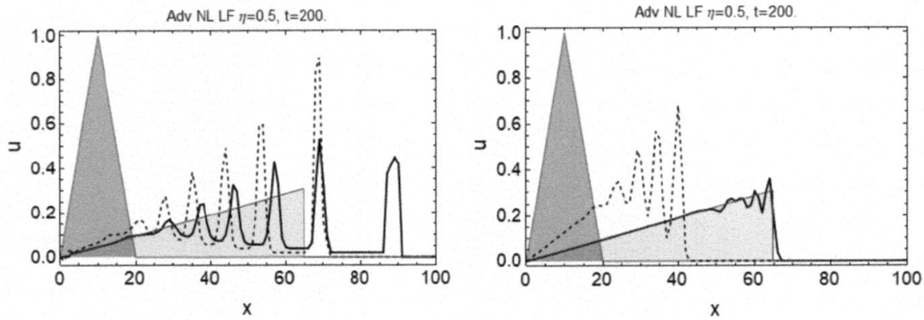

Fig. 5.10. Burgers equation: snapshot of the profile after propagation at times 0 (dark grey triangle), 67 (dotted) and 200 (black line). $u(x, 0)$ is a tent function. The profile of the exact solution is shown with the light-grey-filled square triangle. Leapfrog scheme with added linear diffusion term with β = (left) −0.005 and (right) −0.02

nature of the artefacts generated by these schemes on shock profiles has been clarified. Moreover, a correction drastically diminishing the amplitude of the artefacts of second-order schemes has been suggested. It is also possible to derive a correction to compensate dissipation. Although the results are not as accurate as in the linear case, because an exact solution for the equivalent equation is lacking, from the practical point of view, artefacts generated by the schemes are rather well understood, and efficient means of reducing them are available. Further work is however needed to predict their shape accurately, as has been achieved in the linear case.

5.4 Navier–Stokes System

The approach presented above has been generalized to the Navier–Stokes system [7], focusing mainly on simple wave solutions, namely shocks and contact discontinuities. The main aim of the work was, like for the Burgers equation, to predict the shapes of discrete shocks or contact discontinuities for representative numerical schemes. A pioneering paper on this subject is Ref. [8], where Lerat and Peyret analysed such profiles for the so-called $S_\alpha{}^\beta$ family of three-points predictor corrector schemes.

5.4.1 Shock Profiles

To study the shock profiles, we used specific initial conditions, leading to a single propagating shock. We wrote the Euler system in 1D geometry along a space axis denoted x, within the conservative form:

$$\frac{\partial \vec{W}}{\partial t} + \frac{\partial \vec{F}}{\partial x} = \vec{0}, \tag{5.29}$$

where $\vec{W}$ denotes the vector $(\rho, \rho u, E)$ and $\vec{F}$ the flux vector $\left[\rho u, \ \rho u^2 + p, (E + p)u\right]$, where ρ stands for the fluid density, E for the total energy and p the pressure. The total energy E is the sum of a kinetic and a potential part, and is therefore given by the relation $E = \rho u^2/2 + p/(\gamma - 1)$, where γ stands for the ratio of the constant pressure to constant volume specific heats. The relation between the internal energy $E - \rho v^2/2$ and the pressure p is the so-called equation of state. The flux can be considered as a function of $\vec{W} = (W_j)_{j=1,3}$ via the relation:

$$\vec{F}(\vec{W}) = \left[W_2, W_2^2/W_1 + p, (W_3 + p)W_2/W_1\right] \text{ with } p = (\gamma - 1)\left(W_3 - W_2^2/2W_1\right). \tag{5.30}$$

We defined the local acoustic (or sound) velocity by $c_s = \sqrt{\gamma p/\rho}$. To stabilize the shock profiles, we added into the above system the additional dissipation flux vector

$$\vec{H} = \left[0, \nu\frac{\partial}{\partial x}\left(\rho\frac{\partial u}{\partial x}\right), \nu\frac{\partial}{\partial x}\left(\rho u\frac{\partial u}{\partial x}\right)\right], \tag{5.31}$$

which transforms the Euler equations into the Navier–Stokes system written in divergence form. We sought for a travelling wave solution, depending only on the variable x-Vt, and with limit values at infinity $(\rho_{-\infty}, u_{-\infty}, E_{-\infty})$, $(\rho_\infty, \ 0, \ E_\infty)$. The system can be reduced to a single equation on the velocity u, which turns out, once integrated, to be the ordinary differential reduction of the Burgers equation (5.7)

$$2V \ \nu\frac{\partial u}{\partial \theta} - V(1+\gamma)u^2 + 2\left[V^2 + (1-\gamma)\gamma E_\infty\right]u = 0. \tag{5.32}$$

The analytic solution is given by Eq. (16) of Ref. [7]:

$$U = \frac{u_{-\infty}}{2}\{1 - \text{th}[k(x - Vt)]\}, \tag{5.33}$$

where

$$k = \frac{(1+\gamma)u_{-\infty}}{4\nu}, \quad V = \frac{(1+\gamma)u_{-\infty} + \sqrt{16(\gamma-1)\gamma E_\infty + (1+\gamma)^2 u_{-\infty}^2}}{4}. \tag{5.34}$$

5.4.2 Method of Analysis

To study a specific scheme, we used in Ref. [7] the same method as previously. Namely, we first computed the equivalent equation of the scheme, which turned out to be rather complicated. It is however possible, as in the continuous case, to reduce the system to a differential equation for u. Once integrated, its left-hand side is the ordinary differential reduction of the Burgers equation, and its right-hand side is a perturbation term.

Of course, this equation is not exactly solvable, but, by using the same perturbation method as before, it is possible to find an approximate solution. The solution is as previously searched in the form, $u(\theta) = u_0(\theta) + su_1(\theta) + \ldots,$ where $\theta = x - Vt$. The leading order term $u_0(\theta)$ is the shock solution of the Navier–Stokes equation. The first perturbation term $su_1(\theta)$ is the solution of a first order differential equation and can be determined explicitly. More specifically, s is given by Eq. (29) of Ref. [7], which we refrain from reproducing because of its complexity, and u_1 by Eq. (31) of Ref. [7], which is given below:

$$u_1 = \frac{8}{\nu\left(u_{-\infty}(1+\gamma) + \sqrt{16(-1+\gamma)\gamma E_\infty + (1+\gamma)^2 u_{-\infty}^2}\right)} \log[\cosh[k\theta)]. \tag{5.35}$$

The perturbation generates, as in the case of Burgers equation (5.19), a bump on the top of the shock.

5.4.3 Comparison to Numerical Simulation for Isolated Shocks

We have chosen to isolate the shock wave and so we have modified the initial conditions on the left side:

$$\Delta x = 0.001,\ N_x = 400,\ N_t\,\Delta t = 0.3,\ (\rho,\ u,\ p)_L = (1.419,\ 0.68,\ 1.172),$$
$$(\rho,\ u,\ p)_R = (1,\ 0,\ 1/\gamma)$$

In Fig. 5.11, we use a CFL parameter $\eta = 0.2$. Without viscosity, oscillations appears on the top of the shock. Increasing the viscosity first damps the oscillations, then suppress them, and the artefact becomes a bump, as predicted by the perturbation method. Increasing further the viscosity, the profile becomes monotonous (Fig. 5.11).

Increasing Δt makes the CFL parameter increase to $\eta = 0.6$ and consequently, the dissipation increases. As a result, oscillations are reduced and a lower viscosity can be used to get rid of these oscillations (Figs. 5.12 and 5.13).

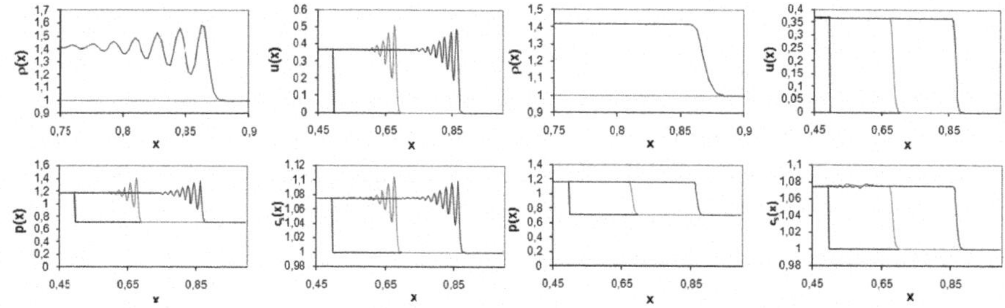

Fig. 5.11. Adiabatic Euler system: solution at times $t = 0.25$ and 0.3 provided by simulation using the Lax–Wendroff method. Viscosity $\nu =$ (left) 0, (right) $1.25\ \ 10^{-3}$. $\Delta t = 0.003$

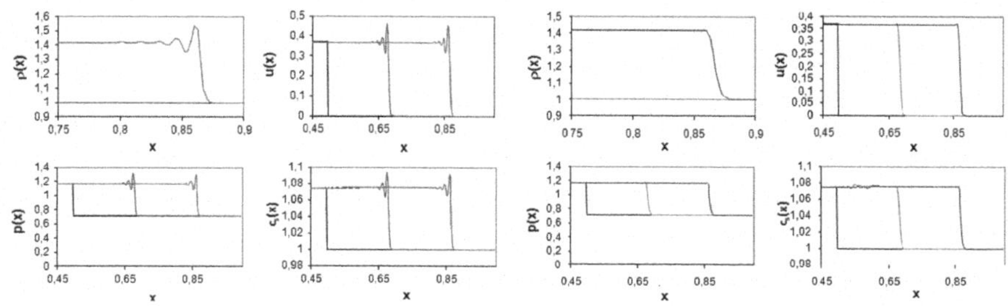

Fig. 5.12. Adiabatic Euler system: solution at times $t = 0.25$ and 0.3 provided by simulation using the Lax–Wendroff method. Viscosity $\nu =$ (left) 0, (right) $9.4\ \ 10^{-4}$. $\Delta t = 0.001$

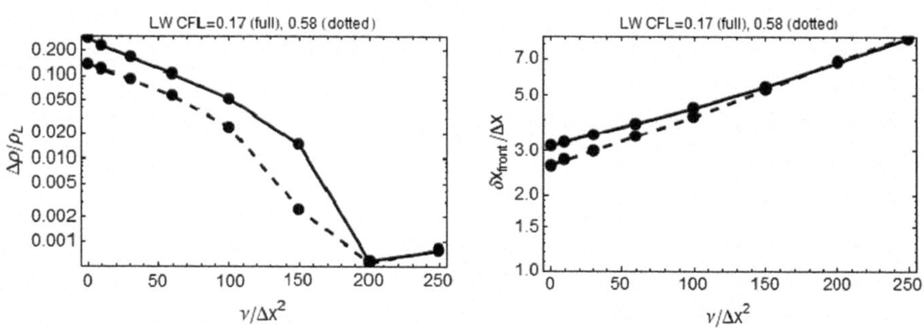

Fig. 5.13. Adiabatic Euler system: influence of the viscosity coefficient on (left) the amplitude of ρ oscillations, (right) the front width. Simulation results using the Lax–Wendroff method. The CFL values are given just above the figure frames. $\Delta t = 0.001$

5.4.4 Disintegrating Discontinuities

In the previous section, we explained the artefacts generated by Lax–Wendroff scheme on the shock profile. We now turn to a more general discontinuity, which disintegrates into three simple waves, as explained below. For higher density and pressure on the left side, a shock propagates rightwards, a rarefaction wave leftwards, and in between them, a contact discontinuity appears.

We simulated the adiabatic Euler system with different values of Δx and ν. As initial condition, the x-axis is divided into two uniform areas with different triplet (ρ_L, u_L, p_L) of the left side and (ρ_R, u_R, p_R) on the right side. First we have chosen an immobile fluid $u_L = u_R = 0$, and higher density and pressure on left side compared to the right area associated to the so-called Sod tube [9]. The analytical solution exhibits three waves: one contact discontinuity (called entropy wave where u and p are kept constant across the discontinuity) and two nonlinear waves (called acoustic waves), each of which being either a shock or a rarefaction wave. As a consequence, the two initial states in close contact become separated by three new states at $t > 0$ denoted from the left to the right (ρ_d, u_d, p_d), (ρ_2, u_2, p_2), (ρ_1, u_1, p_1), the last two triplets being independent of time.

Across the rarefaction wave (case $p \leq p_L$), a Riemann invariant is used:

$$u_L + \frac{2}{\gamma - 1} c_{sL} = u + \frac{2}{\gamma - 1} c_s. \tag{5.36}$$

By use of the polytropic law p/ρ^γ = constant, we are let with $\rho = (p/p_L)^{1/\gamma} \rho_L$ and $c_s = \sqrt{\gamma p/\rho}$, leading to a relation between c_s and p:

$$c_s = c_{sL} \left(\frac{p}{p_L} \right)^{(\gamma-1)/2\gamma}. \tag{5.37}$$

For the contact discontinuity parting two domains labelled 2 and 1, we have $u_2 = u_1$, $p_2 = p_1$.

Across the shock wave (case $p_1 \geq p_R$), the Rankine–Hugoniot relations are used with $u_R = 0$:

$$\text{Density jump:} \quad \frac{\rho_1}{\rho_R} = \frac{(\gamma - 1) + (\gamma + 1) p_1/p_R}{(\gamma + 1) + (\gamma - 1) p_1/p_R}. \tag{5.38}$$

$$\text{Shock velocity:} \quad u_s = \left(\frac{1}{\rho_R} \frac{p_1 - p_R}{1 - \rho_R/\rho_1} \right)^{\frac{1}{2}} \Rightarrow \frac{p_1}{p_R} = \frac{2\gamma}{\gamma + 1} \frac{u_s^2}{c_{sR}^2} - \frac{\gamma - 1}{\gamma + 1}. \tag{5.39}$$

$$\text{Material velocity and shock velocity:} \quad u_1^2 = \frac{1}{\rho_R} (p_1 - p_R)(1 - \rho_R/\rho_1). \tag{5.40}$$

$$\Rightarrow \frac{u_1^2}{u_s^2} = \left(1 - \frac{\rho_R}{\rho_1} \right)^2 \Rightarrow \frac{u_1}{u_s} = 1 - \frac{\rho_R}{\rho_1} = \frac{-2/(\gamma + 1)(1 - p_1/p_R)}{(\gamma - 1)/(\gamma + 1) + p_1/p_R}. \tag{5.41}$$

With use of discontinuity contact relations, we are left with

$$u_1 = \frac{2}{\gamma - 1} c_{sL} \left[1 - \left(\frac{p_1}{p_R} \frac{p_R}{p_L} \right)^{(\gamma-1)/2\gamma} \right] = \frac{2}{\gamma - 1} c_{sL} \left[1 - \left(\frac{p_1}{p_R} \frac{p_R}{p_L} \right)^{(\gamma-1)/2\gamma} \right]. \tag{5.42}$$

With use of the above relation between p_1 and u_s, we find

$$\frac{u_1}{u_s} = \frac{2}{\gamma + 1} \left(1 - \frac{c_{sR}^2}{u_s^2} \right) \Rightarrow u_1 = \frac{2}{\gamma + 1} \left(u_s - \frac{c_{sR}^2}{u_s} \right). \tag{5.43}$$

To summarize, the analytical solution can be cast into five regions (Table 5.2).

Here above, M_s stands for the Mach number u_s/c_{sR} as the solution of the implicit expression:

$$M_s - \frac{1}{M_s} - \frac{c_{sL}}{c_{sR}} \frac{\gamma + 1}{\gamma - 1} \left\{ 1 - \left[\frac{p_R}{p_L} \left(\frac{2\gamma}{\gamma + 1} M_s^2 - \frac{\gamma - 1}{\gamma + 1} \right) \right]^{(\gamma-1)/2\gamma} \right\} = 0. \tag{5.44}$$

If the left-side triplet is chosen to correspond to (ρ_1,u_1,p_1) given above, the solution will consist in only one isolated shock: the left and right sides are connected by one shock only, instead of three waves. Here below, we have chosen $\gamma = 1.4$ and imposed $c_{sL}/c_{sR} = 1$ to have a one-to-one relation between M_s and the pressure ratio p_L/p_R.

5.4.5 Comparison to Numerical Simulations

The simulation space is limited to the segment [0, 1] and initially the fluid division is located at $x = 0.5$. In the following, we examine the influence of two parameters, namely the time step Δt and the viscosity amplitude added to the numerical method. We have used the following conditions exhibiting the three different flows with a strong shock: $\Delta x = 0.0025$, $N_x = 400$, $(\rho, u, p)_L = (8, 0, 10/\gamma)$, $(\rho, u, p)_R = (1, 0, 1/\gamma)$.

The density profile within the figure shown below exhibits from the left to right a descending ramp associated to the expansion wave, the contact discontinuity and then the shock wave. Expansion and shock can be located in the velocity and pressure profiles as well (Fig. 5.14).

Figure 5.15 presents the results of the simulations performed with the LW scheme. The CFL, defined by $\max(u_i + c_{si})\Delta t/\Delta x$, i being the mesh node index, stays around 0.55 during the whole simulation. Compared to the analytical solution, we can observe a deviation in the profiles, which takes the form of a bump or an overshoot on the shock wave profile, as explained above. Oscillations are also present on the contact discontinuity, and on the bottom of the rarefaction wave. For a given value of Δx, the size of the bump decreases when the viscosity ν increases, and it disappears above a threshold value of ν. But it also decreases with Δx (Fig. 5.15).

Table 5.2. Characteristics of the five areas associated to the an initial difference of the (density, velocity, pressure) at $x = 0$

Area	Location $x - x_0$	Density	Velocity	Pressure	Acoustic velocity
L: Left side	$(-\infty, \ -c_{sL}t]$	ρ_L	$uL = 0$	p_L	$c_{sL} = \sqrt{\gamma \frac{p_L}{\rho_L}}$
d: Expansion	$[-c_{sL}t, \ (u_2 - c_{s2})t]$	$\gamma \frac{p_d}{c_{sd}^2}$	$u_d = \frac{2}{\gamma+1}\left(c_{sL} + \frac{x-x_0}{t}\right)$	$p_d = p_L \left(\frac{c_{sd}}{c_{sL}}\right)^{2\gamma/(\gamma-1)}$	$c_{sd} = \frac{2}{\gamma+1} \left(c_{sL} - \frac{\gamma-1}{2}\frac{x-x_0}{t}\right)$
2: Slipping line downstream	$[(u_2 - c_{s2})t, \ u_2 t]$	$\rho_2 = \rho_L \left(\frac{p_1}{p_L}\right)^{1/\gamma}$	$u_2 = u_1$	$p_2 = p1$	$c_{s2} = c_{sL} - \frac{\gamma-1}{2} u_1$
1: Shock downstream	$[u_2 t, M_s c_{sR} t]$	$\rho_1 = \rho_R \left(\frac{2}{\gamma+1}\frac{1}{M_s^2} + \frac{\gamma-1}{\gamma+1}\right)^{-1}$	$u_1 = \frac{2c_{sR}}{\gamma+1}\left(M_s - \frac{1}{M_s}\right)$	$p_1 = p_R \left(\frac{2\gamma}{\gamma+1} M_s^2 - \frac{\gamma-1}{\gamma+1}\right)$	$c_{s1} = \sqrt{\gamma \frac{p_1}{\rho_1}}$
R: Right side	$[M_s c_{sR} t, +\infty)$	ρ_R	$uR = 0$	p_R	$c_{sR} = \sqrt{\gamma \frac{p_R}{\rho_R}}$

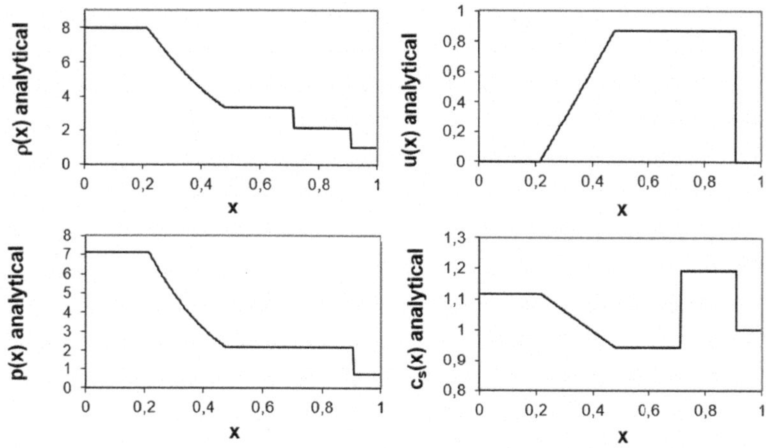

Fig. 5.14. Adiabatic Euler system: analytical solution at time $t = 0.25$ for strong shock conditions

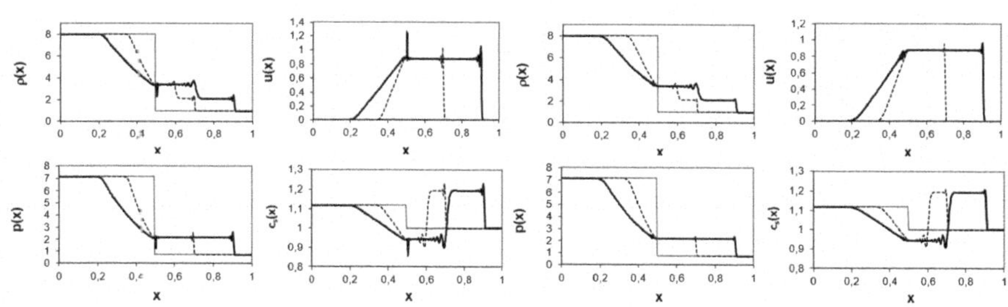

Fig. 5.15. Adiabatic Euler system: solution at times 0.125 (dotted line) and 0.25 (full line) provided by simulation using the Lax–Wendroff method. (left) no viscosity, (right) $\nu = 8 \cdot 10^{-4}$. Time step $\Delta t = 6 \cdot 10^{-4}$

We report now on other simulations performed with conditions of weak shock by choosing $(\rho, u, p)_L = (2.8, 0, 2.8/\gamma)$ and keeping the other parameters. The analytical profile is shown in Fig. 5.16.

Here below, the simulation results got with CFL numbers 0.4 (left) and 1 (right). The viscosity is observed to rule out the oscillations on both velocity and pressure profiles but is less efficient on the density profile.

In this section, we focus on how the shape of the contact discontinuity is predicted, following the analysis presented in Ref. [10]. The equivalent equation is written for this simple wave. As the field in this case linearly degenerates, the left-hand side is simply the linear advection equation. The right-hand side is composed of a lot of terms, but we kept only two terms, the (main) third-order dispersive, and the fourth-order dissipative. The results of the prediction are in good agreement with the numerical simulations: the neglected terms have only minor influence [10]. One can also see, at least qualitatively, in Figs. 5.17 and 5.18 that the dissipative third-order Airy function correctly describes the profiles of the contact discontinuity.

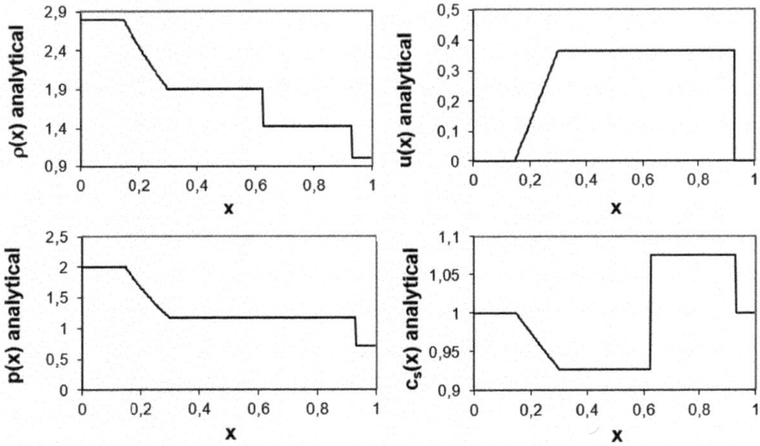

Fig. 5.16. Adiabatic Euler system: analytical solution at time $t = 0.4$ for weak shock conditions

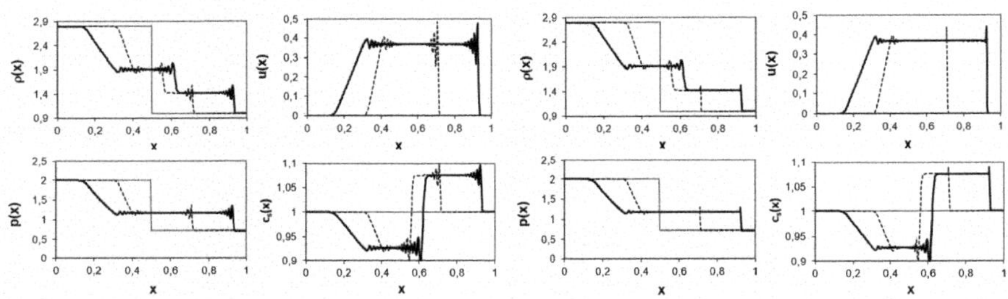

Fig. 5.17. Adiabatic Euler system: density profile at times $t = 0.175$ (dotted), 0.35 (full line) provided by simulation using the Lax–Wendroff method. Time step $\Delta t =$ (left) $6 \cdot 10^{-4}$, (right) $1.6 \cdot 10^{-3}$. No viscosity

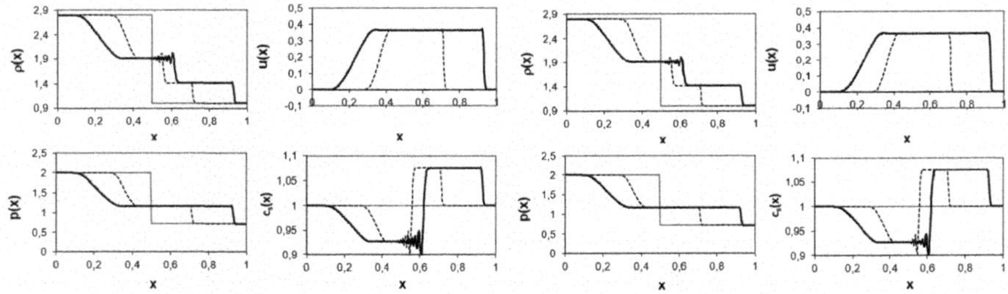

Fig. 5.18. Adiabatic Euler system: density profile in same conditions than Fig. 5.17 but with added viscosity $\nu = 10^{-3}$

5.4.6 *Conclusion on the Navier–Stokes System*

As for the Burgers equation, the perturbation method of the solution of the equivalent equation was quite efficient to understand the nature of artefacts generated by dissipative second-order schemes on the shock profiles. They generate bumps on the top or the bottom of the shock according to the sign of the dispersive term. The artefacts generated on the contact discontinuities are similar to that obtained for linear advection.

5.5 Another Method of Suppressing the Oscillations Generated by Dispersive Schemes

The presence of oscillations generated by dispersive numerical schemes is central in this paper. For the Burgers equation as well as the Navier–Stokes system, the oldest way of suppressing or at least reducing them is to add an appropriate amount of artificial viscosity. More recently, many methods have used limiters. The literature on this subject is huge [11]. We have seen in Sec. III another way achieving this goal, by adding a nonlinear cubic correction.

But an original approach was proposed in Ref. [12]. The idea comes from control theory. As the real profile for a shock is known, a way of suppressing parasitic oscillations is to consider this profile as the target function in the control algorithm. Different implementations of the method for isothermal Euler equation has been tested in Ref. [12]. The method is quite effective, although the structure of the control function results in very small addition to the scheme. We refer to Figs. 5 and 6 of Ref. [12] for illustration of the efficiency of the method, respectively, without and with viscosity.

The method was successfully applied to two-step shock waves propagation for isothermal Euler equations in Ref. [13].

5.6 Conclusion

The present paper gives an overview of a large amount of results on the artefacts generated by numerical schemes discretising the linear and nonlinear advection equation, the Burgers equation with linear viscosity and the Euler and Navier–Stokes system. In close collaboration with Professor Alexei Porubov, we have analysed the shock profiles provided by a range of numerical schemes, with a special emphasis on the parasitic oscillations generated by second-order schemes. These oscillations have a practical importance, as they can seriously spoil the numerical results, for example by generating negative values of physically positive quantities, and drive a great theoretical interest. As a result of the collaboration, the position and magnitude of these oscillations were thoroughly investigated and analytically predicted. Moreover, different original ways of reducing or even suppressing them have been proposed, by appropriate addition of compensation terms, as well as by control. Results on oscillations generated by third order schemes have also been obtained. For the sake of brevity, we have only given a brief account of this topic. We hope that this contribution dedicated to the 60th birthday of Professor Alexey Porubov has somehow given essential results and will induce the interested reader to have a closer look at the references, that contain more detailed material.

References

1. Shokin Y (1983) The method of differential approximation. Springer, Berlin
2. Bouche D, Bonnaud G, Ramos D (2003) Comparison of numerical schemes for solving the advection equation. Appl Math Lett 16:147–154
3. Bouche D, Weens W (2024) Quantitative analysis of numerical schemes of PDE. EDP Science, Paris
4. Arbia G, Bouche D (2014) Closed form solution and equivalent equation approximation of linear advection by a non dissipative second order scheme for step initial conditions. Acta Applicandae Mathematicae 130:151–162
5. Porubov AV, Bouche D, Bonnaud G (2008) Description of numerical shock profiles of non-linear Burgers'equation. Int J Finite Vols 5:1–16
6. Porubov AV, Bouche D, Bonnaud G (2010) Compensation of the scheme dispersion and dissipation by artificial non-linear additions. In: LNCS transactions on computational science, vol 5890, pp 122–131
7. Porubov AV, Bouche D, Bonnaud G (2013) Analytical solutions to detect the scheme dispersion for the coupled nonlinear equations. Commun Nonlinear Sci Num Simulat 18:2679–2688
8. Lerat A, Peyret R (1975) Propriétés dispersives et dissipatives d'une classe de schémas aux différences pour les systèmes hyperboliques non-linéaires. La Recherche Aérospatiale 2:61–79
9. Sod GA (1978) A survey of several finite difference methods for systems of nonlinear hyperbolic conservation laws. J Comp Phys 27:1–31
10. Bonnaud G, Bouche D, Porubov AV (2008) Numerical oscillations close to discontinuities. In: International conference on fluid dynamics—Cargèse
11. Kriel AJ (2017) Error analysis of flux limiter schemes at extrema. J Comp Phys 328:371–386
12. Porubov AV, Bondarenkov RS, Bouche D, Fradkov AL (2017) Control of nonlinear shock waves propagation for isothermic Euler equations. ZAMM 98-3:448–453
13. Porubov AV, Bondarenkov RS, Bouche D, Fradkov AL (2018) Two-step shock waves propagation for isothermal Euler equations. Appl Math Comput 332:160–166

Chapter 6
Time-Resolved Estimation of Multifractal Spectra of Ferroelectric/ Piezoelectric Polymer Dynamics and Neuromimetic Fiber Orientation: Towards Electric-Field- and Electron-Beam-Controllable Scaffolds and Tissue-Engineering Constructs with Dynamic Beads

Eugenia L. Buryanskaya, Oleg V. Gradov, Margarita A. Gradova, Alexey L. Iordanskii, Valentin V. Kochervinskii, Irina A. Maklakova, Anatoly A. Olkhov, and Anna V. Ratnovskaya

Abstract This paper describes a novel approach for monitoring the efficiency of control over ferroelectric fibrous nonwoven materials (and scaffolds based on them), which are sensitive to the electric field and irradiation using charged particle beams. The method proposed is based on the dynamic time-resolved analysis of multifractal spectra of the fiber networks and time-resolved estimation of the fractal dimensions of such networks in the external fields or under the electron beam irradiation conditions. The introduction also provides a bibliographic background for the use of fractal and multifractal approaches to the analysis of "intelligent scaffolds" and networks of electrospun polarized fibers with dynamically regulated porosity as modern structured materials for the cell growth guidance and tissue-engineering constructs. The possibility of electrodynamic analysis of the behavior of fibers and scaffolds with (multi)fractal properties in an external field using mathematical models of fractal

E. L. Buryanskaya · O. V. Gradov (✉) · M. A. Gradova · A. L. Iordanskii · I. A. Maklakova · A. A. Olkhov
N.N. Semenov Federal Research Center for Chemical Physics RAS, Kosygina Str. 4, Moscow 119991, Russia
e-mail: o.v.gradov@gmail.com

E. L. Buryanskaya · V. V. Kochervinskii
Bauman Moscow State Technical University, 2nd Baumanskaya Str. 5/1, Moscow 105005, Russia

O. V. Gradov · I. A. Maklakova · A. A. Olkhov · A. V. Ratnovskaya
Emanuel Institute of Biochemical Physics RAS, Kosygina Str. 4, Moscow 119334, Russia

H. Altenbach and V. A. Eremeyev (eds.), *Dynamics of Discrete and Continuum Structures and Media*, Advanced Structured Materials 221,
https://doi.org/10.1007/978-3-031-75626-9_6

antennas is suggested. The reference list includes 77 items and the paper contains 7 tables with various multifractal graphs.

Keywords Nonwoven structured materials · Field-controllable scaffolds · Beam-controlled scaffolds · Tissue-engineering constructs · Neuromorphic fibers · Beads · Electrospinning · Piezoelectric polymers · Ferroelectric polymers · Multifractality · Multifractal spectra · Fractal scaffolds · Fractal antenna models · Fractal surface

6.1 Introduction

Recent reports on the piezoelectric properties of polylactide films, micro- and nanofibrils, which are quite sufficient for their application in sensors and actuators [1–7], are a promising physical basis for a new branch of nanoindustry—the industry of biodegradable and biocompatible piezoelectric nanofibers. Such systems, by definition, can be used in nanobiotechnologies of the future, including the control upon tissue engineering structures using both electrical and biomechanical feedbacks. It should be noted that previously the most abundant piezoelectric polymer was PVDF, which is biocompatible but not biodegradable, so it could be used in biomedicine only as a non-bioresorbable substance, for example, as a component of biocompatible nanoimplants [8–10].

At the same time, the physical foundations of the biomimetic approach to the development of implants and tissue engineering structures based on ferroelectric systems initially also provided for the design of biodegradable ferroelectrics that are bioresorbable and functionally substituted by the living tissues controlled by the same signals/stimuli (mechanical and electrical ones) as the polymer ferroelectrics [11]. It is quite obvious that such approaches in the future could have a significant impact on nanobiomedical technologies associated with regeneration of the excitable tissues, such as muscle and nerve tissues.

It is noteworthy that, in contrast to the conventional methods of nanomedicine and regenerative bioengineering of the nervous tissue, which usually imply biochemical and molecular genetic agents, including RNA expression [12–15], in piezoelectrically active nanoimplants, the agent and stimulator of the tissue regeneration is the electrically activated implant itself, which can be considered not only as a passive substrate/support/scaffold for the cell growth, but also as a nanomedical "electroceutical" agent [16–18]. In this case, which is typical for convergent technologies, the following properties combine and give rise to a new emergent quality of the material: a nanodevice that stimulates tissue regeneration; an endogenous bioelectric generator in the plane of the interface of the excitable tissue membrane and the stimulating material (fibers or films); transmission of information over a regenerated fiber in the fields generated by the polymer ferroelectric and the subsequent axonal path finding, axonal guidance under the conditions of galvanotaxis in the above fields [19]. In this case, the inevitable interference between the signals of the polymer ferroelectric at the interface with the tissue and the membrane bioelectric generator of the nervous tissue at the interface with the polymer ferroelectric leads to the indivisibility of the properties of such ferroelectric structures as implantable electrical/electromechanical sensors and actuators. In other words, due to the effectively coupled (at the level of the interface between the living tissue and nanomaterial) feedback loops, implantation of such nanomaterials into the tissue automatically leads not only to a prolonged therapeutic or sensory effect, but also to the complex nanotheranostics, where it is impossible to separate the effect and its diagnostic registration.

Unlike other methods of nanotheranostics of the nervous tissue, for example, those applied for the treatment of brain tumors [20–25], which mostly imply pharmaceutical or biochemical theranostic agents, ferroelectric polymers provide biophysical/electrophysical (electrophysiological) form of action and signal detection by the tissue instead of the external device, due to which therapy can be carried out internally, without deciphering/interpreting the diagnostic signals of the theranostic agents, in accordance with the tissue response to their stimulation with the same frequencies and signal amplitudes that they can support. This is due to the crucial role of the impedance matching between the living tissue and ferroelectric polymers [28, 36]. However, this aspect of nanobiocompatibility for polylactide and scaffolds from its nanofibers has not been extensively studied, especially in the high-frequency (MHz and higher) range, due to the technical and methodological limitations. This engineering and computational problem can be solved in the nearest future, which will allow to design theranostic scaffolds based not on the arbitrary empirical observations, but on the theoretical modeling results and predictions, minimizing the problems of biocompatibility and rejection or encapsulation of a nanoscaffold by the tissue resulting from the incompatibility in their conductive, hydrophobic, and electrogenic characteristics.

In the previous work [29] we proposed to consider the scaffold simultaneously as a sensor and a sonar-actuator that implements both electrical (electrophysiological) and acoustic and (or) electroacoustic tissue stimulation, as well as recording of its own signals, which transfers controlled tissue regeneration to a section of theranostics where the scaffold itself is the descriptor source for the tissue regenerated. This approach turned out to be progressive and is still relevant [30, 31]. However, firstly, the properties of such generators and sonars depend on the structure, size of the active elements, distances between such elements (like polymer fibers), resonant characteristics and quality factor for a given frequency range. Secondly, it is obvious that for electrospun fibers, the thicknesses and distances between the scaffold fibers will depend on the rates and modes of the polymer melt/solution supply during the fiber production, its concentration and viscosity, voltage applied, static or dynamic nature of the substrate, etc. [32–34].

Therefore, it is necessary to provide a multilevel analysis of the entire hierarchy of the structures responsible for the generation of potential or acoustic signals, antenna and electret behavior of the system which is supposed to be used to modulate the effect on biological tissue. In addition, tissue regeneration in the scaffold depends on the scaffold structure even if it is not an "intelligent scaffold", i.e. if it does not respond to electrical, acoustic, or chemical signals of the surrounding medium. Of particular interest are the fractal grids/meshes/lattices, on which the cells with different sizes of filopodia/protrusions grow and differentiate. (Such scaffolds allow building both tissue engineering structures for osteoregeneration, and artificial neural networks of the novel architecture including those based on microacoustofluidics controlled by the electrical signals, or to regenerate nervous tissue using acoustic and electrical signals propagating along the fractal scaffold and distributed in it according to its impedances [35]).

The need to analyze the fractal parameters of potential scaffolds from the point of view of regenerative biomedicine and biomedical physics is associated with the difference in the regenerative parameters of the scaffold depending on its surface morphology/topology and the different efficiency of filopodia growth and cell protrusions into the scaffold depending on its pore size distribution and the mechanical properties of the scaffold itself. From the viewpoint of piezotronics and high-frequency (or radiofrequency) electroacoustics and electromechanics, the feasibility of using field-controlled biomedical structures with multifractal properties can be intuitively illustrated by the analogy of a fractal antenna, which makes it possible to simultaneously receive many signals with different frequencies and modulations, leading to different responses of the multimodal "receiver"/controllable object.

Currently, there are many works on the design of fractal scaffolds and fractal characterization of scaffolds from capillary morphogenesis to neuromorphogenesis and the design of artificial neural networks [36–50]. Thus, our approach including dynamic electron microscopic study of such structures with time-resolved characterization of their fractal properties is quite reasonable. In [47], it is proposed to use the method of box counting for multi-scale porous scaffold characterization, which is a part of the multifractal processing and multifractal image analysis techniques. Therefore, in this paper we will address not conventional calculation of the Hurst exponents and Hausdorff dimension, but the multifractal method of multilevel analysis of the material structure.

Fractal properties of PLA-based structures degradable by both biological and physicochemical factors have been studied since the early 1990s [51, 52]. The application of fractal descriptors to the analysis and characterization of the structure of PLA-based composites for engineering purposes (both for construction problems and for electronics) has become common since the early 2010s [53–56]. However, until the second half of the 2010s application of the fractal method in biological studies using PLA and its composites was, as a rule, reduced to the analysis of biodegradation and soil degradation of the polymer/composite [57]. At the same time, it should be noted that fabrication and fractal study of PLA gradient scaffolds for tissue engineering has been performed already in the mid-2000s, but was not demanded, since it has been published not in the top edition [49].

We believe that, having proved in our present work the feasibility of dynamic analysis of the multifractal parameters of PLA-based structures during activation by an electron beam, or during transduction of the acoustic/mechanical waves through the scaffold material for regenerative medical applications, we will also renew the interest in dynamic analysis of multifractal/fractal characteristics of PLA fibers for biomedical applications. This may be due to the possibility of using fractal dimension metrics for both normal cells and malignant cells [58–64].

6.2 Materials and Methods

The polymer fiber samples were obtained from Poly-L-lactic acid (PLA) solution in chloroform by electrospinning method in accordance with the procedure described elsewhere [65–76]. SEM studies were performed using JEOL T330-A scanning electron microscope with the image digitization developed by Gradov and Alexandrov [33, 34]. The data was processed using HARFA software for a series of video recordings obtained under the electron beam. Processing of the dynamical SEM data was performed using the box counting method, similar to the technique applied earlier for multifractal analysis of piezoelectric polymer thread-based microfluidics or acoustofludics [35].

The following multifractal parameters were calculated:

D(q)—a multifractal spectrum where D_q is a generalized dimension for a dataset and Q is an arbitrary set of exponents;
$f(\alpha)$—another type of a multifractal spectrum allowing to determine the fractal dimension at $Q = 0$;
M(q)—the "total mass" or sum of pixels in all boxes for the arbitrary scale ("box size" in box counting method).

We studied the following states of PLA electrospun polymer fiber structures (Fig. 6.1), which are important for understanding and modeling developmental mechanics of neuromorphogenesis in regenerative medicine:

(a) The states with uncharged or positively charged (and, accordingly, poorly visualized under the electron beam, forming dark spots in the image upon charging) fibers and charged contrasting beads—products of varicose instability of the electrospinning processes. Such separated contrasting structures, which are not prone to charge drainage, behave in the electromagnetic field as multiple point sources and often form aggregates and temporary clusters/modules depending on the field configuration, which is also influenced by their own charging. In this case, the influence of dielectric thin PLA fibers on electromechanics can be neglected—it cannot be demonstrated in the experiment since they are poorly visualized (unlike points *b, c, d, f*).
(b) The states with rarer beads and well-charged fibers (with greater thickness relative to the diameter of the beads), influencing the course of beads clustering and beam-induced electromechanics of the fiber system. In contrast to the above case, in this case an important role is played by the geometry of beads, which, as a rule, are not approximated by spheroids, and the orientation of beads with a given geometry (ellipsoidal or even more complex shape of the solid of revolution) either along the fiber or across it, along or perpendicular to the field. In this case, the charge drainage phenomena, charge exchange between the fiber and the beads and vice versa, and (re)polarization are observed, resulting in the sharp contrast changes over time, in which orientation of the beads to the fibers often changes. In this case, contacts between beads play an important role, accompanied by a charge transfer at the contacts (which is similar to the charge

Fig. 6.1 Various configurations of electrospun PLA fiber networks with electrohydrodynamic instabilities (beads) and without them

exchange at electrical synapses/ephapses). Figure 6.1b illustrates an example of this state, in which one bead is aligned orthogonally/perpendicularly relative to the other one.

(c) Figure 6.1c shows an example of a similar state with the beads arranged in a chain with the same non-orthogonal orientation. In developmental mechanics, such circuits are usually arranged during the formation of signal chains of neurons or simple reflex arcs, and in bioelectrophysical experiments in vitro during the assembly of a neuron chain in galvanotaxis when oriented along the electric field lines.

(d) Another state of the circuit with beads missing or masked by the charging effects and well-visualized and contrasted fibers that can be effectively charged, which are characterized by wandering/redistribution of the charge over the surface or the polymer fiber network, along the equivalent circuit of the network. Breathing of the network in this case is determined by the charge wandering along the fibers induced by the electron gun and their discharging onto the neighboring fibers. In such cases, as a rule, no exotic processes are observed in the network dynamics, in contrast to the networks with beads.

(e) The states with pronounced single PLA fibers mechanically supporting the beads but not electrically interacting with them (i.e., not electrostatically adhering to the fibers). These states can be considered as a simple case of type (a) states, in which the PLA fibers are charged and contrasted under the electron beam. Often such states are induced by the electron beam from the states of (a) type when the specific electron flux per the sample area is increased, for example, when moving from a lower magnification to a higher one using the hardware of the electron beam instrument/setup.

(f) The states with pronounced fibers mechanically supporting the beads and electrically and electromechanically interacting with them (electrostatically adhering to the fibers until completely enveloping them and forming contacts through which the charge flows). An example of such a mechanically interacting network is given in Fig. 6.1f. Often similar phenomena occur on the polymer fibers that under an electron beam desorb intercalated compounds into the vacuum of a column or the relatively rarefied gas phase of an LVEM/ASEM microchamber. Some of the phenomena observed on such fiber structures are described only by electrohydrodynamics, taking into account electrohydrodynamic instabilities.

6.3 Results and Discussion

The main results obtained are represented in Figs. 6.2, 6.3, 6.4, 6.5, 6.6, 6.7 and 6.8 and summarized below:

- $D(q)$ calculated with geometric box counting method (Fig. 6.2) and with arithmetic box counting method (Fig. 6.7) are non-linear and indicate the multifractal nature of the structures/textures in the images within the series. In non-fractal circle case (according to the empirical box counting method) at dimension = 1.0 the spectrum would be linear, and in the case of mono-fractal structure it would be more nonlinear than in the first case, but much less nonlinear than can be observed in our series. Changes in the nature of the multifractal spectrum curves from N-shaped (as in the current–voltage characteristics of a tunnel diode) to sigmoidal ones (logistic curves and Gudermannian functions, i.e. "hyperbolic amplitudes") and vice versa during dynamic fluctuations (transient processes in an ungrounded circuit) and electromechanical optimization of the network of neuromimetic fibers in external fields evidence the exit and entry of the neuromimetic guiding fibers from the ROI (region of interest) field when adjusting the network topology, as well as changes in connectivity within the "connectome" of the neuromimetic network of the fiber material during "reception" of the electrical signal.

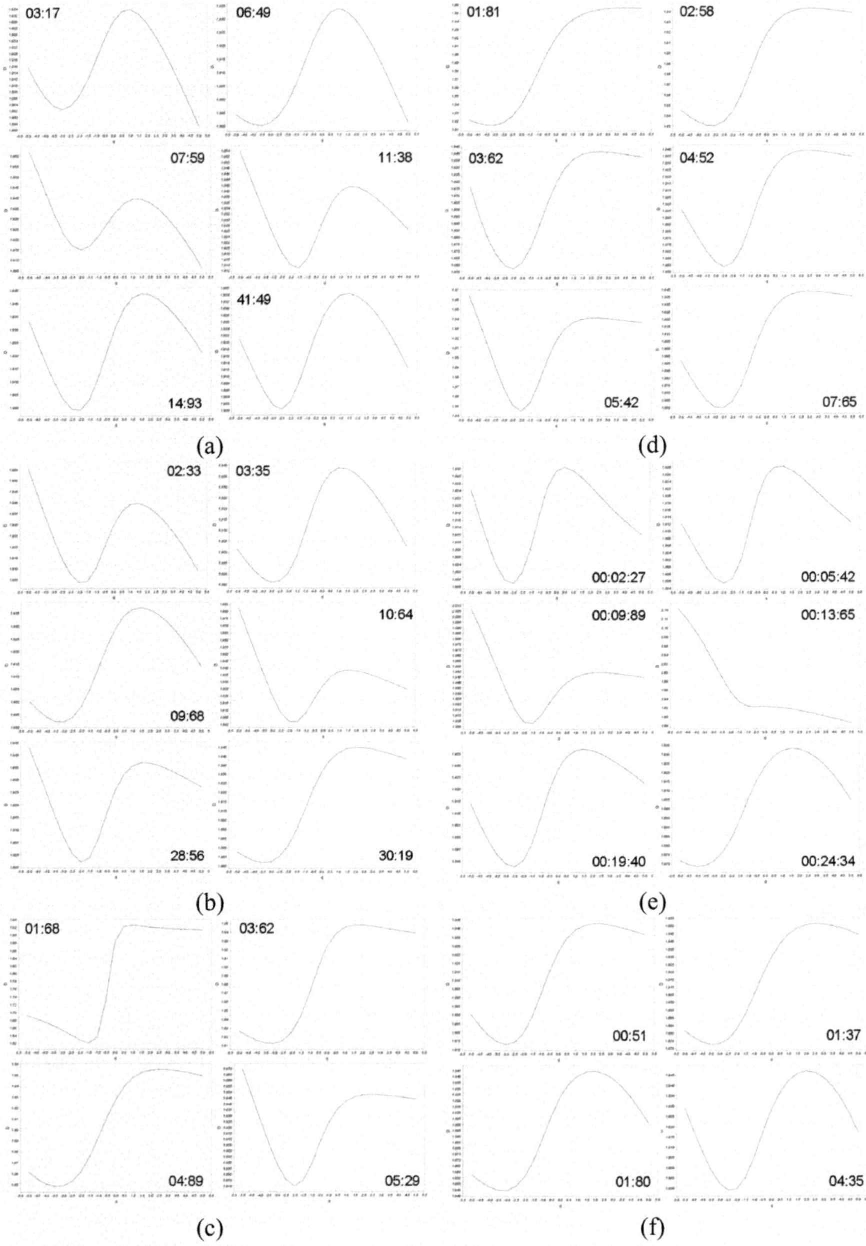

Fig. 6.2 $D(q)$ for the structure dynamics observed in Fig. 6.1a–f calculated with geometric box counting method

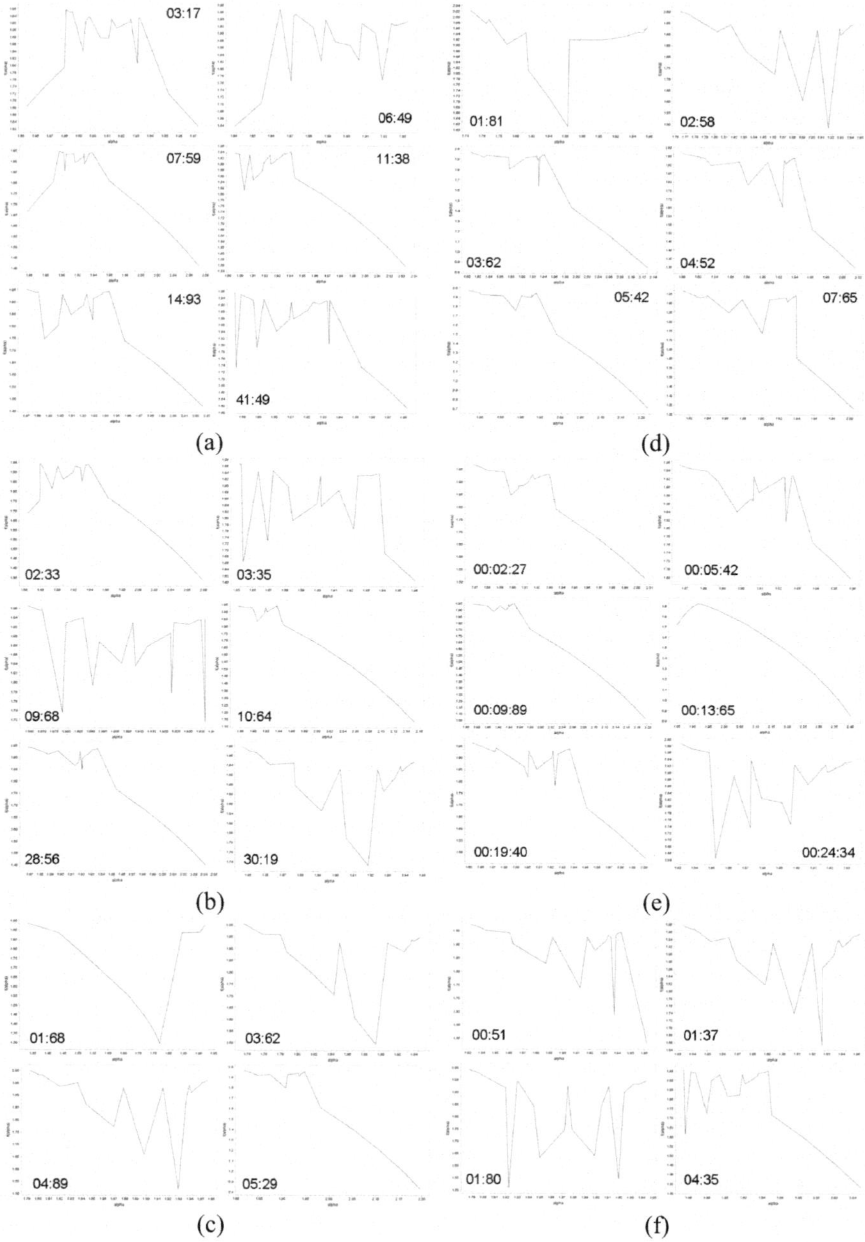

Fig. 6.3 f(α) for the structure dynamics observed in Fig. 6.1a–f calculated with geometric box counting method

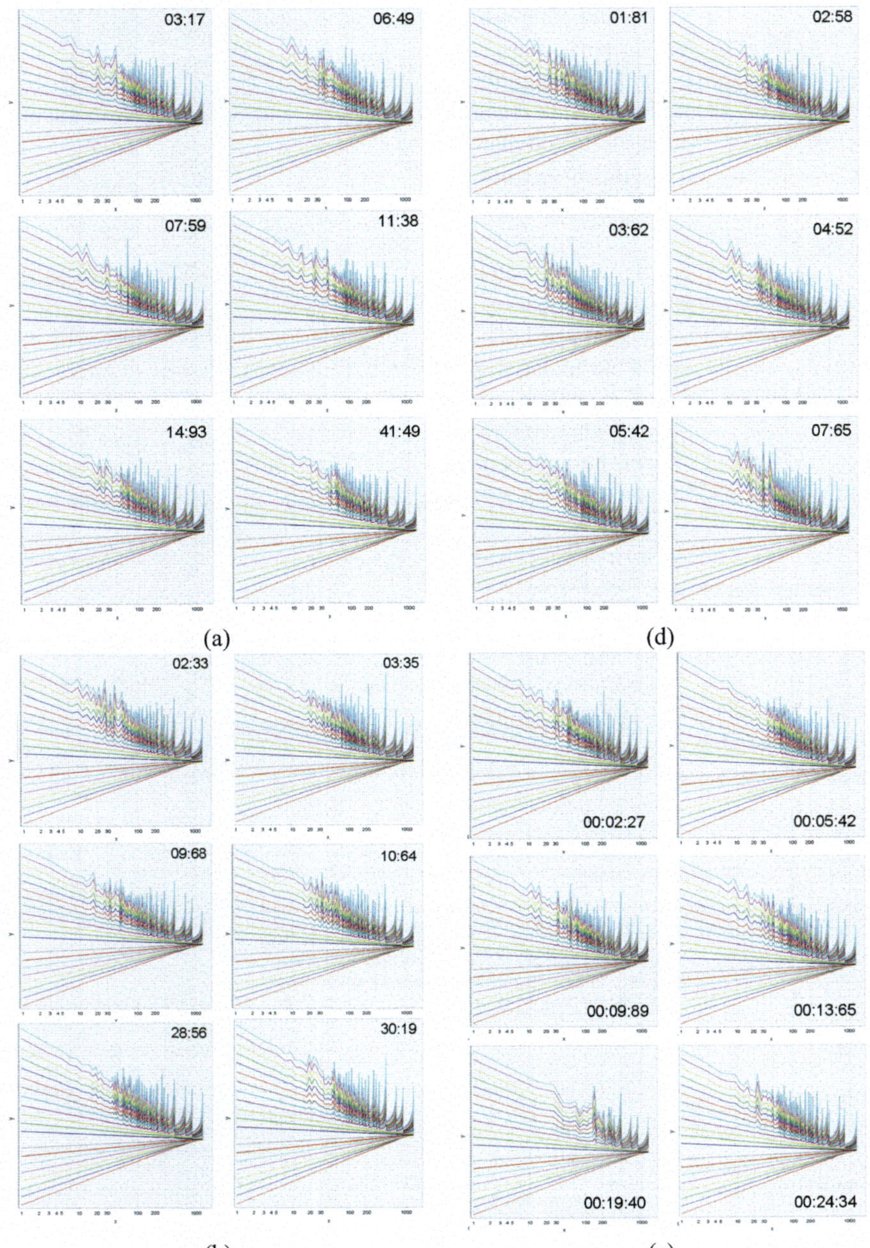

Fig. 6.4 $M(q)$ for the structure dynamics observed in Fig. 6.1a–f

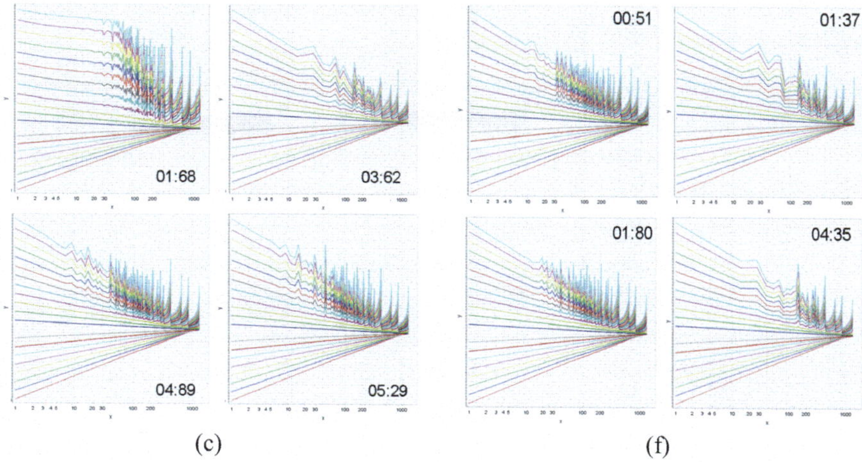

Fig. 6.4 (continued)

- The shift of $f(\alpha)$ peaks along the abscissa (Figs. 6.3 and 6.8) during the adjustment of the network topology up to the linearization of individual fragments and signal inversion (the amplitude goes into the negative range where only positive peaks were observed earlier) indicates the dynamics of the network reconfiguration with the removal of individual neuromimetic and control fibers from the network (among the charged fibers conducting the electrophysical signal). It can also be seen that many fiber networks go through similar stages in their evolution according to the f(α) peak criterion. Such peak shapes can be observed in Fig. 6.3b, c, as well as in Fig. 6.3d, e. Similar stages can be observed in the graphical representations of the arithmetic box counting results in Fig. 6.8b, c, as well as in Fig. 6.8d, e.
- Categorical differences in the peak distribution in the graphs of the "total mass" or sum of pixels in all boxes (see Fig. 6.4) for this arbitrary scale in different groups confirm the original hypothesis about the difference in the network topology, observed in the EM photo (Fig. 6.1). A change in the distribution over time within one group and the similarity of the graphs for some groups by the end of the experiment indicates the presence of some structural optima of the habitus/ "fractal circuit diagram" of the network (achieved when optimizing for the local boundary conditions, but unattainable in general due to the non-stationary nature of the process during "galvanotactic" signal transduction in a self-organizing neuromimetic network).
- Empirical and estimated box counting graphs calculated by the arithmetic box counting method (Fig. 6.5) show that in a number of cases (for example, in the structures at the images (e), (c), and (f)) there is a depletion of the graph of empirical points by the end of monitoring or a temporary change with relaxation of the point configuration during electromechanical or other consistent/matched

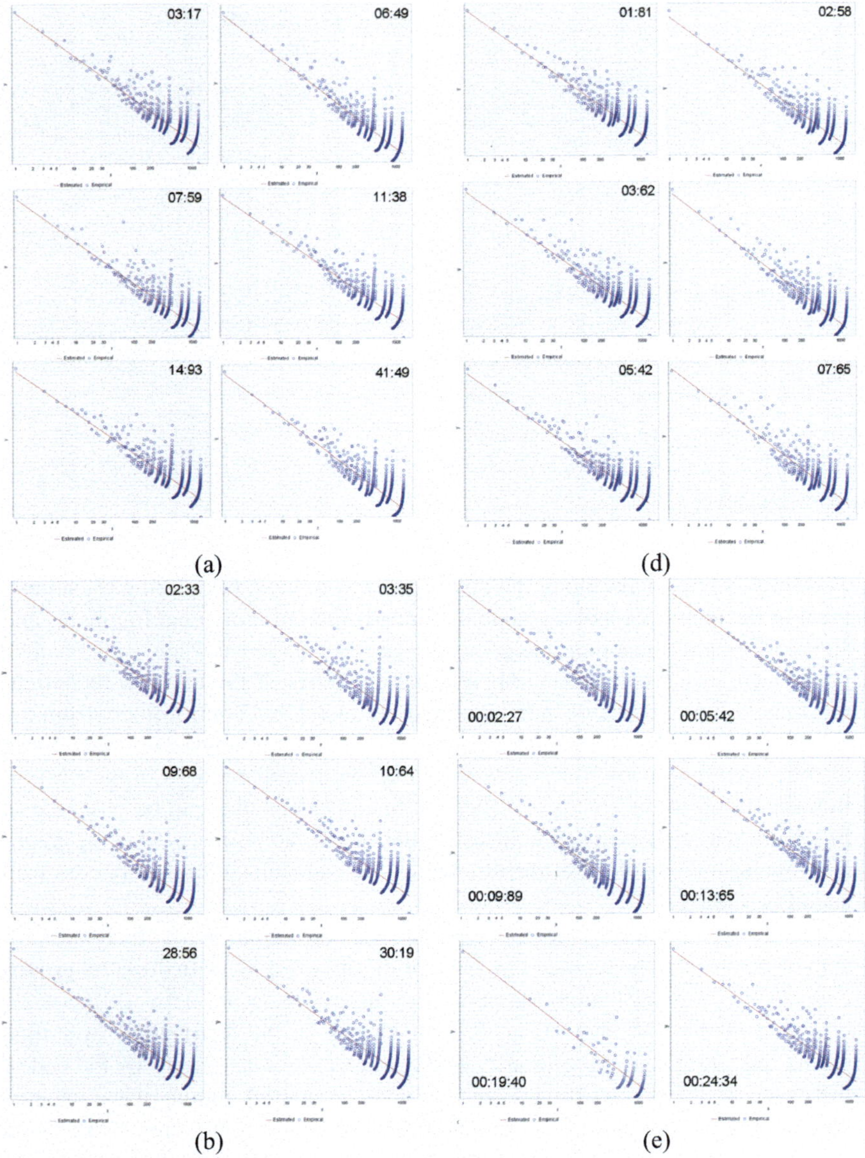

Fig. 6.5 Empirical (blue points) and estimated (red line) box counting graphs calculated with arithmetic box counting method for the structure dynamics in Fig. 6.1a–f

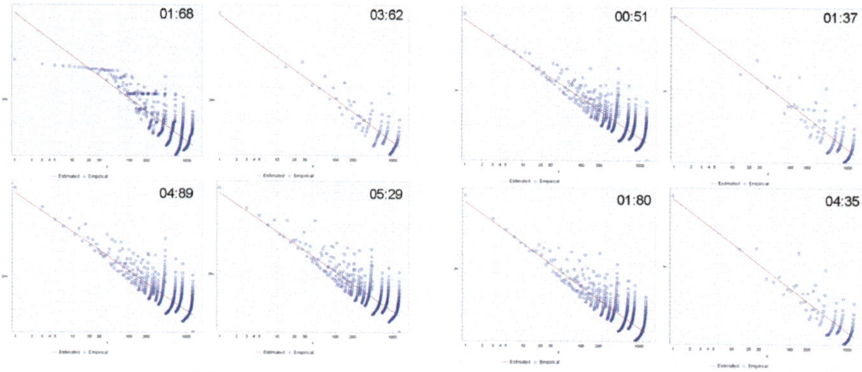

(c) (f)

Fig. 6.5 (continued)

oscillations (with a constant field and the electron source fluence) of the network when reconfiguring its topology.

- Comparison of the scaling behavior for different groups (see Fig. 6.6) shows that in electromechanical dynamics there are areas and groups of structures that generally preserve fractal properties, and there are those that change unstably (up to the selection and degeneration of the spectrum—the elimination of many peaks as the network develops). At the same time, the most neuromimetic fibers in habit are characterized by relaxation (in the simple mechanical sense of this term), while for the fibers with alternative topologies and, in particular, buckling, there is a risk/high probability of transition into a completely stable and degenerate state, according to the criteria of scaling invariance.
- All the transitions between habits of individual microfiber structures under the influence of the electromagnetic field affect the network topology and the fractal behavior of the irradiated system. All the transitions in the network topology influence the fractal behavior of the system and can be indicated using specific graphs shown above. For all the electromechanically controlled (or secondarily responsive to the changes in the network topology) fibers, transition points can be found using multifractal spectra and adjacent graphs obtained during calculations. We provide a detailed description of this approach and method in other works that are less focused on the role of network electromechanics and the aspects of materials science and biotechnical applications, and mostly focused on the fundamental aspects, but are beyond the scope of this special issue.

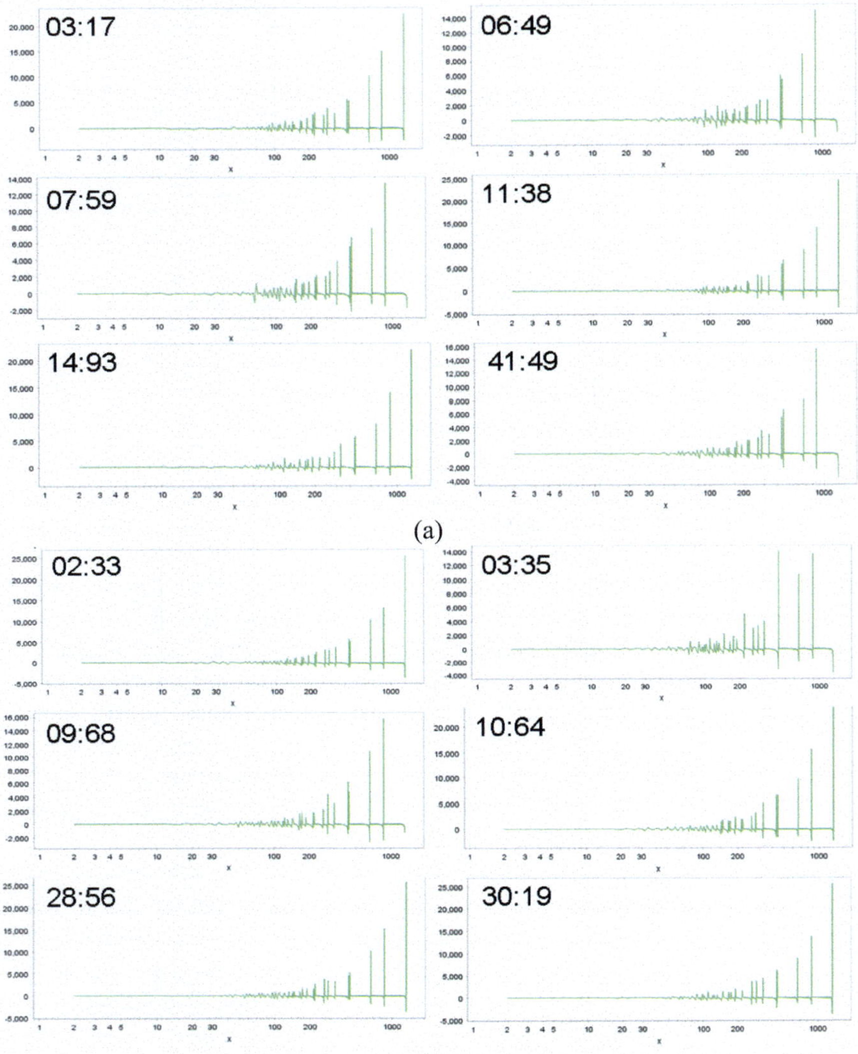

Fig. 6.6 Scaling behavior “fingerprints” for the structure dynamics observed in Fig. 6.1a–f calculated with arithmetic box counting method

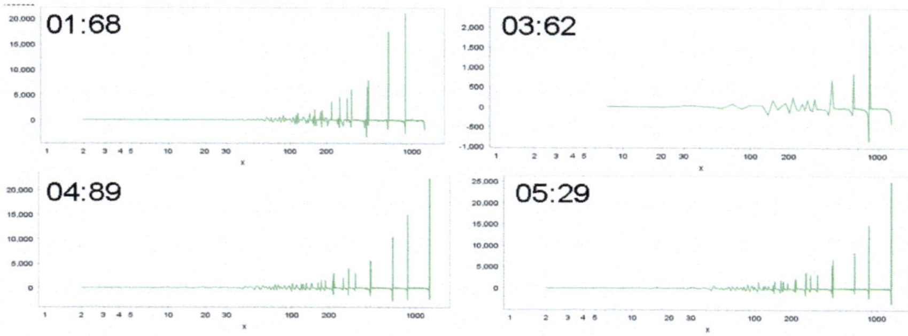

(c)

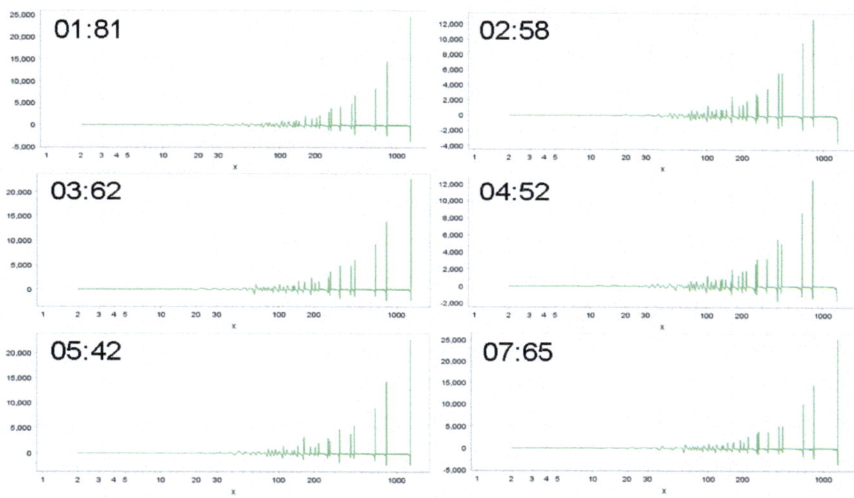

(d)

Fig. 6.6 (continued)

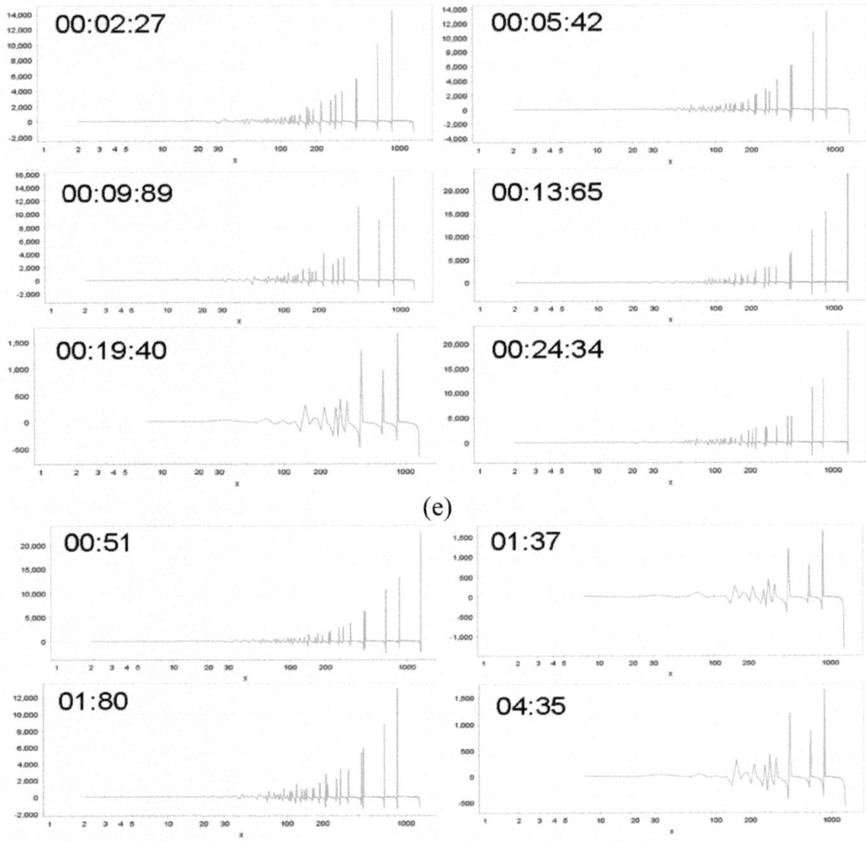

(f)

Fig. 6.6 (continued)

Acknowledgements The authors express their gratitude to:

• Mikhail K. Filippov for reconstructing the old SEM device for its use in electrophysical and electrophysiological experiments (since 2018);

• Valentina N. Seliverstova, Alexander V. Kholyavko, Vladimir S. Kalinin, and Evgeniy N. Kulpin (Department of Experimental Technology of the Federal Research Center for Chemical Physics of the Russian Academy of Sciences) for the production of unique components / details for electron microscopes according to the drawings of Mikhail K. Filippov and Oleg V. Gradov;

• Pavel L. Alexandrov for automation and digitization of data collection from the electron microscope (since 2018).

Author Contributions BEL—multifractal analysis and formal mathematical interpretation;

GOV—conceptualization and bibliographic analysis; development of a system for visualization and dynamic registration of the behavior of nonwoven fibers;

GMA—electron microscopic study of the samples; text editing and translation;

IAL—conceptualization of fundamental ideas about the applicability of polymer fibers in prosthetics of neural and neuromimetic structures and, in the future, the brain; founder of the scientific

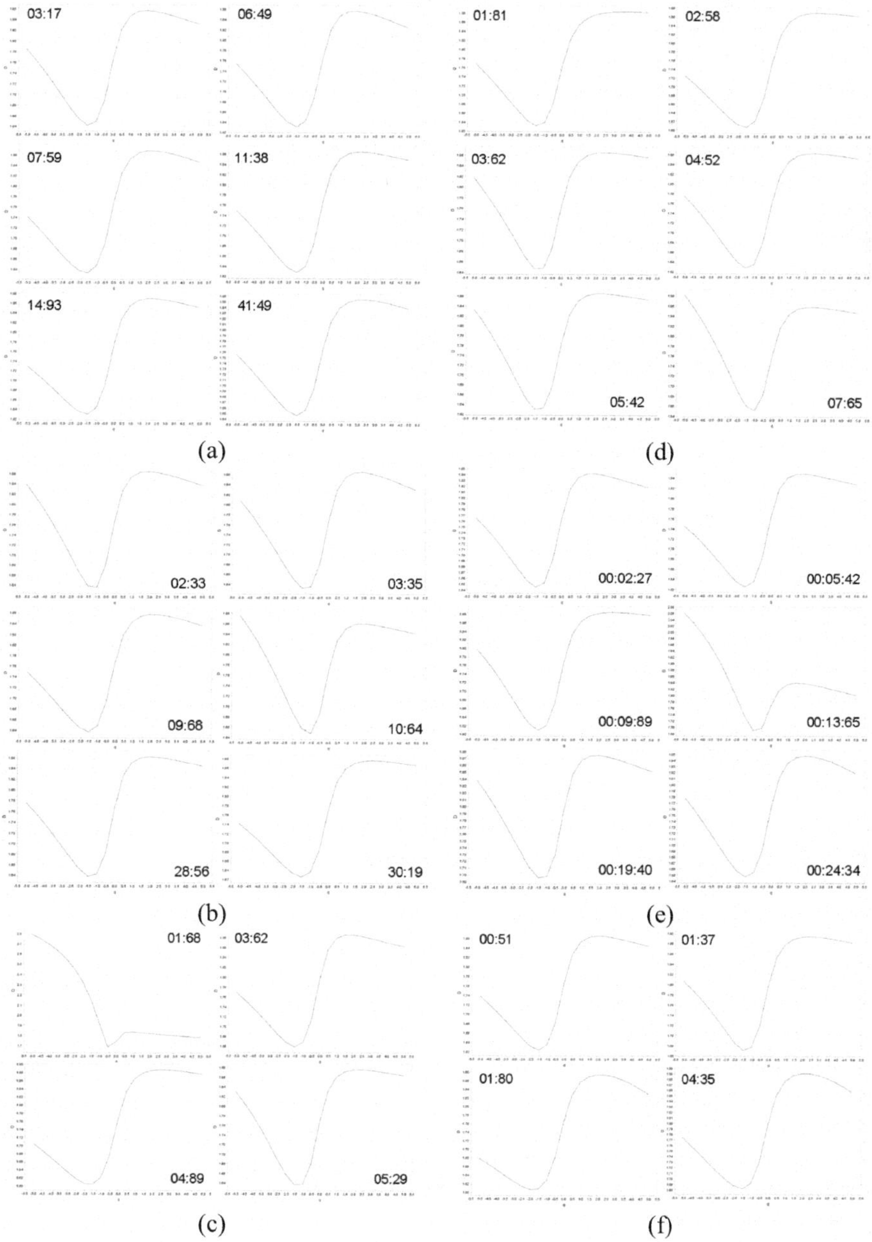

Fig. 6.7 $D(q)$ for the structure dynamics observed in Fig. 6.1a–f calculated with arithmetic box counting method

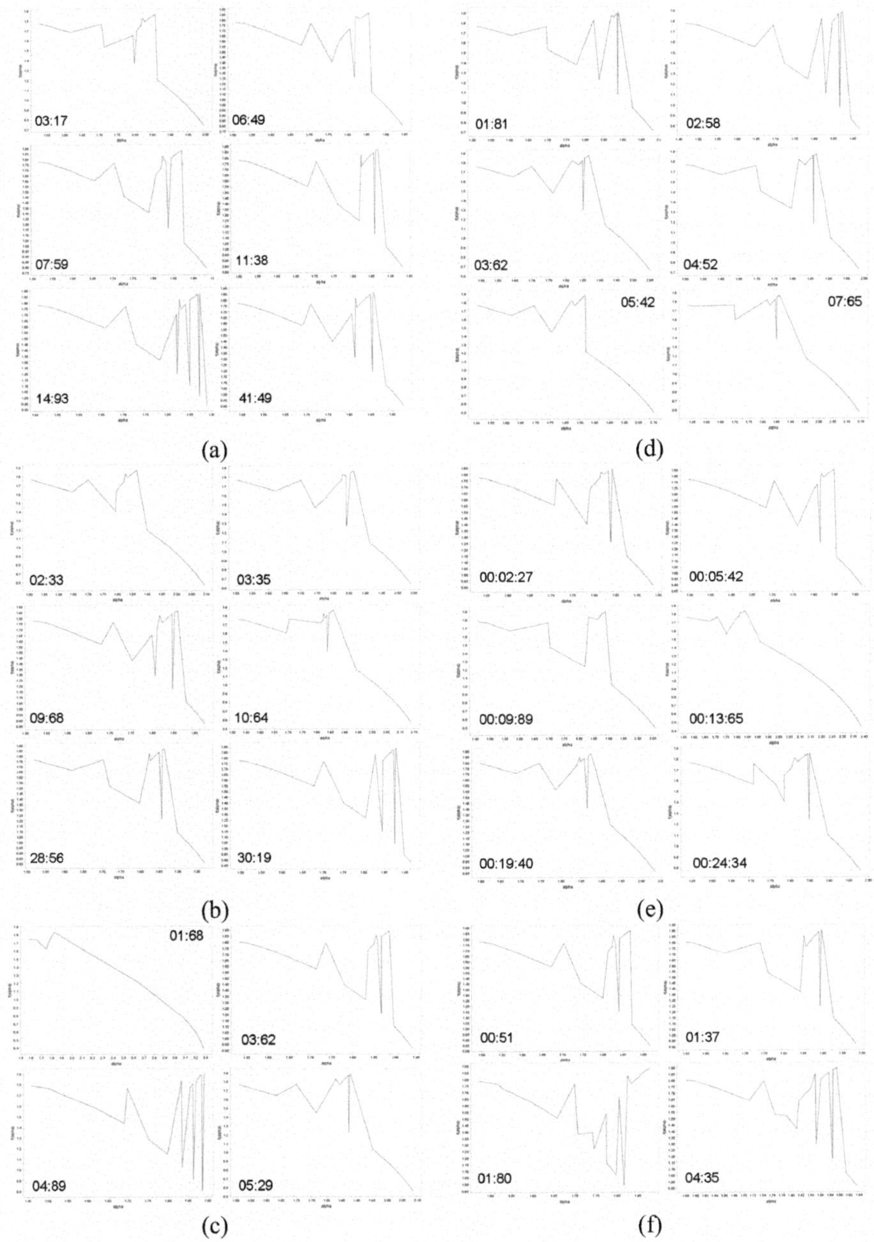

Fig. 6.8 $f(\alpha)$ for the structure dynamics observed in Fig. 6.1a–f calculated with arithmetic box counting method

school ("virtual college") at the Emanuel Institute of Biochemical Physics of the Russian Academy of Sciences;

MIA—dynamic electron microscopy of PLA fiber samples;

KVV—electrophysical studies of piezoelectric and ferroelectric polymers and fundamental contribution to their physical interpretation; founder of a scientific school ("virtual college") at Karpov Institute of Physical Chemistry and Bauman Moscow State Technical University;

OAA—synthesis of PLA fibers by electrospinning method and contribution to the analysis of the sources of electrohydrodynamic instabilities and "beads" in the nonwoven fiber geometry.

RAV—electron microscopy of neuromorphic fibers (2019–2021).

All the authors have made an equal contribution to this research and are listed in alphabetical order without prioritization.

References

1. Ando M, Kawamura H, Kageyama K, Tajitsu Y (2012) Film sensor device fabricated by a piezoelectric poly (L-lactic acid) film. Jpn J Appl Phys 51(9S1):09LD14. https://doi.org/10.1143/JJAP.51.09LD14
2. Sencadas V, Ribeiro C, Heredia A, Bdikin IK, Kholkin AL, Lanceros-Méndez S (2012) Local piezoelectric activity of single poly (L-lactic acid)(PLLA) microfibers. Appl Phys A 109(1):51–55. https://doi.org/10.1007/s00339-012-7095-z
3. Ando M, Kawamura H, Kitada H, Sekimoto Y, Inoue T, Tajitsu Y (2013) Pressure-sensitive touch panel based on piezoelectric poly (L-lactic acid) film. Jpn J Appl Phys 52(9S1):09KD17. https://doi.org/10.7567/JJAP.52.09KD17
4. Varga M, Morvan J, Diorio N, Buyuktanir E, Harden J, West JL, Jákli A (2013) Direct piezoelectric responses of soft composite fiber mats. Appl Phys Lett 102(15):153903. https://doi.org/10.1063/1.4802593
5. Lee SJ, Arun AP, Kim KJ (2015) Piezoelectric properties of electrospun poly (l-lactic acid) nanofiber web. Mater Lett 148:58–62. https://doi.org/10.1016/j.matlet.2015.02.038
6. Zheng T, Yue Z, Wallace GG, Du Y, Higgins MJ (2020) Nanoscale piezoelectric effect of biodegradable PLA-based composite fibers by piezoresponse force microscopy. Nanotechnology 31(37):375708. https://doi.org/10.1088/1361-6528/ab96e3
7. Zhukov S, Ma X, Seggern HV, Sessler GM, Dali OB, Kupnik M, Zhang X (2020) Biodegradable cellular polylactic acid ferroelectrets with strong longitudinal and transverse piezoelectricity. Appl Phys Lett 117(11):112901. https://doi.org/10.1063/5.0023153
8. Kochervinskii VV, Gradov OV, Gradova MA (2022) Fluorine-containing ferroelectric polymers and their application in engineering and biomedicine. Russian Chem Rev 91(11):RCR5037. https://doi.org/10.57634/RCR5037
9. Gradov OV, Gradova MA, Kochervinskii VV (2023) Biocompatible biomimetic polymer structures with an active response for implantology and regenerative medicine. Part I: Basic principles of the active implant's biocompatibility. Siberian J Life Sci Agric 15(1):346–377. https://doi.org/10.12731/2658-6649-2023-15-1-346-377
10. Kochervinskii VV, Gradov OV, Gradova MA (2021) Applications of ferroelectric polymers in technology and medicine. Palmarium Academic Publishing, Kishinew
11. Gradov OV, Gradova MA, Kochervinskij VV (2021) Biomimetic biocompatible ferroelectric polymer materials with an active response for implantology and regenerative medicine. In: Organic ferroelectric materials and applications. United Kingdom, pp 571–619. https://doi.org/10.1016/B978-0-12-821551-7.00012-9
12. Santos T, Boto C, Saraiva CM, Bernardino L, Ferreira L (2016) Nanomedicine approaches to modulate neural stem cells in brain repair. Trends Biotechnol 34(6):437–439. https://doi.org/10.1016/j.tibtech.2016.02.003

13. Lu L, Wang Y, Zhang F, Chen M, Lin B, Duan X, Cao M, Zheng C, Mao J, Shuai X, Shen J (2018) MRI-Visible siRNA nanomedicine directing neuronal differentiation of neural stem cells in stroke. Adv Func Mater 28(14):1706769. https://doi.org/10.1002/adfm.201706769
14. Lin B, Lu L, Wang Y, Zhang Q, Wang Z, Cheng G, Duan X, Zhang F, Xie M, Le H, Shuai X (2021) Nanomedicine directs neuronal differentiation of neural stem cells via silencing long noncoding RNA for stroke therapy. Nano Lett 21(1):806–815. https://doi.org/10.1021/acs.nanolett.0c04560
15. Wang Z, Cai Y, Zhang Q, Li J, Lin B, Zhao J, Zhang F, Li Y, Yang X, Lu L, Shuai X (2022) Upregulating HIF-1α to boost the survival of neural stem cells via functional peptides-complexed MRI-visible nanomedicine for stroke therapy. Adv Healthcare Mater 11(23):2201630. https://doi.org/10.1002/adhm.202201630
16. Tomaskovic-Crook E, Higginbottom SL, James EC, Rathbone SJ, Crook JM (2020) Electroceuticals for neural regenerative nanomedicine. In: Neural regenerative nanomedicine. Academic, pp 213–257. https://doi.org/10.1016/B978-0-12-820223-4.00007-3
17. Long Y, Li J, Yang F, Wang J, Wang X (2021) Wearable and implantable electroceuticals for therapeutic electrostimulations. Adv Sci 8(8):2004023. https://doi.org/10.1002/advs.202004023
18. Maeng WY, Tseng WL, Li S, Koo J, Hsueh YY (2022) Electroceuticals for peripheral nerve regeneration. Biofabrication 14(4):042002. https://doi.org/10.1088/1758-5090/ac8baa
19. Levin M, Selberg J, Rolandi M (2019) Endogenous bioelectrics in development, cancer, and regeneration: drugs and bioelectronic devices as electroceuticals for regenerative medicine. Iscience 22:519–533. https://doi.org/10.1016/j.isci.2019.11.023
20. Wang L, Habib AA, Mintz A, Li KC, Zhao D (2017) Phosphatidylserine-targeted nanotheranostics for brain tumor imaging and therapeutic potential. Mol Imaging 16:1536012117708722. https://doi.org/10.1177/1536012117708722
21. Viswanadh MK, Singh RP, Agrawal P, Mehata AK, Pawde DM, Sonkar R, Muthu MS (2018) Nanotheranostics: emerging strategies for early diagnosis and therapy of brain cancer. Nanotheranostics. 2(1):70. https://doi.org/10.7150/ntno.21638
22. Wu H, Lu H, Xiao W, Yang J, Du H, Shen Y, Qu H, Jia B, Manna SK, Ramachandran M, Xue X (2020) Sequential targeting in crosslinking nanotheranostics for tackling the multibarriers of brain tumors. Adv Mater 32(14):1903759. https://doi.org/10.1002/adma.201903759
23. Gao H, Chu C, Cheng Y, Zhang Y, Pang X, Li D, Wang X, Ren E, Xie F, Bai Y, Chen L (2020) In situ formation of nanotheranostics to overcome the blood–brain barrier and enhance treatment of orthotopic glioma. ACS Appl Mater Interfaces 12(24):26880–26892. https://doi.org/10.1021/acsami.0c03873
24. Carvalho IC, Mansur AA, Carvalho SM, Mansur HS (2021) Nanotheranostics through mitochondria-targeted delivery with fluorescent peptidomimetic nanohybrids for apoptosis induction of brain cancer cells. Nanotheranostics 5(2):213. https://doi.org/10.7150/ntno.54491
25. Mansur AA, Paiva MR, Cotta OA, Silva LM, Carvalho IC, Capanema NS, Carvalho SM, Costa ÉA, Martin NR, Ecco R, Santos BS (2022) Carboxymethylcellulose biofunctionalized ternary quantum dots for subcellular-targeted brain cancer nanotheranostics. Int J Biol Macromol 210:530–544. https://doi.org/10.1016/j.ijbiomac.2022.04.207
26. Portosi V, Loconsole AM, Prudenzano F (2020) A split ring resonator-based metamaterial for microwave impedance matching with biological tissue. Appl Sci 10(19):6740. https://doi.org/10.3390/app10196740
27. Vrbová B, Vrba J (2014) Theory and design of microwave applicator coupled to biological tissue based on impedance matching. In: 2014 24th international conference radioelektronika. IEEE, pp 1–4. https://doi.org/10.1109/Radioelek.2014.6828492
28. Hu YC, Liao PL, Shih WP., Wang XY, Chang PZ (2009) Study on the acoustic impedance matching of human tissue for power transmitting/charging system of implanted biochip. In: 2009 IEEE 3rd international conference on nano/molecular medicine and engineering, pp 201–205. https://doi.org/10.1109/NANOMED.2009.5559087
29. Kochervinskii VV, Gradov OV, Gradova MA (2019) Ferroelectric polymers in regenerative medicine. Genes Cells 14(S1):122–123. https://doi.org/10.23868/gc122911

30. Adamovich ED, Buryanskaya EL, Gradova MA, Gradov OV (2023) Reaction-diffusion effects and spatiotemporal oscillations under SEM, STM and AFM-assisted charging in fiber-like and wire-like systems: From molecular and quantum wires to cooperative ferroelectric nanofibers and microfibers (Invited Review). Mater Technol Rep 1:135. https://doi.org/10.59400/mtr.v1i1.135
31. Buryanskaya EL, Gradov OV, Gradova MA, Kiselev DA, Kochervinsky VV (2022) Biomedical applications of ferroelectric polymers based on vinylidene fluoride in regenerative medicine. Genes Cells 17(3):39. https://doi.org/10.23868/gc122163
32. Gradov OV (2019) Multiangle laser porosimetry and electron beam porosimetry for scaffolds, decellularized extracellular matrix samples and tissue-like models, including ESEM-based & CLEM-based realizations. Genes Cells 14(S1):69–70. https://doi.org/10.23868/gc122421
33. Gradov OV, Gradova MA, Kholuiskaya SN, Olkhov AA (2022) Electron plasma charging effects on the biocompatible electrospun dielectric fibers. IEEE Trans Plasma Sci 50(1):178–186. https://doi.org/10.1109/TPS.2021.3130854
34. Gradov OV, Gradova MA, Iordanskii AL, Olkhov AA, Kholuiskaya SN (2020) Isopotential mapping of electron beam induced dielectric charging of the phb nonwoven fabric structures using Sobel-Feldman gradient operator. In: 2020 7th international congress on energy fluxes and radiation effects (EFRE). IEEE, pp 662–665. https://doi.org/10.1109/EFRE47760.2020.9242099
35. Buryanskaya EL, Gradov OV, Gradova MA, Kochervinskii VV, Maklakova IA (2023) Time-resolved multifractal analysis of electron beam induced piezoelectric polymer fiber dynamics: towards multiscale thread-based microfluidics or acoustofludics. Adv Struct Mater 195:35–58. https://doi.org/10.1007/978-3-031-28744-2_3
36. Pandithevan P, Saravana KG (2009) Personalised bone tissue engineering scaffold with controlled architecture using fractal tool paths in layered manufacturing. Virtual Phys Prototyping 4(3):165–180. https://doi.org/10.1080/17452750903055512
37. Warner J, Soman P, Zhu W, Tom M, Chen S (2016) Design and 3D printing of hydrogel scaffolds with fractal geometries. ACS Biomater Sci Eng 2(10):1763–1770. https://doi.org/10.1021/acsbiomaterials.6b00140
38. Zhao Z, Chen X (2019) Fractal algorithm for 3D-printed continuous porous scaffold design. J Eng 2019(14):485–489. https://doi.org/10.1049/joe.2018.9404
39. Qu H, Han Z, Chen Z, Tang L, Gao C, Liu K, Pan H, Fu H, Ruan C (2021) Fractal design boosts extrusion-based 3D printing of bone-mimicking radial-gradient scaffolds. Research (in press). https://doi.org/10.34133/2021/9892689
40. Ciaravella G, Vozzi G, Bianchi F, Rosi M, Costanza C, Madeddu P, Ahluwalia A (2002) Microfabricated fractal trees as scaffolds for capillary morphogenesis: applications in therapeutic angiogenesis and vascular tissue engineering. In: Proceedings of the IEEE-EMBS special topic conference on molecular, cellular and tissue engineering. IEEE, pp 46–47. https://doi.org/10.1109/MCTE.2002.1174996
41. Spaletta G, Della CC, Bodria M, Ravera S, Dallatana D, Castorina S, Martorella A, Toni R (2006) A fractal model for bioengineering of the stromal/vascular scaffold of a bionic human thyroid gland. Ital J Anat Embryol 111:75–75
42. Yanen W, Xiutian Y (2010) Applying fractal methodology to establishing bone scaffold microstructure models. J Northwestern Polytechnical Univ 28(4):541–546
43. Kumar GS, Pandithevan P (2009) Fractal tool paths for layered manufacturing of scaffolds with matched bone properties. In: Innovative developments in design and manufacturing: advanced research in virtual and rapid prototyping–proceedings of VRP4, p 59. https://doi.org/10.1201/9780203859476-14
44. Derimay F, Souteyrand G, Motreff P, Guerin P, Pilet P, Ohayon J, Darremont O, Rioufol G, Finet G (2016) Sequential proximal optimizing technique in provisional bifurcation stenting with everolimus-eluting bioresorbable vascular scaffold: fractal coronary bifurcation bench for comparative test between absorb and XIENCE Xpedition. JACC: Cardiovasc Interv 9(13):1397–1406. https://doi.org/10.1016/j.jcin.2016.04.021

45. Broadbent B (2015) Evaluation of MEMS fabricated fractal based free standing scaffolds for the purposes of developing a brain bioreactor. Wayne State University
46. Van Kampen K, Mota C, Moroni L (2022) Mechanical stimulation of fractal based scaffolds to steer stem cell activity. Tissue Eng Part A 28:S47–S48
47. Guarino V, Guaccio A, Netti PA, Ambrosio L (2010) Image processing and fractal box counting: user-assisted method for multi-scale porous scaffold characterization. J Mater Sci Mater Med 21(12):3109–3118. https://doi.org/10.1007/s10856-010-4163-9
48. Bassoli E, Denti L, Gatto A, Spaletta G, Sofroniou M, Parrilli A, Fini M, Giardino R, Zamparelli A, Zini N, Barbaro F (2013) A planar fractal analysis of the arterial tree of the human thyroid gland: Implications for additive manufacturing of 3D ramified scaffolds. In: High value manufacturing: advanced research in virtual and rapid prototyping: proceedings of the 6th international conference on advanced research in virtual and rapid prototyping. CRC Press, Leiria, Portugal, p 423
49. Liu WJ, Xu RW, Xue R, Li MW, Yu DS (2006) The fabrication and fractal study of poly (l-lactic acid) gradient scaffolds for tissue engineering. J Beijing Univ Chem Technol 33(3):56–60
50. Bassoli E, Denti L, Gatto A, Spaletta G, Sofroniou M, Parrilli A, Fini M, Giardino R, Sanduzzi ZA, Zini N, Barbaro F (2018) Towards additive manufacturing of ramified scaffolds of the thyroid vascular system: a preliminary fractal analysis. Int J Mech Eng Technol 9(8):429–437
51. Inoue K, Kikuchi H, Asai S, Sumita M (1996) Fractal and fine structure of poly (L-lactic acid) in degradation. Sen'i Gakkaishi 52(9):473–480. https://doi.org/10.2115/fiber.52.9_473
52. Inoue K, Masato S, Asai S, Sumita M (1997) Fractal and structure of poly (L-lactic acid) in fragmentation. Sen'i Gakkaishi 53(7):265–271. https://doi.org/10.2115/fiber.53.7_265
53. Qiang T, Yu DM, Gao HH (2012) Impact strength and fractal characteristic of PLA-based wood plastic composites. Appl Mech Mater 174:683–686. https://doi.org/10.4028/www.scientific.net/AMM.174-177.683
54. Qiang T, Yu DM, Gao HH (2012) Fractal characteristic and impact strength of PLA-based wood plastic composites. Adv Mater Res 532:49–52. https://doi.org/10.4028/www.scientific.net/AMR.532-533.49
55. Takeda M, Yoshino K, Mizukami Y, Sato Y, Ito A, Bao Y, Munakata F (2019) Fractal characters of material texture of self-assembled $BaTiO_3$/Poly-L-lactic-acid composites. J Jpn Soc Powder Powder Metall 66:122–127
56. Mwema FM, Adediran AA, Akinlabi ET, Adeleke AA, Olayanju TM (2021) Fractal characterization of conductive Ag/flexible 3D printed PLA. IOP Conf Ser Mater Sci Eng 1107:012050. https://doi.org/10.1088/1757-899X/1107/1/012050
57. Meng Z, Ren X, Gao Y (2014) Effects of PLA sand barriers on soil fractal dimension. Chin J Soil Sci 45(3):613–617
58. Takeda T, Ishikawa A, Ohtomo K, Kobayashi Y, Matsuoka T (1992) Fractal dimension of dendritic tree of cerebellar Purkinje cell during onto-and phylogenetic development. Neurosci Res 13(1):19–31. https://doi.org/10.1016/0168-0102(92)90031-7
59. Dokukin ME, Guz NV, Gaikwad RM, Woodworth CD, Sokolov I (2011) Cell surface as a fractal: normal and cancerous cervical cells demonstrate different fractal behavior of surface adhesion maps at the nanoscale. Phys Rev Lett 107(2):028101. https://doi.org/10.1103/PhysRevLett.107.028101
60. Dokukin ME, Guz NV, Woodworth CD, Sokolov I (2015) Emergence of fractal geometry on the surface of human cervical epithelial cells during progression towards cancer. New J Phys 17(3):033019. https://doi.org/10.1088/1367-2630/17/3/033019
61. Guz NV, Dokukin ME, Woodworth CD, Cardin A, Sokolov I (2015) Towards early detection of cervical cancer: Fractal dimension of AFM images of human cervical epithelial cells at different stages of progression to cancer. Nanomed Nanotechnol Biol Med 11(7):1667–1675. https://doi.org/10.1016/j.nano.2015.04.012
62. Sokolov I, Dokukin ME (2017) Fractal analysis of cancer cell surface. Methods Mol Biol 1530:229–245. https://doi.org/10.1007/978-1-4939-6646-2_13
63. Dokukin M, Guz N, Woodworth C, Sokolov I (2012) Emergence of fractal behavior and other changes of cell surface during malignant transformation: AFM study of human cervical epithelial cells. Bull Am Phys Soc 57(1):D42.00005

64. Kim YJ, Kim HD, Kim HH, Shin SM, Kang CJ, Lee KH (2015) Fractal analysis of cell boundary ultrastructure imaged by atomic force microscopy. Animal Cells Syst 19(3):161–167. https://doi.org/10.1080/19768354.2015.1037347
65. Cao K, Liu Y, Olkhov AA, Siracusa V, Iordanskii AL (2018) PLLA-PHB fiber membranes obtained by solvent-free electrospinning for short-time drug delivery. Drug Deliv Transl Res 8:291–302. https://doi.org/10.1007/s13346-017-0463-7
66. Iordanskii A, Karpova S, Olkhov A, Borovikov P, Kildeeva N, Liu Y (2019) Structure-morphology impact upon segmental dynamics and diffusion in the biodegradable ultrafine fibers of polyhydroxybutyrate-polylactide blends. Eur Polym J 117:208–216. https://doi.org/10.1016/j.eurpolymj.2019.05.012
67. Olkhov AA, Karpova SG, Tyubaeva PM, Zhulkina AL, Zernova YN, Iordanskii AL (2020) Effect of ozone and ultraviolet radiation on structure of fibrous materials based on poly (3-hydroxybutyrate) and polylactide. Inorg Mater Appl Res 11:1130–1136. https://doi.org/10.1134/S2075113320050251
68. Tyubaeva P, Varyan I, Lobanov A, Olkhov A, Popov A (2021) Effect of the hemin molecular complexes on the structure and properties of the composite electrospun materials based on poly (3-Hydroxybutyrate). Polymers 13(22):4024. https://doi.org/10.3390/polym13224024
69. Tyubaeva P, Zykova A, Podmasteriev V, Olkhov A, Popov A, Iordanskii A (2021) The investigation of the structure and properties of ozone-sterilized nonwoven biopolymer materials for medical applications. Polymers 13(8):1268. https://doi.org/10.3390/polym13081268
70. Olkhov A, Alexeeva O, Konstantinova M, Podmasterev V, Tyubaeva P, Borunova A, Siracusa V, Iordanskii AL (2021) Effect of Glycero-(9, 10-trioxolane)-trialeate on the physicochemical properties of non-woven polylactic acid fiber materials. Polymers 13(15):2517. https://doi.org/10.3390/polym13152517
71. Olkhov AA, Tyubaeva PM, Vetcher AA, Karpova SG, Kurnosov AS, Rogovina SZ, Iordanskii AL, Berlin AA (2021) Aggressive impacts affecting the biodegradable ultrathin fibers based on poly (3-hydroxybutyrate), polylactide and their blends: water sorption, hydrolysis and ozonolysis. Polymers 13(6):941. https://doi.org/10.3390/polym13060941
72. Rogovina SZ, Zhorina LA, Olkhov AA, Yakhina AR, Kucherenko EL, Iordanskiy AL, Berlin AA (2022) Hydrolysis of biodegradable fibrous and film materials based on polylactide and Poly (3-hydroxybutyrate) polyesters and their compositions with chitosan. Polym Sci, Ser D 15(3):447–451. https://doi.org/10.1134/S1995421222030224
73. Tyubaeva PM, Varyan IA, Zykova AK, Yarysheva AY, Ivchenko PV, Olkhov AA, Arzhakova OV (2022) Bioinspired electropun fibrous materials based on Poly-3-hydroxybutyrate and hemin: preparation, physicochemical properties, and weathering. Polymers 14(22):4878. https://doi.org/10.3390/polym14224878
74. Olkhov AA, Mastalygina EE, Ovchinnikov VA, Kurnosov AS, Popov AA, Iordanskii AL (2023) Biological and oxidative degradation of ultrathin-fibrous nonwovens based on Poly (lactic Acid)/Poly (3-Hydroxybutyrate) blends. Int J Mol Sci 24(9):7979. https://doi.org/10.3390/ijms24097979
75. Tyubaeva PM, Gasparyan KG, Fedotov AY, Lobzhanidze PV, Baranov OV, Egorov AA, Sirotinkin VP, Komlev VS, Olkhov AA (2023) Development of Nonwoven fibrous materials based on Poly-3-hydroxybutyrate with a high content of α-tricalcium phosphate. Polymers 15(15):3167. https://doi.org/10.3390/polym15153167
76. Karpova SG, Olkhov AA, Varyan IA, Shilkina NG, Berlin AA, Popov AA, Iordanskii AL (2023) Biocomposites based on electrospun fibers of Poly (3-hydroxybutyrate) and nanoplatelets of graphene oxide: thermal characteristics and segmental dynamics at hydrothermal and ozonation impact. Polymers 15(20):4171. https://doi.org/10.3390/polym15204171

Chapter 7
Meta-Truss and Generalized Elastic Lattices with Long-Range Interactions

Noël Challamel, H. P. Nguyen, Giuseppe Ruta, C. M. Wang, and Y. P. Zhang

Abstract This paper aims to enhance comprehension of two-dimensional (2D) lattice elasticity by analyzing the intricate interplay between short-range and long-range interactions. The 2D square elastic lattices considered herein asymptotically converge towards linear isotropic elasticity. Pioneered by (Gazis et al. in Phys Rev 119:533–544, 1960), the lattice comprising central and angular interactions is generalized to include short- and long-range interactions. This generalized 2D lattice can be also viewed as a particular case of a Molecular Dynamic System (MDS) with discrete short- and long-range interactions. It is shown herein that the associated mixed differential-difference equations of this generalized lattice asymptotically converge towards Navier's elastodynamic partial differential equations. Owing to the long-range interactions, the mixed differential-difference equation takes on a higher-order type for spatial dependence. The short- and long-range lattice parameters may be calibrated to enforce the asymptotic continuum to behave as a homogeneous linear elastic isotropic medium with free Poisson's ratio. A higher-order expansion of the mixed differential-difference equation provides gradient elasticity. The wave dispersive behavior of this generalized lattice with long-range interactions is investigated in an exact form, with a particular emphasis on the short- to long-range interaction ratio. Additionally, static homogeneous compression and shear of square specimens exhibiting short- and long-range interactions are analytically and numerically investigated. It is shown that the homogeneous response of the higher-order lattice strongly

N. Challamel (✉)
Université Bretagne Sud, IRDL (CNRS UMR 6027), Centre de Recherche, Rue de Saint Maudé, BP92116, 56321 Lorient Cedex, France
e-mail: noel.challamel@univ-ubs.fr

H. P. Nguyen · C. M. Wang · Y. P. Zhang
School of Civil Engineering, University of Queensland, St Lucia, QLD 4072, Australia
e-mail: huuphu.nguyen@uq.edu.au

C. M. Wang
e-mail: cm.wang@uq.edu.au

G. Ruta
Dipartimento d'ingegneria strutturale e geotecnica, Università "La Sapienza", Roma, Italy
e-mail: giuseppe.ruta@uniroma1.it

H. Altenbach and V. A. Eremeyev (eds.), *Dynamics of Discrete and Continuum Structures and Media*, Advanced Structured Materials 221,
https://doi.org/10.1007/978-3-031-75626-9_7

depends on the calibration of the higher-order stiffness in the vicinity of the boundary, as previously commented in the literature for one-dimensional higher-order lattices.

Keywords Lattice/discrete elasticity · Continuum elasticity · Elastodynamics · Discrete elasticity · Difference equations · Partial differential equations · Microstructured solids · Long-range interactions · Molecular dynamics

7.1 Introduction

Lattice mechanics, grounded in elementary interactions, is a key physical theory at the foundation of continuum mechanics [3, 6, 7, 31]. One-dimensional lattices based on direct interactions (i.e., with closest neighbors) have been widely studied in the literature [5, 27, 28]. Lagrange lattice [27, 28], also dubbed as Born-Kármán lattice, consists of elementary particles with direct interactions (also called short-range interaction), that may classify this lattice as a simple lattice. It asymptotically converges towards a 1D continuum ruled by the classical wave equation in a first-order approximation. References [27, 28] obtained the relevant governing mixed differential-difference equation, where the difference operator in space is a linear second-order difference operator in the case of direct interaction. Lagrange [27, 28] derived the exact eigenfrequency of this simple finite lattice when both ends are fixed; the wave dispersion equation was obtained by [5] for such an infinite lattice (still with direct interactions). Generalized lattices exhibiting also long-range interactions were studied during the 20th century [7, 15, 18, 39]. Brillouin [7] and later [18] obtained the wave dispersive equation of a generalized Lagrange lattice (one-dimensional lattice with short- and long-range interactions with p interactions). As shown by Brillouin [7] and [18], the generalized lattice is ruled by a mixed differential-difference equation, where the difference operator in space is linear of order $2p$, where p is the number of interactions ($p = 1$ for Lagrange lattice). The eigenfrequencies of this fixed–fixed higher-order lattice with p interactions were determined by [18] and more recently by [15]. Challamel et al. [13] pointed out the role of calibrating the higher-order stiffnesses near the boundaries, with strong scale dependence response, if truncated higher-order stiffnesses are considered there. The calibration of the higher-order stiffness parameters in the vicinity of the boundaries should respect some symmetric or anti-symmetric properties of the discrete displacement field at the boundary, to fulfill the homogeneity of the response under elementary loading (see also [16] for the static analysis of generalized Lagrange 1D lattice with direct and indirect interactions). Otherwise, as also confirmed by the present paper devoted to 2D lattices, the homogeneity of the response under elementary loading may be lost, with some localization phenomena close to the boundaries (see also [16]).

The generalization of Lagrange mixed differential-difference equation to 2D or 3D dates to the early 20th century with the pioneer contribution of [5] (see also the historical review of [14]). This attempt to bridge between molecular and continuum elasticity attracted a lot of researchers during the 19th century [8, 9]. Born and vonKármán [5] formulated the mixed differential-difference equations for a simple cubic lattice with central and non-central interactions. They obtained Navier's partial differential equations of linear cubic elasticity at the asymptotic limit and calibrated the lattice parameters with respect to the macroscopic ones. Born and von Kármán's cubic lattice also converges, as a particular case, towards an isotropic linear elastic solid in the asymptotic limit and, in the absence of non-central forces, converges towards Navier's rare-constant elastodynamics [36] with equal Lamé coefficients (and a Poisson's ratio equal to ¼). In this last case, the lattice with pure central forces coincides with a simple truss exhibiting central interactions with first and second neighbors. The non-central forces considered by [5] are frame-dependent or can be viewed as angular elastic interactions (or shear interaction) connected to the axis of symmetry of the cubic solid. Consequently, it can be easily shown that these non-central forces do not fulfill the invariance principle of the associated potential energy under rigid body motions (in particular, under rotations). Note that a consistent lattice theory should fulfill both translational and rotational invariance principles [1, 20, 24, 25, 29]. A consistent and alternative lattice model was developed in the 1960s by [19] for a simple cubic lattice having central and non-central interactions, with frame independent angular interactions. The model of [19] asymptotically converges towards linear cubic continuum elasticity at the asymptotic limit, which includes also linear isotropic continuum elasticity. Gazis et al. [19] gave a correct answer to the question of bridging cubic lattice elasticity with continuum elasticity for various values of the elasticity parameters, by considering central interactions and objective non-central interactions. The lattice model of [19] is, however, valid for sufficiently small Poisson's ratios as compared to that of the rare-constant theory (for Poisson's ratio smaller than *1/4* for 3D linear isotropic elasticity or for plane strain linear isotropic elasticity, and for Poisson's ratio smaller than *1/3* for plane stress linear isotropic elasticity). An alternative lattice model (or complementary model) for 2D analysis is the surface interaction lattice model that permits to calibrate macroscopic Poisson's ratio larger than the rare-constant limit (see [12]).

In this paper, we investigate the behavior of 2D lattices with short- and long-range interactions, thus generalizing [19] lattice initially restricted to short-range interactions. This can also be understood as the extension of the generalized one-dimensional lattice (Brillouin lattice) with short- and long-range interactions to 2D lattice elasticity with short- and long-range interactions. 2D or 3D generalized lattices have already been studied based on short- and long-range interactions. For instance, [37] studied the wave dispersive properties in a meta-truss with one-cell and two-cell interactions ($p = 2$) (lattice equivalent to a meta-truss with pure central interactions). Tuna et al. [41] investigated the plane static behavior of a two-cell system ($p = 2$) with both central and rotational interactions. [23] studied the static behavior of a meta-truss with one-cell and two-cell interactions, i.e., for $p = 2$ (based on pure central interactions). Here, we derived an exact solution for the wave dispersive properties of a consistent generalized elastic infinite lattice, with $p = 2$ cell interactions. Additionally, we investigate the static behavior of such generalized finite lattice under elementary homogeneous loadings.

7.2 Generalized Lattices with Angular Interactions and Short- and Long-Range Interactions

Gazis et al. model in 2D with one-cell interaction $p = 1$ is a monoatomic, simple square lattice, composed of particles of equal mass M and lattice spacing a. This lattice is governed by the following mixed differential-difference equation [19]:

$$\begin{cases} \alpha_1(u_{i+1,j} - 2u_{i,j} + u_{i-1,j}) + \dfrac{\beta_1}{2}(u_{i+1,j+1} + u_{i-1,j-1} + u_{i-1,j+1} + u_{i+1,j-1} - 4u_{i,j}) + \\ \left(\dfrac{\beta_1}{2} + \gamma_1\right)(v_{i+1,j+1} - v_{i-1,j+1} - v_{i+1,j-1} + v_{i-1,j-1}) + 4\gamma_1(u_{i,j+1} - 2u_{i,j} + u_{i,j-1}) = M\ddot{u}_{i,j} \\ \alpha_1(v_{i,j+1} - 2v_{i,j} + v_{i,j-1}) + \dfrac{\beta_1}{2}(v_{i+1,j+1} + v_{i-1,j-1} + v_{i-1,j+1} + v_{i+1,j-1} - 4v_{i,j}) + \\ \left(\dfrac{\beta_1}{2} + \gamma_1\right)(u_{i+1,j+1} - u_{i-1,j+1} - u_{i+1,j-1} + u_{i-1,j-1}) + 4\gamma_1(v_{i+1,j} - 2v_{i,j} + v_{i-1,j}) = M\ddot{v}_{i,j} \end{cases} \tag{7.1}$$

where M is the mass of each particle, α_1 the stiffness of axial springs (horizontal or vertical springs) and β_1 that of the diagonal springs, both modelling central interactions. The noncentral interactions are modeled by rotational springs of stiffness $C_1 = \gamma_1 a^2 \cdot u_{i,j}$ and $v_{i,j}$ are the inplane horizontal and vertical displacements of each particle at nodes (i, j). The mechanical interactions of such a lattice are represented in Fig. 7.1 for a 2D version of Gazis et al. lattice, i.e.a square lattice.

For the sake of simplicity, and due to symmetry arguments, we only consider Eq. (7.1) with $p = 1$ cell interactions:

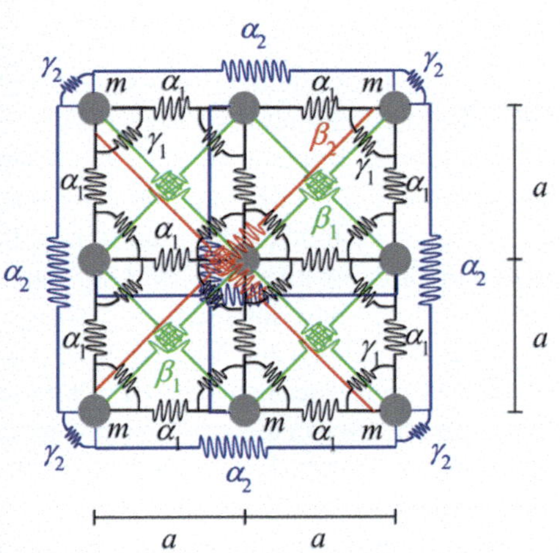

Fig. 7.1 2 * 2 cell problem with central and angular interactions $(\alpha_1, \beta_1, \gamma_1, \alpha_2, \beta_2, \gamma_2) - p = 2$ interactions

$$\alpha_1\left(u_{i+1,j}-2u_{i,j}+u_{i-1,j}\right)+\frac{\beta_1}{2}\left(u_{i+1,j+1}+u_{i-1,j-1}+u_{i-1,j+1}+u_{i+1,j-1}-4u_{i,j}\right)$$
$$+\left(\frac{\beta_1}{2}+\gamma_1\right)\left(v_{i+1,j+1}-v_{i-1,j+1}-v_{i+1,j-1}+v_{i-1,j-1}\right)$$
$$+4\gamma_1\left(u_{i,j+1}-2u_{i,j}+u_{i,j-1}\right)=M\,\ddot{u}_{i,j} \tag{7.2}$$

By displaying in terms of elastic (or material) constants E Young's modulus, ν Poisson's ratio, λ, μ Lamé's constants of usual linear elasticity, the micro–macro identifications in plane strain are (see [14]):

$$\begin{cases}\alpha_1+\beta_1=(\lambda+2\mu)h\\ 2\beta_1+4\gamma_1=(\lambda+\mu)h\\ 4\gamma_1+\beta_1=\mu h\end{cases}\Rightarrow\begin{cases}\alpha_1=2\mu h=\dfrac{Eh}{1+\upsilon}\\ \beta_1=\lambda h=\dfrac{E\upsilon h}{(1+\upsilon)(1-2\upsilon)}\\ \gamma_1=\dfrac{\mu-\lambda}{4}h=\dfrac{E(1-4\upsilon)}{8(1+\upsilon)(1-2\upsilon)}h\end{cases} \tag{7.3}$$

We may also perform the micro–macro identifications in plane stress [14]:

$$\begin{cases}\alpha_1+\beta_1=\dfrac{Eh}{1-\upsilon^2}\\ 2\beta_1+4\gamma_1=\dfrac{Eh}{2(1-\upsilon)}\\ 4\gamma_1+\beta_1=\dfrac{Eh}{2(1+\upsilon)}\end{cases}\Rightarrow\begin{cases}\alpha_1=\dfrac{Eh}{1+\upsilon}\\ \beta_1=\dfrac{Eh\upsilon}{1-\upsilon^2}\\ \gamma_1=\dfrac{Eh(1-3\upsilon)}{8\left(1-\upsilon^2\right)}\end{cases} \tag{7.4}$$

The lattice model of [19] coincides with the simple cubic lattice of Born-Karman without shear interaction, in case of pure central forces (2D lattice with pure central interactions, both for plane stress or for plane strain analyses). In this last case, the angular interaction is neglected, and the two-central interaction parameters α_1 and β_1 are constrained to fulfill the elastic asymptotic property:

$$\gamma_1=0\quad\Rightarrow\quad\alpha_1=2\beta_1 \tag{7.5}$$

The generalization of Eq. (7.2) with $p=2$ interactions is given by (see Fig. 7.1):

$$\begin{aligned}&\alpha_1\left(u_{i+1,j}-2u_{i,j}+u_{i-1,j}\right)+\alpha_2\left(u_{i+2,j}-2u_{i,j}+u_{i-2,j}\right)\\&+\frac{\beta_1}{2}\left(u_{i+1,j+1}+u_{i-1,j-1}+u_{i-1,j+1}+u_{i+1,j-1}-4u_{i,j}\right)\\&+\frac{\beta_2}{2}\left(u_{i+2,j+2}+u_{i-2,j-2}+u_{i-2,j+2}+u_{i+2,j-2}-4u_{i,j}\right)\\&+\left(\frac{\beta_1}{2}+\gamma_1\right)\left(v_{i+1,j+1}-v_{i-1,j+1}-v_{i+1,j-1}+v_{i-1,j-1}\right)\\&+\left(\frac{\beta_2}{2}+\gamma_2\right)\left(v_{i+2,j+2}-v_{i-2,j+2}-v_{i+2,j-2}+v_{i-2,j-2}\right)\\&+4\gamma_1\left(u_{i,j+1}-2u_{i,j}+u_{i,j-1}\right)+4\gamma_2\left(u_{i,j+2}-2u_{i,j}+u_{i,j-2}\right)=M\ddot{u}_{i,j}\end{aligned}\tag{7.6}$$

Equation (7.6) are the mixed differential-difference equations of the generalized Gazis et al. lattice that account for short- and long-range interactions (with $p = 2$ cells). Equation (7.6) can also be understood as the superposition of one-cell interaction and two-cell interactions (homothetic short- and long-range interactions). We implicitly assume that short- and long-range interactions have similar form.

If we pose ξ_1, ξ_2 the (positive) mixture parameters ratios (or, equivalently, weighted coefficients) of short- and long-range interactions, respectively, we may then perform the micro–macro identifications in for plane strain:

$$\begin{cases}\alpha_1=2\mu h\xi_1\\ \beta_1=\lambda h\xi_1\\ \gamma_1=\dfrac{\mu-\lambda}{4}h\xi_1\end{cases}\quad\text{and}\quad\begin{cases}\alpha_2=2\mu h\dfrac{\xi_2}{4}\\ \beta_2=\lambda h\dfrac{\xi_2}{4}\\ \gamma_2=\dfrac{\mu-\lambda}{4}h\dfrac{\xi_2}{4}\end{cases}\quad\text{and}\quad \xi_1+\xi_2=1;\ \xi_1\geq 0;\ \xi_2\geq 0\tag{7.7}$$

For plane stress analysis, the micro–macro relations are identified in the same manner:

$$\begin{cases}\alpha_1=\dfrac{Eh}{1+\upsilon}\xi_1\\ \beta_1=\dfrac{Eh\upsilon}{1-\upsilon^2}\xi_1\\ \gamma_1=\dfrac{Eh(1-3\upsilon)}{8\left(1-\upsilon^2\right)}\xi_1\end{cases}\quad\text{and}\quad\begin{cases}\alpha_2=\dfrac{Eh}{1+\upsilon}\dfrac{\xi_2}{4}\\ \beta_2=\dfrac{Eh\upsilon}{1-\upsilon^2}\dfrac{\xi_2}{4}\\ \gamma_2=\dfrac{Eh(1-3\upsilon)}{8\left(1-\upsilon^2\right)}\dfrac{\xi_2}{4}\end{cases}\tag{7.8}$$
$$\text{and}\ \xi_1+\xi_2=1;\ \xi_1\geq 0;\ \xi_2\geq 0$$

This generalizes the two-cell 1D interaction problem considered by [39] or [15] to two-cell 2D elastic interactions.

The two-cell problem with pure central interactions is reduced to a generalized truss, also called the meta-truss, when the angular interactions are neglected (see Fig. 7.2) both for plane stress or plane strain analysis:

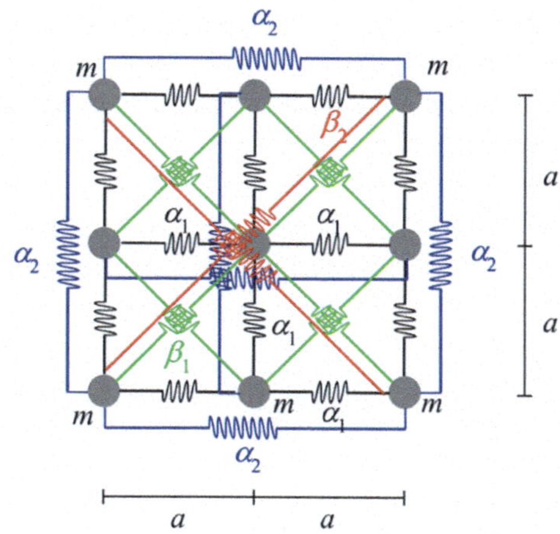

Fig. 7.2 2*2 cell problem with pure central forces $(\alpha_1, \beta_1, \alpha_2, \beta_2) - p = 2$ interactions

$$\gamma_1 = \gamma_2 = 0 \qquad \Rightarrow \quad \alpha_1 = 2\beta_1 \quad \text{and} \quad \alpha_2 = 2\beta_2 \tag{7.9}$$

The static behavior of the generalized truss under various loading conditions will be treated in the last part of this paper.

First, the asymptotic properties of the generalized lattice (with both central and non-central short- and long-range interactions) will be investigated. The mixed differential-difference equations of this generalized square lattice asymptotically converge towards the partial differential equations of Navier's elastodynamics for sufficiently small spacing. Between the discrete and the equivalent continuum connected at the nodes by the equivalent conditions $u_{i,j} = u(x = ai, y = aj)$ and $v_{i,j} = v(x = ai, y = aj)$ on the displacement components, the following relation holds for sufficiently smooth functions:

$$u_{i+1,j+1} = u(x + a, y + a) = e^{a(\partial_x + \partial_y)} u(x, y) \tag{7.10}$$

We use [10] technique, which allows the expansion of some difference operators in Taylor series of differential operators by assuming a sufficiently smooth evolution function (see also [2, 26, 38]). Thus, the expansion of the difference operators in the governing Eq. (7.6) up to the first terms yields Navier's type partial differential equations:

$$\begin{aligned} &(\alpha_1 + 4\alpha_2)(a\partial_x)^2 u + (\beta_1 + 4\beta_2)\left[(a\partial_x)^2 + \left(a\partial_y\right)^2\right] u \\ &+ \left(\frac{\beta_1}{2} + 2\beta_2 + \gamma_1 + 4\gamma_2\right) 4a^2 \partial_x \partial_y v + 4(\gamma_1 + 4\gamma_2)\left(a\partial_y\right)^2 u = M\ddot{u} \end{aligned} \tag{7.11}$$

Using the stiffness parameters calibrated in Eq. (7.7) for the plane strain analysis, Eq. (7.11) can be identified with the corresponding continuous linear isotropic Navier's elastodynamics equation:

$$(\lambda + 2\mu)\partial_x^2 u + (\lambda + \mu)\partial_x \partial_y v + \mu \partial_y^2 u = \rho \ddot{u} \tag{7.12}$$

Using the stiffness parameters calibrated in Eq. (7.8) for the plane stress, Eq. (7.11) can be identified with the corresponding continuous linear isotropic Navier's elastodynamics equation:

$$\frac{E}{1-\upsilon^2}\partial_x^2 u + \frac{E}{2(1-\upsilon)}\partial_x \partial_y v + \frac{E}{2(1+\upsilon)}\partial_y^2 u = \rho \ddot{u} \tag{7.13}$$

If now Eq. (7.6) is expanded up to some higher terms, the gradient elasticity continuum can be obtained for plane strain analysis:

$$\begin{aligned}
&(\lambda + 2\mu)\partial_x^2 u + (\lambda + \mu)\partial_x \partial_y v + \mu \partial_y^2 u \\
&+\frac{a^2}{12}(\xi_1 + 4\xi_2)(\lambda + 2\mu)\partial_x^4 u \\
&+\frac{a^2}{12}(\xi_1 + 4\xi_2)\mu \partial_y^4 u + \frac{a^2}{2}(\xi_1 + 4\xi_2)\lambda \partial_x^2 \partial_y^2 u \\
&+\frac{a^2}{6}(\xi_1 + 4\xi_2)(\lambda + \mu)\left(\partial_x^3 \partial_y + \partial_x \partial_y^3\right) v = \rho \ddot{u}
\end{aligned} \tag{7.14}$$

In the case $p = 1$, associated with $(\xi_1, \xi_2) = (1, 0)$, we recover the gradient elasticity model derived by Mindlin [34] from the lattice model of [19] with only direct interactions:

$$\begin{aligned}
&(\lambda + 2\mu)\partial_x^2 u + (\lambda + \mu)\partial_x \partial_y v + \mu \partial_y^2 u \\
&+\frac{a^2}{12}(\lambda + 2\mu)\partial_x^4 u + \frac{a^2}{12}\mu \partial_y^4 u + \frac{a^2}{2}\lambda \partial_x^2 \partial_y^2 u + \frac{a^2}{6}(\lambda + \mu)\left(\partial_x^3 \partial_y + \partial_x \partial_y^3\right) v = \rho \ddot{u}
\end{aligned} \tag{7.15}$$

Equation (7.15) was also recently obtained by [11] or [14].

It appears that the two-cell problem $p = 2$ is equivalent to the one-cell problem $p = 1$, when the higher-order terms are multiplied by a factor $\xi_1 + 4\xi_2 = 1 + 3\xi_2$ with $\xi_2 \in [0; 1]$, resulting in

$$\begin{aligned}
&(\lambda + 2\mu)\partial_x^2 u + \mu \partial_y^2 u + (\lambda + \mu)\partial_x \partial_y v \\
&+\frac{a^2}{12}(1 + 3\xi_2)\left[(\lambda + 2\mu)\partial_x^4 u + \mu \partial_y^4 u + 6\lambda \partial_x^2 \partial_y^2 u\right] \\
&+\frac{a^2}{6}(1 + 3\xi_2)(\lambda + \mu)\left(\partial_x^3 \partial_y + \partial_x \partial_y^3\right) v = \rho \ddot{u}
\end{aligned} \tag{7.16}$$

This result, that generalizes the gradient elasticity law derived by Mindlin [34] from the lattice of [19] with $p = 1$ interaction, is consistent with available results derived for 1D lattices exhibiting short- and long-range interactions. Rosenau [39] or [15] obtained a similar correction of the gradient elasticity law derived from

Lagrange lattice with the already quoted factor $\xi_1 + 4\xi_2 = 1 + 3\xi_2$ for the higher-order gradient terms. As detailed in Appendix A, the partial differential Eq. (7.16) shows that the zero-th order medium can be enforced to be isotropic whereas the second-order medium (gradient elasticity medium) cannot be isotropic and possesses the symmetry of the initial discrete lattice.

The wave dispersion equation of this two-dimensional elastic lattice with short- and long-range interactions can be obtained from the following harmonic wave discrete displacement fields:

$$\begin{aligned} &u_{i,j}(t) = U\exp\left[J\left(\omega t - k_x x_i - k_y y_j\right)\right] \quad \text{and} \\ &v_{i,j}(t) = V\exp\left[J\left(\omega t - k_x x_i - k_y y_j\right)\right] \quad \text{with} \\ &x_i = ai \quad \text{and} \quad y_j = aj \end{aligned} \tag{7.17}$$

with ω is the propagating angular frequency, $J = \sqrt{-1}$ is the imaginary unit, U, V the wave amplitudes, and k_x, k_y the wave numbers in the x, y directions respectively.

In view of Eq. (7.17), the equation of the 2D generalized Gazis et al.'s lattice, i.e., the two mixed differential-difference equations induced by Eq. (7.6), furnishes the following 2D wave dispersion equation:

$$\begin{vmatrix} D_{11} & D_{12} \\ D_{21} & D_{22} \end{vmatrix} = 0 \quad \text{with}$$
$$\begin{cases} D_{11} = 2\alpha_1\left(1 - c_{1,x}\right) + 2\alpha_2\left(1 - c_{2,x}\right) + 2\beta_1\left(1 - c_{1,x}c_{1,y}\right) + 2\beta_2\left(1 - c_{2,x}c_{2,y}\right) \\ \quad +8\gamma_1\left(1 - c_{1,y}\right) + 8\gamma_2\left(1 - c_{2,y}\right) - M\omega^2 \\ D_{12} = D_{21} = (2\beta_1 + 4\gamma_1)s_{1,x}s_{1,y} + (2\beta_2 + 4\gamma_2)s_{2,x}s_{2,y} \\ D_{22} = 2\alpha_1\left(1 - c_{1,y}\right) + 2\alpha_2\left(1 - c_{2,y}\right) + 2\beta_1\left(1 - c_{1,x}c_{1,y}\right) + 2\beta_2\left(1 - c_{2,x}c_{2,y}\right) \\ \quad +8\gamma_1\left(1 - c_{1,x}\right) + 8\gamma_2\left(1 - c_{2,x}\right) - M\omega^2 \end{cases} \tag{7.18}$$

where the dimensionless parameters c_i and s_i for $i \in \{1; 2\}$ are defined as

$$\begin{aligned} &c_{1,x} = \cos(k_x a);\ c_{1,y} = \cos\left(k_y a\right);\ s_{1,x} = \sin(k_x a);\ s_{1,y} = \sin\left(k_y a\right) \\ &c_{2,x} = \cos(2k_x a);\ c_{2,y} = \cos\left(2k_y a\right);\ s_{2,x} = \sin(2k_x a) \text{ and } s_{2,y} = \sin\left(2k_y a\right) \end{aligned} \tag{7.19}$$

For one-cell interactions ($p = 1$), i.e., for $\alpha_2 = \beta_2 = \gamma_2 = 0$, the wave dispersive equation of the 2D lattice of Gazis et al. is obtained [14, 19]:

$$\begin{vmatrix} \begin{matrix} 2\alpha_1\left(1 - c_{1,x}\right) + 2\beta_1\left(1 - c_{1,x}c_{1,y}\right) \\ +8\gamma_1\left(1 - c_{1,y}\right) - M\omega^2 \end{matrix} & (2\beta_1 + 4\gamma_1)s_{1,x}s_{1,y} \\ (2\beta_1 + 4\gamma_1)s_{1,x}s_{1,y} & \begin{matrix} 2\alpha_1\left(1 - c_{1,y}\right) + 2\beta_1\left(1 - c_{1,x}c_{1,y}\right) \\ +8\gamma_1\left(1 - c_{1,x}\right) - M\omega^2 \end{matrix} \end{vmatrix} = 0 \tag{7.20}$$

For pure truss interactions (or central interactions) which corresponds to a meta-truss with $p = 2$ interactions, the problem is reduced due to the following set of parameters:

$$\gamma_1 = \gamma_2 = 0; \alpha_1 = 2\beta_1; \alpha_2 = 2\beta_2; \alpha_1 = \alpha\xi_1\alpha_2 = \alpha\frac{\xi_2}{4}; \xi_1 \in [0;\, 1] \text{ and } \xi_2 = 1 - \xi_1 \tag{7.21}$$

The wave dispersive equation is then simplified for the meta-truss problem (with $p = 2$ interactions) according to

$$\begin{vmatrix} D_{11} & D_{12} \\ D_{21} & D_{22} \end{vmatrix} = 0 \quad \text{with}$$

$$\begin{cases} D_{11} = 2\alpha_1(1 - c_{1,x}) + 2\alpha_2(1 - c_{2,x}) + \alpha_1(1 - c_{1,x}c_{1,y}) + \alpha_2(1 - c_{2,x}c_{2,y}) - M\omega^2 \\ D_{12} = D_{21} = \alpha_1 s_{1,x}s_{1,y} + \alpha_2 s_{2,x}s_{2,y} \\ D_{22} = 2\alpha_1(1 - c_{1,y}) + 2\alpha_2(1 - c_{2,y}) + \alpha_1(1 - c_{1,x}c_{1,y}) + \alpha_2(1 - c_{2,x}c_{2,y}) - M\omega^2 \end{cases} \tag{7.22}$$

The wave dispersive equation in the generalized truss can be written in terms of dimensionless square frequency parameter $\Omega^2 = 2M\omega^2/\alpha$, i.e.,

$$\begin{vmatrix} \tilde{D}_{11} & \tilde{D}_{12} \\ \tilde{D}_{21} & \tilde{D}_{22} \end{vmatrix} = 0 \quad \text{with}$$

$$\begin{cases} \tilde{D}_{11} = \xi_1(3 - 2c_{1,x} - c_{1,x}c_{1,y}) + \frac{\xi_2}{4}(3 - 2c_{2,x} - c_{2,x}c_{2,y}) - \frac{\Omega^2}{2} \\ \tilde{D}_{12} = \tilde{D}_{21} = \xi_1 s_{1,x}s_{1,y} + \frac{\xi_2}{4} s_{2,x}s_{2,y} \\ \tilde{D}_{22} = \xi_1(3 - 2c_{1,y} - c_{1,x}c_{1,y}) + \frac{\xi_2}{4}(3 - 2c_{2,y} - c_{2,x}c_{2,y}) - \frac{\Omega^2}{2} \end{cases} \tag{7.23}$$

which may be expressed in a single quartic equation of the dimensionless frequency:

$$\begin{aligned} &\frac{1}{4}\Omega^4 - \left[\xi_1(3 - c_{1,x} - c_{1,y} - c_{1,x}c_{1,y}) + \frac{\xi_2}{4}(3 - c_{2,x} - c_{2,y} - c_{2,x}c_{2,y})\right]\Omega^2 \\ &+\left[\xi_1(3 - 2c_{1,x} - c_{1,x}c_{1,y}) + \frac{\xi_2}{4}(3 - 2c_{2,x} - c_{2,x}c_{2,y})\right] \\ &\left[\xi_1(3 - 2c_{1,y} - c_{1,x}c_{1,y}) + \frac{\xi_2}{4}(3 - 2c_{2,y} - c_{2,x}c_{2,y})\right] \\ &-\left(\xi_1 s_{1,x}s_{1,y} + \frac{\xi_2}{4} s_{2,x}s_{2,y}\right)^2 = 0 \end{aligned} \tag{7.24}$$

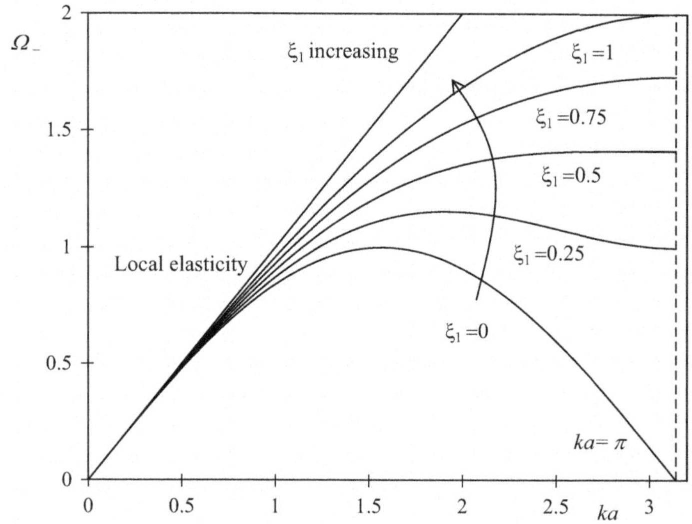

Fig. 7.3 Shear wave dispersive properties in the generalized two-dimensional lattice—$p = 2$ interactions; Influence of the long-range to short-range influence ratio $\xi_1 \in \{0; 0.25; 0.5; 0.75; 1\}$

Equation (7.24) is a quartic equation in frequency, which admits two branches of solutions. Figures 7.3 and 7.4 show the wave dispersive curve in one direction (for instance direction x with $k_y = 0$), associated with the lattice wave dispersive equations:

$$k_y = 0 \quad \Rightarrow \quad \Omega^2 = (4 \pm 2)\left[\xi_1\left(1 - c_{1,x}\right) + \frac{\xi_2}{4}\left(1 - c_{2,x}\right)\right] \tag{7.25}$$

The wave dispersive properties of the generalized truss can be reformulated in terms of dimensionless coefficient $\xi_1 \in [0; 1]$:

$$k_y = 0 \quad \Rightarrow \quad \Omega^2 = (2 \pm 1)\left[4\xi_1 \sin^2\left(\frac{k_x a}{2}\right) + (1 - \xi_1)\sin^2(k_x a)\right] \tag{7.26}$$

This equation resembles the wave dispersive equation of a generalized Lagrange's lattice (one-dimensional lattice) with direct and indirect interactions, as previously obtained by Brillouin [7], and later by [18] or [39], with $p = 2$ cell interactions. A similar dispersive equation was obtained by [37] starting from a 2D truss with $p = 2$ interactions and investigating the propagation of a longitudinal plane wave.

The two solutions derived from the wave dispersion analysis correspond to the shear wave and the compression wave of the 2D lattice with direct and indirect interactions:

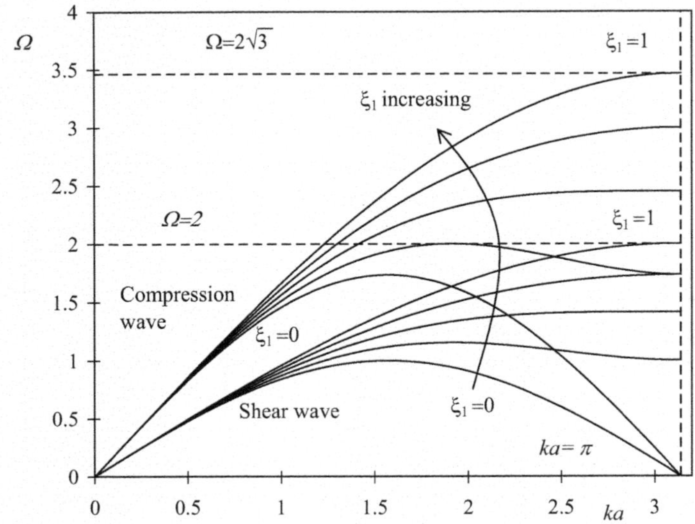

Fig. 7.4 Wave dispersive properties in the generalized two-dimensional lattice—$p=2$ interactions; Influence of the long-range to short-range influence ratio $\xi_1 \in \{0; 0.25; 0.5; 0.75; 1\}$ for both the shear and the compression waves

$$k_y = 0 \quad \Rightarrow \quad \Omega_-^2 = \left[4\xi_1 \sin^2\left(\frac{k_x a}{2}\right) + (1-\xi_1)\sin^2(k_x a)\right] \text{ and}$$
$$\Omega_+^2 = 3\left[4\xi_1 \sin^2\left(\frac{k_x a}{2}\right) + (1-\xi_1)\sin^2(k_x a)\right] \tag{7.27}$$

The dimensionless frequency solution of the shear wave is denoted by Ω_- while the dimensionless frequency solution of the compression wave is denoted by Ω_+. The shear wave dispersive equation is highlighted in Fig. 7.3, whereas Fig. 7.4 compares both dispersive wave solutions, the shear wave and the similar (in terms of equation formulation) compression wave solutions.

From symmetry reasons, we have the same behavior in the orthogonal direction (direction y with $k_x = 0$), i.e.,

$$k_x = 0 \quad \Rightarrow \quad \Omega_-^2 = \left[4\xi_1 \sin^2\left(\frac{k_y a}{2}\right) + (1-\xi_1)\sin^2(k_y a)\right] \text{ and}$$
$$\Omega_+^2 = 3\left[4\xi_1 \sin^2\left(\frac{k_y a}{2}\right) + (1-\xi_1)\sin^2(k_y a)\right] \tag{7.28}$$

Interestingly, as remarked by [37], the optimal (or critical) frequency values of the dispersive curves depend on the short- to long-range interaction ratio ξ_1. For the shear wave solution, with $k_y = 0$, we obtain the critical values for predominant short-range interactions as

$$\xi_1 \in \left[\frac{1}{2}; 1\right] \quad \Rightarrow \quad (k_x a)_{cr} = \pi \quad \text{and} \, (\Omega_-)_{cr} = 2\sqrt{\xi_1}. \tag{7.29}$$

It is worth mentioning that the critical wave number is not dependent in this case to the short to long-range interaction parameter ξ_1. For predominant long-range interactions, the critical wave number of the shear wave solution depends on this ratio, ξ_1, as expected:

$$\begin{aligned} &\xi_1 \in \left[0; \frac{1}{2}\right] \quad \Rightarrow \quad (k_x a)_{cr} = \arccos\left(\frac{-\xi_1}{1-\xi_1}\right) \text{ and} \\ &(\Omega_-)_{cr} = \frac{1}{\sqrt{1-\xi_1}} \end{aligned} \tag{7.30}$$

The same conclusion can be reached for the compression wave where the critical frequency appears to be ξ_1-dependent (i.e., dependent on the short to long-range ratio ξ_1):

$$\begin{aligned} &\xi_1 \in \left[0; \frac{1}{2}\right] \quad \Rightarrow \quad (\Omega_+)_{cr} = \sqrt{\frac{3}{1-\xi_1}} \text{ and} \\ &\xi_1 \in \left[\frac{1}{2}; 1\right] \quad \Rightarrow \quad (\Omega_+)_{cr} = 2\sqrt{3\xi_1} \end{aligned} \tag{7.31}$$

A 3D representation of the wave dispersive surfaces in the space of dimensionless wave numbers are shown in Figs. 7.5, 7.6, and 7.7. Both the shear wave and the compression wave solutions are compared in this spatial representation. Figure 7.5 is the classical solution with pure short-range interaction with $\xi_1 = 1$. This case was investigated in detail by [4, 35] or [17] for 2D lattices with central force (edge and diagonal) interactions among first and second neighboring nodes. The effect of long-range forces is highlighted in Fig. 7.6 with $\xi_1 = 1/2$, and Fig. 7.7 with $\xi_1 = 0$ (this asymptotic case corresponds to the case of a pure long-range interaction). Figures 7.8 and 7.9 compare the distortion of both wave surfaces (shear wave surface and compression wave surface) as a function of the short- to long-range ratios $\xi_1 = 0, \xi_1 = 1/2$ and $\xi_1 = 1$. As expected, the predominant short-range interaction gives higher frequency regime, when compared to the predominant long-range interaction. In the next part, the static response of this 2D generalized lattice will be studied for some academic loading modes.

7.3 Finite Generalized Lattice

The static response of the meta-truss (generalized lattice with $p = 2$ interactions and without angular interactions, i.e., $\gamma_1 = \gamma_2 = 0$) is studied. We consider a meta-truss under several loading modes, including the uniaxial compression as shown

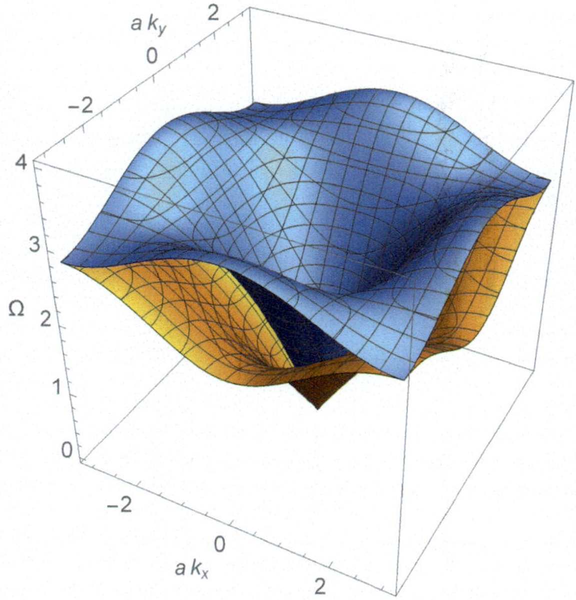

Fig. 7.5 $\Omega \in [0; 4]$; shear wave dispersive surface (orange surface) and compression wave dispersive surface (blue surface); pure central interactions; pure short interactions; $\xi_1 = 1$ and $\xi_2 = 0$

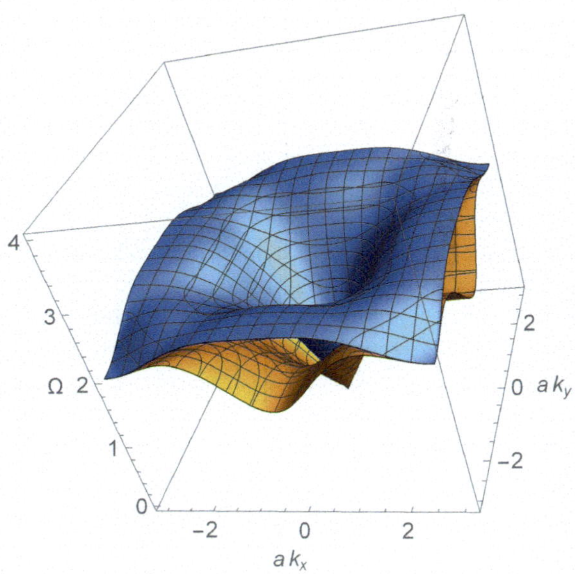

Fig. 7.6 $\Omega \in [0; 4]$; shear wave dispersive surface (orange surface) and compression wave dispersive surface (blue surface); pure central interactions; short and long-range interactions; $\xi_1 = \xi_2 = 0.5$

in Fig. 7.10. In the following, n is related to the number of cells in the finite 2D lattice solid, whereas p is related to the number of interactions (short- and long-range interactions). Only $p = 2$ will be theoretically and numerically studied. The numerical response is computed using a Finite Element truss model based on bar elements, which, however, can produce analytical results as well as numerical ones.

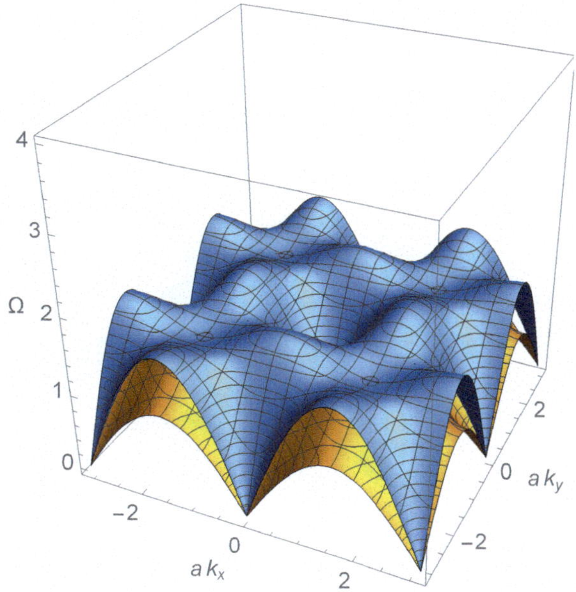

Fig. 7.7 $\Omega \in [0; 4]$; shear wave dispersive surface (orange surface) and compression wave dispersive surface (blue surface); pure central interactions; pure long interactions; $\xi_1 = 0$ and $\xi_2 = 1$

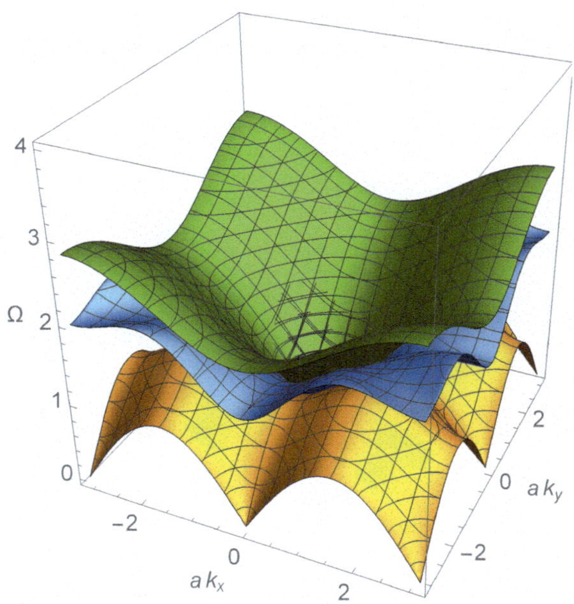

Fig. 7.8 $\Omega \in [0; 4]$; Comparison of the shear wave dispersive surfaces for several values of the long-range interaction parameter (orange surface for $\xi_1 = 0$, blue surface for $\xi_1 = 0.5$, and green surface for $\xi_1 = 1$)

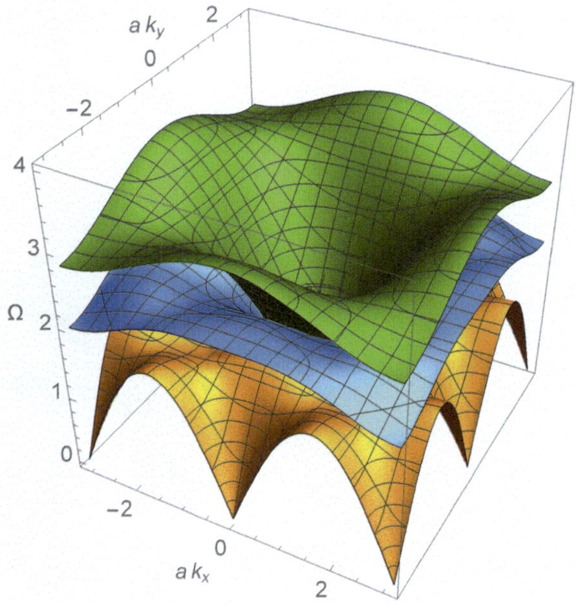

Fig. 7.9 $\Omega \in [0; 4]$; Comparison of the compression wave dispersive surfaces for several values of the long-range interaction parameter (orange surface for $\xi_1 = 0$, blue surface for $\xi_1 = 0.5$ and green surface for $\xi_1 = 1$)

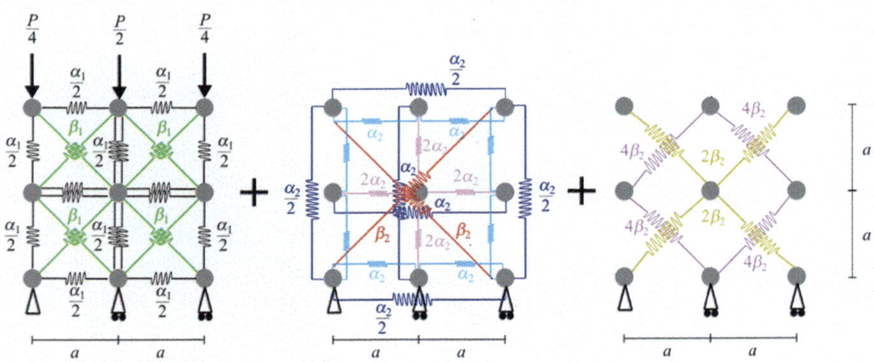

Fig. 7.10 Compression of 2 * 2 cell problem with pure central forces $(\alpha_1, \beta_1, \alpha_2, \beta_2) - n = 2; p = 2$ interaction; Equivalent Poisson's ratio $\upsilon = 1/3$

For the considered plane stress analysis, the following micro/macro scaling parameters are introduced:

$$\alpha_1 = 2\beta_1; \alpha_2 = 2\beta_2; \alpha_1 = \frac{3Eh}{4}\xi_1; \alpha_2 = \frac{3Eh}{4}\frac{\xi_2}{4}; \xi_1 \in [0; 1] \text{ and } \xi_2 = 1 - \xi_1 \tag{7.32}$$

Also, we adopt the following parameters for the combination of the short- and long-range interactions:

$$\frac{\alpha_1}{\xi_1} = \frac{4\alpha_2}{\xi_2} \text{ with } \xi_1 \in [0;\, 1] \text{ and } \xi_2 = 1 - \xi_1 \in [0;\, 1] \tag{7.33}$$

This leads to the meta-truss calibration associated with an equivalent Poisson's ratio $\upsilon = 1/3$ in plane stress:

$$\frac{\alpha_1}{\xi_1} = \frac{4\alpha_2}{\xi_2} = \frac{3Eh}{4} \tag{7.34}$$

For the static analysis of the equivalent generalized truss with $\gamma_1 = \gamma_2 = 0$, the governing difference equations have the following form:

$$\begin{aligned}
&\alpha_1\left(u_{i+1,j} - 2u_{i,j} + u_{i-1,j}\right) + \alpha_2\left(u_{i+2,j} - 2u_{i,j} + u_{i-2,j}\right)\\
&+\frac{\beta_1}{2}\left(u_{i+1,j+1} + u_{i-1,j-1} + u_{i-1,j+1} + u_{i+1,j-1} - 4u_{i,j}\right)\\
&+\frac{\beta_2}{2}\left(u_{i+2,j+2} + u_{i-2,j-2} + u_{i-2,j+2} + u_{i+2,j-2} - 4u_{i,j}\right)\\
&+\frac{\beta_1}{2}\left(v_{i+1,j+1} - v_{i-1,j+1} - v_{i+1,j-1} + v_{i-1,j-1}\right)\\
&+\frac{\beta_2}{2}\left(v_{i+2,j+2} - v_{i-2,j+2} - v_{i+2,j-2} + v_{i-2,j-2}\right) = 0
\end{aligned} \tag{7.35}$$

In an initial analysis, the truss is loaded under pure compression with a prescribed displacement at the border nodes:

$$i \in \{0, 1, \ldots, n\} \quad \Rightarrow \quad v_{i,0} = 0 \text{ and } v_{i,n} = \overline{v} \tag{7.36}$$

For this loading mode, the homogeneous discrete field is obtained, which is the solution of both the static difference equations and the boundary conditions:

$$\begin{cases} u_{i,j} = -\dfrac{1}{3}\dfrac{i}{n}\overline{v} \\[2ex] v_{i,j} = \dfrac{j}{n}\overline{v} \end{cases} \tag{7.37}$$

The displacement field is independent on the short- or long-range interactions. We also have the size-independent load–displacement relationship:

$$P = -Eh\overline{v} \tag{7.38}$$

where we have used half of the spring stiffness for the border springs, as recommended by [22] or [32] (see also [14]). The methodology has been generalized here to include long-range lattice interactions. Also, the long-range stiffness (also dubbed higher-order connections) was doubled when virtual nodes outside the considered cell are included, a methodology that is very similar to the correction of the stiffness at the clamped section in Hencky bar-chain model; i.e., by doubling the rotational spring stiffness for the clamped end [21] or more recently [42]).

For plane stress analysis, we theoretically and numerically found that the ratio between the lateral displacement to the axial displacement to be equal to, for $n = 2$ (2 * 2 cell problem):

$$\frac{u_{n,n}}{v_{n,n}} = -\upsilon = -\frac{1}{3} \text{ in plane stress} \tag{7.39}$$

Figure 7.10 highlights the calibration of the higher-order stiffnesses in the vicinity of the finite lattice for $n = 2$ (2 * 2 cell problem), in order to preserve the homogeneous response of the lattice loaded in pure compression. The numerical response is shown in Fig. 7.11 which confirms the exact theoretical solution Eq. (7.37) of the difference equation Eq. (7.35) with the equivalent Poisson's ratio $\upsilon = 1/3$ in plane stress. The case $n = 3$ (3 * 3 cell problem) was studied in Figs. 7.12 and 7.13, still with $p = 2$ interactions. The generalized lattice is the sum of short-range interactions (Fig. 7.12) and long-range interactions (Fig. 7.13). The process to account for all interactions is described by Fig. 7.13, where the higher-order stiffnesses are corrected at the border to take into account symmetry considerations of the virtual nodes (or external nodes). The numerical response of the pure compression test with $n = 3$ (3 * 3 cell problem) is highlighted in Fig. 7.14, where the homogeneous response is numerically observed as theoretically predicted by Eq. (7.37) with the equivalent Poisson's ratio $\upsilon = 1/3$. The homogeneity of the lattice response with short- and long-range interactions is very sensitive to the calibrated higher-order stiffnesses. For instance, we have performed some static tests without the specific calibration of the higher-order stiffnesses to highlight the loss of homogeneity of the lattice response. In Fig. 7.15, the compression of a 2 * 2 truss problem with $p = 2$ interactions, is considered and some higher-order stiffnesses are purposely omitted (this truss differs from the one considered in Fig. 7.10 due to the omission of the diagonal higher-order stiffnesses). In this last case, the equivalent Poisson's ratio υ clearly differs from 1/3, as numerically indicated by Fig. 7.16 (also with loss of homogeneity of the displacement solution).

Exact solutions may be also obtained for the pure shear test of the generalized lattice with short- and long-range interactions, as considered in Fig. 7.17 (for $n = 2$ and $p = 2$) and Fig. 7.18 (for $n = 3$ and $p = 2$). The meta-truss (or lattice composed of short- and long-range central interactions) is loaded by horizontal and vertical forces parallel to the boundaries of the lattice, to perform a pure shear test. The loading is corrected at the boundary nodes (as already considered for the pure compression case, following the recommendations of [22, 32] for the definition of equivalent nodal forces). The truss is under pure shear with a prescribed displacement at the boundary nodes $u_{i,n} = \overline{u}$:

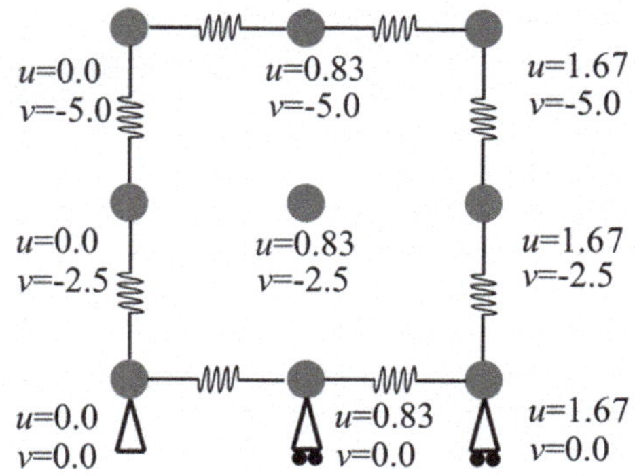

Fig. 7.11 Numerical response of the response of the 2 * 2 cell problem in compression with pure central forces $(\alpha_1, \beta_1, \alpha_2, \beta_2) - p = 2$ interactions; $\overline{v} = -5$ in the numerical test; Equivalent Poisson's ratio $\upsilon = 1/3$

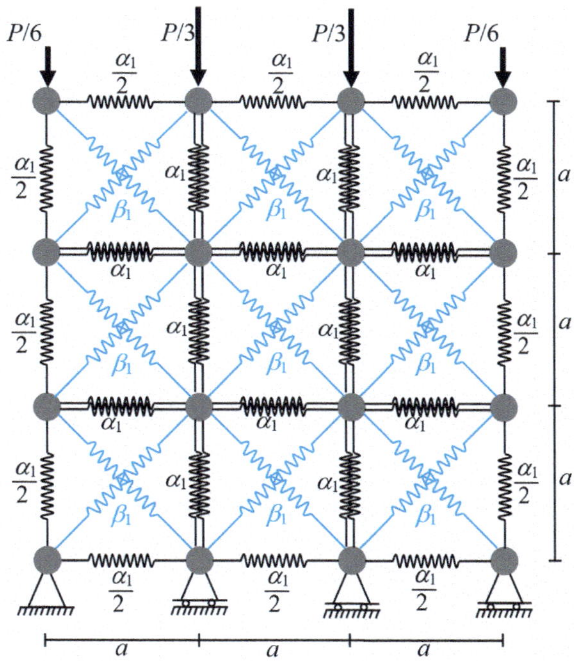

Fig. 7.12 Compression of *3*3* cell problem with pure central forces $(\alpha_1, \beta_1, \alpha_2, \beta_2) - n = 3$; $p = 2$ interaction; Equivalent Poisson's ratio $\upsilon = 1/3$; Representation of the short-range interaction

$$i \in \{0, 1, ..., n\} \quad \Rightarrow \quad u_{i,0} = v_{i,0} = 0 \text{ and } u_{i,n} = \overline{u} \geq 0 \tag{7.40}$$

One recognizes the pure shear mode, which is characterized by the following kinematics for the $n*n$ cells lattice:

$$\begin{cases} u_{i,j} = \dfrac{j}{n}\overline{u} = \dfrac{j}{n}\dfrac{Q}{Gh} \\ v_{i,j} = 0 \end{cases} \tag{7.41}$$

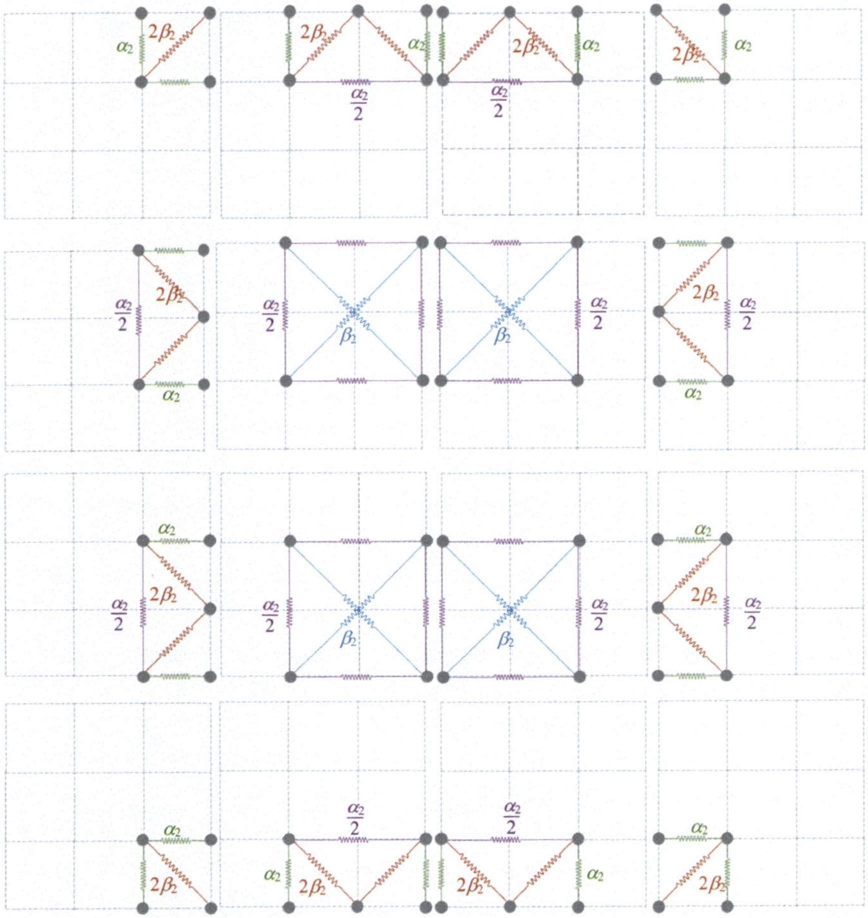

Fig. 7.13 Compression of 3 * 3 cell problem with pure central forces $(\alpha_1, \beta_1, \alpha_2, \beta_2) - n = 3; p = 2$ interaction; Equivalent Poisson's ratio $\upsilon = 1/3$; Representation of the long-range interaction

where the shear modulus $G = E/[2(1+\upsilon)]$ has been used (and $\upsilon = 1/3$ is considered for the pure truss model in plane stress). In particular, the discrete shear angle for this homogeneous test is uniform and is equal to:

$$\frac{u_{i,j+1} - u_{i,j} + v_{i+1,j} - v_{i,j}}{a} = \frac{\overline{u}}{na} = \frac{1}{na}\frac{Q}{Gh} \tag{7.42}$$

Numerical values are reported in Fig. 7.19 (for $n = 2$ and $p = 2$) or Fig. 7.20 (for $n = 3$ and $p = 2$), which confirms the pure shear discrete solution given by Eq. (7.41).

A rhombus shear test is also performed, as shown by the loading mode described in Fig. 7.21: this is identical to the previous one, except that the kinematical constraints are modified in such a way that the nodes along the first diagonal are free to move

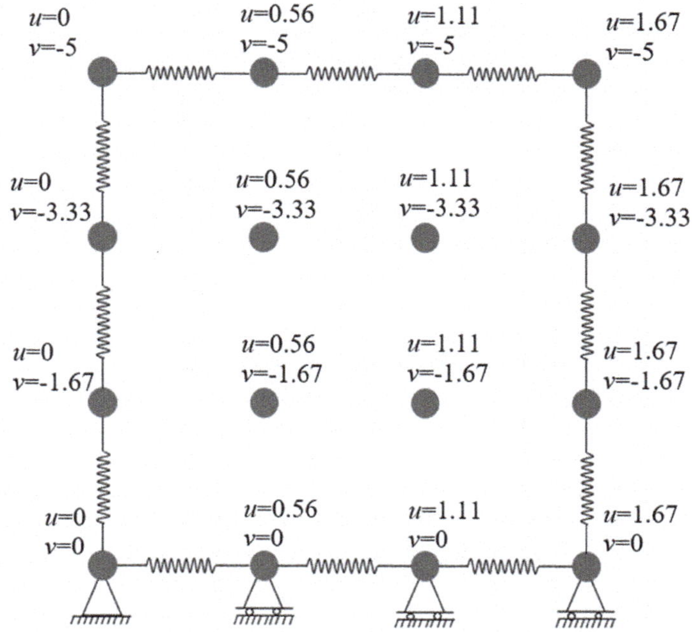

Fig. 7.14 Numerical response of the response of the 3 * 3 cell problem in compression with pure central forces (alpha1, beta1, alpha2, beta2) – p = 2 interactions; $\overline{v} = -5$ in the numerical test; Equivalent Poisson's ratio $\upsilon = 1/3$

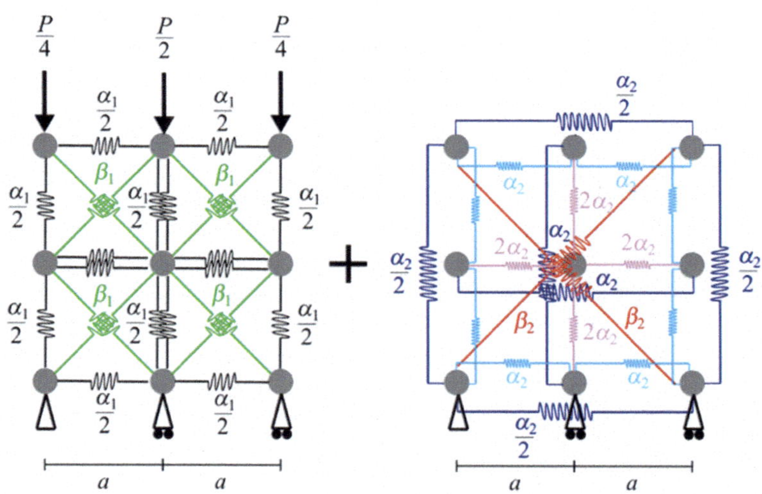

Fig. 7.15 Compression of 2 * 2 cell problem with pure central forces $(\alpha_1, \beta_1, \alpha_2, \beta_2) - n = 2$; $p = 2$ interaction; some higher-order stiffnesses are voluntarily omitted; Equivalent Poisson's ratio υ different from *1/3*

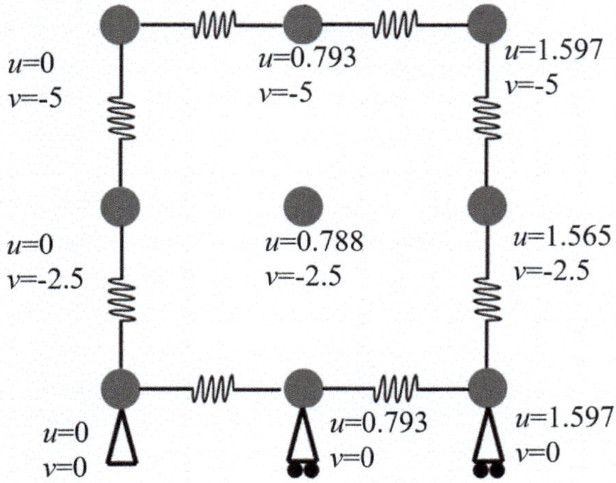

Fig. 7.16 This figure shows the numerical results for the problem in Fig. 7.15; Some higher-order stiffnesses are voluntarily omitted; Equivalent Poisson's ratio υ different from *1/3*

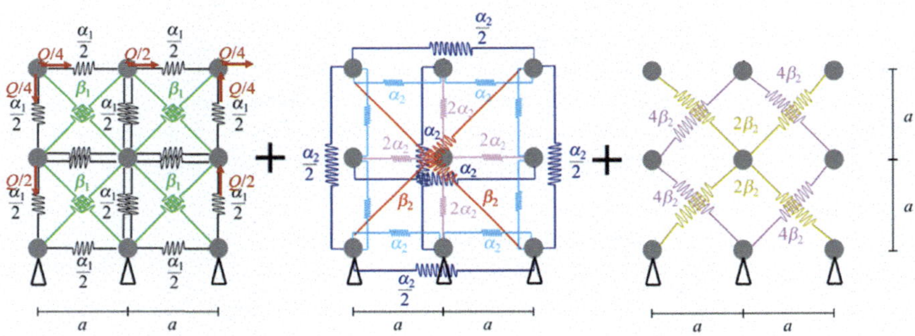

Fig. 7.17 Pure shear test model gives a parallelogram deformed configuration; $n = 2$, $p = 2$; Equivalent Poisson's ratio $\upsilon = 1/3$

along the first diagonal:

$$u_{i,i} = v_{i,i} \text{ for } i \in \{1.2....; n-1\} \tag{7.43}$$

The case $n = 2$ and $p = 2$ is first considered, as shown in Fig. 7.21, which includes the corrections of higher-order stiffnesses in the vicinity of the boundary. The discrete exact solution of this problem for the rhombus shear test is obtained:

$$\begin{cases} u_{i,j} = \dfrac{j}{n}\dfrac{Q}{2Gh} \\ v_{i,j} = \dfrac{i}{n}\dfrac{Q}{2Gh} \end{cases} \tag{7.44}$$

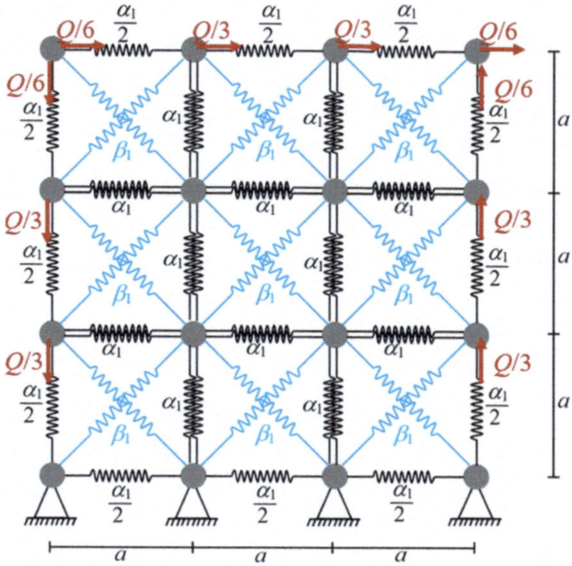

Fig. 7.18 Pure shear test model gives a parallelogram deformed configuration; $n = 3$, $p = 2$; Equivalent Poisson's ratio $\upsilon = 1/3$; Higher-order stiffnesses are shown in Fig. 7.13

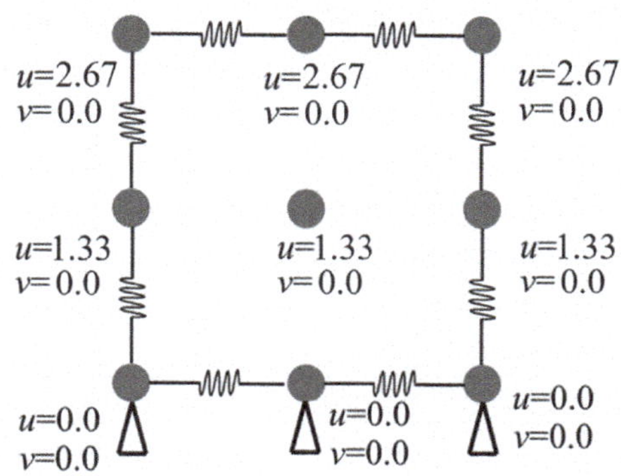

Fig. 7.19 Numerical response of the 2 * 2 cell problem in pure (parallelogram) shear test $(\alpha_1, \beta_1, \alpha_2, \beta_2) - p = 2$ interactions; $Q = 1$, $E = 1$, $h = 1$ in the numerical test; Equivalent Poisson's ratio $\upsilon = 1/3$

In particular, the discrete shear angle for this rhombus shear test is also uniform and is equal to:

$$\frac{u_{i,j+1} - u_{i,j} + v_{i+1,j} - v_{i,j}}{a} = \frac{1}{na}\frac{Q}{Gh} \tag{7.45}$$

as considered for the pure shear test. The numerical values in Fig. 7.22 confirm the expected theoretical solution expressed by Eq. (7.44) for the case $n = 2$ and $p = 2$. The same test was performed for $n = 3$ and $p = 2$, as shown by Fig. 7.23. Higher-order interactions are the same as those represented in Fig. 7.13, and, again, the numerical results of Fig. 7.24 corroborate the theoretical solution of Eq. (7.44). The deformed generalized truss under pure compression, under pure traction, under pure shear and under pure rhombus shear tests is represented in Figs. 7.25, 7.26, 7.27, and

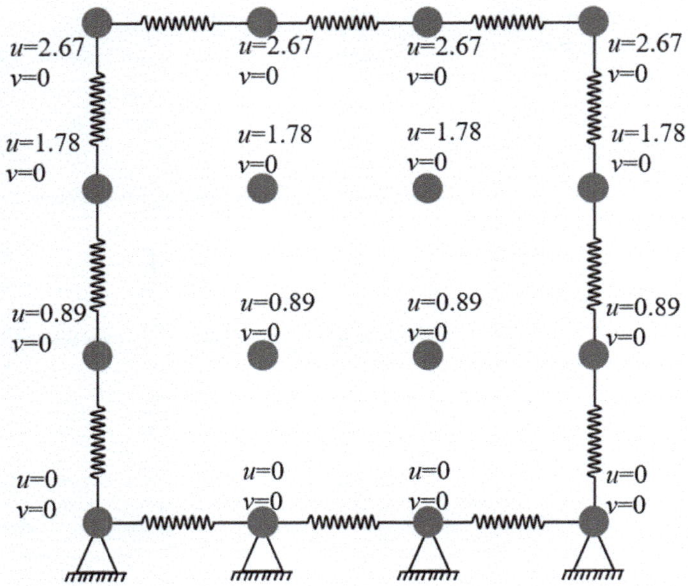

Fig. 7.20 Numerical response of the 3 * 3 cell problem in shear test $(\alpha_1, \beta_1, \alpha_2, \beta_2) - p = 2$ interactions; $Q = 1, E = 1, h = 1$ in the numerical test; Equivalent Poisson's ratio $\upsilon = 1/3$

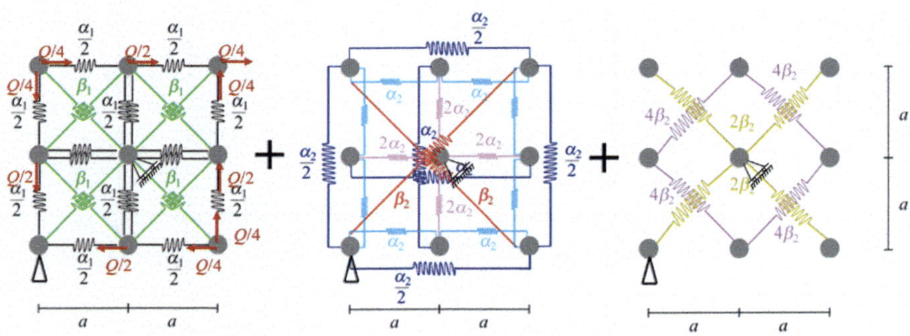

Fig. 7.21 2*2 cell problem in (rhombus) shear test $(\alpha_1, \beta_1, \alpha_2, \beta_2) - p = 2$ interactions; $Q = 1$, $E = 1, h = 1, \xi_1 = 1, u_{0,0} = v_{0,0} = 0$ and $u_{1,1} = v_{1,1}$ in the numerical test; Equivalent Poisson's ratio $\upsilon = 1/3$

7.28 for $n = 2$ and $p = 2$. The homogeneity of the associated discrete displacement field is firmly related to the higher-order stiffness calibration in the vicinity of the boundaries. The equivalent Poisson's ratio $\upsilon = 1/3$ characterizes this generalized truss, as a combination of short- and long-range truss (which may be called meta-truss). The same response is observed in Figs. 7.29, 7.30, 7.31, and 7.32 for $n = 3$ and $p = 2$, under the same academic loading modes.

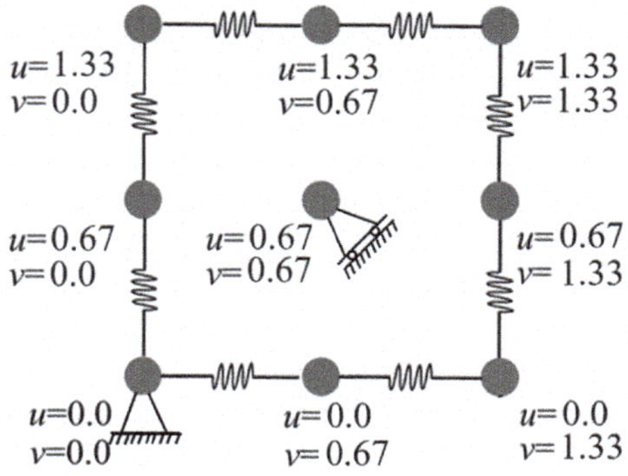

Fig. 7.22 Numerical response of the response of the 2*2 cell problem in rhombus shear test $(\alpha_1, \beta_1, \alpha_2, \beta_2) - p = 2$ interactions; $Q = 1$, $E = 1$, $h = 1$ in the numerical test; Equivalent Poisson's ratio $\upsilon = 1/3$

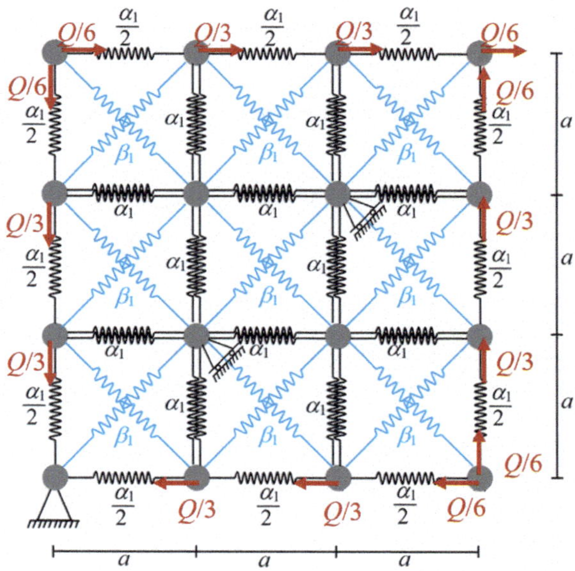

Fig. 7.23 3*3 cell problem in rhombus shear test $(\alpha_1, \beta_1, \alpha_2, \beta_2) - p = 2$ interactions; $Q = 1$, $E = 1$, $h = 1$, $\xi_1 = 1$, $u_{0,0} = v_{0,0} = 0$, $u_{1,1} = v_{1,1}$ and $u_{2,2} = v_{2,2}$ in the numerical test; Equivalent Poisson's ratio $\upsilon = 1/3$; Higher-order stiffnesses are shown in Fig. 7.13

7.4 Conclusions

This paper deals with the static and the dynamic analysis of a 2D elastic lattice exhibiting short-range and long-range interactions. This lattice can be also referred as a generalized lattice, which includes the simple one with only short-range interactions (one-cell interaction model as a particular case). It is shown that the generalized lattice which accounts for central and non-central interactions (of angular type) asymptotically converges towards linear isotropic elasticity, and can also be considered as the generalization of Gazis et al. lattice [19] for multi-cell interaction analysis.

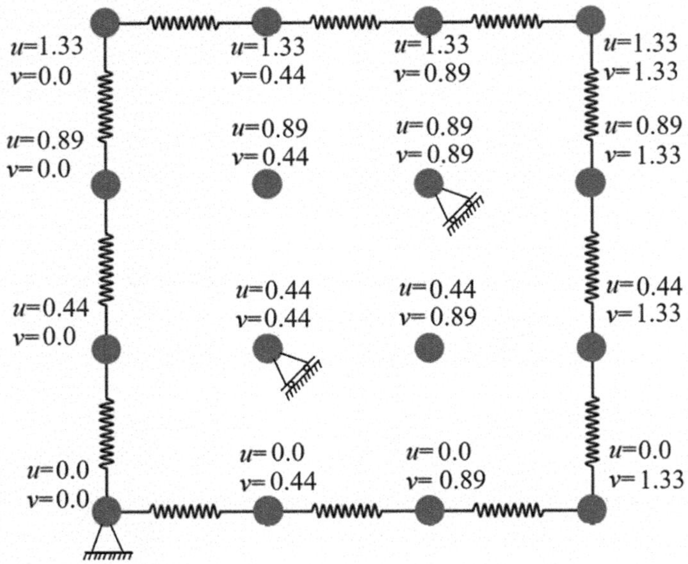

Fig. 7.24 Numerical response of the response of the 3 * 3 cell problem in rhombus shear test $(\alpha_1, \beta_1, \alpha_2, \beta_2) - p = 2$ interactions; $Q = 1$, $E = 1$, $h = 1$ in the numerical test; Equivalent Poisson's ratio $\upsilon = 1/3$

Fig. 7.25 Compression test; $n = 2$ and $p = 2$; black: original shape; grey: deformed shape (suitably rescaled)

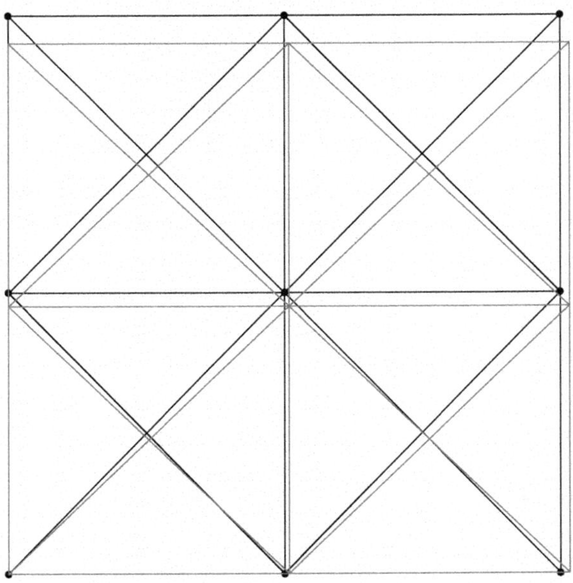

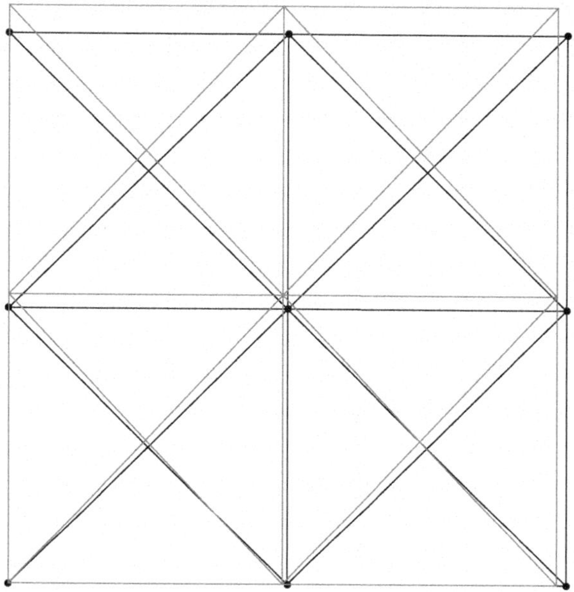

Fig. 7.26 Traction test; $n = 2$ and $p = 2$; black: original shape; grey: deformed shape (suitably rescaled)

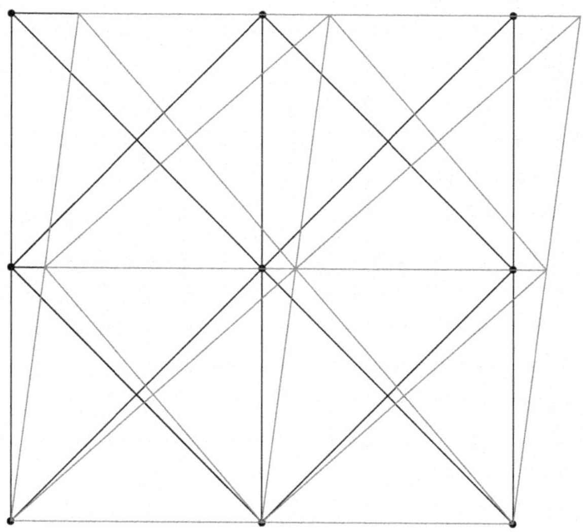

Fig. 7.27 Parallelogram shear test; $n = 2$ and $p = 2$; black: original shape; grey: deformed shape (suitably rescaled)

An exact dispersive wave equation was derived for such generalized lattice (with central, non-central, direct and indirect interactions). The lattice parameters are calibrated so to fit Navier's elastodynamic partial differential equations in the asymptotic limit. A cubic elastic linear continuum may be obtained in the asymptotic limit, which can also degenerate into the linear isotropic elastic medium. A gradient elasticity approach is also derived for this generalized lattice with short- and long-range interactions. The wave dispersive properties of this generalized lattice show an isotropic

Fig. 7.28 Rhombus shear test; $n = 2$ and $p = 2$; black: original shape; grey: deformed shape (suitably rescaled)

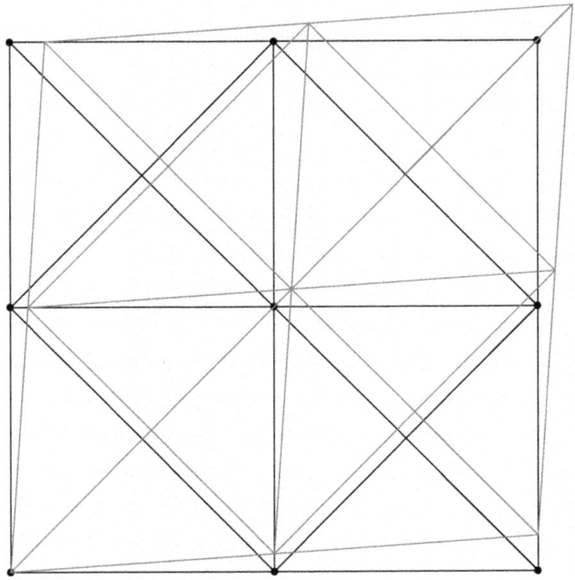

Fig. 7.29 Compression test; $n = 3$ and $p = 2$; black: original shape; grey: deformed shape

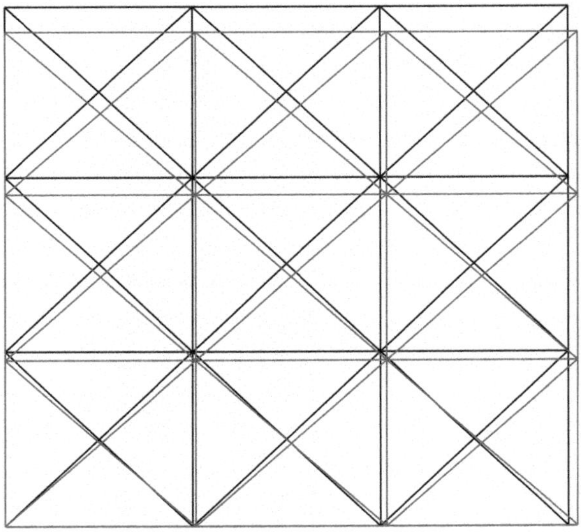

regime for low frequencies, and a loss of isotropy in the high frequency regime, as already commented by [11] or [14] for the lattice with only $p = 1$ interaction (simple lattice).

Finally, a static approach to characterize the finite generalized lattice is performed, which highlights the need of specific calibration of the higher-order stiffness in the vicinity of the boundaries. When considering only central interaction, the lattice degenerates into a pure truss with direct and indirect connections (meta-truss model).

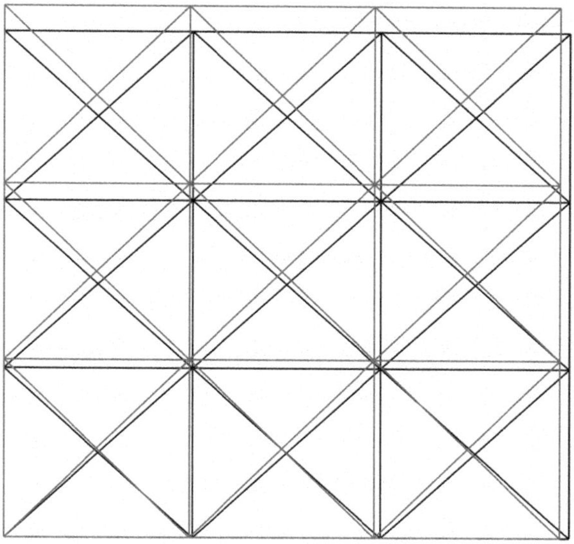

Fig. 7.30 Traction test; $n = 3$ and $p = 2$; black: original shape; grey: deformed shape

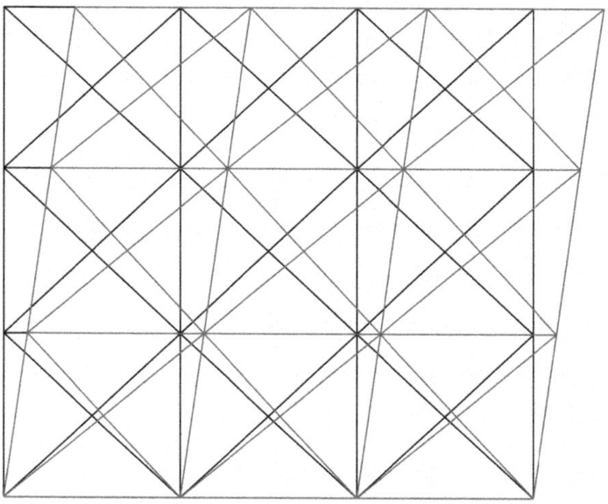

Fig. 7.31 Parallelogram shear test; $n = 3$ and $p = 2$; black: original shape; grey: deformed shape

The response of this finite truss with $p = 2$ interactions is computed by using a truss Finite Element model. It is shown that the response is very sensitive to the calibration of the higher-order stiffnesses close to the boundaries, as previously highlighted by [15, 13] or [16] for static or dynamic of one-dimensional generalized lattices (with direct and indirect interactions). In the future, more interactions with p larger than 3 will be considered, for a complete understanding of the statics and the dynamics of generalized lattices.

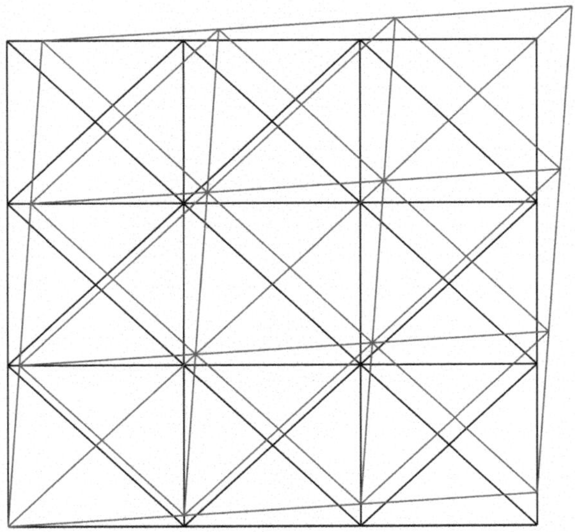

Fig. 7.32 Rhombus shear test; $n = 3$ and $p = 2$; black: original shape; grey: deformed shape

Appendix A—Lattice-Based Gradient Elasticity Model Based on Angular Interactions

In this Appendix A, we show that the gradient elasticity model derived from the generalized lattice of [19] (with both short- and long-range interactions) cannot be isotropic for the gradient elasticity contribution. The lattice-based gradient elasticity equations are obtained for the generalized 2D Gazis et al. lattice model in plane strain, from Eq. (7.16):

$$\begin{cases} (\lambda+2\mu)\dfrac{\partial^2 u}{\partial x^2}+\mu\dfrac{\partial^2 u}{\partial y^2}+(\lambda+\mu)\dfrac{\partial^2 v}{\partial x\partial y} \\ +\dfrac{a^2}{12}(1+3\xi_2) \\ \left[(\lambda+2\mu)\dfrac{\partial^4 u}{\partial x^4}+\mu\dfrac{\partial^4 u}{\partial y^4}+6\lambda\dfrac{\partial^4 u}{\partial x^2\partial y^2}+2(\lambda+\mu)\left(\dfrac{\partial^4 v}{\partial x\partial y^3}+\dfrac{\partial^4 v}{\partial y\partial x^3}\right)\right]=\rho\dfrac{\partial^2 u}{\partial t^2} \\ (\lambda+2\mu)\dfrac{\partial^2 v}{\partial y^2}+\mu\dfrac{\partial^2 v}{\partial x^2}+(\lambda+\mu)\dfrac{\partial^2 u}{\partial x\partial y} \\ +\dfrac{a^2}{12}(1+3\xi_2) \\ \left[(\lambda+2\mu)\dfrac{\partial^4 v}{\partial y^4}+\mu\dfrac{\partial^4 v}{\partial x^4}+6\lambda\dfrac{\partial^4 v}{\partial x^2\partial y^2}+2(\lambda+\mu)\left(\dfrac{\partial^4 u}{\partial x\partial y^3}+\dfrac{\partial^4 u}{\partial y\partial x^3}\right)\right]=\rho\dfrac{\partial^2 v}{\partial t^2} \end{cases} \tag{7.46}$$

An interchange of the variables x and y with the displacement field u and v in the first equation gives the second equation, due to the symmetry $D4$, and vice versa.

Equation (7.46) cannot be expressed as an isotropic linear gradient elasticity model. In other words, it is not possible from Eq. (7.46) to express this system of partial differential equations from an isotropic linear gradient elasticity model that depends on two additional length scales of the strain gradient medium, as expressed by Mindlin [33] and given by

$$(\lambda + 2\mu)(1 - l_1^2\Delta)(\partial_x^2 u + \partial_x\partial_y v) + \mu(1 - l_2^2\Delta)\left(\partial_y^2 u - \partial_x\partial_y v\right) = \rho\, \ddot{u} \tag{7.47}$$

For symmetry reasons, only the first equation is explicitly written. By comparing Eq. (7.47) of the linear isotropic strain gradient elasticity medium with Eq. (7.46) issued of the generalized 2D Gazis et al. lattice, one obtains a system of identification:

$$\begin{cases} -l_1^2(\lambda + 2\mu) = (1 + 3\xi_2)\dfrac{a^2}{12}(\lambda + 2\mu) \quad \Rightarrow \quad l_1^2 = -\dfrac{a^2}{12}(1 + 3\xi_2) \\ -\mu l_2^2 = \mu\dfrac{a^2}{12}(1 + 3\xi_2) \quad \Rightarrow \quad l_2^2 = -\dfrac{a^2}{12}(1 + 3\xi_2) \\ -l_1^2(\lambda + 2\mu) - \mu l_2^2 = 6\lambda\dfrac{a^2}{12}(1 + 3\xi_2) \\ -l_1^2(\lambda + 2\mu) + \mu l_2^2 = \dfrac{a^2}{6}(\lambda + \mu)(1 + 3\xi_2) \end{cases} \tag{7.48}$$

Equation (7.48) is over-constrained and has no solution. It means that the gradient elasticity medium obtained from the lattice equations of [19] with short- and long-range interactions is not isotropic, even if the zero-th order medium may be enforced to be a linear isotropic elastic medium. This conclusion agrees with the one already formulated for the simple lattice of [19] which cannot be isotropic for the second-order medium, for the same reasons (as already outlined by [14]).

The 3D cubic linear gradient elasticity model has 14 material parameters (3 classical and 11 additional gradient elasticity parameters), as already mentioned by Mindlin [34] (see more recently [30, 40]). References [30, 40] showed that the elastodynamic equations of the linear strain gradient model with cubic symmetry can be reduced to 9 parameters, i.e., 3 classical parameters and 6 additional length scale parameters, as opposed to the linear strain gradient isotropic model (associated with two classical parameters and 2 additional length scale parameters). The 2D plane strain equations of a linear strain gradient medium with cubic symmetry are obtained from (see for instance [30, 40]):

$$\begin{aligned} &(c_{12} + 2c_{44})(1 - l_1^2\Delta)(\partial_x^2 u + \partial_x\partial_y v) + c_{44}(1 - l_2^2\Delta)\left(\partial_y^2 u - \partial_x\partial_y v\right) \\ &-c_{44}l_3^2\partial_x^4 u - c_{44}l_4^2\left(2\partial_x^4 u + \partial_x\partial_y^3 v + \partial_x^3\partial_y v\right) - c_{44}l_5^2\left(\partial_x^4 u + \partial_y^4 u\right) \\ &+(c_{11} - c_{12} - 2c_{44})(1 - l_6^2\Delta)\partial_x^2 u = \rho\ddot{u} \end{aligned} \tag{7.49}$$

where (c_{11}, c_{12}, c_{44}) are the three classical parameters and $(l_1, l_2, l_3, l_4, l_5, l_6)$ are the six additional length scales of the gradient elasticity model. For an isotropic material, the number of characteristic lengths reduces to the two length scales (l_1, l_2) already introduced, and the three classical elasticity parameters reduce to the two Lamé parameters from $c_{11} = \lambda + 2\mu$, $c_{44} = \mu$ and $c_{12} = \lambda$.

Equation (7.49) can be equivalently reformulated as

$$\begin{aligned} &c_{11}\partial_x^2 u + c_{44}\partial_y^2 u + (c_{12} + c_{44})\partial_x\partial_y v \\ &+(C_1 + C_3 + 2C_4 + C_5 + C_6)\partial_x^4 u + (C_2 + C_5)\partial_y^4 u \\ &+(C_1 + C_2 + C_6)\partial_x^2\partial_y^2 u + (C_1 - C_2 + C_4)\left(\partial_x\partial_y^3 v + \partial_x^3\partial_y v\right) = \rho\, \ddot{u} \end{aligned} \tag{7.50}$$

where the 6 gradient elasticity parameters identified are

$$\begin{aligned} &C_1 = -l_1^2(c_{12} + 2c_{44});\ C_2 = -l_2^2 c_{44};\ C_3 = -l_3^2 c_{44};\ C_4 = -l_4^2 c_{44};\ C_5 = -l_5^2 c_{44} \text{ and} \\ &C_6 = -l_6^2(c_{11} - c_{12} - 2c_{44}) \end{aligned} \tag{7.51}$$

By comparing Eq. (7.50) with Eq. (7.46), the macroscopic gradient elasticity parameters can be calibrated from the 2D generalized square lattice with short- and long-range interactions:

$$\begin{cases} c_{11} = \lambda + 2\mu \\ c_{44} = \mu \\ c_{12} = \lambda \end{cases} \text{and} \begin{cases} C_1 + C_3 + 2C_4 + C_5 + C_6 = (1 + 3\xi_2)\dfrac{a^2}{12}(\lambda + 2\mu) \\ C_2 + C_5 = (1 + 3\xi_2)\dfrac{a^2}{12}\mu \\ C_1 + C_2 + C_6 = (1 + 3\xi_2)\dfrac{a^2}{2}\lambda \\ C_1 - C_2 + C_4 = (1 + 3\xi_2)\dfrac{a^2}{6}(\lambda + \mu) \end{cases} \tag{7.52}$$

As remarked by Mindlin [34] for the identification of the gradient elasticity cubic parameters from the simple lattice model of [19] (with one-cell interaction), the identification procedure is not unique. One possibility is to assume that the isotropic gradient elasticity parameters vanish, as assumed by Mindlin [34], i.e., $C_1 = C_2 = 0$, leading to the following macroscopic gradient elasticity parameters which account for both short- and long-range interactions:

$$C_1 = 0;\ C_2 = 0;\ C_3 = -(1+3\xi_2)\frac{a^2}{4}(3\lambda+\mu);\ C_4 = (1+3\xi_2)\frac{a^2}{6}(\lambda+\mu);$$
$$C_5 = (1+3\xi_2)\frac{a^2}{12}\mu \text{ and } C_6 = (1+3\xi_2)\frac{a^2}{2}\lambda \tag{7.53}$$

With such an identification, the classical continuous medium (0th order medium) is isotropic and the strain gradient continuous medium associated with the generalized 2D square lattice conserves the square symmetry, and cannot be enforced to be isotropic.

References

1. Alexander S (1998) Amorphous solids: their structure, lattice dynamics and elasticity. Phys Rep 296:65–236
2. Andrianov IV, Awrejcewicz J, Danishevskyy VV (2021) Linear and nonlinear waves in microstructured solids–Homogenization and asymptotic approaches. Springer
3. Askar A (1986) Lattice dynamical foundations of continuum theories: elasticity, piezoelectricity, viscoelasticity, plasticity. World Scientific, Singapore
4. Blackman M (1935) Contributions to the theory of the specific heat of crystals. II. On the vibrational spectrum of cubical lattices and its application to the specific heat of crystals. Proc R Soc Lond A 148:384–406
5. Born M, von Kármán T (1912) On fluctuations in spatial grids. Physikalishe Zeitschrift 13:297–309
6. Born M, Huang K (1954) Dynamical theory of crystal lattices. Oxford University Press
7. Brillouin L (1946) Wave propagation in periodic structures: electric filters and crystal lattices. Mc Graw-Hill
8. Capecchi D, Ruta G, Trovalusci P (2010) From classical to Voigt's molecular models in elasticity. Arch Hist Exact Sci 64:525–559
9. Capecchi D, Ruta G, Trovalusci P (2011) Voigt and Poincaré's mechanistic-energetic approaches to linear elasticity and suggestions for multiscale modelling. Arch Appl Mech 81:1573–1584
10. Cauchy A (1828) Sur les différences finies et les intégrales aux différences des fonctions entières d'une on de plusieurs variables. Exercices de mathématiques, pp 155–159
11. Challamel N, Atanackovic T, Zhang YP, Wang CM (2022) A fractional nonlocal elastic model for lattice wave analysis. Mech Res Comm 126(103999):1–9
12. Challamel N, Ruta G, Nguyen HP, Wang CM, Zhang YP (2024) On angular and surface interactions in two-dimensional elastic lattices. Eur J Mech A/Solids 104(105176):1–25
13. Challamel N, Zhang H, Wang CM, Kaplunov J (2019) Scale effect and higher-order boundary conditions for generalized lattices, with direct and indirect interactions. Mech Res Commun 97:1–7
14. Challamel N, Zhang YP, Wang CM, Ruta G, Dell'Isola F (2023) Discrete and continuous models of linear elasticity: history and connections. Continuum Mech Thermodyn 35(2):347–391
15. Challamel N, Wang CM, Zhang H, Kitipornchai S (2018) Exact and nonlocal solutions for vibration of axial lattices with direct and indirect neighbouring interactions. J Eng Mech 144(5):04018025, 1–9
16. Challamel N, Picandet V (2021) Localization in the static response of higher-order lattices with long range interactions. In: Challamel N, Kaplunov J, Takewaki I (eds) Modern trends in structural and solid mechanics 1.—statics and stability, ISTE-Wiley, pp 67–89

17. De Launay JD (1956) Solid state physics, vol 2, Seitz F, Turnbull D (eds). Academic, pp 219–303
18. Eaton HC, Peddieson J Jr (1973) On continuum description of one-dimensional lattice mechanics. J Tenesse Acad Sci 18(3):96–100
19. Gazis DC, Herman R, Wallis RF (1960) Surface elastic waves in cubic crystals. Phys Rev 119:533–544
20. Gazis DC, Wallis RF (1966) Conditions for rotational invariance of a harmonic lattice. Phys Rev 151(2):578, 580–see also Gazis and Wallis RF, Erratum, Phys Rev 156:1038 (1967)
21. Hencky H (1920) Über die angenäherte Lösung von Stabilitätsproblemen im Raummittels der elastischen Gelenkkette. Der Eisenbau 11:437–452
22. Hrennikoff A (1941) Solutions of problems of elasticity by the framework method. J Appl Mech 8:A169–A175
23. Karpov EG, Saha D (2022) Collective elastic properties of mechanical metamaterials. Extrem Mech Lett 52:101661
24. Keating PN (1966a) Effect of invariance requirements on the elastic strain energy of crystals with application to the diamond structure. Phys Rev 145(2):637–645
25. Keating PN (1966b) Relationship between the macroscopic and microscopic theory of crystal elasticity. I. Primitive crystals. Phys Rev 152(2):774–779
26. Kruskal MD, Zabusky NJ (1964) Stroboscopic-perturbation procedure for treating a class of nonlinear wave equations. J Math Phys 5:231
27. Lagrange JL (1759) Recherches sur la nature et la propagation du son, Miscellanea Philosophico-Mathematica Societatis Privatae Taurinensis I, 2rd Pagination, i–112 (see also Œuvres, Tome 1, 39–148)
28. Lagrange JL (1853) Mécanique analytique, Paris, 1788–3rd Edition, Mallet-Bachelier, Gendre et successeur de Bachelier, Imprimeur-Libraire du bureau des longitudes, de l'école Polytechnique, de l'école centrale des arts et manufactures, Paris, p 367
29. Lax M (1965) E1. The relation between microscopic and macroscopic theories of elasticity. In: Wallis RF (ed) Lattice dynamics. Pergamon Press, New-York, pp 583–596
30. Lazar M, Agiasofitou E, Böhlke T (2022) Mathematical modeling of the elastic properties of cubic crystals at small scales based on the Toupin-Mindlin anisotropic first strain gradient elasticity. Cont Mech Thermodyn 34:107–136
31. Maugin GA (1999) Nonlinear waves in elastic crystals. Oxford University Press
32. McHenry D (1943) A lattice analogy for the solution of stress problems. J Inst Civil Eng Lond 2(5350):59–82
33. Mindlin RD (1964) Micro-structure in linear elasticity. Arch Rat Mech Anal 16:51–78
34. Mindlin RD (1968) Theories of elastic continua and crystal lattice theories. In: Kröner E (ed) Mechanics of generalized continua, proceedings of the IUTAM-symposium on the generalized cosserat continuum and the continuum theory of dislocations with applications. Freudenstadt and Stuttgart, Germany. Published by Springer, pp 312–320
35. Montroll EW (1947) Dynamics of a square lattice. J Chem Phys 15(8):575–591
36. Navier L (1823) Sur les lois de l'équilibre et du mouvement des corps solides élastiques. In: Bulletin des sciences par la Société Philomatique de Paris, pp 177–181 (in French)
37. Porubov AV, Osokina AE, Michelitsch T (2018) Nonlocal approach to square lattice dynamics, Chapter 34, 639–652, Special Issue in honour of Prof. Maugin. In: Altenbach H, Pouget J, Rousseau M, Collet B, Michelitsch T (eds) Generalized models and non-classical approaches in complex materials, Advanced structured materials. Springer
38. Rosenau P (1986) Dynamics of nonlinear mass-spring chains near the continuum limit. Phys Lett A 118(5):222–227
39. Rosenau P (1987) Dynamics of dense lattices. Phys Rev B 36(11):5868–5876

40. Shodja HM, Moosavian H, Ojaghnezhad F (2018) Toupin-Mindlin first strain gradient theory revisited for cubic crystals of hexoctahedral class: analytical expression of the material parameters in terms of the atomic force constants and evaluation via ab initio DFT. Mech Mater 123:19–29
41. Tuna M, Kirca M, Trovalusci P (2019) Deformation of atomic models and their equivalent continuum counterparts using Eringen's two-phase local/nonlocal model. Mech Res Commun 97:26–32
42. Wang CM, Zhang H, Challamel N, Pan W (2020) Hencky-Bar-Chain/Net for structural analysis. World Scientific

Chapter 8
Field-Driven and Electron Beam-Driven Discrete Multi-stable Microrotators Based on Modified HPLC Sorbents

Anton B. Elfimov, Oleg V. Gradov, Margarita A. Gradova, Irina A. Maklakova, and Andrey I. Sergeev

Abstract Different chromatographic sorbents are characterized by different size, charge, and exchange properties, as well as surface groups. In HPLC modified Silasorbs with chemically bonded polar phases (usually 3-glycerolpropyl, 3-aminopropyl, 3-cyanopropyl derivatives of Silasorb 300) and non-polar phases (for reversed-phase HPLC, usually n-octyl, n-octadecyl, or 2-phenethyl derivatives of Silasorb 300, as well as dimethylsilyl derivatives of Silasorb 600) are widely used. For example, Silasorb DEA and Silasorb DEA 300 Polygosil $N(CH_3)_2$ are weak anion exchangers, while Silasorb S and Silasorb S 300 Nucleosil 100 SA (SCX) are strong cation exchangers. Therefore, their response to the electron and ion beams will differ and it is possible to select such combinations of sorbents and irradiation conditions under which the surface charge caused by the specific functional groups will lead to the exotic electrostatic effects, and after sorption the surface charge will change and the nature of the effect observed will also change (depending on the sorbent grain/particle position in the field). This can be demonstrated by time-resolved SEM analysis with millisecond resolution sufficient to observe the dynamics of microparticles. Such effects can be observed on a number of ion exchangers, but here we illustrate this approach using the simplest example of Silasorb C18 (n-octadecyl derivative of Silasorb 300). The particle rotation and the presence of several discrete stationary positions of the Silasorb particle under the electron beam are shown. For the pairs of particles, the bistability of the charge switching and particle dynamics is observed, and for the ensembles of several particles, the presence of several stable states is possible—multistability. In both cases, the presence of stable states can also be proved using oscilloscopic monitoring.

A. B. Elfimov
Emanuel Institute of Biochemical Physics, Russian Academy of Sciences, Kosygina Str., 4, 119334 Moscow, Russia

O. V. Gradov (✉) · M. A. Gradova · I. A. Maklakova · A. I. Sergeev
N.N. Semenov Federal Research Center for Chemical Physics, Russian Academy of Sciences, Kosygina Str., 4, 119991 Moscow, Russia
e-mail: o.v.gradov@gmail.com

H. Altenbach and V. A. Eremeyev (eds.), *Dynamics of Discrete and Continuum Structures and Media*, Advanced Structured Materials 221,
https://doi.org/10.1007/978-3-031-75626-9_8

Keywords Smart materials · Active materials · Su-field analysis · Microrotators · Nanorotators · Microrotors · Nanorotors · Micropropellers · Nanopropellers · HPLC sorbents · Weak anion exchangers · Strong cation exchangers · Chemically bonded phases · Silasorb · Field-driven microactuators · Bistability · Multistability

8.1 The Problem of the Rotating Element Microminiaturization and Its Solution in the XX–XXI Centuries: From the Industrial Restrictions of the XX Century to the Terminological Problems of the XXI Century

The problem of microminiaturization of the rotating elements, such as propellers, rotators, rotors, and gyroscopes, is a part of the global trend of the technical systems' microminiaturization existing since the late 1950s in electronics (transition to the solid-state microelectronics, lithographic techniques [1–4], which was recorded back in 1958 in the works of "Symposium on Microminiaturization of Electronic Assemblies"). Microminiaturization came to the field of electromechanics in the 1960s, but for a long time, it did not affect rotational dynamics due to the manufacturing complexity of the corresponding systems. Initially, microminiaturized electromechanical systems, such as resonant gate transistors and resonistors [5, 6], as follows from the definition, used the principles of resonance. The term MEMS (MicroElectroMechanical Systems) itself was proposed only in the 1980s and formalized by DARPA and IEEE. Until the mid-1980s this term did not designate the corresponding class of devices, although the first devices included in it had already been in development since the late 1970s [7]. Many of the principles of microminiaturization used in MEMS design can be attributed to the application of su-field analysis (SFA) or vepol methodology (SFM), since they explicitly use the principles of interaction between a substance and a field or several fields that control its dynamics (because "su-field" term = "substance and field(s)" or "surface and field(s)" in other interpretations, used in microelectronics, engineering/technical physics of solids and surfaces) [8–12].

Nevertheless, since the 1960s, the area of microminiaturization and miniaturization of purely mechanical rotary elements has been developing independently of electromechanics. The works on the design and study of the dynamics of small passive rotational elements, as well as the study of longitudinal velocity fluctuations in uniform and steady open channel flow over rough beds studied by measuring the statistical variation of velocity with tiny propellers, are examples of purely engineering works on the so-called "micropropellers" of the 1960s, finding a target audience of readers and citation sources in publications on the ocean engineering or coastal engineering, civil engineering and engineering mechanics, applied hydraulics, hydro-sedimentology, and hydro-seismology [13–22]. Such works do not find applications in microengineering for the design of controlled actuators, but are inspiratory for a variety of mechanical/hydraulic/(aero-)hydrodynamic sensors. This trend has prevailed since the late 1960s through the 1980s, spreading from the individual laboratories (Dutch Waterloopkundig Laboratorium) to the wide industrial use of such devices, the leaders in the development and manufacture of which are Dutch and Danish specialists [23–26]. Many of these works are, unfortunately, not currently accessible to the English-speaking reader due to their publication in national languages and the industry-specific nature of publications in the local information whitepapers, newsletters, and reports, which are rarely digitized and made publicly available. However, references to these works are widely found in English periodicals (in the last quarter of the 20th century and the beginning of the 21st century as well). Specialists in the history of science and technology usually associate the active progress of these studies in the Netherlands and Denmark with hydraulic engineering and hydroclimatic factors of these territories, since the construction of dams (in the Netherlands the total length of protective dams is about 3500 km) for flood protection and the use of water energy—from mechanical mills to the simple hydroelectric power stations of the early 20th century (for example, like Gudenaacentralen in Denmark between Kongensbro and Bjerringbro, launched in 1921) formed a "hydraulic engineering culture" that required improvement of both the means of hydrodynamic/hydraulic measurements and the mathematical apparatus that allows one to interpret these measurements and to model and predict the dynamics of natural hydraulic systems.

The issue of using propellers "for micro-hydro-applications" is still relevant today [27–29], but its emphasis has moved to the area of microfluidic applications [30], and also, due to the shift in the "technological poles" of development from Europe to Asia, particularly to Korea and China, and from the university science to the corporate research within spin-offs. For example, one of the most recognizable specialists in this field is Kim Byung-Kon, the member of the Korean Society for Fluid Machinery, who publishes with the affiliation of DSK Engineering Co. He has performed a simulation of micropropeller-type hydro turbines using computational fluid dynamic (CFD) methods [31–33]. Taiwan can be considered as another new center for this work. However, as the terms move deeper into the continent, their meaning is lost (the well-known mechanisms of metonymy and reification). Therefore, by now, terms like "micropopeller" are interpreted very broadly, including both autonomous and non-autonomous agents, both active and passive systems, sensor, transducer, and actuator elements, both micron and decamicron structures and almost centimeter-sized elements for unmanned aerial vehicles (in other words, the use of such terminology is transferred from a strict engineering and metrological vocabulary to the field of advertising "hot labels"). Therefore, despite the large amount of work in this area, in our review, we do not consider a significant part of the available bibliographic pool, which is becoming an area of the "information noise". In addition, we deliberately exclude from consideration the areas of applied applicability that are not related to the use of the properties of the structured materials or their dynamics (such as the development of Micro Air Vehicles (MAV) powered by electric motor-driven micro propellers and micro-DC-motor-driven propeller–thrusters [34–36]). If some application area goes beyond the laboratory and becomes engineering, then the noise that arises in it and extremely expands the meaning of the initial terms impedes the development of the entire direction, since the criteria of the scientific novelty are replaced by the criteria of economics and practice. Below we will consider physical directions that possess promising novelty, set by the principles of su-field analysis (SFA) or vepol methodology (SFM), since only they take full advantage of the micro-miniaturization of rotational elements that arise due to the comparability of the scale of the effective field action with the size of the controlled rotational element.

8.2 SFA-Based Classification of Micropropellers/Nanopropellers and Microrotors/Nanorotors by the Field Control Principles

Within the framework of su-field analysis (SFA), controlled technical systems can be classified according to the physical fields that govern them. Thus, for micropropellers, the following physical and chemical forms of control can be distinguished:

(a) Optical control, which can be used in different forms, such as laser light activation ("the high propulsion of micro-torpedoes under the illumination of near-infrared light is owing to the photothermal effect of gold clusters, generating a thermal gradient inside the micro-torpedoes" [37], while "speed of micro-torpedoes is dependent on the laser powers and media") or light harvesting [30, 38]. The mechanisms for the indirect implementation of this control principle may be much more complex than assumed by the classical theory used to describe the conventional experiment of P.N. Lebedev with a rotator responding to the radiation pressure/light pressure, or what is provided by the classical model with coupled dipole method for the computation of optical torque, taking into account the influence of geometry and relative permittivity on the optical torque [39]. Thus, calculations with Matlab using a Metallic Nano-Particle Boundary Element Method (MNPBEM) toolbox, which considers geometry for plasmonically active micropropellers (thesis [30]), as well as COMSOL models that take into account natural convection currents as results of the nano/micro/particle heating, allow one to simulate indeterminism/the causes of chaotic dynamics in the rotation of micropropellers based on such metal particles. The experimental part of the work (by Elwood [38]) shows that "a combination of surface plasmon resonance, thermophoretic and convective forces enable rotation of the micropropeller in an aqueous environment"; "forces from the Poynting vector caused by electric field generation and the force from the convective current were found to be possible in a 50 °C aqueous environment". Thus, in part, micropropellers controlled by the optical coherent sources can also be considered as thermally controlled, convection-controlled, or plasmon-controlled ones. However, as a rule, such structures are considered as "light driven" and as "non-reactive" micropropellers and nanopropellers or microrotors and nanorotors (see the paper [40], where "thermophoretic migration and active colloid dynamics of light propelled nanostructures" and "circular polarized light mediated notation of plasmonic nanostructure have been reviewed", but the corresponding micropropellers and nanopropellers are considered as "non-reactive light driven" ones).

(b) Acoustically-controlled/acoustically-actuated bubble-powered rotational micropropellers [41]. In fact, in this case, not only the driving signals of the acoustic field can be considered as the driving forces (although a variety of acoustically driven or acoustic(ally)-powered active dynamical agents such as microswimmers [42–49], actuators and microrobots [50, 51] are well known and widely applicable not only for heterogeneous, but also for soft matter environments (for example, in mesophase/liquid crystalline environments), and the variety of hybrid systems, including control or power supply with acoustic field energy, is extremely large and includes acoustomagnetic and optoacoustic, as well as electroacoustic variations [52–55]), but mostly the bubbles emerging within the cavitation phenomena. It should be noted that bubble-powered micropropellers and bubble-powered microrotors have been known since the second half of the 2000s [56–58], but, in fact, the very concept of the bubble-powered microactuators originated in the first half of the 1990s [59], and in the work of L. Lin in 1998 [60] it was logically stated that "characterization of microbubble formation mechanisms leads to new scientific research in microscale heat transfer and provides design guidelines for bubble-powered micromachines." Since then, many devices have been designed based on the above principles of control and energy supply, in particular, bubble-powered actuators [61], bubble-powered motors [62, 63], and micropumps [64], as well as the above mentioned "swimmers" or microswimmers [65, 66]. Acoustically-actuated bubble-powered rotational micropropellers are only a particular case of these bubble-powered/bubble-actuated microstructures or "microdevices".

(c) Magnetic control of dynamics of microrotors/microrotators/micropropellers. This is one of the simplest forms of rotational dynamics control, which does not require additional explanation (except for the cases of photomagnetic, acoustomagnetic, and other hybrid mechanisms and forms of control), known since the times of Faraday's experiments. Magnetically-controlled rotors/propellers can be of arbitrary shape and dynamic symmetry to ensure switching between the directions/trajectories of movement [67–70]. At the same time, there are nonlinear magnetic micropropellers that can be used in practice (for example, the developments at the University of Potsdam [71]). The fundamental and distinctive feature of magnetic control, due to the long history of its use in science and technology (even Heinrich Altshuler used to describe the operation of "su-fields" based on magnetic fields in the 1970s), is its readiness for industrial implementation and the highest (compared to many other field forms of the actuator control) readiness for integration into the engineering structures [72]. Based on the magnetically-controlled systems, it is possible not only to provide regulation of movement in the environment, but also to provide targeted drug delivery in biomedicine, theranostics, and personalized pharmacology, using the principle of magnetic field-controlled microswimmers and nanoswimmers, including those based on self-assembled magnetic micropropellers [73]. Due to the emergence of a hydrodynamic bound state between magnetically powered micropropellers, it is possible to create "swarm robotics"/"swarm intelligence" systems based on autonomous collective agents with magnetic control principles [74].

(d) Various chemical methods for dynamics control, from ion micropropellers of the 1980s [75] (which, in the case of using ion beams, already become physical, but not chemical in nature) to the modern multifunctional chiral chemically-powered micropropellers applicable for cargo transport and manipulation [76]. According to the terminology, these systems belong to micropropellers, but, according to the principle of organization and controllability, they belong to nanotechnology in the futuristic Drexler's interpretation. A special case of such systems is catalytic, in particular, enzymatic micropropellers [77]. In the field of phenomenological and epiphenomenological modeling of enzymatic catalysis, it is possible to use both organic and inorganic carriers and modeling structures [78–80]. In this regard, the models of catalytically driven micropropellers, even with an enzymological prototype, can be either inorganic in the chemical composition or the products of synthetic organic/organometallic chemistry, extremely far from the biological origin. At the same time, in a number of cases, the basis of microscopic and submicroscopic particle mobility is nanoscale mobility of the inorganic clusters and metal ligands, due to which, in essence, when observing physical mobility at the microlevel, we are actually faced with the underlying and supporting mechanisms at the levels of femtochemistry (by time) to the dynamics of nanostructures (in the characteristic scales) [81–83].

It is noteworthy that despite the abundance of biomedical applications [84, 85], biological (by the control methods) microrotors and micropropellers are not specifically distinguished, since they are based on the reductionistically interpreted principles of physical chemistry/chemical physics/conventional physics and biophysics. This enables bionic/biomimetic reproduction of their activity without using specific forms of the movement initiation and control of As an example of such reductionist developments, one can consider the recent article "Synthetic and living micropropellers for convection-enhanced nanoparticle transport" in Science Advances [86]. Strictly speaking, what is fundamental, from the point of view of the charge and size effects and similarity criteria, is the scale at which controlled (or controlling) phenomena occur, due to which the phenomena, essentially occurring at the nanoscale, do not make sense to be reduced to the microscales, even if the devices implementing them are microscale or even larger systems. Thus, speaking about catalytic micropropellers, one can, appealing to the scale of the catalytic mechanisms' manifestation, immediately reduce this problem to the nanoscale phenomenology [87], just as in the case of chiral micropropellers, we must operate at the level of chirality manifestation, that is, starting from the nanolevel according to the characteristic scales of self-assembly [88]. Accordingly, moving to the bionic criteria for reproducing the mechanisms of dynamics of such molecular nanomachines and supramolecular machines, we will avoid logical errors of extrapolation and essentially false reification, but will reproduce the primary catalytic and nonlinear (bio)physical mechanisms, and through this we will obtain a sufficient degree of mechanical similarities in the action of artificial and natural biological machines or micropropellers with the same su-field driving stimuli. In this case, it will be necessary to take into account the entire complexity of the similarity criteria and dimensional theory applicable in modeling. Thus, in the case of modeling bacterial flagellation movement, which ensures the action of nanopropellers in bacterial cells [89–91], it is advisable to take into account mechanics and hydrodynamics, bioenergetic electrogenic machinery, teinochemical phenomena, and conformational mobility of the corresponding biomacromolecules or their models. Otherwise, constructed without considering similarity at the level of mechanisms, the models will not only lack the proper efficiency of the prototype, but will also not fully provide all the phenomena of the prototype that are inevitably associated with the generation of the simulated effect, taking into account the similarity criteria and the effectiveness of action in the specific environment (i.e., hydrodynamics and rheology [92–94]).

And, if in micropropeller manufacturing in industry it is possible to use conventional methods of the parts' manufacturing and processing (achieving minimal microroughness by grinding and using ultraprecision milling machines [95–97]), then in the case of nanopropellers and nanorotors, the classical problems of field-induced and field-controlled self-organization/self-assembly arise, and the emergence of such parts is usually associated with the natural or laboratory-reproducible physical processes at the level of aggregation, adhesion and cohesion, agglomeration, agglutination of the single particles of different shapes. Standard approaches to the static and dynamic testing of such structures (micropropellers or rotors), borrowed from the corresponding branches of technology, appear to be not applicable to the naturally self-organizing propellers and rotors at the decamicron, micron, and submicron scales [98, 99].

The design principles of 3D propellers at the level of "microdevices" (in fact, as an advertising term based on the use of micro- and, occasionally, nanotechnology principles during manufacturing) are not applicable in the assembly and functioning of the real micropropellers, due to the limitations of microhydrodynamics (including those arising according to the similarity criteria) and the size effects, as well as a number of some other specific reasons [100, 101].

8.3 "Dialectics of the Collective and Individual Conditions" in the Manifestations of Short-Range and Long-Range Control Factors in the Dynamics of Microrotors and Nanorotors

Localized controlling fields (for example, surface plasmons under local irradiation [102]) are usually used to control single nano- and micropropellers and rotors, while the long-range distributed fields (e.g., magnetic field, with its power decreasing according to a quadratic law with the distance from the field source) are used in determining the microdynamics of the agent flows and swarm intelligence-like systems (such as "smart dust"), which include not only the active orientable microdevices aligned along field lines ("artificial plankton" in magnetic fields), but also the passive systems of particles without physical separation of the parts/"organs" receiving and detecting the field, oriented in it due to their own physical properties (magnetic nanoparticles and microparticles, including iron oxide or ferrite-based ones [103]). Nanopropellers based on such passive magnetic principles are currently widespread and are being developed/selected to perform a number of the applied functions in specific media (including rheologically complex ones) and soft material environments [104, 105], up to the creation of biocompatible magnetic microdevices or nanodevices that perform various functions up to the cell transfections [106, 107]. Their performance in rheologically complex, including non-Newtonian, environments makes it possible to use them as metrological probes, in particular, for analyzing blood (and other biofluids) [108–110]. The potential for introducing such agents into clinical practice is determined or limited by the dispersity and coercivity of the materials used in their manufacturing (from nano- to micropropellers or magnetic nanorotors [111]). In some cases, a clear tendency can be traced: collective effects are preferable for the distributed nanoscale systems, while the individually addressable field forms of the response and control are more effective at microscales. For example, a laser beam can be focused on several microns, in contrast to the magnetic or microwave field (acting at the nanoscale and synchronizing or collinearizing the dynamics of the field-controlled nanoparticles, regardless of their number or concentration in the field). Significant effects of self-assembly and photoinduced self-organization of the dynamic nanostructures in an optical field only confirm this rule, since in such situations, when the particles are precursors for the assembly of the higher-level functional ensembles, irradiation is carried out at the scales significantly exceeding the size of the single nanoparticles.

In the case of the above mentioned nanorotors, the situation is similar to that for the so-called "nanopropellers". In particular, among them it is also possible to distinguish (in addition to the above mentioned "magnetic field-controlled") a number of forms of the su-field control, both provided by the action of the field on the substance, and during the field formation by the controlled substance itself (that is, the feedback principle of control works in this case):

- Optically driven/light driven nanorotors [112–114], including levitated ones [115, 116]. Such systems can self-assemble and synchronize under the optical beam/light source [117] or can be passively captured by it, as in the case of optical laser tweezers [118]. The criteria for their control can be not only the beam energy, but also polarization, coherence, and interference patterns, and the response criteria depend on the refractive index, absorption, and reflection of their material in the spectral region used. Therefore, it is possible to control and detect (for example, microinterferometrically in the time-resolved sounding/probing mode with simultaneous response registration by the modulation interference microscopy in the "phase dialogue" regime) the activity of single particles and their ensembles using a single beam [119]. The effect on them can be not only thermodynamically positive, but also negative, as is in the special case of the optically driven coherently controlled nanorotors [120], in which the effect of cooling of the nanorotors by elliptic coherent scattering is possible [121, 122]. According to the classical concepts of thermodynamics, bringing the systems out of the equilibrium can be inspired not only by an increase in their thermodynamic level (like in the "negative luminescence" [123–127]: "Negative luminescence is the deficit of the radiation of a body which is in thermodynamic equilibrium with its environment in regard to thermal radiation" [128]; "Negative luminescence is the emission of less radiation than a body emits in thermal equilibrium" [129]). Photothermal and photothermomechanical nanorotors can be considered as another example of such phenomena (although they can be based on nanomagnetic particles and are, in fact, photothermomagnetic or magnetothermal actuators of various geometries and orientations in the field [130–132]).

- Electrokinetic nanorotors and nanorotors with different types of electric trapping [133–135].
- Nanorotors, which are due to the quantum and dimensional effects controlled by the quantum and quantum-electrodynamic factors [136, 137], as well as the exotic spin and spintronic (including those controlled by the spin chemistry principles) nanorotors [138, 139]
- Tunnel nanorotors can be classified as a separate class of the controlled systems. The work [140] (for some reason often cited as: Eisfeld (2016). Nanorotor driven by single-electron tunneling. New J. Phys, 18, 063001, although the title of the paper sounds like "Rotational directionality via symmetry-breaking in an electrostatic motor", there are 4 times more authors in it, both in the paper itself and in the poster report "Nanorotor driven by single-electron tunneling" at the "Trends in Nanotechnology International Conference" at TU Dresden where the first author was A. Celestino describes a "motor... composed by a rotor and two electronic reservoirs", for which "the charge dynamics and the electrostatic field drive rotations of the rotor". The authors of the cited paper use two different approaches: "The hopping process approach takes into account charge fluctuations, which can appear along Coulomb blockade effects in nanoscale systems" and "The mean-field approach neglects the charge fluctuations on the islands, which is typically suitable for larger systems". The authors draw conclusions that are mathematically provable and illustrated with fairly simple modeling: for a system described by the mean-field equations one can in principle prepare initial conditions to obtain a desired rotational direction. In contrast, this is not possible in the stochastic description. However, for both cases, one can achieve rotational directionality by changing the geometry of the rotor. Simple verification tests carried out on the basis of SFA showed the possibility of using this control scheme in various nanorotors, including biomimetic ones (taking into account electrostatic methods of realizing enzymatic catalytic acts [141–148]).
- Catalytic nanorotors and nanomotors controlled by the same factors as the corresponding catalytic acts (including, probably, photocatalytic, electrocatalytic,

and microwave catalytic nanorotors), including those using various types of non-covalent interactions or the field agents affecting electrostatics [149–156].

- Other molecular, but in most cases supramolecular nanorotors, to which the same non-covalent/electrostatic methods of controlling interactions, (self)assembly, repulsion, and dynamics under specific environmental conditions are applicable (as it has been repeatedly shown for both synthetic and biogenic systems since the 1990s [157–170]). Therefore, it is not surprising that works have appeared not only on molecular chemically-propelled nanorotors [171], but also on various supramolecular systems with properties similar to those of nanorotors, such as supramolecular rotors/supramolecular nanorotors [172–175] (including multi-component ones operating based on the same forces that assemble supramolecular synthons together).

Note that, as in the cases cited in the above section, the class of biomimetic, bionic, and bioinspired nanorotors [176–178] is not specifically distinguished, since they are also based on the reductionist physical or (physico)chemical principles. Even the difference in the possible self-organization of such structures, characteristic of microrotors and micropropellers, is not fundamental in the case of nanorotors, especially supramolecular nanorotors, since bacterial self-organizing (supra)molecular machines operating on the principles of nanorotors are self-organizing or self-assembling according to the same principles, as synthetic supramolecular machines based on the above mechanisms for controlling supramolecular assembly [179].

It is also noteworthy that since the size of the nanosystems approaches the minimum sizes for implementation of the corresponding physical features (for example, they can work as single domains in magnetic field effects, or as single actors for surface plasmon resonance, or as single nanodipoles in electrical interactions), they satisfy the definition of su-fields as the minimal technical systems. So, to ensure multifunctionality, agglomeration or self-organization of many nanoparticles may be required, due to which, even with the monofunctionality of certain types of particles, the multiphysical activity of their agglomerates or assemblies as a whole will be achieved. Therefore, it is not surprising that nanorotors with a large number of parts/components will have a more wide range of the controllability options by different physical fields (including several fields simultaneously). Currently, there are two-, three-, four-, five-component, and even more multicomponent nanorotors (and nanosliders) [180–185]. One can imagine nanorotors of various shapes/geometry and multicomponent microrotors (combined from various nanoparticles or nanostructures), since the materials from which nanorotors can be made (or the particles of which can be interpreted as single nanorotors) are very different: photosensitive, luminescent and scintillation active at a wide wavelength range semiconductor oxides and sulfides (for example, zinc-containing ones [186, 187], photoactive and photosensitive porphyrins and phthalocyanines [188], nanotubes differing in the electrophysical properties (including ferroelectric and piezoelectric ones in the future [189–193]) [194, 195], polymer dielectric systems and structures [196] and other macromolecular and biomacromolecular systems (up to the DNA-based nanorotors, which are referred to in [197] and a number of inaccessible Eastern/Asian works).

It is obvious that in such variable combinations, non-conventional heterostructures, multipoles, and translation-diffusion effects will arise (that is, there will be a transition from mass transfer to a reaction with diffusion, according to A. Turing, as a form of morphogenesis of the structures based on similar nanorotors and their surrounding media) [198, 199]. In such systems, which, in our opinion, are of purely scientific interest (since in the case of multiparametric control it is often impossible to reduce the number of variables taken into account at the stage of selecting the control factors, and instead of controllability one obtains adaptability to a wide range of the action factors or cross-adjustability of the system components in multiphysical response and interactions), elements of regular (in strictly controllable units) and chaotic (in multiparametric units according to the controllability criteria) dynamics will have to coexist, which will inevitably lead to the different trajectories of movement in combined external fields [200–203].

Repeating the thesis from the last section, it should be noted that conventional industrial approaches to the analysis of such systems in static conditions (positioned as applicable for nano-scale rotors [204–206]) for self-organizing nanorotors, especially multiparametric in the nature of control, are almost inapplicable. Moreover, in a number of cases it is impossible to state that the system is in an equilibrium state, since at this moment it can be supported (or pumped up) by the other field sources not taken into account in the simplified analysis, and hence, be in a multi-stable state or in a state of dynamic equilibrium. Accordingly, both discrete states of punctuated equilibrium and multi-stable constructs with the field-induced switching of states between several particles will be possible. This, however, is a fundamental possibility for the manifestation of nonlinear dynamics or dynamical systems theory, which is not limited to the nanoparticles, and hence, it will manifest itself at the microlevel, subject to the above control/regulation principles.

We will consider a number of such examples in the following experimental sections.

8.4 Materials and Methods

We illustrate the above approach using the simplest example of a sorbent—Silasorb C18 (n-octadecyl derivative of Silasorb 300).

The dynamics and rotation of the particles were initiated by a 10 kV electron beam from the SEM cathode with simultaneous visualization at 2000x–3500x magnification. The studies were carried out using a setup for time-resolved scanning electron microscopy with the real-time correlation-spectral analysis based on JEOL JSM-T330A (Resolution: 5 nm (at 30 kV; WD = 10 nm); Magnification: LGS 15x (WD = 48 mm) – 200 000x, SGZ 35x (WD = 38 mm) –200 000x; Accelerating Voltage: 0.5–30 kV).

The output video signal was recorded using a video signal digitization system synchronized using a time-lapse videorecorder, providing digitization with a millisecond time resolution. This approach allowed the recording of high-speed rotation processes of micropropellers and microrotators. Image recording with the same time resolution was also realized using a specialized camera mounted in a 3D-printed holding case directly opposite the phosphor of the cathode ray tube of an electron microscope operating in a TV mode.

In the latest versions of the setup, the possibility of installing a camera phone instead of the camera was implemented for the purpose of recording and broadcasting the measurement results in a stream form, but the main disadvantage of this technology is the impossibility of the operator's control over the objective focus, framerates, "sensitometric", photometric/exponometric and colorimetric parameters of the video signal. Data from such stream recordings possess unstable image/signal quality, and therefore cannot be used for the subsequent correlation and comparative analysis of the shape and dynamics of the sorbent particles.

Therefore, one should choose which data sources to use for correlation-spectral analysis, which ones for displaying in the oscilloscopic mode on the SEM display, and which ones for digitization with a high accuracy and time resolution.

8.5 Software

Real-time correlation-spectral analysis was carried out using QAVIS software based on the FFTW library, which allows to work directly with the display memory with comparison intervals of up to 0.4 s.

8.6 Open Access Data

Experimental results obtained are available at the Center for Open Science platform (VA, Charlottesville) and deposited on the servers in Sydney (Australia) at the following addresses: jumping and rotation of Silasorb C18 particles [207] (positive contrast film and negative contrast film images; Sobel-Feldman operator assisted edge detection); bistability and reorientation of the Silasorb C18 particles (oscilloscopic indication) [208]. The data is also available at the Scanning Electron Microscopy Facility Dataport servers at the Federal Research Center for Chemical Physics of the Russian Academy of Sciences (Semenov Institute of Chemical Physics) and the Institute of Biochemical Physics of the Russian Academy of Sciences and are available at the channel https://www.youtube.com/@scanningelectronmicroscopy8023 (Scanning Electron Microscopy Facility ICP—IBCP).

8.7 Results and Discussion

An example of a discrete rotation of a single Silasorb C18 particle on the surface of another Silasorb C18 particle, recorded with millisecond time resolution, is given in Fig. 8.1a, b. It can be seen that the particles actually rotate, and statistically reproducible discrete quasi-stationary states or the states of dynamic equilibrium are recorded, in which the rotating particle stops for some time. Electrostatic forces can involve the nearby particles into the "game" of rotation, approaching up to the point of contact with the rotating particle and transferring an electrostatic charge to it (or removing an electrostatic charge from it).

Direct identification of the contact and semi-contact states of interaction is available using 2D FFT method with the calculation of integral spatial characteristics showing at what angles the interaction occurs. We implemented this using QAVIS software. A system of two interacting particles can be seen in Fig. 8.2a, and the corresponding FFT spectra to these frames are shown in Fig. 8.2b. In this case, according to the real-time optical diffractometry data (emulated by the FFT analysis), it is possible to find the most frequently occurring angular combinations and to relate the retention times of certain particles at certain positions with the electrostatic interactions of these particles with the other particles.

Using the same FFT-assisted technique, one can obtain information on the discrete states of the rotator. The discrete states of the rotator—a sorbent particle—can be most clearly represented by the time-resolved analysis of the integral spatial characteristics superimposed on the images of the corresponding stages of the particle rotation. In this case, it is also useful to pre-process the images in order to contrast the particle edges for FFT analysis (the edges on which scattering is emulated). Figure 8.3a, b show integral spatial characteristics obtained by QAVIS in a real-time mode when analyzing the images from a video stream pre-processed by a Sobel filter (discrete differential Sobel-Feldman operator). It can be seen that for a number of the angular positions of the edges of the Silasorb C18 microparticles integral spatial characteristics are repeated or reproducible and extremely similar (at the same time, there are several states with very sharp angle distributions of the integral spatial characteristics determined by the peculiarly scattering edges).

From the above data it follows that the system of Silasorb C18 particles works:

(a) as a microrotator / microrotor or a micropropeller;
(b) discretely with the presence of a number of steady/equilibrium states, which, according to the scattering emulation of an optical laser beam, are often orthogonal/perpendicular;
(c) is controlled by the electrostatic field (which corresponds to the behavior of such a sorbent particles in the assemblies of the charged microparticles of the same composition—see Fig. 8.2a, b).

The hypothesis on the presence of stationary electrostatic states can be tested using the built-in oscilloscope (for the analysis of cross-sections and median profiles in YMD mode). It turns out that a single particle (Fig. 8.4a, at the top) in an external field

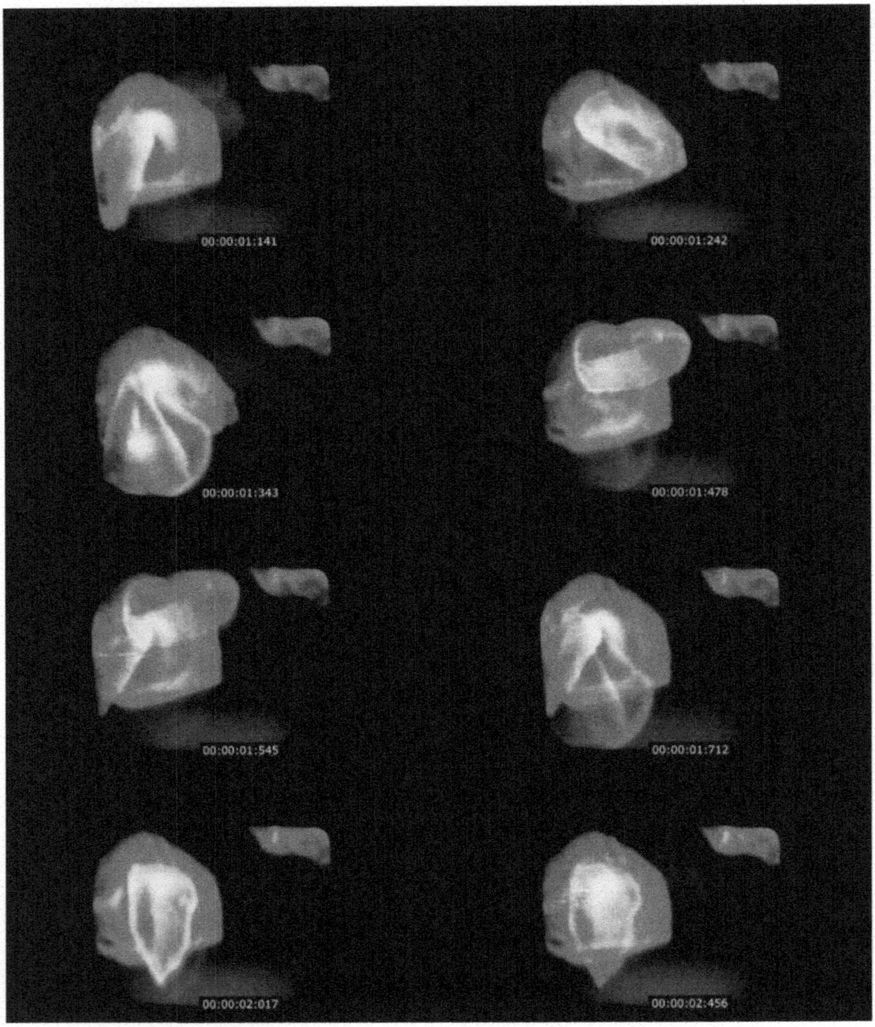

Fig. 8.1 **a** Discrete rotation of a single Silasorb C18 particle on the surface of another Silasorb C18 particle for different time instants. **b** Discrete rotation of a single Silasorb C18 particle on the surface of another Silasorb C18 particle for different time instants

is also characterized by the regular phase switching from the longitudinal direction to the transverse direction, as well as the frequent "jumps" in space (during the transient processes). It can also be seen that the areas of high brightnesses on the particle surface move during the observation time of approximately 10 s. In another case (Fig. 8.4b), one can see that the Silasorb particle (in a system with another particle flaring up nearby—often asynchronously or in antiphase with it) switches from one state to another several times during the same observation time, which

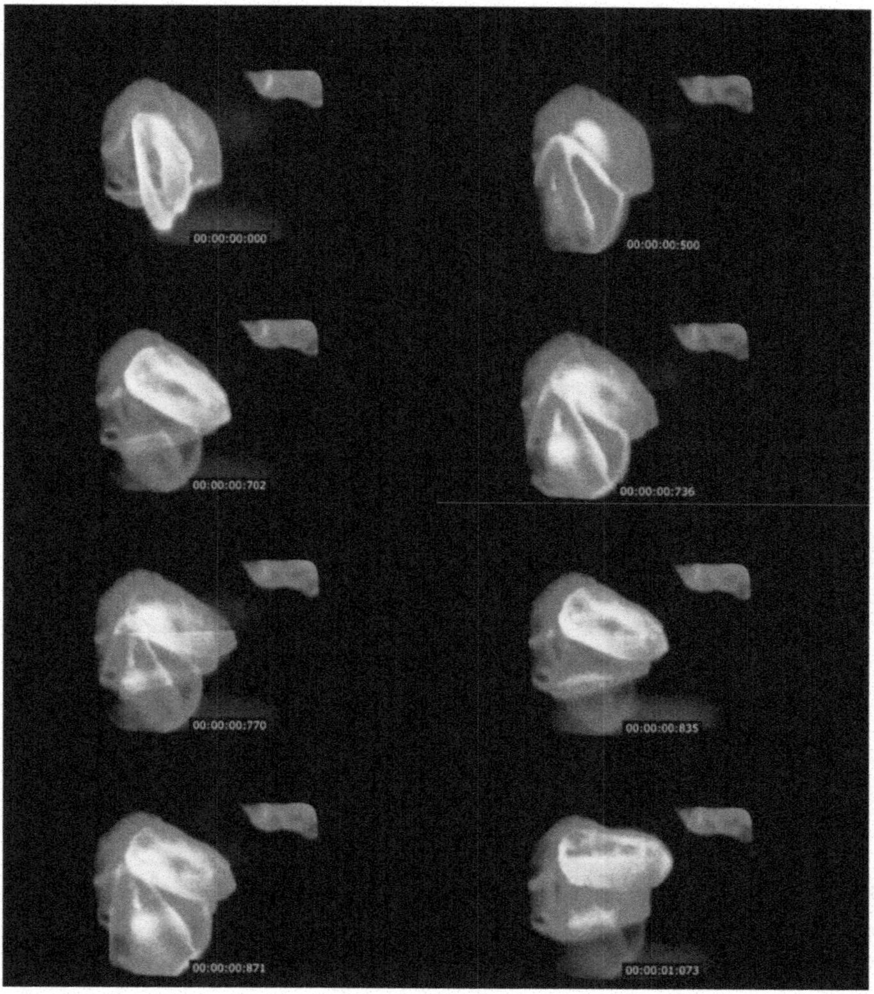

b

Fig. 8.1 (continued)

indicates a possible charge exchange during charging-discharging/charge exchange in a system of two particles.

In other words, synchronization of the mechanical movements of the particles can occur together with their switching between the discrete charge states, being determined by them or determining them (didactically speaking, as in an electric motor or a dynamo). This observation makes the microparticles studied in this work even more similar to the field-driven microrotor/microrotators described from the standpoint of the Su-field analysis in the first part of this work.

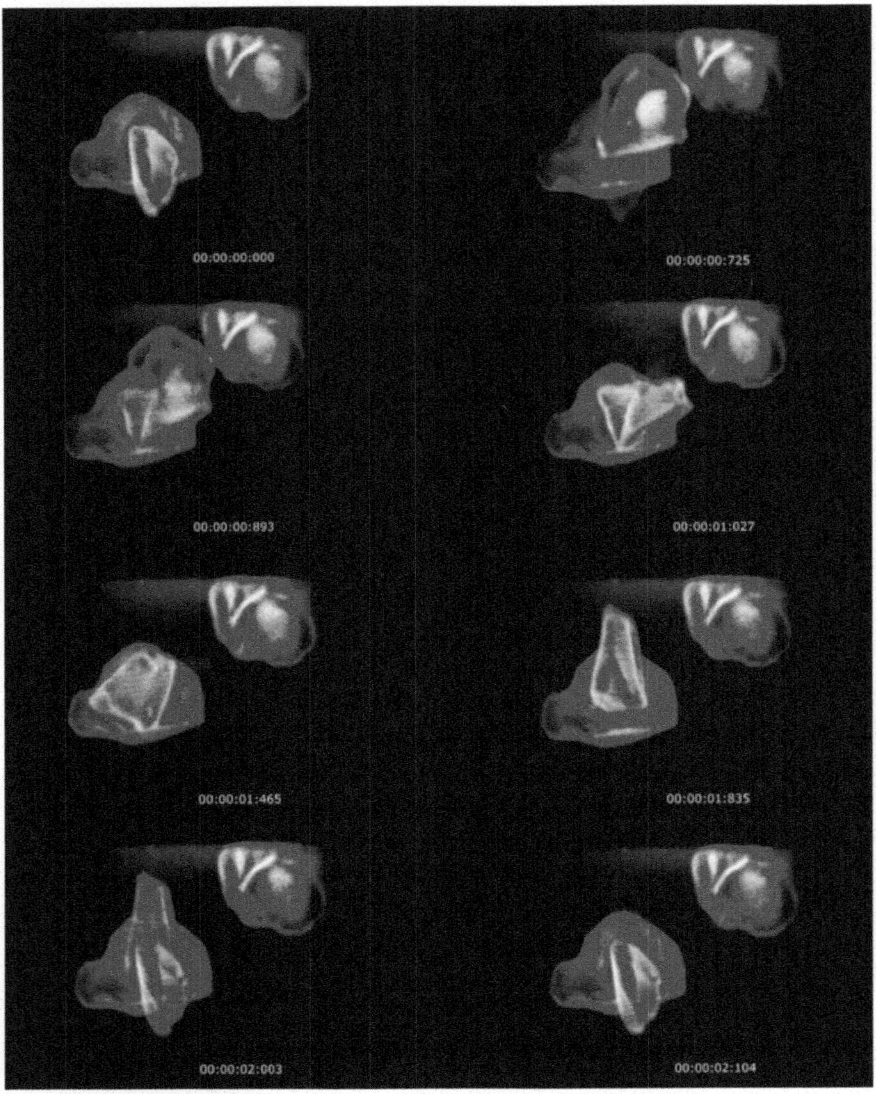

a

Fig. 8.2 **a** A system of two interacting particles. **b** FFT spectra

8.8 Conclusions

Based on the above considerations, one can imagine a whole range of applications of chemomechanical and electrochemomechanical systems based on sorbents and catalytic actors in the field of low-current high-voltage sources (including those with modulation and repolarization in the electrophysical control link), not only

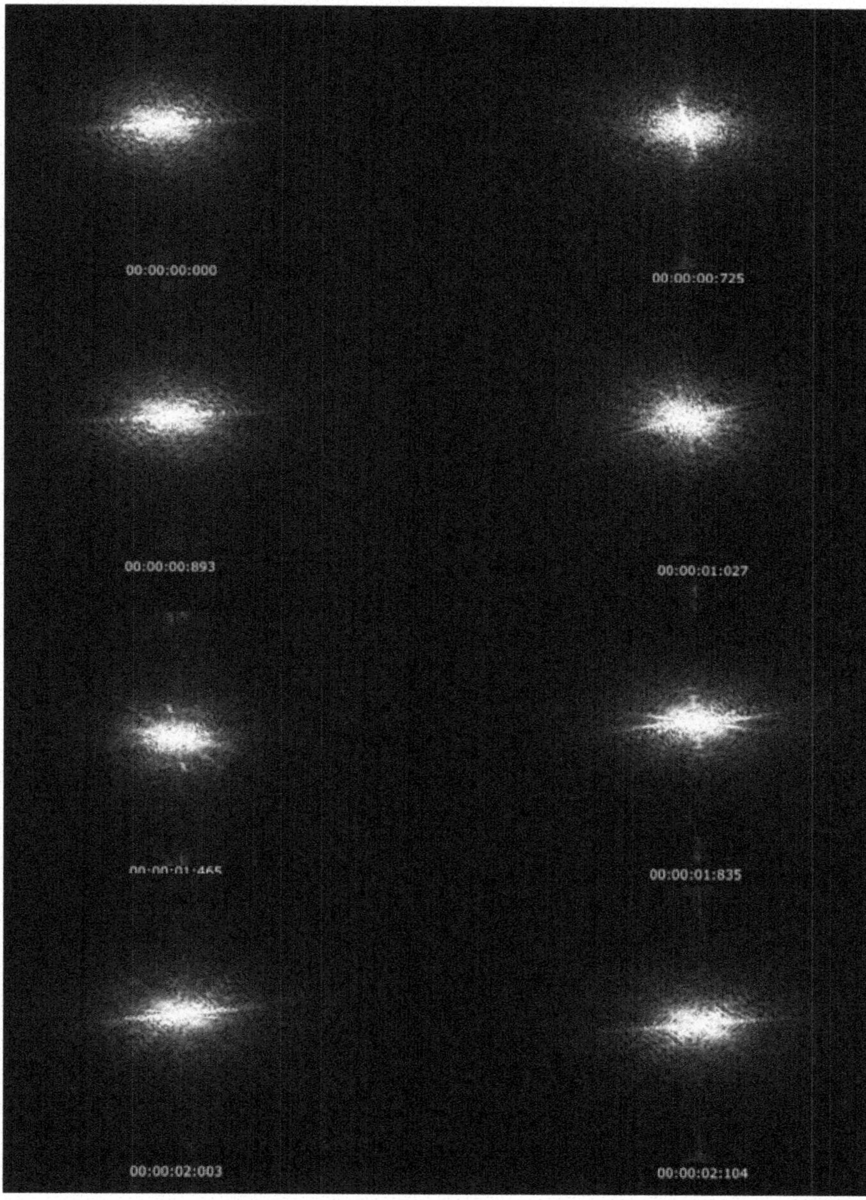

b

Fig. 8.2 (continued)

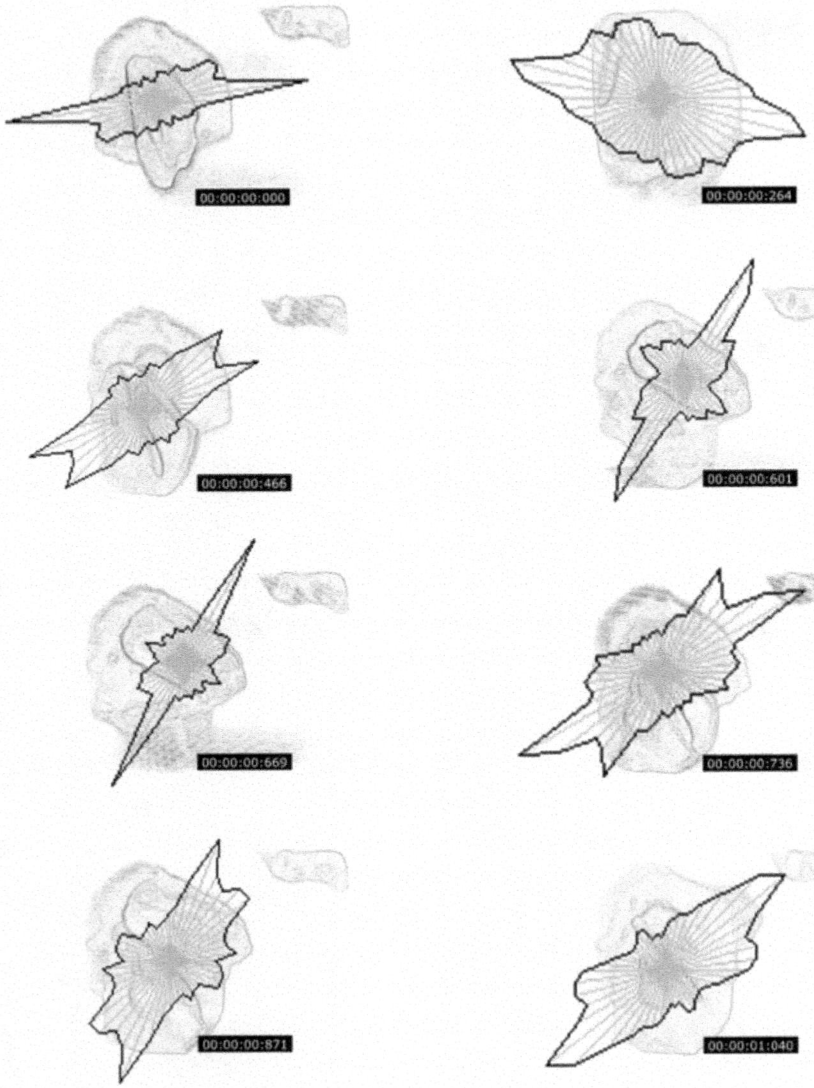

a

Fig. 8.3 **a** Integral spatial characteristics obtained by QAVIS in a real-time mode. **b** Integral spatial characteristics obtained by QAVIS in a real-time mode

converting the field, due to their polar properties, into the mechanics of electrostatically induced motion, but also transferring a charge during this motion to the other particles, depending on the surface parameters of these particles. In the combination of chemically (and/or sorption)-induced dynamics, electrostatic dynamics, and mechanically provided charge transfer (from the particles with some electrostatic

Fig. 8.3 (continued)

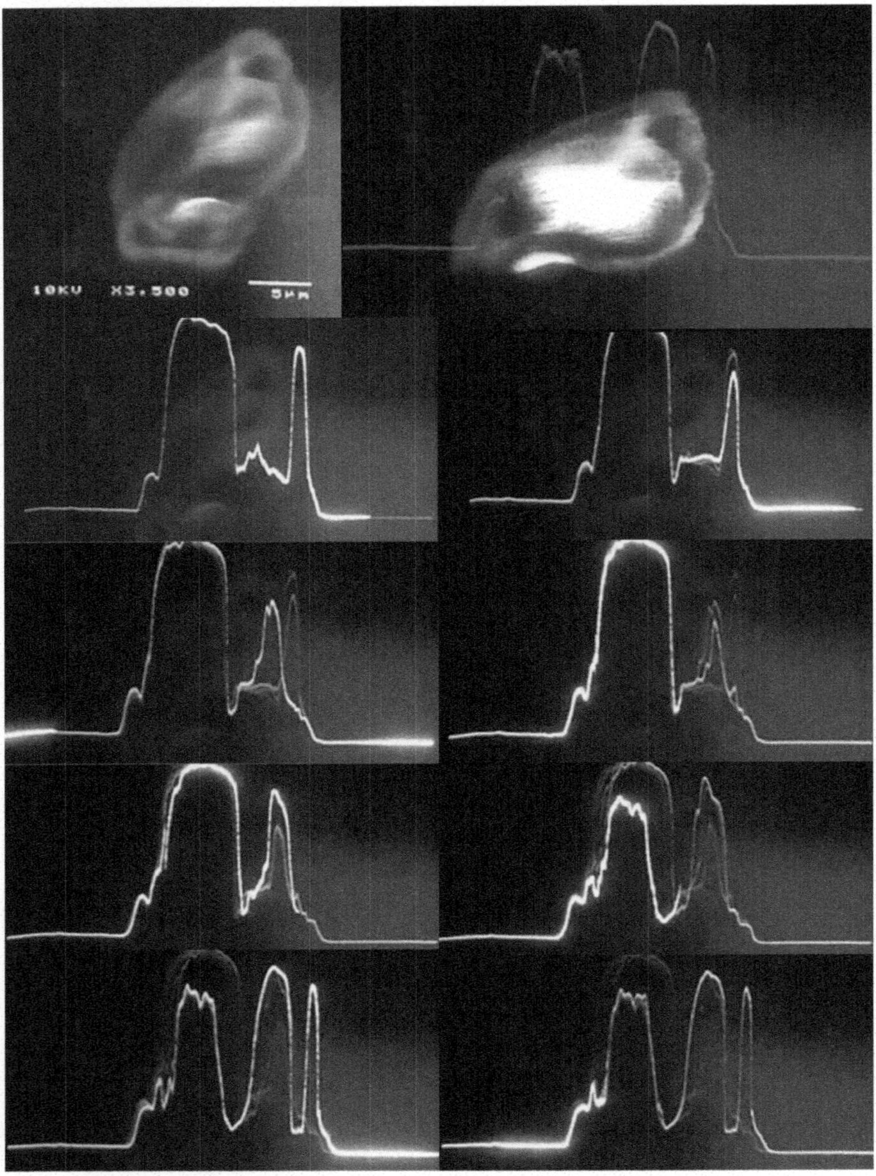

a

Fig. 8.4 a Single particle behavior. b Silasorb particle and the corresponding response

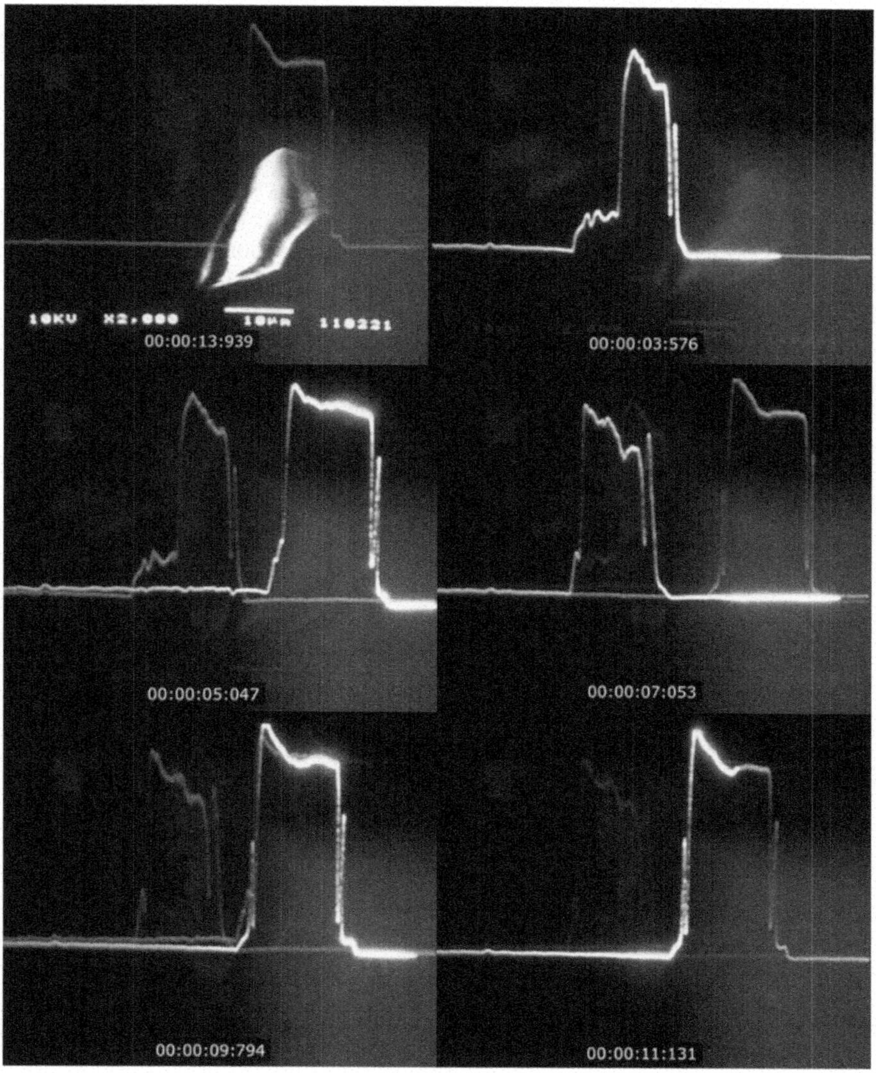

b

Fig. 8.4 (continued)

and surface chemical properties to the particles with the other corresponding properties), the Su-field form of control and regulation of the particle dynamics through the multiple feedbacks is manifested. In this regard, considering these filed-regulated rotary microgranules of sorbents as the prototypes of technical systems, one can consider them as su-fields in the engineering Altshuller's sense, that is, minimal technical systems regulated and controlled by the field. In this system, various forms of movement will transform into each other, providing technical flexibility of the design due to the multifactorial response of the structured (including the application of polar phases) material.

To compensate the purely technical nature of this conclusion, we will end it with a quote: "We see a series of forms of motion, mechanical motion, heat, light, electricity, magnetism, chemical union and decomposition, transitions of states of aggregation, organic life, all of which, if at present we still make an exception of organic life, pass into one another, mutually determine one another, are in one place cause and in another effect... In order to understand the separate phenomena, we have to tear them out of the general inter-connection and consider them in isolation, and then the changing motions appear, one as cause and the other as effect" [208]. Having learned to manage these transitions between causes and effects in advanced structured materials, humanity is approaching the creation of truly smart and adaptive materials, the ultimate form of which will be the biomimetic form of a complex exchange of regulatory signals and switching between causes and effects in the form in which it exists in biological systems (and if a century ago this required the exception of organic life), now, as it is evidence from the literature review, there are biomimetic structures such as artificial bacterial nanorotors, on which these transitions from bioenergetics and cytoelectrophysiology to biomechanics are reliably simulated).

Authors' Contributions Elfimov A. B.—glass chemistry; inspiration of glass chemistry and silicate chemistry approaches for microrotator engineering; measurements of pH-induced transitions and conductivity-induced transitions in such systems.

Gradov O. V.—TRIZ-/SFA-based analysis of microrotator vs. nanorotator engineering; foresight review writing (80%); SEM/ESEM instrumentation engineering and automation; conceptualization.

Gradova M. A.—assistance in SEM / CLEM microscopy of microrotators; text translation and literature quality improvement.

Maklakova I. A.—scanning electron microscopy and dynamic stroboscopic scanning electron microscopy of HPLC sorbents and micro-/nanorotators; granulometric measurements of above mentioned Silasorb particles.

Sergeev A. I.—providing of different samples of HPLC sorbents and their characterization by NMR techniques.

References

1. Finkelstein M (1959) Microminiaturization techniques. IRE Trans Prod Tech 5(1):50–50. https://doi.org/10.1109/TPGPT.1959.1136050
2. Danko SF, Doxey WL, McNaul JP (1959) The micro-module: a logical approach to microminiaturization. Proc IRE 47(5):894–903. https://doi.org/10.1109/JRPROC.1959.287285
3. Liimatainen TM (1959) Recent advances in the application of photo-lithographic techniques to semiconductor devices and microminiaturization. Proc IEE-Part B: Electron Commun Eng 106(15S):385–390. https://doi.org/10.1049/pi-b-2.1959.0086
4. Herwald SW, Angello SJ (1960) Integration of circuit functions into solids: the trend in electronic circuit construction is toward microminiaturization and molecular electronics. Science 132(3434):1127–1133. https://doi.org/10.1126/science.132.3434.1127
5. Nathanson HC, Wickstrom RA (1965) A resonant-gate silicon surface transistor with high-Q band-pass properties. Appl Phys Lett 7(4):84–86. https://doi.org/10.1063/1.1754323

6. Wilfinger RJ, Bardell PH, Chhabra DS (1968) The resonistor: a frequency selective device utilizing the mechanical resonance of a silicon substrate. IBM J Res Dev 12(1):113–118. https://doi.org/10.1147/rd.121.0113
7. Muro H (2013) History and recent progress of MEMS physical sensors. Adv Sci Technol 81:1–8. https://doi.org/10.4028/www.scientific.net/AST.81.1
8. Dung P (1995) TRIZ: inventive creativity based on the laws of systems development. Creativity Innov Manag 4(1):19–29. https://doi.org/10.1111/j.1467-8691.1995.tb00198.x
9. Dung P (1997) Dialectical systems thinking for problem solving and decision making. In: The 7th international conference on thinking (Singapore International Convention and Exhibition Centre, June 1–6, 1997; National Institute of Education, Nanyang Technological University, Singapore). https://books.google.ru/books/about/7th_International_Conference_on_Thinking.html?id=GrvlXwAACAAJ&redir_esc=y
10. Heller JE, van der Beek J, Dittmann C, Feldhusen J (2013) A methodical approach for designing innovative products based on computer aided functional modelling. In: DS 75–1: Proceedings of the 19th international conference on engineering design (ICED13), design for harmonies, vol. 1: Design Processes. Seoul, Korea, 19–22.08. 2013. https://www.designsociety.org/download-publication/34838/a_methodical_approach_for_designing_innovative_products_based_on_computer_aided_functional_modelling
11. Souchkov V (2016) A glossary of essential TRIZ terms. Research and practice on the theory of inventive problem solving (TRIZ) linking creativity, engineering and innovation, pp 265–281. https://doi.org/10.1007/978-3-319-31782-3_17
12. Kaliteevskii V, Bryksin M, Chechurin L (2021) TRIZ application for digital product design and management. IFIP Adv Inf Commun Technol 635:245–255. https://doi.org/10.1007/978-3-030-86614-3_20
13. Jonsson IG (1965) On turbulence in open channel flow. Statistical theory applied to micropropeller measurements. Acta Polytechnica Scandinavica: Civil Eng Build Constr Ser (Finnish Academy of Technical Sciences) 31:3–49
14. Yokosi S (1967) The structure of river turbulence. Bull Disaster Prev Res Inst 17(2):1–29. https://cir.nii.ac.jp/crid/1050001201940072704
15. Plate EJ, Bennett JP (1969) Rotary flow meter as turbulence transducer. J Eng Mech Div 95(6):1307–1329. https://doi.org/10.1061/JMCEA3.0001184
16. Cederwall K (1971) A float diffusion study. Water Res 5(10):889–907. https://doi.org/10.1016/0043-1354(71)90025-X
17. Imamoto H (1972) On the basic characteristics of turbulence in free surface shear flows. Proc Jpn Soc Civil Eng 197:83–91. https://doi.org/10.2208/jscej1969.1972.197_83
18. Hansen E (1972) Lagrangian characteristics of surface turbulence. J Hydraul Div 98(7):1255–1273. https://doi.org/10.1061/JYCEAJ.0003354
19. Jonsson IG, Carlsen NA (1976) Experimental and theoretical investigations in an oscillatory turbulent boundary layer. J Hydraul Res 14(1):45–60. https://doi.org/10.1080/00221687609499687
20. Jonsson IG (1980) A new approach to oscillatory rough turbulent boundary layers. Ocean Eng 7(1):109–152. https://doi.org/10.1016/0029-8018(80)90034-7
21. Bartholdy J (1982) On measuring time-averaged velocities in open channel flow. Sedimentology 29(1):91–94. https://doi.org/10.1111/j.1365-3091.1982.tb01711.x
22. Justesen P (1988) Prediction of turbulent oscillatory flow over rough beds. Coast Eng 12(3):257–284. https://doi.org/10.1016/0378-3839(88)90008-7
23. Pedersen FB (1969) Turbulence measurement in water by means of micro-propellers. Educ Fluid Mech Bull 7 (cited by secondary source)
24. Brolsma P (1973) Temperature correction for micro propeller measurements. Dutch Waterloopkundig Laboratorium Informatie H, 18 (cited by secondary source)
25. Jensen P (1973) [Experimental Investigations of Breaking Waves Using a Micro-Propeller] (Doctoral dissertation, M. Sc thesis (in Danish), Institute of Hydrodynamic and Hydraulic Engineering, Techn. Univ., Denmark)

26. Basco DR, Svendsen IA, Christensen J (1982) 3. Measurements with a bi-directional micro-propeller current meter. Institute of Hydrodynamic and Hydraulic Engineering (ISVA), Tech. Univ. Denmark, Prog Rep 57:25–32
27. Singh P, Nestmann F (2009) Experimental optimization of a free vortex propeller runner for micro hydro application. Exp Thermal Fluid Sci 33(6):991–1002. https://doi.org/10.1016/j.expthermflusci.2009.04.007
28. Kim BK (2014) Optimization of micro hydro propeller turbine blade using NSGA-II. KSFM J Fluid Mach (한국유체기계학회 논문집) 17(4):19–29. https://doi.org/10.5293/kfma.2014.17.4.019
29. Kim BK (2015) Development of a submerged propeller turbine for micro hydro power. KSFM J Fluid Mach (한국유체기계학회 논문집) 18(6):45–56. https://doi.org/10.5293/kfma.2015.18.6.045
30. Shih J (2016) Optimization and modeling of an energy harvesting micro-propeller for microfluidic applications (Honor Thesis; College of Nanoscale Science and Engineering, University at Albany, State University of New York & SUNY Polytechnic Institute). https://scholarsarchive.library.albany.edu/honorscollege_nano/19
31. Park J, Kim Y, Kim B, Lee Y (2012) Variable flow rate characteristics of a low micro propeller-type hydro turbine by CFD. In: Abstracts of the Korean society for new and renewable energy conference (한국신· 재생에너지학회 학술대회 초록집), pp 145–145. https://www.dbpia.co.kr/Journal/articleDetail?nodeId=NODE06235523
32. Park JH, Kim YT, Kim BG, Lee YH (2012) CFD analysis of a micro propeller-type hydro Turbine by varying the number of runner and guide vanes. In: Proc AFORE—2012 237. https://www.dbpia.co.kr/Journal/articleDetail?nodeId=NODE06236013
33. Kim BK (2014) Performance characteristics of micro propeller turbine. KSFM J Fluid Mach (한국유체기계학회 논문집) 17(3):25–32. https://doi.org/10.5293/kfma.2014.17.3.025
34. Boscoianu M, Cîrciu I (2011) An analysis of the efficiency of the functional matching between a flying wing MAV airframe and different types of micro propellers. Incas Bull 3(1):23–29. https://bulletin.incas.ro/files/mircea_boscoianu_v3is1_full.pdf
35. Ranjanag S, Nagpurwala QH, Subbaramu S (2013) Numerical studies on the effect of design trim on aerodynamic performance of a micro propeller for MAV application. Proc ASME 35161:V001T01A015. https://doi.org/10.1115/gtindia2013-3671
36. Yang SF, Chou JH (2009) A mechatronic positioning system actuated using a micro DC-motor-driven propeller–thruster. Mechatronics 19(6):912–926. https://doi.org/10.1016/j.mechatronics.2009.05.005
37. Rao Q, Si T, Wu Z, Xuan M, He Q (2017) A light-activated explosive micropropeller. Sci Rep 7(1):4621. https://doi.org/10.1038/s41598-017-04908-x
38. Elwood J (2016) Optimization and modeling of an energy harvesting optical micropropeller for microfluidic applications. (Honor Thesis; College of Nanoscale Science and Engineering, University at Albany, State University of New York & SUNY Polytechnic Institute). https://scholarsarchive.library.albany.edu/honorscollege_nano/11
39. Chaumet PC, Billaudeau C (2007) Coupled dipole method to compute optical torque: application to a micropropeller. J Appl Phys 101(2):023106. https://doi.org/10.1063/1.2409490
40. Mitra S, Basak M (2023) Recent trends in non-reactive light driven Micro/-nano propellers and rotors. Appl Mater Today 31:101748. https://doi.org/10.1016/j.apmt.2023.101748
41. Mohanty S, Zhang J, McNeill JM, Kuenen T, Linde FP, Rouwkema J, Misra S (2021) Acoustically-actuated bubble-powered rotational micro-propellers. Sens Actuators, B Chem 347:130589. https://doi.org/10.1016/j.snb.2021.130589
42. Ahmed D, Lu M, Nourhani A, Lammert PE, Stratton Z, Muddana HS, Crespi VH, Huang TJ (2015) Selectively manipulable acoustic-powered microswimmers. Sci Rep 5(1):9744. https://doi.org/10.1038/srep09744
43. Ahmed D, Baasch T, Jang B, Pane S, Dual J, Nelson BJ (2016) Artificial swimmers propelled by acoustically activated flagella. Nano Lett 16(8):4968–4974. https://doi.org/10.1021/acs.nanolett.6b01601

44. Ren L, Nama N, McNeill JM, Soto F, Yan Z, Liu W, Wang W, Wang J, Mallouk TE (2019) 3D steerable, acoustically powered microswimmers for single-particle manipulation. Sci Adv 5(10):eaax3084. https://doi.org/10.1126/sciadv.aax3084
45. Chen Q, Liu FW, Xiao Z, Sharma N, Cho SK, Kim K (2019) Ultrasound tracking of the acoustically actuated microswimmer. IEEE Trans Biomed Eng 66(11):3231–3237. https://doi.org/10.1109/TBME.2019.2902523
46. Liu J, Ruan H (2020) Modeling of an acoustically actuated artificial micro-swimmer. Bioinspir Biomim 15(3):036002. https://doi.org/10.1088/1748-3190/ab6a61
47. Cui M, Kim M, Weisensee PB, Meacham JM (2021) Thermal considerations for microswimmer trap-and-release using standing surface acoustic waves. Lab Chip 21(13):2534–2543. https://doi.org/10.1039/D1LC00257K
48. Katuri J, Snezhko A, Sokolov A (2022) Motility of acoustically powered micro-swimmers in a liquid crystalline environment. Soft Matter 18(45):8641–8646. https://doi.org/10.1039/D2SM01171A
49. Liu B, Zhang S, Yang J, Li D, Liu H (2023) Acoustically driven microswimmer with real-time controllable motion direction. Sens Actuators, B Chem 396:134603. https://doi.org/10.1016/j.snb.2023.134603
50. Aghakhani A, Yasa O, Wrede P, Sitti M (2020) Acoustically powered surface-slipping mobile microrobots. Proc Natl Acad Sci 117(7):3469–3477. https://doi.org/10.1073/pnas.1920099117
51. Xiao Y, Zhang J, Fang B, Zhao X, Hao N (2022) Acoustics-actuated microrobots. Micromachines 13(3):481. https://doi.org/10.3390/mi13030481
52. Ahmed D, Dillinger C, Hong A, Nelson BJ (2017) Artificial acousto-magnetic soft microswimmers. Adv Mater Technol 2(7):1700050. https://doi.org/10.1002/admt.201700050
53. Ahmed D, Sukhov A, Hauri D, Rodrigue D, Maranta G, Harting J, Nelson BJ (2021) Bioinspired acousto-magnetic microswarm robots with upstream motility. Nat Mach Intell 3(2):116–124. https://doi.org/10.1038/s42256-020-00275-x
54. Chen XZ, Liu JH, Dong M, Müller L, Chatzipirpiridis G, Hu C, Terzopoulou A, Torlakcik H, Wang X, Mushtaq F, Puigmartí-Luis J (2019) Magnetically driven piezoelectric soft microswimmers for neuron-like cell delivery and neuronal differentiation. Mater Horiz 6(7):1512–1516. https://doi.org/10.1039/C9MH00279K
55. Tang S, Zhang F, Zhao J, Talaat W, Soto F, Karshalev E, Chen C, Hu Z, Lu X, Li J, Lin Z (2019) Structure-dependent optical modulation of propulsion and collective behavior of acoustic/light-driven hybrid microbowls. Adv Func Mater 29(23):1809003. https://doi.org/10.1002/adfm.201809003
56. Kao J, Warren J, Xu J, Attinger D (2006) A bubble-powered micro-rotor: manufacturing, assembly and characterization. Proc ASME 47756:611–615. https://doi.org/10.1115/IMECE2006-14675
57. Kao J, Wang X, Warren J, Xu J, Attinger D (2007) A bubble-powered micro-rotor: conception, manufacturing, assembly and characterization. J Micromech Microeng 17(12):2454. https://doi.org/10.1088/0960-1317/17/12/010
58. Jang D, Jeon J, Chung SK (2018) Acoustic bubble-powered miniature rotor for wireless energy harvesting in a liquid medium. Sens Actuators, A 276:296–303. https://doi.org/10.1016/j.sna.2018.04.023
59. Lin L, Pisano AP (1994) Thermal bubble powered microactuators. Microsyst Technol 1:51–58. https://doi.org/10.1007/BF01367761
60. Lin L (1998) Microscale thermal bubble formation: thermophysical phenomena and applications. Microscale Thermophys Eng 2(2):71–85. https://doi.org/10.1080/108939598199991
61. Dijkink RJ, Van Der Dennen JP, Ohl CD, Prosperetti A (2006) The 'acoustic scallop': a bubble-powered actuator. J Micromech Microeng 16(8):1653. https://doi.org/10.1088/0960-1317/16/8/029
62. Zhou C, Gao C, Lin Z, Wang D, Li Y, Yuan Y, Zhu B, He Q (2020) Autonomous motion of bubble-powered carbonaceous nanoflask motors. Langmuir 36(25):7039–7045. https://doi.org/10.1021/acs.langmuir.9b03398

63. Dincel O, Ueta T, Kameoka J (2019) Acoustic driven microbubble motor device. Sens Actuators, A 295:343–347. https://doi.org/10.1016/j.sna.2019.05.013
64. Jung JY, Kwak HY (2007) Fabrication and testing of bubble powered micropumps using embedded microheater. Microfluid Nanofluid 3:161–169. https://doi.org/10.1007/s10404-006-0116-5
65. Feng J, Yuan J, Cho SK (2016) 2-D steering and propelling of acoustic bubble-powered microswimmers. Lab Chip 16(12):2317–2325. https://doi.org/10.1039/C6LC00431H
66. Liu FW, Zhan Y, Cho SK (2021) Propulsion reversal in oscillating-bubble powered micro swimmer. J Micromech Microeng 31(8):084001. https://doi.org/10.1088/1361-6439/ac0e7f
67. Vach PJ, Fratzl P, Klumpp S, Faivre D (2015) Fast magnetic micropropellers with random shapes. Nano Lett 15(10):7064–7070. https://doi.org/10.1021/acs.nanolett.5b03131
68. Vach PJ, Klumpp S, Faivre D (2015) Steering magnetic micropropellers along independent trajectories. J Phys D Appl Phys 49(6):065003. https://doi.org/10.1088/0022-3727/49/6/065003
69. Codutti A, Bachmann F, Faivre D, Klumpp S (2018) Bead-based hydrodynamic simulations of rigid magnetic micropropellers. Front Robot AI 5:109. https://doi.org/10.3389/frobt.2018.00109
70. Bachmann F, Bente K, Codutti A, Faivre D (2019) Using shape diversity on the way to structure-function designs for magnetic micropropellers. Phys Rev Appl 11(3):034039. https://doi.org/10.1103/PhysRevApplied.11.034039
71. Bachmann F (2019) Non-linearity of magnetic micropropellers: chances and challenges of complex dynamics at the microscale (Doctoral dissertation, Universität Potsdam). https://publishup.uni-potsdam.de/frontdoor/index/index/docId/43603 https://hdl.handle.net/21.11116/0000-0005-6454-C
72. Schwarz L, Medina-Sánchez M, Schmidt OG (2016) Easily scalable high speed magnetic micropropellers. In: 2016 international conference on manipulation, automation and robotics at small scales (18–22 July 2016; Paris, France). IEEE, pp 1–4. https://doi.org/10.1109/MARSS.2016.7561727
73. Calero C, García Torres JM, Sagués Mestres F, Pagonabarraga I, Tierno P (2018) Swimming modes of self-assembled magnetic micropropeller. In: BIMS 2018: Bio-inspired magnetic systems, p 18. http://hdl.handle.net/2117/169139
74. Martinez-Pedrero F, Navarro-Argemí E, Ortiz-Ambriz A, Pagonabarraga I, Tierno P (2018) Emergent hydrodynamic bound states between magnetically powered micropropellers. Sci Adv 4(1):eaap9379. https://doi.org/10.1126/sciadv.aap9379
75. Sandonato GM, Ferreira JL, Ferreira JG, Ludwig GO (1986) [Preliminary results of a ion micropropeller]. Ciencia e Cultura. Suplemento 38(7):392. [Language: Portuguese]. https://inis.iaea.org/search/search.aspx?orig_q=RN:17084741
76. McGovern AD, Huang MJ, Wang J, Kapral R, Aranson IS (2023) Multifunctional chiral chemically—powered micropropellers for cargo transport and manipulation. Small 2304773. https://doi.org/10.1002/smll.202304773
77. Walker D, Käsdorf BT, Jeong HH, Lieleg O, Fischer P (2015) Enzymatically active biomimetic micropropellers for the penetration of mucin gels. Sci Adv 1(11):e1500501. https://doi.org/10.1126/sciadv.1500501
78. Purmal AP, Nikolaev LA (1985) The modelling of biological catalysts. Russian Chem Rev 54(5):466–475. https://www.mathnet.ru/rus/rcr3634
79. Nikolaev LA (1964) The principles of biocatalyst modelling. Russian Chem Rev 33(5):275–286. https://www.mathnet.ru/rus/rcr1824
80. Nikolaev LA (1961) The evolution of biocatalysts. Russian Chem Rev 30(3):117–133. https://www.mathnet.ru/rus/rcr1628
81. Gao PX, Wang ZL (2004) Nanopropeller arrays of zinc oxide. Appl Phys Lett 84(15):2883–2885. https://doi.org/10.1063/1.1702137
82. Fortenberry RC (2016) Analysis of a bicyclic, triple disulphide molecular nanopropeller. RSC Adv 6(49):43509–43517. https://doi.org/10.1039/C6RA08520B

83. Kephart JA, Boggiano AC, Kaminsky W, Velian A (2020) Inorganic clusters as metalloligands: ligand effects on the synthesis and properties of ternary nanopropeller clusters. Dalton Trans 49(45):16464–16473. https://doi.org/10.1039/D0DT02416C
84. Wu Z, Troll J, Jeong HH, Wei Q, Stang M, Ziemssen F, Wang Z, Dong M, Schnichels S, Qiu T, Fischer P (2018) A swarm of slippery micropropellers penetrates the vitreous body of the eye. Sci Adv 4(11):eaat4388. https://doi.org/10.1126/sciadv.aat4388
85. Schnichels S, Peter F, Kadiri VM, Muehl H, Löscher M, Fischer P, Hurst J (2022) Customized magnetic nanopropellers grant targeted delivery to the retina. Invest Ophthalmol Vis Sci 63(7):4150–F0142. https://iovs.arvojournals.org/article.aspx?articleid=2781836
86. Schuerle S, Soleimany AP, Yeh T, Anand GM, Häberli M, Fleming HE, Mirkhani N, Qiu F, Hauert S, Wang X, Nelson BJ (2019) Synthetic and living micropropellers for convection-enhanced nanoparticle transport. Sci Adv 5(4):eaav4803. https://doi.org/10.1126/sciadv.aav4803
87. Kephart JA, Mitchell BS, Chirila A, Anderton KJ, Rogers D, Kaminsky W, Velian A (2019) Atomically defined nanopropeller $Fe_3Co_6Se_8$ (Ph2PNTol) 6: Functional model for the electronic metal–support interaction effect and high catalytic activity for carbodiimide formation. J Am Chem Soc 141(50):19605–19610. https://doi.org/10.1021/jacs.9b12473
88. Zheng Y, Li X, Huang L, Li X, Yang S, Wang Q, Du J, Wang Y, Ding W, Gao B, Chen H (2023) Homochiral nanopropeller via chiral active surface growth. J Am Chem Soc 146(1):410–418. https://doi.org/10.1021/jacs.3c09652
89. Mittal A, Hasija V, Bhat V, Arora S, Sharma M, Gupta R (2011) Bacterial Nanopropellers. In: Encyclopedia of Nanoscience and Nanotechnology, vol 11. American Scientific Publishers, pp 439–448. https://www.ingentaconnect.com/contentone/asp/enn2/2011/00000011/00000001/art00013
90. Gupta R, Sharma M, Mittal A (2006) Effects of membrane tension on nanopropeller driven bacterial motion. J Nanosci Nanotechnol 6(12):3854–3862. https://doi.org/10.1166/jnn.2006.630
91. Soong RK, Neves HP, Schmidt JJ, Montemagno CD (2001) Engineering issues in the fabrication of a hybrid nano-propeller system powered by F 1-ATPase. Biomed Microdevice 3:71–73. https://doi.org/10.1023/A:1011441825422
92. Walker D, Kubler M, Morozov KI, Fischer P, Leshansky AM (2015) Optimal length of low Reynolds number nanopropellers. Nano Lett 15(7):4412–4416. https://doi.org/10.1021/acs.nanolett.5b01925
93. Schamel D, Mark AG, Gibbs JG, Miksch C, Morozov KI, Leshansky AM, Fischer P (2014) Nanopropellers and their actuation in complex viscoelastic media. ACS Nano 8(9):8794–8801. https://doi.org/10.1021/nn502360t
94. Wada H (2009) A semiflexible polymer ring acting as a nano-propeller. Euro Phys J E 28:11–16. https://doi.org/10.1140/epje/i2008-10396-1
95. Cooke JP, Campbell MF, Steager EB, Bargatin I, Yim MH, Park GI (2023) Numerical and experimental study on the addition of surface roughness to micro-propellers. Phys Fluids 35(11):113607. https://doi.org/10.1063/5.0176690
96. Cîrciu I, Pahonie R, Boscoianu M (2009) Considerations about the micro propeller design and it's interaction with the aerodynamic surfaces. Rev Air Force Acad (1):49–54. https://ns.afahc.ro/ro/revista/Nr_1_2009/rev_1_2009.pdf#page=50
97. Takeuchi Y, Sawada K, Sata T (1997) Manufacture of micropropellers by means of ultraprecision milling machine. J Robot Mechatron (Fuji Technology Press Ltd.) 9(6):475–479. https://cir.nii.ac.jp/crid/1360574093771171584
98. Deters R, Selig M (2008) Static testing of micro propellers. In: 26th AIAA applied aerodynamics conference (paper 2008–6246). https://doi.org/10.2514/6.2008-6246
99. Smedresman A, Yeo D, Shyy W (2011) Design, fabrication, analysis, and dynamic testing of a micro air vehicle propeller. In: 29th AIAA applied aerodynamics conference (paper 2011–3817). https://doi.org/10.2514/6.2011-3817
100. Lee KH, Jeon YH, Kim KH, Lee DH, Lee KT (2007) Three-dimensional micro propeller design by using efficient two step optimization. J Fluid Sci Technol 2(2):334–345. https://doi.org/10.1299/jfst.2.334

101. Zhang Y, Tan X, Li X, Lv P, Huang TY, Yang J, Duan H (2021) 3D propulsions of rod-shaped micropropellers. Adv Intell Syst 3(11):2100083. https://doi.org/10.1002/aisy.202100083
102. Jiao J, Lin E, Liang G, Zhao Q (2017) Nano optical propeller based on localized field intensity enhancement of surface plasmons. Proc SPIE 10173:455–460. https://doi.org/10.1117/12.2268294
103. Yang WH, Lee CF, Tang HY, Shieh DB, Yeh CS (2006) Iron oxide nanopropellers prepared by a low-temperature solution approach. J Phys Chem B 110(29):14087–14091. https://doi.org/10.1021/jp062371t
104. Vach PJ, Brun N, Bennet M, Bertinetti L, Widdrat M, Baumgartner J, Klumpp S, Fratzl P, Faivre D (2013) Selecting for function: solution synthesis of magnetic nanopropellers. Nano Lett 13(11):5373–5378. https://doi.org/10.1021/nl402897x
105. Ghosh A, Ghosh A (2022) Correction to: mapping viscoelastic properties using helical magnetic nanopropellers. Trans Indian Natl Acad Eng 7(4):1409–1409. https://doi.org/10.1007/s41403-022-00357-9
106. Kadiri VM, Bussi C, Holle AW, Son K, Kwon H, Schütz G, Gutierrez MG, Fischer P (2020) Biocompatible magnetic micro-and nanodevices: fabrication of FePt nanopropellers and cell transfection. Adv Mater 32(25):2001114. https://doi.org/10.1002/adma.202001114
107. Qiu T, Jeong M, Goyal R, Kadiri VM, Sachs J, Fischer P (2022) Magnetic Micro-/Nanopropellers for biomedicine. In: Sun Y, Wang X, Yu J (eds) Field-driven micro and nanorobots for biology and medicine. Springer, Cham, pp 389–411. https://doi.org/10.1007/978-3-030-80197-7_16
108. Ghosh A (2018) Dynamics, Fluctuations and Rheological Applications of Magnetic Nanopropellers (Doctoral dissertation). https://etd.iisc.ac.in/handle/2005/2984
109. Venugopalan PL, Ghosh A (2015) Application of ferrite coated nanopropellers for measurement of rheological properties of human blood. In: 2015 IEEE 15th international conference on nanotechnology (IEEE-NANO). IEEE, pp 428–431. https://doi.org/10.1109/NANO.2015.7388628
110. Ghosh A, Ghosh A (2021) Mapping viscoelastic properties using helical magnetic nanopropellers. Trans Indian Natl Acad Eng 6:429–438. https://doi.org/10.1007/s41403-021-00212-3
111. Feyen M, Heim E, Ludwig F, Schmidt AM (2008) Magnetic nanorotors with tailored field-induced dynamics. Chem Mater 20(9):2942–2948. https://doi.org/10.1021/cm703419t
112. Khan M, Sood AK, Deepak FL, Rao CNR (2007) Optically driven nanorotors: experiments and model calculations. J Nanosci Nanotechnol 7(6):1800–1803. https://doi.org/10.1166/jnn.2007.719
113. Jones PH, Palmisano F, Bonaccorso F, Gucciardi PG, Calogero G, Ferrari AC, Maragó OM (2009) Rotation detection in light-driven nanorotors. ACS Nano 3(10):3077–3084. https://doi.org/10.1021/nn900818n
114. Ding H, Kollipara PS, Kotnala A, Chen Z, Zheng Y (2021) Symmetric and isotropic micro/nanorotors driven by a plane-polarized gaussian laser beam. Proc SPIE 11798:153–157. https://doi.org/10.1117/12.2593464
115. Xu Z, Li T (2020) Detecting rotational vacuum friction by optically levitated nanorotor. In: Laser science (OSA Technical Digest; Optica Publishing Group, 2020). paper LTh2G.2. https://doi.org/10.1364/LS.2020.LTh2G.2
116. Ahn J, Xu Z, Bang J, Ju P, Gao X, Li T (2020) Ultrasensitive torque detection with an optically levitated nanorotor. Nat Nanotechnol 15(2):89–93. https://doi.org/10.1038/s41565-019-0605-9
117. Cui X, Mylnikov V, Johansson P, Käll M (2024) Synchronization of optically self-assembled nanorotors. Sci Adv 10(10):eadn3485. https://doi.org/10.1126/sciadv.adn3485
118. Khan M, Sood AK, Deepak FL, Rao CNR (2006) Nanorotors using asymmetric inorganic nanorods in an optical trap. Nanotechnology 17(11):S287. https://doi.org/10.1088/0957-4484/17/11/S10
119. Schrinski B, Stickler BA, Hornberger K (2022) Interferometric control of nanorotor alignment. Phys Rev A 105(2):L021502. https://doi.org/10.1103/PhysRevA.105.L021502

120. Stockman MI, Li K, Brasselet S, Zyss J (2006) Octupolar metal nanoparticles as optically driven, coherently controlled nanorotors. Chem Phys Lett 433(1–3):130–135. https://doi.org/10.1016/j.cplett.2006.11.015
121. Schäfer J, Rudolph H, Hornberger K, Stickler BA (2020) Cooling nanorotors by elliptic coherent scattering. arXiv:2006.04090. https://doi.org/10.48550/arXiv.2006.04090
122. Schäfer J, Rudolph H, Hornberger K, Stickler BA (2021) Cooling nanorotors by elliptic coherent scattering. Phys Rev Lett 126(16):163603. https://doi.org/10.1103/PhysRevLett.126.163603
123. De Wilde Y, Haidar R (2019) Optical cooling without lasers. Nature 566(7743):186–187. https://doi.org/10.1038/d41586-019-00526-x
124. Liptuga AI, Malyutenko VK, Pipa VI, Levash LV (1997) Radiative cooling under the conditions of magnetoconcentration. Semiconductors 31:423–426. https://doi.org/10.1134/1.1187176
125. Pipa VI, Liptuga AI (2002) Parameter analysis and optimization for the radiative cooling effect due to negative luminescence. J Appl Phys 92(9):5053–5059. https://doi.org/10.1063/1.1509104
126. Pipa VI (1995) Cooling effect caused by a negative luminescence. Semiconductors 29(3):272–274
127. Chen K, Santhanam P, Fan S (2016) Near-field enhanced negative luminescent refrigeration. Phys Rev Appl 6(2):024014. https://doi.org/10.1103/PhysRevApplied.6.024014
128. Kollyukh AG, Malyutenko VK, Rybak AM (1990) Spectra of the negative luminescence of semiconductors under the condition of charge carrier exclusion. J Appl Spectrosc 53(2):848–851. https://doi.org/10.1007/BF00659397
129. Berdahl P, Malyutenko V, Morimoto T (1989) Negative luminescence of semiconductors. Infrared Phys 29(2–4):667–672. https://doi.org/10.1016/0020-0891(89)90107-3
130. Cuba-Supanta G, Fernández-Escamilla HN, Guerrero-Sanchez J, Rojas-Tapia J, Takeuchi N (2020) Structural properties and thermal stability of multi-walled black phosphorene nanotubes and their operation as temperature driven nanorotors. Nanoscale 12(35):18313–18321. https://doi.org/10.1039/D0NR03815F
131. Wang D, Jia Y, He Y, Wang L, Xie H, Yu W (2019) Photothermal efficiency enhancement of a nanofluid-based direct absorption solar collector utilizing magnetic nano-rotor. Energy Convers Manag 199:111996. https://doi.org/10.1016/j.enconman.2019.111996
132. Kiani K (2016) Thermo-mechanical analysis of functionally graded plate-like nanorotors: a surface elasticity model. Int J Mech Sci 106:39–49. https://doi.org/10.1016/j.ijmecsci.2015.11.029
133. Martinetz L, Hornberger K, Stickler BA (2021) Electric trapping and circuit cooling of charged nanorotors. New J Phys 23(9):093001. https://doi.org/10.1088/1367-2630/ac1c82
134. Hoang MT, Aziz D, Feldman L, Filler M, Shan J (2021) A simple electrokinetic nanorotor. Bull Am Phys Soc 66(17), Abstract P09.00003. https://meetings.aps.org/Meeting/DFD21/Session/P09.3
135. Durán R, Ramírez A, Zehe A (2005) A nanorotor driven by the electrokinetic effect acting on a bioparticle. Trans Biol Biomed 2(3):294–298
136. Ma Y, Khosla KE, Stickler BA, Kim MS (2020) Quantum persistent tennis racket dynamics of nanorotors. Phys Rev Lett 125(5):053604. https://doi.org/10.1103/PhysRevLett.125.053604
137. Stickler BA, Papendell B, Kuhn S, Schrinski B, Millen J, Arndt M, Hornberger K (2018) Probing macroscopic quantum superpositions with nanorotors. New J Phys 20(12):122001. https://doi.org/10.1088/1367-2630/aaece4
138. Izumida W, Okuyama R, Sato K, Kato T, Matsuo M (2022) Einstein–de Haas nanorotor. Phys Rev Lett 128(1):017701. https://doi.org/10.1103/PhysRevLett.128.017701
139. Rusconi CC, Perdriat M, Hétet G, Romero-Isart O, Stickler BA (2022) Spin-controlled quantum interference of levitated nanorotors. Phys Rev Lett 129(9):093605. https://doi.org/10.1103/PhysRevLett.129.093605
140. Celestino A, Croy A, Beims MW, Eisfeld A (2016) Rotational directionality via symmetry-breaking in an electrostatic motor. New J Phys 18(6):063001. https://doi.org/10.1088/1367-2630/18/6/063001

141. Warshel A (1998) Electrostatic origin of the catalytic power of enzymes and the role of preorganized active sites. J Biol Chem 273(42):27035–27038. https://doi.org/10.1074/jbc.273.42.27035
142. Warshel A, Sharma PK, Kato M, Xiang Y, Liu H, Olsson MH (2006) Electrostatic basis for enzyme catalysis. Chem Rev 106(8):3210–3235. https://doi.org/10.1021/cr0503106
143. Yahashiri A, Nimrod G, Ben-Tal N, Howell EE, Kohen A (2009) Effect of electrostatic shielding on H-tunneling in R67 dihydrofolate reductase. Chembiochem: Euro J Chem Biol 10(16):2620. https://doi.org/10.1002/cbic.200900451
144. Hammes-Schiffer S (2014) Hydrogen tunneling, electrostatics, and conformational motions in enzyme catalysis. Biophys J 106(2):445a–446a. 10.1016%2Fj.bpj.2013.11.2522
145. Delgado M, Gorlich S, Longbotham JE, Scrutton NS, Hay S, Moliner V, Tuñón I (2017) Convergence of theory and experiment on the role of preorganization, quantum tunneling, and enzyme motions into flavoenzyme-catalyzed hydride transfer. ACS Catal 7(5):3190–3198. https://doi.org/10.1021/acscatal.7b00201
146. Zheng H, Yang SJ, Zheng YC, Cui Y, Zhang Z, Zhong JY, Zhou J (2020) Electrostatic effect of functional surfaces on the activity of adsorbed enzymes: simulations and experiments. ACS Appl Mater Interfaces 12(31):35676–35687. https://doi.org/10.1021/acsami.0c08080
147. Lin X, Wei ZQ, Mo QL, Hou S, Xu S, Fu XY, Xiao FX (2021) Electron tunneling through interim ligand layers towards photoredox selective organic transformation. J Catal 400:28–39. https://doi.org/10.1016/j.jcat.2021.05.019
148. Ma Z, Yan Z, Li X, Chung LW (2023) Quantum tunneling in reactions modulated by external electric fields: reactivity and selectivity. J Phys Chem Lett 14(5):1124–1132. https://doi.org/10.1021/acs.jpclett.2c03461
149. Qin L, Banholzer MJ, Xu X, Huang L, Mirkin CA (2007) Rational design and synthesis of catalytically driven nanorotors. J Am Chem Soc 129(48):14870–14871. https://doi.org/10.1021/ja0772391
150. Mirkovic T, Zacharia NS, Scholes GD, Ozin GA (2010) Nanolocomotion—catalytic nanomotors and nanorotors. Small 6(2):159–167. https://doi.org/10.1002/smll.200901340
151. Chen Y, Shi Y (2013) Computational design of chemically propelled catalytic nanorotors. J Chem Phys 139(6):064707. https://doi.org/10.1063/1.4817343
152. Ghosh A, Paul I, Saha S, Paululat T, Schmittel M (2018) Machine metathesis: thermal and catalyzed exchange of piston rods in multicomponent nanorotor/nanoslider ensemble. Org Lett 20(24):7973–7976. https://doi.org/10.1021/acs.orglett.8b03541
153. Goswami A, Pramanik S, Schmittel M (2018) Catalytically active nanorotor reversibly self-assembled by chemical signaling within an eight-component network. Chem Commun 54(32):3955–3958. https://doi.org/10.1039/c8cc01496e
154. Biswas PK, Goswami A, Saha S, Schmittel M (2020) Dynamics of hydrogen bonding in three-component nanorotors. Chem Euro J 26(62):14095–14099. https://doi.org/10.1002/chem.202002877
155. Biswas PK (2020) From stand-alone multicomponent nanorotors as catalysts to networking catalytic machinery through logic gate operation (Doctoral dissertation, Universität-Gesamthochschule-Siegen)
156. Goswami A, Özer MS, Paul I, Schmittel M (2021) Evolution of catalytic machinery: three-component nanorotor catalyzes formation of four-component catalytic machinery. Chem Commun 57(58):7180–7183. https://doi.org/10.1039/D1CC02805G
157. Raymo FM, Houk KN, Stoddart JF (1998) Origins of selectivity in molecular and supramolecular entities: solvent and electrostatic control of the translational isomerism in [2] catenanes. J Org Chem 63(19):6523–6528. https://doi.org/10.1021/jo980543f
158. Crowley JD, Goshe AJ, Bosnich B (2003) Molecular recognition. Electrostatic effects in supramolecular self-assembly. Chem Commun 3:392–393. https://doi.org/10.1039/B210957C
159. Gröhn F (2008) Electrostatic self-assembly as route to supramolecular structures. Macromol Chem Phys 209(22):2295–2301. https://doi.org/10.1002/macp.200800290

160. Schneider HJ (2009) Binding mechanisms in supramolecular complexes. Angew Chem Int Ed 48(22):3924–3977. https://doi.org/10.1002/anie.200802947
161. Tomba G, Stengel M, Schneider WD, Baldereschi A, De Vita A (2010) Supramolecular self-assembly driven by electrostatic repulsion: the 1D aggregation of rubrene pentagons on Au (111). ACS Nano 4(12):7545–7551. https://doi.org/10.1021/nn101884p
162. Willerich I, Gröhn F (2011) Molecular structure encodes nanoscale assemblies: understanding driving forces in electrostatic self-assembly. J Am Chem Soc 133(50):20341–20356. https://doi.org/10.1021/ja207565m
163. Zhuang YZ, Gu WX, Yang JJ, Chen X, Gao H (2015) Supramolecular nanoparticles constructed by balancing the forces between attractive host–guest and repulsive electrostatic interactions in two positively charged polymers. RSC Adv 5(117):96464–96471. https://doi.org/10.1039/C5RA18031G
164. Xie Y, Wang X, Huang R, Qi W, Wang Y, Su R, He Z (2015) Electrostatic and aromatic interaction-directed supramolecular self-assembly of a designed Fmoc-tripeptide into helical nanoribbons. Langmuir 31(9):2885–2894. https://doi.org/10.1021/la504757c
165. Tantakitti F, Boekhoven J, Wang X, Kazantsev RV, Yu T, Li J, Zhuang E, Zandi R, Ortony JH, Newcomb CJ, Palmer LC (2016) Energy landscapes and functions of supramolecular systems. Nat Mater 15(4):469–476. https://doi.org/10.1038/nmat4538
166. Bai Y, Luo Q, Liu J (2016) Protein self-assembly via supramolecular strategies. Chem Soc Rev 45(10):2756–2767
167. Haynes DA, Rawson JM (2018) Molecular electrostatic potential as a predictor of supramolecular synthons in non-hydrogen-bonded systems: application to heavier p-block systems. Eur J Inorg Chem 2018(31):3554–3564. https://doi.org/10.1002/ejic.201800357
168. Zhang Q, Liu C, Ju G, Cheng M, Shi F (2018) Macroscopic supramolecular assembly through electrostatic interactions based on a flexible spacing coating. Macromol Rapid Commun 39(20):1800180. https://doi.org/10.1002/marc.201800180
169. Baer A, Hänsch S, Mayer G, Harrington MJ, Schmidt S (2018) Reversible supramolecular assembly of velvet worm adhesive fibers via electrostatic interactions of charged phosphoproteins. Biomacromol 19(10):4034–4043. https://doi.org/10.1021/acs.biomac.8b01017
170. Zapata F, González L, Bastida A, Bautista D, Caballero A (2020) Formation of self-assembled supramolecular polymers by anti-electrostatic anion–anion and halogen bonding interactions. Chem Commun 56(52):7084–7087. https://doi.org/10.1039/d0cc02831b
171. Robertson B, Stark H, Kapral R (2018) Collective orientational dynamics of pinned chemically-propelled nanorotors. Chaos: Interdisc J Nonlinear Sci 28(4):045109. https://doi.org/10.1063/1.5018297
172. Jiang Y, Mattioli EJ, Calvaresi M, Wang Z (2020) Theoretical design of an ultrafast supramolecular rotor composed of carbon nano-rings. Chem Commun 56(79):11835–11838
173. Samanta SK, Schmittel M (2013) Four-component supramolecular nanorotors. J Am Chem Soc 135(50):18794–18797. https://doi.org/10.1021/ja411011a
174. Tang L (2019) A nanorotor for amplifying molecular movements. Nat Methods 16(9):802–802. https://doi.org/10.1038/s41592-019-0556-y
175. Burtch NC, Torres-Knoop A, Foo GS, Leisen J, Sievers C, Ensing B, Dubbeldam D, Walton KS (2015) Understanding DABCO nanorotor dynamics in isostructural metal–organic frameworks. J Phys Chem Lett 6(5):812–816. https://doi.org/10.1021/jz502653y
176. Zhao S, Liu Z, Che P, Li B, Dang T, Bu C (2020) Propulsion performance investigation of bio-inspired nano rotor base on fluid–structure interaction. Int J Micro Air Veh 12:1756829319900886. https://doi.org/10.1177/1756829319900886
177. Shanyong Z, Zhen L, Yachuan S, Tianjiao D, Shiqi L (2020) Bio-inspired nano rotor investigation based on UVLM. Biosurface Biotribology 6(1):20–24. https://doi.org/10.1049/bsbt.2019.0019
178. Zhao S, Liu Z, Lu K, Su D, Wu S (2021) Investigation on aerodynamic characteristics of bionic membrane nano rotor. Int J Micro Air Veh 13:17568293211043304. https://doi.org/10.1177/17568293211043304

179. Stock D, Namba K, Lee LK (2012) Nanorotors and self-assembling macromolecular machines: the torque ring of the bacterial flagellar motor. Curr Opin Biotechnol 23(4):545–554. https://doi.org/10.1016/j.copbio.2012.01.008
180. Goswami A, Paul I, Schmittel M (2017) Three-component nanorotors generated from fusion of complexes and post-fusion metal–metal exchange. Chem Commun 53(37):5186–5189. https://doi.org/10.1039/c7cc01977g
181. Biswas PK, Saha S, Nanaji Y, Rana A, Schmittel M (2017) Influence of rotator design on the speed of self-assembled four-component nanorotors: coordinative versus dispersive interactions. Inorg Chem 56(11):6662–6670. https://doi.org/10.1021/acs.inorgchem.7b00740
182. Samanta SK, Bats JW, Schmittel M (2014) A five-component nanorotor with speed regulation. Chem Commun 50(18):2364–2366. https://doi.org/10.1039/C3CC49476D
183. Samanta SK, Rana A, Schmittel M (2016) Conformational slippage determines rotational frequency in five-component nanorotors. Angew Chem Int Ed 55(6):2267–2272. https://doi.org/10.1002/anie.201509108
184. Samanta SK (2013) From Supramolecular Multi Porphyrin Assemblies to a 5-component Adjustable Speed Nanorotor (Doctoral dissertation)
185. Schmittel M, Biswas P, Goswami A, Paul I, Saha S (2017) Dynamic multi-component machinery: nanorotors and nanosliders in action. Abstr Pap Am Chem Soc 254
186. Hussain SG, Liu D, Huang X, Ali S, Sayyad MH (2008) A facile route to heterostructured ZnO: S/ZnO nanorotors: structural and optical properties. J Phys Chem C 112(30):11162–11168. https://doi.org/10.1021/jp803323a
187. Hussain SG, Liu D, Wang T, Huang X, Sulieman KM, Chi Q (2008) Synthesis and optical properties of S-doped ZnO nanorotors. In: 2008 international conference on smart manufacturing application. IEEE, pp 272–275. https://doi.org/10.1109/ICSMA.2008.4505656
188. Özer MS, Rana A, Biswas PK, Schmittel M (2017) Four-component zinc-porphyrin/zinc-salphen nanorotor. Dalton Trans 46(29):9491–9497. https://doi.org/10.1039/c7dt01323j
189. Morozovska AN, Glinchuk MD, Eliseev EA (2007) Ferroelectricity enhancement in ferroelectric nanotubes. Phase Transitions 80(1–2):71–77. https://doi.org/10.1080/01411590601092746
190. Nonnenmann SS, Gallo EM, Coster MT, Soja GR, Johnson CL, Joseph RS, Spanier JE (2009) Piezoresponse through a ferroelectric nanotube wall. Appl Phys Lett 95(23):232903. https://doi.org/10.1063/1.3263714
191. Bernal A, Tselev A, Kalinin S, Bassiri-Gharb N (2012) Free-standing ferroelectric nanotubes processed via soft-template infiltration. Adv Mater 24(9):1160–1165. https://doi.org/10.1002/adma.201103993
192. Kim J, Yang SA, Choi YC, Han JK, Jeong KO, Yun YJ, Kim DJ, Yang SM, Yoon D, Cheong H, Chang KS (2008) Ferroelectricity in highly ordered arrays of ultra-thin-walled Pb (Zr, Ti) O_3 nanotubes composed of nanometer-sized perovskite crystallites. Nano Lett 8(7):1813–1818. https://doi.org/10.1021/nl080240t
193. Yan C, Nikolova L, Dadvand A, Harnagea C, Sarkissian A, Perepichka DF, Xue D, Rosei F (2010) Multiple $NaNbO_3/Nb_2O_5$ heterostructure nanotubes: a new class of ferroelectric/semiconductor nanomaterials. Adv Mater 22(15):1741–1745. https://doi.org/10.1002/adma.200903589
194. Ouakad HM, Sedighi HM, Al-Qahtani HM (2020) Forward and backward whirling of a spinning nanotube nano-rotor assuming gyroscopic effects. Adv Nano Res 8(3):245–254. https://doi.org/10.12989/anr.2020.8.3.245
195. Zhang B, Li R, Peng Q (2022) Controlling CNT-based nanorotors via hydroxyl groups. Nanomaterials 12(19):3363. https://doi.org/10.3390/nano12193363
196. Brus J, Zhigunov A, Czernek J, Kobera L, Uchman M, Matejicek P (2014) Control over the self-assembly and dynamics of metallacarborane nanorotors by the nature of the polymer matrix: a solid-state NMR study. Macromolecules 47(18):6343–6354. https://doi.org/10.1021/ma501117a
197. Huo S, Kwak M, Qin J, Dittrichn B, Herrmann A (2021) Dynamic DNA-based biomaterials interacting with external, macroscopic, and molecular stimuli. Mater Today 49:378–390. https://doi.org/10.1016/j.mattod.2021.04.010

198. Nourhani A, Lammert P, Borhan A, Crespi V (2012) Dimensional transitions for coupled rotational/translational diffusion in powered nanorotors. Bull Am Phys Soc 57(1), ID: BAPS.2012.MAR.T41.13 https://meetings.aps.org/Meeting/MAR12/Session/T41.13
199. Nourhani A, Lammert P, Borhan A, Crespi V (2012) Effective translational diffusion of nanorotors as rotary powered random walkers. Bull Am Phys Soc 57(17), ID: BAPS.2012.DFD.D30.10. https://meetings.aps.org/Meeting/DFD12/Session/D30.10
200. Zhao SY, Liu Z, Bu C, Che PL, Dang TJ (2019) Unsteady flow mechanism investigation on pitching-blade rigid nano rotor. Lect Notes Electri Eng 459:836–853
201. Kiani K (2017) In-plane vibration and instability of nanorotors made from functionally graded materials accounting for surface energy effect. Microsyst Technol 23(10):4853–4869. https://doi.org/10.1007/s00542-017-3321-0
202. Kiani K, Soltani S (2018) Three-dimensional dynamics of beam-like nanorotors on the basis of newly developed nonlocal shear deformable mode shapes. Euro Phys J Plus 133:1–21. https://doi.org/10.1140/epjp/i2018-12197-4
203. Nourhani A, Byun YM, Lammert PE, Borhan A, Crespi VH (2013) Nanomotor mechanisms and motive force distributions from nanorotor trajectories. Phys Rev E 88(6):062317. https://doi.org/10.1103/PhysRevE.88.062317
204. Liu Z, Bu C, Kong X, Yang D, Li B (2018) Computational investigation of noise interaction for a nano counter-rotating rotor in a static condition. Int J Comput Mater Sci Eng 7(01n02):1850004. https://doi.org/10.1142/S2047684118500045
205. Liu Z, Dong L, Moschetta JM, Zhao J, Yan G (2013) Evaluation of Nano-scale rotors and motors at static condition. Int J Micro Air Veh 5(3):193–206. https://doi.org/10.1260/1756-8293.5.3.193
206. Liu Z, Moschetta JM, Chapman N, Barènes R, Xu M (2010) Design of test benches for the hovering performance of nano-rotors. Int J Micro Air Veh 2(1):17–32. https://doi.org/10.1260/1756-8293.2.1.17
207. Gradov OV, Maklakova IA, Sergeev AI (2021) Jumping and rotation of Silasorb C18 particles (Sobel-Feldman filter). Scanning Electron Microscopy Facility Dataport (ICP RAS—IBCP RAS). https://doi.org/10.17605/OSF.IO/9ACSD
208. Gradov OV, Maklakova IA, Sergeev AI (2021) Bistability of Silasorb C18 microparticles (SEM—oscilloscopic indication). Scanning Electron Microscopy Facility Dataport (ICP RAS—IBCP RAS). https://doi.org/10.17605/OSF.IO/VRTFX
209. Engels F (2012) Dialectics of nature. Wellred Books, p 410
210. Engels F (1972) Dialectics of nature. Progress Publishers, p 403

Chapter 9
Neural Network-Assisted Sedimentography Based on Discrete Angular Rotation Algorithms and Real-Time Correlation-Spectral Analysis: From Physics of Natural Structured Geomaterials to Multilayer Nanomaterials

Anton Elfimov, Oleg Gradov, Margaret Gradova, and Nikolai Marnautov

Abstract The paper proposes a number of methods for neural network-assisted sedimentography based on discrete angular rotation algorithms and real-time correlation-spectral analysis, including those using synthetic data to identify the layered structures. Their applicability to various layered materials is shown, from geomineral textures formed by cyclic sedimentation to the layered graphene oxide-based membranes fabricated either by deposition onto a substrate or by other methods. It is pointed out that the main physical properties, in particular, permeability of such layered structured materials strongly depend on the layered structure, while the latter depends on the preparation method, which makes it possible to predict the main functional properties of membranes based on the results of the neural network-assisted sedimentography and real-time correlation-spectral analysis of the self-assembled multilayers and/or multilayer nanomaterials (both synthetic and natural, formed under geological conditions or during biomineralization). To prove the universal nature of the approach proposed, a description of the methodological aspects of the study and presentation of the experimental data are preceded by an analytical literature review in five parts (containing more than 200 references), in which, from the standpoint of physical reductionism, various forms of sedimentation are analyzed based on the unified physical background—from soft matter physics (colloids, polymers, mesophases), biochemistry, cell biology, and biomedicine, to nanomineralogy,

A. Elfimov · N. Marnautov
Emanuel Institute of Biochemical Physics, Russian Academy of Sciences, Kosygina 4, 119334 Moscow, Russia

O. Gradov (✉) · M. Gradova
N.N. Semenov Federal Research Center for Chemical Physics, Russian Academy of Sciences, Kosygina 4, 119991 Moscow, Russia
e-mail: o.v.gradov@gmail.com

H. Altenbach and V. A. Eremeyev (eds.), *Dynamics of Discrete and Continuum Structures and Media*, Advanced Structured Materials 221,
https://doi.org/10.1007/978-3-031-75626-9_9

hydrogeology and geology of sedimentary rocks, including pyroclastic ones and rhythmic sedimentation products. The proposed methodological approaches allow, regardless of the scale of the structured materials, to analyze the uniformity of the layers and angular distributions (angular anisotropy) for different stages of their formation, as well as the chemical substitution of such textures or rhythmic sedimentation products on the arbitrary time scales (from minutes of the laboratory experiments to the geological eons in the case of analysis of the natural sedimentary multilayers). In the experimental part, we showed that the order parameters of the synthetic and natural layered textures can be quite close, and also illustrated the scale dependence of the order parameters calculated using Legendre and Chebyshev polynomials. Both methods developed on the basis of discrete angular rotation algorithms and correlation-spectral techniques based on 2D FFT are suitable for identifying the orientational disorganization of the structural elements in multilayer structured materials and their anisotropy. In the future, such techniques can be used in a wide range of scientific and applied problems from the analysis of nanomaterials (such as carbon multilayer nanostructures and MXenes) to experimental mineralogy and interpretation of geophysical sounding results.

Keywords Sedimentography · Sedimentometry · Multilayer structures materials · Graphene oxide-based membrane · MXenes · Discrete angular rotation algorithm · Periodic sedimentation · Rhythmic sedimentation · Order parameter · Legendre and Chebyshev polynomials · Time-resolved sedimentation measurements · Synthetic data · Neural networks · UNet · Correlation-spectral analysis · 2D FFT

9.1 Main Problems of Sedimentography of Self-organized Layers: From General Physics to Earth Sciences and Materials Sciences

For granulometric analysis of the deposition of organic and inorganic particles and modeling of geo-hydrological sedimentation processes, sedimentographs of various designs are used [1–6], the earliest of which were developed about a century ago for colloid chemistry [7]. Since the 1950s, there are simple and affordable optical sedimentographs and photosedimentographs for analyzing sedimentation rate of erythrocytes and some other cells [8–11]. Until now, the principles of quantitative sedimentography are based on the optical detection of particles by optical density, usually with time resolution (in ESR analysis) or time-lapse recording (in photosedimentographs) or with both spatial and temporal resolution (as in the case of analysis of hydrological sedimentation, in which more or less regular dense layers are formed).

Figure 9.1 shows a "Seishin" (two-channel correlating/two-module comparative) sedimentographic complex operating at the authors' laboratory, serving the Institute of Biochemical Physics of the Russian Academy of Sciences and the former Institute of Chemical Physics of the Russian Academy of Sciences. This model has been repeatedly used in works ranging from metallography to mineralogy and analysis of the products of pulsed shock wave synthesis [12, 13]. This complex includes a centrifuge and two optical reading devices with analog-to-digital converters, from which digital data is received via the GPIB/HPIB protocol (IEEE-488, Standard Digital Interface for Programmable Instrumentation). To control the particles deposited on the cuvette walls, we use a horizontal optical microscope (similar to those used in tensiometers for contact angle measurements) with a modernized and robotized rotation table (Fig. 9.2). Such measurements do not require application of advanced mathematical methods, since the experiment usually fits into the theoretical framework of the simplified models of the 1950s if we ignore the spatial distribution of sedimented particles taking into account diffusion, convection, and advection [14–16].

The problem of sedimentation dynamics and formation of organized structures during/as a result of sedimentation is still of significant fundamental and applied interest for various sciences—from soft matter physics (colloids, polymers, mesophases, etc.), working at micro- and nanoscales [17], and cell biology/biomedicine with micron and decamicron order of discreteness [18, 19], to hydrogeology and geology of sedimentary rocks, operating at macroscopic scales [20, 21].

Fig. 9.1 Two-module/two-channel comparative "Seishin" (MICRON PHOTO SIZER) sedimentographic complex with analog-to-digital converters, from which digital data is received via the GPIB/HPIB protocol (upper illustration) and "Seishin" M.P.S. centrifuge for sample preparation (below). Such instrumentation is optimal for routine sedimentographic analysis based on the optical principles

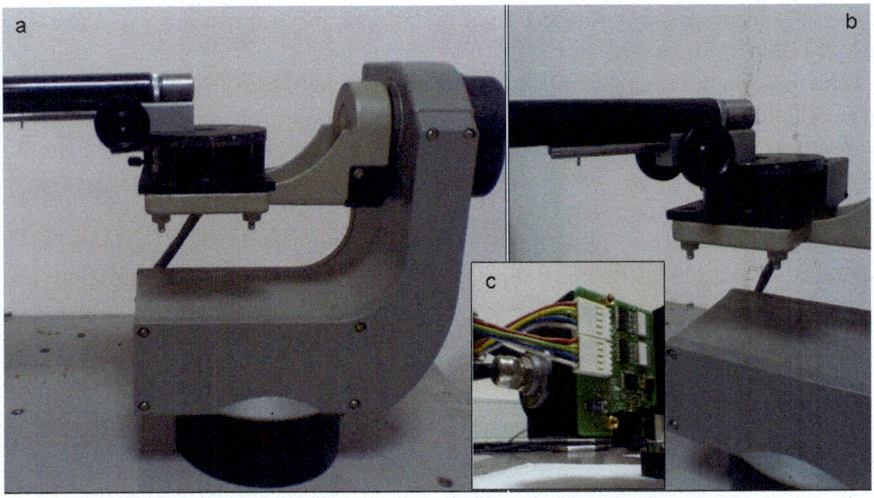

Fig. 9.2 Horizontal rotation microscope based on goniometric horizontal table (**a**, **b**—different rotation positioning angles) and a robotized angular positioner, which can be modernized with a specially developed PCB circuit (**c**—a well-known Oganessian–Notchenko scheme, constructed in 2011–2016 and controlled by a G-code program)

Despite the long history of development, many issues in the statistical mechanics of steady sedimentation [22] (not to mention the intriguing fluctuations and instabilities in sedimentation [23]) are not fully understood. In sedimentation simulation most authors prefer to use simple types of figures approximating sedimented particles (e.g., spherical/spheroidal [24, 25] or rod-like particles [26]), and to reduce the problems to laminar flows or two-dimensional channels, which is good for microfluidics, but does not solve the problems of spatial morphopoiesis in liquids [27] and cannot be applied to construct the full models of three-dimensional layered or percollatively organized structures that arise during natural sedimentation processes. The old models taking into account polydispersity (while modern authors, striving for beauty, try to reduce everything to the case of monodisperse particles both in experimental practice and in theory) also focus on spherical or spheroidal particles [28, 29], for which packing (pertiling) parameters and angular characteristics can be easily calculated (since, being isotropic particles, they do not have additional maxima in the integral spatial characteristic—ISC). At the same time, real sedimentation processes in hydrogeology and formation of sedimentary rocks resemble filling of the field in Tetris game, in which, at a high "sedimentation" rate of the complex shaped particles (similar to the variety of constituents in sedimentary rocks—not only geological fragments but also calcareous shells of nannoplankton, fractal dendrite-like aggregates, etc.) "clusters of squares" are often built not into the empty spaces (according to the "key-lock" principle), but into the arbitrary points, leaving "pores" or "through percolation regions" (by analogy with the physics of sedimentary rocks). As a consequence, layers of more and less ordered sedimentation of "Tetris clusters" arise (depending on the sedimentation rate) to which the 2D analogy of dense and less dense layers in sedimentology is applicable.

Application of modern methods of computational fluid dynamics [30] only partly solves the problem of the adequacy of modeling in sedimentography, being a "niche solution," as well as the study of sedimentation in jet fuels due to watering using methods of molecular mechanics [31]. Any models of this kind do not describe the order formation in sedimented layers. The algorithm for coarse particle sedimentation simulation by Stokesian Dynamics [32] is more progressive from the point of view of computer technology. Stokesian Dynamics, by definition, is a solution technique for the Langevin equation, which is the relevant form of Newton's 2nd law for a Brownian particle, and, according to [33], this method treats the suspended particles in a discrete sense while the continuum approximation remains valid for the surrounding fluid, i.e., the suspended particles are generally assumed to be significantly larger than the molecules of the solvent. This brings us closer to objective physical modeling, but this approach is not scalable for understanding "self-organizing criticality" and stratification of layers in sedimentation that arise according to the above mechanism [34–38].

9.2 From the Mechanics of Pyroclastic and Sedimentary Rocks to Geostratigraphy and Hydrostratigraphy

Meanwhile, layers are a common form of conservative self-organization (self-assembly) in volcanology [39–41] and geology of sedimentary rocks [42–44] at different scales—from microscale to gigascale. Layering is a criterion for time-resolved sedimentation dynamics, and the properties of the layers correlate with the conditions of sedimentation and the properties of the sedimentary and inclusive rocks composing the layers. The minimum level of layered sedimentation discreteness for geochemical and biogeochemical systems corresponds to the level of template deposition of single particles or cells undergoing biomineralization (as in the case of some biofilms involving chemoautotrophs), while the maximum level is determined by the mass transfer values of global geomechanics and hydrophysical parameters of river and sea transport of microparticles and fluvial processes in their regular phenological and chaotic catastrophic dynamics. In contrast to the simplest and most widespread models such as layered sedimentation in suspensions of monodisperse spherical particles [45], multilayer shallow water models are calculated for polydisperse sedimentation [46–50].

In the broad sense of the word, layer-by-layer regularity and causality of sedimentation from the most ancient deposits of pyroclastic rocks to the weathering products and aeolian systems underlies modern geostratigraphy [51–56]. Sedimentation mechanics can influence geomorphogenesis and hence, the resulting geomorphology of sedimentary deposits. Hydrostratigraphy as a discipline also inevitably develops under the influence of the modern concepts about sedimentation in lakes and seas, river estuaries, and also in complex meteorological processes or natural disasters, accompanied by different thermal and pressure conditions and characterized by different thermobaroclines [57–61]. Resuspension for such systems can be regarded as a manifestation of thermodynamics of reversible processes, which prevents self-organization of one type or another. The role of mass transfer mechanisms, including alluvial and eluvial components, can be considered within the framework of geomorphogenesis using a reaction with diffusion and advection. At the same time, the settling of plankton in accordance with the phenological patterns, and of inorganic/colloidal mineral suspensions with the known seasonal dynamics should lead to heterogeneous (spatially modulated) stratified macrostructures, composed of the layers [62–64] different in chemical compositions and the presence of biomarkers [65–67] (or biologically specific isotope markers/isotope tracers in carbon geochemistry [68–77]). Therefore, both for the purpose of analyzing seasonal heterogeneity and for the purpose of analyzing the mechanisms and physicochemical components of geomorphogenesis of sedimentary rocks, it is necessary to analyze the geometry and dimensions of such layers in their natural alternation. Diagenetic, metamorphic processes and other processes of transformation of hydrogeological particulated matter [78] within the framework of this approach can be considered as reaction–diffusion processes with the substitution of individual layers while maintaining the overall structural regularity of the layers. Later we will encounter reaction–diffusion mechanisms in the analysis of macromolecular sedimentation in which they also play a crucial role [79].

9.3 Towards Physical Reductionism: Analysis of Reaction–Diffusion Mechanisms and the Influence of Convection and Advection in Sedimentation Processes

Within the framework of the physical reductionism, all geophysical and hydrophysical phenomena of sedimentation and concomitant aggregation of particles are based on the basic principles of mechanics, which are common for sedimented particles of different compositions [80–82], and in stratified media/stratified liquids this is also influenced by the medium parameters and the medium stratification [83–85]. Usually, stably stratified environments are considered, since with stratification oscillations and nonlinear perturbations it is extremely difficult to understand the system behavior. However, even stably stratified environments may also be characterized by instabilities, such as sedimentation-induced instabilities, affecting the evolution of concentration profiles in self-organizing stratified structures at different levels of organization and regularity periods (of the layer sequences) [86]. In the case of oscillations during the formation of layers, waviness is formed along at least one of the coordinates or orientations of the liquid flows and flow-guided particles (which is clearly observed in the model experiments using particle image velocimetry).

Superposition of oscillations at different scales leads to the phenomena which has been well known in physics since the 19th century—macroscale analogs of Bragg's scattering of light by sound, reproducing the scattering patterns on diffraction gratings of the corresponding periodicity. Therefore, decoding of such "barcodes" of processes with different periodicities in sedimentology should be carried out considering all the influencing factors that determine this periodicity. The role of some of them can be compared with the role of oscillating grids during sedimentation in dense layers of suspended particles stirred by oscillating grids [87]. Undoubtedly, the nature of regulation depends on the coordinate, since in some cases it will affect horizontal mass transfer, in other cases—vertical mass transfer, in some cases it will affect convection, while in others—advection. It is obvious that in complex reaction–diffusion processes, accompanied by stratification and deformation of layers, neither neglecting convection nor neglecting advection is acceptable. Many effects in the real geosphere, hydrosphere and atmosphere, as well as in the body (including the Beenakker–Mazur effect [88]) are based on the integral account of the role of convection in sedimentation, and the corresponding modeling capabilities are almost unlimited. Arguments in favor of this thesis are numerous mathematical and computational works that discuss:

- different schemes for convection–diffusion reaction system modeling continuous sedimentation [89];
- Runge–Kutta methods for convection–diffusion problems modeling polydisperse sedimentation [90];
- formal mathematical analogies between colloidal sedimentation and turbulent convection at high Prandtl numbers [91];
- numerical simulation of sedimentation in the presence of convection with reconstruction of the particle-radius distribution functions [92];
- numerical simulation of free convection in sedimentation transports using immersed boundary method [93].

Unfortunately, such works often use unacceptable simplifications; in particular, they exclusively consider degenerating convection–diffusion reaction systems or strongly degenerate convection–diffusion problems in the modeling of sedimentation–consolidation processes (morphogenesis and self-organization of sedimented layers and agglomerates) [89, 94]. Attempts to neglect convection end up even worse, although, even in soft matter physics, it is known that convection-free boundary sedimentation is possible only for dilute protein solutions (since the 1970s [95]), since during sedimentation of undiluted macromolecular solutions supramolecular structures and facies layers inevitably form, in some cases with self-organization of boundaries, often due to the self-pinning effect on the surface [96–98] or syneresis and tractographic phenomena on the surface [99–108]. The latter phenomena also actively involve the mechanisms of convection and advection. Even in such a simple case as in evaporating layers of sucrose aqueous solutions, the mechanisms of free and guided convection [109] participate in transport and sedimentation of the solid particles (and in a number of experimental conditions they coexist and influence each other [110]).

The problem of the contribution of convection to sedimentation has been actively discussed since the 1970s [111]. The relevance of such research is evident in a wide variety of classical and exotic cases, such as:

1. Centrifugation of macromolecules in self-generating density gradients, in which the simplest case is convection-free band sedimentation, but in fact, it is a simplification/idealization [112];
2. Natural processes of geological sedimentography/sedimentology, for example, such as particle sedimentation with diffusive convection in volcanic ash clouds [113];
3. Convective ferrofluidics caused by barometrical sedimentation and gravitational sedimentation of magnetic particles on ferrofluid convection [114, 115];
4. Aerosol sedimentation and bioaerosol sedimentation by convection (including diffusion-containing ones measurable with diffusion aerosol spectrometers) [116–127];
5. Pulmonary acinar particle transport with interplayed convection, sedimentation, and diffusion [128];
6. Haemocytometry and measurements of the erythrocyte sedimentation rate [129, 130];
7. Thermophoresis, thermal diffusion, and the Soret effect in various applications—from hydrophysics and subsurface geophysics to medicine and materials science [131–133];
8. Solidification with melt convection and grain sedimentation in material sciences and industry [134];
9. Machinery of applied hydrodynamics in the intensification of industrial processes—from convection induced by hydrostatic pressure in sedimentation velocity experiments to convection and mixing driven by sedimentation [135, 136].

Regarding the last point, it should be added that currently the mechanisms of layer-by-layer sedimentation of particles are often used for the fabrication of functionally graded and multilayer composites [137, 138]. In this case, the mechanisms of mixed columnar-equiaxed solidification with melt convection and grain sedimentation are often involved [139, 140]. At the same time, many similar sedimentographic mechanisms often used in technology using reactions with diffusion and convection or advection are common components of the dynamics of processes in geology and planetology (for example, the formation of compositional structures by sedimentation in vigorous convection [141]). In fact, it is impossible to draw a clear line between the natural and technical fields in which these approaches can be used. This is an additional argument in favor of physical reductionism adopted as the basis for universalization in the developed neural network sedimentography and other computer-optical methods of sedimentographic analysis using reactions with diffusion and convection/advection.

9.4 On the Role of Geometry and Mechanical Properties of Particles in Sedimentation and Self-assembly Processes

In accordance with the above-mentioned discrete mathematical analogy of the particle sedimentation as the Tetris-type cellular automata (if we were talking about dynamically transforming cluster shapes, then the analogy with Conway's game "Life" would be more appropriate), the role of their geometry and proportions (such as the ellipticity coefficient, Feret diameters, etc.) in ensuring either uniform or non-uniform layer sedimentation is extremely significant and determines the course of the whole process. As can be seen from the simple experiments, sedimentation of particles of different shapes and orientations and of crystals of different systems in geological processes forms layers of different friability or compactness, completely different in porometric and permeametric properties and percolation indicators, affecting the geological exploration qualities of the corresponding horizons. The periodicity of layers depends on the type of crystals and the possibility of tiling the surface of the layer with them (in this regard, the work on the formation of periodic layers of a dodecagonal quasicrystal and a hexagonal crystal in sedimentation–diffusion equilibria of colloids [142] is worth mentioning). On the other hand, in addition to the tessellation/tiling ("paving the parquet") in such situations, phenomena of superstructure along a fractal tree or Sierpinski carpet and similar phenomena with scale invariance can be observed, which can affect both the nano-micro level (in the formation of structures, starting from atomic clusters in sedimentation with self-assembly in equilibrium conditions), and at the level of mesomechanics (including in conditions of self-organization of systems, starting with dendrites arising due to diffusion-limited aggregation) and macrosystems, up to the fractal properties of different layer sedimentation behind dams that are significant in geoengineering [143]). Under conditions of significant plastic and elastic deformations in soft mater physics, these phenomena do not usually occur (except for the cases of crystallization of the stratified films of proteins and colloids, in which the solidified part works in favor of this scenario). Therefore, one can conclude that the diversity of sedimentation products is determined by the geometric physics of the individual fractions and "populations" of the particles. It is inappropriate to ignore and simplify these aspects of the phenomenon. Often, for the sake of simplicity, sedimentation of spheroidal particles in sratified fluids [144] is studied, however problems become more complex and interesting when taking into account the entire range of the particle shapes, such as rods and disks [145], as well as porous and solid particles in stratified fluids [146]. In addition, sedimentation of a stratified suspension in a quiescent fluid [147] is usually considered, while the real liquid medium (an elementary example is the sea in the processes of littoral/sublittoral sedimentation) may not be quiescent, and therefore the mechanism of sedimentation of the individual particles or fractions may not be direct or spontaneously smooth, but may be regulated by convection/advection or turbulence of the medium and thermal or density (e.g., thermohaline) stratification, and hence, can be forced. In such processes, the sequence of layers does not always fully correspond to the linear deposition of particles ranked by mass, and the contribution of geometric parameters/form factor in determining their deposition sequence greatly increases (some particles can be removed due to sailing properties, while others can be physically selected due to instabilities, etc.) [148–150]. The interaction with optical, thermal and radio frequency radiation of particles of different shapes also differs significantly (as anyone interested can see using programs like COMSOL MULTIPHYSICS).

The geometry of particles and agglomerates of particles during sedimentation, accordingly, also affects the signals of their detection/registration in the optical methods (and other devices based on the principles of wave optics and quasi-optics), from photon correlation spectrometers to simple sedimentographs built on the basis of the modified and automated measuring units of Seishin Micron Photo Sizers (see Fig. 9.1), including the cells containing electrical and magnetic components for measurements of sedimentation–diffusion profiles and layered sedimentation of charged colloids [151] in extrinsic control fields. One of the notable works in this area is the work "Study on phosphor sedimentation effect in white light-emitting diode packages by modeling multi-layer phosphors with the modified Kubelka–Munk theory" [152]. Back in the 1960s, methods of particle sizing using turbidimetric layered sedimentation [153, 154] were actively developed, and in the 1970s, with the development of laser technology, in addition to laser aerosol counters or laser aerosol spectrometers [155–158], automatic particle sizing devices based on gas laser and layered sedimentation principles were developed [159]. The possibility of laser sedimentation analysis in dynamics was proven, taking into account the kinetics of reaction–diffusion processes and the mechanisms of agglomeration of layers/in layers. It has been shown that for both hard particles and soft granules (from soft gauge latexes to polyelectrolyte granules/coacervates and proteinoid microspheres) sedimentation processes can be described perfectly (and the degree of contribution of regular and chaotic dynamics to sedimentation varies greatly). If we take as an example the natural geological process with sedimentation, erosion, and terrain inversion [160], then its components will be described by the mechanics of a deformable solid (in its bulk version), and in geomimetic processes the concept of describing sedimentation mechanisms will be similar. For obvious reasons, layer porosity will be taken into account, and the porosity of a sedimentation layer will have to be considered as a function of the particle size, the layer thickness, and the adhesive-force level [161]. Deformation can occur under the influence of various stresses (including tangential stresses at the boundaries between layers [162]) and the interlayer interactions induced by them; however, in general, it is impossible to fill all the percolation areas and all pores with the fragments of various sizes, even purely mechanically, not to mention tiling restrictions and limits of applicability of the power laws.

If we consider a viscous and fluid medium, then, ultimately, there will be no pores at all. For obvious reasons, pores/voids in the pitch drop experiment (and its multiphase modifications, such as an experiment with several different pitches and natural diffusion and mixing) quickly heal. Microheterogeneous medium (bubble fluid) sooner or later loses all its "pores" during effervescence. Therefore, outside the condensed medium, there is no self-organization of layers, and in the field of mechanical weathering in sedimentology, layers are an extremely simple (for research and interpretations—since 1930–1960 years) and regular object with different periodicity formation mechanisms [163–183]. Intermediate levels, where both regular and chaotic dynamics coexist in sedimentography, including those described by the wave equations, are at the level where the colloidal properties of dispersed systems (such as clays in natural sedimentology/sedimentography or oriented nanoparticles in the engineering of structured materials, layers and multilayers of nanostructures, etc.) are included in the field of soft matter physics, and the flows are described by elementary and easily approximated hydrodynamics. Thus, a compromise is achieved between the boundaries in solid mechanics and the mobility necessary for the implementation of convection, advection, diffusion, and chaos (micro)hydrodynamics [184–194], which is necessary for reaction–diffusion and transport processes, multilayer self-organization of soft matter systems, including the processes occurring in a density gradient.

The most irreproducible and morphologically variable, due to the microrheological viscoelastic properties of the medium, is the formation of layers and multilayers in soft matter physics, which by definition includes liquid crystals, colloids, polymers, and biomacromolecular/biosupramolecular structures. Thus, electrostatic interactions in the adsorption of proteins can be analyzed by a sedimentation technique [195], while methods of sedimentation of red blood cells, depending on the degree of maturity, underlie hematocrite centrifugation or "ellocrite" centrifugation (see, for example, classical biomedical papers with cellocrite microhematocrite tubes [196–204]) or, as well as the methods for analyzing the erythrocyte sedimentation rate which are relevant for the clinical applications [204, 205]. In this regard, we do not focus on the above details in this article which has an emphasis on the mechanics and structuring of materials. In this paper, we will consider two examples of application of neural network sedimentography based on generative adversarial networks: to graphite oxides (fairly regular layers that arise during a controlled technological process in the laboratory) and to siltstones (sedimentation products of the clay layers in sedimentology, which are disordered).

9.5 An Example of the Layer Formation and Their Structural Disturbances is Graphene Oxide

One of the obvious and most noteworthy areas in which our approach can be applied is the self-assembly of graphene and graphite oxide multilayers. Sedimentography or sedimentology of graphite structures originates in geology and tectonics [206]. Studies on graphite sedimentation under various electrochemical conditions/ at different pH values have been known for more than 20 years [207]. Also since the 2010s, the effect of size fractionation of graphene oxide sheets by pH-assisted selective sedimentation is known [208]. The effects of specific electrical conductivity of the medium and salinity (sometimes almost along the Hofmeister series) determine the efficiency of sedimentation of the individual fractions and aggregation of graphene oxide particles; the particle size influences aggregation and sedimentation differently from the conventional abstract models with significant simplifications [209]. Differences in the sedimented particle size and their structure (for example, in the case of heteroaggregation of graphene oxide with nanometer- and micrometer-sized colloids [210, 211]) lead to the distortions of the layer surface, the appearance of micro-roughness of the layers and influence the sorption and adhesive properties. Accordingly, each layer of graphene or graphite oxide on the previous layer is formed not as a perfect layering of sheets in a stack, but as waves or heterogeneous surfaces with nanoroughness or microroughness (depending on how far the layering process has progressed in the areas of heterogeneity, nucleation and "epitaxy"). Different surface geometries or roughness provide different forms of more highly organized microscopic and mesoscopic structures using layers with different regularity thicknesses and different angular orientation distributions of the constituent leaflets.

It is clear that maximally linear distribution of the sheets is possible only for the first layer on an initially planar and physically straightened substrate (classical metallographic methods since the 1970s have associated such forms with planar sedimentation on highly polished steels [212], but this approach produces mistakes and inconsistencies in the case of pitting or other corrosion types even at the initial stages, during which inhomogeneities of the order of tens to hundreds of nanometers appear in the system). In real conditions, such approaches do not work, due to the reasons described in this section, including the following paragraphs. For graphene sheets, they do not work due to the thickness of the sheets, which can be reduced to a differentiable form of infinitesimal, due to which the possibilities of introducing chaotic dynamics into the morphology of sedimented layers are virtually limitless. In hydrodynamic approximations, a system of such small sheets can be represented as a flow of particles.

The introduction of additional disturbances into the system, such as chemical interactions and catalysis at hot spots, heating, irradiation, and photochemical reactions [213, 214] change dramatically the morphology of these nanosystems (if we imagine natural sedimentation in an equilibrium context [215]). If you introduce flows (hydrodynamics) into the system, then heat transfer and mass transfer will work as heat and mass transfer, in which it is necessary to take sedimentation into account in the study of heat transfer performance and thermo-hydraulic characteristics of graphene nanofluids [216]. The situation will be even more complicated in the case of reaction–diffusion dynamics with gas bubble formation of filaments from sedimented graphene oxide [217], especially if this gas, being in reactive conditions, participates in the reaction with which diffusion is associated during the processes of the layered structure morphogenesis.

The introduction of external fields—mechanical (as in the case of centrifugation and rotation, in the broad sense of the term [218, 219]) or electromagnetic (as in the case of electrosynthesis or plasma treatment and electrophoretic sedimentation [220]) further distorts the picture of field-induced morphogenesis in the system of deposited graphene-oxide or graphite-oxide sheets. The introduction of rheological or hydrodynamic variables, again, additionally complicates the situation. For example, the use of a fluid magnetocomposite with field-programmable fluidity properties and ferrohydrodynamic instabilities as a magnetorheological working fluid with enhanced sedimentation stability leads to the fact that, instead of stable sedimentation, we will observe the effects of ferrohydrodynamic and magnetoconvective instability [22, 221]. If we are talking about the adsorption and desorption phases of the process in an external field (and the nature of sedimentation of the graphene sheets depends on the quality of desorption—for example, those heavier due to effective sorption will settle faster), then we can refer to the technologies such as ISSMI (In Situ Sludge Magnetic Impregnation), using nitric acid treated graphene oxide for dye removal using enhancement of sludge sedimentation [223]. In such cases, the formation of the so-called "nano-sludge-composite layers" is regulated at the same time by the field and the sorption history ("memory") of graphene/graphite oxide sheets. In this case, the field, depending on the configuration, can be either an agent that resets memory or that ensures the stationary state of the system (controlling transitions between Markov and non-Markov processes). We do not specifically focus here on the limitless possibilities of introducing an electric field into the system and sedimentation of graphene oxide directly on the wire/electrode [224]. Focusing on this issue would lead to a twofold increase in the number of references, based on the work experience of the third author (Oleg Gradov) in 2014–2015, and at the laboratory dealing with the biological effects of nanostructures in 2016–2018. However, the possibilities of using modulated electromagnetic fields to form graphene layers and regulate their local properties seem to be almost inexhaustible.

From the above, it follows that the possibilities of graphene and graphite oxide texture morphogenesis during their sedimentation are extremely variable, and the control parameters and fields can be applied in any arbitrarily complex combination, allowing to achieve any arbitrarily complex effect. At the same time, the means of controlling micro-roughness, curvature, waviness, and other key morphological parameters of these structures are extremely underdeveloped. Therefore, software tools are needed for analyzing data on the surface peculiarities from layer to layer or on a cross section/fracture of a layered graphene or graphite oxide sample, allowing one to identify the production conditions based on the sample geometry or to reconstruct the sample geometry in a model based on the known production conditions (in the future). At the same time, for such a task it is inevitable to solve the problem of studying the distribution of layers with different regularity thicknesses of the sheets and different angle distributions of the sheet's constituent orientation. In this work, we present the results which have first been shown at a conference on graphene in 2023 [225] and later partially described in an article published in 2024 [226]. They show that the proposed approach is reliable and productive as a data source for such beautiful aims (see next section).

9.6 Methods of Sedimentographic and Sedimentometric Analytics Based on Discrete Angular Rotation Algorithm and Correlation-Spectral Validation Approach

The current versions of the free software developed for analysis, including LayerNet Segmenter and OrientLayer Analyser, as well as other image analysis tools, are available at: ImageMorphLab [https://nmar33.github.io/ImageMorphLab]. Using synthetic data, a model (U-Net) was trained to segment layered structures. The synthetic data was a dataset containing images simulating layered structures and a mask by which the layer boundaries can be segmented (see Fig. 9.3). The dataset was generated in such a way that the neural network did not respond to the color changes and gradients, but readily responded to the thin interlayer boundaries (an example of such a dataset is available: https://www.kaggle.com/datasets/nikolaymarnautov/synthetic-layered-images-segmentation-small; https://doi.org/10.34740/kaggle/dsv/6851202].

The image analysis was performed as follows:

(1) First, the initial micrograph is processed by a neural network, resulting in an image that represents the network's confidence level regarding the presence of a lay-ered boundary in specific areas of the original image (see Fig. 9.4).

(2) Then the expert manually selects a binarization threshold. Manual selection of the binarization threshold is due to the fact that the data on which the model was trained (synthetically generated dataset) and the real data for analysis of which the model is used (micrographs of membranes) are of a different nature. The expert participation ensures the correct identification of the interlayer boundaries by the neural network. Examples of selectable binarization thresholds are shown in Fig. 9.5.

(3) Padding is added to the neural network mask so that the mask does not go beyond the image boundaries when it is rotated. Then a set of 360 images is obtained, where each image is a neural network mask rotated from 0 to 359° with a step of 1°, and a further binarization is performed according to the selected threshold (see Fig. 9.6).

(4) Then a search for horizontal boundaries for each binarized mask in the set is performed. For this purpose, a sliding window, which is a rectangle, is passing

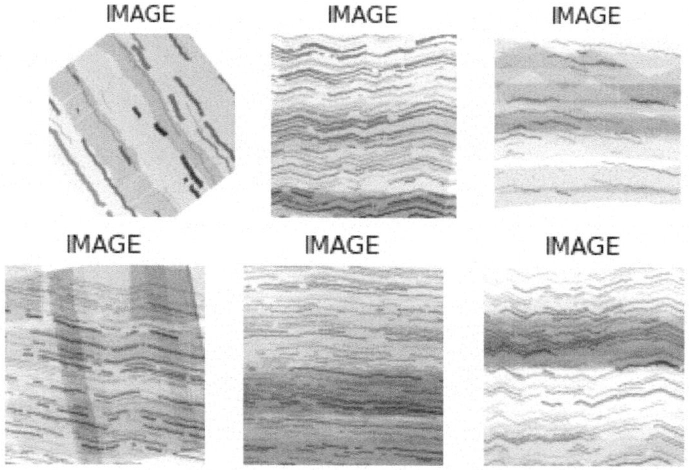

a) Images simulating layered structures

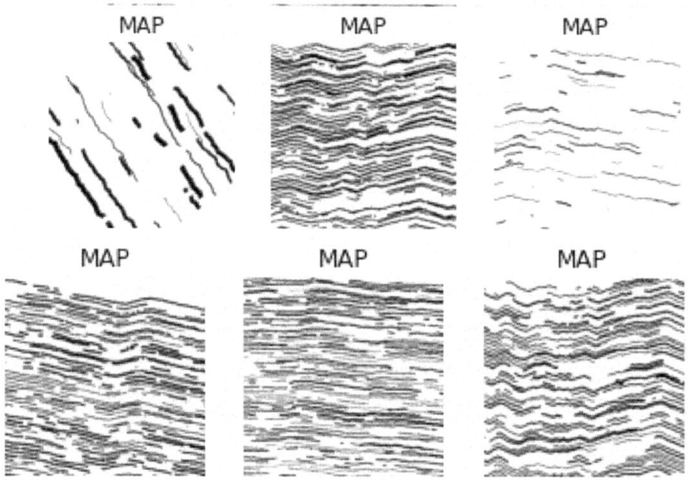

b) Mask

Fig. 9.3 An example of using synthetic data simulating layered structures (**a**) and a mask by which the layer boundaries can be highlighted/selected (**b**)

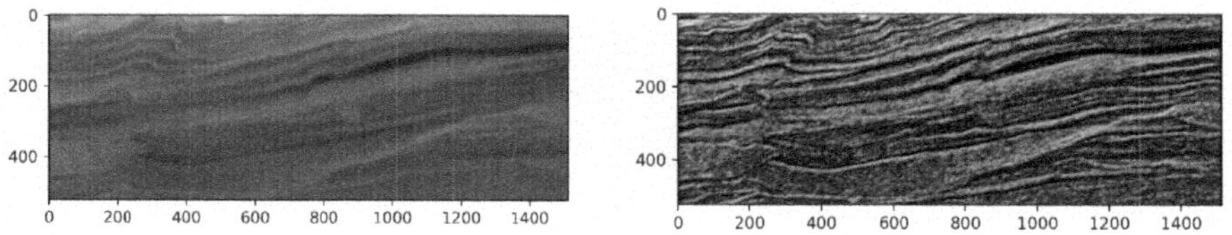

a b

Fig. 9.4 Original EM micrograph (**a**) and a neural network mask obtained from this micrograph (**b**)

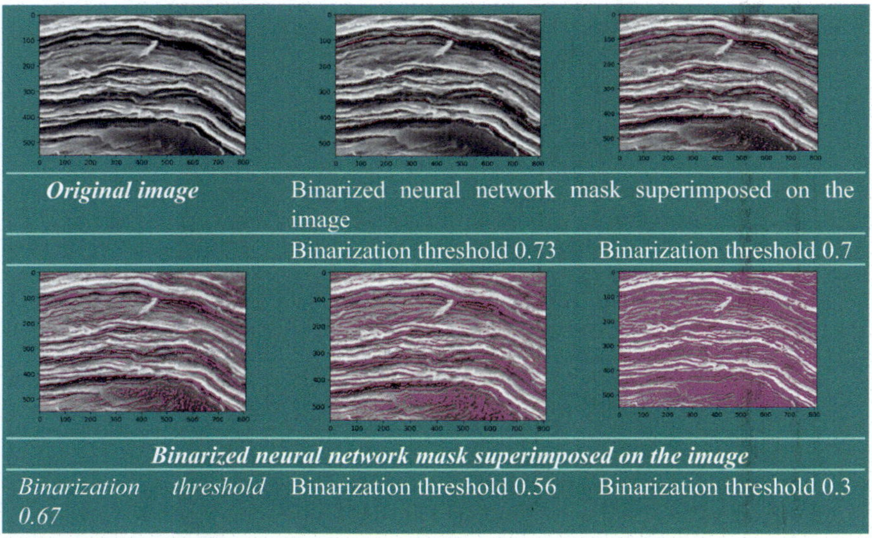

Fig. 9.5 Manual expert selection of a binarization threshold and visualization of a binarized neural network mask superimposed on the original image

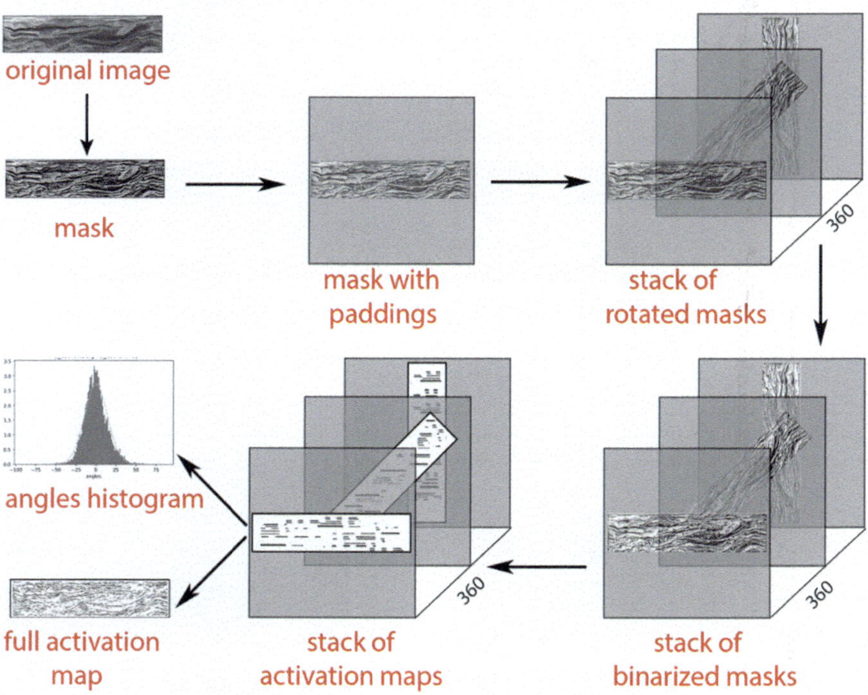

Fig. 9.6 Scheme of a discrete angular rotation algorithm (OrientLayer Algorithm developed by N. A. Marnautov)

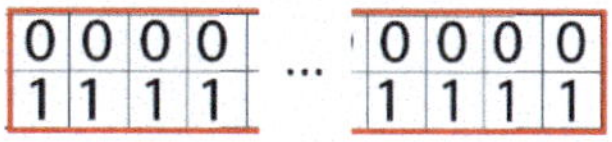

Fig. 9.7 A horizontal filter for cosine similarity determination

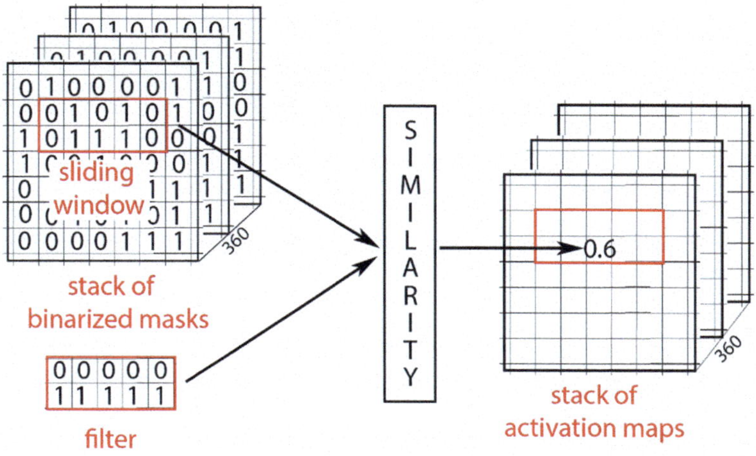

Fig. 9.8 A set of the activation maps

through the image and the similarity coefficient is calculated with a "horizontal filter" (see Fig. 9.7).

Thus, a set of activation maps is obtained, where each pixel is assigned a value corresponding to the coefficient between the horizontal filter and the sliding window centered on this pixel (see Fig. 9.8).

Then non-maximum suppression (NMS) for each pixel is performed, i.e., for each specific pixel an activation map with the highest value is found, and this value does not change, while on the other activation maps the value of this pixel is equated to zero (Fig. 9.9).

Fig. 9.9 Non-maximum Suppression scheme

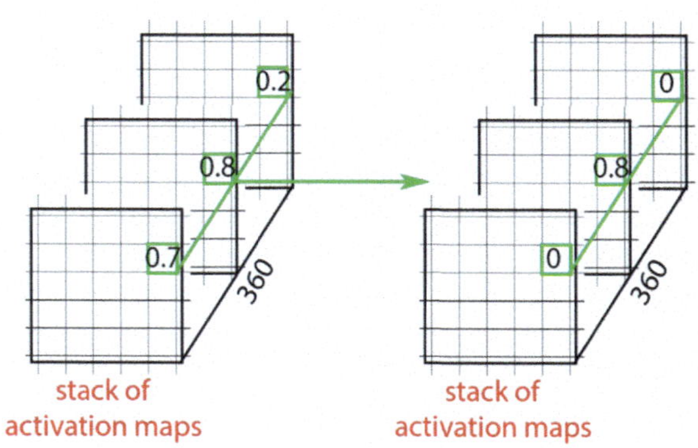

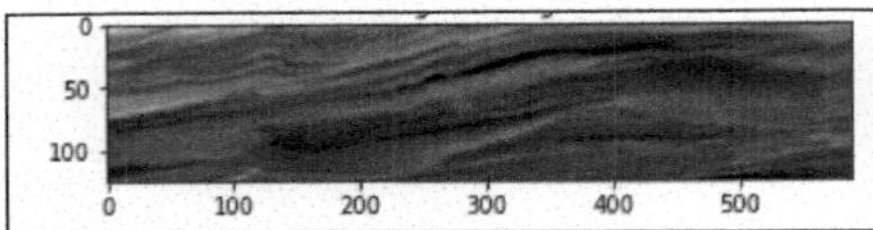	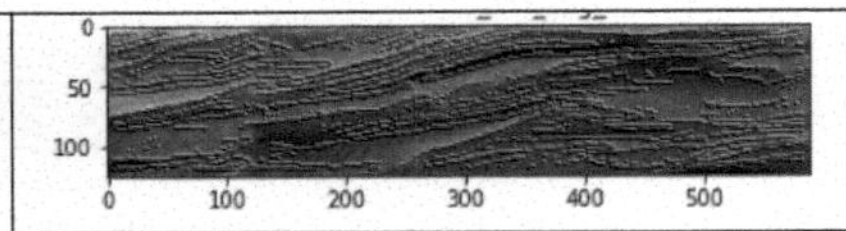
Original micrograph	Complete activation map superimposed on the original image
	Binarization threshold 0.75
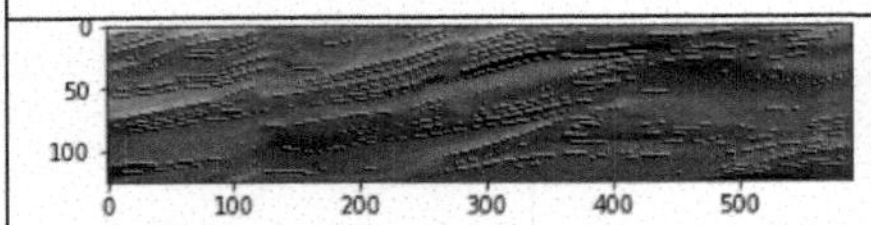	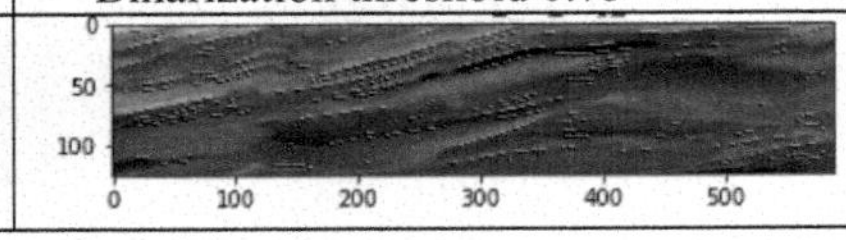
Complete activation map superimposed on the original image	
Binarization threshold 0.8	Binarization threshold 0.85

Fig. 9.10 Binarization of the activation maps with a binarization threshold selected by an expert and visualization of a complete (integrated) activation map by adding individual binarized activation maps

Finally, the activation maps obtained are binarized with the binarization threshold being selected by an expert. For ease of visualization, a complete activation map is obtained by adding the binarized activation maps (Fig. 9.10).

Angular distribution histogram is plotted using the data on binarized activation maps through the sum calculation on each activation map. The resulting histogram is centered relative to the main peak (examples of such histograms are given in the sections below). As a control, one can use the analysis of the image spatial characteristics using the Fourier transform, which simulates radiation diffraction by the periodic structures. The difference in visualization will include the radiation scattering being perpendicular to the periodic structure (for example, a diffraction grating), and the diffracting structure orientation in this case occurs using a secondary transformation based on the integral spatial characteristic (ISC) using QAVIS software [227–232].

9.7 Examples of Heterogeneous Layer Formation in Some Siltstones (Aleurolites)

As the first example, let us consider the analysis of sedimentary rocks—siltstones (photos are taken from the authors' collection based on field, laboratory, and museum sources). Figures 9.11and 9.12 demonstrate two examples of layered sedimentary rocks—the products of rhythmic sedimentation. As the algorithm is applied, the following sequence of images is passed through: a—original images; b—binarized masks; c—angular activation maps. At each step, as a control, "diffraction" analysis of the structure regularity/periodicity can be applied, as shown in Figs. d, e, and f located under a (original image), b (binarized mask), and c (angular activation maps), respectively. In this case, both the image and the masks and maps can be binarized, as shown in Figs. g and h. A "diffractometric" approach can also be applied to the binarized images, as it is shown in Figs. i and j. For comparison, the corresponding orthogonal mappings of the angular orientation of the real structures ("orientation histograms") in polar coordinates, obtained using the discrete angular rotation algorithm developed by Nikolai Marnautov, are shown in insert k (they will be discussed in more detail below). The actual angular distribution histograms for the layers in the sedimentary rock samples analyzed are given below.

Figures 9.13 and 9.14a, b show angle distribution histograms in Cartesian and polar coordinates, respectively, for the images prepared for analysis, as shown in Figs. 9.11, 9.12. Figures 9.13 and 9.14c, d illustrate the attempts to apply Gaussian approximation to the distribution histograms in Cartesian and polar coordinates shown in Figs. 9.13and 9.14a, b. As one can see, the distribution is not completely

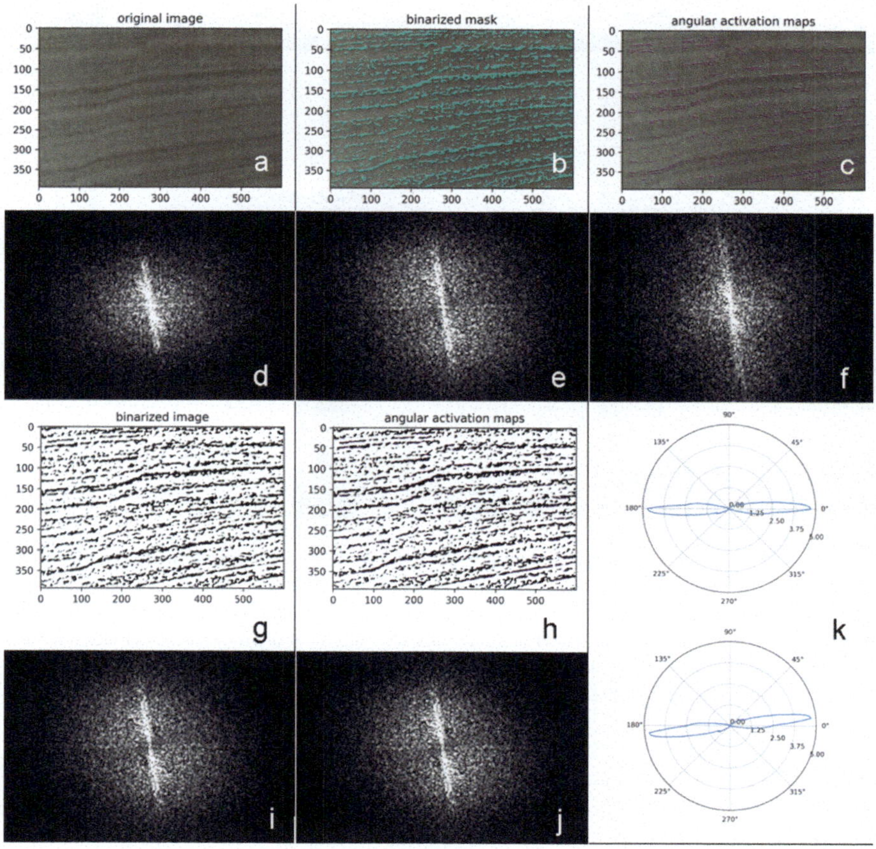

Fig. 9.11 An example of the siltstone image processing from binarization to obtaining angular activation masks and a comparative series of 2D FFT controls of the layer regularity, indicating the contribution of the non-parallel/inclined structural elements to the final signal, which should appear in the form of additional modes in the distribution on the statistical histogram of the angular distribution

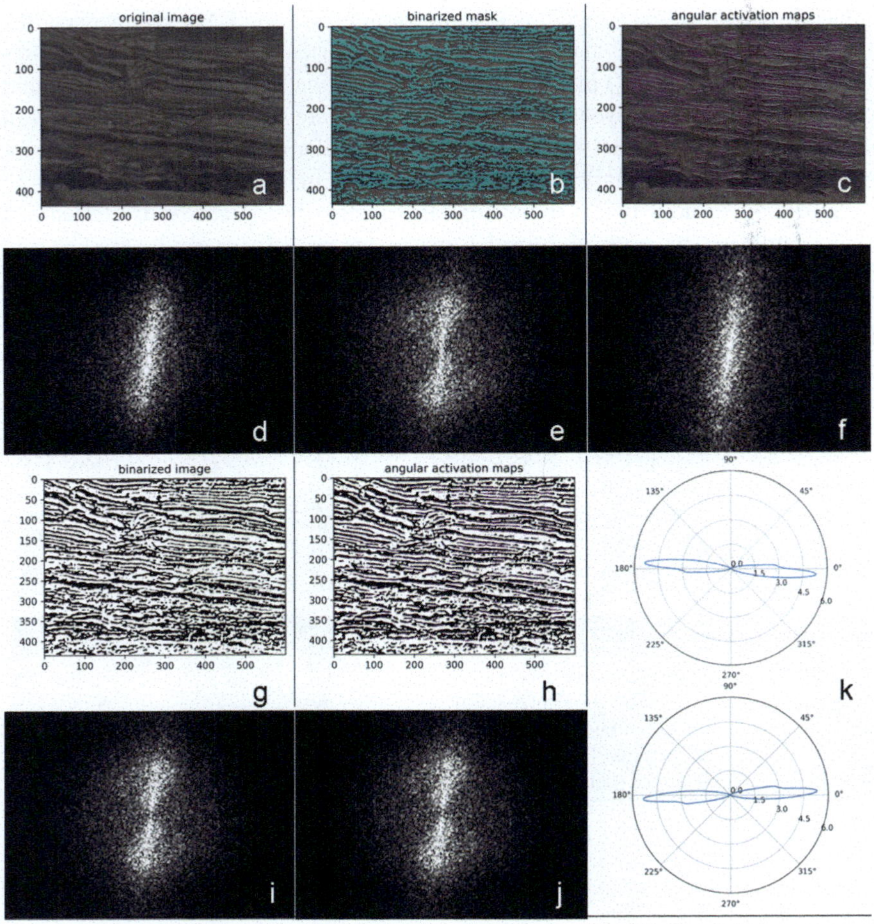

Fig. 9.12 An example of the image processing of siltstones from binarization to obtaining angular activation masks and a comparative series of 2D FFT controls of the layer regularity, indicating the contribution of partial disorder (in the layer organization) to the final signal (scattering broadening and the lateral shift), which should correlatively appear in the form of Gaussian distribution broadening (in the "tails" of the distribution and the distribution asymmetry or identification of several unresolved narrow peaks against the background of a wider one, possibly with a slight shift)

unimodal and fails to be described by a simple Gaussian approximation, which indicates the presence of additional orientations of the elements within the layers and their incomplete parallelism. This observation corresponds to the presence of the additional reflections in 2D FFT (Fig. 9.11d, f–j) and scattering broadening on the sedimentary rock structure with the appearance of additional diffuse components of the texture orientation (Fig. 9.12d, f–j). Thus, layered (rhythmic) sedimentary rocks contain both a regular component, containing several prevailing directions of

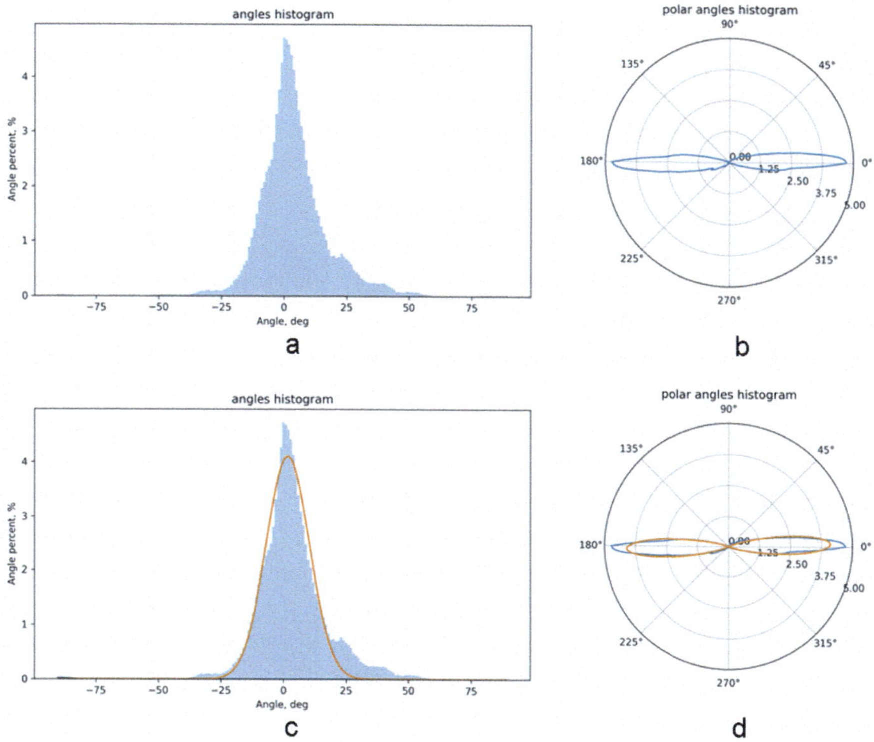

Fig. 9.13 Statistical distribution histograms (angular orientation histograms) in Cartesian (**a**) and polar (**b**) coordinates and their approximations for the (meso)texture of a sedimentary rock shown in Fig. 9.11

orientation (translating into each other as sedimentation occurs), and the elements of anisotropy and chaos (observed in a scattering halo).

9.8 Results of Sedimentographic Analysis of Graphene Oxides Produced by Sedimentation on the Carrier Surface

The main idea in the report [225] was that discrete angular rotation algorithms provide the possibility to distinguish between the real graphene oxide membrane (which, being obtained under the appropriate conditions, possesses valuable functional properties which can be applied in industry) and the loose "piled flakes" (or lamellas) of the same chemical composition, resulting from the ineffective sedimentation under suboptimal conditions. That is, the real task may be a quality control of graphene oxide and graphene membranes, the physicochemical properties of which depend on

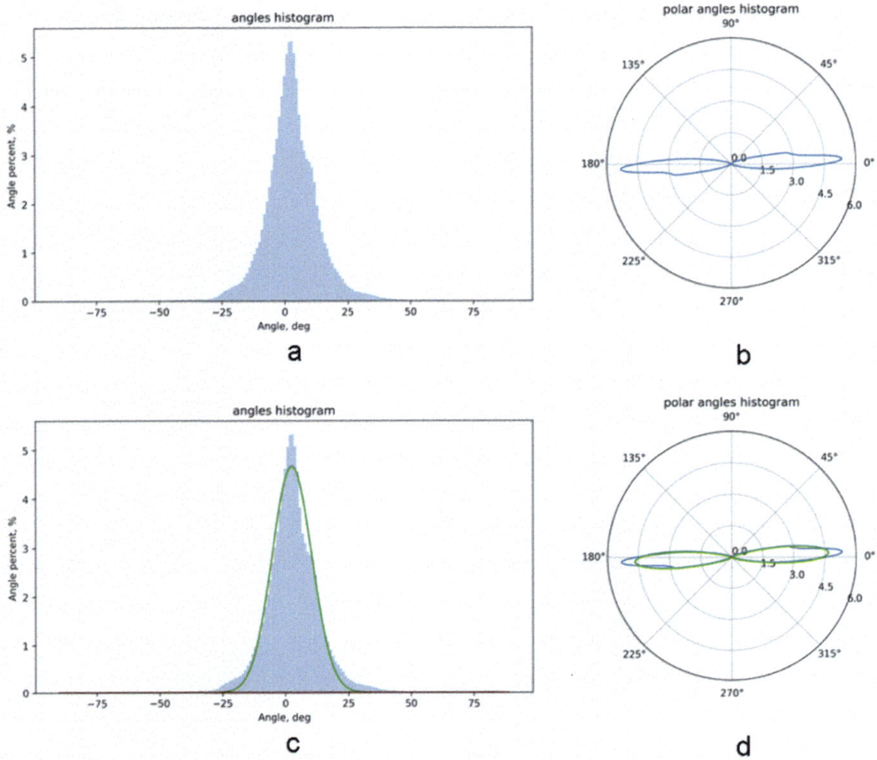

Fig. 9.14 Statistical distribution histograms (angular orientation histograms) in Cartesian (**a**) and polar (**b**) coordinates and their approximations for the (meso)texture of a sedimentary rock shown in Fig. 9.12

their compactness (otherwise, problems arise with transport and filtration of various agents, which we do not dwell on due to the non-chemical nature of this paper). It is obvious that different forms of structurally ordered graphene oxide lamellas, with different angular ordering, depending on the production procedure during deposition on the substrate under specific conditions, lead to non-equivalent values of the target parameters, depending on the physical conditions of sedimentation. In granular and non-lamellar materials in most cases, the level of ordering is lower, since the failures between the layers are not limited by the mechanics of the lamellae/sheets, and turbulence and shear flows lead to different geometries as a consequence of chaos (up to the "baker's transformation").

Generally speaking, the order in the structure of the sheets/lamellas determines the nature of permeability and the possibility of considering the corresponding carbon nanostructures as membrane mimetic materials [233–235] with the conductivity ordering or simply as materials with porosity and any computable percolation parameters (similar to the clay particles in the problems of the oil reservoir porosity). It can be shown that the difference between chaos and order in the conditions of synthesis and production of the layered structures fundamentally determines the order in the resulting structures. For this purpose, one should apply the sedimentometry method with the analysis of order and angular distributions based on the Marnautov algorithm. In this case, one should move away from the a priori idea that graphene sheets always lie compactly and parallelly, since, as it has been shown in the paper by Marnautov et al. [226], this idealized idea is not true. In graphene oxide, the layers or lamellas can also be directed non-parallel and at an angle to each other (if we are talking about lamellas), so we must always take into account their "emergent" production conditions. Under extreme conditions, clays in a banded form are compact and collinear/parallel, while poorly prepared graphene oxide lamella membranes may in some cases be much less ordered in organization than clays or have biomodal, trimodal, and higher order (see Table 9.1 with the distributions of different modalities for clays and graphene oxide, where the scores with an even number of the data lines refer to siltstones, and the scores with an odd number of the data lines refer to graphene oxide) distributions of the structural elements by angles, explicitly determined by the method of discrete rotational angular analysis (Marnautov's algorithm).

The main experimental data was presented in the work by Marnautov et al. [226] and its supplement, so here we will show that similar results can be obtained from the arbitrary images of the layered carbon nanostructures. For this purpose, we will take an arbitrary micrograph from the open-access published paper [236] and process it in the same way as we processed the images of sedimentary rocks above, simultaneously with 2D FFT analysis in real time using the QAVIS software (Pacific Oceanological Institute—POI FEB RAS) based on the FFTW library [227–232].

Figures 9.15 and 9.16 show two micrographs (at different magnifications) of graphene oxide layers taken from the paper [236]. Same as in Figs. 9.11 and 9.12,

Table 9.1 Examples of distributions of different modalities calculated using Marnautov's discrete rotation algorithm

	Angle	Sigma	Amplitude	Area	Percent
0	2,03	10,0733	0,0387	0,9776	100
	Angle	Sigma	Amplitude	Area	Percent
0	−89,3	3,4051	0,0002	0,0017	0,2
1	1,68	8,6532	0,0411	0,8913	99,8
	Angle	Sigma	Amplitude	Area	Percent
0	−58,35	4,1749	0,0003	0,0028	0,3
1	1,2	10,0122	0,0379	0,9518	99,7
2	64	1,6559	0,0001	0,0002	0
	Angle	Sigma	Amplitude	Area	Percent
0	−87,12	1,7518	0	0,0001	0
1	−75,98	2,2284	0	0	0
2	2,61	7,9507	0,0467	0,9303	99,9
3	81,36	2,8798	0,0001	0,0005	0,1

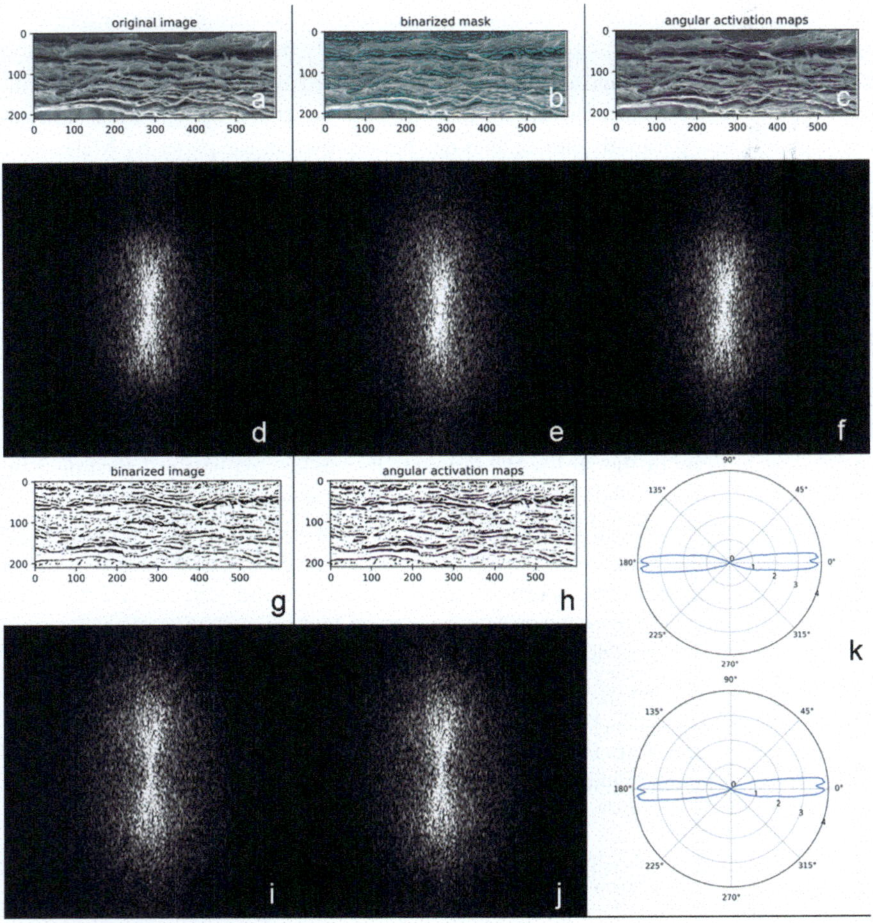

Fig. 9.15 An example of image processing of graphene oxide sheets/lamellas from binarization to obtaining angular activation masks and a comparative series of 2D FFT controls of the layer regularity, indicating a significant contribution of non-parallel/inclined structural elements to the final signal, which should manifest itself in the form of additional modes in the statistical histogram of the angular distribution

the sequence of images is given: a—original images; b—binarized masks; c—angular activation maps. 2D FFT data obtained using QAVIS algorithms and FFTW library are also shown in Figs. 9.11 and 9.12d, e, f. The same is true for segmented and binarized data (Figs. g and h correspond to 2D FFTs in Figs. i and j). Visualization of the angle distribution of graphene oxide sheets/lamellas is shown in the insert k, as well as in Figs. 9.11 and 9.12. Figures 9.17 and 9.18 show angle distribution histograms in Cartesian and polar coordinates, respectively, for the images prepared for analysis, as shown in Figs. 9.15 and 9.16. Figures 9.17 and 9.18c, d illustrate the attempts to apply Gaussian approximation to the angle distribution histograms in Cartesian and polar coordinates shown in Figs. 9.17 and 9.18a, b. As can be seen, the angle distribution for the multilayers shown in Fig. 9.15 is clearly bimodal at the vertex, although it is relatively approximated by a Gaussian (see Fig. 9.17). The above bimodality could be predicted from the FFT in Fig. 9.15, where splitting/bimodality of the FFT pattern is clearly observed. Figure 9.18 shows angle distributions in Cartesian and polar coordinates for graphene oxide layers given in Fig. 9.16 and its Gaussian approximation, proving that the angle distribution is also not completely unimodal. Nevertheless, as in the work [226], to a certain approximation, it makes sense to calculate the order parameters using Legendre and Chebyshev polynomials, as it is shown in Tables 9.2 and 9.3 (for the images of graphene oxide given in Figs. 9.15 and 9.16).

In the case of graphene oxide membranes, calculation of the order parameters has a clear physical meaning. This issue is discussed in more detail in the paper [226]. In another work inaccessible in English [225], it is stated that:

> "About ten years ago, the unique selectivity of permeability of graphene oxide membranes for both liquids and gases was discovered [237, 238]. It was shown that the transport rate of water through such membranes exceeds the transport rate of alcohols by many orders of magnitude. In addition, it has been established that graphene oxide membranes are effective molecular sieves that pass through hydrated ions and molecules with the radius less than 4.5 Å [239]. For this reason, graphene oxide-based membranes are currently considered as promising materials for water purification and separation of various liquids and dissolved molecules [240]. The passage mechanism of various liquids through graphene oxide membranes has not currently been established. However, experimental data indicate that this process is significantly influenced by the degree of orientational ordering of the oxygraphene layers, which, in turn, is determined by the membrane manufacturing method. Thus, in work [241] it was shown that more ordered membranes produced by vacuum filtration are more efficient in dehydrating 1-butanol/water mixtures compared to the less ordered membranes obtained by spontaneous evaporation of a graphene oxide suspension. The authors of [242] showed that the flow of water through highly ordered membranes obtained by shear alignment of the liquid crystalline graphene oxide/water phase significantly exceeds the flow of water through the membranes fabricated by vacuum filtration. In the work [243] it was noted that the curvature degree of the oxygraphene layers also has a great influence on the resulting membrane permeability."

Describing the approach used in Marnautov's program [244, 245] customized for graphene-based samples, the authors write: "the procedure for numerical analysis of SEM micrographs of the membrane lateral surface is based on machine learning technologies. This approach allows us to establish orientation parameters of the order of the membrane-forming lamellas."

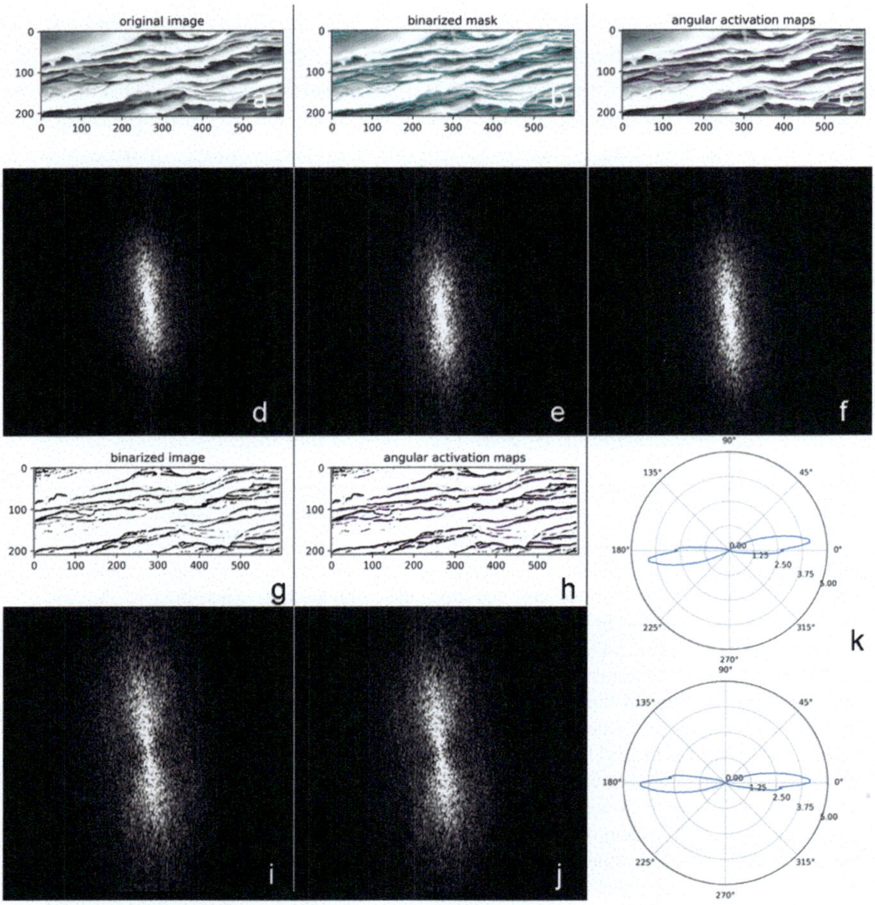

Fig. 9.16 An example of the image processing of graphene oxide sheets/lamellas from binarization to obtaining the angular activation masks and a comparative series of 2D FFT controls of the layer regularity (the higher magnification is the result of the magnification dependence analysis of the algorithm operation), indicating higher relative ordering at the local scales: despite the waviness of the sheets, no new modes are observed in their 2D FFT pattern and in the diagram in polar coordinates (k), in contrast to the case shown in Fig. 9.15. The contribution of horizontal waviness is also visible in the binarized images below (i, j)

9.9 Concluding Remarks

Note that, for layered clay minerals and other geological structures, it is also possible to apply the order parameter analysis using discrete angular rotation algorithm and computations of Legendre and Chebyshev polynomials, but unification of the above parameters for all geological structures is impossible due to different mechanisms of their formation. In the coming years, the authors set the task of describing and publishing results on the order parameters of such structures and their application to

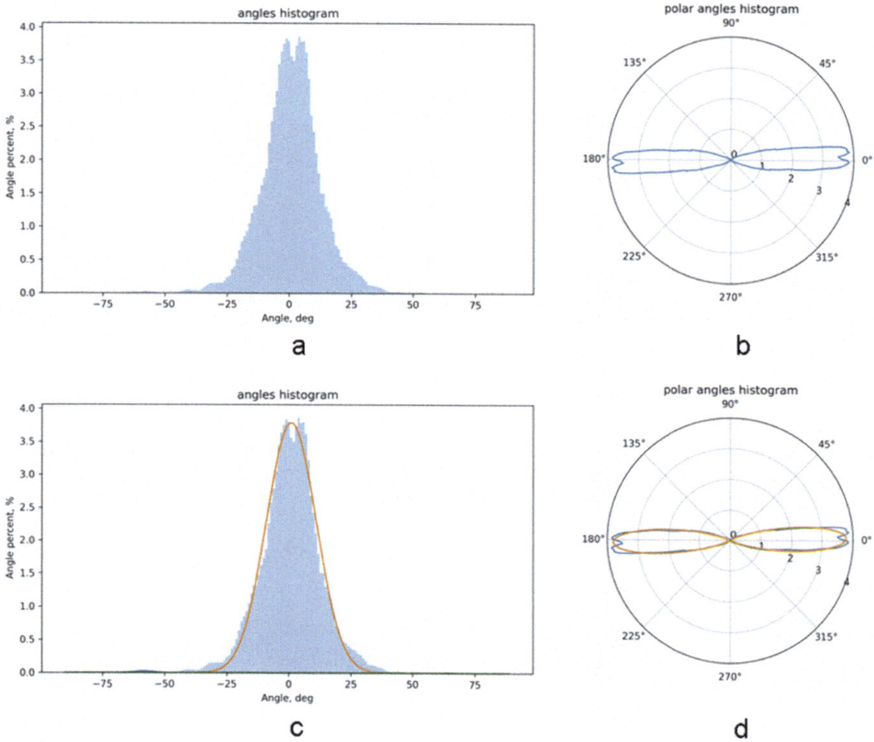

Fig. 9.17 Angle statistical distribution histograms (angular orientation histograms) in Cartesian (**a**) and polar (**b**) coordinates and their approximation for graphene oxide layers shown in Fig. 9.15

the formation of geological, geophysical, and geomineral structures, which have been insufficiently and fragmentarily studied using Legendre polynomials and Chebyshev polynomials in recent decades (previous works more often were related to the problems of seismology, magnetometry, and geophysical sounding/probing as a whole) [246–249].

As can be seen from Tables 9.4 and 9.5, mineral structures can have order indices no lower than those of graphene membranes with an average compactness, up to the numerical overlap in the region of the high orders (the above tables show data for the second, fourth, sixth, eighth, and tenth orders). It is possible that, during compaction of geomaterials and thermobarogeochemical processes in them (interpreted in the broad sense of the term, taking into account the low-temperature region in sedimentology), it is possible to observe the ordering processes in the layered minerals in real time. Such a problem, from our point of view, may be of interest to the laboratories dealing with experimental mineralogy and modeling of geophysical processes, as well as the groups involved in the field of application of nonlinear physics, synergetics, and various approaches of the theory of self-organization to geophysics and geochemistry of sedimentary rocks.

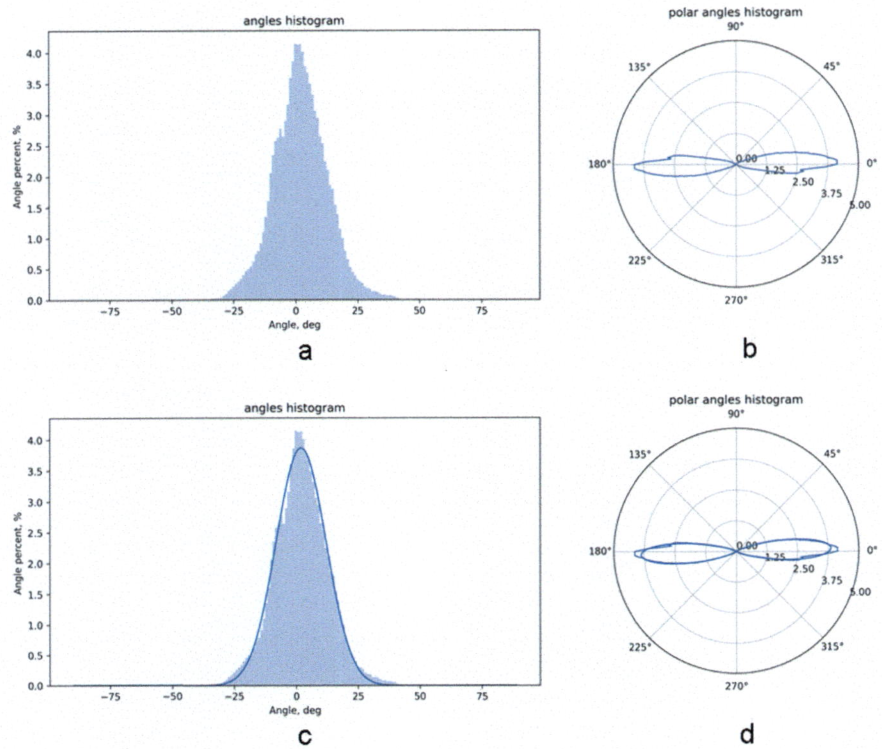

Fig. 9.18 Angle statistical distribution histograms (angular orientation histograms) in Cartesian (**a**) and polar (**b**) coordinates and their approximation for graphene oxide layers shown in Fig. 9.16

Table 9.2 Order parameter calculation data using Legendre and Chebyshev polynomials, calculated by Marnautov's software with GUI (OrientLayer Analyser) using discrete angular rotation algorithm for the images of graphene oxide structures shown in Fig. 9.15

Legendre polynomials	LgP: 2	4	6	8	10
LgP normalized angles	0,936	0,821	0,696	0,581	0,488
LgP angles wo norm	0,936	0,821	0,696	0,581	0,488
Chebyshev polynomials	ChP: 2	4	6	8	10
ChP normalized angles	0,915	0,722	0,517	0,345	0,226
ChP angles wo norm	0,915	0,721	0,516	0,345	0,226

Table 9.3 Order parameter calculation data using Legendre and Chebyshev polynomials, calculated by Marnautov's software with GUI (OrientLayer Analyser) using discrete angular rotation algorithm for the images of graphene oxide structures shown in Fig. 9.16

Legendre polynomials	LgP: 2	4	6	8	10
LgP normalized angles	0,946	0,842	0,719	0,599	0,497
LgP angles wo norm	0,928	0,792	0,641	0,505	0,398
Chebyshev polynomials	ChP: 2	4	6	8	10
ChP normalized angles	0,928	0,752	0,544	0,358	0,219
ChP angles wo norm	0,904	0,675	0,425	0,224	0,094

Table 9.4 Order parameter calculation data using Legendre and Chebyshev polynomials calculated by Marnautov's software with GUI (OrientLayer Analyser) using discrete angular rotation algorithm for the images of the layered mineral sedimentary structures presented in Fig. 9.11

Legendre polynomials	LgP: 2	4	6	8	10
LgP normalized angles	0,92	0,8	0,685	0,594	0,515
LgP angles wo norm	0,896	0,745	0,61	0,507	0,418
Chebyshev polynomials	ChP: 2	4	6	8	10
ChP normalized angles	0,894	0,694	0,517	0,397	0,293
ChP angles wo norm	0,862	0,612	0,411	0,28	0,168

Table 9.5 Order parameter calculation data using Legendre and Chebyshev polynomials calculated by Marnautov's software with GUI (OrientLayer Analyser) using discrete angular rotation algorithm for the images of the layered mineral sedimentary structures presented in Fig. 9.12

Legendre polynomials	LgP: 2	4	6	8	10
LgP normalized angles	0,948	0,856	0,752	0,653	0,567
LgP angles wo norm	0,941	0,834	0,712	0,596	0,497
Chebyshev polynomials	ChP: 2	4	6	8	10
ChP normalized angles	0,931	0,776	0,604	0,452	0,331
ChP angles wo norm	0,921	0,741	0,538	0,36	0,226

Acknowledgements The authors are grateful to their co-authors from the papers [225, 226] for the fruitful discussions and for the permissions for using of the graphical materials from the above mentioned papers for both review and educational aims. Quantitative determination of the orientational ordering of graphene oxide membranes by spin probe techniques developed by above mentioned co-authors was described in [250–254].

O. V. Gradov is also grateful to V. A. Oganessian and A. V. Notchenko for the common successful experience in multi-coordinate goniometric instrumentation development in 2010–2012 at RAMS and BMSTU. This experience was an inspiration for the development of the ideas on multi-angle sedimentography.

Authors' Contributions Elfimov Anton B.—image selection; data processing and analysis; testing of different modifications of algorithms and software versions with GUI (OrientLayer Analyser, 2023–2024);

Gradov Oleg V.—analytical review (70%) and conceptualization; ideas on the application of the above algorithms for the sedimentographic purposes;

Gradova Margaret A.—the final text translation, proofreading, and editing;

Marnautov Nikolai A.—Software development (LayerNet Segmenter, Ori-entLayer Analyser, SynthiLayer); creation and detailed description of the OrientLayer algorithm (2022–2023).

References

1. Kolomeitsev VV, Kolomeitseva EF (1980) Analysis in a liquid sedimentograph of the dispersion of suspensions of vitreous silica. Refractories 21(11–12):658–661. https://doi.org/10.1007/BF01398343
2. Berezin PN, Voronin AD (1981) [Use of the sedimentograph for particle size analysis of soils and subsoils]. Pochvovedeniye 5:56–61 (in Russian)
3. Mazumdar DH, Blum WE, Elahi SF (1994) Stratigraphic and sedimentographic interpretation of particle size distribution in two well drained pedons developed from Madhupur clay, Bangladesh. Arch Agron Soil Sci 38(3):203–214. https://doi.org/10.1080/03650349409365852
4. Kalin L, Govindaraju RS, Hantush MM (2003) Effect of geomorphologic resolution on modeling of Runoff hydrograph and sedimentograph over small watersheds. J Hydrol 276(1–4):89–111. https://doi.org/10.1016/S0022-1694(03)00072-6
5. Kalin L, Govindaraju RS, Hantush MM (2004) Development and application of a methodology for sediment source identification. I: Modified unit sedimentograph approach. J Hydrol Eng 9(3):184–193. https://doi.org/10.1061/(ASCE)1084-0699(2004)9:3(184)
6. Stadnichenko SM, Podoba VO (2008) [Technique of using sedimentograph MasterSizer 2000 for granulometric analysis and results interpretation question] Proc Inst Geol Sci Natl Acad Sci Ukraine (6):269–277. (in Ukrainian) https://paperity.org/p/292233917/technique-of-using-sedimentograph-mastersizer-2000-for-granulometric-analysis-and-results
7. Sauer H (1937) Der Sedimentograph. Kolloid-Zeitschrift 80(3):287–291. https://doi.org/10.1007/BF01518554
8. Wyss F (1955) Sedimentograph, a new apparatus for measurement and registration of the changing stratification limit of mixed fluids, especially for blood sedimentation reaction. Schweizerische Medizinische Wochenschrift 85(5):113–115. https://pubmed.ncbi.nlm.nih.gov/14358617/
9. Val'dman VA (1955) [Sedimentograph]. Fiziologicheskii zhurnal SSSR imeni IM. Sechenova 41(3):430–432. (in Russian) https://pubmed.ncbi.nlm.nih.gov/13241619/
10. Sapinski A, Sapinski W (1959) Sedimentograph-apparatus for automatic registration of erythrocyte sedimentation. Polski tygodnik lekarski 14:2202–2203. https://pubmed.ncbi.nlm.nih.gov/14441675/ (in Poland)
11. Hasznos P, Almasi F, Hete S (1988) Examining separation with hydrocyclone used in wheat-starch making using a photo-sedimentograph Fritsch Analysette 20. Hung J Ind Chem 16(2):151–155
12. Pietriková A, Búgel M, Golja M (2004) Pripremanje praha SiO_2 luženjem serpentina. Metalurgija 43(4):299–304. https://hrcak.srce.hr/128112
13. Kozyrev NV, Larionov BV, Sakovich GV (2008) Influence of HMX particle size on the synthesis of nanodiamonds in detonation waves. Combust Explosion Shock Waves 44:193–197. https://doi.org/10.1007/s10573-008-0026-9
14. Kynch GJ (1952) A theory of sedimentation. Trans Faraday Soc 48:166–176. https://doi.org/10.1039/TF9524800166
15. Goldberg RJ (1953) Sedimentation in the ultracentrifuge. J Phys Chem 57(2):194–202. https://doi.org/10.1021/j150503a014

16. Williams JW, Van Holde KE, Baldwin RL, Fujita H (1958) The theory of sedimentation analysis. Chem Rev 58(4):715–744. https://doi.org/10.1021/cr50022a005
17. Jamie EAG, Wensink HH, Aarts DGAL (2010) Sedimentation–diffusion dynamics in colloid–polymer mixtures. J Stat Mech: Theory Exp 2010(11):P11008. https://doi.org/10.1088/1742-5468/2010/11/P11008
18. Whittington RB (1942) Blood sedimentation: a study in haemo-mechanics. Proc R Soc Lond Ser B-Biol Sci 131(863):183–190. https://doi.org/10.1098/rspb.1942.0024
19. Trowbridge EA (1982) The fluid mechanics of blood: equilibrium and sedimentation. Clin Phys Physiol Meas 3(4):249–265. https://doi.org/10.1088/0143-0815/3/4/001
20. Ashmore P, Conly FM, DeBoer D, Martin Y, Petticrew E, Roy A (2000) Recent (1995–1998) Canadian research on contemporary processes of river erosion and sedimentation, and river mechanics. Hydrol Process 14(9):1687–1706. https://doi.org/10.1002/1099-1085(20000630)14:9%3C1687::AID-HYP72%3E3.0.CO;2-9
21. De Boer DH, Hassan MA, MacVicar B, Stone M (2005) Recent (1999–2003) Canadian research on contemporary processes of river erosion and sedimentation, and river mechanics. Hydrol Process Int J 19(1):265–283. https://doi.org/10.1002/hyp.5767
22. Ramaswamy S (2001) Issues in the statistical mechanics of steady sedimentation. Adv Phys 50(3):297–341. https://doi.org/10.1080/00018730110050617
23. Guazzelli É, Hinch J (2011) Fluctuations and instability in sedimentation. Annu Rev Fluid Mech 43:97–116. https://doi.org/10.1146/annurev-fluid-122109-160736
24. Nissanka K, Ma X, Burton JC (2023) Dynamics of mass polar spheroids during sedimentation. J Fluid Mech 956:A28. https://doi.org/10.1017/jfm.2023.32
25. Yazdi AA, D'Avino G (2023) Sedimentation and migration dynamics of a spherical particle in an elastoviscoplastic fluid near a wall. J Nonnewton Fluid Mech 314:105004. https://doi.org/10.1016/j.jnnfm.2023.105004
26. Okada K, Satoh A (2020) Sedimentation characteristics of spherical and rod-like particles in the gravitational field by Brownian dynamics simulations: for the improvement of the visibility of small lakes and ponds. Environ Fluid Mech 20:765–790. https://doi.org/10.1007/s10652-019-09722-9
27. Pan TW, Glowinski R (2017) On the dynamics of particle sedimentation in viscoelastic fluids: A numerical study on particle chaining in two-dimensional narrow channels. J Nonnewton Fluid Mech 244:44–56. https://doi.org/10.1016/j.jnnfm.2017.04.001
28. Batchelor GK (1982) Sedimentation in a dilute polydisperse system of interacting spheres. Part 1. General theory. J Fluid Mech 119:379–408. https://doi.org/10.1017/S0022112082001402
29. Batchelor GK, Wen CS (1982) Sedimentation in a dilute polydisperse system of interacting spheres. Part 2. Numerical results. J Fluid Mech 124:495–528. https://doi.org/10.1017/S0022112082002602
30. Anuar NSS, Talib SHA, Radzaly NFI, Hashim SINS, Abustan MS (2020) The effect of sedimentation to the pump sump system by using computational fluid dynamics (CFD) model. J Adv Res Fluid Mech Thermal Sci 68(1):86–97. https://doi.org/10.37934/arfmts.68.1.8697
31. Oreshenkov AV (2020) Study of sedimentation in jet fuels due to watering using methods of molecular mechanics. Chem Technol Fuels Oils 56(5):763–772. https://doi.org/10.1007/s10553-020-01189-5
32. Wang L, Li JC, Zhou JF (2009) An algorithm for coarse particle sedimentation simulation by Stokesian Dynamics. In: New trends in fluid mechanics research: proceedings of the fifth international conference on fluid mechanics (Shanghai, 2007). Springer, Berlin, pp 432–435. https://doi.org/10.1007/978-3-540-75995-9_141
33. Sierou A (2005) stokesian dynamics simulations for particle laden flows. In: Yip S (ed) Handbook of materials modeling. Springer, Dordrecht. https://doi.org/10.1007/978-1-4020-3286-8_139
34. Gomez B, Page M, Bak P, Trustrum N (2002) Self-organized criticality in layered, lacustrine sediments formed by landsliding. Geology 30(6): 519–522. https://doi.org/10.1130/0091-7613(2002)030%3C0519:SOCILL%3E2.0.CO;2

35. Fonstad M, Marcus WA (2003) Self-organized criticality in riverbank systems. Ann Assoc Am Geogr 93(2):281–296. https://doi.org/10.1111/1467-8306.9302002
36. Sidorchuk A (2006) Stages in gully evolution and self—organized criticality. Earth Surf Process Landforms J Br Geomorphol Res Group 31(11):1329–1344. https://doi.org/10.1002/esp.1334
37. Corté L, Gerbode SJ, Man W, Pine DJ (2009) Self-organized criticality in sheared suspensions. Phys Rev Lett 103(24):248301. https://doi.org/10.1103/PhysRevLett.103.248301
38. Van De Wiel MJ, Coulthard TJ (2010) Self-organized criticality in river basins: challenging sedimentary records of environmental change. Geology 38(1):87–90. https://doi.org/10.1130/G30490.1
39. Carey S (1997) Influence of convective sedimentation on the formation of widespread tephra fall layers in the deep sea. Geology 25(9):839–842. https://doi.org/10.1130/0091-7613(1997)025%3C0839:IOCSOT%3E2.3.CO;2
40. Carey S (1998) Influence of convective sedimentation on the formation of widespread tephra fall layers in the deep sea. Oceanogr Lit Rev 1(45):70. https://www.infona.pl/resource/bwmeta1.element.elsevier-ec894fa6-d786-3764-929e-72bbf1ce801f
41. Hardardóttir J, Geirsdóttir A, Thordarson T (2001) Tephra layers in a sediment core from Lake Hestvatn, southern Iceland: implications for evaluating sedimentation processes and environmental impacts on a lacustrine system caused by tephra fall deposits in the surrounding watershed. Volcaniclastic Sed Lacustrine Settings 223–246. https://doi.org/10.1002/9781444304251.ch11
42. Pozhidayev SD, Babenko VP (1976) Origin of clay layers in Carboniferous coal seams of the Donbass and their relationships with fluctuating movements in regions of supply and sedimentation. Int Geol Rev 18(7):829–833. https://doi.org/10.1080/00206817609471285
43. Hieke W (2000) Transparent layers in seismic reflection records from the central Ionian Sea (Mediterranean)—evidence for repeated catastrophic turbidite sedimentation during the Quaternary. Sed Geol 135(1–4):89–98. https://doi.org/10.1016/S0037-0738(00)00065-8
44. Wheatcroft RA, Drake DE (2003) Post-depositional alteration and preservation of sedimentary event layers on continental margins, I. The role of episodic sedimentation. Mar Geol 199(1–2):123–137. https://doi.org/10.1016/S0025-3227(03)00146-4
45. Siano DB (1979) Layered sedimentation in suspensions of monodisperse spherical colloidal particles. J Colloid Interface Sci 68(1):111–127. https://doi.org/10.1016/0021-9797(79)90263-7
46. Fernández-Nieto ED, Koné EH, De Luna TM, Bürger R (2013) A multilayer shallow water system for polydisperse sedimentation. J Comput Phys 238:281–314. https://doi.org/10.1016/j.jcp.2012.12.008
47. Bürger R, Fernández-Nieto ED, Osores V (2019) A dynamic multilayer shallow water model for polydisperse sedimentation. ESAIM: Math Model Numer Anal 53(4):1391–1432. https://doi.org/10.1051/m2an/2019032
48. Bürger R, Fernández-Nieto ED, Osores V (2020) A multilayer shallow water approach for polydisperse sedimentation with sediment compressibility and mixture viscosity. J Sci Comput 85:1–40. https://doi.org/10.1007/s10915-020-01334-6
49. Posedel N, Faganeli J (1991) Nature and sedimentation of suspended particulate matter during density stratification in shallow coastal waters(Gulf of Trieste, northern Adriatic). Mar Ecol Progress Ser Oldendorf 77(2):135–145. https://www.int-res.com/articles/meps/77/m077p135.pdf
50. Choux C, Druitt T, Thomas N (2004) Stratification and particle segregation in flowing polydisperse suspensions, with applications to the transport and sedimentation of pyroclastic density currents. J Volcanol Geoth Res 138(3–4):223–241. https://doi.org/10.1016/j.jvolgeores.2004.07.004
51. Dellino P, Dioguardi F, Doronzo DM, Mele D (2020) A discriminatory diagram of massive versus stratified deposits based on the sedimentation and bedload transportation rates. Experimental investigation and application to pyroclastic density currents. Sedimentology 67(4):2013–2039. https://doi.org/10.1111/sed.12693

52. Johnson SM, Dashtgard SE (2014) Inclined heterolithic stratification in a mixed tidal–fluvial channel: differentiating tidal versus fluvial controls on sedimentation. Sed Geol 301:41–53. https://doi.org/10.1016/j.sedgeo.2013.12.004
53. Elçi Ş, Work PA, Hayter EJ (2007) Influence of stratification and shoreline erosion on reservoir sedimentation patterns. J Hydraul Eng 133(3):255–266. https://doi.org/10.1061/(ASCE)0733-9429(2007)133:3(255)
54. Thorne JA, Grace E, Swift DJP, Niedoroda A (1992) Sedimentation on continental margins, III: the depositional fabric—an analytical approach to stratification and facies identification. In: Shelf sand and sandstone bodies: geometry, facies and sequence stratigraphy (Special Publication Number 14 of the International Association of Sedimentologists). Blackwell Scientific Publications, Oxford, London, pp 59–87. https://doi.org/10.1002/9781444303933.ch3
55. Cotter E, Graham JR (1991) Coastal plain sedimentation in the late Devonian of southern Ireland; hummocky cross-stratification in fluvial deposits? Sed Geol 72(3–4):201–224. https://doi.org/10.1016/0037-0738(91)90012-3
56. Byrne ML, McCann SB (1990) Stratification and sedimentation in complex vegetated coastal dunes, Sable Island, Nova Scotia. Sed Geol 66(3–4):165–179. https://doi.org/10.1016/0037-0738(90)90058-2
57. Jaillard E, Chihaoui A, Latil JL, Zghal I (2021) Sequences, discontinuities and water stratification in a low-energy ramp: the Early Albian sedimentation in central Tunisia. Int J Earth Sci 110:263–285. https://doi.org/10.1007/s00531-020-01951-4
58. Mondal MS, Kouchi Y, Okubo K (2009) Sedimentation and thermal stratification during floods: a case study on Hail Haor of North-East Bangladesh. J Hydrol Meteorol 6(1):15–25. http://soham.org.np/wp-content/uploads/2009/08/vol6-2.pdf
59. Koren N, Ostrovsky I (2000) Sedimentation in a stratified subtropical lake. Internationale Vereinigung für theoretische und angewandte Limnologie: Verhandlungen 27(5):2636–2639. https://doi.org/10.1080/03680770.1998.11898144
60. Lundhansen L, Pejrup M, Valeur J, Jensen A (1993) Gross sedimentation-rates in the north sea-baltic sea transition-effects of stratification, wind energy-transfer, and resuspension. Oceanol Acta 16(3):205–212
61. Kanari S (1966) Some experiments on the sedimentation in estuaries with density stratification, vol 6. Special Contributions of the Geophysical Institute, Kyoto University, pp 127–133. http://hdl.handle.net/2433/178511
62. Sandgren CD, Robinson JV (1984) A stratified sampling approach to compensating for non-random sedimentation of phytoplankton cells in inverted microscope settling chambers. Brit Phycol J 19(1):67–72. https://doi.org/10.1080/00071618400650071
63. Bhosle NB, Sawant SS, Sankaran PD, Wagh AB (1989) Sedimentation of particulate material in stratified and nonstratified water columns in the Bombay High area of the Arabian Sea. Mar Ecol Progress Ser Oldendorf 57(3):225–236. https://doi.org/10.3354/meps057225
64. Ott I, Rakko A, Sarik D, Nõges P, Ott K (2005) Sedimentation rate of seston during the formation of temperature stratification after ice break-up in the partly meromictic Lake Verevi. Hydrobiologia 547:51–61. https://doi.org/10.1007/s10750-005-4143-0
65. Barlow RG, Mantoura RFC, Peinert RD, Miller AEJ, Fileman TW (1996) Distribution, sedimentation and fate of pigment biomarkers following thermal stratification in the western Alboran Sea. Oceanogr Lit Rev 5(43):478. https://www.infona.pl/resource/bwmeta1.element.elsevier-e8ebb375-60f7-399e-a38e-98e122dbe727/
66. Barlow RG, Mantoura RFC, Peinert RD, Miller AEJ, Fileman TW (1995) Distribution, sedimentation and fate of pigment biomarkers following thermal stratification in the western Alboran Sea. Mar Ecol Prog Ser 125:279–291. https://doi.org/10.3354/meps125279
67. Guy M, Taylor WD, Carter JC (1994) Decline in total phosphorus in the surface waters of lakes during summer stratification, and its relationship to size distribution of particles and sedimentation. Can J Fish Aquat Sci 51(6):1330–1337. https://doi.org/10.1139/f94-132
68. Morales-Pineda M, Ubeda B, Cozar A, Obrador B, Galvez JA (2016) Organic carbon sedimentation dominates over CO_2 emission in two net heterotrophic mediterranean reservoirs during stratification. Aquat Sci 78:279–290. https://doi.org/10.1007/s00027-015-0423-6

69. Kodina LA (2010) Carbon isotope fractionation in various forms of biogenic organic matter: I. Partitioning of carbon isotopes between the main polymers of higher plant biomass. Geochem Int 48:1157–1165. https://doi.org/10.1134/S0016702910120013
70. Galimov EM (2006) Isotope organic geochemistry. Org Geochem 37(10):1200–1262. https://doi.org/10.1016/j.orggeochem.2006.04.009
71. Galimov E (1981) The biological fractionation of isotopes. Academic, London. https://doi.org/10.1016/B978-0-12-273970-5.X5001-4
72. Schidlowski M (2001) Carbon isotopes as biogeochemical recorders of life over 3.8 Ga of Earth history: evolution of a concept. Precambrian Res 106(1–2):117–134. https://doi.org/10.1016/S0301-9268(00)00128-5
73. Bykovskii YA, Timoshin VT, Laptev ID, Manykin EA (1986) Correlation between neutron-binding energy and anomalous fractionation of isotopes in medical-biological objects. Sov Phys J 29:827–831. https://doi.org/10.1007/BF00900749
74. Horita J (2005) Some perspectives on isotope biosignatures for early life. Chem Geol 218(1–2):171–186. https://doi.org/10.1016/j.chemgeo.2005.01.017
75. Schmidt HL, Robins RJ, Werner RA (2015) Multi-factorial in vivo stable isotope fractionation: causes, correlations, consequences and applications. Isot Environ Health Stud 51(1):155–199. https://doi.org/10.1080/10256016.2015.1014355
76. Schidlowski M (1987) Application of stable carbon isotopes to early biochemical evolution on Earth. Annu Rev Earth Planet Sci 15(1):47–72. https://doi.org/10.1146/annurev.ea.15.050187.000403
77. Galimov EM (2003) Isotopic criteria for identification of organic carbon on Earth and meteorites. Space Sci Rev 106:249–262. https://doi.org/10.1023/A:1024697923188
78. Hupfer M, Herzog C, Schmieder P (2010) Sedimentation and diagenetic transformation of particulate organic phosphorus in a stratified eutrophic lake. Internationale Vereinigung für theoretische und angewandte Limnologie: Verhandlungen 30(9):1389–1392. https://doi.org/10.1080/03680770.2009.11902338
79. Schuck P (2010) Diffusion of the reaction boundary of rapidly interacting macromolecules in sedimentation velocity. Biophys J 98(11):2741–2751. https://doi.org/10.1016/j.bpj.2010.03.004
80. Dance SL, Maxey MR (2003) Particle density stratification in transient sedimentation. Phys Rev E 68(3):031403. https://doi.org/10.1103/PhysRevE.68.031403
81. González AE (2006) Stratification of colloidal aggregation coupled with sedimentation. Phys Rev E 74(6):061403. https://doi.org/10.1103/PhysRevE.74.061403
82. Wadell H (1934) The coefficient of resistance as a function of Reynolds number for solids of various shapes. J Franklin Inst 217(4):459–490. https://doi.org/10.1016/S0016-0032(34)90508-1
83. Hoyal DC, Bursik MI, Atkinson JF (1999) Settling-driven convection: a mechanism of sedimentation from stratified fluids. J Geophys Res Oceans 104(C4):7953–7966. https://doi.org/10.1029/1998JC900065
84. Sutherland BR, Hong YSD (2016) Sedimentation from particle-bearing plumes in a stratified ambient. Phys Rev Fluids 1(7):074302. https://doi.org/10.1103/PhysRevFluids.1.074302
85. Lee H, Fouxon I, Lee C (2019) Sedimentation of a small sphere in stratified fluid. Phys Rev Fluids 4(10):104101. https://doi.org/10.1103/PhysRevFluids.4.104101
86. Blanchette F, Bush JW (2005) Particle concentration evolution and sedimentation-induced instabilities in a stably stratified environment. Phys Fluids 17(7):073302. https://doi.org/10.1063/1.1947987
87. Huppert HE, Turner JS, Hallworth MA (1995) Sedimentation and entrainment in dense layers of suspended particles stirred by an oscillating grid. J Fluid Mech 289:263–293. https://doi.org/10.1017/S0022112095001339
88. Nozieres P (1987) A local coupling between sedimentation and convection: application to the Beenakker-Mazur effect. Physica A 147(1–2):219–237. https://doi.org/10.1016/0378-4371(87)90107-5

89. Bürger R, Diehl S, Mejías C (2018) A difference scheme for a degenerating convection-diffusion-reaction system modelling continuous sedimentation. ESAIM: Math Model Numer Anal 52(2):365–392. https://doi.org/10.1051/m2an/2017038
90. Boscarino S, Bürger R, Mulet P, Russo G, Villada LM (2016) On linearly implicit IMEX Runge-Kutta methods for degenerate convection-diffusion problems modeling polydisperse sedimentation. Bull Brazilian Math Soc New Ser 47:171–185. https://doi.org/10.1007/s00574-016-0130-5
91. Tong P, Ackerson BJ (1998) Analogies between colloidal sedimentation and turbulent convection at high Prandtl numbers. Phys Rev E 58(6):R6931. https://doi.org/10.1103/PhysRevE.58.R6931
92. Parchevsky KV (2001) Numerical simulation of sedimentation in the presence of 2D compressible convection and reconstruction of the particle-radius distribution function. J Eng Math 41:203–220. https://doi.org/10.1023/A:1011944412466
93. Majlesara M, Abouali O, Kamali R (2021) Fully resolved numerical simulation of free convection of falling spherical particles in sedimentation transports using immersed boundary method. Iranian J Sci Technol Trans Mech Eng 45(4):961–976. https://doi.org/10.1007/s40997-020-00348-7
94. Bürger R, Evje S, Karlsen KH (2000) On strongly degenerate convection–diffusion problems modeling sedimentation–consolidation processes. J Math Anal Appl 247(2):517–556. https://doi.org/10.1006/jmaa.2000.6872
95. Noel JF, Schumaker VN (1973) Convection-free boundary sedimentation of dilute protein solutions. Anal Biochem 54(2):586–591. https://doi.org/10.1016/0003-2697(73)90390-4
96. Weon BM, Je JH (2013) Self-pinning by colloids confined at a contact line. Phys Rev Lett 110(2):028303. https://doi.org/10.1103/PhysRevLett.110.028303
97. Mehta S, Bahadur J, Sen D (2023) Understanding the self-pinning driven jamming behavior of colloids in drying droplets. Colloids Surf, A 676:132284. https://doi.org/10.1016/j.colsurfa.2023.132284
98. Yang KC, Wang C, Hu TY, Lin HP, Cho KH, Chen LJ (2020) Self-pinning of silica suspension droplets on hydrophobic surfaces. J Colloid Interface Sci 579:212–220. https://doi.org/10.1016/j.jcis.2020.06.059
99. Rapis EG (2000) Self-assembly of cluster protein films (allotropic nonequilibrium noncrystalline modification) during the process or their condensation. Tech Phys 45:121–131. https://doi.org/10.1134/1.1259582
100. Rapis E (2004) Self-organization and supramolecular chemistry of protein films from the nano-to the macroscale. Tech Phys 49:494–498. https://doi.org/10.1134/1.1736921
101. Golbraikh E, Rapis EG, Moiseev SS (2003) On the crack pattern formation in a freely drying protein film. Tech Phys 48:1333–1337. https://doi.org/10.1134/1.1620131
102. Rapis E (2001) Properties and symmetry of the solid cluster phase of protein. Tech Phys 46:1307–1313. https://doi.org/10.1134/1.1412069
103. Rapis E (2007) On the nonequilibrium phase transition in protein. Tech Phys 52:787–792. https://doi.org/10.1134/S1063784207060199
104. Rapis E (2006) Nonequilibrium state of self-organized protein nanostructures. Tech Phys 51:268–273. https://doi.org/10.1134/S1063784206020198
105. Rapis E (2008) Evolutionary aspect of protein self-organization. Tech Phys 53:783–788. https://doi.org/10.1134/S1063784208060182
106. Rapis E (2005) Relaxation of the energy of the protein colloidal solution arising at drying in open and closed systems. Tech Phys 50:1236–1238. https://doi.org/10.1134/1.2051470
107. Rapis E (2005) On the problem of nucleation (cell formation) in self-organization of protein nanostructures in vitro and in vivo. Tech Phys 50:780–786. https://doi.org/10.1134/1.1947357
108. Orekhov TK, Gradov OV (2023) From desolvation-induced self-organization on the MALDI anchor target chip surfaces to laser-induced self-organization in MALDI techniques: Correlation-spectral analysis and complex wavelet analysis of tesiographic spots on the anchor chips. Invited Paper. Mater Technol Rep 1:124. https://doi.org/10.59400/mtr.v1i1.124

109. Simon B, Pomeau Y (1991) Free and guided convection in evaporating layers of aqueous solutions of sucrose. Transport and sedimentation of solid particles. Phys Fluids A: Fluid Dyn 3(3):380–384. https://doi.org/10.1063/1.858147
110. Kouloulias K, Sergis A, Hardalupas Y (2016) Sedimentation in nanofluids during a natural convection experiment. Int J Heat Mass Transf 101:1193–1203. https://doi.org/10.1016/j.ijheatmasstransfer.2016.05.113
111. Hill WD, Rothfus RR, Li K (1977) Boundary-enhanced sedimentation due to settling convection. Int J Multiph Flow 3(6):561–583. https://doi.org/10.1016/0301-9322(77)90030-1
112. Vinograd J, Bruner R (1966).Band centrifugation of macromolecules in self-generating density gradients. III. Conditions for convection-free band sedimentation. Biopolym Original Res Biomole 4(2):157–170. https://doi.org/10.1002/bip.1966.360040203
113. Carazzo G, Jellinek AM (2013) Particle sedimentation and diffusive convection in volcanic ash-clouds. J Geophys Res Solid Earth 118(4):1420–1437. https://doi.org/10.1002/jgrb.50155
114. Bozhko A, Tynjälä T (2005) Influence of gravitational sedimentation of magnetic particles on ferrofluid convection in experiments and numerical simulations. J Magn Magn Mater 289:281–284. https://doi.org/10.1016/j.jmmm.2004.11.080
115. Tynjälä T, Bozhko A, Bulychev P, Putin G, Sarkomaa P (2006) On features of ferrofluid convection caused by barometrical sedimentation. J Magn Magn Mater 300(1):e195–e198. https://doi.org/10.1016/j.jmmm.2005.10.078
116. Ryazanov VV (2001) Aerosol sedimentation by convection. J Aerosol Sci 32:543–544. https://doi.org/10.1016/S0021-8502(21)00250-0
117. Kirsch AA, Zhulanov UV (1978) Measurement of aerosol penetration through high efficiency filters. J Aerosol Sci 9(4):291–298. https://doi.org/10.1016/0021-8502(78)90031-9
118. Zhulanov IV, Petrianov IV, Sadovskii BF (1978) Photoelectric laser spectrometer for large condensation nuclei. Fizika Atmosfery i Okeana (Akademiia Nauk SSSR, Izvestiia) 14:520–526 (in Russian)
119. Zhulanov I (1985) An investigation of the size spectrum transformation of submicron mist particles. Fizika Atmosfery i Okeana (Akademiia Nauk SSSR, Izvestiia) 21:629–637 (in Russian)
120. Julanov YV, Lushnikov AA, Zagaynov VA (2002) Diffusion aerosol spectrometer. Atmos Res 62(3–4):295–302. https://doi.org/10.1016/S0169-8095(02)00015-7
121. Ankilov A, Baklanov A, Colhoun M, Enderle KH, Gras J, Julanov Y, Kaller D, Lindner A, Lushnikov AA, Mavliev R, McGovern F (2002) Intercomparison of number concentration measurements by various aerosol particle counters. Atmos Res 62(3–4):177–207. https://doi.org/10.1016/S0169-8095(02)00010-8
122. Ankilov A, Baklanov A, Colhoun M, Enderle KH, Gras J, Julanov Y, Kaller D, Lindner A, Lushnikov AA, Mavliev R, McGovern F (2002) Particle size dependent response of aerosol counters. Atmos Res 62(3–4):209–237. https://doi.org/10.1016/S0169-8095(02)00011-X
123. Dubtsov S, Ovchinnikova T, Valiulin S, Chen X, Manninen HE, Aalto PP, Petäjä T (2017) Laboratory verification of aerosol diffusion spectrometer and the application to ambient measurements of new particle formation. J Aerosol Sci 105:10–23. https://doi.org/10.1016/j.jaerosci.2016.10.015
124. Zhulanov Y, Makaveev P, Gradov O (2018) Towards the aerosol cytometry and hydrosol cytometry based on laser aerosol spectrometers. IEEE DataPort (Datasets). https://ieee-dataport.org/documents/towards-aerosol-cytometry-and-hydrosol-cytometry-based-laser-aerosol-spectrometers, https://doi.org/10.21227/njak-qr29
125. Gradov O, Zhulanov Y, Makaveev P (2020) Optical ultrastructural virometry and its limitations. Photonics Russia 14(6):542–549. https://doi.org/10.22184/1993-7296.FRos.2020.14.6.542.549
126. Gradov O, Zhulanov Y, Makaveev P (2021) Optical ultrastructural virometry using optoelectronic aerosol counters and laser aerosol spectrometers. Is it possible to pose the problem correctly? Nanosyst Nanomater Nanotechnol 19(3):487–512. https://doi.org/10.15407/nnn.19.03.487

127. Gradov OV, Zhulanov YuV, Makaveev PYu, Marnautov NA (2022) [Integration of lensless holographic microscopy and optical aerosol particle counting for the analysis of spores on a chip and in projection aerodynamic schlieren system: from identification to monitoring of species]. Curr Mycol Russia 9(2):109–110. https://doi.org/10.14427/cmr.2022.ix.04
128. Hofemeier P, Sznitman J (2015) Revisiting pulmonary acinar particle transport: convection, sedimentation, diffusion, and their interplay. J Appl Physiol 118(11):1375–1385. https://doi.org/10.1152/japplphysiol.01117.2014
129. Hocking MB, O'Brien RN (1987) Importance of convection to the enhancement of erythrocyte sedimentation rates in inclined tubes. Biorheology 24(5):473–482. https://doi.org/10.3233/bir-1987-24505
130. Gradov OV, Krukowskikh VV, Nasirov PA, Jablokov AG (2018) Multi-axis goniometric 3d-visualization of vector diagrams of optical characteristics of dispersed and biological structures on a chip using different laser scanning regimes and trajectories. Photonics Russia 4(72):448–458. https://doi.org/10.22184/1993-7296.2018.72.4.448.458
131. Gan H, Chang J, Feng JJ, Hu HH (2003) Direct numerical simulation of the sedimentation of solid particles with thermal convection. J Fluid Mech 481:385–411. https://doi.org/10.1017/S0022112003003938
132. Hadji L, DarAssi M (2014) Influence of sedimentation on the threshold for Soret-driven convection in colloidal suspensions. Phys Rev E 89(1):013014. https://doi.org/10.1103/PhysRevE.89.013014
133. DarAssi M, Hadji L (2014) Analysis of the interplay between sedimentation and thermophoresis in the presence of convection in colloidal suspensions. Proc ASME (American Society of Mechanical Engineers) 46247:V01DT32A001. https://doi.org/10.1115/FEDSM2014-21078
134. Wu M, Ludwig A (2009) Modeling equiaxed solidification with melt convection and grain sedimentation—I: model description. Acta Mater 57(19):5621–5631. https://doi.org/10.1016/j.actamat.2009.07.056
135. Kegeles G (1969) Convection induced by hydrostatic pressure in sedimentation velocity experiments. Biopolym Original Res Biomole 7(1):83–86. https://doi.org/10.1002/bip.1969.360070108
136. Cardoso SSS, Woods AW (1996) On convection and mixing driven by sedimentation. J Fluid Mech 285:165–180. https://doi.org/10.1017/S0022112095000498
137. Chen F, Yu L, Shen Q, Zhang L (2008) Fabrication of TZ-3Y20A/Mo multilayer composites by particle sedimentation method. J Mater Process Technol 199(1–3):37–40. https://doi.org/10.1016/j.jmatprotec.2007.07.042
138. Pourmajidian M, Akhlaghi F (2014) Fabrication and characterization of functionally graded Al/SiC p composites produced by remelting and sedimentation process. J Mater Eng Perform 23:444–450. https://doi.org/10.1007/s11665-013-0798-2
139. Wu M, Fjeld A, Ludwig A (2010) Modelling mixed columnar-equiaxed solidification with melt convection and grain sedimentation–Part I: Model description. Comput Mater Sci 50(1):32–42. https://doi.org/10.1016/j.commatsci.2010.07.005
140. Wu M, Ludwig A, Fjeld A (2010) Modelling mixed columnar-equiaxed solidification with melt convection and grain sedimentation–Part II: Illustrative modelling results and parameter studies. Comput Mater Sci 50(1):43–58. https://doi.org/10.1016/j.commatsci.2010.07.006
141. Höink T, Schmalzl J, Hansen U (2005) Formation of compositional structures by sedimentation in vigorous convection. Phys Earth Planet Inter 153(1–3):11–20. https://doi.org/10.1016/j.pepi.2005.03.014
142. Pattabhiraman H, Dijkstra M (2017) Periodic layers of a dodecagonal quasicrystal and a floating hexagonal crystal in sedimentation-diffusion equilibria of colloids. J Chem Phys 147(10):104902. https://doi.org/10.1063/1.4993521
143. Hassanzadeh A, Seyedian M, Rouhani H, Farasati M (2020) Evaluation of fractal properties of different layers sedimentation behind the check dam. Iranian J Watershed Manag Sci Eng 14(49):59–69. http://dorl.net/dor/20.1001.1.20089554.1399.14.49.10.9

144. More R, Ardekani M, Brandt L, Ardekani A (2020) Sedimentation of spheroidal particles in density stratified fluid (APS Division of Fluid Dynamics Meeting Abstracts). Bull Am Phys Soc 65(13), ID: Y05.00008. https://meetings.aps.org/Meeting/DFD20/Session/Y05.8
145. Mercier M, Pemeja J, Ern P (2014) Sedimentation dynamics of disks in a linearly stratified fluid (APS Division of Fluid Dynamics Meeting Abstracts). Bull Am Phys Soc 59(20), ID: BAPS.2014.DFD.R17.4. https://meetings.aps.org/Meeting/DFD14/Event/229319
146. Khatri S, Arnosti C, Camassa R, Falcon C, He X, McLaughlin R, Praine J, White B, Yu S, Ziervogel K (2012) Sedimentation of porous and solid particles in stratified fluids (APS Division of Fluid Dynamics Meeting Abstracts). Bull Am Phys Soc 57(17), ID: BAPS.2012.DFD.E1.3. https://meetings.aps.org/Meeting/DFD12/Session/E1.3
147. Mitsui T, Harada S, Asakura K (2007) Sedimentation of a stratified suspension in a quiescent fluid (Fluids Engineering Division Summer Meeting). Proc ASME (American Society of Mechanical Engineers) 42886:379–383. https://doi.org/10.1115/FEDSM2007-37149
148. Gómez DC, Bergougnoux L, Hinch J, Guazzelli É (2007) On stratification control of the velocity fluctuations in sedimentation. Phys Fluids 19(9):098102. https://doi.org/10.1063/1.2772902
149. Sutherland B, Hong YS (2015) Sedimentation from Particle-Laden Plumes in Stratified Fluid (APS Division of Fluid Dynamics Meeting Abstracts—2015). Bull Am Phys Soc 60(21), ID: BAPS.2015.DFD.A11.5. https://meetings.aps.org/Meeting/DFD15/Session/A11.5
150. Hung CY, Niu YY, Chou YJ (2020) Numerical study of double-diffusive sedimentation in thermally stratified fluid. J Fluid Mech 893:A27. https://doi.org/10.1017/jfm.2020.223
151. Philipse AP, Koenderink GH (2003) Sedimentation–diffusion profiles and layered sedimentation of charged colloids at low ionic strength. Adv Coll Interface Sci 100:613–639. https://doi.org/10.1016/S0001-8686(02)00078-7
152. Hu R, Wang Y, Zou Y, Chen X, Liu S, Luo X (2013) Study on phosphor sedimentation effect in white light-emitting diode packages by modeling multi-layer phosphors with the modified Kubelka-Munk theory. J Appl Phys 113(6):063108. https://doi.org/10.1063/1.4792051
153. Schleusener SA, Read AA (1967) Small particle sizing using turbidimetric layered sedimentation. Rev Sci Instrum 38(8):1152–1156. https://doi.org/10.1063/1.1721040
154. Schleusener SA (1969) Automatic high-speed particle sizing using a gas laser and nuclear counter. Powder Technol 2(6):365–367. https://doi.org/10.1016/0032-5910(69)80030-6
155. Zhulanov YV, Sadovskii BF, Petryanov IV (1975) Use of a laser resonator for raising the sensitivity of laser-type aerosol spectrometers. Soviet Phys Doklady 20(6):437
156. Zhulanov Y, Sadovskii B, Nevskii I, Petryanov I (1977) Laser spectrometer for submicronic aerosols. Colloid J USSR 39(6):936–941
157. Zhulanov YV, Nikitin O, Saprykin K, Nevskii I, Sedovskii B, Petryanov I (1983) Laser particle counter. Instrum Exp Tech 26(3):686–690
158. Gebhart J (1988) Laser submicronic particle-counter for gases-technical status and perspectives. Chem Ing Tec 60(3):225
159. Anderson SR, JA (1971) Automatic particle sizing using a gas laser and layered sedimentation. New Mexico State University. Proquest page short https://gclnk.com/eTHb1eqM
160. Fassett CI, Head III JW (2007) Layered mantling deposits in northeast Arabia Terra, Mars: Noachian-Hesperian sedimentation, erosion, and terrain inversion. J Geophys Res Planets 112(E8):E08002.1–E08002.19. https://doi.org/10.1029/2006JE002875
161. Dueck JG, Purevjav D, Kilimnik DY, A contribution to the theory of porosity of fine-grained sediments. J Eng Phys Thermophys 77(1):93–102. https://doi.org/10.1023/B:JOEP.0000020722.04941.c5
162. Shiryaeva EV, Zhukov MY (2018) On the tangential stresses at the boundary between the layers for two-layer sedimentation models. Polarforschung 87(2):211–214. https://doi.org/10.2312/polarforschung.87.2.211
163. Docherty P (1992) Solving for the thickness and velocity of the weathering layer using 2-D refraction tomography. Geophysics 57(10):1307–1318. https://doi.org/10.1190/1.1443198
164. Brueckner WD (1953) Cyclic calcareous sedimentation as an index of climatic variations in the past. J Sediment Res 23(4):235–237. https://doi.org/10.1306/D4269620-2B26-11D7-8648000102C1865D

165. Nott JF, Young RW, Idnurm M (1991) Sedimentology, weathering, age and geomorphological significance of Tertiary sediments on the far south coast of New South Wales. Aust J Earth Sci 38(3):357–373. https://doi.org/10.1080/08120099108727978
166. Weller JM (1930) Cyclical sedimentation of the Pennsylvanian period and its significance. J Geol 38(2):97–135. https://doi.org/10.1086/623695
167. Wells AJ (1960) Cyclic sedimentation: a review. Geol Mag 97(5):389–403. https://doi.org/10.1017/S0016756800061744
168. Sarin DD (1962) Cyclic sedimentation of primary dolomite and limestone. J Sediment Res 32(3):451–471. https://doi.org/10.1306/74D70CEA-2B21-11D7-8648000102C1865D
169. Beerbower JR (1964) Cyclothems and cyclic depositional mechanisms in alluvial plain sedimentation. Kansas Geol Surv Bull 169(1):31–32. https://www.kgs.ku.edu/Publications/Bulletins/169/Beerbower/
170. Wanless HR (1964) Local and regional factors in Pennsylvanian cyclic sedimentation. Kansas Geol Surv Bull 169:593–606. https://www.kgs.ku.edu/Publications/Bulletins/169/Wanless/
171. De Raaf JFM, Reading HG, Walker RG (1965) Cyclic sedimentation in the Lower Westphalian of no Devon, England. Sedimentology 4(1–2):1–52. https://doi.org/10.1111/j.1365-3091.1965.tb01282.x
172. Wilson JL (1967) Cyclic and reciprocal sedimentation in Virgilian strata of southern New Mexico. Geol Soc Am Bull 78(7):805–818. https://doi.org/10.1130/0016-7606(1967)78[805:CARSIV]2.0.CO;2
173. Duff PMD (1967) Cyclic sedimentation in the Permian coal measures of New South Wales. J Geol Soc Aust 14(2):293–307. https://doi.org/10.1080/00167616708728667
174. Carr TR (1982) Log-linear models, Markov chains and cyclic sedimentation. J Sediment Res 52(3):905–912. https://doi.org/10.1306/212F808A-2B24-11D7-8648000102C1865D
175. Oschmann W (1988) Kimmeridge Clay sedimentation—a new cyclic model. Palaeogeogr Palaeoclimatol Palaeoecol 65(3–4):217–251. https://doi.org/10.1016/0031-0182(88)90025-9
176. Van Buchem FSP, McCave IN (1989) Cyclic sedimentation patterns in Lower Lias mudstones of Yorkshire (GB). Terra Nova 1(5):461–467. https://doi.org/10.1111/j.1365-3121.1989.tb00411.x
177. Humphrey NF, Heller PL (1995) Natural oscillations in coupled geomorphic systems: an alternative origin for cyclic sedimentation. Geology 23(6):499–502. https://doi.org/10.1130/0091-7613(1995)023%3C0499:NOICGS%3E2.3.CO;2
178. Weedon GP, Read WA (1995) Orbital-climatic forcing of Namurian cyclic sedimentation from spectral analysis of the limestone coal formation, central Scotland. Geol Soc Lond Special Publ 85(1):51–66. https://doi.org/10.1144/GSL.SP.1995.085.01.04
179. Rouchy JM, Noel D, Wali AMA, Aref MAM (1995) Evaporitic and biosiliceous cyclic sedimentation in the Miocene of the Gulf of Suez—depositional and diagenetic aspects. Sed Geol 94(3–4):277–297. https://doi.org/10.1016/0037-0738(94)00095-C
180. Trichet J, Défarge C, Tribble J, Tribble G, Sansone F (2001) Christmas Island lagoonal lakes, models for the deposition of carbonate–evaporite–organic laminated sediments. Sed Geol 140(1–2):177–189. https://doi.org/10.1016/S0037-0738(00)00177-9
181. Kim W, Paola C (2007) Long-period cyclic sedimentation with constant tectonic forcing in an experimental relay ramp. Geology 35(4):331–334. https://doi.org/10.1130/G23194A.1
182. Auler AS, Smart PL, Wang X, Piló LB, Edwards RL, Cheng H (2009) Cyclic sedimentation in Brazilian caves: mechanisms and palaeoenvironmental significance. Geomorphology 106(1–2):142–153. https://doi.org/10.1016/j.geomorph.2008.09.020
183. Palma RM, Kietzmann DA, Bressan GS, Martín-Chivelet J, López-Gómez J, Farias ME, Llanos MPI (2013) Peritidal cyclic sedimentation from La Manga Formation (Oxfordian), Neuquén Basin, Mendoza, Argentina. J S Am Earth Sci 47:1–11. https://doi.org/10.1016/j.jsames.2013.05.008
184. Niebuhr B, Prokoph A (1997) Periodic-cyclic and chaotic successions of Upper Cretaceous (Cenomanian to Campanian) pelagic sediments in the North German Basin. Cretac Res 18(5):731–750. https://doi.org/10.1006/cres.1997.0083

185. Sivakumar B (2002) A phase-space reconstruction approach to prediction of suspended sediment concentration in rivers. J Hydrol 258(1–4):149–162. https://doi.org/10.1016/S0022-1694(01)00573-X
186. Sivakumar B, Jayawardena AW (2002) An investigation of the presence of low-dimensional chaotic behaviour in the sediment transport phenomenon. Hydrol Sci J 47(3):405–416. https://doi.org/10.1080/02626660209492943
187. Aidun CK, Ding EJ (2003) Dynamics of particle sedimentation in a vertical channel: period-doubling bifurcation and chaotic state. Phys Fluids 15(6):1612–1621. https://doi.org/10.1063/1.1571825
188. Vaalgamaa S, Korhola A (2004) Searching for order in chaos: a sediment stratigraphical study of a multiple-impacted bay of the Baltic Sea. Estuar Coast Shelf Sci 59(2):319–332. https://doi.org/10.1016/j.ecss.2003.09.005
189. Sivakumar B (2004) Chaos theory in geophysics: past, present and future. Chaos, Solitons Fractals 19(2):441–462. https://doi.org/10.1016/S0960-0779(03)00055-9
190. Shang P, Kamae S (2005) Fractal nature of time series in the sediment transport phenomenon. Chaos, Solitons Fractals 26(3):997–1007. https://doi.org/10.1016/j.chaos.2005.01.051
191. Angilella JR (2008) Chaotic particle sedimentation in a rotating flow with time-periodic strength. Phys Rev E 78(6):066310. https://doi.org/10.1103/PhysRevE.78.066310
192. Pemberton SG, MacEachern JA, Gingras MK, Saunders TD (2008) Biogenic chaos: cryptobioturbation and the work of sedimentologically friendly organisms. Palaeogeogr Palaeoclimatol Palaeoecol 270(3–4):273–279. https://doi.org/10.1016/j.palaeo.2008.01.036
193. Shang P, Na X, Kamae S (2009) Chaotic analysis of time series in the sediment transport phenomenon. Chaos, Solitons Fractals 41(1):368–379. https://doi.org/10.1016/j.chaos.2008.01.014
194. Verjus R, Guillou S, Ezersky A, Angilella JR (2016) Chaotic sedimentation of particle pairs in a vertical channel at low Reynolds number: multiple states and routes to chaos. Phys Fluids 28(12):123303. https://doi.org/10.1063/1.4968559
195. Hemar Y, Horne DS (1998) Electrostatic interactions in adsorbed protein layers probed by a sedimentation technique. J Colloid Interface Sci 206(1):138–145. https://doi.org/10.1006/jcis.1998.5711
196. Vecchio F, Schettini F, Di Francesco L, Meloni T, Russino G (1966) Electrophoretic studies of erythrocyte giucose-6-phosphate dehydrogenase in normal and enzyme-deficient sardinian subjects. Acta Haematol 35(1):46–52. https://doi.org/10.1159/000209107
197. Schettini F (1969) Diagnosis of thalassemia major in the first months of life. Ann N Y Acad Sci 165(1):387–393. https://doi.org/10.1111/j.1749-6632.1969.tb27808.x
198. Schettini F, di Bitonto G, Mautone A (1971) Rate of synthesis of haemoglobin A 2 in thalassaemia. Z Kinderheilkd 109:271–278. https://doi.org/10.1007/BF00438970
199. Meloni T, Cagnazzo G, Dore A, Cutillo S (1973) Phenobarbital for prevention of hyperbilirubinemia in glucose-6-phosphate dehydrogenase-deficient newborn infants. J Pediatr 82(6):1048–1051. https://doi.org/10.1016/S0022-3476(73)80442-1
200. Cutillo S, Meloni T, Dore A (1974) Effect of orotic acid upon serum bilirubin in newborn infants with erythrocyte G-6-PD deficiency. Acta Pædiatrica 63(1):143–146. https://doi.org/10.1111/j.1651-2227.1974.tb04363.x
201. Meloni T, Costa S, Cutillo S (1975) Haptoglobin, hemopexin, hemoglobin and hematocrit in newborns with erythrocyte glucose-6-phosphate dehydrogenase deficiency. Acta Haematol 54(5):284–288. https://doi.org/10.1159/000208087
202. Meloni T, Costa S, Cutillo S (1976) Three years experience in preventing severe hyperbilirubinemia in newborn infants with erythrocyte G-6-PD deficiency. Neonatology 28(5–6):370–374. https://doi.org/10.1159/000240839
203. Meloni T, Pilo G, Gallisai D, Dore A (1979) Serum glutamic oxalacetic transaminase, glutamic pyruvic transaminase, gamma-glutamyl transpeptidase and glutamic dehydrogenase levels in favism. Acta Haematol 62(2):71–73. https://doi.org/10.1159/000207541
204. Stephens JG (1938) Stratified blood sedimentation-isolation of immature red cells. Nature 141(3580):1058. https://doi.org/10.1038/1411058a0

205. Onyilagha JC, Omenyi SN, Snyder RS (1989) Application of some droplet sedimentation theories to layered erythrocyte suspensions. Cell Biophys 14:217–230. https://doi.org/10.1007/BF02797269
206. Galvez ME, Beyssac O, Martinez I, Benzerara K, Chaduteau C, Malvoisin B, Malavieille J (2013) Graphite formation by carbonate reduction during subduction. Nat Geosci 6(6):473–477. https://doi.org/10.1038/ngeo1827
207. Starchenko VY, Bulavin LA, Boiko YP, Moraru VN, Lebovka NI (2005) Conductometric and gravimetric studies of the kinetics of graphite sedimentation in aqueous dispersions. Colloid J 67:755–759. https://doi.org/10.1007/s10595-005-0160-y
208. Wang X, Bai H, Shi G (2011) Size fractionation of graphene oxide sheets by pH-assisted selective sedimentation. J Am Chem Soc 133(16):6338–6342. https://doi.org/10.1021/ja200218y
209. Ding G, Zhang N, Wang C, Li X, Zhang J, Li W, Li R, Yang Z (2018) Effect of the size on the aggregation and sedimentation of graphene oxide in seawaters with different salinities. J Nanopart Res 20:1–10. https://doi.org/10.1007/s11051-018-4421-1
210. Feng Y, Liu X, Huynh KA, McCaffery JM, Mao L, Gao S, Chen KL (2017) Heteroaggregation of graphene oxide with nanometer-and micrometer-sized hematite colloids: influence on nanohybrid aggregation and microparticle sedimentation. Environ Sci Technol 51(12):6821–6828. https://doi.org/10.1021/acs.est.7b00132
211. Feng Y, Huynh KA, Xie Z, Liu G, Gao S (2019) Heteroaggregation and sedimentation of graphene oxide with hematite colloids: influence of water constituents and impact on tetracycline adsorption. Sci Total Environ 647:708–715. https://doi.org/10.1016/j.scitotenv.2018.08.046
212. Traverse JP, Roux P, Bressolles JC, Snoeck E, Roucau C (1993) $YBa_2Cu_3O_7$-x/Ag composite films layered on stainless steel by sedimentation. Supercond Sci Technol 6(8):573. https://doi.org/10.1088/0953-2048/6/8/003
213. Kozur A, Burk L, Thomann R, Lutz PJ, Mülhaupt R (2019) Graphene oxide grafted with polyoxazoline as thermoresponsive support for facile catalyst recycling by reversible thermal switching between dispersion and sedimentation. Polymer 178:121553. https://doi.org/10.1016/j.polymer.2019.121553
214. Li Y, Zhang M, Hao R, Wang H, He X, Du T, Yang X, Wang Y (2019) Influence of (photo) bromination on the transformation, aggregation and sedimentation of graphene oxide. Chem Eng J 355:487–497. https://doi.org/10.1016/j.cej.2018.08.199
215. Zhang X, Shi Z, Song B, Feng J, Niu Y (2021) Reduction, structures, and electrical properties of multi-layer reduced graphene oxide papers by natural sedimentation. Chem Phys Lett 785:139141. https://doi.org/10.1016/j.cplett.2021.139141
216. Lakshmanan P, Periyasamy S, Dhairiyasamy R (2023) Analysis of heat transfer performance and thermo-hydraulic characteristics of graphene nanofluids: impact of sedimentation effects. Mater Res Express 10(9):095004. https://doi.org/10.1088/2053-1591/acf7b2
217. Song Z, Lin ES, Abid HA, Ong JW, Ng TW (2023) Gas bubble formation of filaments from sedimented graphene oxide. Mater Chem Phys 309:128325. https://doi.org/10.1016/j.matchemphys.2023.128325
218. Babakhani P, Bridge J, Phenrat T, Doong RA, Whittle KR (2018) Aggregation and sedimentation of shattered graphene oxide nanoparticles in dynamic environments: a solid-body rotational approach. Environ Sci Nano 5(8):1859–1872. https://doi.org/10.1039/C8EN00443A
219. Huang JJ, Yuan YJ (2016) A sedimentation study of graphene oxide in aqueous solution using gradient differential centrifugation. Phys Chem Chem Phys 18(17):12312–12322. https://doi.org/10.1039/C6CP00167J
220. Dada OJ (2020) Free standing porous graphene and reduced graphene papers manufactured from sedimentation of arc discharge of graphite rods. In: Electrochemical society meeting abstracts prime, vol 7, pp 1088–1088. https://doi.org/10.1149/MA2020-0271088mtgabs

221. Kumar S, Sehgal R, Wani MF, Sharma MD (2023) Magnetorheological study of core– shell structured carbonyl iron/2-D graphene oxide microparticles suspensions with improved sedimentation stability. J Magn Magn Mater 579:170852. https://doi.org/10.1016/j.jmmm.2023.170852
222. Wang G, Ma Y, Tong Y, Dong X (2017) Development of manganese ferrite/graphene oxide nanocomposites for magnetorheological fluid with enhanced sedimentation stability. J Ind Eng Chem 48:142–150. https://doi.org/10.1016/j.jiec.2016.12.032
223. Moharramzadeh S, Baghdadi M (2016) In situ sludge magnetic impregnation (ISSMI) as an efficient technology for enhancement of sludge sedimentation: Removal of methylene blue using nitric acid treated graphene oxide as a test process. J Environ Chem Eng 4(2):2090–2102. https://doi.org/10.1016/j.jece.2016.03.039
224. Chuprov PN, Zaitsev EV, Buslaeva EY, Gubin SP (2018) Electrophoretic sedimentation of graphene oxide on cylindrical surface of microwires. Radioelectronics Nanosystems Inf Technol RENSIT 10(1):65–70
225. Chumakova N, Gradov O, Marnautov N, Matveev M (2023) Orientational ordering of graphene oxide membranes according to the data of spin probe technique and scanning electron microscopy. In: Proceedings of V International scientific-practical conference graphene and related structures: synthesis, production, and application, pp 212–216
226. Marnautov NA, Matveev MV, Gulin AA, Kálai T, Bognár B, Rebrikova AT, Chumakova NA (2024) Orientational ordering of graphene oxide membranes by a spin probe technique and SEM image analysis. J Phys Chem C 128(6):2543–2550. https://doi.org/10.1021/acs.jpcc.3c07127
227. Goncharova AA, Fischenko VK (2011) QAVIS—program for quick image and video analysis. In: Proceedings of 8th open German-Russian workshop on pattern recognition and image understanding (OGRW-8 2011), p 85
228. Goncharova AA, Fischenko VK, Dubina VA (2012) Use the express-analysis program QAVIS for the satellite monitoring. Curr Problems Re-Mote Sens Earth Space 9(3):293–298
229. Fishchenko VK, Dolgikh GI, Zimin PS, Subote AE (2018) Some results of oceanological video monitoring. Dokl Earth Sci 482:1244–1247. https://doi.org/10.1134/S1028334X18090283
230. Dolgikh GI, Fishchenko VK, Goncharova AA (2019) Potential for recording of waves and sea level fluctuations in the world ocean coastal areas by internet video analysis. Dokl Earth Sci 488:1264–1267. https://doi.org/10.1134/S1028334X19100209
231. Fischenko VK, Goncharova AA, Dolgikh GI, Zimin PS, Subote AE, Klescheva NA, Golik AV (2021) Express image and video analysis technology QAVIS: application in system for video monitoring of Peter the Great Bay (Sea of Japan/East Sea). J Mar Sci Eng 9(10):1073. https://doi.org/10.3390/jmse9101073
232. Fischenko V, Mitnik L, Dolgikh G, Goncharova A, Zimin P, Klescheva N, Golik A (2022) QAVIS technology: Measuring wave processes in coastal zones based on the analysis of internet video broadcast. In: IGARSS 2022–2022 IEEE international geoscience and remote sensing symposium. IEEE, pp 6801–6804
233. Gradov OV (2016) Can graphene bilayers be the membrane mimetic materials? Radioelectronics Nanosystems Inf Technol 8(1):25–38. https://doi.org/10.17725/rensit.2016.08.025
234. Gradov OV, Gradova MA (2016) Can graphene bilayers be the membrane mimetic materials? ion channels in graphene-based nanostructures. Radioelectronics Nanosystems Inf Technol 8(2):154–170. https://doi.org/10.17725/rensit.2016.08.154
235. Gradov OV (2018) Can graphene bilayers be the membrane mimetic materials? arXiv:1808.00109. https://doi.org/10.48550/arXiv.1808.00109
236. Aixalà-Perelló A, Pedico A, Laurenti M, Fontananova E, Bocchini S, Ferrari IV, Lamberti A (2023) Scalable and highly selective graphene-based ion-exchange membranes with tunable permselectivity. NPJ 2D Mater Appl 7(1):46. https://doi.org/10.1038/s41699-023-00399-9
237. Nair RR, Wu HA, Jayaram PN, Grigorieva IV, Geim AK (2012) Unimpeded permeation of water through helium-leak–tight graphene-based membranes. Science 335(6067):442–444. https://doi.org/10.1126/science.1211694

238. Joshi RK, Carbone P, Wang FC, Kravets VG, Su Y, Grigorieva IV, Wu HA, Geim AK, Nair RR (2014) Precise and ultrafast molecular sieving through graphene oxide membranes. Science 343(6172):752–754. https://doi.org/10.1126/science.1245711
239. Abraham J, Vasu KS, Williams CD, Gopinadhan K, Su Y, Cherian CT, Dix J, Prestat E, Haigh SJ, Grigorieva IV, Carbone P (2017) Tunable sieving of ions using graphene oxide membranes. Nat Nanotechnol 12(6):546–550. https://doi.org/10.1038/nnano.2017.21
240. Ma J, Ping D, Dong X (2017) Recent developments of graphene oxide-based membranes: a review. Membranes 7(3):52. https://doi.org/10.3390/membranes7030052
241. Tsou CH, An QF, Lo SC, De Guzman M, Hung WS, Hu CC, Lee KR, Lai JY (2015) Effect of microstructure of graphene oxide fabricated through different self-assembly techniques on 1-butanol dehydration. J Membr Sci 477:93–100. https://doi.org/10.1016/j.memsci.2014.12.039
242. Akbari A, Sheath P, Martin ST, Shinde DB, Shaibani M, Banerjee PC, Tkacz R, Bhattacharyya D, Majumder M (2016) Large-area graphene-based nanofiltration membranes by shear alignment of discotic nematic liquid crystals of graphene oxide. Nat Commun 7(1):10891. https://doi.org/10.1038/ncomms10891
243. Huang H, Song Z, Wei N, Shi L, Mao Y, Ying Y, Sun L, Xu Z, Peng X (2013) Ultrafast viscous water flow through nanostrand-channelled graphene oxide membranes. Nat Commun 4(1):2979. https://doi.org/10.1038/ncomms3979
244. Marnautov NA (2024) LayerNet Segmenter, version 1.0.0. https://colab.research.google.com/drive/1UcfWkLQdwx8XpMW1X4i2CoFAAep_ftNI?usp=sharin.
245. Marnautov NA (2024) OrientLayer Analyser, version 1.0.0. https://colab.research.google.com/drive/1yl9UGMGGjR3cFTzAqibZsER3trWyC4h_?usp=sharing
246. Holmes SA, Featherstone WE (2002) A unified approach to the Clenshaw summation and the recursive computation of very high degree and order normalised associated Legendre functions. J Geodesy 76:279–299. https://doi.org/10.1007/s00190-002-0216-2
247. Dimitrakopoulos R, Yao L (2020).High-order spatial stochastic models. In: Daya Sagar BS, Cheng Q, McKinley J, Agterberg F (eds) Encyclopedia of mathematical geosciences. Springer, Cham, pp 1–10. https://doi.org/10.1007/978-3-030-26050-7_16-1
248. Yao L, Dimitrakopoulos R, Gamache M (2020) High-order sequential simulation via statistical learning in reproducing kernel Hilbert space. Math Geosci 52(5):693–723. https://doi.org/10.1007/s11004-019-09843-3
249. Fathy A, Ghamry E (2021) A two-dimensional lithospheric magnetic anomaly field model of Egypt using the measurements from Swarm satellites. Geodesy Geodyn 12(3):229–238. https://doi.org/10.1016/j.geog.2021.03.004
250. Vorobiev AK, Yankova TS, Chumakova NA (2012) Orientation distribution function and order parameters of oriented spin probe as determined by EPR spectroscopy. Chem Phys 409:61–73. https://doi.org/10.1016/j.chemphys.2012.10.006
251. Chumakova NA, Rebrikova AT, Talyzin AV, Paramonov NA, Vorobiev AK, Korobov MV (2018) Properties of graphite oxide powders and membranes as revealed by electron paramagnetic resonance spectroscopy. J Phys Chem C 122(39):22750–22759. https://doi.org/10.1021/acs.jpcc.8b07221
252. Chumakova NA, Kalai T, Rebrikova AT, Sár C, Kokorin AI (2022) Novel orientation-sensitive spin probes for graphene oxide membranes study. Membranes 12(12):1241. https://doi.org/10.3390/membranes12121241
253. Chumakova NA, Kh VA, Astvatsaturov DA, Yankova TS, Rebrikova AT, Matveev MV (2022) Quantitative determination of the orientational ordering of graphene oxide membranes by the spin probe technique. Phys Chem Chem Phys 24(8):4881–4890. https://doi.org/10.1039/D1CP05417A
254. Astvatsaturov DA, Kokorin AI, Melnikov MY, Chumakova NA (2023) Spin exchange between paramagnetic probes inside graphite oxide. Chem Phys Lett 833:140946. https://doi.org/10.1016/j.cplett.2023.140946

Chapter 10
Uncovering an Overlooked Natural Frequency in Free Vibrations of a Plate with Attached Resonators

Isaac Elishakoff and Matthieu Zawadzki

Abstract In this paper, we study the free vibrations of the plate with metamaterial, i.e., with an arbitrary number of masses and springs. The free vibration frequencies and the mode shapes are determined as a prelude of studying forced and random vibrations. The comparison is made with previous studies devoted to a plate with a single mass, or with five resonators. Numerical results are reported for the plate supported with numerous resonators.

10.1 Introduction

Liao et al. [1] state that "Metamaterials, which are ingeniously designed artificial structures, are characterized by supernormal physical properties that natural materials do not possess. In the field of electromagnetism, the concept of meta-materials was first theoretically proposed by Veselago in 1968 [2]. Acoustic metamaterials (AMs), one of the most significant embranchments of metamaterials, composed of a novel type of artificial periodic superatom with subwavelength thickness, were first recognized and presented by Liu et al. [3, 4]." Since then, several books have been written. The interested reader can consult the books by Romero-Garcia and Hladky-Hennion [5], Jiménez et al. [6], Gao and Ding [7], and Hu and Zhou [8]. Numerous review papers have been composed by several authors, namely by Fok et al. [9], Chen et al. [10], Zangeneh-Nejad and Fleury [11], Kumar and Pueh Lee [12], Liu et al. [13], Barchiesi et al. [14], Suresh Kumar et al. [15], Liao et al. [1], Song et al. [16], and Dalela et al. [17].

I. Elishakoff (✉)
Department of Ocean and Mechanical Engineering, Florida Atlantic University, Boca Raton, FL 33431-0991, USA
e-mail: elishako@fau.edu

M. Zawadzki
SIGMA Clermont, 27 Rue Roche Genès, 63170 Aubière, France

H. Altenbach and V. A. Eremeyev (eds.), *Dynamics of Discrete and Continuum Structures and Media*, Advanced Structured Materials 221,
https://doi.org/10.1007/978-3-031-75626-9_10

It should be stressed that the study of the rectangular plate with concentrated masses has a long history. The research was apparently originated by Dana Young [18]. It was followed by works by Avalos et al. [19], Avalos and Larrondo [20], Wu and Luo [21], Laura et al. [22], Avalos et al. [23], Wu et al. [24], Wu and Luo [26, 28], Wu and Huang [28], Amabili [29], Amabili and Cara [30], and Amabili et al. [31]. Papers [18–28] dealt with the linear problem, whereas Amabili and his collaborators [29–31] considered the problem in a more realistic, nonlinear setting, with attendant experiments [30].

This study is a revisit of the problem in simpler, linear setting to grasp the problem, as the necessary prelude to the nonlinear treatment that is underway, and will be submitted elsewhere.

10.2 Formulation of the Problem

This article deals with a simply supported at its boundaries plate coupled with a multi-mass system. This multi-mass system is composed by N mass-spring resonators. The configuration of the plate and the mass-spring system is represented in Fig. 10.1. The length, width, and thickness of the plate are denoted as a, b, and h, respectively.

There are N resonators on the plate. Fig. 10.2 represents one resonator, where m_j represents the mass of the tip-mass and k_j is the stiffness of the spring of the jth resonator.

Fig. 10.3 represents the system of coordinates. The jth resonator is situated at $\mathbf{x_j} = (x_j, y_j)$. One point on the plate is denoted as $\mathbf{x} = (x, y)$.

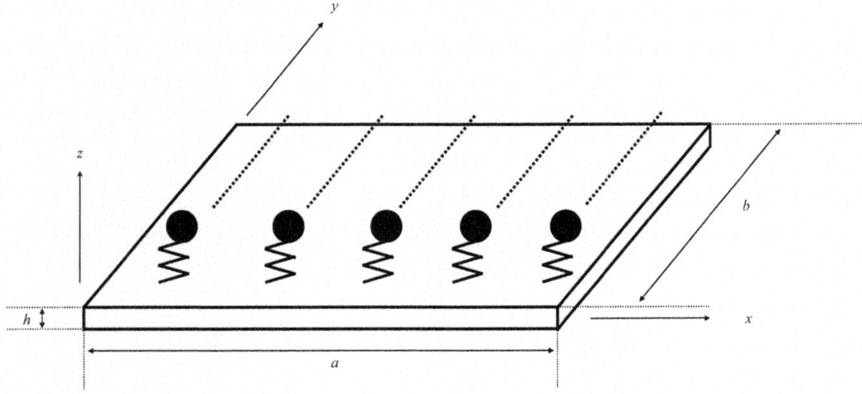

Fig. 10.1 Representation of the plate and the multi-mass-spring system

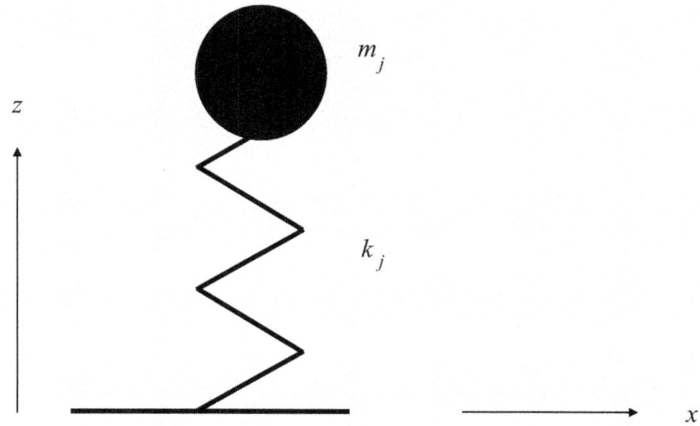

Fig. 10.2 Representation of one resonator

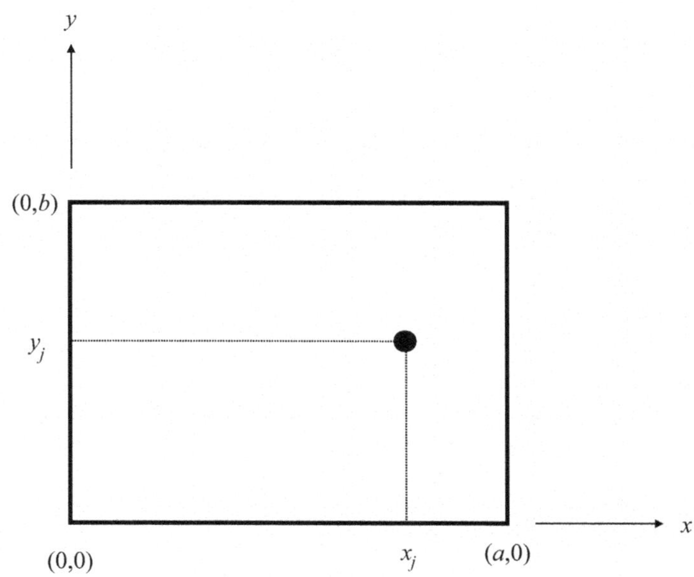

Fig. 10.3 Representation of the system of coordinates

10.3 Solution of the Problem

One starts from the governing equation of the plate:

$$D\nabla^2 w + \rho h \ddot{w} = \sum_{j=1}^{N} k_j (z_j - w_j)\delta(\mathbf{x} - \mathbf{x_j}) \tag{10.1}$$

where ρ is the density of the plate, w denotes $w(\mathbf{x}, t)$, and w_j denotes $w(\mathbf{x_j}, t)$. D is the flexural rigidity of the plate whose expression is given below:

$$D = \frac{Eh^3}{12(1-\upsilon^2)} \tag{10.2}$$

with E, Young's modulus of the material of the plate, and υ, Poisson's ratio. The expression of the biharmonic operator ∇^4 is given as follows:

$$\nabla^4 = \frac{\partial^4}{\partial x^4} + 2\frac{\partial^4}{\partial x^2 \partial y^2} + \frac{\partial^4}{\partial y^4} \tag{10.3}$$

δ is the Dirac-δ function.

One may also write the equation of motion of the *jth* resonator.

$$m_j \frac{d^2 z_j}{dt^2} + k_j\left(z_j - w_j\right) = 0 \tag{10.4}$$

One notes

$$w(\mathbf{x}, t) = W(\mathbf{x})e^{i\omega t} \tag{10.5a}$$

$$z_j(t) = Z_j e^{i\omega t} \tag{10.5b}$$

Substituting Eqs. (10.5a) and (10.5b) in Eq. 10.1 yields

$$D\left[\frac{d^4W}{dx^4} + 2\frac{d^4W}{dx^2dy^2} + \frac{d^4W}{dy^4}\right] - \omega^2 \rho h W = \sum_{j=1}^{N} k_j\left(Z_j - W_j\right)\delta\left(\mathbf{x} - \mathbf{x_j}\right) \tag{10.6}$$

with $W_j = W(\mathbf{x_j})$. For a simply supported plate at its boundaries, one writes

$$W(\mathbf{x}) = \sum_{k=1}^{\infty}\sum_{l=1}^{\infty} W_{kl} \sin\left(\frac{k\pi x}{a}\right) \sin\left(\frac{l\pi y}{b}\right) \tag{10.7}$$

Substituting Eqs. (10.7) in (10.6) yields

$$\begin{aligned} &\sum_{k=1}^{\infty}\sum_{l=1}^{\infty}\left[D\left[\left(\frac{k\pi}{a}\right)^2 + \left(\frac{l\pi}{b}\right)^2\right]^2 - \omega^2 \rho h\right] W_{kl}\alpha_{kl} \\ &= \sum_{j=1}^{N} k_j\left[Z_j - \sum_{k=1}^{\infty}\sum_{l=1}^{\infty} W_{kl}\alpha_{kl,j}\right]\delta\left(x - x_j\right) \end{aligned} \tag{10.8}$$

where

$$\alpha_{kl} = \sin\left(\frac{k\pi x}{a}\right)\sin\left(\frac{l\pi y}{b}\right) \tag{10.9a}$$

$$\alpha_{kl,j} = \sin\left(\frac{k\pi x_j}{a}\right)\sin\left(\frac{l\pi y_j}{b}\right) \tag{10.9b}$$

By multiplying Eq. (10.8) by α_{mn} and integrating over the plate area, one obtains

$$\begin{gathered}\left[D\left[\left(\frac{m\pi}{a}\right)^2 + \left(\frac{n\pi}{b}\right)^2\right]^2 - \omega^2\rho h\right]W_{mn}\int\alpha_{mn}^2 d\Omega \\ = \sum_{j=1}^{N} k_j\left[Z_j - \sum_{k=1}^{\infty}\sum_{l=1}^{\infty} W_{kl}\alpha_{kl,j}\right]\int\delta(\mathbf{x}-\mathbf{x_j})\alpha_{mn}d\Omega\end{gathered} \tag{10.10}$$

Using the orthogonality condition and the property of the Dirac-δ function, one rewrites Eq. (10.10) as follows:

$$\frac{W_{mn}ab}{4}\left[D\left[\left(\frac{m\pi}{a}\right)^2 + \left(\frac{n\pi}{b}\right)^2\right]^2 - \omega^2\rho h\right] = \sum_{j=1}^{N} k_j\left[Z_j - \sum_{k=1}^{\infty}\sum_{l=1}^{\infty} W_{kl}\alpha_{kl,j}\right]\alpha_{mn,j} \tag{10.11}$$

One may now solve Eq. (10.4) and find the following relations:

$$k_j\left[Z_j - \sum_{k=1}^{\infty}\sum_{l=1}^{\infty} W_{kl}\alpha_{kl,j}\right] = \omega^2 m_j Z_j \tag{10.12}$$

10.4 Particular Case $k_r - m_r\omega^2 \neq 0$

When $k_j \neq m_j\omega^2$, one writes Eq. (10.12) as follows:

$$Z_j = \frac{k_j}{k_j - m_j\omega^2}\sum_{k=1}^{\infty}\sum_{l=1}^{\infty} W_{kl}\alpha_{kl,j} \tag{10.13}$$

Substituting Eq. (10.13) yields

$$\begin{gathered}\frac{ab}{4}W_{mn}\left[\left[\left(\frac{m\pi}{a}\right)^2 + \left(\frac{n\pi}{b}\right)^2\right]^2 - \omega^2\rho h\right] \\ = \sum_{j=1}^{N} k_j\left[\frac{k_j}{k_j - \omega^2 m_j}\sum_{k=1}^{\infty}\sum_{l=1}^{\infty} W_{kl}\alpha_{kl,j} - \sum_{k=1}^{\infty}\sum_{l=1}^{\infty} W_{kl}\alpha_{kl,j}\right]\#\alpha_{mn,j}\end{gathered} \tag{10.14}$$

Using Eq. (10.13), one may rewrite Eq. (10.14) as follows:

$$W_{mn}\left[D\left[\left(\frac{m\pi}{a}\right)^2+\left(\frac{n\pi}{b}\right)^2\right]^2-\omega^2\rho h\right]=\frac{4}{ab}\sum_{j=1}^{N}k_j\left(\frac{k_j}{k_j-m_j\omega^2}-1\right)\alpha_{mn,j}\sum_{k=1}^{\infty}\sum_{l=1}^{\infty}W_{kl}\alpha_{kl,j} \tag{10.15}$$

From Eq. (10.15), we need a numerical approach to find natural frequency of the plate with N resonators.

If one chooses that each resonator has the same stiffness k_r and the same mass m_r, Eq. (10.15) becomes

$$\left(D\left[\left(\frac{m\pi}{a}\right)^2+\left(\frac{n\pi}{b}\right)^2\right]^2-\omega^2\rho h\right)W_{mn}=\frac{4}{ab}k_r\left(\frac{k_r}{k_r-m_r\omega^2}-1\right)\sum_{j=1}^{N}\alpha_{mn,j}\left[\sum_{k=1}^{\infty}\sum_{l=1}^{\infty}W_{kl}\alpha_{kl,j}\right] \tag{10.16}$$

This is equivalent to

$$\left(D\left[\left(\frac{m\pi}{a}\right)^2+\left(\frac{n\pi}{b}\right)^2\right]^2-\omega^2\rho h\right)W_{mn}=\frac{4}{ab}k_r\left(\frac{k_r}{k_r-m_r\omega^2}-1\right)\sum_{j=1}^{N}\alpha_{mn,j}\left[\sum_{k=1}^{\infty}\sum_{l=1}^{\infty}W_{kl}\alpha_{kl,j}\right] \tag{10.17}$$

Developing this equation yields

$$\omega^4W_{mn}-\omega^2\left[\frac{D}{\rho h}\left[\left(\frac{m\pi}{a}\right)^2+\left(\frac{n\pi}{b}\right)^2\right]^2W_{mn}+\frac{k_r}{m_r}W_{mn}+\frac{4k_r}{ab\rho h}\sum_{j=1}^{N}\alpha_{mn,j}\sum_{k=1}^{\infty}\sum_{l=1}^{\infty}W_{kl}\alpha_{kl,j}\right]$$
$$+\frac{k_rD}{m_r\rho h}\left[\left(\frac{m\pi}{a}\right)^2+\left(\frac{n\pi}{b}\right)^2\right]W_{mn}=0 \tag{10.18}$$

For the numerical approach, one introduces M as the number of retained terms and rewrites Eq. (10.18) as follows:

$$\omega^4W_{mn}-\omega^2\left[\frac{D}{\rho h}\left[\left(\frac{m\pi}{a}\right)^2+\left(\frac{n\pi}{b}\right)^2\right]^2W_{mn}+\frac{k_r}{m_r}W_{mn}+\frac{4k_r}{ab\rho h}\sum_{j=1}^{N}\alpha_{mn,j}\sum_{k=1}^{M}\sum_{l=1}^{M}W_{kl}\alpha_{kl,j}\right]$$
$$+\frac{k_rD}{m_r\rho h}\left[\left(\frac{m\pi}{a}\right)^2+\left(\frac{n\pi}{b}\right)^2\right]^2W_{mn}=0 \tag{10.19}$$

10.5 Free Vibration of a Plate with One Resonator: Comparison with the Works by Avalos et al. and Wu et al.

In this section, we study a plate with a single spring-mass system situated at x_j and y_j. We keep only the frequency whose values are under 400 Hz.

$$\omega^4 W_{mn} - \omega^2\left[\frac{D}{\rho h}\left(\left(\frac{m\pi}{a}\right)^2+\left(\frac{n\pi}{b}\right)^2\right)^2 W_{mn} + \frac{k_r}{m_r}W_{mn} + \frac{4k_r}{ab\rho h}\alpha_{mn,1}\sum_{k=1}^{M}\sum_{l=1}^{M} W_{kl}\alpha_{kl,1}\right]$$
$$+\frac{k_r}{m_r\rho h}D\left(\left(\frac{m\pi}{a}\right)^2+\left(\frac{n\pi}{b}\right)^2\right)^2 W_{mn} = 0 \tag{10.20}$$

For $m = 1$ and $n = 1$, we obtain

$$\begin{pmatrix} m_{1,1} & m_{1,2} & m_{1,3} & m_{1,4} & m_{1,5} \end{pmatrix}\begin{pmatrix} W_{11} \\ W_{12} \\ W_{21} \\ W_{22} \\ W_{31} \end{pmatrix} = 0 \tag{10.21}$$

with

$$m_{1,1} = \left[\omega^4 - \omega^2\left[\tfrac{D}{\rho}\left(\left(\tfrac{\pi}{a}\right)^2+\left(\tfrac{\pi}{b}\right)^2\right)^2 + \tfrac{k_r}{m_r} + \tfrac{4k_r}{ab\rho h}\alpha_{11,1}^2\right] + \tfrac{k_r}{m_r\rho h}D\left(\left(\tfrac{\pi}{a}\right)^2+\left(\tfrac{\pi}{b}\right)^2\right)^2\right] \tag{10.22a}$$

$$m_{1,2} = -\omega^2\tfrac{4k_r}{ab\rho h}\alpha_{11,1}\alpha_{12,1} \tag{10.22b}$$

$$m_{1,3} = -\omega^2\tfrac{4k_r}{ab\rho h}\alpha_{11,1}\alpha_{21,1} \tag{10.22c}$$

$$m_{1,4} = -\omega^2\tfrac{4k_r}{ab\rho h}\alpha_{11,1}\alpha_{22,1} \tag{10.22d}$$

$$m_{1,5} = -\tfrac{\omega^2 4k_r}{ab\rho h}\alpha_{11,1}\alpha_{31,1} \tag{10.22e}$$

For $m = 1$ and $n = 2$, we obtain

$$\begin{pmatrix} m_{2,1} & m_{2,2} & m_{2,3} & m_{2,4} & m_{2,5} \end{pmatrix}\begin{pmatrix} W_{11} \\ W_{12} \\ W_{21} \\ W_{22} \\ W_{31} \end{pmatrix} = 0 \tag{10.23}$$

with

$$m_{2,1} = m_{1,2} \tag{10.24a}$$

$$m_{2,2} = \omega^4 - \omega^2\left[\frac{D}{\rho}\left(\left(\frac{\pi}{a}\right)^2 + \left(\frac{2\pi}{b}\right)^2\right)^2 + \frac{k_r}{m_r} + \frac{4k_r}{ab\rho h}\alpha_{12,1}^2\right] + \frac{k_r}{m_r\rho h}D\left(\left(\frac{\pi}{a}\right)^2 + \left(\frac{2\pi}{b}\right)^2\right)^2 \tag{10.24b}$$

$$m_{2,3} = -\omega^2\frac{4k_r}{ab\rho h}\alpha_{12,1}\alpha_{21,1} \tag{10.24c}$$

$$m_{2,4} = -\omega^2\frac{4k_r}{ab\rho h}\alpha_{12,1}\alpha_{22,1} \tag{10.24d}$$

$$m_{2,5} = -\frac{\omega^2 4k_r}{ab\rho h}\alpha_{12,1}\alpha_{31,1} \tag{10.24e}$$

For $m = 2$ and $n = 1$, we obtain

$$\begin{pmatrix} m_{3,1} & m_{3,2} & m_{3,3} & m_{3,4} & m_{3,5} \end{pmatrix}\begin{pmatrix} W_{11} \\ W_{12} \\ W_{21} \\ W_{22} \\ W_{31} \end{pmatrix} = 0 \tag{10.25}$$

with

$$m_{3,1} = m_{1,3} \tag{10.26a}$$

$$m_{3,2} = m_{2,3} \tag{10.26b}$$

$$m_{3,3} = \omega^4 - \omega^2\left[\frac{D}{\rho}\left(\left(\frac{2\pi}{a}\right)^2 + \left(\frac{\pi}{b}\right)^2\right)^2 + \frac{k_r}{m_r} + \frac{4k_r}{ab\rho h}\alpha_{21,1}^2\right] + \frac{k_r}{m_r\rho h}D\left(\left(\frac{2\pi}{a}\right)^2 + \left(\frac{\pi}{b}\right)^2\right)^2 \tag{10.26c}$$

$$m_{3,4} = -\omega^2\frac{4k_r}{ab\rho h}\alpha_{21,1}\alpha_{22,1} \tag{10.26d}$$

$$m_{3,5} = -\frac{\omega^2 4k_r}{ab\rho h}\alpha_{21,1}\alpha_{31,1} \tag{10.26e}$$

For $m = 2$ and $n = 2$, we obtain

$$\begin{pmatrix} m_{4,1} & m_{4,2} & m_{4,3} & m_{4,4} & m_{4,5} \end{pmatrix} \begin{pmatrix} W_{11} \\ W_{12} \\ W_{21} \\ W_{22} \\ W_{31} \end{pmatrix} = 0 \tag{10.27}$$

with

$$m_{4,1} = m_{1,4} \tag{10.28a}$$

$$m_{4,2} = m_{2,4} \tag{10.28b}$$

$$m_{4,3} = m_{3,4} \tag{10.28c}$$

$$\begin{aligned} m_{4,4} = \omega^4 - \omega^2 \left[\frac{D}{\rho} \left(\left(\frac{2\pi}{a} \right)^2 + \left(\frac{2\pi}{b} \right)^2 \right)^2 + \frac{k_r}{m_r} + \frac{4k_r}{ab\rho h} \alpha_{22,1}^2 \right] \\ + \frac{k_r}{m_r \rho h} D \left(\left(\frac{2\pi}{a} \right)^2 + \left(\frac{2\pi}{b} \right)^2 \right)^2 \end{aligned} \tag{10.28d}$$

$$m_{4,5} = -\frac{\omega^2 4k_r}{ab\rho h} \alpha_{22,1} \alpha_{31,1} \tag{10.28e}$$

For $m = 3$ and $n = 1$, we obtain

$$\begin{pmatrix} m_{5,1} & m_{5,2} & m_{5,3} & m_{5,4} & m_{5,5} \end{pmatrix} \begin{pmatrix} W_{11} \\ W_{12} \\ W_{21} \\ W_{22} \\ W_{31} \end{pmatrix} = 0 \tag{10.29}$$

with

$$m_{5,1} = m_{1,5} \tag{10.30a}$$

$$m_{5,2} = m_{2,5} \tag{10.30b}$$

$$m_{5,3} = m_{3,5} \tag{10.30c}$$

$$m_{5,4} = m_{4,5} \tag{10.30d}$$

$$m_{5,5} = \omega^4 - \omega^2\left[\frac{D}{\rho}\left(\left(\frac{3\pi}{a}\right)^2 + \left(\frac{\pi}{b}\right)^2\right)^2 + \frac{k_r}{m_r} + \frac{4k_r}{ab\rho h}\alpha_{31,1}^2\right] + \frac{k_r}{m_r\rho h}D\left(\left(\frac{3\pi}{a}\right)^2 + \left(\frac{\pi}{b}\right)^2\right)^2 \tag{10.30e}$$

Then we can write the following equation:

$$\begin{pmatrix} m_{1,1} & m_{1,2} & m_{1,3} & m_{1,4} & m_{1,5} \\ m_{2,1} & m_{2,2} & m_{2,3} & m_{2,4} & m_{2,5} \\ m_{3,1} & m_{3,2} & m_{3,3} & m_{3,4} & m_{3,5} \\ m_{4,1} & m_{4,2} & m_{4,3} & m_{4,4} & m_{4,5} \\ m_{5,1} & m_{5,2} & m_{5,3} & m_{5,4} & m_{5,5} \end{pmatrix}\begin{pmatrix} W_{11} \\ W_{12} \\ W_{21} \\ W_{22} \\ W_{31} \end{pmatrix} = M \times \begin{pmatrix} W_{11} \\ W_{12} \\ W_{21} \\ W_{22} \\ W_{31} \end{pmatrix} = 0 \tag{10.31}$$

We are looking for the non-trivial solution then we write the following equation:

$$\det(M) = 0 \tag{10.32}$$

We introduce the characteristics of the plate that are given in Table 10.1.

The numerical results for the case of the plate carrying a single spring-mass system placed at $x_j = 0.75a$ and $y_j = 0.75b$ are presented and compared with some references in Table 10.2.

To compare the results of Table 10.2 with frequencies of "bare" plate, we list frequencies of the latter plate in Table 10.3, where frequencies are evaluated by classical formula

Table 10.1 Characteristic of the plate and a resonator

a (m)	2.0
b (m)	1.0
h (m)	0.005
E (GPa)	205.1
ν	0.3
ρ (kg/m^3)	7850
k_r (N/m)	2641.23
m_r (kg)	39.25
$M_p = ab\rho h$ (kg)	78.5

Table 10.2 Comparison of the first five natural frequencies between numerical approach developed above and some references

Methods	ω_1 (Hz)	ω_2 (Hz)	ω_3 (Hz)	ω_4 (Hz)	ω_5 (Hz)
Avalos et al.	3.8651	95.4346	152.6884	248.0864	324.4224
J-S Wu et al.	3.8650	97.8775	160.0768	261.5045	327.2913
Present method	3.8652	95.4346	152.6884	248.0864	324.4224

Table 10.3 First five natural frequencies of a SSSS plate without resonator

ω_{11} (Hz)	ω_{21} (Hz)	ω_{31} (Hz)	ω_{12} (Hz)	ω_{22} (Hz)
95.4150	152.6639	248.0789	324.4108	381.6598

Table 10.4 First five Natural frequencies of a SSSS plate carrying a single spring-mass system at its middle

ω_{11} *(Hz)*	ω_{21} *(Hz)*	ω_{31} *(Hz)*	ω_{12} *(Hz)*	ω_{22} *(Hz)*
3.8634	95.4934	152.6639	248.1090	324.4108

$$\omega_{mn} = \left[\left(\frac{m\pi}{a}\right)^2 + \left(\frac{n\pi}{b}\right)^2\right]\sqrt{\frac{D}{\rho h}} \tag{10.33}$$

As is seen from Table 10.2, the single added resonator introduces addition natural frequency given in the first column of Table 10.2, whereas "bare" plate frequencies are close to the frequencies listed in columns numbered 2, 3, 4, and 5, respectively.

The partial natural frequency of the resonator that is suspended at the location of a nonmoving point on the plate is given with the following expression:

$$\sqrt{\frac{k_r}{m_r}} = 3.8670 \tag{10.34}$$

We observe that the first natural frequencies listed in Table 10.2, are extremely close to this partial frequency. We arrive at a remarkable conclusion that the inclusion of the suspended oscillator introduces an additional natural frequency in the combined "plate-oscillator" system.

We now report results for the case of a square plate carrying a single spring-mass system placed at $x_j = 0.5a$ and $y_j = 0.5b$. With $k_r = 0.5D/a^2$ and $m_r = 0.25\rho abh$, the following results are obtained (Table 10.4).

In the particular case, $k_r - m_r\omega^2 = 0$, we have a coincidence of the partial frequency ω_r and a frequency of a SSSS plate having a point support in the middle.

10.6 Uncovering Additional Frequency: Particular Case $k_r - m_r\omega^2 = 0$

In that case, Eq. (10.12) becomes

$$k_r \sum_{k=1}^{\infty} \sum_{l=1}^{\infty} W_{kl}(t)\alpha_{kl,j} = 0 \tag{10.35}$$

Then, our problem is now reduced to the study of a plate which is fixed at prespecified N point supports, namely at the positions where the spring-mass systems are attached.

Note that the free vibrations of plates with point supports was investigated in the literature widely. For example, Nowacki states that the natural frequency parameter for fundamental frequency of a plate

$$\omega a^2 \sqrt{\frac{\rho}{D}} \tag{10.36}$$

Equals, respectively, 52.6 for a square plate, 73.1 for a rectangular plate with side ratio $a/b = 1.5$, and 91.1 for the case $a/b = 2$. This implies that if the oscillator is suspended at the center of the rectangular plate the system "plate plus oscillator" might have an additional frequency equal to partial frequency of the oscillator, in our specific case, it is equal to 3.8670, if either of following equalities hold, for various side ratios:

$$52.6\sqrt{\frac{\rho}{D}\frac{1}{a^2}} = \sqrt{\frac{k_r}{m_r}}(a = b) \tag{10.37}$$

$$73.1\sqrt{\frac{\rho}{D}\frac{1}{a^2}} = \sqrt{\frac{k_r}{m_r}}\left(\frac{a}{b} = 1.5\right) \tag{10.38}$$

$$91.1\sqrt{\frac{\rho}{D}\frac{1}{a^2}} = \sqrt{\frac{k_r}{m_r}}\left(\frac{a}{b} = 2.0\right) \tag{10.39}$$

This possible natural frequency of the hybrid system appears to be largely overlooked.

Another case of free vibrations of isotropic plates supported at certain points has been studied by Huang et al. For a square plate having a point support at the center, they give the first natural frequency parameter $\lambda_1 = 7.272$. Using their formula,

$$\lambda^4 = \frac{\rho h \omega^2 a^4}{D(1-\nu^2)} \tag{10.40}$$

we give the fundamental $\omega_1 = 78.031$.

10.7 Free Vibrations of a Square Plate with Multiple Spring-Mass Systems

We study a SSSS square plate carrying 2, 3, 4 oscillators represented on Figs. 10.4, 10.5, and 10.6, respectively. The characteristics of the plate that in studied in this section are presented Table 10.5.

The numerical results for all those cases are reported in the following Table 10.6. We also give the value of the partial frequency:

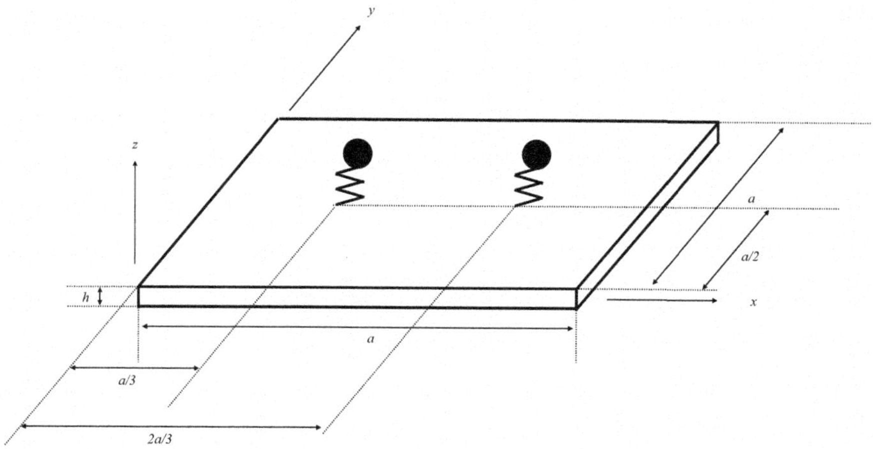

Fig. 10.4 Representation of the plate carrying two spring-mass systems

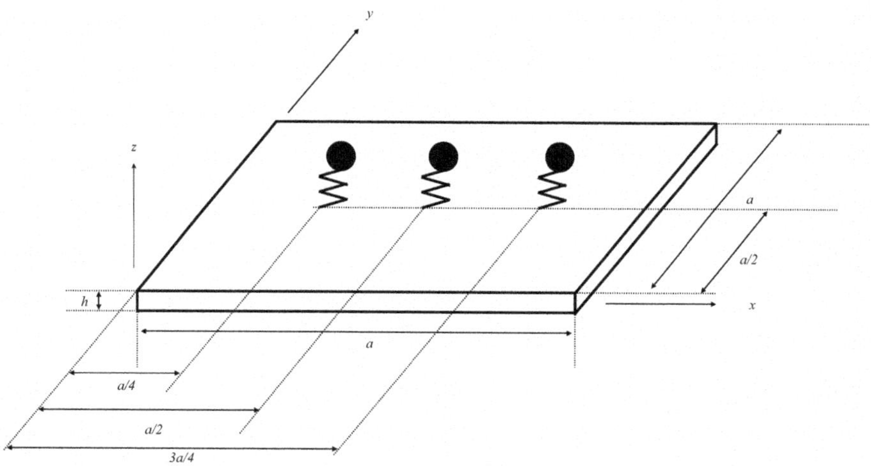

Fig. 10.5 Representation of the plate carrying three spring-mass systems

$$\omega_r = \sqrt{\frac{k_r}{m_r}} \approx 2.18752 \tag{10.37}$$

We make the following observation, for N resonators attached to the plate, N natural frequencies close to ω_r are added to the spectrum of the combined system. Nevertheless, if one wants to obtain those N additional natural frequencies, one must build a $(N \times N)$ matrix M.

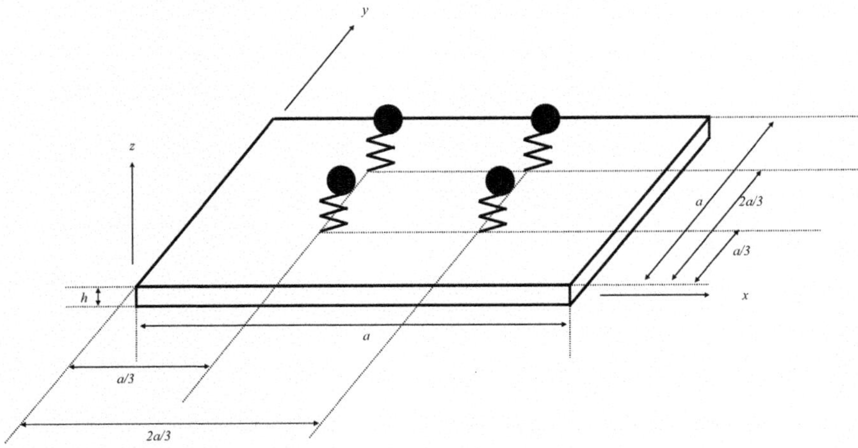

Fig. 10.6 Representation of the plate carrying four spring-mass systems

Table 10.5 Characteristics of the plate

a (m)	1.0
b (m)	1.0
h (m)	0.001
E (GPa)	205.1
ν	0.3
ρ (kg/m^3)	7850
k_r (N/m)	$0.5D/a^2$
m_r (kg)	$0.25\rho a^2 h$

Table 10.6 Numerical results for the plate with 2, 3, and 4 resonators

	ω_1(Hz)	ω_2(Hz)	ω_3(Hz)	ω_4(Hz)	ω_5(Hz)	ω_6(Hz)
2 Resonators	2.17910	2.18752	30.65070	76.33196	122.13114	152.66393
3 Resonators	2.17632	2.18707	2.18752	30.68991	76.33196	122.13114
4 Resonators	2.17493	2.18550	2.18673	2.18752	30.70949	76.40251

10.8 Conclusion

According to Romero-Garcia and Hladky-Hennion [5], "During the recent decades, metamaterials have revolutionized the way waves are controlled in the broad field of wave physics due to extraordinary physical properties they present. Their locally resonant structure, introducing deep wavelength band gaps, regions of frequencies where propagation is forbidden, among other properties, have motivated a plethora of applications not available up to now and creating an inflection point in the material science conception."

References

1. Liao G, Luan C, Wang Z, Liu J, Yao X, Fu J (2021) Acoustic metamaterials: a review of theories, structures, fabrication approaches, and applications. Adv Mater Technol 6(5), article 2000787
2. Veselago VG (1968) Electrodynamics of substances with simultaneously negative electrical and magnetic permeabilities. Soviet Physics Uspekhi 10(4):504–509
3. Liu Z, Zhang X, Mao Y, Zhu YY, Yang Z, Chan CT, Sheng P (2000) Locally resonant sonic materials. Science 289(5485):1734–1736
4. Liu Z, Chan CT, Sheng P (2002) Three-component elastic wave band-gap material. Phys Rev B 65(16), article165116
5. Romero-Garcia V, Hladky-Hennion A-C (2019) Fundamentals and applications of acoustic metamaterials: from seismic to radio frequencies. ISTE-Wiley, London
6. Jiménez N, Umnova O, Groby J-P (2021) Acoustic waves in periodic structures, metamaterials, and porous media: from fundamentals to industrial applications. Springer Nature Switzerland, Cham
7. Gao N, Ding J (2022) Dynamic equivalent modeling of acoustic metamaterials: solving problem of noise and vibration. Springer Nature Singapore
8. Hu G-K, Zhou X-M (2024) Acoustic metamaterials and wave control. World Scientific, Singapore
9. Fok L, Ambati M, Zhang X (2008) Acoustic metamaterials. MRS Bull 33(10):931–934
10. Chen S, Fan Y, Fu Q, Wu H, Jin Y, Zheng J, Zhang F (2018) A review of tunable acoustic metamaterials. Appl Sci 8(9), article 1480
11. Zangeneh-Nejad F, Fleury R (2019) Active times for acoustic metamaterials. Rev Phys 4, article 100031
12. Kumar S, Pueh Lee H (2021) Recent advances in active acoustic metamaterials. Int J Appl Mech 11(8), article 1950081
13. Liu J, Guo H, Wang T (2020) A review of acoustic metamaterials and phononic crystals. Crystals 10(4), article 305
14. Barchiesi E, Di Cosmo F, Laudato M (2020) A review of some selected examples of mechanical and acoustic metamaterials. Discrete Continuous Models Complex Metamater 2020:52–102
15. Suresh Kumar N, Naidu KC, Banerjee P, Anil Babu TV, Venkata Shiva Reddy B (2021) A review on metamaterials for device applications. Crystals 11(5), article 518
16. Song H, Ding X, Cui Z, Hu H (2021) Research progress and development trends of acoustic metamaterials. Molecules 26(13), article 4018
17. Dalela S, Balaji PS, Jena DP (2022) A review on application of mechanical metamaterials for vibration control. Mech Adv Mater Struct 29(22):3237–3262
18. Young D (1952) Theory of dynamic vibration absorbers for beams. In: Proceedings of the First U.S. National Congress of Applied Mechanics. ASME Press, New York, pp 91–96
19. Avalos DR, Larrondo H, Laura PAA (1993) Vibrations of a simply supported plate carrying an elastically mounted concentrated mass. Ocean Eng 20(2):195–205
20. Avalos DR, Larrondo HA (1997) Comments on "Use of the analytical and numerical combine method in the free vibration analysis of a rectangular plate with any number of point masses and translational springs" (with authors' reply). J Sound Vib 207(4):591–592
21. Wu JS, Luo SS (1997) Use of the analytical-and-numerical-combined method in the free vibration analysis of a rectangular plate with any number of point masses and translational springs. J Sound Vib 200(2):179–194
22. Laura PAA, Filipich CP, Cortinez VH (1987) Vibrations of beams and plates carrying concentrated masses. J Sound Vib 117(3):459–465
23. Avalos DR, Larrondo HA, Laura PAA (1994) Transverse vibrations of a circular plate carrying an elastically mounted mass. J Sound Vib 177(2):251–258
24. Wu JS, Chou H-M, Chen D-W (2003) Free vibration analysis of a rectangular plate carrying multiple various concentrated elements. Proc Inst Mech Eng Part K: J Multi-body Dyn 217(2):171–183(13)

25. Wu JS, Luo SS (1997) Vibration analysis of a rectangular plate carrying any number of point masses and translational springs by using the modified and quasi-analytical and numerical combined methods. Int J Numer Meth Eng 40:2171–2193
26. Wu JS, Luo SS (1997b) Use of the analytical and numerical combined method in the free vibration analysis of a rectangular plate with any number of points masses and translational springs. J Sound Vib 200(2):179–194
27. Wu JS, Luo SS (1997) Free vibration analysis of a rectangular plate carrying any number of point masses and translational springs by using the modified and quasi analytical and numerical combined methods. Int J Numer Meth Eng 40:2171–2193
28. Wu JS, Huang CG (1995) Free and forced vibrations of a Timoshenko beam with any number of translational and rotational springs and lumped masses. Commun Numer Methods Eng 11(9):743–756
29. Amabili M (2010) Geometrically nonlinear vibrations of rectangular plates carrying a concentrated mass. J Sound Vib 329(21):4501–4514
30. Amabili M, Carra S (2012) Experiments and simulations for large-amplitude vibrations of rectangular plates carrying concentrated masses. J Sound Vib 331(1):155–166
31. Amabili M, Pellegrini M, Righi F, Vinci F (2006) Effect of concentrated masses with rotary inertia on vibrations of rectangular plates. J Sound Vib 295(1–2):1–12

Chapter 11
Dispersion, Attenuation and Nonlinear Spatial Locality of Longitudinal Waves Propagating in Materials with Point Defects

Vladimir I. Erofeev, Anna V. Leonteva, and A. V. Shekoyan

Abstract This paper deals with the study of the propagation of longitudinal waves in materials with point defects. The problem is described by the system of differential equations comprising the dynamic equation of the elasticity theory and kinetic equations of defect density allowing for the mutual influence of defects and a propagating wave, as well as the mutual recombination of defects. There are considered herein both the limiting cases presenting materials with one type of point defect (vacancies, interstitials) and the general case, if the material contains both types of point defects (vacancies and interstitials). A self-consistent mathematical model under conditions of the static deformation is reduced to a nonlinear evolution equation which combines the well-known wave dynamics equations of Korteweg-de Vries-Burgers and Klein-Gordon. Exact analytical solutions to the resulting evolution equation were found and analyzed. There was analyzed the influence of point defect parameters, which characterize the diffusion of defects, the rate of their recombination at sinks and the change in the volume of the material, if a single point defect is formed therein, on the harmonic wave amplitude and velocity. It was shown that in vacancy-containing media the low-frequency longitudinal waves have a greater amplitude and velocity than in interstitials-containing media. And in vacancy-containing media the low-frequency perturbation velocities reach larger values, but in interstitials-containing media they reach lower values as compared to high-frequency perturbations. A frequency range was identified wherein the dispersion of longitudinal waves was significant; in vacancy-containing media it was normal, and in interstitials-containing

V. I. Erofeev (✉) · A. V. Leonteva
Mechanical Engineering Research Institute of the Russian Academy of Sciences, Nizhny Novgorod, Russia
e-mail: erof.vi@yandex.ru

A. V. Leonteva
e-mail: aleonav@mail.ru

A. V. Shekoyan
Institute of Mechanics of the National Academy of Sciences of the Republic of Armenia, Yerevan, Republic of Armenia

H. Altenbach and V. A. Eremeyev (eds.), *Dynamics of Discrete and Continuum Structures and Media*, Advanced Structured Materials 221,
https://doi.org/10.1007/978-3-031-75626-9_11

media, it was anomalous. An increase in the diffusion coefficient or a decrease in the dilatation parameter value contributed to the lower grade dispersion. For high-frequency waves, media with vacancies and interstitials were barely distinguishable; the availability of any point defects had almost no effect on the propagation velocity of high-frequency perturbations and their amplitude. It was noted that the defect diffusion coefficients did not affect the existence of an additional low-frequency wave in media containing two types of point defects. In the limiting case associated with the Korteweg-de Vries-Burgers equation it was shown that there were stationary waves with a decreasing or increasing profile for the function of the volume concentration of point defects. In the limiting case associated with the Klein-Gordon equation, with a negative value of the coefficient for the linear term with a zero derivative the propagation of stationary compression waves was possible with any sign of the coefficient for the nonlinear term; with a positive value of the coefficient for the linear term with a zero derivative, the propagation of a rarefaction wave was possible. In media with single type of point defect, the propagation of nonlinear stationary deformation waves is possible. Their velocity is fixed and its values differ for vacancy-containing media and interstitials-containing media, i.e. velocities pertain to different intervals. The type of material defects affects the type of the stationary wave, so in vacancy-containing media a compression wave is formed and in interstitials-containing media a rarefaction wave is formed. In vacancy-containing media, a nonlinear wave has the higher velocity but the smaller amplitude and width as compared to a wave propagating in interstitials-containing media.

Keywords Point defects · Vacancies · Interstitials · Longitudinal wave · Dispersion · Nonlinear stationary wave

11.1 Introduction

It is known [1] that in the course of the ion implantation and the nanosecond laser alloying of metallic materials, the anomalous mass transfer occurs. It is largely due to the formation of numerous point defects (vacancies, interstitial atoms), both under the exposure to a laser beam generating the stress-strain state in the material surface layer [2–10] and during the ion implantation [11–25].

The formation of point defect clusters in iron crystallites under radiation exposure was studied in [26]. It was found that atomic displacement cascades could cause the radiation-stimulated migration of grain boundaries.

The effect of point defects on the crack resistance of the joined material boundaries in composite structures was studied in [27]. It was ascertained how the crack propagation regime changed with increasing vacancy concentration.

In [28] the adsorption of lithium was simulated on a carbon nanotube used as anode of a lithium-ion battery and containing vacancy-type defects. It was shown that any available defects could create more efficient adsorption centers as compared to the surface of a perfect tube and the energy of adsorption on centers within the defect region was up to twice higher than that on the perfect tube center. Furthermore, the effect of increasing sorption activity had a long-range nature as it was identified for adsorption centers in the first and second entourage of the defect and even for centers located on the tube opposite to the defect. The impact of a vacancy edge defect (namely, the breaking of bonds between carbon atoms) of a graphene sheet on the Raman spectrum was studied in [29].

To construct both the said and any other effects associated with available point defects, it is relevant to identify the peculiar features for the propagation of elastic waves in condensed media and their interaction with the defect structure.

It was shown in [7] that the problem of the elastic wave propagation in material with point defects should be considered as a self-consistent one comprising along with the dynamic equation of the elasticity theory a kinetic equation for the density of defects.

The transmission of a longitudinal (especially nonlinear, higher-intensity) wave contributes to some change in the activation energy for the formation of point defects in areas of tension and compression, thus causing their spatial redistribution. While migrating through the material, defects recombine at various types of centers. As such centers there can be dislocations, interstitial impurities, etc.

Wave effects in dislocation ensembles were studied in papers [30–38]. In [39] the interaction of a nonlinear deformation wave with the field for concentration of point defects (vacancies, interstitials) was studied which caused both the wave scattering and a change in the activation energy for the formation of defects and their spatial redistribution. It was thereat assumed that main processes determining the defect behavior were the generation, recombination and diffusion processes. The volume mutual recombination of unlike defects was not regarded.

In [40, 41] the effect of an external nonstationary inhomogeneous temperature field was analyzed on the nonlinear deformation waves propagating in medium containing one type of point defect.

The propagation of nonlinear longitudinal waves in a plate was studied in [42] with regard to the interaction of the longitudinal component of the medium displacements with the thermal field and the concentration of nonequilibrium atomic point defects.

In [43], in the one-dimensional case for elastic medium with moving defects an equation was obtained from the main system of equations consisting of coupled equations for concentrations of various types of point defects and for nonlinear longitudinal displacements, taking into account the minor effect of dissipation, and its important particular cases were analyzed. The mutual volume recombination of various types of defects was not regarded. As a result, it was found that the recombination at sinks and the diffusion of defects affected the possibility for the formation of shock waves and quasi-stationary load/unload waves in the shape of a kink with a "cap". It was noted that the generation of a bell-shaped solitary wave was impossible.

In this paper the dispersion and attenuation are studied in the linear problem statement, and in the nonlinear statement the spatial localization of longitudinal waves propagating in materials with point defects is studied with regard to the mutual volume recombination of unlike defects.

11.2 The Problem Statement

A self-consistent problem comprising the dynamic equation of the elasticity theory and the kinetic equation for the density of defects was proposed in [7] and is written as

$$\frac{\partial^2 U}{\partial t^2} - c_l^2 \frac{\partial^2 U}{\partial x^2} - \frac{\beta_N}{\rho} \frac{\partial U}{\partial x} \frac{\partial^2 U}{\partial x^2} = -\left(\lambda + \frac{2}{3}\mu\right) \frac{\Omega_V}{\rho} \frac{\partial n_V}{\partial x} - \left(\lambda + \frac{2}{3}\mu\right) \frac{\Omega_i}{\rho} \frac{\partial n_i}{\partial x}, \tag{11.1}$$

$$\frac{\partial n_V}{\partial t} = q_0 + q_\varepsilon \frac{\partial U}{\partial x} + D_V \frac{\partial^2 n_V}{\partial x^2} - \beta_V n_V - \beta_{iV} n_i, \tag{11.2}$$

$$\frac{\partial n_i}{\partial t} = q_0 + q_\varepsilon \frac{\partial U}{\partial x} + D_i \frac{\partial^2 n_i}{\partial x^2} - \beta_i n_i - \beta_{Vi} n_V. \tag{11.3}$$

Here, $U(x, t)$ is the longitudinal movement of material particles (the wave is considered as flat); $n_j(x, t)$ is the volume concentration of point defects ($j = V$ is for vacancies, $j = i$ is for interstitials); $c_l = \sqrt{(\lambda + 2\mu)/\rho}$ is the velocity of the longitudinal wave propagation in the material, if there are no defects therein; λ, μ are Lamé's constants; ρ is the material density; β_N is the nonlinearity factor, $\beta_N = 3\lambda + 6\mu + 2A + 6B + 2C$; A, B, C are Landau third-order modules; Ω_j is the dilatation parameter characterizing the change in the material volume when a single point defect is formed therein; for vacancies $\Omega_j < 0$, for interstitials $\Omega_j > 0$. The rate of the point defect generation under no deformation is denoted by q_0; the second term on the right-hand side (11.2) represents the deformation correction to the generation of defects; D_j is the diffusion coefficient for the j type of defect; β_j is the recombination rate at sinks. The rates of the mutual recombination of the "interstitial-vacancy" type and the "vacancy-interstitial" type are respectively denoted by β_{iV} and β_{Vi}.

11.3 Harmonic Wave

The propagation of a harmonic wave under no-constraining force ($q_0 = 0$) is described by the following linear system:

$$\frac{\partial^2 U}{\partial t^2} - c_l^2 \frac{\partial^2 U}{\partial x^2} = -\left(\lambda + \frac{2}{3}\mu\right) \frac{\Omega_V}{\rho} \frac{\partial n_V}{\partial x} - \left(\lambda + \frac{2}{3}\mu\right) \frac{\Omega_i}{\rho} \frac{\partial n_i}{\partial x}, \tag{11.4}$$

$$\frac{\partial n_V}{\partial t} = q_\varepsilon \frac{\partial U}{\partial x} + D_V \frac{\partial^2 n_V}{\partial x^2} - \beta_V n_V - \beta_{iV} n_i, \tag{11.5}$$

$$\frac{\partial n_i}{\partial t} = q_\varepsilon \frac{\partial U}{\partial x} + D_i \frac{\partial^2 n_i}{\partial x^2} - \beta_i n_i - \beta_{Vi} n_V. \tag{11.6}$$

By excluding n_i

$$n_i = \frac{1}{\beta_{iV}}\left(-\frac{\partial n_V}{\partial t} + q_\varepsilon \frac{\partial U}{\partial x} + D_V \frac{\partial^2 n_V}{\partial x^2} - \beta_V n_V\right),$$

the system (11.4)–(11.6) is reduced to the system of two equations

$$\frac{\partial^2 U}{\partial t^2} - \left(c_l^2 - \left(\lambda + \frac{2}{3}\mu\right)\frac{q_\varepsilon \Omega_i}{\rho \beta_{iV}}\right)\frac{\partial^2 U}{\partial x^2} + \left(\lambda + \frac{2}{3}\mu\right)\frac{1}{\rho}\left(\Omega_V - \frac{\beta_V \Omega_i}{\beta_{iV}}\right)\frac{\partial n_V}{\partial x} - \\ -\left(\lambda + \frac{2}{3}\mu\right)\frac{\Omega_i}{\rho \beta_{iV}}\frac{\partial^2 n_V}{\partial x \partial t} + \left(\lambda + \frac{2}{3}\mu\right)\frac{\Omega_i D_V}{\rho \beta_{iV}}\frac{\partial^3 n_V}{\partial x^3} = 0 \tag{11.7}$$

$$\frac{\partial^2 n_V}{\partial t^2} - (\beta_V D_i + \beta_i D_V)\frac{\partial^2 n_V}{\partial x^2} + (\beta_i + \beta_V)\frac{\partial n_V}{\partial t} - (D_V + D_i)\frac{\partial^3 n_V}{\partial x^2 \partial t} + D_i D_V \frac{\partial^4 n_V}{\partial x^4} + \\ + q_\varepsilon(\beta_{iV} - \beta_i)\frac{\partial U}{\partial x} - q_\varepsilon \frac{\partial^2 U}{\partial x \partial t} + q_\varepsilon D_i \frac{\partial^3 U}{\partial x^3} + (\beta_i \beta_V - \beta_{iV}\beta_{Vi})n_V = 0. \tag{11.8}$$

We substitute to the system of differential equations (11.7), (11.8) the solution in the form of monochromatic waves with random amplitudes U^0, n_V^0:

$$\begin{cases} U = U^0 \exp(i\omega t - ikx) \\ n_V = n_V^0 \exp(i\omega t - ikx) \end{cases}$$

where ω and k are the wave frequency and wave number. We compose the system determinant and equate it to zero.

Dispersion equation in non-dimensional variables

$$\overline{\omega} = \frac{\omega}{\beta_i + \beta_V}, \ \overline{k} = \frac{k}{(\beta_i + \beta_V)/c_l}$$

($\overline{\omega}$, $\overline{k}$ are non-dimensional values of frequency and wave number, respectively) are obtained as follows (the bar is omitted):

$$\omega^4 + i[-q_7(1+q_2)(1+q_3)k^2\omega^3 - \omega^3] - q_3 q_7^2(1+q_2)^2 k^4 \omega^2 - \\ -(1 + q_7(q_2+q_3))k^2\omega^2 - \frac{(q_2 - q_4 q_5)}{(1+q_2)^2}\omega^2 + \\ + i\left[q_7(1+q_2)(1+q_3)k^4\omega - \left(\frac{q_1(1+q_6)}{1+q_2} - 1\right)k^2\omega\right] + q_3 q_7^2 (1+q_2)^2 k^6 + \\ + q_7(q_2 + q_3 - q_1(q_3+q_6))k^4 + \frac{(q_2 - q_4 q_5 - q_1(q_2 + q_6 - q_4 - q_5 q_6))}{(1+q_2)^2}k_2 = 0 \tag{11.9}$$

where non-dimensional parameters are

$$q_1 = \left(\lambda + \frac{2}{3}\mu\right)\frac{q_\varepsilon}{\rho c_l^2}\frac{\Omega_V}{\beta_V},\ q_2 = \frac{\beta_i}{\beta_V},\ q_3 = \frac{D_i}{D_V},\ q_4 = \frac{\beta_{iV}}{\beta_V},\ q_5 = \frac{\beta_{Vi}}{\beta_V},\ q_6 = \frac{\Omega_i}{\Omega_V},\ q_7 = \frac{\beta_V D_V}{c_l^2}.$$

All parameters are positive, except for these two $q_1 < 0, q_6 < 0$.

In the complex dispersion equation (11.9) the wave number must be complex $k = k_1 + ik_2$ [44], where k_1 and k_2 are the real and imaginary parts of the wave number. Allowing for this circumstance, the longitudinal displacement will be written as follows:

$$U = U^0 \exp(k_2 x)\exp(i\omega t - ik_1 x),$$

where k_1 and k_2 characterize the wave propagation and attenuation, respectively, at a fixed frequency value.

11.3.1 Medium with One Type of Point Defect

Let us first consider the following two simpler particular cases, if there is a single type of point defect in the material: either vacancies or interstitials. The system (11.4)–(11.6) will be as follows:

$$\frac{\partial^2 U}{\partial t^2} - c_l^2\frac{\partial^2 U}{\partial x^2} = -\left(\lambda + \frac{2}{3}\mu\right)\frac{\Omega_j}{\rho}\frac{\partial n_j}{\partial x},$$

$$\frac{\partial n_j}{\partial t} = q_\varepsilon\frac{\partial U}{\partial x} + D_j\frac{\partial^2 n_j}{\partial x^2} - \beta_j n_j$$

($j = V$ and $j = i$ in systems with vacancies and interstitials, respectively).

The complex dispersion ratio (11.9) will be simplified ($q_2 = 0, q_3 = 0, q_4 = 0, q_6 = 0$):

$$\omega^3 - i(1 + q_7 k^2)\omega^7 - k^2\omega + i(q_7 k^4 + (1 - q_1)k^2) = 0, \tag{11.10}$$

where $q_1 = \left(\lambda + \frac{2}{3}\mu\right)\frac{q_\varepsilon}{\rho c_l^2}\frac{\Omega_j}{\beta_j},\ q_7 = \frac{\beta_j D_j}{c_l^2}\ (j = V, i)$.

By expanding the wave number and separating the real and imaginary parts of the dispersion equation (11.10), we obtain the following system:

$$\begin{cases}\omega^3 + 2q_7 k_1 k_2 \omega^2 - (k_2^2 - k_2^2)\omega - 4q_7 k_1^3 k_2 + 4q_7 k_1 k_2^3 - 2(1 - q_1)k_1 k_2 = 0\\ (1 + q_7(k_1^2 - k_2^2))\omega^2 + 2k_1 k_2\omega - q_7(k_1^4 + k_2^4) + 6q_7 k_1^2 k_2^2 - (1 - q_1)(k_1^2 - k_2^2) = 0.\end{cases} \tag{11.11}$$

When solving the system (11.11), we limit ourselves to considering wave numbers with a positive real part. Dispersion curves $k_1(\omega)$, $k_2(\omega)$ are shown in Fig. 11.1. One pair of branches (k_1, k_2) ("1", Fig. 11.1) has asymptotes in the form of an oblique line $k_1 = \omega$ emanating from the origin of coordinates and a horizontal straight line $k_2 = 0$. The other pair of branches ("2", Fig. 11.1) is limited by parts of a quadratic parabola $k_{1,2} = \pm\sqrt{\dfrac{\omega}{2q_7}}$ (short dashed line, Fig. 11.1).

The attenuation factor $\gamma = \dfrac{k_2}{k_1}$ within the interval $-1 < \gamma < 1$ corresponds to the propagating wave. This factor within the said ranges ($\gamma_1 \to 0$, $|\gamma_2| > 1$) is for the only pair of branches (k_1, k_2). The attenuation factor is positive ($\gamma_1 > 0$) for vacancies, and it is negative ($\gamma_1 < 0$) for interstitials (Fig. 11.2).

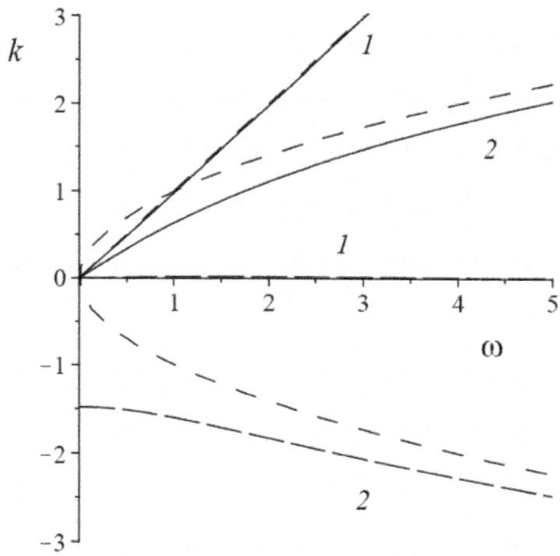

Fig. 11.1 Dependences $k_1(\omega)$ (solid line), $k_2(\omega)$ (long dashed line)

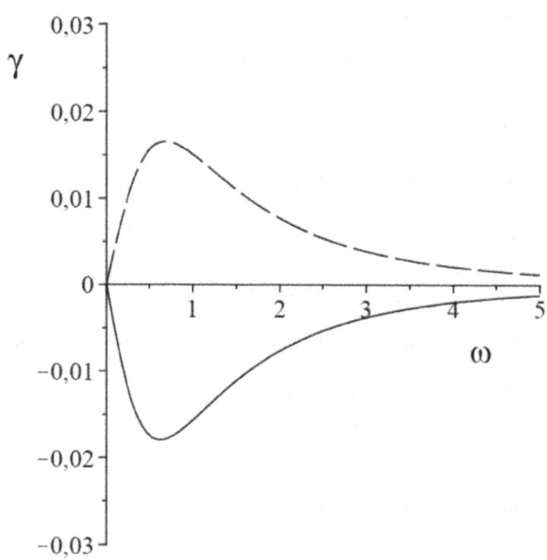

Fig. 11.2 Dependences $\gamma(\omega)$ at different values of parameter q_1 : $q_1 < 0$ (dashed line), $q_1 > 0$ (solid line)

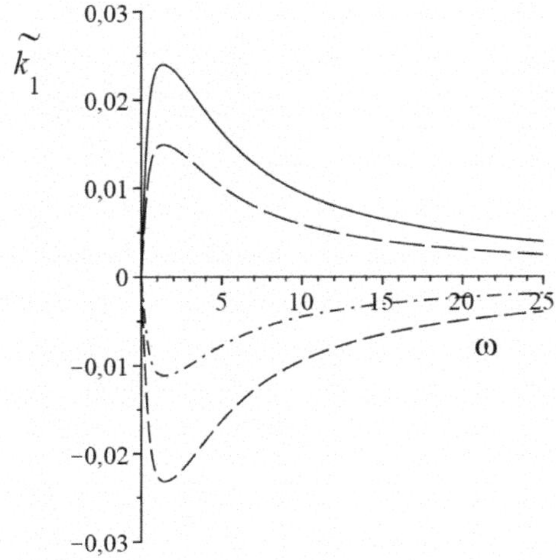

Fig. 11.3 Dependence $\widetilde{k_1}(\omega)$ for different values of parameter q_1 : $q_1^{(1)}$ (short dashed line), $q_1^{(2)}$ (dash-dot line), $q_1^{(3)}$ (long dashed line), $q_1^{(4)}$ (solid line); $q_1^{(1)} < q_1^{(2)} < 0 < q_1^{(3)} < q_1^{(4)}$, $q_1^{(4)} = |q_1^{(1)}|$, $q_1^{(3)} > |q_1^{(2)}|$

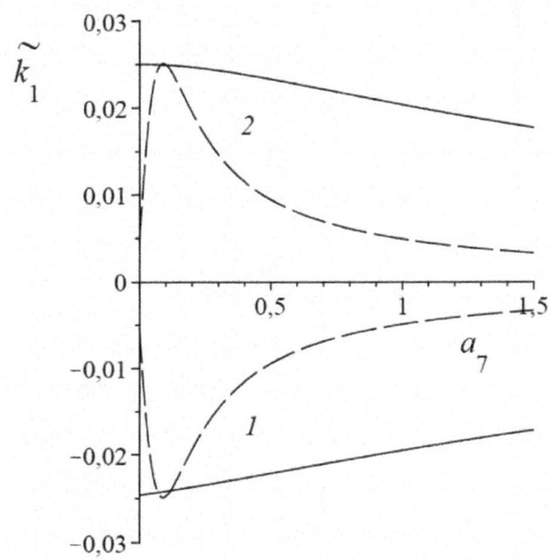

Fig. 11.4 Dependence $\widetilde{k_1}(q_7)$ for the fixed value of frequency ω and different values of parameter q_1: $\omega < 1$ (solid line), $\omega >> 1$ (dashed line) and $q_1 < 0$ ("1"), $q_1 > 0$ ("2")

The dynamics of one branch (corresponding to the propagating wave) of the dispersion curve $k_1(\omega)$ with changing parameters q_1, q_7 can be traced through the difference of this branch of the curve and its asymptote. Let us consider the function $\widetilde{k_1}(\omega) = k_1(\omega) - \omega$. The effect of the parameter q_1 on the behavior of the curve $\widetilde{k_1}(\omega)$ is shown in Fig. 11.3. Figure 11.4 shows the dependences $\widetilde{k_1}(q_7)$ typical for low and high frequencies.

The clear-cut dispersion is visible in the low-frequency region (Fig. 11.3). At any frequency values: $\widetilde{k_1} < 0$ at $q_1 < 0$ and $\widetilde{k_1} > 0$ at $q_1 > 0$.

The dependence $\widetilde{k_1}(\omega)$ on the parameter q_1 is monotonic: as the absolute value of the parameter increases, the curve shifts toward increasing absolute values of $\widetilde{k_1}$. The dependence $\widetilde{k_1}(\omega)$ on the parameter q_7 is monotonic (it decreases at $q_1 > 0$ and increases at $q_1 < 0$) in the low-frequency region and non-monotonic in the

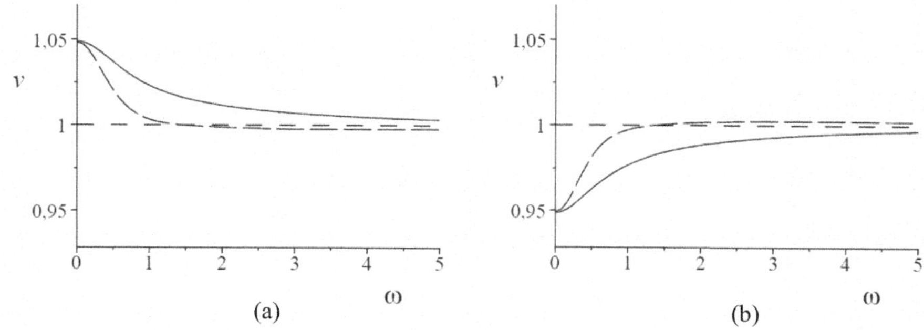

Fig. 11.5 Dependences $v_{ph}(\omega)$ (solid line) and $v_{gr}(\omega)$ (long dashed line) at different values of the parameter q_1: $q_1 < 0$ (**a**), $q_1 > 0$ (**b**)

high-frequency region (Fig. 11.4). As the value of parameter q_7 increases, the curve $\tilde{k}_1(\omega)$ shifts first toward increasing, then toward decreasing of absolute values of $\tilde{k}_1$ $(\omega >> 1)$.

The dependences of the phase $v_{ph} = \dfrac{\omega}{k_1}$ and group $v_{gr} = \dfrac{d\omega}{dk_1}$ velocities are shown in Fig. 11.5. For velocities in media with vacancies and interstitials, the inequalities $v_{ph} > v_{gr}$ and $v_{ph} < v_{gr}$ are satisfied, respectively, over the entire frequency interval. The velocities are close to the value $v_0 = \sqrt{1 - q_1}$ at low frequencies, and they are close to unity ($v_\infty \to 1$) at high frequencies. Furthermore, if defects are vacancies, $v_0 > v_\infty$, and if defects are interstitials, then $v_0 < v_\infty$.

The dispersion is obvious at low frequencies (Fig. 11.5) just the same as in Fig. 11.3. The normal dispersion is observed in media with vacancies, and the anomalous dispersion is observed in media with interstitials.

11.3.2 Medium with Two Types of Point Defects

Let us now consider a more complex general case, when both types of point defects: vacancies and interstices simultaneously occur in the material. We still assume that the waves travel in the direction of increase of the x axis ($k_1 > 0$).

As compared to the previous case, where the dispersion curve has two pairs of branches (and only one of them corresponds to a propagating wave with weak attenuation, $k_2 \to 0$), the curve herein has another pair of branches $k_1(\omega)$, $k_2(\omega)$, which can correspond at certain values of the system parameters to a propagating wave with the constant non-zero attenuation.

The real branches of the dispersion curve (11.9) have three asymptotes: one straight line and two quadratic parabolas with vertices at the origin of the coordinates

$$\omega = k_1, \ \omega = 2q_7(1 + q_2)k_1^2, \ \omega = 2q_3q_7(1 + q_2)k_1^2.$$

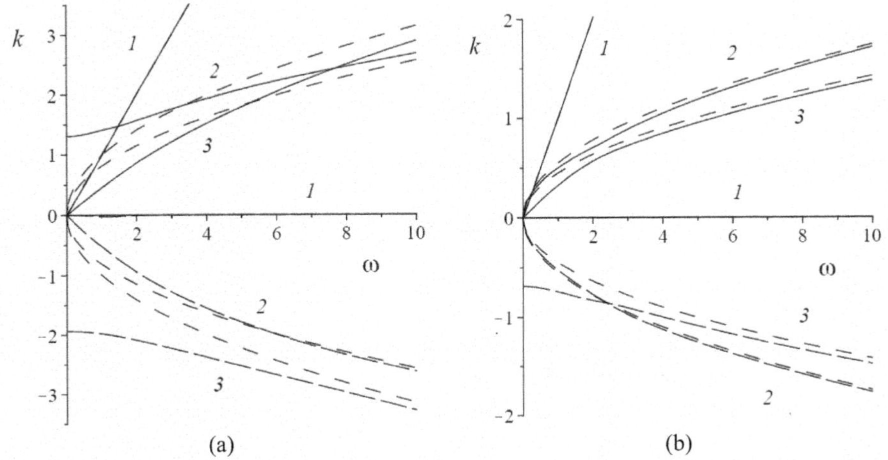

Fig. 11.6 Dependences $k_1(\omega)$ (solid line) and $k_2(\omega)$ (long dashed line) under the fulfillment (**a**) and non-fulfillment (**b**) of the condition (11.12)

One of three branches of the dependence $k_1(\omega)$ has a wave number cutoff, if the system parameters satisfy the inequality

$$-q_1q_4 - q_2 - q_1q_5q_6 + q_1q_6 + q_1q_2 + q_4q_5 > 0, \qquad (11.12)$$

the left-hand side thereof does not depend on the parameters q_3, q_7.

The system parameter ratio has a qualitative effect on the type of the dispersion curve (Fig. 11.6). Pairs of branches (k_1, k_2) are conditionally numbered with numerals in the figure. The corresponding dependences of the attenuation factor $\gamma_1 - \gamma_3$ are shown in Fig. 11.7. Asymptotes in the figures are marked as before with short dashed line.

The k_1 wave number cutoff is observed in Fig. 11.6a, but it is not shown in Fig. 11.6b.

If inequality (11.12) is true, at low frequencies there may be another longitudinal wave $(-1 < \gamma_2 < 1)$, which quickly decays in the course of its propagation. If the system parameters do not satisfy the inequality (11.12), only one longitudinal wave $(\gamma_1 \to 0)$ propagates in the medium over the entire frequency range as in the limiting cases (with one type of point defect).

The parameters q_3 and q_7 do not affect the existence of the second low-frequency wave.

Dependences $\widetilde{k_1}(\omega)$, $\widetilde{k_2}(\omega)$ and $\gamma(\omega)$ of the propagating wave are presented in Fig. 11.8.

It follows from the analysis of $\widetilde{k_1}(\omega)$ that the curve $k_1(\omega)$ mainly lies below its asymptote (Fig. 11.8a). Curves $k_2(\omega)$ and $\gamma(\omega)$ change the sign. Values k_1 and k_2 have different signs at low frequencies and the same sign at high frequencies. As the wave propagates, its amplitude decreases (the wave attenuates) at low frequencies;

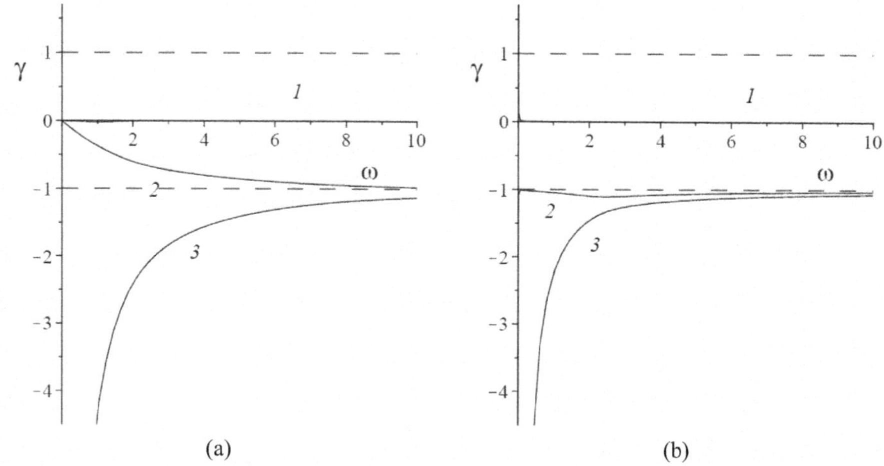

Fig. 11.7 Dependences $\gamma(\omega)$ (solid line) under the fulfillment (**a**) and non-fulfillment (**b**) of the condition (11.12)

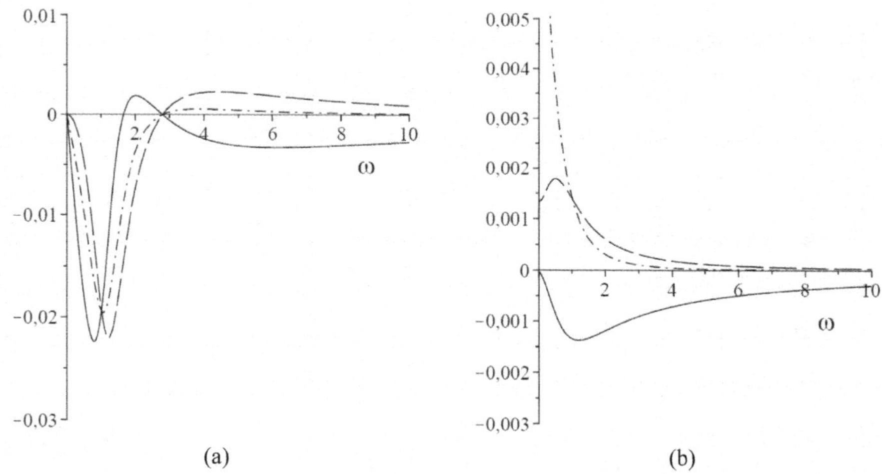

Fig. 11.8 Dependences $\widetilde{k_1}(\omega)$ (solid line), $\widetilde{k_2}(\omega)$ (long dashed line) and $\gamma(\omega)$ (dash-and-dot line) under the fulfillment (**a**) and non-fulfillment (**b**) of the condition (11.12)

a similar behavior is observed in the same-type media with interstitials. At high frequencies the amplitude of the wave increases (the wave intensifies). An increase in the wave amplitude is typical for media with vacancies.

If the inequality (11.12) is violated, k_1 and k_2 have the same sign over the entire frequency range.

Curves of the phase and group velocities are shown in Fig. 11.9.

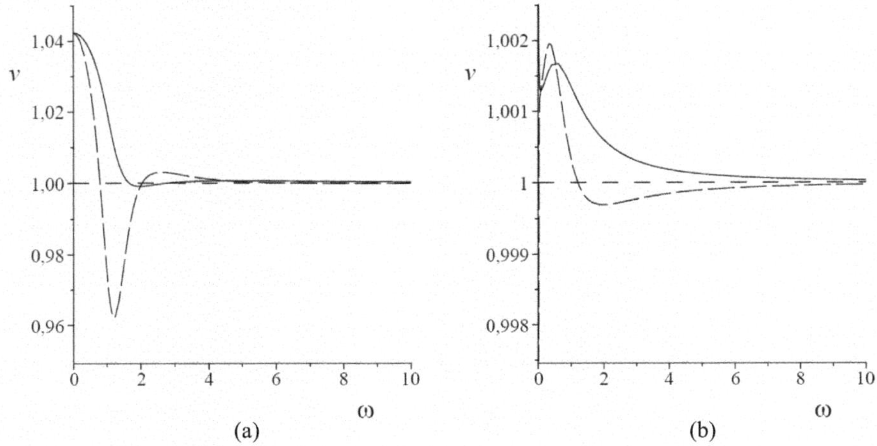

Fig. 11.9 Dependences $v_{ph}(\omega)$ (solid line) and $v_{gr}(\omega)$ (long dashed line) under the fulfillment (**a**) and non-fulfillment (**b**) of the condition (11.12)

The group velocity curve is located both below and above the phase velocity curve (Fig. 11.9a). At low and high frequencies, there are areas of the normal dispersion enclosing an area of the anomalous dispersion therebetween.

At high frequencies the normal dispersion is in both options.

It is obvious that if the defect diffusion coefficients are equal $q_3 \to 1$, the asymptotes-parabolas of the dispersion curve (11.9) coincide. The curves of the phase and group velocities have one intersection point. At low frequencies the normal dispersion is, at high frequencies the anomalous one emerges.

11.4 Evolution Equation and Nonlinear Stationary Wave

If there is only the static deformation $U = U(x)$, the system of Eqs. (11.1)–(11.3) will be reduced to one equation describing the dynamics of point defects:

$$\frac{\partial^2 n_V}{\partial t^2} - (\beta_V D_i + \beta_i D_V)\frac{\partial^2 n_V}{\partial x^2} + (\beta_i + \beta_V)\frac{\partial n_v}{\partial t} - (D_i + D_V)\frac{\partial^3 n_V}{\partial x^2 \partial t} + D_i D_V \frac{\partial^4 n_V}{\partial x^4} +$$
$$+(\beta_i\beta_V - \beta_{iV}\beta_{Vi})n_V \mp \frac{1}{2}\left(\lambda + \frac{2}{3}\mu\right)^2 \frac{\beta_N}{\rho^3 c_l^6} q_\varepsilon D_i \left(\frac{\Omega_i \beta_V}{\beta_{iV}} - \Omega_V\right)^2 \frac{\partial^2}{\partial x^2}\left(n_V^2\right) \pm$$
$$\pm\frac{1}{2}\left(\lambda + \frac{2}{3}\mu\right)^2 \frac{\beta_N}{\rho^3 c_l^6} q_\varepsilon(\beta_i - \beta_{iV})\left(\frac{\Omega_i \beta_V}{\beta_{iV}} - \Omega_V\right)^2 n_V^2 = (\beta_i - \beta_{iV})(q_0 - Q_\pm) \tag{11.13}$$

$(Q_+ = 0,\ Q_- = \dfrac{2q_\varepsilon \rho c_l^2}{\beta_N})$. The following expressions are thereat obtained for the static deformation:

$$\left(\frac{dU}{dx}\right)_\pm = \pm c_1 \frac{\partial^2 n_V}{\partial x^2} \mp \left(c_2 \frac{\partial n_V}{\partial t} + c_3 n_V^2 + c_4 n_V\right) - C_\pm^{(1)}, \tag{11.14}$$

where

$$c_1 = \left(\lambda + \frac{2}{3}\mu\right) \frac{1}{\rho c_l^2} \frac{\Omega_i D_V}{\beta_{iV}},\quad c_2 = \left(\lambda + \frac{2}{3}\mu\right) \frac{1}{\rho c_l^2} \frac{\Omega_i}{\beta_{iV}},$$

$$c_3 = \frac{1}{2}\left(\lambda + \frac{2}{3}\mu\right)^2 \frac{\beta_N}{\rho^3 c_l^6} \left(\frac{\Omega_i \beta_V}{\beta_{iV}} - \Omega_V\right)^2,$$

$$c_4 = \left(\lambda + \frac{2}{3}\mu\right) \frac{1}{\rho c_l^2} \left(\frac{\Omega_i \beta_V}{\beta_{iV}} - \Omega_V\right),\quad C_+^{(1)} = 0,\ C_-^{(1)} = \frac{2}{\beta_N}(\lambda + 2\mu).$$

The upper sign in the last expression determines the upper sign in Eq. (11.13) and subsequent equations in the text.

Equation (11.13) and dependence (11.14) were obtained based on the fact that the recombination rates at sinks and the mutual recombination rates of defects are of the same order and subject to the minor deformation correction ($\beta_i \sim \beta_V,\ \beta_{iV} \sim \beta_{Vi},\ q_\varepsilon \to 0$).

We look for a solution to Eq. (11.13) in the form of the asymptotic expansion for a small parameter: $n_V = n_0 + \tilde{\varepsilon} n_1 + \ldots\ (\tilde{\varepsilon} << 1)$. Let us introduce herein new variables: $\xi = x - ct$. This choice of variables is explained by the fact that the perturbation propagating at velocity along the axis slowly evolves in space because of the nonlinearity, dispersion and dissipation.

In the zero approximation for $\tilde{\varepsilon}$, we obtain an expression for velocity: $c = \sqrt{\beta_V D_i + \beta_i D_V}$. The first approximation for $\tilde{\varepsilon}$ leads to the evolution equation with respect to n_0:

$$\frac{\partial}{\partial \xi}\left(\frac{\partial n_0}{\partial \eta} \pm a_1 n_0 \frac{\partial n_0}{\partial \xi} - a_2 \frac{\partial^3 n_0}{\partial \xi^3} - a_3 \frac{\partial^3 n_0}{\partial \xi^3} + a_4 n_0\right) = a_5 n_0 \pm a_6 n_0^2 - a_0. \tag{11.15}$$

Here the following notations are agreed:

$$a_1 = \frac{q_\varepsilon}{2\tilde{\varepsilon}} \frac{\beta_N D_i \left(\lambda + \frac{2}{3}\mu\right)^2 \left(\frac{\Omega_i \beta_V}{\beta_{iV}} - \Omega_V\right)^2}{\rho^3 c_l^6 (\beta_V D_i + \beta_i D_V)},\quad a_2 = \frac{D_i + D_V}{2\tilde{\varepsilon}\sqrt{\beta_V D_i + \beta_i D_V}},$$

$$a_3 = \frac{D_i D_V}{2\tilde{\varepsilon}(\beta_V D_i + \beta_i D_V)}, \quad a_4 = \frac{\beta_i + \beta_V}{2\tilde{\varepsilon}\sqrt{\beta_V D_i + \beta_i D_V}}, \quad a_5 = \frac{\beta_i \beta_V - \beta_{iV}\beta_{Vi}}{2\tilde{\varepsilon}(\beta_V D_i + \beta_i D_V)},$$

$$a_6 = \frac{q_\varepsilon}{4\tilde{\varepsilon}} \frac{\beta_N(\beta_i - \beta_{iV})\left(\lambda + \frac{2}{3}\mu\right)^2 \left(\frac{\Omega_i \beta_V}{\beta_{iV}} - \Omega_V\right)^2}{\rho^3 c_l^6 (\beta_V D_i + \beta_i D_V)}, \quad a_0 = \frac{(\beta_i - \beta_{iV})q_0}{2\tilde{\varepsilon}(\beta_V D_i + \beta_i D_V)}.$$

Note that in Eq. (11.15) the terms with coefficients a_1 and a_6 are determined by the available nonlinearity, the terms with coefficients a_2 and a_4 are responsible for dissipative processes, and the terms with coefficients a_3 and a_5 are responsible for dispersion processes.

If the recombination of defects and diffusion processes in the system occur equally strong, the terms in Eq. (11.15) cannot be neglected; it is necessary to consider the entire equation. We assume that there is no constraining force ($a_0 = 0$).

Equation (11.15) in the traveling wave variables $n_0(\xi, \eta) = n_0(z)$, $z = \xi - v\eta$ (z is the traveling coordinate, v is a non-dimensional value meaningful as the nonlinear wave velocity) will be as follows:

$$\frac{d}{dz}\left(-v\frac{dn_0}{dz} \pm a_1 n_0 \frac{dn_0}{dz} - a_2 \frac{d^2 n_0}{dz^2} - a_3 \frac{d^3 n_0}{dz^3} + a_4 n_0\right) = a_5 n_0 \pm a_6 n_0^2. \quad (11.16)$$

The solution to this equation can be found using the method of simple equations [45, 46]. The method was proposed by N. A. Kudryashov. It is used to search for exact solutions of nonlinear ordinary differential equations and enables one to construct solutions to the original equation based on known solutions to the simplest equations. The simplest equations can be the Riccati, Jacobi and Weierstrass equations. Here the simplest equation is the Riccati equation with constant coefficients.

We find a partial solution to Eq. (11.16) as follows:

$$n_0(z) = b_0 Y^2 + b_1 Y + b_2, \quad (11.17)$$

where $Y(z)$ is the solution to the Riccati equation $Y_z = -Y^2 + B_0$, the constants b_0, b_1, b_2 and B_0 are determined upon substituting (11.17) into Eq. (11.16) and have the form

$$b_0 = \pm\frac{12 a_3}{a_1}, \quad b_1 = \mp\frac{12}{5}\frac{a_2}{a_1}, \quad b_2 = \mp\frac{20 a_3 a_6}{a_1^2} \pm \frac{10 a_3 a_4}{a_1 a_2} \mp \frac{8}{25}\frac{a_2^2}{a_1 a_3} \pm \frac{1}{10}\frac{a_2 a_4}{a_3 a_6} \mp \frac{a_5}{1 a_6},$$

$$B_0 = -\frac{5}{4}\frac{a_4}{a_2} + \frac{5}{2}\frac{a_6}{a_1} + \frac{1}{100}\frac{a_2^2}{a_3^2}.$$

In Eq. (11.16), a constraint is imposed on the dimensionless velocity of a nonlinear wave

$$v = \frac{1}{10}\frac{a_1 a_2 a_4}{a_3 a_6} - \frac{a_1 a_5}{2a_6} - \frac{a_2^2}{5a_3} - \frac{2a_3 a_6}{a_1}.$$

Furthermore, there are two more nonlinear algebraic equations of the system; therefrom we obtain three pairs of relations binding the parameters of Eq. (11.16). Two parameters of the equation appear here to be dependent on the remaining four ones, provided that there is no constraining force.

Additional conditions for the parameters thereunder Eq. (11.16) are integrated

$$\left[\begin{array}{l} a_6 = \frac{1}{50}\frac{a_1 a_2^2}{a_3^2},\ a_4 = \frac{6}{125}\frac{a_2^3}{a_3^2} - \frac{3125}{36}\frac{a_3^4 a_5^2}{a_2^5} \\ a_6 = \frac{25}{12}\frac{a_1 a_3 a_5}{a_2^2},\ a_4 = \frac{25}{6}\frac{a_3 a_5}{a_2} \\ a_6 = -\frac{25}{12}\frac{a_1 a_3 a_5}{a_2^2},\ a_4 = -\frac{25}{6}\frac{a_3 a_5}{a_2}. \end{array}\right. \tag{11.18}$$

In each of the cases (11.18) the solution (11.17) is as follows:

$$\bar{n}_{0,1}^{(\pm)}(\bar{z}_1) = \pm\frac{1}{4\tilde{a}}(th(\bar{z}_1) - \tilde{a})^2 \mp \frac{1}{2}sign(a_5) \mp \frac{1}{4}\left(\frac{1}{\tilde{a}} + \tilde{a}\right), \tag{11.19}$$

$$\bar{n}_{0,2}^{(\pm)}(\bar{z}_2) = \pm\frac{1}{4}(th(\bar{z}_2) - 1)^2 \mp 1, \tag{11.20}$$

$$\bar{n}_{0,3}^{(\pm)}(\bar{z}_3) = \pm\frac{1}{4}(th(\bar{z}_3) - 1)^2, \tag{11.21}$$

where $\tilde{a} = \frac{6}{625}\frac{a_2^4}{a_3^3|a_5|}$; $\bar{n}_{o,i}$, $\bar{z}_i$ are non-dimensional variables, $\bar{n}_{o,i} = n_{o,i}/A_i$, $\bar{z}_i = z/\Delta_i$; $A_i > 0$, $\Delta_i > 0$ are specific values of the volume concentration and coordinates

$$A_1 = \frac{50a_3^2|a_5|}{a_1 a_2^2},\ \Delta_1 = \frac{12}{125}\frac{a_2^3}{a_3^2|a_5|}, \tag{11.22}$$

$$A_2 = A_3 = \frac{12}{25}\frac{a_2^2}{a_1 a_3} = \frac{|a_5|}{|a_6|},\ \Delta_2 = \Delta_3 = \frac{10a_3}{a_2} = \frac{12}{5}\frac{a_4}{|a_5|}. \tag{11.23}$$

The amplitude and width of stationary waves (11.19)–(11.21) are determined for each case by dependences (11.22), (11.23). Taking into account (11.18), the wave velocities assume the form

$$v_1 = -\frac{25^3}{36}\frac{a_3^5 a_5^2}{a_2^6}\left(1 + \frac{36}{625}\frac{a_2^4}{a_3^3 a_5}\right), \tag{11.24}$$

$$v_2 = -v_3 = -\frac{25}{6}\frac{a_3^2 a_5}{a_2^2}\left(1 + \frac{36}{625}\frac{a_2^4}{a_3^3 a_5}\right). \tag{11.25}$$

Solutions (11.19) degenerate into solutions (11.20) or (11.21), under the fulfillment of the condition

$$\widetilde{a} = 1 \tag{11.26}$$

and, accordingly, at $a_5 > 0$ or $a_5 < 0$. The amplitude and width dependences (11.22) transform into (11.23). The velocity (11.24) turns thereat into (11.25) and is equal to $v_1 = -\dfrac{7}{25}\dfrac{a_2^2}{a_3} = v_2$ or $v_2 = \dfrac{a_2^2}{5a_3} = v_3$, respectively.

Solutions (11.20), (11.21) shall be considered as special cases of the solution (11.19) obtained with a certain ratio of the initial parameters (11.26).

Figures 11.10, 11.11 and 11.12 show the profiles of waves corresponding to solutions (11.19)–(11.21); each of the profiles has the shape of a kink—a smooth drop between two function values. The solution $\bar{n}_{0,1}^{(+)}$ appears to be a non-monotonic kink in the region of the jump, which at certain parameter values is transformed into a monotonic decreasing kink with a symmetrical profile. The derivative of the nonmonotonic dependence has two extrema. Solutions $\bar{n}_{0,2}^{(+)}$, $\bar{n}_{0,3}^{(+)}$ have the shape of monotonic decreasing kinks which are asymmetrical about the inflection point (Fig. 11.11). Dependences $\bar{n}_{0,1}^{(-)}$ for different parameter $\widetilde{a}$ values are shown in Fig. 11.12. Asymptotes are marked in the figures with short dashed lines with spaces.

It is clear from (11.19)–(11.21) that $\bar{n}_{0,i}^{(+)}$ and $\bar{n}_{0,i}^{(-)}$ $(i = \overline{1,3})$ are mirror images of each other relative to the horizontal axis. For any values of parameters (a_1, a_2, a_3, a_5) for monotonic and non-monotonic kinks $n_{0,i}^{\pm(z)}$ is fulfilled

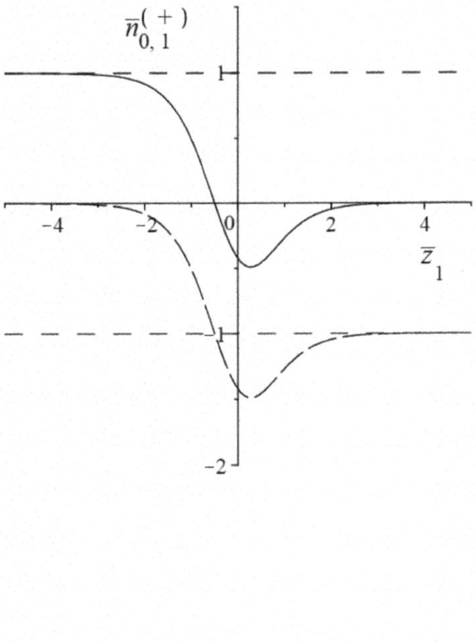

Fig. 11.10 Wave profile $\bar{n}_{0,1}^{(+)}(\bar{z}_1)$ at $a_5 < 0$ (solid line) and at $a_5 > 0$ (long dashed line)

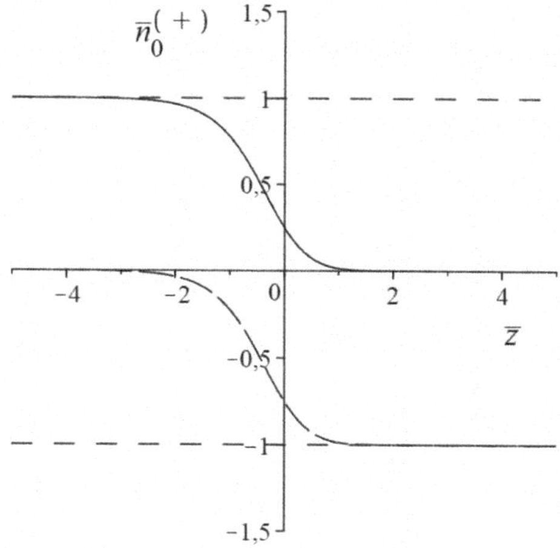

Fig. 11.11 Wave profile $\bar{n}_{0,2}^{(+)}(\bar{z}_2)$ (long dashed line), $\bar{n}_{0,3}^{(+)}(\bar{z}_3)$ (solid line)

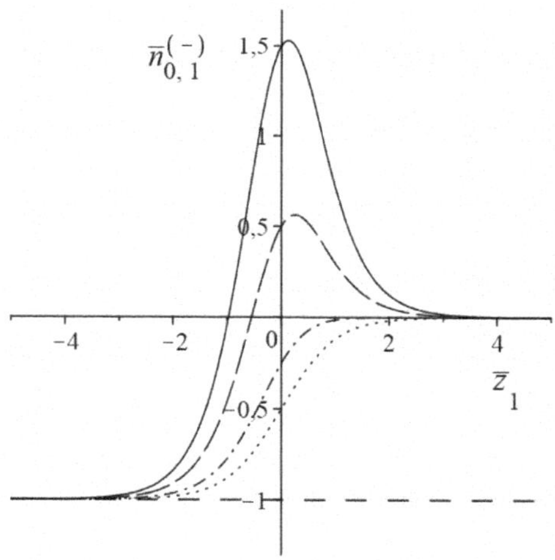

Fig. 11.12 Wave profile $\bar{n}_{0,1}^{(-)}(\bar{z}_1)$ under changing parameter $\widetilde{a}$: $\widetilde{a} = \widetilde{a}^{(1)}$ (solid line), $\widetilde{a} = \widetilde{a}^{(2)}$ (long dashed line), $\widetilde{a} = 1$ (dash-and-dot line), $\widetilde{a} = \widetilde{a}^{(3)}$ (dotted line) $0 < \widetilde{a}^{(1)} < \widetilde{a}^{(2)} < 1$, $\widetilde{a}^{(3)} >> 1$ $(a_5 < 0)$

$$\left[\begin{matrix} n_{0,1}^{(\pm)}\Big|_{a_5>0} \\ n_{0,2}^{(\pm)} \end{matrix}\right]\Bigg|_{z\to-\infty} \to \mp 0, \quad \left[\begin{matrix} n_{0,1}^{(\pm)}\Big|_{a_5>0} \\ n_{0,2}^{(\pm)} \end{matrix}\right]\Bigg|_{z\to+\infty} \to \mp A_i,$$

$$\left[\begin{matrix} n_{0,1}^{(\pm)}\Big|_{a_5<0} \\ n_{0,3}^{(\pm)} \end{matrix}\right]\Bigg|_{z\to-\infty} \to \pm A_i, \quad \left[\begin{matrix} n_{0,1}^{(\pm)}\Big|_{a_5<0} \\ n_{0,3}^{(\pm)} \end{matrix}\right]\Bigg|_{z\to+\infty} \to \pm 0.$$

Dependences $n_{0,i}^{(+)}(z)$ are decreasing $\left(n_{0,i}^{(+)}\Big|_{z\to-\infty} > n_{0,i}^{(+)}\Big|_{z\to+\infty}\right)$, and $n_{0,i}^{(-)}(z)$ are increasing $\left(n_{0,i}^{(-)}\Big|_{z\to-\infty} < n_{0,i}^{(-)}\Big|_{z\to+\infty}\right)$.

The parameter $\widetilde{a}$ affects the type of curve (11.19) and its monotonicity. At parameter values $\widetilde{a} < 1$ the decreasing/increasing kink is non-monotonic, and at $\widetilde{a} > 1$ it is monotonic. At parameter values $\widetilde{a} >> 1$, the kink acquires symmetry about the inflection point. The critical value of the parameter $a_2 = a_2^*$ ($a_3 = a_3^*$, $|a_5| = a_5^*$) is determined from the condition (11.26), thereunder the kink monotonicity is violated.

Increasing values of the parameter a_1 reduces the amplitude (A_1) but does not change the shape of the curve $n_{0,1}^{(\pm)}(z)$. The kink speed (v_1) and width (Δ_1) do not depend on the parameter a_1.

With increasing values of the parameter or decreasing values of the parameter a_3 ($|a_5|$): the amplitude (A_1) decreases, the width (Δ_1) increases, the absolute values of the velocity (v_1) decrease or increase ($a_5 < 0$) within a small interval of the parameter a_2 ($a_{3,5}$).

The sign of the parameter a_5 affects the displacement of the curve $n_{0,1}^{(\pm)}(z)$ along the vertical axis. In case of the parameter positive values the kink differs in its displacement along the vertical axis down/up ($n_{0,1}^{(+)}/n_{0,1}^{(-)}$) by the amplitude value (A_1).

The effect of parameters a_1, a_2, a_3 on dependences $n_{0,2}^{(\pm)}(z)$, $n_{0,3}^{(\pm)}(z)$ is other than on $n_{0,1}^{(\pm)}(z)$. An increase in values of the parameter a_1 reduces the amplitude of the decreasing/increasing kink. When increasing values of the parameter a_2 or decreasing values of the parameter a_3: the amplitude increases, the width decreases and the absolute values of velocity (v_2, v_3) decrease and then increase. Intervals of positive and negative values of the velocity (v_2, v_3) emerge at $a_5 < 0$; at $a_5 > 0$: $v_2 < 0$, $v_3 > 0$.

Positive and negative values of velocity v designate the relative direction of the wave movement. Values $v > 0$ increase the wave propagation velocity, and $v < 0$ decrease thereof.

Regarding the positivity of the volume concentration function, dependences $n_{0,1}^{(+)}\Big|_{a_5<0}$, $n_{0,1}^{(-)}\Big|_{a_5>0}$, $n_{0,2}^{(-)}$ and $n_{(0,3)}^{(+)}$ shall be the most significant for the analysis.

The amplitude, width and velocity (11.22), (11.23), (11.24), (11.25) of a nonlinear wave are bound with the initial parameters of the system by the following relations:

$$A_1 = \frac{50}{q_\varepsilon}\frac{(\lambda+2\mu)^3}{\beta_N\left(\lambda+\dfrac{2}{3}\mu\right)^2}\frac{D_i D_V^2}{(D_i+D_V)^2}\frac{1}{\left(\dfrac{\beta_V}{\beta_{iV}}\Omega_i-\Omega_V\right)^2}\frac{|\beta_i\beta_V-\beta_{iV}\beta_{Vi}|}{(\beta_V D_i+\beta_i D_V)},$$

$$\Delta_1 = \frac{12}{125}\frac{(D_i+D_V)^3}{D_i^2 D_V^2}\frac{(\beta_V D_i+\beta_i D_V)^{3/2}}{|\beta_i\beta_V-\beta_{iV}\beta_{Vi}|},$$

$$v_1 = -\frac{25}{2\varepsilon}\frac{D_i^2 D_V^2}{(D_i+D_V)^2}\frac{(\beta_i\beta_V-\beta_{iV}\beta_{Vi})}{(\beta_V D_i+\beta_i D_V)^2}\left(\frac{625}{36}\frac{D_i^3 D_V^3}{(D_i+D_V)^4}\frac{(\beta_i\beta_V-\beta_{iV}\beta_{Vi})}{(\beta_V D_i+\beta_i D_V)^2}+1\right)$$

$$A_2 = A_3 = \frac{12}{25q_\varepsilon}\frac{(\lambda+2\mu)^3}{\beta_N\left(\lambda+\dfrac{2}{3}\mu\right)^2}\frac{(D_i+D_V)^2}{D_V D_i^2}\frac{(\beta_V D_i+\beta_i D_V)}{\left(\dfrac{\beta_V}{\beta_{iV}}\Omega_i-\Omega_V\right)^2},$$

$$\Delta_2 = \Delta_3 = \frac{10}{\sqrt{\beta_V D_i + \beta_i D_V}} \frac{D_i D_V}{(D_i + D_V)},$$

$$v_2 = -v_3 = -\frac{1}{\varepsilon}\left(\frac{25}{12}\frac{D_i^2 D_V^2}{(D_i + D_V)^2}\frac{(\beta_i \beta_V - \beta_{iV}\beta_{Vi})}{(\beta_V D_i + \beta_i D_V)^2} + \frac{3}{25}\frac{(D_i + D_V)^2}{D_i D_V}\right).$$

When analyzing them, we note that the width and velocity of stationary waves only depend on the diffusion and recombination rate of defects.

Longitudinal deformation (11.14) in the zero approximation for $\widetilde{\varepsilon}$ ($U = U_0 + \widetilde{\varepsilon}U_1 + ..., n_V = n_0 + \widetilde{\varepsilon}n_1 + ..., \xi = x - ct, \eta = \widetilde{\varepsilon}x$)

$$\left(\frac{\partial U}{\partial \xi}\right)_\pm = \pm\left(c_1\frac{\partial^2 n_0}{\partial \xi^2} + c_2 c\frac{\partial n_0}{\partial \xi}\right) \mp (c_3 n_0^2 + c_4 n_0) - C_\pm^{(1)}$$

and in traveling wave variables ($U_0(\xi, \eta) = U_0(z), n_0(\xi, \eta) = n_0(z), z = \xi - v\eta$)

$$\left(\frac{dU_0}{dz}\right)_\pm = \pm\left(c_1\frac{d^2 n_0}{dz^2} + c_2 c\frac{dn_0}{dz}\right) \mp (c_3 n_0^2 + c_4 n_0) - C_\pm^{(1)}, \tag{11.27}$$

where $c = \sqrt{\beta_V D_i + \beta_i D_V}$.

It is here useful to introduce non-dimensional parameter complexes q_i ($i = \overline{1, 7}$) as follows:

$$\begin{gathered} q_1 = \left(\lambda + \frac{2}{3}\mu\right)\frac{q_\varepsilon}{\rho c_l^2}\frac{\Omega_V}{\beta_V}, \quad q_2 = \frac{\beta_i}{\beta_V}, \quad q_3 = \frac{D_i}{D_V}, \quad q_4 = \frac{\beta_{iV}}{\beta_V}, \\ q_5 = \frac{\beta_{Vi}}{\beta_V}, \quad q_6 = \frac{\Omega_i}{\Omega_V}, \quad q_7 = \frac{\beta_V D_V}{c_l^2}; \end{gathered} \tag{11.28}$$

all the parameters are positive, except for $q_1 < 0, q_6 < 0$.

Deformation (11.27) will be written as

$$\left(\frac{dU_0}{dz}\right)_\pm = c_3 A_i^2 (W_0)_\pm,$$

$$(W_0)_\pm = \pm\left(\widetilde{c}_1\frac{d^2\overline{n}_{0,i}}{d\overline{z}_i^2} + \widetilde{c}_2\frac{d\overline{n}_{0,i}}{d\overline{z}_i}\right) \mp (\widetilde{c}_3\overline{n}_{0,i}^2 + \widetilde{c}_3\overline{n}_{0,i}) - C_\pm^{(2)}, \tag{11.29}$$

where $\widetilde{c}_1 = \left(\frac{125}{12}\right)^2 \frac{q_3^4 (q_2 - q_4 q_5)^2}{q_4 (1 + q_3)^6 (q_2 + q_3)^3}$, $\widetilde{c}_2 = \frac{125}{12}\frac{q_3^2 |q_2 - q_4 q_5|}{q_4 (1 + q_3)^3 (q_2 + q_3)}$, $\widetilde{c}_3 = \frac{25 q_3 |q_2 - q_4 q_5|}{q_1 q_6 (1 + q_3)^2 (q_2 + q_3)}$, $\widetilde{c}_4 = -\frac{q_4 - q_6}{q_4 q_6}$, $C_+^{(2)} = 0$, $C_-^{(2)} = \frac{1}{25}\frac{q_1 (1 + q_3)^2 (q_2 + q_3)(q_4 - q_6)^2}{q_3 q_4^2 q_6 |q_2 - q_4 q_5|}$.

In non-dimensional parameters conditions (11.18) assume the following form:

$$\begin{gathered}\frac{1}{25}\frac{(q_2+q_3)(1+q_3)^2}{q_3(q_2-q_4)}=1,\\ \frac{6}{125}\frac{(q_2+q_3)(1+q_3)^3}{q_3^2(1+q_2)}-\frac{3125}{36}\frac{q_3^4(q_2-q_4q_5)^2}{(1+q_2)(1+q_3)^5(q_2+q_3)^3}=1,\end{gathered} \tag{11.30}$$

$$\frac{(q_2-q_4)(1+q_3)}{q_3(1+q_2)}=1,\ \frac{25}{6}\frac{q_3(q_2-q_4q_5)}{(1+q_2)(1+q_3)(q_2+q_3)}=1,$$

$$\frac{(q_2-q_4)(1+q_3)}{q_3(1+q_2)}=1,\ \frac{25}{6}\frac{q_3(q_2-q_4q_5)}{(1+q_2)(1+q_3)(q_2+q_3)}=-1.$$

We turn now in (11.19) to parameters q_i

$$\overline{n}_{0,1}^{(\pm)}(\overline{z}_1)=\pm\frac{1}{4\widetilde{q}}(th(\overline{z}_1)-\widetilde{q})^2\pm\frac{1}{2}\mp\frac{1}{4}\left(\frac{1}{\widetilde{q}}+\widetilde{q}\right), \tag{11.31}$$

where $\widetilde{q}=\dfrac{6}{625}\dfrac{(1+q_3)^4(q_2+q_3)^2}{q_3^3|q_2-q_4q_5|}$. Excluding parameters q_4, q_5 based on (11.30), we obtain the expression for the volume defect concentration (11.31) depending on two parameters q_2 and q_3, which characterize the relative values, respectively, of the rate of the defect recombination at sinks and the diffusion coefficient of defects. Expressions for volume concentrations $\overline{n}_{0,2}^{(\pm)}(\overline{z}_2)$, $\overline{n}_{0,3}^{(\pm)}(\overline{z}_3)$ do not depend on parameters.

The positivity of parameters q_4, q_5 determines the range of possible values for parameters q_2, q_3 in the first ($\overline{n}_{0,1}^{(+)}$) and third ($\overline{n}_{0,3}^{(\pm)}$) cases, respectively,

$$\begin{cases}q_2(1-23q_3+q_3^2)+q_3(1+q_3)^2<0\\ q_2(1+6q_3)(6-18q_3+q_3^2)+q_3(6+q_3)(1-18q_3+6q_3^2)>0\end{cases}\quad\text{and } q_3<q_2. \tag{11.32}$$

The regions (“A”, “B”, “E”) of acceptable values for parameters q_2, q_3 (11.32) are shown in Fig. 11.13. The dotted line marks here the condition (11.26), which looks in the new parameters like this

$$q_2(q_3^3-22q_3^2+3q_3+1)+q_3(q_3^3+3q_3^2-22q_3+1)=0. \tag{11.33}$$

In the first case, within the boundaries of the “A” and “B” regions along the upper and lower edges, respectively, there are narrow stripes—the regions “C” and “D”, there within the dependence $\overline{n}_{0,1}^{(-)}(\overline{z}_1)$ makes sense.

In the second case ($\overline{n}_{0,2}^{(\pm)}$), under the simultaneous fulfillment of inequalities $q_4>0, q_5>0$, there is no acceptable region of admissible values in the plane of parameters (q_2, q_3).

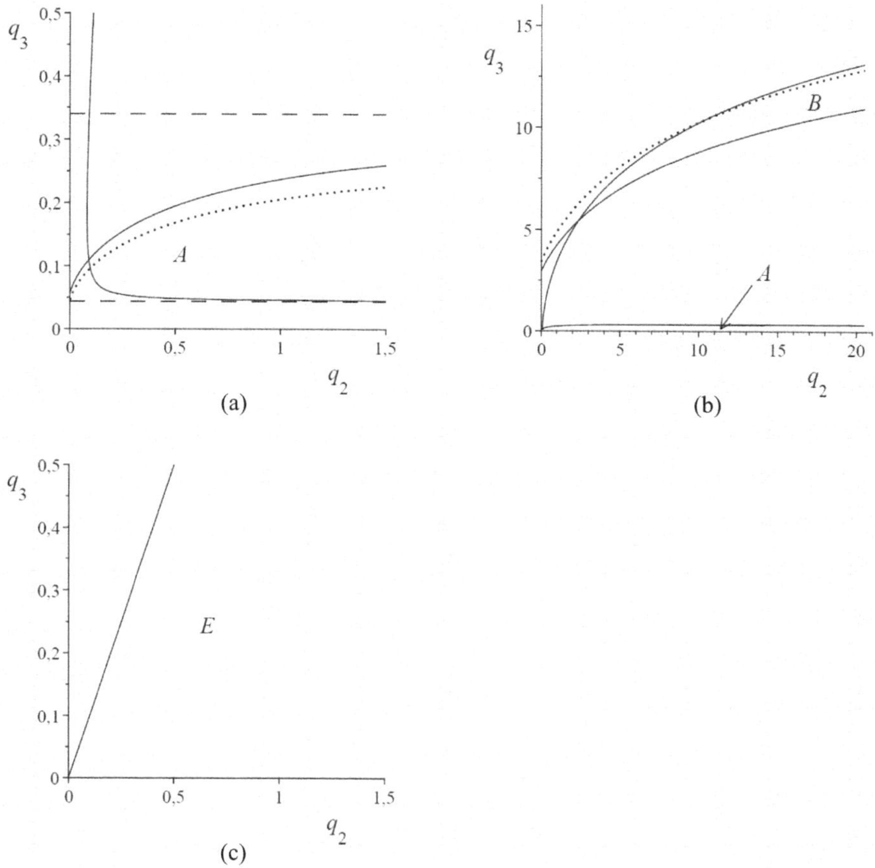

Fig. 11.13 Ranges of admissible values “A”, “B” ($\overline{n}_{0,1}^{(+)}$) and “E” ($\overline{n}_{0,3}^{(\pm)}$) in the plane of parameters (q_1, q_3)

Profiles of volume concentration dependences $\overline{n}_{0,1}^{(\pm)}(\overline{z}_1)$ in the “A” and “C” regions at a fixed value of q_2 and different values of q_3 are shown in Fig. 11.14. In the “B” and “D” regions, the $\overline{n}_{0,1}^{(\pm)}(\overline{z}_1)$ dependence profiles have a similar appearance. The relevant deformation dependences (11.29) are shown below in Fig. 11.15.

Condition (11.33) limits the region of existence of dependences $\overline{n}_{0,1}^{(\pm)}(\overline{z}_1)$ with a monotonic profile and determines the critical value of the parameter: q_2^* or q_3^*.

At parameter values $q_2 < q_2^*/q_3 > q_3^*$ (“A”) or $q_2 > q_2^*/q_3 < q_3^*$ (“B”), the dependence $\overline{n}_{0,1}^{(+)}(\overline{z}_1)$ has a monotonic decreasing profile. The dependence $\overline{n}_{0,1}^{(-)}(\overline{z}_1)$ in the “C” or “D” region has a monotonic increasing profile. Both dependences are located in the upper half-plane.

The increasing (need not be monotonic) dependence characterizing the deformation $(W_0)_+$ is mainly in the lower half-plane, and the upper limit is zero. The monotonic increasing dependence characterizing the deformation $(W_0)_-$ shifts from the

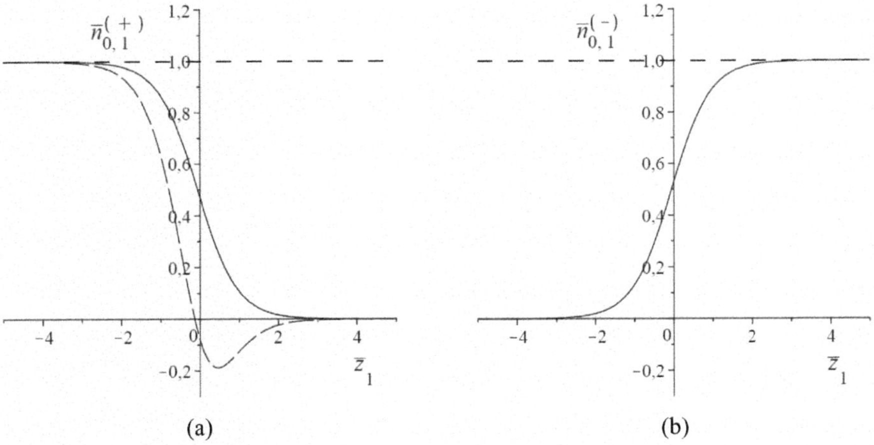

Fig. 11.14 Wave profiles $\overline{n}_{0,1}^{(+)}(\overline{z}_1)$ (a) and $\overline{n}_{0,1}^{(-)}(\overline{z}_1)$ (b) when changing the parameter q_3: $q_3^{(1)}$ (long dashed line), $q_3^{(2)}$ (solid line), $q_3^{(1)} < q_3^* < q_3^{(2)}$

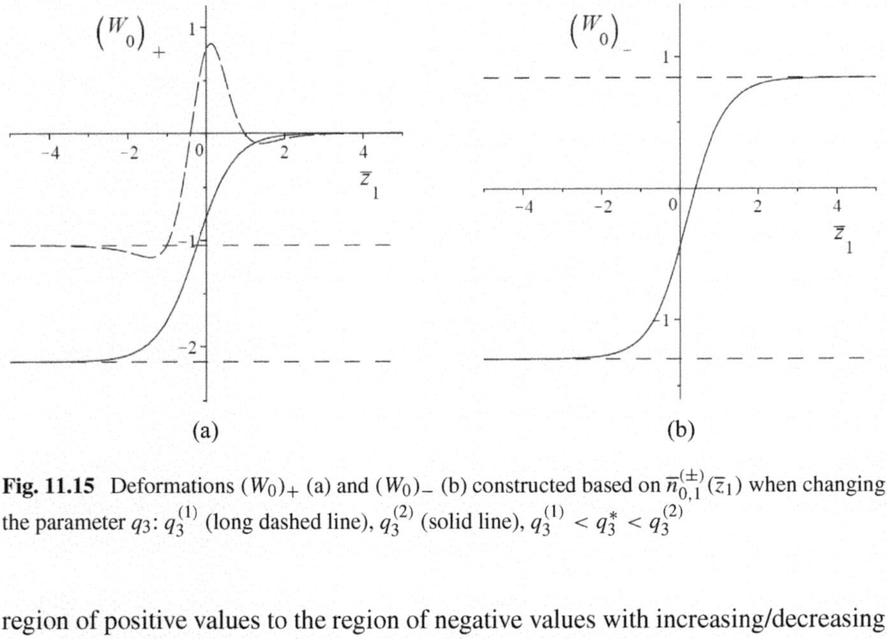

Fig. 11.15 Deformations $(W_0)_+$ (a) and $(W_0)_-$ (b) constructed based on $\overline{n}_{0,1}^{(\pm)}(\overline{z}_1)$ when changing the parameter q_3: $q_3^{(1)}$ (long dashed line), $q_3^{(2)}$ (solid line), $q_3^{(1)} < q_3^* < q_3^{(2)}$

region of positive values to the region of negative values with increasing/decreasing values of the parameter q_3 from the "C"/"D" region. At a certain parameter value from the acceptable region the upper limit tends to zero.

Dependences characterizing the deformation $(W_0)_+$ (11.29) in the third case at a fixed value of q_3 and different values of q_2 are shown in Fig. 11.16.

The deformation $(W_0)_+$ takes on positive values starting from a certain value $\overline{z}_3 = \overline{z}_3^*$ $((W_0)_+|_{\overline{z}_3=\overline{z}_3^*})$. In this case, the critical parameter value $q_2 = q_2^*$ points to the minimum value of the coordinate $\overline{z}_3^* > 0$ therewith the deformation changes the sign.

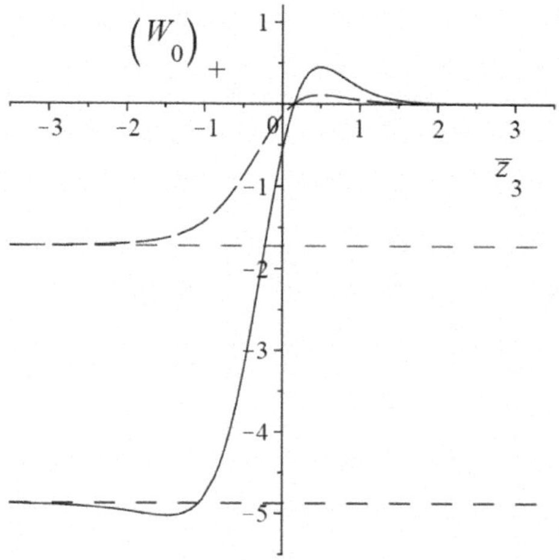

Fig. 11.16 Deformation $(W_0)_+$ constructed based on $\overline{n}_{0,3}^{(+)}(\overline{z}_3)$ when changing the parameter q_2: $q_2^{(1)}$ (solid line), $q_2^{(2)}$ (long dashed line), $q_3 < q_2^{(1)} < q_2^* < q_2^{(2)}$

When increasing/decreasing the parameter value in the interval $q_2 > q_2^*/q_2 < q_2^*$ within the "E" region, the interception point moves to the right.

11.5 Special Cases of the Evolution Equation

Let us analyze the evolution Eq. (11.15) in limiting cases to be valid when diffusion or recombination processes dominate.

11.5.1 Korteweg-De Vries-Burgers Equation

If in the system under consideration the diffusion prevails over the defect recombination ($a_4 = a_5 = a_6 = 0$), Eq. (11.15) is transformed into the Korteweg-de Vries-Burgers equation

$$\frac{\partial n_0}{\partial \eta} \pm a_1 n_0 \frac{\partial n_0}{\partial \xi} - a_2 \frac{\partial^2 n_0}{\partial \xi^2} - a_3 \frac{\partial^3 n_0}{\partial \xi^3} = 0 \tag{11.34}$$

We consider now solutions to the equation in variables of the traveling wave $n_0(\xi, \eta) = n_0(z)$, where $z = \xi - v\eta$ is the traveling coordinate; $v = const$ is a non-dimensional value, and has the meaning of the stationary wave velocity. By substituting into (11.34) and integrating once on z, we obtain an ordinary second-order differential equation

$$-vn_0 \pm \frac{1}{2} a_1 n_0^2 - a_2 \frac{dn_0}{dz} - a_3 \frac{d^2 n_0}{dz^2} = 0, \tag{11.35}$$

where the integration constant is assumed to be equal to zero.

Equation (11.35) may be classified as an oscillator equation with quadratic nonlinearity and light damping. There are two equilibrium states in its phase portrait; one of them is a “saddle” and the other is either a “focus” or a “node” depending on the equation parameter ratio. The phase portrait analysis shows that the equation does not have any periodic solutions.

Equation (11.35) enables one to find an exact solution using the method of simple equations [45, 46] earlier applied herein to Eq. (11.16).

The unknown coefficients and dimensionless velocity of the nonlinear wave are determined by the following expressions:

$$b_0 = \pm\frac{12a_3}{a_1},\ b_1 = \mp\frac{12}{5}\frac{a_2}{a_1},\ b_2 = \pm\frac{v}{a_1} \mp \frac{3}{25}\frac{a_2^2}{a_1 a_3},\ B_0 = \frac{1}{100}\frac{a_2^2}{a_3^2},\ v^2 = \frac{36}{625}\frac{a_2^4}{a_3^2}.$$

For velocities $v = \frac{6}{25}\frac{a_2^2}{a_3}$ ($v \to v_3$ at $a_5 \to 0$) and $v = -\frac{6}{25}\frac{a_2^2}{a_3}$ ($v \to v_2$ at $a_5 \to 0$) solutions (11.17) take the form (11.21) ($\overline{n}_{0,3}^{(\pm)}(\overline{z}_3)$) and (11.20) ($\overline{n}_{0,2}^{(\pm)}(\overline{z}_2)$), respectively:

$$\overline{n}_{0,11}^{(\pm)}(\overline{z}) = \pm\frac{1}{4}(th(\overline{z}) - 1)^2 = \overline{n}_{0,3}^{(\pm)}(\overline{z}_3)\ \text{and}\ \overline{n}_{0,12}^{(\pm)}(\overline{z}) = \pm\frac{1}{4}(th(\overline{z}) - 1)^2 \mp 1 = \overline{n}_{0,2}^{(\pm)}(\overline{z}_2) \tag{11.36}$$

where $\overline{n}_{0,1i} = n_0/A$, $\overline{z} = z/\Delta$ are non-dimensional values, $A = A_2 = A_3$, $\Delta = \Delta_2 = \Delta_3$—the amplitude and width coincide with dependences (11.23).

Solutions $\overline{n}_{0,11}^{(+)}, \overline{n}_{0,12}^{(-)}$ make sense; each of them corresponds to a stationary shock wave. The leading edge of the wave is steeper than the trailing edge. In the initial parameters, the stationary wave width does not depend on the mutual recombination rates but depends on the rates of recombination at the sinks and the diffusion of defects. The stationary wave velocity depends only on the diffusion of defects

$$v = \pm\frac{6}{25}\frac{a_2^2}{a_3} = \pm\frac{3}{25}\frac{(D_i + D_V)^2}{\tilde{\varepsilon} D_i D_V}.$$

11.5.2 Klein-Gordon Equation

And, on the contrary, if in the system the defect recombination prevails over diffusion processes ($a_1 = a_2 = a_3 = 0$), evolution equation (11.15) is transformed into the Klein-Gordon equation with a dissipative term:

$$\frac{\partial}{\partial\xi}\left(\frac{\partial n_0}{\partial\eta} + a_4 n_0\right) = a_5 n_0 \pm a_6 n_0^2. \tag{11.37}$$

In the traveling wave variables Eq. (11.37) is as follows:

$$-v\frac{d^2 n_0}{dz^2} + a_4\frac{dn_0}{dz} - a_5 n_0 \mp a_6 n_0^2 = 0. \tag{11.38}$$

Equation (11.38), with accuracy up to coefficients, is similar to Eq. (11.35) and has solutions only in the form of stationary shock waves.

The unknown values of expansion (11.17) are as follows:

$$b_0 = \mp\frac{6v}{a_6},\ b_1 = \mp\frac{6}{5}\frac{a_4}{a_6},\ b_2 = \pm\frac{3}{50}\frac{a_4^2}{a_6 v} \mp \frac{1}{2}\frac{a_5}{a_6},\ B_0 = \frac{25}{144}\frac{a_5^2}{a_4^2},\ v_2 = \frac{36}{625}\frac{a_4^4}{a_5^2}.$$

Let us consider the positive values of the coefficient a_6 for the nonlinear term in Eq. (11.38). At negative coefficient values curves $n_0^{(\pm)}$ are symmetrically reflected about the horizontal axis. The solutions to Eq. (11.38) with an upper sign are $n_0^{(+)}$ and with a lower sign are $n_0^{(-)}$.

For different values of the parameter a_5 and velocity $v = \pm\frac{6}{25}\frac{a_4^2}{a_5}$, we obtain four exact solutions satisfying Eq. (11.38).

At parameter values $a_5 > 0$ and velocities $v > 0$ and $v < 0$ ($v \to v_2$ at $a_2 \to 0$) solutions $n_0^{(\pm)}$, respectively, are as follows:

$$n_{0,21}^{(\pm)}(z) = \mp\frac{1}{4}A\left(th\left(\frac{z}{\Delta}\right)+1\right)^2 \text{ and } n_{0,22}^{(\pm)}(z) = \pm\frac{1}{4}A\left(th\left(\frac{z}{\Delta}\right)-1\right)^2 \mp A = n_{0,2}^{(\pm)}(z), \tag{11.39}$$

where $A = A_2$, $\Delta = \Delta_2$.

At parameter values $a_5 < 0$ and velocities $v > 0$ and $v < 0$ ($v \to v_3$ at $a_2 \to 0$) solutions $n_0^{(\pm)}$, respectively, are as follows:

$$n_{0,23}^{(\pm)}(z) = \mp\frac{1}{4}A\left(th\left(\frac{z}{\Delta}\right)+1\right)^2 \pm A \text{ and } n_{0,24}^{(\pm)}(z) = \pm\frac{1}{4}A(th(z)-1)^2 = n_{0,3}^{(\pm)}(z), \tag{11.40}$$

where $A = A_3$, $\Delta = \Delta_3$.

Dependences (11.39), (11.40) in non-dimensional variables take the following form:

$$\overline{n}_{0,21}^{(\pm)}(\overline{z}) = \mp\frac{1}{4}(th(\overline{z})+1)^2,\ \overline{n}_{0,22}^{(\pm)}(\overline{z}) = \pm\frac{1}{4}(th(\overline{z})-1)^2 \mp 1 = \overline{n}_{0,2}^{(\pm)}(\overline{z}_2),$$

$$\overline{n}_{0,23}^{(\pm)}(\overline{z}) = \mp\frac{1}{4}(th(\overline{z})+1)^2 \pm 1,\ \overline{n}_{0,24}^{(\pm)}(\overline{z}) = \pm\frac{1}{4}(th(\overline{z})-1)^2 = \overline{n}_{0,3}^{(\pm)}(\overline{z}_3),$$

where $\overline{n}_{0,2i} = n_0/A$, $\overline{z} = z/\Delta$.

Among obtained solutions (11.39), (11.40), there are both those previously found when solving Eq. (11.16) and those different therefrom, which are not solutions to the general equation. Note that formally solutions $\overline{n}_{0,21}^{(\pm)}$ and $\overline{n}_{0,23}^{(\pm)}$ result from (11.19) under the fulfillment of the condition, which does not make sense for Eq. (11.15).

Dependences characterizing $n_0^{(+)}(z)$ decrease and $n_0^{(-)}(z)$ increase, moreover

$$\begin{bmatrix} n_{0,21}^{(\pm)} \\ n_{0,22}^{(\pm)} \end{bmatrix}\Bigg|_{z\to-\infty} \to \mp 0, \quad \begin{bmatrix} n_{0,21}^{(\pm)} \\ n_{0,22}^{(\pm)} \end{bmatrix}\Bigg|_{z\to+\infty} \to \mp A,$$

$$\begin{bmatrix} n_{0,23}^{(\pm)} \\ n_{0,24}^{(\pm)} \end{bmatrix}\Bigg|_{z\to-\infty} \to \pm A, \quad \begin{bmatrix} n_{0,23}^{(\pm)} \\ n_{0,24}^{(\pm)} \end{bmatrix}\Bigg|_{z\to+\infty} \to \pm 0.$$

At fixed values of parameters, $a_4 - a_6$ and velocity v curves characterizing $n_0^{(+)}$ and $n_0^{(-)}$ are symmetrical about the horizontal axis.

Only solutions (11.39) with a lower sign and solutions (11.40) with an upper sign have a physical significance provided that $a_6 > 0$. If $a_6 < 0$, on the contrary, the solutions $n_{0,21}^{(+)}$, $n_{0,22}^{(+)}$ and $n_{0,23}^{(-)}$, $n_{0,24}^{(-)}$ make sense.

In this particular case, the sign of velocity v affects the steepness of the leading and trailing wave fronts. So, at positive values of the velocity ($v > 0$) the trailing edge is steeper than the leading one, and at negative values thereof the opposite is true. The stationary shock wave velocity does not depend on the diffusion but depends on the recombination rates of defects

$$v = \pm\frac{6}{25}\frac{a_4^2}{a_5} = \pm\frac{3}{25}\frac{(\beta_i + \beta_V)^2}{\widetilde{\varepsilon}(\beta_i\beta_V - \beta_{iV}\beta_{Vi})}.$$

11.6 Nonlinear Stationary Deformation Wave

If the material contains one type of point defect, the nonlinear system of Eqs. (11.1)–(11.3) becomes as follows:

$$\frac{\partial^2 U}{\partial t^2} - c_l^2\frac{\partial^2 U}{\partial x^2} - \frac{\beta_N}{\rho}\frac{\partial U}{\partial x}\frac{\partial^2 U}{\partial x^2} = -\left(\lambda + \frac{2}{3}\mu\right)\frac{\Omega_i}{\rho}\frac{\partial n_j}{\partial x}, \tag{11.41}$$

$$\frac{\partial n_j}{\partial t} = q_\varepsilon\frac{\partial U}{\partial x} + D_j\frac{\partial^2 n_j}{\partial x^2} - \beta_j n_j \tag{11.42}$$

($j = V$ and $j = i$ in systems with vacancies and interstitials, respectively).

The system (11.41), (11.42) is reduced to one fourth-order equation with respect to the longitudinal displacement, which in non-dimensional variables

$$z = \frac{x}{c_l/\beta_j}, \quad \tau = \frac{t}{1/\beta_j}, \quad U = u/u_0$$

(z, τ and U are non-dimensional values of coordinate, time and displacement, respectively, u_0 is some characteristic value of displacement, for example, the maximum displacement thereunder the deformation remains elastic) will become as follows:

$$\frac{\partial^2 U}{\partial \tau^2} - (1 - q_1)\frac{\partial^2 U}{\partial z^2} + \frac{\partial^3 U}{\partial \tau^3} - \frac{\partial^3 U}{\partial z^2 \partial \tau} - q_7 \left(\frac{\partial^4 U}{\partial z^2 \partial \tau^2} - \frac{\partial^4 U}{\partial z^4} \right) - \\ -q_8 \frac{\partial U}{\partial z}\frac{\partial^2 U}{\partial z^2} - q_8 \frac{\partial}{\partial \tau}\left(\frac{\partial U}{\partial z}\frac{\partial^2 U}{\partial z^2} \right) + q_7 q_8 \frac{\partial^2}{\partial z^2}\left(\frac{\partial U}{\partial z}\frac{\partial^2 U}{\partial z^2} \right) = 0 \tag{11.43}$$

where $q_1 = \left(\lambda + \frac{2}{3}\mu\right)\frac{q_\varepsilon}{\rho c_l^2}\frac{\Omega_i}{\beta_j}, q_7 = \frac{\beta_j D_j}{c_l^2}, q_8 = \frac{u_0 \beta_N \beta_j}{\rho c_l^3}$ $(j = V, i)$, q_1, q_7 coincide with (11.28) at $j = V$. The parameter q_1 characterizes here the type of point defect in the material, q_7 characterizes the diffusion of defects ($q_7 > 0$), and q_8 specifies the nonlinearity of the system. For media with vacancies there is $q_1 < 0$, for media with interstitials there is $q_1 > 0$.

As before, we look for a solution to Eq. (11.43) as an asymptotic expansion for the small parameter $U = U_0 + \widetilde{\varepsilon} U_1 + \ldots$ ($\widetilde{\varepsilon} << 1$) and introduce slow variables of coordinates and time: $\xi = \widetilde{\varepsilon}^{1/2}(z - c\tau)$ and $\eta = \widetilde{\varepsilon}^{3/2}\tau$. Perturbation propagating at velocity c along the z axis slowly evolves in space and time because of nonlinearity, dispersion and dissipation.

In the first approximation for $\widetilde{\varepsilon}$, we find an expression for the velocity $c = \sqrt{1 - q_1}$ (the value of the phase and group velocities at low frequencies). The second approximation for the small parameter enables us to obtain an evolution equation for the longitudinal deformation function $W = \frac{\partial U_0}{\partial \xi}$:

$$\frac{\partial W}{\partial \eta} - \frac{q_1}{2\sqrt{\varepsilon}}\frac{\partial^2 W}{\partial \xi^2} - \frac{q_1 q_7}{2\sqrt{1 - q_1}}\frac{\partial^3 W}{\partial \xi^3} + \frac{q_8}{2\sqrt{\varepsilon}\sqrt{1 - q_1}} W \frac{\partial W}{\partial \xi} - \frac{1}{2} q_8 \frac{\partial}{\partial \xi}\left(W \frac{\partial W}{\partial \xi} \right) = 0. \tag{11.44}$$

In variables of the traveling wave $W(\xi, \eta) = W(\chi)$, $\chi = \xi - v\eta$, where χ is the traveling coordinate and v is the nonlinear stationary wave velocity, Eq. (11.44) is as follows:

$$-\frac{q_1 q_7}{2\sqrt{1 - q_1}} W'' - \frac{q_1}{2\sqrt{\varepsilon}} W' - vW - \frac{1}{2} q_8 W W' + \frac{q_8}{4\sqrt{\varepsilon}\sqrt{1 - q_1}} W^2 = 0. \tag{11.45}$$

The exact particular solution to nonlinear Eq. (11.45) is the dependence

$$W(\chi) = -\frac{sign(q_1)}{2} A \left(th\left(-\frac{\chi}{\Delta} \right) + 1 \right), \tag{11.46}$$

where the constants

$$A = \frac{2(1 - q_1 + q_7)|q_1|}{\sqrt{\varepsilon}(1 - q_1)q_8}, \quad \Delta = \frac{2q_7\sqrt{\varepsilon}\sqrt{1 - q_1}}{1 - q_1 + q_7}$$

determine the wave parameters (amplitude, width). The wave velocity is assumed to be a fixed value equal to

$$v = -\frac{(1 - q_1 + q_7)q_1}{2\sqrt{\varepsilon}\sqrt{(1 - q_1)^3}}.$$

The profile of solution (11.46) is a kink symmetrical about the inflection point. At negative values of the parameter q_1, we have a monotonic decreasing kink located in the upper half-plane; at positive values of the parameter, we have a monotonic increasing kink in the lower half-plane.

The evolution Eq. (11.44) contains dispersive (the third derivative), dissipative (the second derivative) and two different nonlinear terms. We consider now special cases of this equation.

If the dispersion term in the evolution equation is small ($q_1 \sim \sqrt{\varepsilon}$ or $q_7 \sim \sqrt{\varepsilon}$) as compared to the other terms of the equation, so, the generalized Burgers equation is applicable. The generalization is due to an additional nonlinear term—the last term of the equation. In this case, in Eq. (11.45) there is no highest derivative

$$-\frac{q_1}{2\sqrt{\varepsilon}}W' - vW - \frac{1}{2}q_8WW' + \frac{q_8}{4\sqrt{\varepsilon}\sqrt{1 - q_1}}W^2 = 0. \tag{11.47}$$

Taking into account the boundary conditions

$$W(\chi) = \begin{cases} W_1, & \chi \to +\infty \\ W_2, & \chi \to -\infty. \end{cases}$$

Equation (11.47) allows the integration. The solution is presented as follows:

$$\chi = \frac{2\sqrt{1 - q_1}}{q_8(W_2 - W_1)}\left((q_1 + \sqrt{\varepsilon}q_8W_2)\ln(W_2 - W) - (q_1 + \sqrt{\varepsilon}q_8W_1)\ln(W - W_1)\right), \tag{11.48}$$

the velocity is defined by the expression

$$v = \frac{q_8}{4\sqrt{\varepsilon}\sqrt{1 - q_1}}(W_1 + W_2).$$

Dependence (11.48) is represented as a kink, which is asymmetrical about the inflection point (Fig. 11.17). In the figure horizontal, asymptotes are marked with a short dashed line. The leading edge of the wave is steeper than its trailing edge.

If in the evolutionary equation only one type of nonlinearity ($q_8 \sim \sqrt{\varepsilon}$) is regarded, i.e. if the last term of the equation is far smaller than its other terms, we obtain the

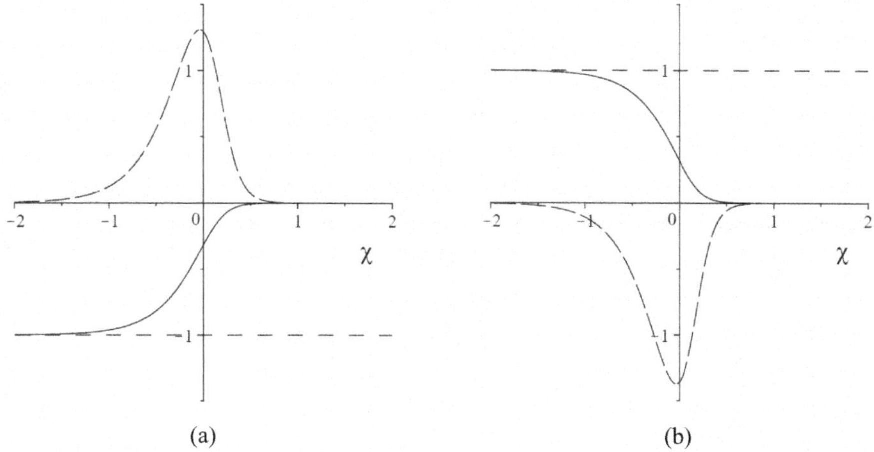

Fig. 11.17 Profiles of the stationary deformation waves $W(\chi)$ (solid line) and the first derivative $W'(\chi)$ (long dashed line) at $q_1 < 0$ (**a**) and $q_1 > 0$ (**b**)

Korteweg-de Vries-Burgers equation, which is similar (with an accuracy to variables and coefficients) to Eq. (11.34).

In this case Eq. (11.45) and its exact solution will be as follows:

$$-\frac{q_1 q_7}{2\sqrt{1-q_1}} W'' - \frac{q_1}{2\sqrt{\varepsilon}} W' - vW + \frac{q_8}{4\sqrt{\varepsilon}\sqrt{1-q_1}} W^2 = 0, \tag{11.49}$$

$$W(\chi) = \frac{sign(q_1)}{4} A\left(\left(th\left(\frac{\chi}{\Delta}\right) - 1\right)^2 - 2\right) + \frac{2\sqrt{\varepsilon}\sqrt{1-q_1}}{q_8} v, \tag{11.50}$$

where the amplitude, width and velocity of the wave are equal

$$A = \frac{12}{25} \frac{(1-q_1)|q_1|}{\sqrt{\varepsilon} q_7 q_8}, \quad \Delta = \frac{10 q_7 \sqrt{\varepsilon}}{\sqrt{1-q_1}}, \quad v = \pm \frac{3}{25} \frac{q_1 \sqrt{1-q_1}}{\varepsilon q_7}.$$

Solutions (11.50) have profiles of asymmetric increasing (at $q_1 < 0$) and decreasing (at $q_1 > 0$) kinks located both above (at $v > 0$) and below (at $v < 0$) the horizontal axis.

If in the evolution equation the dispersion and one nonlinear term ($q_1 \sim \sqrt{\varepsilon}$ and $q_8 \sim \sqrt{\varepsilon}$ or $q_7 \sim \sqrt{\varepsilon}$ and $q_8 \sim \sqrt{\varepsilon}$) are simultaneously negligible, Eq. (11.44) is reduced to the Burgers equation, which in ordinary derivatives is as follows:

$$-\frac{q_1}{2\sqrt{\varepsilon}} W' - vW + \frac{q_8}{4\sqrt{\varepsilon}\sqrt{1-q_1}} W^2 = 0. \tag{11.51}$$

The solution of the Burgers equation in the form of a stationary shock wave resulting from the balance of nonlinear and dissipative effects is known [47]. In the case of Eq. (11.51) the solution of the Burgers equation in the form of a stationary wave will be written as follows:

$$W(\chi) = \frac{1}{2}(W_1 + W_2) - \frac{1}{2}Ath\left(\frac{\chi}{\Delta}\right), \tag{11.52}$$

where W_1, W_2 are values at the boundaries such that $W_1 < W_2$. The parameters characterizing the amplitude, width and velocity of the wave are equal

$$A = W_2 - W,\ \Delta = \frac{4q_1\sqrt{1-q_1}}{q_8(W_2 - W_1)},\ v = \frac{(W_1 + W_2)q_8}{4\sqrt{\varepsilon}\sqrt{1-q_1}}.$$

11.7 Conclusion

The simulated medium is that for lack of any defects we have a classical elastic medium wherein a longitudinal wave propagates without dispersion and attenuation. Point defects contribute to the emergence of its dispersion and attenuation within quite a wide frequency range. Low-frequency and high-frequency perturbations thereat propagate nearly without dispersion and losses.

The type of point defect available in an elastic medium exerts an opposite effect on a longitudinal wave: vacancies promote an increase in the wave amplitude and interstitials contribute to a decrease in the amplitude of the wave during its propagation. Probably, in vacancy formation regions some neighboring lattice eigenmodes occur (vacancies "pulsate"), i.e. vacancies behave like oscillators.

In medium containing two types of point defects (vacancies and interstitials), one or two longitudinal waves propagate at low frequencies depending on the system parameter ratio. The only longitudinal wave propagates over the entire frequency range, as in media containing one type of point defect (vacancies, interstitials). At high frequencies this wave has some little normal dispersion. In the defect containing medium the propagation velocity and the amplitude of high-frequency oscillations are close, respectively, to the velocity of the longitudinal wave propagation in a solid and its amplitude.

Based on the analysis of exact analytical solutions to the evolution equation for the volume concentration of point defects in a material, it is shown that vacancies and interstitials can contribute to the formation of spatially localized nonlinear waves.

When the diffusion predominates over the recombination of defects, the evolution equation is transformed into the Korteweg-de Vries-Burgers equation. If defect recombination processes dominate over diffusion processes, the evolution equation is reduced to the Klein-Gordon equation containing the quadratic nonlinearity.

In particular cases both the solutions were obtained which were limiting solutions for solutions already found in the general case, as well as new solutions lacking in the general case. Therefore, it is necessary to consider in addition to the general evolution equation regarding the contributions of the processes of diffusion and defect recombination, also the limiting cases of the evolution equation, which emerge when one or another factor prevails.

In two limiting cases, we have stationary shock waves emerging as a result of the compensation of processes of dispersion, dissipation and nonlinearity. It is natural to assume that in the case generalizing two limiting cases, a similar reasoning is valid and, as a result, the existence of this type of wave.

Acknowledgements The work was supported by the Russian Science Foundation (project No 20-19-00613).

References

1. Bykovsky YA, Nevolin VN, Fominsk VY (1991) Ionic and laser implantation of metal materials. Energoatomizdat, Moscow, p 235
2. Mirzoev FK, Panchenko VY, Shelepin LA (1996) Laser control processes in solids. Phys Uspekhi 39(1):1–29. https://doi.org/10.1070/PU1996v039n01ABEH000125
3. Emel'yanov VI, Uvarova IF (1988) Laser-induced electron-deformation-thermal instability and a semiconductor-metal phase transition involving superstructure formation. Sov Phys J Exp Theor Phys 67(2)(8):1662–1670
4. Emel'yanov VI, Sumbatov AA (1988) Crystallization-deformation-thermal instability and formation of ordered structures during laser crystallization. Phys Chem Mech Surf (7):122–127
5. Emel'ianov VI, Uvarova IF (1989) Vacancy-deformation instability with formation of ordered structures under laser action on thin metal plates. Metallofizika 11(5):101–106
6. Emel'yanov VI, Mirzoev FK, Shelepin LA (1994) Mechanisms of the formation of ordered defect structures by the action of concentrated energy fluxes. Quantum Electron 24(8):714–717. https://doi.org/10.1070/QE1994v024n08ABEH000173
7. Mirzoev F, Shelepin LA (2001) Nonlinear strain waves and densities of defects induced in metal plates by external energy fluxes. Tech Phys 46(8):952–955. https://doi.org/10.1134/1.1395114
8. Khomenko MD, Mirzade FK, Makoana NW, Pityana S (2021) Coupled heat transfer, fluid flow and solidification kinetics for laser additive manufacturing applications. J Manuf Process 67:611–618. https://doi.org/10.1016/j.jmapro.2021.05.019
9. Seydgazov RD, Mirzade FK (2021) Threshold conditions for thermocapillary formation of a deep cavity in the additive process for selective laser melting of a metal powder layer. Weld Int 35(7–9):359–368. https://doi.org/10.1080/09507116.2021.1979829
10. Seidgazov RD, Mirzade FK (2023) Excitation of a capillary wave as a mechanism for the formation of pores in the process of deep penetration by laser radiation. Tech Phys Lett 49(6):65–67. https://doi.org/10.61011/TPL.2023.06.56384.19553
11. Ryssel H, Ruge I (1978) Ion implantation. B.G. Teubner Verlag, Stuttgart, p 366. https://doi.org/10.1007/978-3-663-05668-3
12. Ohtsuki Y-H (1983) Charged beam interaction with solids. Taylor & Francis Publication, London, p 248
13. Anishchik VM, Uglov VV (2003) Modification of instrumental materials by ion and plasma beams. Izd-vo BGU, Minsk, p 190
14. Wesch W, Wendler E, Gotz G, Kekelidse NP (1989) Defect production during ion implantation of various AIIIBV semiconductors. J Appl Phys 65(2):519–526. https://doi.org/10.1063/1.343134
15. Liu X, Prasad A, Nishio J, Weber ER, Liliental-Weber Z, Walukiewicz W (1995) Native point defects in low-temperature-grown GaAs. Appl Phys Lett 67(2):279–281. https://doi.org/10.1063/1.114782

16. Dymova NN, Kunitsyn AE, Chaldyshev VV, Markov AV (1997) Silicon and phosphorus co-implantation into undoped and indium-doped GaAs substrates. Semiconductors 31(12):1217–1220. https://doi.org/10.1134/1.1187296
17. Chaldyshev VV, Dymova NN, Kunitsyn AE, Markov AV (1997) Formation of n+-layers in undoped and indium-doped GaAs wafers by Si and Si+P ion implantation. Phys Status Solidi (A) 163(1):81–86. https://doi.org/10.1002/1521-396X(199709)163:1<81::AID-PSSA81>3.0.CO;2-J
18. Napolitani E, Impellizzeri G (2015) Ion implantation defects and shallow junctions in Si and Ge. Semicond Semimet 91:93–122. https://doi.org/10.1016/bs.semsem.2015.01.001
19. Okulich EV, Okulich VI, Tetelbaum DI (2018) Calculation of the influence of the ion current density and temperature on the accumulation kinetics of point defects under the irradiation of Si with light ions. Semiconductors 52(9):1091–1096. https://doi.org/10.1134/S1063782618090105
20. Nikolskaya AA, Korolev DS, Mikhaylov AN, Belov AI, Sushkov AA, Krivulin NO, Muhamatchin KR, Elizarova AA, Marychev MO, Konakov AA, Tetelbaum DI, Pavlov DA (2018) Light-emitting 9R-Si phase formed by Kr+ ion implantation into SiO2/Si substrate. Appl Phys Lett 113:182103. https://doi.org/10.1063/1.5052605
21. Okulich EV, Okulich VI, Tetelbaum DI (2020) Calculating silicon-amorphization doses under medium-energy light-ion irradiation. Semiconductors 54(8):916–922. https://doi.org/10.1134/S1063782620080175
22. Nikolskaya A, Belov A, Mikhaylov A, Konakov A, Tetelbaum D, Korolev D (2021) Photoluminescence of ion-synthesized 9R-Si inclusions in SiO2/Si structure: effect of irradiation dose and oxide film thickness. Appl Phys Lett 118(21):212101. https://doi.org/10.1063/5.0052243
23. Nikolskaya A, Korolev D, Belov A, Konakov A, Pavlov D, Mikhaylov A, Tetelbaum D (2022) Influence of chemical nature of implanted atoms on photoluminescence of ion-synthesized 9R-Si hexagonal silicon. Mater Lett 308:131103. https://doi.org/10.1016/j.matlet.2021.131103
24. Nikolskaya A, Korolev D, Mikhaylov A, Konakov A, Okhapkin A, Kraev S, Sushkov A, Pavlov D, Tetelbaum D (2023) Luminescence of modified W-centers arising in silicon upon irradiation of the SiO2/Si system by Kr+ ions. Mater Lett 342:134302. https://doi.org/10.1016/j.matlet.2023.134302
25. Nikolskaya AA, Korolev DS, Mikhaylov AN, Konakov AA, Okhapkin AI, Kraev SA, Andrianov AI, Moiseev AD, Sushkov AA, Pavlov DA, Tetelbaum DI (2023) Formation of hexagonal phase 9R-Si in SiO2/Si system upon Kr+ ion implantation. Mosc Univ Phys Bull 78(3):361–367. https://doi.org/10.3103/S0027134923030153
26. Zolnikov KP, Korchuganov AV, Kryzhevich DS, Chernov VM, Psakhie SG (2019) Formation of point defect clusters in metals with grain boundaries under irradiation. Phys Mesomech 22(5):355–364. https://doi.org/10.1134/S1029959919050023
27. Goldstein RV, Makhviladze TM, Sarychev ME (2019) Effect of point defects on the interface strength of joint materials. Phys Mesomech 22(4):269–274. https://doi.org/10.1134/S1029959919040027
28. Sozykin SA, Beskachko VP (2022) The structure and properties of a carbon nanotube (7, 7) with a vacancy defect. Lett Mater 12(1):32–36. https://doi.org/10.22226/2410-3535-2022-1-32-36
29. Ten GN, Gerasimenko AY, Savelyev MS, Kuksin AV, Vasilevsky PN, Kitsyuk EP, Baranov VI (2020) Influence of edge defects on Raman spectra of graphene. Lett Mater 10(1):89–93. https://doi.org/10.22226/2410-3535-2020-1-89-93
30. Burlak GN, Ostrovskii IV (1997) Acoustic hysteresis phenomena due to the dislocation nonlinearity in crystals. Tech Phys Lett 23(9):725–726. https://doi.org/10.1134/1.1261669
31. Erofeev VI, Romashov VP (2002) Effect of dislocations on the dispersion and damping of ultrasound in solids. Tech Phys Lett 28(3):218–220. https://doi.org/10.1134/1.1467280
32. Erofeev VI (2008) Self-modulation of an acoustic wave in a dislocated solid. Tech Phys Lett 34(2):150–152. https://doi.org/10.1134/S1063785008020193
33. Shekoyan AV (2009) Wave beams in crystals with dislocations and quadratic nonlinearity. Tech Phys Lett 35(4):337–339. https://doi.org/10.1134/S1063785009040142

34. Bagdoev AG, Erofeyev V, Shekoyan AV (2016) Wave dynamics of generalized continua. Springer, Berlin, p 274. https://doi.org/10.1007/978-3-642-37267-4
35. Bagdoev AG, Erofeyev V, Shekoyan AV (2016) Wave dynamics of generalized continua. Springer, Berlin, p 274
36. Erofeev VI, Kazhaev VV (2010) Nonlinear stationary acoustic wave in a solid with dislocations. Tech Phys 55(4):580–583. https://doi.org/10.1134/S1063784210040262
37. Sarafanov GF (2010) Collective and wave effects in the ensemble of dislocations in plastic deformation of metals. Litera, Nizhny Novgorod, p 359
38. Erofeev VI, Malkhanov AO (2019) Dispersion and self-modulation of waves propagating in a solid with dislocations. Phys Mesomech 22(3):173–180. https://doi.org/10.1134/S1029959919030019
39. Erofeev VI, Artamonova OA (2010) Effect of point defects in the material on the propagation of a nonlinear acoustic wave. In: Trudy XXII Sessii Rossiiskogo akusticheskogo obshchestva i Nauchnogo soveta po akustike RAN, vol 1. GEOS, Moscow, pp 158–159
40. Erofeev VI, Leont'eva AV, Shekoyan AV (2020) Elastic waves in a thermoelastic medium with point defects. Tech Phys 65(1):22–28. https://doi.org/10.1134/S1063784220010053
41. Erofeev VI, Leonteva AV, Shekoyan AV (2021) Dispersion, attenuation and spatial localization of thermoelastic waves in a medium with point defects. In: Altenbach H, Eremeyev VA, Igumnov LA (eds) Multiscale solid mechanics. Advanced structured materials, vol 141. Springer, Cham, STRUCTMAT, pp 123–144. https://doi.org/10.1007/978-3-030-54928-2_10
42. Mirzade FK, Shelepin LA (2007) Nonlinear longitudinal waves in the presence of the interaction between the strain and temperature fields and the field of nonequilibrium atomic defects. Acoust Phys 53(6):676–682. https://doi.org/10.1134/S1063771007060048
43. Porubov AV (2003) Amplification of nonlinear strain waves in solids. Series on stability, vibration and control of systems, series a, vol 9. World Scientific, p 228. https://doi.org/10.1142/5238
44. Erofeev VI, Kazhaev VV, Semerikova NP (2002) Waves in the rods. Dispersion. Dissipation. Nonlinearity. Fizmatlit, Moscow, p 208
45. Kudryashov NA (2004) Analytical theory of nonlinear differential equations. Institute of Computer Research, Moscow-Izhevsk, p 360
46. Kudryashov NA (2010) Methods of nonlinear mathematical physics: textbook. Publishing House, House Intellect, Dolgoprudny, p 368
47. Ryskin NM, Trubetskov DI (2017) Nonlinear waves. Lenand, Moscow, p 312

Chapter 12
Towards the Design of Adhesives and Binders with Adhesive Ability Programmed Through Morphogenetic Rearrangement of Their Particle Geometry: Transgenic Adhesives and Binders from Genetically Modified Starch Producing Plants

Anna G. Filatova, Oleg V. Gradov, Margarita A. Gradova, Irina A. Maklakova, Irina G. Plashchina, and Lubov A. Wasserman

Abstract The literature review in the introduction of this paper provides information on the relationship between the adhesion mechanisms and viscosity, which is in good agreement with the classical works demonstrating the relationship between rheology, adhesion, and surface energy. It is indicated that for sufficiently thick binder layers, in which redistribution of particles and their dense (or not) packing during compression is possible, adhesion itself, as well as friction or compression, can influence the packing geometry in starch-containing dispersed systems of the binder particles. It is underlined that when the particle shapes are not spherical, this simplified model approach has no predictive power. In this regard, it is necessary to take into account the non-spherical particles' morphology and the size classes of the structures characterized not only by a granulometric diameter/radius but also by a number of additional descriptors, such as elongation, ellipticity, moment of inertia, convexity, etc. It is also necessary to account for the dynamic shape changes of the particles/vesicles or micelles in connection with their rheological behavior and response to the effects of medium, in addition to their static characteristics. Adhesion of the emulsion components and polyelectrolytes can be characteristically related to the particle morphology that determines the rheology of the starch-containing mixtures. At the same time, it is

A. G. Filatova · O. V. Gradov (✉) · M. A. Gradova · I. A. Maklakova
N.N. Semenov Federal Research Center for Chemical Physics, Russian Academy of Sciences, Kosygina Str., 4, 119991 Moscow, Russia
e-mail: o.v.gradov@gmail.com

O. V. Gradov · I. G. Plashchina · L. A. Wasserman
Emanuel Institute of Biochemical Physics, Russian Academy of Sciences 119334, Kosygina Str., 4, Moscow, Russia

H. Altenbach and V. A. Eremeyev (eds.), *Dynamics of Discrete and Continuum Structures and Media*, Advanced Structured Materials 221,
https://doi.org/10.1007/978-3-031-75626-9_12

possible to create controllable starch-containing adhesives and binders or adhesives with the adhesive ability that is programmable by rearranging their particle geometry. In addition to the currently known methods of changing the already formed starch granules (ultrasound, pH, chemical treatment and encapsulation), it is also possible to control the starch granule shape in vivo or in situ—at the granule formation stage, using transgenic varieties of the cultivated plants that give programmable size distributions or geometry of the starch granules during cultivation under the programmed environmental conditions. But such granules require a characterization method. In this paper, multiparametric granulometry methods for analyzing the starch granules' geometry are tested. The compatibility of this approach with modeling of rheological and hydrodynamic characteristics of the adhesive particles is illustrated using an elementary model didactic example.

Keywords Adhesives · Binders · Hydrodynamics of non-Newtonian fluids · Starch morphometry · Starch granulometry · Soft matter hydrodynamics · Microrheology · Transgenic adhesives · Genetically modified starch-producing plants

12.1 Introduction

It is well known that starch-based binders and glues/adhesives are used in a variety of industries, from conventional wallpaper pastes and plywood adhesives (the development of which peaked in the 1970s and 1980s [1–3], although in Asian catch-up economies and a more extensive approach to industrialization, this trend was still relevant in the 2000s [4]) and decorative panels [5] to modern conductive mounting binders for supercapacitors [6]. The above systems, depending on the fixation mechanism, can be designed for a wide range of bond strengths: from water-soluble and bioresorbable (for example, in the case of starch binders in the tablets of pharmaceuticals [7]) or biodegradable [8, 9] adhesives for soft matter, often associated with the application in green chemistry, to starch-based adhesives for gluing and interlayering thin metal films and foils [10, 11]. According to [10], the adhesion mechanism in this case is associated with the starch glue viscosity, which is in good agreement with the conventional works [12, 13] demonstrating the relationships between rheology, adhesion, and surface energy. However, in this case, it is necessary to distinguish between the different physical mechanisms of adhesion and, sensu lato, fixation using starch-containing binders, based on the types of surface interactions and the weighting coefficients of the contribution of the factors involved in the contact formation, including rheological characteristics of the corresponding media. Indeed, on the one hand, the bonding processes involve molecular mechanisms and forces (oligomers and rheological modifiers with different energy [14]; the chemistry of grafted surface compounds or cross-linked polymer networks that provide different rheological and surface properties [15]; non-branching and branching molecular chains with different branching ratios, affecting both micromechanics and rheology of the binder [16]; the presence of points of supramolecular interactions, including cross-linking points, non-linearly affecting rheology, adhesion, and pour point [17]) affecting rheology of the adhesive (and, accordingly, adhesion). On the other hand, the rheology of biocolloidal starch-containing mixture in the corresponding adhesives is associated with the particle morphology (however, this is observed not only in the case of nanodispersed polysaccharide and polysaccharide-containing systems and hydrogels [18–20] but also in many other micro- and ultradisperse systems, including carbon nanostructures [21–21], which illustrates the universal colloidal nature of this phenomenon). Thus, separate consideration of the contribution of physical (in particular, rheological) and chemical (in particular, supramolecular) factors is not always possible or accessible for the experimenter. However, an emergent approach to analyzing the connection/bonding properties (using binders) based on the ideas about their formation mechanisms under specific environmental conditions (which satisfies both the criteria of supramolecular chemistry and the well-known dependence of the properties of colloidal systems on their preparation procedure) suggests the need to take into account both chemical and physical bonding factors.

This is the most important task, unsolvable without analyzing the contact morphology during its formation, and therefore, without analyzing the morphology of starch-containing raw materials, since, in addition to the recently actively studied coordination of molecular geometry and adhesion forces at the stage of the interface formation [25], the effects of the combined action of adhesion and dispersity, as well as the influence of the glue particle shape or geometry on the interaction efficiency with the surface have been known since the 1990s. [26]. Moreover, the surface energy involved in adhesion is associated with the topography of a given surface type [27, 28], and various geometric components of this process, both from the binder particles [29] and from the contact surfaces, significantly contribute to the adhesion [30] (the so-called "geometry-controlled adhesion" [31]). On the other hand, it must be taken into account that for sufficiently thick binder layers, in which various distribution of particles and their dense (or not) packing during compression are possible, adhesion itself, as well as friction or compression, can influence the packaging geometry in starch-containing dispersed systems of the binder particles, which is consistent with a simple model analyzed for spherical particles [32]. However, when the particle shapes deviate from the spherical ones and heteroscedastic granulometric statistics appear, unfortunately, this simplified approach ceases to work and fails to predict the properties of binders. To overcome these limitations, it is necessary to take into account the non-spherical particle morphology (and size classes of the structures, characterized not only by granulometric diameter/radius, but also by a number of additional descriptors, such as elongation or ellipticity, moment of inertia, convexity, etc.) For adhesion in rheologically complex systems, including emulsified adhesives [33], often based on starch [34–38], it is necessary to take into account dynamic changes in the shape of particles/vesicles or micelles due to their rheological behavior and response to the medium/environmental effects, not just their static parameters determined by the surface tension and leading to the simplifications (such as spherical approximation of the vesicle shape). Adhesion of the emulsion components and polyelectrolytes can be closely related to the particle morphology that determines the rheology of the corresponding mixtures [39, 40]. The relationship between the morphology and adhesion of starch in correlation with the rheological and extrusion properties has been shown on the polymer blends with soil fillers based on coal ash and containing starch as a binder [41, 42]. If we extrapolate this statement to the widely used starch-containing polyelectrolyte mixtures [43–50] based on the general ideas about multiparametric reactivity of the polyelectrolyte systems [51–55], then it seems possible to create controlled starch-containing adhesives and binders or adhesives with adhesive ability programmed by rearranging the particle geometry.

Consequently, by varying the size and geometry of the starch raw material particles, it is possible, with a certain efficiency, to vary the rheological and adhesive parameters of the starch-based adhesives and binders. However, to date, only the methods influencing the already formed starch granules are known (i.e. ultrasound [56, 57], pH [58], chemical treatment, and encapsulation [59, 60]), while there are no methods for controlling such properties in vivo or in situ during the starch granule formation intended for use in the adhesive industry. If such a method existed, it would inevitably account for the phenotype of the plant from which this starch is produced, since any local phytophysiological effect is reversible and unstable in metrological results. The only reliable way to modify the dispersity of starch as a component of the chemical phenotype, from the standpoint of biocolloid chemistry and phytochemistry, is to fix this phenotypic effect in the genome. However, breeding activities of this kind are quite expensive and time-consuming, and adhesive manufacturers are unlikely to be interested in such investments. Therefore, the only rational way to achieve the goal is genetic modification of the plants themselves that produce and store starch [61–63], regardless of the need for its further chemical modification or processing. Methods for achieving such results in starch have been developed by molecular genetics since the 1990s, and have already been tested on almost all agro-relevant plant species: potatoes [64–70], corn [71], rice [72–74] (if we ignore a series of reports on the genetic modification of fungi—biodestructors of polysaccharides [75], dating back to the very beginning of the 1990s, with the goal of optimizing the enzymatic breakdown of starch, which also leads to a change in its granule morphology [76]).

Despite the fact that proposals and pioneering works on the use of starch produced by genetically modified organisms as a raw material for industrial purposes date back to the mid-1990s [77], their introduction into practice was delayed until the second half of the 2000s. At the same time, it should be noted that the delay in the introduction of "genetically modified" starches into practice was due to the associative linking of the starch industry standards to the agro-food application area of its products. As a result, the certified/reference material of genetically modified potatoes—the most common source of starch, with a modified chemical composition and structure—is tested exclusively according to the standards of the agro-food industry (see, for example, [78]), and the standardization of the starch-producing plants and the corresponding cultivars is carried out by EFSA (European Food Safety Authority) [79–83]. This seems unnecessary when using it as a technical raw material outside the food industry, so, of particular interest are the works that consider physical properties of starch obtained from the genetically modified sources, relevant for its use as a raw material for the production of adhesives, binders, etc. (for glued decorative products, wallpaper, planar capacitors, plywood, etc. genetic modification of the source of binders does not pose a serious "life hazard"). In connection with the above formulated rheological or ultramorphological considerations, it is advisable to point out several Russian works in the field of rheological properties of genetically modified starches [84]. Based on this direction and the principles of similarity when analyzing according to the criteria for correlating structures and functions (similar to QSAR and QSPR approaches) in genetically modified starches [85], it will be possible to predict rheological properties, as well as (in the future, as missing correlations for a number of descriptors are identified) the prospects of using such starch raw materials as the components of glues/adhesives/binders. Strictly speaking, an approach to creating adhesives based on a quantitative correlation between the structure and function (both for adhesive components and fillers) possesses the potential for an absolute research novelty, since until now we could speak about the genetically modified sources of adhesives and binders in an unjustifiably expanded interpretation often with a confusion of the concepts of biological specific adhesion and chemical adhesion and autohesion (for example, talking about genetically modified suspensions in artificial morphogenesis in regenerative medicine or gene therapy [86]), and even with confusion in the systematics of intermolecular interactions and their mixing with electroadhesion and mutual diffusion, including its specific biological forms and manifestations. A metrological approach to assessing the quality of genetic modification, mediated by the qualimetry of starches according to the specified rheological and micromorphometric criteria/descriptors, leads to the possibility of controlling the target qualities of the modified starches during their metabolic deposition in the plant. Any other ways to improve the quality of adhesives containing "genetically modified" starch, except for metrological and qualimetric ones, are not capable of providing a qualitative improvement in their target parameters, since the times of W. Thomson (Lord Kelvin) it has been known that "if you cannot measure something, you cannot improve it". However, the state-of-the-art research in the field of genetically modified starches usually does not imply the physical and metrological accuracy of the analysis of the modification results [87].

In this regard, we propose a set of characterization methods for the starch properties, using the results of which, with the correlation of rheological, morphometric, adhesimetric and (bio-)chemical characteristics, it seems possible to synthesize more optimal genetically modified pathways for the formation and deposition of granular starch in plant cells. This approach is based on measuring a wide variety of the geometric parameters in starch granules that affect the flow or fluidity (in rheology) and adhesion of the mixtures containing them (for example, pastes or wallpaper adhesives).

12.2 Materials and Methods

Potato starch granules extracted from potato tubers cultivated in vitro and displaced on the paper were analyzed using SEM (LVEM—Low-Voltage Electron Microscopy). Morphological changes of the starch granule (correlating with in vitro or in vivo cultivation regimes, transgenic or non-transgenic nature of the tubers) were investigated using scanning electron microscopy and multiparametric morphometric analysis. Photos of starch granules were obtained with a scanning electron microscope MIRA3 LMU (Tescan, Brno, Czech Republic) at room temperature in the conditions of high vacuum with accelerating voltage 500 V.

Size and morphology of the potato starch granules were measured or estimated using an extended suite of morphometric algorithms under the "Altami Studio" licensed software GUI. Besides the standard granulometric characteristics, a number of dimensionless shape descriptors (such as ellipticity factor and elongation ratio) were analyzed and compared. Ellipticity factor was calculated by the above-mentioned software as the ratio between the starch granule squares and squares of the ellipses with the equivalent moments of inertia. Elongation ratios were measured and calculated using two different algorithms: as the ratios between the equivalent lengths and the equivalent widths or as the ratios between the maximal and minimal Feret diameters, which can be defined as the distance between the two parallel planes restricting the starch granules perpendicular to their direction. Additionally, Feret diameters of the convex starch granules were calculated as the ratios of the starch granules' perimeters to the π number which is the ratio of the circle's circumference to its diameter.

12.3 Results and Discussion

An example of an electron micrograph of the starch granules on the paper (as a carrier that can potentially be glued using it) is shown in Fig. 12.1. The microphotograph shows granules of different sizes and varying degrees of maturity, differing in aggregative properties and morphology/geometry—usually non-spherical at the later stages. For this reason, the process of starch maturation over time can be considered not only as a simple increase in the granule size but also as an increase in their elongation factors and other descriptors corresponding to the elongation of the granule, no matter what ratios they are based on (Feret diameters, or equal lengths and equal widths, or real lengths and real widths).

Thus, the effectiveness of participation of these fractions or statistical scores of granules in gluing will depend on the time of removal from the plants and the degree of maturity. This approach (multiparametric granulometric fingerprinting) allows us to predict the effectiveness of bonding and the absence of the possibility of coagulation clumping at the level of the individual points or microvolumes of glue-containing fractions/morphometric scores of the corresponding microparticles.

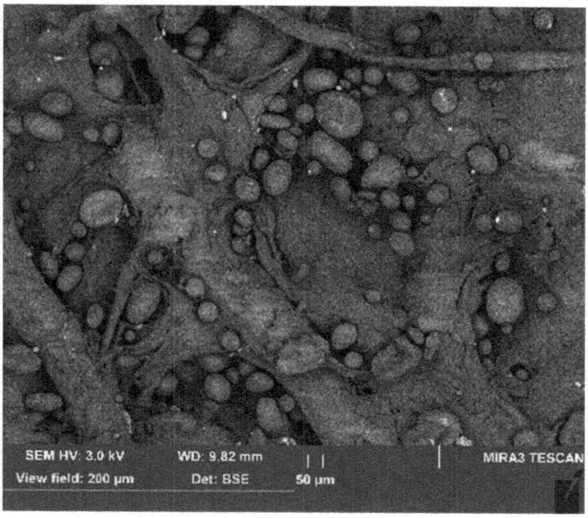

Fig. 12.1 SEM micrograph of potato starch granules (displaced on the paper), extracted from potato tubers cultivated in vitro

An example of multiparametric granulometric fingerprinting of the SEM data on the starch granules is shown in Fig. 12.2. It can be seen that the Feret diameter distributions are significantly different (compare also Tables 12.1 and 12.2), which indicates the non-sphericity of the starch particles and allows one to predict their corresponding behavior in a flow and shear viscosity when gluing with an insufficiently prepared/pretreated paste. A similar effect can be predicted from the comparison of the equal length and equal width distributions or the real length and real width distributions (Tables 12.3, 12.4, 12.5, and 12.6). Direct confirmation of this conclusion can be obtained from the analysis of the elongation factors of the starch granules in multiparametric granulometry data (see Table 12.7), as well as from the gonimetric data comparing major angles and minor angles of the oriented starch structures (Tables 12.8 and 12.9). To predict the behavior of granules in the shear flow and the effectiveness of adhesion when spreading them on paper during gluing, one can also use the calculation of the inertia moments for the same statistical scores of the starch granules (see Table 12.10).

It is possible (and rational) to more accurately assess the morphology, geometry [88] and complex granulometric characteristics of starch using not (semi-)manual but automatic analysis of their geometry based on computer-assisted image segmentation or image edge gradient detection. Examples of automatic starch grain edge detections using Sobel-Feldman operator, Laplacian (Laplace operator), Roberts cross and embossing simulation [89–94] are shown in Fig. 12.3. Obviously, granules with different shapes will behave differently in hydrodynamics, rheology and propagation of the adhesive spot. The effects of particle shape on the hydrodynamics of both active (biological and non-biological man-made) and passive particles and structures have been studied since the nineteenth century and continue to be of interest in the twenty-first century [95–102]. A simple model analogy can be found for starch particles using cross-sections of a cylinder in hydrodynamic models used in biology.

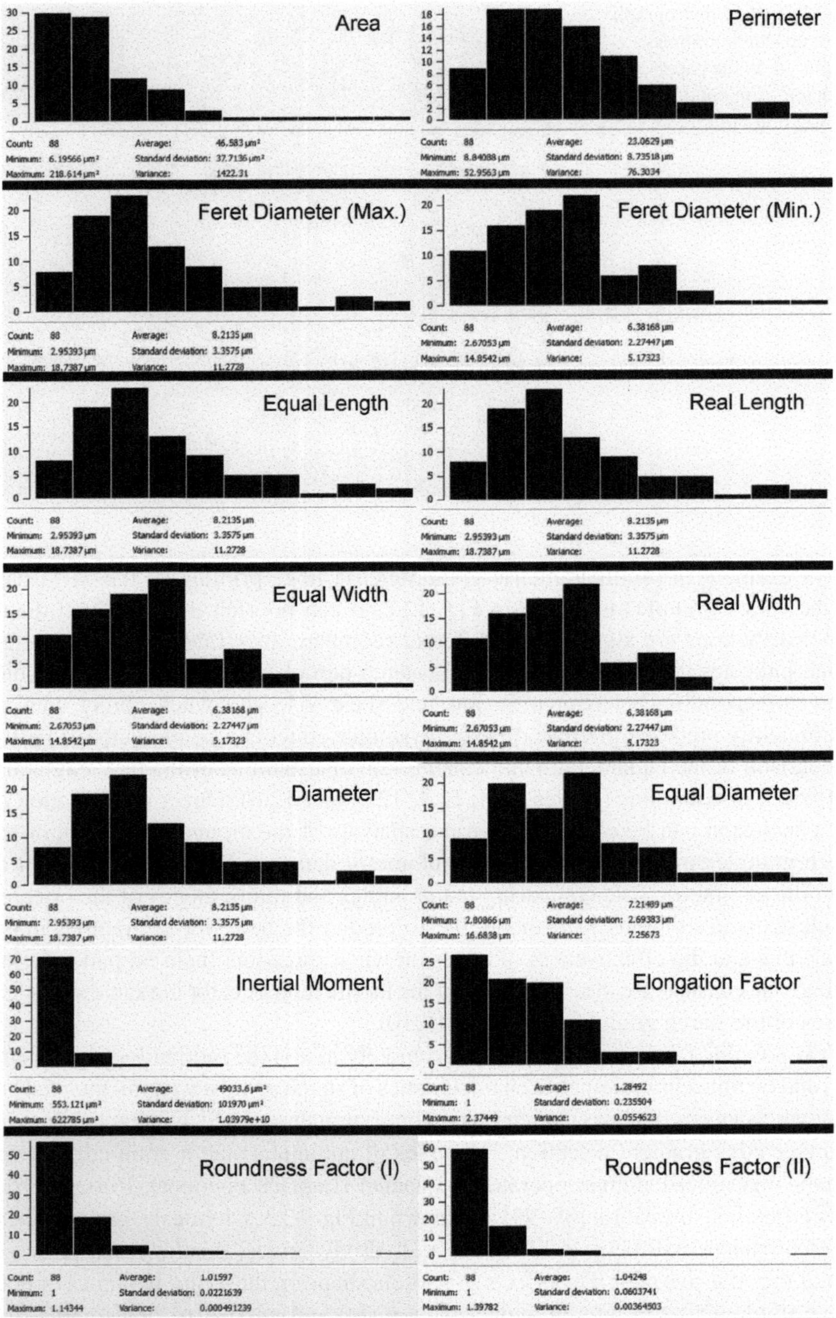

Fig. 12.2 Statistical histograms of multiparametric granulometry of the starch grains

Table 12.1 Feret diameter (max.) distributions for the starch granules on the paper surface

Caption	Amount	Minimum, µm	Maximum, µm	Average, µm	Standard deviation, µm	Variance	Percentage
Range 1	32,0467	2,96,854	4,52,799	4,00,584	0,530,859	0,281,812	9,09,091
Range 2	104,315	4,60,526	6,13,165	5,49,024	0,562,194	0,316,062	21,5909
Range 3	162,991	6,22,598	7,72,105	7,08,658	0,412,974	0,170,547	26,1364
Range 4	110,507	7,73,964	9,08,443	8,50,051	0,485,255	0,235,472	14,7727
Range 5	92,9071	9,75,794	10,7067	10,323	0,323,088	0,104,386	10,2273
Range 6	56,7725	10,9625	11,7999	11,3545	0,363,793	0,132,345	5,68,182
Range 7	65,0688	12,4919	13,584	13,0138	0,45,833	0,210,066	5,68,182
Range 8	15,5242	15,5242	15,5242	15,5242	0	0	1,13,636
Range 9	49,4693	15,9221	16,9824	16,4898	0,43,607	0,190,157	3,40,909
Range 10	36,7627	17,9313	18,8314	18,3814	0,450,039	0,202,535	2,27,273

Table 12.2 Feret diameter (min.) distributions for the starch granules on the paper surface

Caption	Amount	Minimum, µm	Maximum, µm	Average, µm	Standard deviation, µm	Variance	Percentage
Range 1	37,5325	2,68,374	3,86,765	3,41,205	0,333,825	0,111,439	12,5
Range 2	72,7615	3,96,199	5,10,591	4,54,759	0,400,514	0,160,411	18,1818
Range 3	107,575	5,18,832	6,33,131	5,66,182	0,358,871	0,128,789	21,5909
Range 4	153,54	6,38,938	7,57,825	6,97,911	0,366,638	0,134,423	25
Range 5	48,3028	7,59,956	8,44,779	8,05,047	0,320,396	0,102,654	6,81,818
Range 6	73,2011	8,84,163	9,46,467	9,15,013	0,226,635	0,051,364	9,09,091
Range 7	30,8744	10,1686	10,5296	10,2915	0,168,386	0,028,354	3,40,909
Range 8	12,1021	12,1021	12,1021	12,1021	0	0	1,13,636
Range 9	13,5491	13,5491	13,5491	13,5491	0	0	1,13,636
Range 10	14,9277	14,9277	14,9277	14,9277	0	0	1,13,636

Table 12.3 The equal length distributions for the starch granules on the paper surface

Caption	Maximum, µm	Average, µm	Standard deviation, µm	Variance	Percentage
Range 1	4,52,799	4,00,584	0,530,859	0,281,812	9,09,091
Range 2	6,13,165	5,49,024	0,562,194	0,316,062	21,5909
Range 3	7,72,105	7,08,658	0,412,974	0,170,547	26,1364
Range 4	9,08,443	8,50,051	0,485,255	0,235,472	14,7727
Range 5	10,7067	10,323	0,323,088	0,104,386	10,2273
Range 6	11,7999	11,3545	0,363,793	0,132,345	5,68,182
Range 7	13,584	13,0138	0,45,833	0,210,066	5,68,182
Range 8	15,5242	15,5242	0	0	1,13,636
Range 9	16,9824	16,4898	0,43,607	0,190,157	3,40,909
Range 10	18,8314	18,3814	0,450,039	0,202,535	2,27,273

Thus, an elementary example of visualization of flow around sections of particles with different shapes from the paper [97] (in which gratitude is given to one of the authors of this article) is shown in Fig. 12.4 (from author's manuscript; with the permission of the author). The illustration above clearly shows that the vector fields and flow profiles are different for particles of different shapes. The same approach can be applied to the analysis of starch particles and other fillers after precision morphometries and geometrical approximations.

Thus, an elementary example of visualization of flow around sections of particles with different shapes from the paper [97] (in which gratitude is given to one of the authors of this article) is shown in Fig. 12.4 (from author's manuscript; with the permission of the author). The illustration above clearly shows that the vector fields and flow profiles are different for particles of different shapes. The same approach

Table 12.4 The real length distributions for the starch granules on the paper surface

Caption	Amount	Minimum, µm	Maximum, µm	Average, µm	Standard deviation, µm	Variance	Percentage
Range 1	32,0467	2,96,854	4,52,799	4,00,584	0,530,859	0,281,812	9,09,091
Range 2	104,315	4,60,526	6,13,165	5,49,024	0,562,194	0,316,062	21,5909
Range 3	162,991	6,22,598	7,72,105	7,08,658	0,412,974	0,170,547	26,1364
Range 4	110,507	7,73,964	9,08,443	8,50,051	0,485,255	0,235,472	14,7727
Range 5	92,9071	9,75,794	10,7067	10,323	0,323,088	0,104,386	10,2273
Range 6	56,7725	10,9625	11,7999	11,3545	0,363,793	0,132,345	5,68,182
Range 7	65,0688	12,4919	13,584	13,0138	0,45,833	0,210,066	5,68,182
Range 8	15,5242	15,5242	15,5242	15,5242	0	0	1,13,636
Range 9	49,4693	15,9221	16,9824	16,4898	0,43,607	0,190,157	3,40,909
Range 10	36,7627	17,9313	18,8314	18,3814	0,450,039	0,202,535	2,27,273

Table 12.5 The equal width distributions for the starch granules on the paper surface

Caption	Amount	Minimum, µm	Maximum, µm	Average, µm	Standard deviation, µm	Variance	Percentage
Range 1	37,5325	2,68,374	3,86,765	3,41,205	0,333,825	0,111,439	12,5
Range 2	72,7615	3,96,199	5,10,591	4,54,759	0,400,514	0,160,411	18,1818
Range 3	107,575	5,18,832	6,33,131	5,66,182	0,358,871	0,128,789	21,5909
Range 4	153,54	6,38,938	7,57,825	6,97,911	0,366,638	0,134,423	25
Range 5	48,3028	7,59,956	8,44,779	8,05,047	0,320,396	0,102,654	6,81,818
Range 6	73,2011	8,84,163	9,46,467	9,15,013	0,226,635	0,051,364	9,09,091
Range 7	30,8744	10,1686	10,5296	10,2915	0,168,386	0,028,354	3,40,909
Range 8	12,1021	12,1021	12,1021	12,1021	0	0	1,13,636
Range 9	13,5491	13,5491	13,5491	13,5491	0	0	1,13,636
Range 10	14,9277	14,9277	14,9277	14,9277	0	0	1,13,636

Table 12.6 The real width distributions for the starch granules on the paper surface

Caption	Amount	Minimum, μm	Maximum, μm	Average, μm	Standard deviation, μm	Variance	Percentage
Range 1	37,5325	2,68,374	3,86,765	3,41,205	0,333,825	0,111,439	12,5
Range 2	72,7615	3,96,199	5,10,591	4,54,759	0,400,514	0,160,411	18,1818
Range 3	107,575	5,18,832	6,33,131	5,66,182	0,358,871	0,128,789	21,5909
Range 4	153,54	6,38,938	7,57,825	6,97,911	0,366,638	0,134,423	25
Range 5	48,3028	7,59,956	8,44,779	8,05,047	0,320,396	0,102,654	6,81,818
Range 6	73,2011	8,84,163	9,46,467	9,15,013	0,226,635	0,051,364	9,09,091
Range 7	30,8744	10,1686	10,5296	10,2915	0,168,386	0,028,354	3,40,909
Range 8	12,1021	12,1021	12,1021	12,1021	0	0	1,13,636
Range 9	13,5491	13,5491	13,5491	13,5491	0	0	1,13,636
Range 10	14,9277	14,9277	14,9277	14,9277	0	0	1,13,636

Table 12.7 The elongation factor distributions for the starch granules on the paper surface

Caption	Amount	Minimum	Maximum	Average	Standard deviation	Variance	Percentage
Range 1	28,8651	1	1,13,636	1,06,908	0,041,596	0,00,173	30,6818
Range 2	25,3339	1,13,894	1,27,027	1,20,638	0,045,134	0,002,037	23,8636
Range 3	26,5203	1,27,536	1,40,579	1,32,601	0,036,981	0,001,368	22,7273
Range 4	16,0883	1,41,503	1,52,391	1,46,257	0,033,481	0,001,121	12,5
Range 5	4,78,048	1,56,124	1,65,052	1,59,349	0,040,436	0,001,635	3,40,909
Range 6	5,2088	1,70,733	1,76,208	1,73,627	0,022,459	0,000,504	3,40,909
Range 7	1,89,474	1,89,474	1,89,474	1,89,474	0	0	1,13,636
Range 8	2,00,665	2,00,665	2,00,665	2,00,665	0	0	1,13,636
Range 9	–	–	–	–	–	–	0
Range 10	2,37,449	2,37,449	2,37,449	2,37,449	0	0	1,13,636

Table 12.8 The major angle distributions for the starch granules on the paper surface

Caption	Minimum,°	Maximum,°	Amount	Average,°	Standard deviation,°	Variance	Percentage
Range 1	−79,4149	−62,4734	−430,8	−71,8001	5,87,959	34,5696	6,81,818
Range 2	−62,4734	−45,5319	−673,561	−56,1301	5,14,237	26,444	13,6364
Range 3	−45,5319	−28,5904	−388,193	−38,8193	4,29,835	18,4759	11,3636
Range 4	−28,5904	−11,6489	−212,219	−21,2219	4,67,822	21,8858	11,3636
Range 5	−11,6489	5,29,257	-40,3037	−1,61,215	3,38,633	11,4672	28,4091
Range 6	5,29,257	22,2341	73,5913	14,7183	4,58,265	21,0007	5,68,182
Range 7	22,2341	39,1755	87,5592	29,1864	2,6453	6,99,761	3,40,909
Range 8	39,1755	56,117	144,198	48,0662	3,89,106	15,1403	3,40,909
Range 9	56,117	73,0585	185,796	61,9318	4,73,434	22,414	3,40,909
Range 10	73,0585	90	940,302	85,482	5,22,985	27,3514	12,5

Table 12.9 The minor angle distributions for the starch granules on the paper surface

Caption	Amount	Minimum,°	Maximum,°	Average,°	Standard deviation,°	Variance	Percentage
Range 1	109,2	10,5851	26,8781	18,1999	5,87,959	34,5696	6,81,818
Range 2	406,439	27,8146	44,3754	33,8699	5,14,237	26,444	13,6364
Range 3	511,807	45,0563	61,1645	51,1807	4,29,835	18,4759	11,3636
Range 4	687,781	62,6046	75,8329	68,7781	4,67,822	21,8858	11,3636
Range 5	2209,7	79,1643	94,2364	88,3879	3,38,633	11,4672	28,4091
Range 6	523,591	97,219	110,225	104,718	4,58,265	21,0007	5,68,182
Range 7	357,559	115,445	121,068	119,186	2,6453	6,99,761	3,40,909
Range 8	414,198	132,894	142,279	138,066	3,89,106	15,1403	3,40,909
Range 9	455,796	147,703	158,542	151,932	4,73,434	22,414	3,40,909
Range 10	1930,3	166,891	180	175,482	5,22,985	27,3514	12,5

Table 12.10 The calculation results for the inertial moment distributions in the starch granules on the paper surface

Caption	Minimum, μm^2	Maximum, μm^2	Average, μm^2	Standard deviation, μm^2	Percentage
Range 1	558,608	60,685,4	16,560,4	14,980,3	81,8182
Range 2	63,473,1	121,464	77,938,1	18,296,9	10,2273
Range 3	146,371	186,248	166,310	19,938,4	2,27,273
Range 4	–	–	–	–	0
Range 5	285,662	289,791	287,726	2064,15	2,27,273
Range 6	–	–	–	–	0
Range 7	397,985	397,985	397,985	0	1,13,636
Range 8	–	–	–	–	0
Range 9	528,951	528,951	528,951	0	1,13,636
Range 10	628,963	628,963	628,963	0	1,13,636

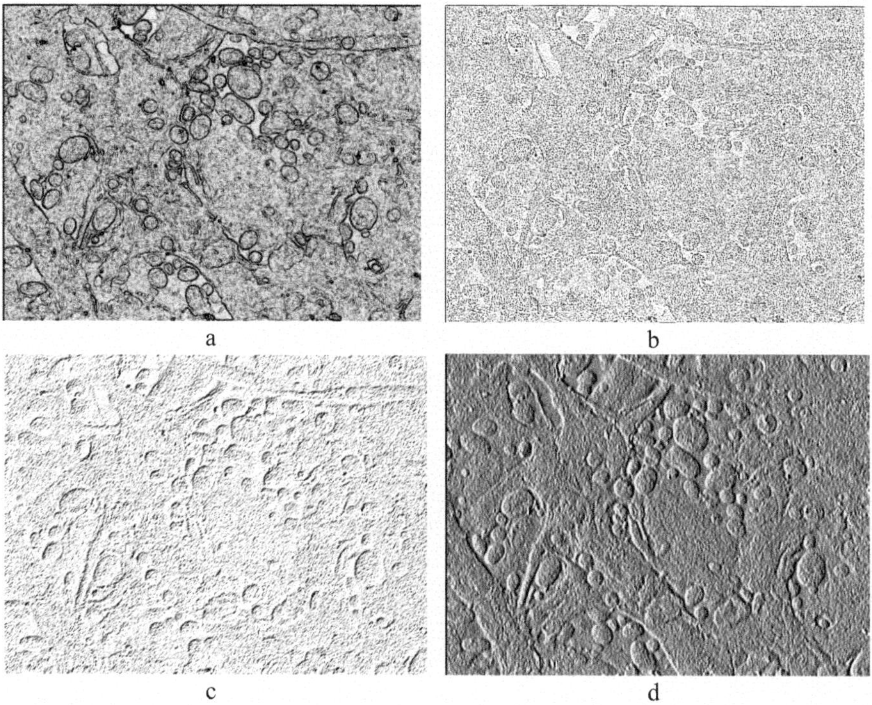

Fig. 12.3 Examples of automatic starch grain edge detections with: **a** Sobel-Feldman operator; **b** Laplacian (Laplace operator); **c** Roberts cross; **d** embossing simulation

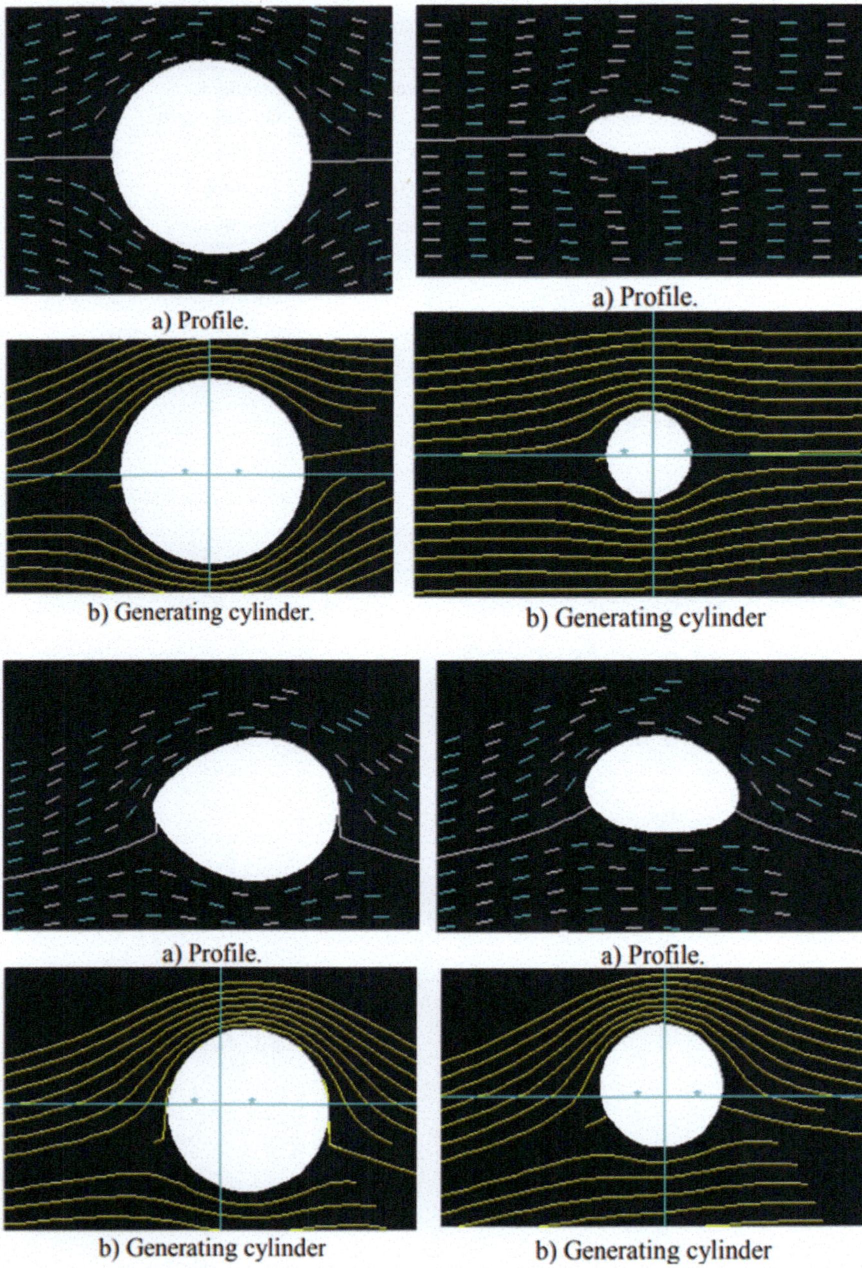

Fig. 12.4 A simple model analogy for starch particles based on cross-sections of profiles and generating cylinders for elementary geometrical-hydrodynamic forms

can be applied to the analysis of starch particles and other fillers after precision morphometries and geometrical approximations. The above example is purely illustrative and didactic, since in real models more complex effects need to be taken into account. In real starch particle cases, it is necessary to take into account the phenomenology of soft matter hydrodynamics and non-Newtonian hydrodynamics [103–109]. The mentioned approaches have been developed for both starch and adhesives in general [110–114], which makes them fully applicable to the case of the physical application of starch to create adhesives described in the article.

12.4 Conclusion

It is possible to create controllable starch-containing adhesives and binders or adhesives with the adhesive ability that is programmable by rearranging their particle geometry. In addition to the currently known methods of changing the already formed starch granules (ultrasound, pH, chemical treatment, and encapsulation), it is also possible to control the starch granule shape in vivo or in situ—at the granule formation stage, using transgenic varieties of the cultivated plants that give programmable size distributions or geometry of the starch granules during cultivation under the programmed environmental conditions. In this paper, multiparametric granulometry methods for analyzing the starch granules' geometry are tested and compared with the elementary "didactic" hydrodynamic model.

References

1. Weakley FB, Roth WB, Mehltretter CL, Rist CE (1971) Protein-dialdehyde starch glue for birch type II plywood. Starch-Stärke 23(2):58–62. https://doi.org/10.1002/star.19710230208
2. Weakley FB, Roth WB, Mehltretter CL, Kennicott JW (1973) Mill evaluation of dialdehyde starch-protein glue for hardwood plywood manufacture. Forest Prod J 23(7):44–48. https://doi.org/10.5555/19730610477
3. Fidel MM (1987) Starch based adhesive for water resistant plywood glue. Philippine Technol J (Philippines) 12(1):55–65
4. Xinbo W, Youshi W, Shuhong Z, Chunfang L (2003) Preparation of a new type urea-dialdehyde starch glue. Chem Ind Eng Progress 22(12):1316–1318
5. Xian-Qing X, Yu-Liang B, Wei-Juan G, Lu F, Zhi-Hui W (2018) Preparation and application of high performance corn starch glue in straw decorative panel. Wood Fiber Sci 50(1):1–8. https://doi.org/10.22382/wfs-2018-009
6. Jeżowski P, Kowalczewski PŁ (2019) Starch as a green binder for the formulation of conducting glue in supercapacitors. Polymers 11(10):1648. https://doi.org/10.3390/polym11101648
7. Doharey V, Sharma N (2010) The permutation role of fenugreek seeds starch and Gunda glue as a binder in Paracetamol tablets. J Pharm Sci Res 2(2):64
8. Jaszczur A, Modzelewska I, Kokoszka A, Żołnowski P (2010) Strength properties and biodegradation of paper products manufactured from broad-leaved bleached kraft pulp supplemented with starch and resin glue additives. Ann Warsaw Univ Life Sci SGGW; Forestry Wood Technol 71:314–318. https://doi.org/10.5555/20113111295

9. Modzelewska I, Sandak A, Sandak J, Jaszczur A (2011) FT-NIR spectrophotometric analysis of paper, made of cellulose sulfate bleached deciduous pulp, with starch and resinous glue additives, subjected to mycological biodegradiation. Acta Scientiarum Polonorum. Silvarum Colendarum Ratio et Industria Lignaria 10(1):29–36. https://doi.org/10.5555/20113202099
10. Zha Z, Jin W, Xiao H (1997) Study on adhesion mechanism and initial viscosity of starch glue for interleaving aluminium foil. Modern Chem Ind 17:26–27
11. Lizhen Y (2005) Preparation of oxidation prevent starch glue for interleaving aluminum foil paper. Chem Ind Times 6 (in Chinese)
12. Kaelble DH (1969) Peel adhesion: influence of surface energies and adhesive rheology. J Adhes 1(2):102–123. https://doi.org/10.1080/00218466908078882
13. Kaelble DH (1971) Rheology of adhesion. J Macromole Sci Rev Macromole Chem C6(1):85–112. https://doi.org/10.1080/15321797108068157
14. Narayan R, Dones M (1997) Energy-curable oligomers, rheology modifiers and adhesion. Paint Coat Ind 13(6):54–57
15. Piau JM, Verdier C, Benyahia L (1997) Influence of rheology and surface properties in the adhesion of uncross-linked pressure sensitive adhesives. Rheol Acta 36(4):449–461. https://doi.org/10.1007/BF00396330
16. Najarzadeh Z, Ajji A, Bruchet JB (2015) Interfacial self-adhesion of polyethylene blends: the role of long chain branching and extensional rheology. Rheol Acta 54(5):377–389. https://doi.org/10.1007/s00397-015-0843-1
17. Callies X, Fonteneau C, Pensec S, Bouteiller L, Ducouret G, Creton C (2016) Adhesion and non-linear rheology of adhesives with supramolecular crosslinking points. Soft Matter 12(34):7174–7185. https://doi.org/10.1039/C6SM01154C
18. Goffin AL, Raquez JM, Duquesne E, Siqueira G, Habibi Y, Dufresne A, Dubois P (2011) Poly (ε-caprolactone) based nanocomposites reinforced by surface-grafted cellulose nanowhiskers via extrusion processing: morphology, rheology, and thermo-mechanical properties. Polymer 52(7):1532–1538. https://doi.org/10.1016/j.polymer.2011.02.004
19. Yang X, Bakaic E, Hoare T, Cranston ED (2013) Injectable polysaccharide hydrogels reinforced with cellulose nanocrystals: morphology, rheology, degradation, and cytotoxicity. Biomacromol 14(12):4447–4455. https://doi.org/10.1021/bm401364z
20. Li MC, Wu Q, Song K, Lee S, Qing Y, Wu Y (2015) Cellulose nanoparticles: structure–morphology–rheology relationships. ACS Sustain Chem Eng 3(5):821–832. https://doi.org/10.1021/acssuschemeng.5b00144
21. Katbab AA, Nazockdast H, Bazgir S (2000) Carbon black—reinforced dynamically cured EPDM/PP thermoplastic elastomers. I. Morphology, rheology, and dynamic mechanical properties. J Appl Polym Sci 75(9):1127–1137. https://doi.org/10.1002/(SICI)1097-4628(20000228)75:9%3C1127::AID-APP5%3E3.0.CO;2-2
22. Otero-Navas I, Arjmand M, Sundararaj U (2017) Carbon nanotube induced double percolation in polymer blends: morphology, rheology and broadband dielectric properties. Polymer 114:122–134. https://doi.org/10.1016/j.polymer.2017.02.082
23. Bai L, He S, Fruehwirth JW, Stein A, Macosko CW, Cheng X (2017) Localizing graphene at the interface of cocontinuous polymer blends: morphology, rheology, and conductivity of cocontinuous conductive polymer composites. J Rheol 61(4):575–587. https://doi.org/10.1122/1.4982702
24. Damodar D, Kumar SK, Martha SK, Deshpande AS (2018) Nitrogen-doped graphene-like carbon nanosheets from commercial glue: morphology, phase evolution and Li-ion battery performance. Dalton Trans 47(35):12218–12227. https://doi.org/10.1039/C8DT01787E
25. Hartwig A, Meissner R, Merten C, Schiffels P, Wand P, Grunwald I (2013) Mutual influence between adhesion and molecular conformation: molecular geometry is a key issue in interphase formation. J Adhes 89(2):77–95. https://doi.org/10.1080/00218464.2013.731363
26. Kim H, Rajagopalan R (1994) Simultaneous effects of adhesion and poly-dispersity on microstructures and geometrical properties in particle deposition. Korean J Chem Eng 11(4):290–292. https://doi.org/10.1007/BF02697396

27. Packham DE (2003) Surface energy, surface topography and adhesion. Int J Adhes Adhes 23(6):437–448. https://doi.org/10.1016/S0143-7496(03)00068-X
28. Violano G, Demelio G, Afferrante L (2019) A note on the effect of surface topography on adhesion of hard elastic rough bodies with low surface energy. J Mech Behav Mater 28(1):8–12. https://doi.org/10.1515/jmbm-2019-0002
29. Mullins ME, Michaels LP, Menon V, Locke B, Ranade MB (1992) Effect of geometry on particle adhesion. Aerosol Sci Technol 17(2):105–118. https://doi.org/10.1080/02786829208959564
30. Varenberg M, Peressadko A, Gorb S, Arzt E (2006) Effect of real contact geometry on adhesion. Appl Phys Lett 89(12):121905. https://doi.org/10.1063/1.2356099
31. Varenberg M, Murarash B, Kligerman Y, Gorb SN (2011) Geometry-controlled adhesion: revisiting the contact splitting hypothesis. Appl Phys A 103(4):933–938. https://doi.org/10.1007/s00339-011-6394-0
32. Martin CL, Bordia RK (2008) Influence of adhesion and friction on the geometry of packings of spherical particles. Phys Rev E 77(3):031307. https://doi.org/10.1103/PhysRevE.77.031307
33. Babak VG (1994) Effect of interfacial tension and droplet size on coagulation, adhesion and rheology of concentrated emulsions. Colloids Surf, A 85(2–3):279–294. https://doi.org/10.1016/0927-7757(94)02873-7
34. Zhou M, Liang B, Huang H (2006) Preparation of PVAc emulsion adhesive modified by dialdehyde starch. China Adhes 15(9):32
35. Shi J, Wei S (2007) Isocyanate adhesive with esterified corn starch emulsion as main agent. Electron J Biol 3(2):35–41
36. Song SH, Kim YS, Cho UR (2009) A study on synthesis of starch-acryl pressure sensitive adhesive by soap-free emulsion polymerization. Elastomers Compos 44(4):429–435. https://koreascience.kr/article/JAKO200911764898669.page
37. He L, Lu W, Hu YA, Wang Y, Zhu Z, Zhong S (2016) The preparation and structure characterization of modified starch adhesive grafted with vinyl acetate-acrylate emulsion. J Forestry Eng 1(3):54–57. https://doi.org/10.5555/20163238727
38. Rodrigues LDA, Guerrini LM, Oliveira MP (2020) Thermal and mechanical properties of cationic starch-graft-poly (butyl acrylate-co-methyl methacrylate) latex film obtained by semi-continuous emulsion polymerization for adhesive application. J Therm Anal Calorim 146(1):143–152. https://doi.org/10.1007/s10973-020-09915-1
39. Cao R, Kumar D, Torres I, Menon N, Dinsmore A (2019) A gel based on vesicle-vesicle adhesion induced by polyelectrolyte: from morphology to rheology. Bull Am Phys Soc 64(2), ID: V63.00002. https://meetings.aps.org/Meeting/MAR19/Session/V63.2
40. Capovilla R, Guven J (2002) Geometry of lipid vesicle adhesion. Phys Rev E 66(4):041604. https://doi.org/10.1103/PhysRevE.66.041604
41. Gyu-Soon J [조규순] (2005) 2P-191: morphology, mechanical properties and biodegradation of PLA/starch blend improved interfacial adhesion by reactive extrusion. 한국공업화학회 연구논문 초록집 (Korean Society of Industrial Chemistry Research Paper Abstracts) 2005(1):228–228. https://kiss.kstudy.com/Detail/Ar?key=3797465
42. Jayasinghe GY, Tokashiki Y, Vidhana Arachchi LP (2005) Study of the production, performance and the micro-morphology of synthetic light weight soil aggregates developed by using coal fly ash treated with starch as an organic binder. In: Proceedings of the international conference on coal science and technology, 2P601, pp 1–12
43. Kovalevskaya EI, Kurilenko OD (1975) Structure-formation in starch paste polyelectrolyte systems. Ukr Khim Zh 41(2):205–206 (in Russian)
44. Mofti A, Hassan EA, Salama JY, El-Sabbah MMB (1982) Activity and activity coefficient of counter monovalent ions in aqueous polyelectrolyte solutions from carboxyalkylated starch. Starch (Stärke) 34(12):406–410. https://doi.org/10.1002/star.19820341204
45. Grote C, Lazik W, Heinze T (2003) Tartaric acid starch ether: a novel biopolymer-based polyelectrolyte. Macromol Rapid Commun 24(16):927–931. https://doi.org/10.1002/marc.200300042

46. Klimavičiūtė R (2004) Starch-cationic polyelectrolyte complexes and their interaction with acid dye. Fibres Textiles Eastern Euro 12(2):74–78. http://www.fibtex.lodz.pl/46_19_74.pdf
47. Pettersson G, Höglund H, Wågberg L (2006) The use of polyelectrolyte multilayers of cationic starch and CMC to enhance strength properties of papers formed from mixtures of unbleached chemical pulp and CTMP: Part I. Nord Pulp Pap Res J 21(1):115–121. https://doi.org/10.3183/npprj-2006-21-01-p115-121
48. Lundström-Hämälä L, Johansson E, Wågberg L (2010) Polyelectrolyte multilayers from cationic and anionic starch: Influence of charge density and salt concentration on the properties of the adsorbed layers. Starch (Stärke) 62(2):102–114. https://doi.org/10.1002/star.200900176
49. Wang YJ, Assaad E, Ispas-Szabo P, Mateescu MA, Zhu XX (2011) NMR imaging of chitosan and carboxymethyl starch tablets: swelling and hydration of the polyelectrolyte complex. Int J Pharm 419(1–2):215–221. https://doi.org/10.1016/j.ijpharm.2011.08.008
50. Fan Y, Bose R, Picchioni F (2020) Highly branched waxy potato starch-based polyelectrolyte: controlled synthesis and the influence of chain composition on solution rheology. Ind Eng Chem Res 59(23):10847–10856. https://doi.org/10.1021/acs.iecr.9b06893
51. Yu L, Xu H, Monro TM, Lancaster DG, Xie Y, Zeng H, Chen GY, Liu X (2017) Ultrafast colorimetric humidity-sensitive polyelectrolyte coating for touchless control. Mater Horiz 4(1):72–82. https://doi.org/10.1039/c6mh00317f
52. del Mercato LL, Abbasi AZ, Parak WJ (2011) Synthesis and characterization of ratiometric ion-sensitive polyelectrolyte capsules. Small 7(3):351–363. https://doi.org/10.1002/smll.201001144
53. Li G, Meng Y, Guo L, Zhang T, Liu J (2014) Formation of thermo-sensitive polyelectrolyte complex micelles from two biocompatible graft copolymers for drug delivery. J Biomed Mater Res, Part A 102(7):2163–2172. https://doi.org/10.1002/jbm.a.34894
54. Ramasamy T, Kim JH, Choi JY, Tran TH, Choi HG, Yong CS, Kim JO (2014) PH sensitive polyelectrolyte complex micelles for highly effective combination chemotherapy. J Mater Chem B 2(37):6324–6333. https://doi.org/10.1039/C4TB00867G
55. Zhulina EB, Borisov OV (2011) Poisson-Boltzmann theory of pH-sensitive (annealing) polyelectrolyte brush. Langmuir 27(17):10615–10633. https://doi.org/10.1021/la201456a
56. Losev NV, Lipatova IM (2009) Prediction of the dispersity of ultrasonically treated starch hydrogels. Russ J Appl Chem 82(6):1070–1073. https://doi.org/10.1134/S1070427209060263
57. Losev NV, Lipatova IM (2010) Prediction of the dispersity of starch hydrogels prepared under hydroacoustic treatment. Russ J Appl Chem 83(7):1309–1313. https://doi.org/10.1134/S1070427210070268
58. Wei B, Hu X, Li H, Wu C, Xu X, Jin Z, Tian Y (2014) Effect of pHs on dispersity of maize starch nanocrystals in aqueous medium. Food Hydrocolloids 36:369–373. https://doi.org/10.1016/j.foodhyd.2013.08.015
59. Wei B, Xu X, Jin Z, Tian Y (2014) Surface chemical compositions and dispersity of starch nanocrystals formed by sulfuric and hydrochloric acid hydrolysis. PLoS ONE 9(2):e86024. https://doi.org/10.1371/journal.pone.0086024
60. Fu Y, Yang J, Jiang L, Ren L, Zhou J (2019) Encapsulation of Lutein into starch nanoparticles to improve its dispersity in water and enhance stability of chemical oxidation. Starch (Stärke) 71(5–6):1800248. https://doi.org/10.1002/star.201800248
61. De Boer ED (1991) Chemically modified derivatives of starch from a new genetic variety of corn. Cereal Foods World 36(8):631–639. https://doi.org/10.5555/19931636037
62. Jacobsen E, Kuipers AGJ, Heeres P, Visser RGF (1997) Genetic modification of starch composition in potato and the biosafety aspects concerning the introduction of these modified potatoes. In: Tewari KK, Singhal GS (eds) Plant molecular biology and biotechnology, pp 196–201. https://library.wur.nl/WebQuery/wurpubs/38165
63. Gao M, Baga M, Chibbar RN (2002) Genetic engineering for modified starch structure in cereals. In: Khachatourians GC, Hui YH, Scorza R, Nip W-K (eds) Transgenic plants and crops. CRC Press, Boca Raton, pp. 169–182. https://doi.org/10.1201/9780203910979-18

64. Kortstee AJ, Suurs LC, Vermeesch AM, Keetels CJ, Jacobsen E, Visser RG (1998) The influence of an increased degree of branching on the physico-chemical properties of starch from genetically modified potato. Carbohyd Polym 37(2):173–184. https://doi.org/10.1016/S0144-8617(98)00045-9
65. Ward I, Chapman JW, Williamson R (1999) New viscosifier for water based muds based on a genetically modified starch. In: SPE international conference on oilfield chemistry? SPE (Society of Petroleum Engineers), pp SPE–50723. https://doi.org/10.2118/50723-MS
66. Wandelt C (2006) General surveillance–Cultivation of genetically modified starch potatoes in a closed-loop system. J Verbr Lebensm 1(1):58–59. https://doi.org/10.1007/s00003-006-0074-6
67. Sadowska J, Budny J, Fornal J (2008) Starch, protein, glycoalkaloids, and l-ascorbic acid content in tubers of genetically modified potato cv. Irga. Euro Food Res Technol 227(1):233–241. https://doi.org/10.1007/s00217-007-0716-7
68. Gschwendtner S (2010) Effects of a genetically modified potato line with altered starch metabolism on carbon fluxes within the plant-soil system and on microbial community structure and function in the rhizosphere (Doctoral dissertation, Technische Universität München). https://mediatum.ub.tum.de/?id=982887
69. Gschwendtner S, Esperschütz J, Buegger F, Reichmann M, Müller M, Munch JC, Schloter M (2011) Effects of genetically modified starch metabolism in potato plants on photosynthate fluxes into the rhizosphere and on microbial degraders of root exudates. FEMS Microbiol Ecol 76(3):564–575. https://doi.org/10.1111/j.1574-6941.2011.01073.x
70. Hannula SE, Boschker HTS, de Boer W, Van Veen JA (2012) ^{13}C pulse-labeling assessment of the community structure of active fungi in the rhizosphere of a genetically starch-modified potato (Solanum tuberosum) cultivar and its parental isoline. New Phytol 194(3):784–799. https://doi.org/10.1111/j.1469-8137.2012.04089.x
71. Okano K, Zhang Q, Shinkawa S, Yoshida S, Tanaka T, Fukuda H, Kondo A (2009) Efficient production of optically pure D-lactic acid from raw corn starch by using a genetically modified L-lactate dehydrogenase gene-deficient and α-amylase-secreting Lactobacillus plantarum strain. Appl Environ Microbiol 75(2):462–467. https://doi.org/10.1128/AEM.01514-08
72. Li M, Piao JH, Yang XG (2009) Comparative analysis of nutritional composition between genetically modified rice enriched with resistant starch and its wild type rice. Sci Technol Eng 9:1847–1849
73. Li M, Piao J, Yang X (2010) Subchronic toxicity test of genetically modified rice with double antisense starch-branching enzyme gene. J Hygiene Res (Wei sheng yan jiu) 39(4):436–439 (in Chinese). https://pubmed.ncbi.nlm.nih.gov/20726232/
74. Min L, Yichun H, Jianhua P, Xiaoguang Y (2011) [Evaluation on the nutritional value of protein in genetically modified rice with double antisense starch-branching enzyme gene]. Chin J Food Hygiene (2):97–99 (in Chinese). http://caod.oriprobe.com/articles/26973801/Evaluation_on_the_nutritional_value_of_protein_in_genetically_modified.htm
75. Joutsjoki V, Fagerström R, Parkkinen E, Home S, Torkkeli T (1993) Synthesis of glucoamylase, limit dextrinase and cellulase activities by genetically modified T. reesei and use of the final growth medium (ALKO-GA) as an effective starch degrading enzyme mix. In: 2nd Tricel symposium on Trichoderma reesei cellulases and other hydrolases (TRICEL93; Finland, Espoo, 2 Jun 1993 - 5 Jun 1993). https://cris.vtt.fi/en/publications/synthesis-of-glucoamylase-limit-dextrinase-and-cellulase-activiti
76. Wasserman LA, Kolachevskaya OO, Krivandin AV, Filatova AG, Gradov OV, Plashchina IG, Romanov GA (2023) Changes in structural and thermodynamic properties of starch during potato tuber dormancy. Int J Mol Sci 24(9):8397. https://doi.org/10.3390/ijms24098397
77. Jansson C, Puthigae S, Sun C, Grenthe C (1995) Genetically modified starch as an industrial raw material. J Macromol Sci Part A Pure Appl Chem 32(4):895–898. https://doi.org/10.1080/10601329508010303
78. Broothaerts W, Corbisier P, Emons H, Emteborg H, Linsinger TP, Trapmann S (2007) Development of a certified reference material for genetically modified potato with altered starch composition. J Agric Food Chem 55(12):4728–4734. https://doi.org/10.1021/jf0701584

79. EFSA (2006) Opinion of the Scientific Panel on Genetically Modified Organisms on an application (Reference EFSA-GMO-UK-2005-14) for the placing on the market of genetically modified potato EH92-527-1 with altered starch composition, for production of starch and food/feed uses, under Regulation (EC) No 1829/2003 from BASF plant science. EFSA J 324:1–20. https://doi.org/10.2903/j.efsa.2006.324
80. EFSA (European Food Safety Authority) (2006) Opinion of the Scientific Panel on Genetically Modified Organisms on a request from the Commission related to the notification (Reference C/SE/96/3501) for the placing on the market of genetically modified potato EH92-527-1 with altered starch composition, for cultivation and production of starch, under Part C of Directive 2001/18/EC from BASF plant science. EFSA J 323:1–20. https://doi.org/10.2903/j.efsa.2006.323
81. European Food Safety Authority (EFSA) (2006) Opinion of the scientific panel on genetically modified organisms [GMO] on an application (Reference EFSA-GMO-UK-2005-14) for the placing on the market of genetically modified potato EH92-527-1 with altered starch composition, for production of starch and food/feed uses, under Regulation (EC) No 1829/2003 from BASF plant science. EFSA J 4(2):324. https://doi.org/10.2903/j.efsa.2006.324
82. European Food Safety Authority (EFSA) (2006) Opinion of the Scientific Panel on genetically modified organisms [GMO] related to the notification (Reference C/SE/96/3501) for the placing on the market of genetically modified potato EH92-527-1 with altered starch composition, for cultivation and production of starch, under Part C of Directive 2001/18/EC from BASF plant science. EFSA J 4(2):323
83. Sørensen IK (2006) Opinion of the scientific panel on genetically modified organisms on an application (Reference EFSA-GMO UK-production of starch and food/feed uses, under Regulation (EC) No 1829/2003 from BASF Plant Science: Question No EFSA-Q-2005-070. https://doi.org/10.2903/j.efsa.2006.323
84. Kanarskii AV, Kanarskaya ZA, Yagofarov DS, Wasserman LA, Vasil'ev VG, Skryabin KG, Kamionskaya AM (2011) Study of structural and rheological properties of starch isolated from genetically modified potato. Appl Biochem Microbiol 47(9):827–832. https://doi.org/10.1134/S000368381109002X
85. Tan I, Halley PJ (2014) "Structure-Property" relationships of genetically modified starch. In: Halley PJ, Avérous L (eds) Starch polymers: from genetic engineering to green applications. Elsevier, pp 31–75. https://doi.org/10.1016/B978-0-444-53730-0.00020-8
86. Tanczos E, Kulmburg P, Andree CH, Kopp J, Jiao XY, Cao L, Flis B, Dodic. T, Stark GB (1998) Genetically modified fibroblasts-fibrin-glue-suspension for in vivo drug delivery of epidermal growth factor (EGF): a surface gene therapy model not only for wounds. In: Biological matrices and tissue reconstruction. Springer, Berlin, pp 47–52. https://doi.org/10.1007/978-3-642-60309-9_6
87. Regina A, Li Z, Morell MK, Jobling SA (2014) Genetically modified starch: state of art and perspectives. In: Halley PJ, Avérous L (eds) Starch polymers. Elsevier, pp 13–29. https://doi.org/10.1016/B978-0-444-53730-0.00019-1
88. Ulusoy U (2023) A review of particle shape effects on material properties for various engineering applications: from macro to nanoscale. Minerals 13(1):91. https://doi.org/10.3390/min13010091
89. Naidu BR, Rao PL, Babu MP, Bhavani KVL (2012) Efficient case study for image edge gradient based detectors- Sobel, Robert cross, Prewitt and Canny. Int J Electron Commun Comput Eng 3(3):561–571
90. Herwindiati DE, Hendryli J (2020) Classification of flower images using SVM method through the colour, texture and shape using Histogram, Haar Wavelet and Robert Cross methods. IOP Conf Ser Mater Sci Eng 1007:012135. https://doi.org/10.1088/1757-899X/1007/1/012135
91. Kojima H, Arai H, Yamazaki M (2007) A web-based system for producing emboss processing-based moving images inducing visual illusion for image interpretation. J Appl Comput Civil Eng 16:43–52. https://doi.org/10.2208/journalac2003.16.0_43
92. Leonardi RM, Giordano D, Maiorana F, Greco M (2010) Accuracy of cephalometric landmarks on monitor-displayed radiographs with and without image emboss enhancement. Euro J Orthod 32(3):242–247. https://doi.org/10.1093/ejo/cjp122

93. Prajapati KA, Shah SM (2017) A review on emboss and deboss features of edge matching. Int J Comput Appl 162(4):22–26. https://doi.org/10.5120/ijca2017913275
94. Jhapate AK, Dronawat R, Saxena M, Chourey R (2022) Improve the detection of retinopathy with Roberts cross edge detection. Lect Notes Netw Syst 400:475–483. https://doi.org/10.1007/978-981-19-0095-2_45
95. Borazjani I, Sotiropoulos F (2010) On the role of form and kinematics on the hydrodynamics of self-propelled body/caudal fin swimming. J Exp Biol 213(1):89–107. https://doi.org/10.1242/jeb.030932
96. Martín M, Galán MA, Cerro RL, Montes FJ (2011) Shape oscillating bubbles: hydrodynamics and mass transfer-a review. Bubble Sci Eng Technol 3(2):48–63. https://doi.org/10.1179/1758897911Y.0000000006
97. Bolkhovitinov AS (2013) Hydrodynamic engineering approaches for modeling of the surface cell morphogenesis as applications of Rosen's optimality theory. J Med Bioeng 2(4):279–289. https://doi.org/10.12720/jomb.2.4.279-289
98. Theers M, Westphal E, Qi K, Winkler RG, Gompper G (2018) Clustering of microswimmers: interplay of shape and hydrodynamics. Soft Matter 14(42):8590–8603. https://doi.org/10.1039/C8SM01390J
99. Qi K, Annepu H, Gompper G, Winkler RG (2020) Rheotaxis of spheroidal squirmers in microchannel flow: interplay of shape, hydrodynamics, active stress, and thermal fluctuations. Phys Rev Res 2(3):033275. https://doi.org/10.1103/PhysRevResearch.2.033275
100. Ma H, Bolster C, Johnson WP, Li K, Pazmino E, Camacho KM, Anselmo AC, Mitragotri S (2020) Coupled influences of particle shape, surface property and flow hydrodynamics on rod-shaped colloid transport in porous media. J Colloid Interface Sci 577:471–480. https://doi.org/10.1016/j.jcis.2020.05.022
101. Dianyu E, Zhou P, Guo S, Zeng J, Cui J, Jiang Y, Lu Y, Jiang Z, Li Z, Kuang S (2022) Particle shape effect on hydrodynamics and heat transfer in spouted bed: a CFD–DEM study. Particuology 69:10–21. https://doi.org/10.1016/j.partic.2021.11.009
102. Eppala VCR, Varghese MM, Vakamalla TR (2023) Effect of particle shape on the hydrodynamics of gas-solid fluidized bed. Chem Eng Res Des 189:461–473. https://doi.org/10.1016/j.cherd.2022.11.025
103. Manghi M, Schlagberger X, Kim YW, Netz RR (2006) Hydrodynamic effects in driven soft matter. Soft Matter 2(8):653–668. https://doi.org/10.1039/B516777A
104. Marchetti MC, Joanny JF, Ramaswamy S, Liverpool TB, Prost J, Rao M, Simha RA (2013) Hydrodynamics of soft active matter. Rev Mod Phys 85(3):1143. https://doi.org/10.1103/RevModPhys.85.1143
105. Czajkowski M, Bi D, Manning ML, Marchetti MC (2018) Hydrodynamics of shape-driven rigidity transitions in motile tissues. Soft Matter 14(27):5628–5642. https://doi.org/10.1039/C8SM00446C
106. Fan XJ, Tanner RI, Zheng R (2010) Smoothed particle hydrodynamics simulation of non-Newtonian moulding flow. J Nonnewton Fluid Mech 165(5–6):219–226. https://doi.org/10.1016/j.jnnfm.2009.12.004
107. Chhabra RP (2020) Hydrodynamics of nonspherical particles in non-Newtonian fluids. Handb Appl Polym Process Technol 1–45. https://doi.org/10.1201/9781003067528-1
108. Hosseini K, Omidvar P, Kheirkhahan M, Farzin S (2019) Smoothed particle hydrodynamics for the interaction of Newtonian and non-Newtonian fluids using the μ (I) model. Powder Technol 351:325–337. https://doi.org/10.1016/j.powtec.2019.02.045
109. Abdolahzadeh M, Tayebi A, Omidvar P (2019) Mixing process of two-phase non-newtonian fluids in 2D using smoothed particle hydrodynamics. Comput Math Appl 78(1):110–122. https://doi.org/10.1016/j.camwa.2019.02.019
110. Chandler HD (2014) An activation energy approach to analysing non-Newtonian slurry viscosities with application to a suspension of starch in a carboxymethylcellulose solution. Powder Technol 268:368–372. https://doi.org/10.1016/j.powtec.2014.08.058
111. Ge Z, Martone R, Brandt L, Minale M (2021) Irreversibility and rate dependence in sheared adhesive suspensions. Phys Rev Fluids 6(10):L101301. https://doi.org/10.1103/PhysRevFluids.6.L101301

112. Müller M, Finke J, Stahl L, Tong Y, Fricke H, Vallée T (2022) Development and validation of a compression flow model of non-Newtonian adhesives. J Adhes 98(9):1260–1297. https://doi.org/10.1080/00218464.2021.1895771
113. Kaufmann M, Flaig F, Müller M, Stahl L, Finke J, Vallée T, Fricke H (2022) Experimental validation of a compression flow model of Non-Newtonian adhesives. J Adhes 98(14):2295–2324. https://doi.org/10.1080/00218464.2021.1971081
114. Ingelsten S, Mark A, Kádár R, Edelvik F (2022) Simulation of viscoelastic squeeze flows for adhesive joining applications. J Nonnewton Fluid Mech 300:104722. https://doi.org/10.1016/j.jnnfm.2021.104722

Chapter 13
Experimental Study of Pine Orthotropy Under Dynamic Compression

Michael E. Gonov, Aleksandr Yu. Konstantinov, and Iuliia V. Konstantinova

Abstract The work is devoted to the experimental study of orthotropy of pine strength characteristics under dynamic loading. Using the Kolsky method, pine samples were tested at normal humidity and room temperature. Loading was carried out at different angles to the fiber direction. Deformation diagrams in the range of strain rates 500–3000 1/s were obtained. The strain rate dependences of the dynamic hardening coefficient (dynamic factor) were determined. A comparison of the strength characteristics of pine under loading at angles of 0, 45, and 90° relative to the fiber direction has been carried out.

Keywords Dynamic strength · Stress · Strain · Strain rate · Orthotropy · Wood · Pine · Dynamic tests · Kolsky method · SHPB · Experiment · Dynamic factor

13.1 Introduction

The dynamic characteristics of pine are of interest due to the use of this material as a damping element in transportation containers, including those for hazardous materials [1–5]. To date, wood is considered to be a material whose properties have orthogonal anisotropy. When calculating wooden structures, the approximation of transversal-isotropic material is usually used, the properties of which differ along and across the fibers. Due to its properties, wood absorbs impact energy well [6, 7]. Systematic studies of dynamic properties of five species of wood (balsa, pine, mahogany, American oak, and yucca) at impact velocities up to 300 m/s are given in

M. E. Gonov · A. Yu. Konstantinov (✉) · I. V. Konstantinova
Research Institute for Mechanics, National Research Lobachevsky State University of Nizhny Novgorod, Gagarin Ave. 23, 603950 Nizhny Novgorod, Russia
e-mail: konstantinov@mech.unn.ru

M. E. Gonov
e-mail: gonov@mech.unn.ru

I. V. Konstantinova
e-mail: ullichka@rambler.ru

H. Altenbach and V. A. Eremeyev (eds.), *Dynamics of Discrete and Continuum Structures and Media*, Advanced Structured Materials 221,
https://doi.org/10.1007/978-3-031-75626-9_13

[8–10]. The growth of ultimate stresses with increasing loading speed was experimentally shown [11]. A large volume of dynamic tests of sequoia, birch, pine, and aspen under loading at different angles to the fiber direction was performed by Novikov and co-workers [12]. Buchar and co-workers [6, 13, 14] investigated the behavior of coniferous and deciduous wood species (spruce, pine, oak, beech, and birch) under intensive dynamic effects during explosive loading of wooden beams and slabs. In their research, the authors used both experimental methods and numerical modeling methods using the LS-DYNA software package. In recent years, the attention of scientists has been focused on the study of the influence of wood density, moisture content, structure, cutting angle, and type of stress–strain state on its mechanical properties [15–21]. In these works, such important characteristics as dynamic deformation diagrams, ultimate strength and deformation characteristics, and fracture energy were obtained for a number of wood species, and the dependence of these characteristics on the strain rate was plotted. The range of investigated strain rates was from 10^{-3} to 10^3 s^{-1}. At strain rates from 10^{-3} to 10° s^{-1} servohydraulic testing machines were used, strain rates of 10^1–10^2 s^{-1} were realized with the help of various pile driving machines, modifications of the Kolsky method were used to obtain dynamic properties at strain rates of ~10^3 s^{-1}. It should be noted that the obtained results on the influence of strain rate on dynamic properties of wood, mainly indicate that the ultimate stresses increase with increasing strain rate. The strongest increase is observed at speeds of about 10^3 s^{-1}. It should be noted that there is no consensus on the interpretation of this effect at present. The study [22] presents experimental results of a series of tests on uniaxial dynamic compression using the Kolsky method of cylindrical samples of five wood species (balsa, yellow pine, mahogany, American oak, and ecchi) selected for the range of densities (from 260 to 1200 kg/m^3) and at impact velocities up to 300 m/s. In [23], the compression behavior of maritime pine in the transverse (radial-tangential) plane under quasi-static and high-velocity deformation regimes is examined. For the high-speed deformation tests, the Kolsky method combined with the digital image correlation method was used to reconstruct the deformation fields. Quasi-static compression tests were also performed to compare the results. In [24], the dynamic compression behavior of cork wood was investigated over a wide range of strain rates. A screw driven testing machine was used for low strain rates (10^{-4} s^{-1}), a servohydraulic testing machine was used for strain rates up to 500 s^{-1}, and for high strain rates (500–4000 s^{-1}) the Kolsky method was used with magnesium alloy rods and semiconductor strain gauges to measure the rod strain. In article [16], the Split Hopkinson Pressure Bar (SHPB) system was used to study the dynamic fracture of dry maple wood. In [25], a comprehensive experimental program was conducted to study the mechanical behavior of maple and ash wood for a range of densities used to make baseball bats. At elevated temperatures, the wood softens significantly. At low strain rates, the yield stress decreases linearly with increasing temperature from 25 to 100 °C [26, 27]. The yield stress is developed at 130–160 °C. At temperatures above 160 °C, any changes in yield stress are due to chemical changes in the wood [26, 28]. In [29], compression tests were carried out on birch plywood samples at negative temperatures and different strain rates (0.004–700 1/s). At −170 °C, the sensitivity to deformation rate is negligible. At room temperature, the temperature dependence of the strain rate becomes significant. The yield strength does not change in the temperature range 0–100 °C. The yield strength drops by one tenth in the temperature range 0–200 °C during quasi-static radial compression [28]. During a series of quasi-static tests [30], Young's modulus for radial compression decreased by 84% when the temperature increased from 22 °C to 140 °C, tangential compression decreased by 82%, and for longitudinal compression by 57%. In high strain rate compression, yield strength and elastic modulus decreased with increasing temperature [31]. In [32], Young's modulus decreased by 93% for radial compression at high strain rate when the temperature was increased from 20 °C to 135 °C. It is important to note that the mechanical properties of wood-sawn timber are highly dependent on the place of growth, its age, the place of cutting of the sample and a number of other factors, so the results obtained by different authors can be quite different from each other; however, the effects obtained are often qualitatively the same.

The present work is aimed at the experimental study of the orthotropy of pine of normal moisture content under dynamic compression.

13.2 Experimental Method

The dynamic deformation and fracture of pine samples were investigated using the classical version of the Kolsky method (Split Hopkinson Pressure Bar method) in compression under uniaxial stress state conditions [33, 34]. The scheme of the method is shown in Fig. 13.1. Numbers denote: 1—striker, 2—loading measuring bar, 3, 6—strain gauges, 4—specimen, and 5—supporting measuring bar. Varying of the loading mode is carried out by changing the speed of the striker. The time dependences of the sample deformation, sample strain rate, and stress in the sample are calculated on the basis of the impulses of elastic strains in the measuring bars registered by the strain gauges according to the Kolsky formulae.

To analyze the influence of direction on the properties of pine, the specimens were manufactured in such a way that the fiber direction made angles 0, 90, and 45° with the loading direction (Fig. 13.2).

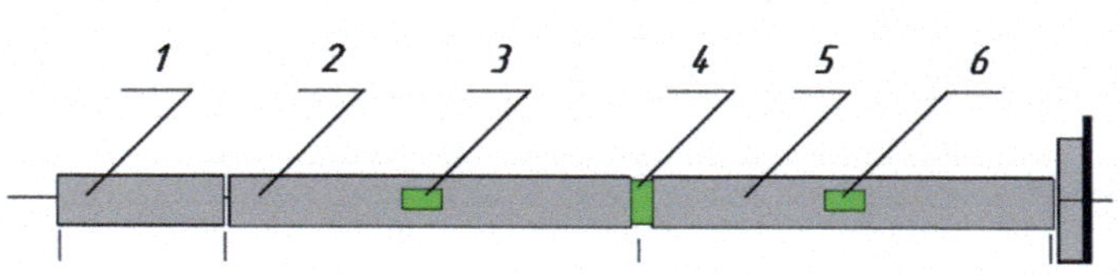

Fig. 13.1 Scheme of the Kolsky method in compression

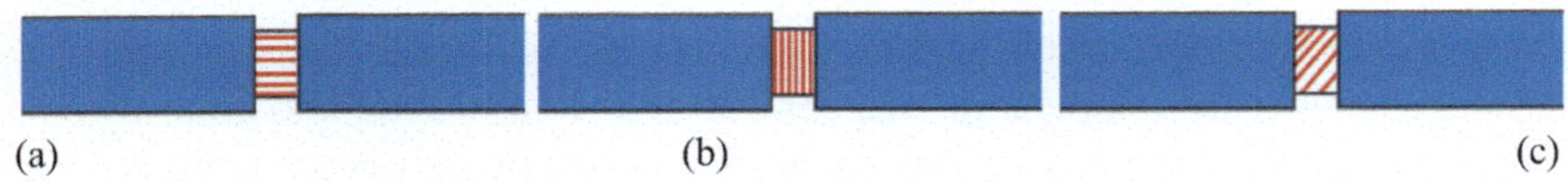

Fig. 13.2 Samples for analysis of directional effects: **a** loading along the fibers, **b** loading across the fibers, **c** loading at an angle of 45°

13.3 Specimens Preparation

To make pine samples from solid wood blanks (usually in the form of beams), plates of the required thickness were cut using a sawing machine with a disk with a diameter of 350 mm, from which samples of the required diameter were drilled using a table-drilling machine NS-12 M and wood drill bits. After that, the working surfaces of the samples were sanded with low-grit sandpaper. Then the diameter and length of each sample were measured, the mass was measured to determine the density, and moisture content was measured with two control meters (needle-type IV 1–1 and surface VIMS-2 moisture meters). Upon completion of the preparatory work, the samples were numbered and packed in polyethylene bags with a lock (Fig. 13.3).

Fig. 13.3 Specimens preparation

13.4 Results of the Experimental Study

High-speed tests of pine samples on one-dimensional compression were carried out in three directions with respect to the direction of fibers: along, across, and at an angle of 45 degrees. The experiments were carried out on the SHPB setup with 20 mm diameter measuring bars on 18 mm diameter and 10 mm length specimens with normal moisture content. When testing specimens with the direction of cutting along the fibers (AF), measuring bars and a striker made of duralumin alloy D16T were used. Loading bar length was 1 m and supporting bar length was 3 m. When testing the specimens across the fibers (CF) and at an angle of 45° (45), a 1 m long D16T duralumin bar was used as a loading bar, and a 2 m long D16T duralumin tube was used as a supporting bar to increase the amplitude of the recorded signal. The length of the impactor made of D16T alloy in all experiments was 250 mm. The loading modes were selected according to the gradual increase in strain rate.

13.5 Compression Along Fibers

The set of dynamic tests of pine with the direction of specimen cutting along the fibers (AF) consisted of four speed modes. At least five repeated tests under the same conditions were performed in each group to evaluate the variation.

The strain pulses recorded in the measuring bars were synchronized in time. Using the formulas of the Kolsky method, stress versus time, strain versus time, and strain rate versus time dependencies were plotted for each experiment. The obtained dependencies were synchronized in time for each velocity regime. After excluding the time parameter, a dynamic deformation diagram $\sigma(\varepsilon)$ was plotted with an additional vertical axis showing the history of strain rate changes for each experiment. Further, averaged diagrams of stress–strain dependence $\sigma(\varepsilon)$ and strain rate–strain dependence, as well as stress–time dependence $\sigma(t)$, and strain rate–time dependence $\sigma(t)$ were plotted for the same-type loading modes.

In mode 1, the average maximum stresses reached 47 MPa, with average strain rates of about 450 s^{-1}. It should be noted that in this mode the specimens retained their integrity, as evidenced by the nature of the diagram, as well as visual inspection of the specimens. In the other experimental modes, almost all the specimens underwent failure. In mode 2, the average maximum stresses reached 100 MPa, with average strain rates of about 1100 s^{-1}. In the third mode, the average maximum stresses reached 97 MPa, at average strain rates of about 1800s^{-1}. In the fourth mode, the average maximum stresses reached 84 MPa, at average strain rates of about 2250 s^{-1}. It can be noted that, depending on the impact velocity, the values of compressive strength values change with a tendency of slight decrease. Analysis of the obtained diagrams of modes 2–4 showed that the initial sections have a character close to linear, declining branches of deformation diagrams and examination of specimens

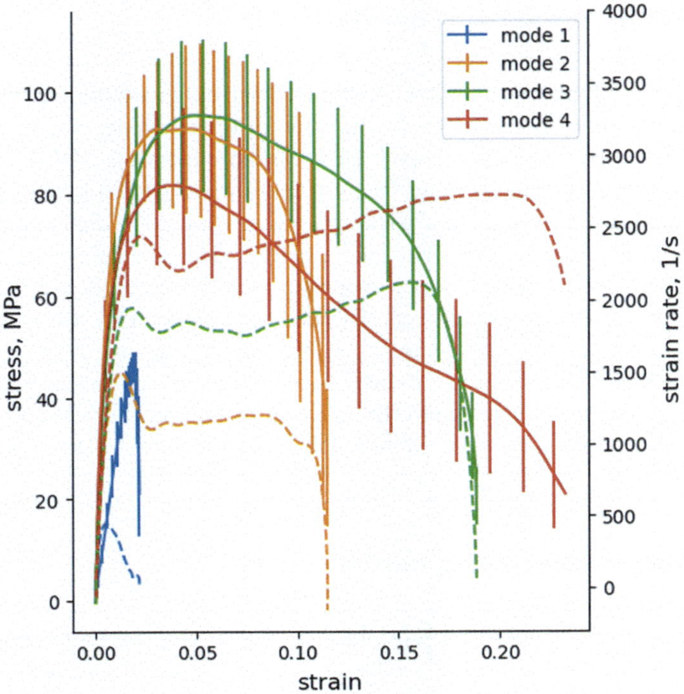

Fig. 13.4 Averaged diagrams of dynamic deformation in compression of pine along fibers

after the experiments indicate that the specimens were destroyed in the process of testing.

Figure 13.4 shows averaged deformation diagrams with chronology of strain rate variation for three rate regimes with confidence intervals.

For the obtained diagrams of individual experiments, characteristic points with maximum achieved stresses are selected, for which the values of time to the onset of fracture are determined. The strain rates were assumed to be maximum before the beginning of failure of the specimens, since they change in the process of their deformation. The results are presented in Fig. 13.5 in the form of strain rate dependencies. Green circles correspond to the unfractured specimens.

The obtained dependences demonstrate that, with the increase of strain rate, there is a tendency of a slight increase of maximum stresses, the corresponding ultimate strains also increase (according to the linear law), and the time to failure decreases (according to the power law).

The dynamic strength of the material is characterized by the dynamic hardening factor (DF), defined as the ratio of the maximum dynamic stress to the static compressive strength of pine (41.2 MPa). The dependence of the DF on the strain rate is shown in Fig. 13.6. In the obtained range of strain rates, the DF varies from 1 to 2.5.

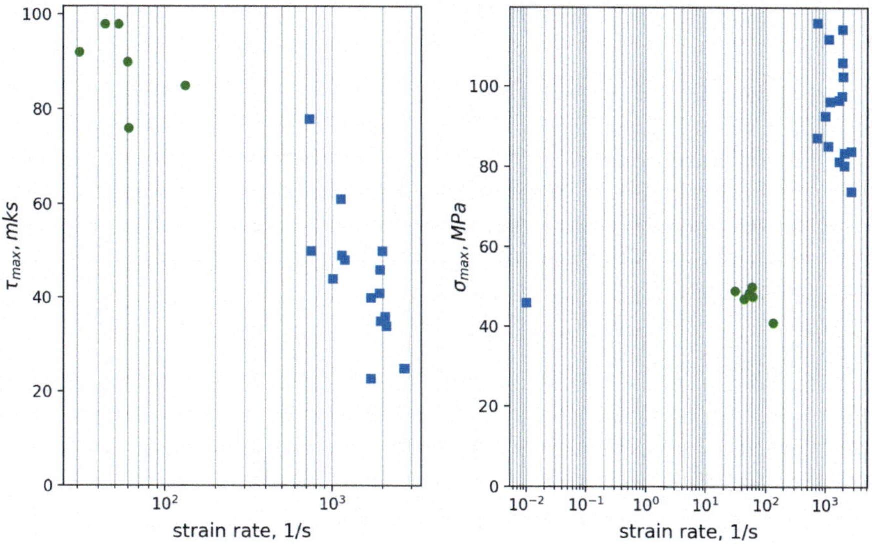

Fig. 13.5 Dependence of ultimate stresses on strain rate in compression of pine along fibers

Fig. 13.6 Dependence of DF on strain rate in compression of pine along fibers

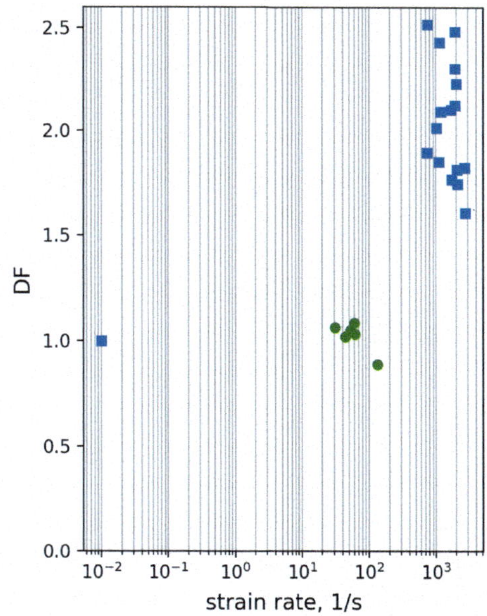

13.6 Compression Across Fibers

The set of dynamic tests of pine with the direction of cutting the specimens across the fibers (CF) consisted of three speed modes with five repeated tests per mode. In mode 1, the average dynamic compressive strength was about 5 MPa, with average strain rates of about 1200 s^{-1}. At mode 2, the average dynamic compressive strength was on the order of 6 MPa, at average strain rates of about 1700s^{-1}. At mode 3, the average dynamic compressive strength also reached about 6 MPa, at average strain rates of about 2100s^{-1}.

Figure 13.7 shows averaged strain diagrams with chronology of strain rate variation for three strain rate modes with confidence intervals.

In the above diagrams in the stress–strain axes ($\sigma \sim \varepsilon$) in the initial loading section, the growth of stress and strain follows a law close to linear, and during further deformation, when the stress limit values are reached, the pine is intensively destroyed,

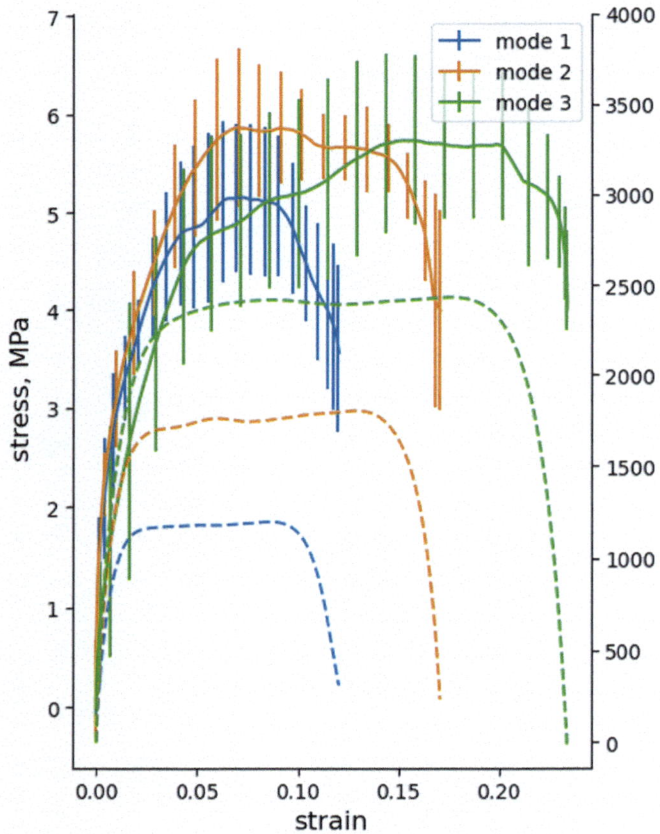

Fig. 13.7 Averaged diagrams of dynamic deformation in compression of pine across fibers

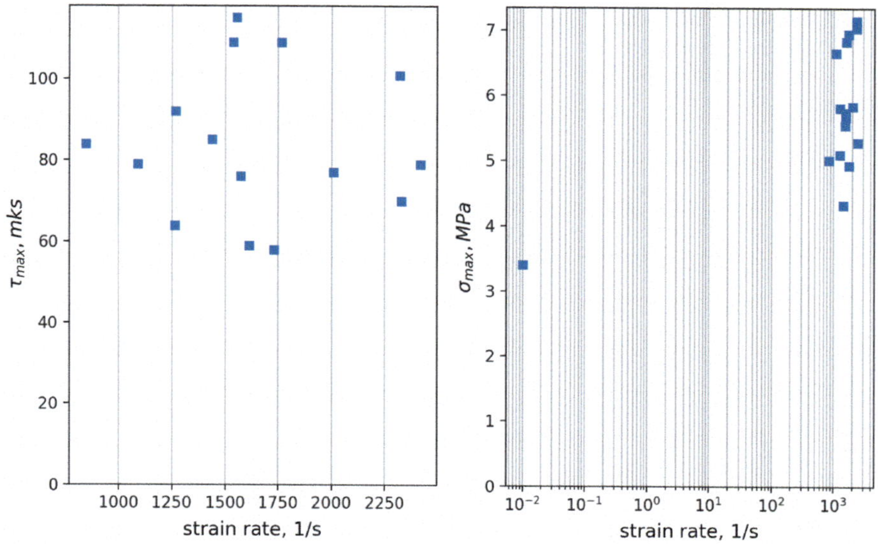

Fig. 13.8 Influence of strain rate on mechanical properties of pine with direction across the fibers under dynamic compression

which is accompanied by a decrease in stress and growth of strain. Dynamic compression tests were conducted to determine the effect of changing the strain rate on the deformation diagram and strength characteristics of pine, shown below (Fig. 13.8).

The obtained dependences demonstrate that with the increase in strain rate according to the linear law, there is a tendency of insignificant growth of maximum stresses, and the limiting strains corresponding to them have a significant growth; the time to the onset of fracture is approximately constant and do not depend on strain rate.

The dependence of the DF on the strain rate is shown in Fig. 13.9. In the obtained range of strain rates, the DF index varies from 1 to 2.1.

13.7 Compression at an Angle of 45°

The set of dynamic tests of pine with the direction of cutting of specimens at an angle of 45° (45) relative to the fibers consisted of three velocity modes with five repeated tests per mode. In mode 1, the average dynamic compressive strength was about 11 MPa, with average strain rates of about 1000 s^{-1}. At mode 2, the average dynamic compressive strength was about 9 MPa, at average strain rates of about 1500 s^{-1}. At mode 3, the average dynamic compressive strength also reached about 18 MPa, at average strain rates of about 1800s^{-1}.

Figure 13.10 shows averaged strain diagrams with chronology of strain rate variation for three strain rate modes with confidence intervals.

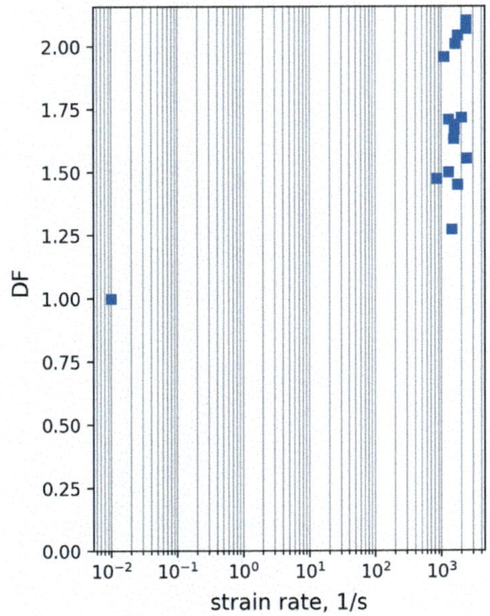

Fig. 13.9 Influence of strain rate on mechanical properties of pine with direction across the fibers under dynamic

In the above diagrams in the stress–strain axes (σ ~ ε) in the initial loading section, the growth of stress and strain follows a law close to linear, and during further deformation, when the stress limit values are reached, the pine is intensively fractured, which is accompanied by a decrease in stress and an increase in strain. The effect of changing the strain rate on the deformation diagram and strength characteristics of pine is shown in Fig. 13.11. The obtained dependencies demonstrate that with the increase in strain rate, the maximum stresses tend to increase according to the linear law, and the limiting strains corresponding to them decrease; the time to the onset of fracture also decreases according to the power law.

The dependence of DF on strain rate is shown in Fig. 13.12. The DF index in the obtained strain rate range varies from 1 to 3.5.

13.8 Orthotropy Analysis

A comparison of the fracture characteristics determined during tests in different directions is presented in Fig. 13.13. It can be noted that the time to failure of the specimens cut in the direction perpendicular to the fiber was slightly longer. The ultimate compressive stresses in the direction perpendicular to the fiber are an order of magnitude lower than the stresses that the specimen withstands when loaded along the fiber. When loaded at an angle of 45 degrees, the specimen fails at higher stresses than when loaded in the direction perpendicular to the fiber. At the same time, the DF was found to be close for the investigated directions as can be seen from Fig. 13.14.

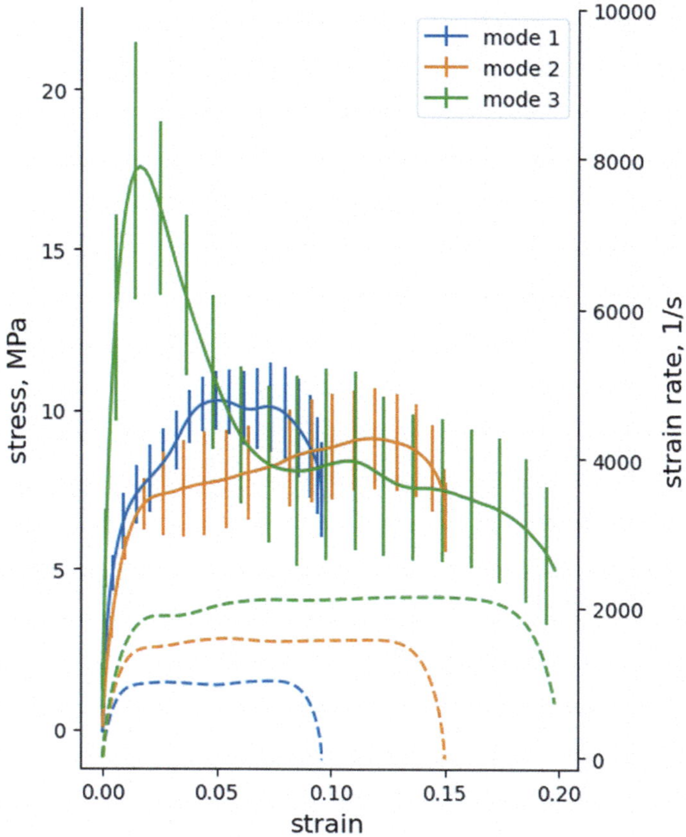

Fig. 13.10 Averaged diagrams of dynamic deformation in compression of pine at an angle of 45^0

13.9 Conclusions

This paper presents the results of dynamic tests of pine under uniaxial compression with respect to three fiber directions. The results of the performed studies showed that the strain rate significantly affects the ultimate fracture stress of pine: the coefficient of dynamic hardening at a strain rate of about 2500 s^{-1} is of the order of 3. This means that the strength of pine at this strain rate is 3 times higher than the static value. Fracture stress under loading in the fiber direction in the dynamic range of strain rates is about 100 MPa, across the fiber—about 6 MPa, at an angle of 45°—about 10 MPa. Destruction of specimens under loading along the fiber occurs at strains of about 5%, which is twice less than for the other two studied directions. The lifetime of the specimens decreases with increasing intensity of loading. Although the ultimate compressive stresses in the direction perpendicular to the fiber are an order of magnitude lower than the stresses that the specimen withstands when loaded along the fiber, the DF was found to be close to the investigated directions. The given data

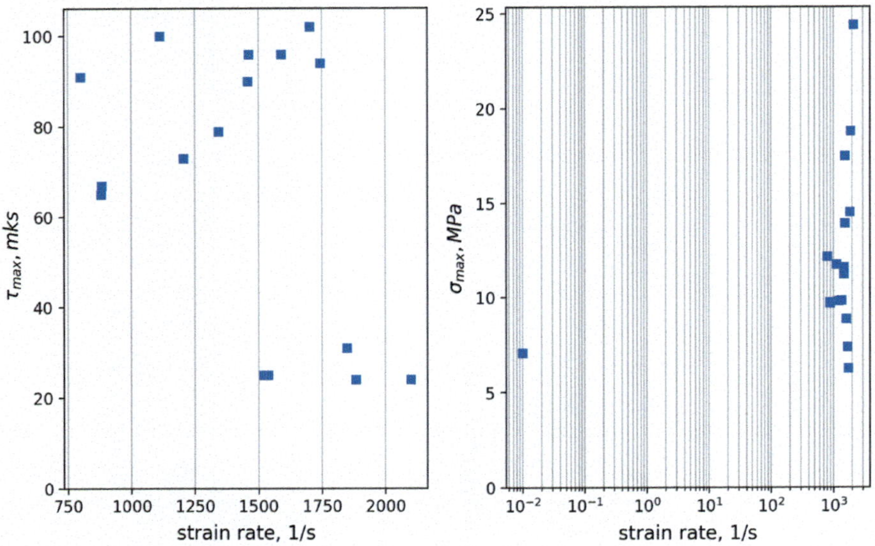

Fig. 13.11 Influence of strain rate on mechanical properties of pine with fiber direction at 45° under dynamic compression

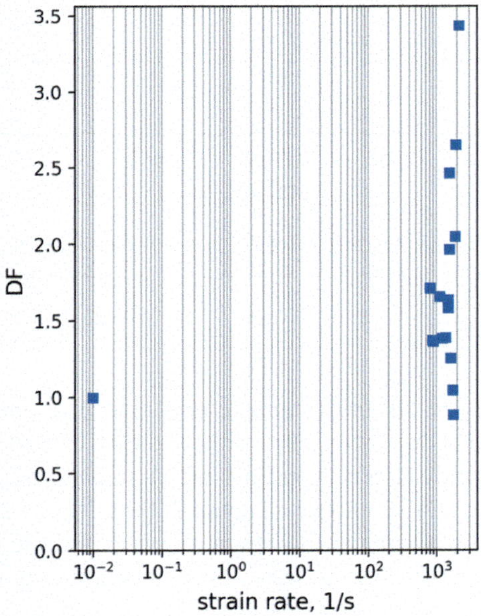

Fig. 13.12 Dependence of DF on strain rate in compression of pine at an angle of 45°

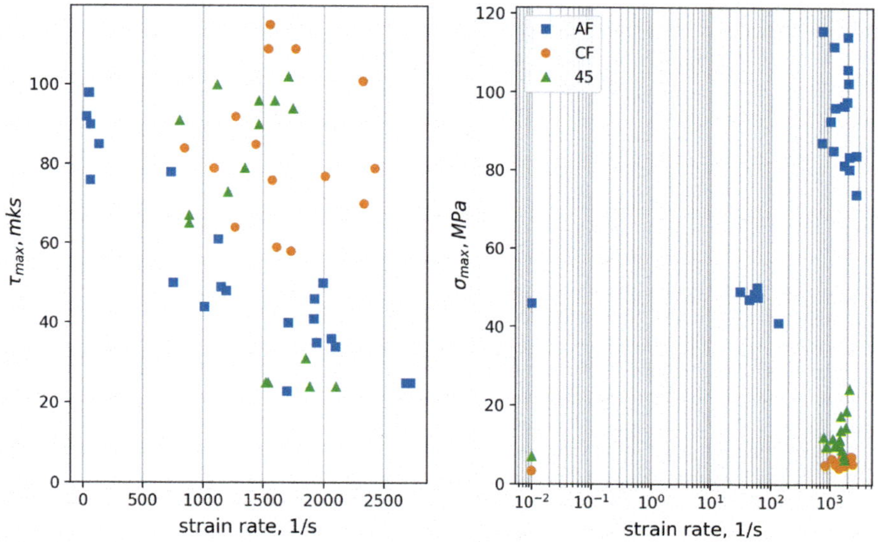

Fig. 13.13 Comparison of data

Fig. 13.14 Comparison of DF in different directions

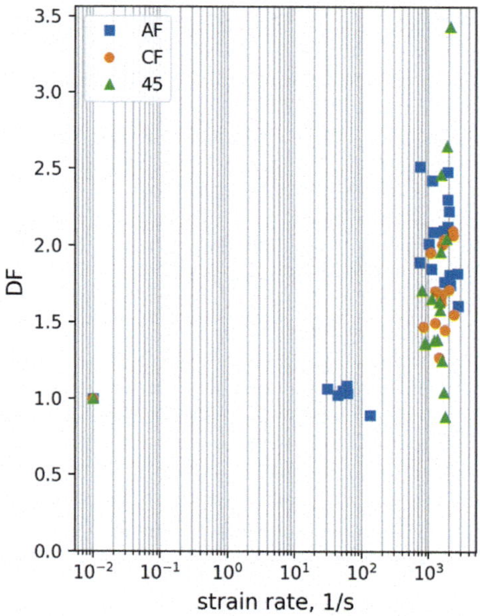

can be used in the problems of strength assessment of dynamically loaded structures with elements made of pine.

Acknowledgements This work was financially supported by the Ministry of Science and Higher Education of the Russian Federation under the basic part of the state assignment (project FSWR-2023-0036).

References

1. Adalian C, Morlier P (1998) Modeling the behaviour of wood during the crash of a cask impact limiter. In: PATRAM'98. Conference proceedings, vol 1. Paris. France
2. Neumann M (2009) Investigation of the behavior of shock-absorbing structural parts of transport casks holding radioactive substances in terms of design testing and risk analysis. PhD thesis. Bergische Universität Wuppertal, Germany
3. Ryabov AA, Romanov VI, Kukanov SS, Skurikhin SG (2006) Numerical simulations of dynamic deformation of air transport package PAT-2 in accidental impacts. In: Proceedings of 9th international LS-DYNA users conference. Dearborn, USA, pp 43–51
4. Eisenacher G et al (2013) Crushing characteristics of spruce wood used in impact limiters of type B packages. In: Packaging and transportation of radioactive materials PATRAM. San Francisco, USA
5. Ryabov AA, Romanov VI, Kukanov SS, Spiridonov VF, Tsiberev KV (2016) Numerical analysis of impact and thermal resistances of air transport package PAT-2. Problems Strength Plastcity J 78(1):101–111 (in Russian)
6. Buchar J, Severa L, Havlicek M, Rolc S (2000) Response of wood to the explosive loading. J Phys IV France 10:529–534
7. Johnson W (1986) Historical and present-day references concerning impact on wood. Int J Imp Eng 4(3):161–174
8. Reid SR, Reddy TY, Peng C (1993) Dynamic compression of cellular structures and materials. In: Jones N, Wierzbicki T (eds) Structural crashworthiness and failure. Taylor & Francis Publ., London, New York, pp 257–294
9. Reid SR, Peng C (1997) Dynamic uniaxial crushing of wood. Int J Imp Eng 19:531–570
10. Harrigan JJ, Reid SR, Tan PJ, Reddy TY (2005) High rate crushing of wood along the grain. Int J Mech Sci 47:521–544
11. Gibson LJ, Ashby MF (1997) Cellular solids-structure and properties, 2nd edn. Cambridge University Press, p 510
12. Bolshakov AP, Gerdyukov NN, Novikov SA et al (2001) Damping properties of Sequoia, Birch, Pine, and Aspen under shock loading. J Appl Mech Tech Phys 42(2):202–210
13. Buchar J, Adamik V (2001) Wood strength evaluation under impact loading. In: 39th international conference on experimental stress analysis. Tabor, Czech Republic
14. Buchar J, Krivanek I, Severa L (2001) High rate behaviour of wood. In: Nowacki WK, Klepaczko JR (eds) New experimental methods in material dynamics and impact, Trends in mechanics of materials. Warsaw, pp 357–362
15. Widehammar S (2004) Stress-strain relationships for spruce wood: influence of strain rate, moisture content and loading direction. Exp Mech 44:44–48
16. Allazadeh MR, Wosu SN (2011) High strain rate compressive tests on wood. Strain 48(2):1–7
17. Zhao S, Zhao JX, Han GZ (2016) Advances in the study of mechanical properties and constitutive law in the field of wood research. IOP Conf Ser Mater Sci Eng 137:012036
18. Gilbertson CG (2011) Dynamic properties of wood using the Split-Hopkinson Pressure Bar, Dissertation, Michigan Technological University
19. Riggio M, Sandak J, Sandak A, Babiński L (2014) Analysis and prediction of selected mechanical/dynamic properties of wood after short and long-term waterlogging. Constr Build Mater 68:444–454

20. Backman AC, Lindberg KAH (2001) Difference in wood material responses for radial and tangential direction as measured by dynamic mechanical thermal analysis. J Mater Sci 36(15):3777–3783
21. Wouts J, Haugou G, Oudjene M, Coutellier D, Morvan H (2016) Strain rate effects on the compressive response of wood and energy absorption capabilities—part a: experimental investigations. Compos Struct 149:315–328
22. Reid SR, Peng C (1997) Dynamic uniaxial crushing of wood. SR Int Z Impact Eng 19(5–6):531–570
23. Gomesa F, Xavierb J, Koerber H (2019) High strain rate compressive behaviour of wood on the transverse plane. Procedia Struct Integrity 17:900–905
24. Tagarielli VL, Deshpande VS, Fleck NA (2008) The high strain rate response of PVC foams and end-grain balsa wood. Compos B 39:83–91
25. Fortin-Smith J, Sherwood J, Drane P, Kretschmann D (2016) Characterization of maple and ash material properties as a function of wood density for bat/ball impact modeling in LS-DYNA. Procedia Eng 147:413–418
26. Muller U, Gindl W, Teischinger A (2003) Effects of cell anatomy on the plastic and elastic behaviour of different wood species loaded perpendicular to grain. IAWA J 24(2):117–128
27. Uhmeier A, Salmen L (1996) Repeated large radial compression of heated spruce. Nord Pulp Pap Res J 11(3):171–176
28. Uhmeier A, Morooka T, Norimoto M (1998) Influence of thermal softening and degradation on the radial compression behavior of wet spruce. Holzforschung 52(1):77–81
29. Caetano L, Grolleau V (2018) High strain rate out-of-plane compression of birch plywood from ambient to cryogenic temperatures. Strain 54(2):12264
30. Law KN, Kokta BV, Mao C (2006) Compression properties of wood and fibre failures. J Pulp Pap Sci 32(4):224–230
31. Salmi A, Salminen LI, Engberg BA, Bjorkqvist T, Hæggström E (2012) Repetitive impact loading causes local plastic deformation in wood. J Appl Phys 111(2):024901
32. Moilanen C, Bjorkqvist T, Engberg BA, Salminen LI, Saarenrinne P (2016) High strain rate radial compression of Norway spruce earlywood and latewood. Cellulose 23:873–889 [JuFo2]
33. Kolsky H (1949) An investigation of the mechanical properties of materials at very high rates of loading. Proc Phys Soc Lond Sect B 62:676–700
34. Kolsky H, Douch LS (1962) Experimental studies in plastic wave propagation. J Mech Phys Solids 10:195–223

Chapter 14
Towards Ferroelectric Smart Polymer Fiber-Based MEMS/MOEMS/NEMS with Laser-Assisted Control and Electron-Beam Addressing of the Fiber Actuators

Oleg V. Gradov, Margarita A. Gradova, Irina A. Maklakova, Anna V. Ratnovskaya, Eugenia L. Buryanskaya, Anatoly A. Olkhov, Alexei L. Iordanskii, and Valentin V. Kochervinskii

Abstract This paper describes MEMS/MOEMS/NEMS based on polymer fibers of polyvinylidene fluoride (PVDF) and other ferroelectric polymers that exhibit piezoelectric effect, that can be spatially directed/oriented and programmably change their shape under the electron beam, respond to the external fields and laser exposure. Original results of our latest experiments in this field are presented, proving that such fibrillar microstructures are capable of performing various mechanical functions and simultaneously acting as the light guides or waveguides, as well as performing the functions of the spatial light modulators or diffraction gratings with an adjustable slit, combining the functions of MEMS and MOEMS. Robustness and reliability of such systems to the mechanical fracture in vacuum are considered with reference to the time-resolved SEM registration of the polymer fibers. The problem of mechanical failures and fuzzy addressing due to the inherent mechanical/elastic characteristics of the fibers is discussed. The possibilities of the smart fiber MEMS/MOEMS and NEMS application in environmental conditions (from ecology and Earth sciences to aerochemical measurements or aerosol sciences) and vacuum space conditions are considered.

O. V. Gradov (✉) · M. A. Gradova · I. A. Maklakova · E. L. Buryanskaya · A. A. Olkhov · A. L. Iordanskii
N.N. Semenov Federal Research Center for Chemical Physics of the Russian Academy of Sciences (FRC CP RAS), Kosygina Str. 4, 119991 Moscow, Russia
e-mail: o.v.gradov@gmail.com

A. V. Ratnovskaya · A. A. Olkhov
Institute of Biochemical Physics of the Russian Academy of Sciences (IBCP RAS), Kosygina Str. 4, 119334 Moscow, Russia

E. L. Buryanskaya · V. V. Kochervinskii
Bauman Moscow State Technical University (BMSTU, Bauman MSTU), 2-Nd Baumanskaya, 5, 105005 Moscow, Russia

H. Altenbach and V. A. Eremeyev (eds.), *Dynamics of Discrete and Continuum Structures and Media*, Advanced Structured Materials 221,
https://doi.org/10.1007/978-3-031-75626-9_14

Keywords MEMS · MOEMS · NEMS · Fiber actuators · Electron beam. HVSEM · SEM · ESEM · ASEM

14.1 Electron Beam Methods for Creating MEMS for Space Vacuum

Electron beam methods for creating MEMS sensors and MEMS actuators are well known, including electron beam lithography and electron beam evaporation technologies [1–6], as well as MEMS-assisted electron beam control methods (including those for electron microscopes and MEMS nanolithography systems under vacuum pumping) [7–13]. On this basis, technologies of accelerated assessment and representation of materials behavior via integrated electron-optical, precision-focused electron or ion beam, and MEMS-based characterization methods [14] are being actively developed. Control of the mechanical activity of the MEMS elements using electron and ion beams, for obvious electrophysical reasons, can be interpreted as an equivalent alternative to the electrostatic MEMS activation known since the end of the last century [15–28]. However, their control by means of the electron beams is seldom implemented due to the atmospheric-chemical conditions in which the electron beam projection is rather problematic. Nevertheless, for cosmic conditions (or high vacuum conditions of the upper rarefied layers of the atmosphere, the exosphere, and the space adjacent to it), design of MEMS controlled by an electron or ion beam seems to be very useful.

The possibilities of creating MEMS for space use have been actively discussed and worked out for more than 20 years [29–36]. Of particular interest is their application in the following areas:

1. In space telescopes, both optical and VUV, X-ray, and corpuscular ones [37–47] (since MEMS have already been used in telescopes and telescopic spectrometers since the beginning of the twenty-first century [48–54]);
2. In space GC-MS [55] (since MEMS-based chromatographs [56–64] and MEMS-based mass spectrometers [55, 65–78] are already known);
3. As the components of microsatellites and nanosatellites (CUBSATs, etc.) [79–83];
4. As microactuators and the micromover components for both suborbital and orbital vehicles [84, 85], gyroscopes that balance the spacecraft movement [86–88], and energy storage systems for microsatellites [89].

Obviously, when modeling the outer space effects on MEMS, one should not ignore cosmic rays—elementary particles of different charges [37, 90–93], causing not only Single Event Upsets but also more complex failures in them. For bi-stable MEMS, the effect of the space charge distortion [94] is fatal, so the exposure to the elementary particles is especially dangerous. In addition, when exposed to different radiations, singlet oxygen is produced, which can shift the functional parameters of a number of special MEMS designs [95–97]. The X-ray component of cosmic rays, which arises during the selective emission of radiation (which can be analyzed by WDX and EDX methods), is also interesting, since MEMS X-ray optics and MEMS detectors are being developed [98–104]. In our opinion, scanning electron microscopy with elemental composition mapping is the only research method that allows position-sensitive examination/observation/monitoring of MEMS under the particle flow and taking into account the emission of characteristic X-ray radiation [105]. Photoacoustic microscopy, as a more common method of dynamic characterization of MEMS, can be used as a part of CLEM complexes for MEMS analysis but cannot be used independently for these purposes, operating in a different (optical) spectral range [106].

14.2 The Possibility of Position-Sensitive Time-Resolved Monitoring and Controlled Experiment with MEMS Under the Electron Beam in Vacuum

In this work, we propose and illustrate a novel approach to 4D (time-resolved position-sensitive) monitoring/controlled study of the vacuum MEMS dynamics under the electron beam, based on the application of stroboscopic TV-SEM, SEM registration of the surface dielectric charging of the active elements of MEMS and the corresponding EDXRS mapping (if possible). For such methods any frequency, "duty cycle" and data acquisition sampling rate compatible with the stroboscopic electron microscopy recording system is allowed: from the order of 100 MHz in the old design versions [107] and up to gigahertz in the new ones [108]; down to subnanosecond and picosecond gating times/strobe times [109, 110]; up to far ultrasonic and radio frequency sub-ranges of frequencies of the observed/induced processes in the chamber [111, 112]; in any range of the electron energies or wavelengths of characteristic X-ray radiation [113, 114]. The latter parameters are of great importance since many types of MEMS are capable to receive signals (like antennas) and operate within RF bands [115–119]. Therefore, stroboscopic microscopy at various modulation rates and probing ranges is, within the framework of this logic, a useful tool not only for the sample registration but also for the sample probing/excitation, selective in activation of the specific functional elements of MEMS. When implementing progressive methods for analyzing and correcting waveform distortion and analyzing the regularity of physical processes (developed since the 1980s [120–123]), it is possible to reconstruct in detail the dynamics of the individual structures (filaments, cantilevers, etc.), moving directly under the beam of the electron microscope which provides registration of their vibrations/oscillations [124].

At the same time, since MEMS are known to be used as the light modulators and programmable "diffractors" (dynamic diffraction gratings, modulators, mono-/polychromators, switches) [125–145], it is possible to monitor their dynamics by changing the beam diffraction patterns on them—two-dimensional Fourier spectra

(2D FFT) and "integral frequency characteristics" and "integral spatial characteristics" calculated from them, as well as by the wavelet digital algorithms [146–149] close to the Fourier-based methods. SEM analysis performed in vacuum only simplifies the problem, since many MEMS are produced at the vacuum lines and are designed to operate in vacuum, and the air evacuation improves controllability of their diffraction characteristics and modulation stability, due to the absence of a damping medium (for high vacuum, medium vacuum and gas-filled evacuated MEMS, see, for example, [150–158]; for evacuation of the medium using MEMS, as well as for measuring the vacuum parameters using MEMS, see [159–164]). Previously such measurements were not carried out using MEMS, although we used diffraction and diffraction emulation methods (2D FFT using FFTW library) to analyze structures and dynamics in SEM and in lensless holographic microscopy [165–169]. The latter, according to the well-known system engineering assumptions, was implemented on CMOS, while CMOS and MEMS are compatible within the sensors [157, 170–172]. Therefore, lensless CMOS arrays can be used for registering lensless dynamic patterns (shadowgrams) of MEMS. At the moment, it seems realistic to integrate lensless CMOS for MEMS and stroboscopic electron microscopy of MEMS in a subchamber with a built-in CMOS sensor (however, this is rather difficult because it does not live long if irradiated with an electron beam) or another similar detector [173], but this is beyond the scope of this article.

14.3 Applicability of Vacuum MEMS in Diffractive Optics. Do the Diffraction Patterns of the Fiber Structures Change Over Time Under an Electron Beam?

To create the above optical diffraction devices, including ptychographic ones, it is necessary to find out if it is possible to control the diffraction patterns of the fiber structures by an electron beam. This can be revealed by modeling laser diffraction using a 2D Fourier transform. We have implemented this using QAVIS software based on the FFTW library [174–178].

As can be seen from Fig. 14.1, after switching on the high voltage, the fibers begin to move rapidly under the beam and "relax to a quasi-stationary state" in half a minute. This is preceded by a complete reconfiguration of the fiber direction depending on their position under the beam, as can be seen from the comparison of 2D Fourier spectra and Integral Spatial Characteristics (ISC) for different time points. After that, reversible oscillations of the fibers also occur, but the network topology remains stable, which is consistent with the stability of the ISC symmetry axes, whose angular ratios vary by only a few degrees (and the full data is also available in the preprint of the article [179]).

In some cases, not the smooth movements are observed, leading to a change in the network configuration under the electron beam, but spontaneous instantaneous changes, indicated by a sharp increase or disappearance of the ISC (Integral Spatial

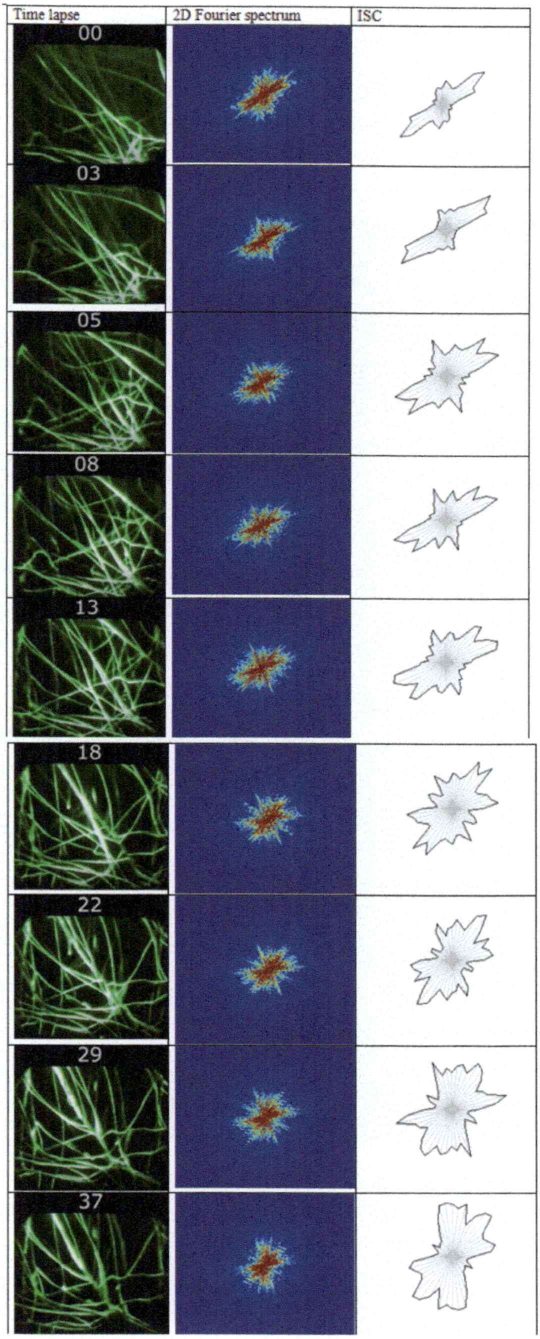

Fig. 14.1 Real-time 2D FFT analysis of a fuzzy fiber MEMS actuator network under the electron beam. After switching on the high voltage regime, ferroelectric fibers begin to move rapidly under the beam due to the local piezoelectric effect

Characteristic) components, as shown in Fig. 14.2. Thus, the processes of movement and relaxation of the network, which are determined by the charge and mechanics of the fibers, are possible, which can be used to control diffraction on such fibers up to their use as spatial light modulators.

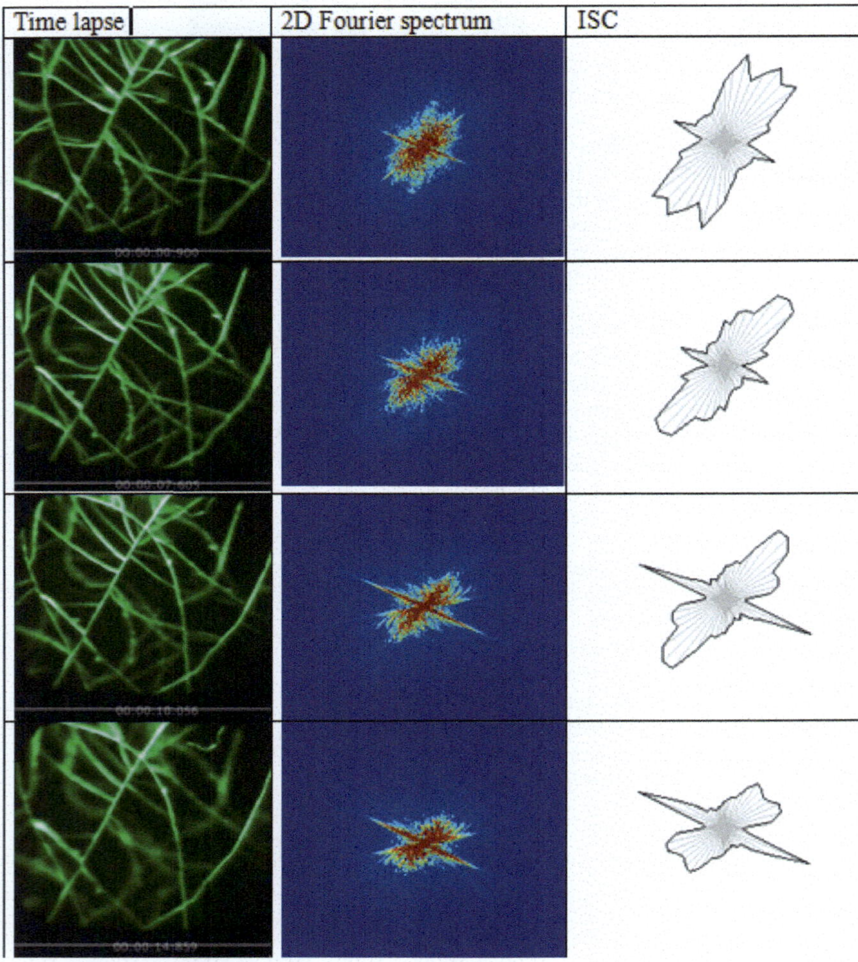

Fig. 14.2 In some cases, not the smooth movements are observed, leading to a change in the network configuration under the electron beam, but spontaneous instantaneous changes, indicated by a sharp increase or disappearance of the ISC components

14.4 Is Programmable Rearrangement of the Structures Based on the Networks of Fuzzy Fiber MEMS/NEMS Possible Under the Electron Beam?

In some cases, rearrangement of the entire networks/assemblies is observed, which can be applied in the electron beam-induced/-controlled microrobotics. Figure 14.3 shows stroboscopic SEM images and correlograms obtained with a time resolution of 0.4 s for an area of such a rearranging network structure. It can be seen that exposure to the electron beam induces mechanical changes in the micromechanical structure as a whole. In this case, different fibers function as "jacks", "drive belts", "springs", "elastic beams", "pushers", etc. The effect of rearrangement of the structures depends on the size of the irradiated area, that is, on the area of the effective cross-section of the focused or scanning electron beam.

It can be confirmed that some (thicker) fibers perform the functions of the load-bearing elements of mechanical strength, while the other (thin) fibers predominantly act as flexible elements controlled by the electron beam. For this, let us turn to the experiment with two such structures. From Fig. 14.4, it can be seen that the movement of a thin bending fiber (nanofilament in the upper part of the micrograph) occurs with the stability of a thicker fiber (in the center). Its features can be tracked according to the data of the wavelet transform (Morlet/Gabor wavelet) in the Re and Im regions (for example, one can observe the drift of the extremum on the "blue" representation).

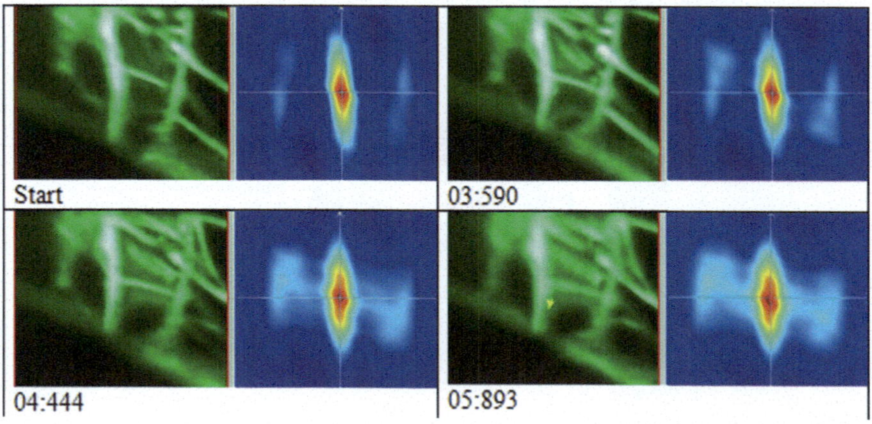

Fig. 14.3 Stroboscopic SEM images and correlograms obtained with a time resolution of 0.4 s for an area of the rearranging network structure with contacted fiber microelements

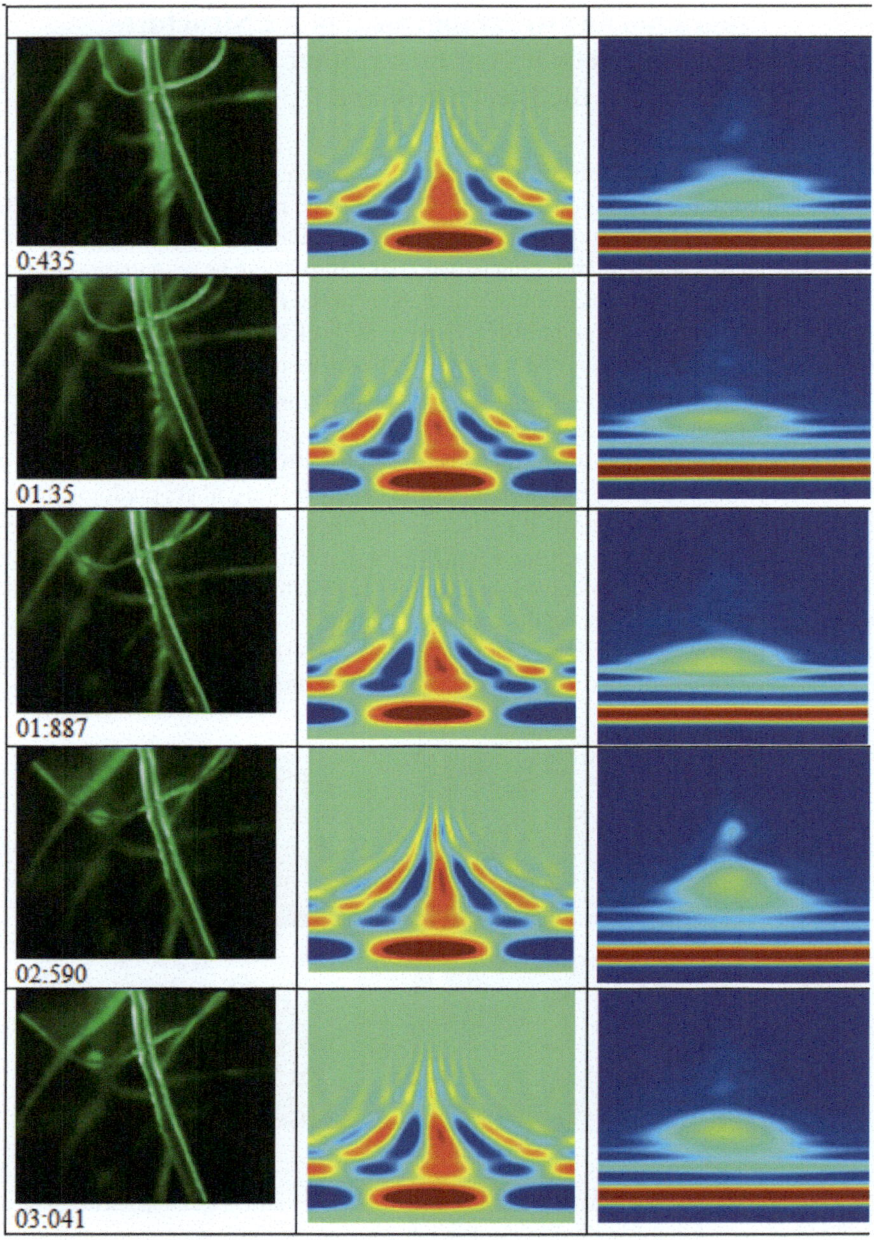

Fig. 14.4 The movement of thin bending nanofilaments occurs with the stability of a thick fiber

14.5 Is It Possible to Perform the Spring Functions by the Fiber MEMS Elements?

In a similar way, it is possible to monitor the vibrations/oscillations of the spring-like elements that roll up, unfold, or change their helicity under the beam, as shown in Fig. 14.5. It can be seen that when a spring-like structure appears in the field of view, changes its geometry (twists) and moves in space, a clearly visible changing in the wavelet pattern (Morlet/Gabor wavelet) is observed, as it can be seen from the part of the frame where the Re button is not pressed.

If some fibers oscillate, move, contract like springs (thin fibers), while others perform a frame function (thick fibers) like an awn in mammalian fur, then the micromechanical properties of the fibers should be evaluated in dynamics. A fiber break is a possible cause of the breakdown/failure of the controlled microelectromechanical structures based on it. Therefore, for each chemical or structural class of fibers, mechanical (bending/fracture) tests should be performed via SEM under the electron beam. In some cases, the fiber mechanics may not satisfy the intended function. Then the use of such fibers of a given thickness and composition or structure and

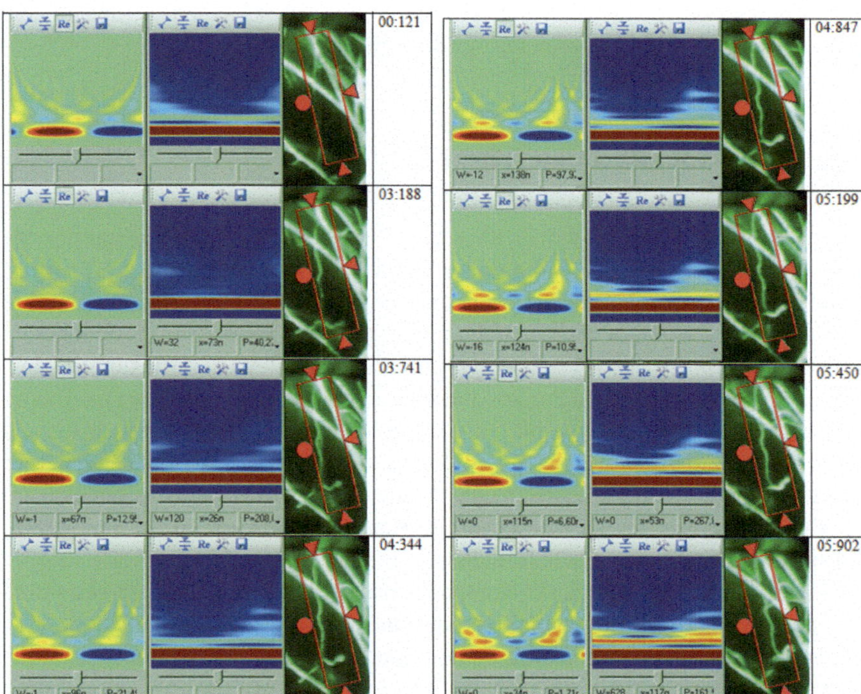

Fig. 14.5 Electron beam-induced vibrations/oscillations of the spring-like elements that roll up, unfold, or change their helicity under the beam

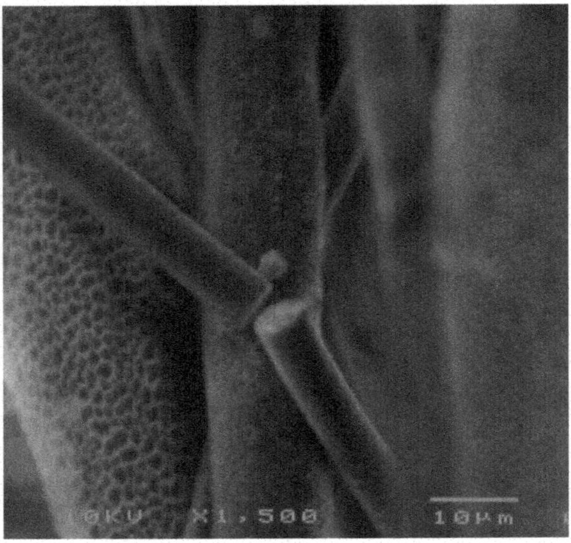

Fig. 14.6 An example of the fiber break in the SEM image

production method for the purposes of robotics and MEMS should be abandoned. An example of the fiber break on the SEM image is given in Fig. 14.6.

14.6 Why Polymer Fiber MEMS?

The general advantage of polymer fiber MEMS in terms of actuation and flexible tunable positioning and pathfinding/fiber guidance is their mechanical properties. Unlike rigid MEMS, where adaptability and tunability, as well as fuzzy logics, are achieved due to the device design, in the case of polymer MEMS, this is an inherent property of the polymer material. Polymers allow to achieve high flexibility, plasticity, elasticity, viscoelastic state, etc., and in terms of the load cycling are much better than metals. Therefore, the possible applications of fiber or thread-based sensors and actuators for MEMS applications (except for vacuum ones) will be considered below.

Usually "fiber MEMS sensors" means "optical fiber MEMS sensors" (which are one of the types of "fiber MEMS sensors"). There are also known infrasonic [180], ultrasonic [181, 182] or, as a whole, acoustic [183–185] optical fiber MEMS sensors or MEMS microphones [186, 187]. Optical fiber MEMS pressure sensors [188–194] have already been commercially produced since the 2000s. They are often very complex in design and require specific conditions that differ significantly from the conventional fiber manufacturing process (as, for example, in MEMS optical fiber pressure sensor fabrication by Au–Au thermal-compression bonding [195, 196], where single crystalline magnesium oxide MEMS-based fiber-optic Fabry–Perot (FP) pressure sensor for 800 °C high-temperature environmental applications was constructed/fabricated by combining wet chemical etching with direct bonding).

Less abundant are optical fiber MEMS displacement sensors [197–199]. However, the above list does not exhaust the full range of even optical MEMS-based sensors, which goes far beyond the boundaries of MEMS optical fiber sensors (for example, there are optical sensors or laser Doppler vibrometers based on MEMS actuators, and not on the optical fiber MEMS sensor principles [200, 201]). If we go beyond sensors towards devices affecting the substance/object/environment (MEMS-based heaters and micro-heaters [202], electrothermal MEMS actuators and scanners [203–206] or other actuators based on fiber MEMS structures [207], then it will turn out that:

- First, the functions of sensor and actuator or sonar in the case of fiber MEMS can be integrated;
- Secondly, fiber MEMS structures are not equivalent to the optical fiber MEMS structures;
- Thirdly, in addition to the optical sensors based on optical fiber MEMS, there are many other types of fiber MEMS sensors and/or fiber-containing MEMS sensors based on different physical principles, and it is possible to integrate different measurement methods within a single fiber component of the device, as well as those different in the signal/energy conversion principles (direct and inverse piezoelectric effect in the field of an external source of a control signal or a detected signal "felt" by the sensor) [208, 209];
- Fourthly, the fundamental task of the theory of control or engineering of controlled sensor-actuator MEMS structures is matching of the incoming signals and the synthesis of communication systems or "connectomes" similar to the fiber MEMS systems in order to ensure their intelligent synchronization and matching in terms of the operating modes in devices [210–212].

14.7 The Relationship Between the Popularity of the Optical Fiber MEMS and the Development of Optical Fiber Technology and Optical Waveguides

Perhaps the popularity of the optical fiber MEMS is associated with the development of the optical fiber technology and the comparability of the fiber size scale in micro-electromechanics with the wavelengths of the optical range. However, NEMS do not contradict the use of the optical spectral range when working with the subwavelength super-resolution fiber optics. For example, in the NSOM development since the late 1990s optical fibers have been used, although NSOM (Near-field Scanning Optical Microscope) or SNOM (Scanning Near-field Optical Microscope) is an instrument for micro-/nanoimaging and manipulation of light at the nanoscale [213–217]. Note that in our experiments with fuzzy MEMS fibers (based on different polymers), it was shown that they can surfacely conduct a CLEM scanning laser pulse and measure variations in the intensity and scattering of the laser beam, depending on the angle and coherence level, as shown in Fig. 14.7a-c.

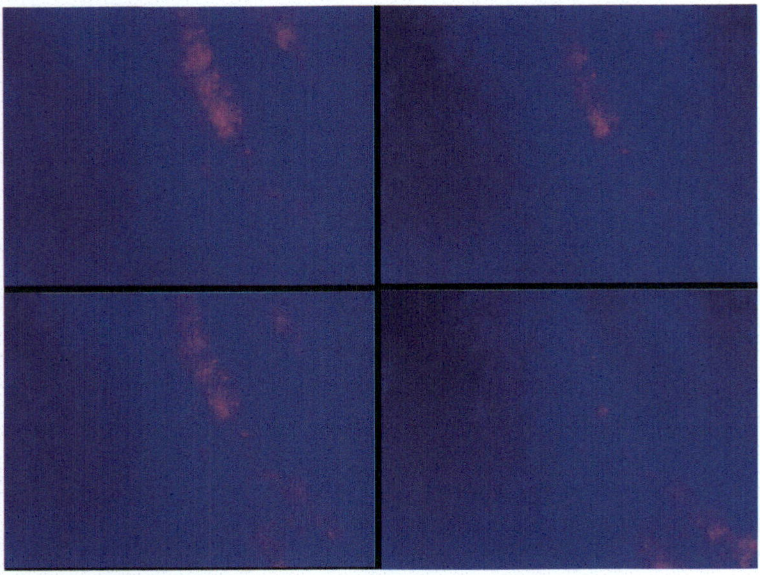

a

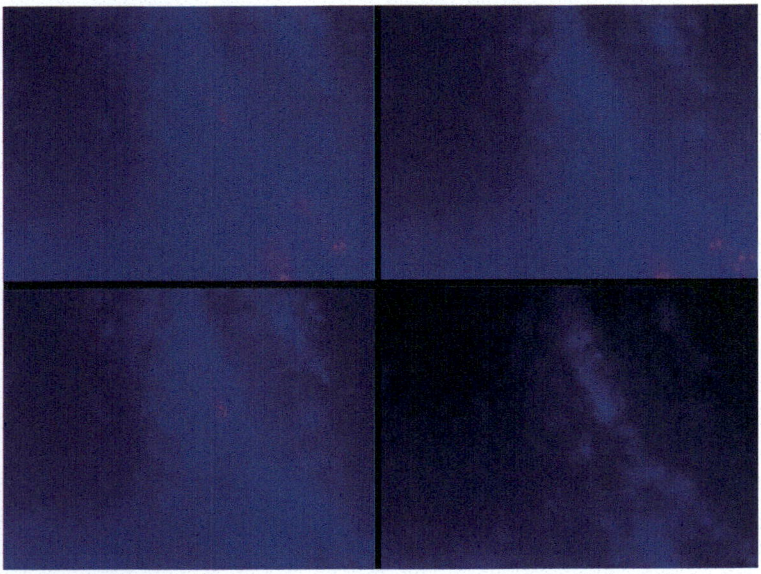

b

Fig. 14.7 **a** Well-defined surface waveguiding/laser optical light guiding ($\lambda = 405$ nm) on the fiber. **b** Low-intensity surface waveguiding/laser optical light guiding ($\lambda = 405$ nm) on the fiber at different laser scanning angles (non-coherent light scattering effect is mostly observed). **c** Comparative analysis of the light intensities in more coherent (λ = 405 nm) and more scattered light on the prototype elastic ferroelectric fiber element of fuzzy MOEMS

c

Fig. 14.7 (continued)

However, at the nanolevel (nanoplasmonics, nanospintronics, etc.), the operation boundaries of nano-/micro-EMS are significantly expanded both in the actuator and sensor modes, including fiber NEMS structures sensitive to the electric and magnetic fields [218–222]. Unfortunately, these works are at the initial stage and have not yet been implemented on the polymer fibers and most of the promising nanocomposites, and their production methods are simpler/not as high-tech as for the optical fiber MEMS [223]. There are also methods based on nanoelectrospinning and spunbond extrusion, and other protocols for the fabrication of nanofibers for MEMS/NEMS [224–227] (transitions between NEMS and MEMS may occur and are not critical, except for the applications in the physics of the size effects [228]). Nanofiber networks are used in hybrid microassemblies and microcircuits that combine nano- and micro-dimensions of not only sensor but also actuator and hybrid (as a physical analog of theranostics, where the impact on the biomaterial and the detection of its results and/or the diagnosis of the medium state are performed simultaneously) type.

The latter analogy may also be appropriate due to the development of trends in biomimetic NEMS [229], inspired by the biomechanics of the cytoskeleton and electrically activated fibrils and neurofilaments) [230]. In this regard, the creation of network and mutually regulated (network-controlled) bionic nanofiber structures and the development of special types of contacts for them become the general problem (by the way, developments in the field of creating contacts for NEMS and nanofiber MEMS were industrially implemented at the end of 2000s—early 2010s [231, 232]). Ideally, one could rely on the principles of organization of the neuromuscular synapse, however, it requires chemical activation (transfer of the neurotransmitter—acetylcholine), and therefore, the possibility of developing and implementing contacts based on the principles of ephapses or electrical synapses seems more suitable [233, 234]. All such systems can be integrated with CMOS structures for various tasks of detection and monitoring, inspection/control of the actuation process [235].

14.8 On the Possibility of Implementing MEMS/NEMS for Analog Mechanical Computing and NEMS Sensors Based on the Fuzzy Logics Principles

As can be seen from the above-cited works (on polyaniline, polyurethane, PVA/PPy/MIP fibers), polymer nanofibers and nanonetworks can also be used for the MEMS/NEMS design purposes. For this reason, it is possible to speak about flexible, physically plastic, and elastic nanoelectromechanical structures, including the structures for analog mechanical calculations and NEMS sensors based on the principles of fuzzy logics (and about the corresponding fuzzy faults/failures of the impulse addressing, but not in the classical sense of the "fuzzy actuator failure" in the control theory or fuzzy logics [236–240]). MEMS containing elastic elements are used for various purposes—from gyroscopes [241, 242] to biomedical electrodes [243, 244] and microfluidics (including microhydraulic actuators) or (bio)analytics [245, 246]. They use various structural elements such as elastic beams, micro-springs and bridge-type elastic structures [247–250]. Composite or multilayered MEMS structures [251] are quite widespread, which allow integrating the properties of several materials into the device design with a large number of the control signals or the sensor pathways, including those designed to perform various tasks. Because of this, firstly, materials differing in mechanical, conductive, converting, thermoreactive and other properties are modeled and empirically selected [252–258]; secondly, various forms of manifestation of these properties in elastic structures are investigated, such as:

- Static friction in elastic adhesion contacts in MEMS [259];
- Visco-elastic properties [260] and elastic-plastic behavior of MEMS [261, 262];
- Thermo-elastic excitation and damping processes; and temperature-dependent elastic constants of MEMS [263–265];
- Elastic-electrostatic characteristics and integrated electro-fluid-elastic models for MEMS [266–269];

The models can be very sophisticated, for example, such as ARX (nonlinear models consist of model regressors and an output function or in other words combination of autoregressive models and exogenous input models) and ARIX (ARX models with integrated noise) [270, 271]. At the same time, such measurements are more accessible due to the elementary nature of the underlying physical principles. The validity of the empirical results is not in doubt, while the models more often remain debatable. Thus, the measurement of nanodisplacement and elastic properties of MEMS via the microscopic hole method [272] is very accurate both in terms of the property analysis and in terms of locality; and there is no need to write a code or draw a graphic image of the model (as it is required in COMSOL Multiphysics, etc.). However, the MEMS itself can be a sensor/probe in various methods of the elastic measurements [273], so it is possible to organize measurements in such a way that one MEMS component will be an actuator, and the second one will be a sensor-encoder of its movements and elastic properties. As a rule, the elastic modulus is measured, but for polymeric MEMS with elastic properties, it is determined differently than for solid-state ones, in which the contribution of the surface carrier concentration effect to the elastic properties is significant [274–278]. Strictly speaking, because of this, highly elastic polymer/composite structures are optimal for the synthesis of fuzzy MEMS. It is noteworthy that the elastic behavior of NEMS has also been actively studied since the 2000s. [279–281], but the possibilities of using polymer and composite systems in NEMS are very limited, since most of the polymers and composites do not satisfy (at least sensu lato) the requirements of dimensional limitations for nanoelectromechanical systems. Therefore, the only structures that somehow correspond to this definition are nanofibers obtained by nanoelectrospinning and similar techniques.

In fact, the term "fuzzy" in relation to MEMS has so far been used in English-language works only in a limited area related to software, regulation theory, and control theory (for example, in the regulation systems in Gyroscopy and Navigation [282–289], including the neuro-fuzzy models and similar casews [290–294]). Our approach, which determines the possibility of fuzzy and neuro-fuzzy behavior of MEMS structures due to the mechanical movement of the fiber and thin beam elements, which provides the possibility of consistent and customizable/adaptive behavior of the elements in a single external field or under the influence of an external stimulus (an electron beam) undoubtedly possesses a fundamental novelty.

14.9 About the Previous Developments in the Field of Polymeric MEMS

It is noteworthy that although the development of polymeric MEMS and polymeric optical MEMS began in the 1990s. References [295, 296] and the main processing methods and technological processes were already established before 2010 [297–304], the most interesting physical problems in this area have not yet been solved, and many of them have not yet been posed. For example, in most cases, film [305, 306] or spring/beam schemes are used, and the full potential of micromechanical components of such mechanisms has not yet been exhausted, despite the fact that polymer springs of micron or submicron size are known in MEMS engineering (but they are little effective in the absence of the effective driving mechanisms or systems of various types that could be driven by them) [307]. To date there are many methods of control and directing of controllable polymer MEMS in different spectral ranges (thermal [308]), radio frequency and field [309, 310], optical [311, 312], etc.), but the particle beam control is not widely used. In practice, there are hybrid metal-polymer [313, 314] and composite [315] MEMS, NEMS, and MOEMS structures, but the advantages of contacts provided by the use of two materials for applied tasks (such as the creation of heterostructures, thermocouples, integral actuators, and indicators-regulators, etc.) are seldom used, as well as the Su-Field type of devices (Su-Field techniques, where the field effects the substance, due to which a physical or a technical contradiction is resolved that prevents the use of this substance for technological purposes in its original state).

On the one hand, there are liquid (including water) actuators and sensors based on MEMS [316, 317]. On the other hand, there are no MEMS that change their characteristics during phase transitions, which, in the case of the correct organization of MEMS as a "hybrid microassembly", can be controlled and regulated due to the internal feedback and regulation circuits or networks. On the one hand, mechanical (and hydrodynamic, in the case of embedding MEMS structures into labs-on-a-chip and BioMEMS [318, 319]) properties of various types of transparent elastic polymers are used, but, on the other hand, they are practically not integrated with the optical properties, which predetermines separation of the ways of technical development of polymer MEMS and MOEMS [320–324], although in this case there should be a convergence (by the way, in our works on microfibers it is achieved in some cases—see Fig. 14.7). In addition, for such systems, where, for the mechanical reasons, a fuzzy response may occur, the very fact of its detection is a kind of diagnosis of their mechanical/robustness properties. Therefore, in essence (as can be seen from Fig. 14.6 with a fiber break), by loading the network of the polymer fibers in MEMS, we test their fuzzy properties as logical or other "network connectome" switches. This direction is formed from the well-known works on the mechanical testing of polymer nanofibers using MEMS [325] and nanotribology of their fibers [326]. Thus, the proposed fuzzy fiber MEMS or elastic polymeric fiber MEMS system based on the ferroelectric fibers is a qualitatively new direction, but it strictly follows the development patterns of the MEMS and polymer sensor/actuator industry in general.

14.10 Conclusion

From the above text, it can be concluded that:

1. Control over the mechanical activity of the MEMS elements using electron beams can be interpreted as an alternative to the electrostatic MEMS actuation known since the end of the last century.
2. It is possible to create flexible polymeric ferroelectric MEMS and NEMS actuated by an electron beam and suitable for operation both in gaseous media and in the conditions of space vacuum and cryovacuum. In the case of operation in the cryoSEM mode and atmospheric ESEM/ASEM [327–329], it is possible to vary the physical parameters of the medium, due to which ferroelectric fibers that are not suitable for the space actuators can be used for the environmental monitoring in the Earth's atmosphere and even in the stratosphere at different heights.
3. It is possible to use stroboscopic SEM modes to analyze the mobility of fiber piezoelectric and ferroelectric elements at different frequencies (up to the megahertz ranges). CLEM modes allow observing the high-frequency activity of MEMS and MOEMS based on the polymer piezoelectrics as spatial (and spatio-temporal) light flux modulators.

4. It is possible to use ferroelectric polymer fuzzy MEMS as diffraction gratings, (semi)stochastic structured illumination systems, tunable coded apertures, and ptychographic coding elements. In the case of sufficiently high transparency and compliance of their optical properties with the requirements of fiber optics, aperture coding of such systems can be combined with the transmission of an optical signal through the individual fiber elements.
5. Exposure to the electron beam induces mechanical changes in the micromechanical structure as a whole. In this case, different fibers function as "jacks", "drive belts", "springs", "elastic beams", "pushers", etc.
6. The effect of rearrangement of the fiber network structures depends on the size of the irradiated area, that is, on the area of the effective cross-section of the focused or scanning electron beam.
7. Tunable but not completely deterministic fuzzy MEMS or fuzzy NEMS in various modes of mobility controlled by an external field or a flow of the charged particles can be considered as the models of pathfinding of the neuromorphic microelectromechanical elements based on the principles of galvanotaxis and its optically controlled microphotoelectromechanical analog. (In the latter case, doping of polymer piezoelectric optical fibers for these purposes may be appropriate). The fuzzy control principle is more consistent with both the mechanisms and formal models of axonal pathfinding (trajectory selection in a multipole system) in an inhomogeneous electric field, and the requirements of the adaptive control and electrostatic self-guidance of microelectromechanical actuators for the purpose of their constructive and physical optimization for operation in serendipity conditions, not (fully) controlled by a person or a control device (for example, in space conditions or in the signal detection in the environment with non-deterministic parameters). Thus, MEMS and NEMS of this kind are both prototypes of neuromorphic sensors (of the medium parameters) in advanced systems of the adaptive biomimetic control for space and terrestrial tasks, and for biomimetic actuators that reproduce fuzzy orientation in the medium, in accordance with the principles of orientation of the fiber biological structures (like the nerve fibers with axons, pollen tubes, root hairs, moss protonemas, fungal hyphae, etc.). However, the latter issues (neuromorphic and biomimetic fuzzy control of MEMS, MOEMS, and NEMS fibers) are not considered in this paper and are the subject of our forthcoming papers.

References

1. Singh K, Joyce R, Varghese S, Akhtar J (2015) Fabrication of electron beam physical vapor deposited polysilicon piezoresistive MEMS pressure sensor. Sens Actuators, A 223:151–158. https://doi.org/10.1016/j.sna.2014.12.033
2. Chase JG, Smith BW (2001) Overview of modern lithography techniques and a MEMS-based approach to high throughput rate electron beam lithography. J Intell Mater Syst Struct 12(12):807–817. https://doi.org/10.1177/104538901400438109

3. Bhat SP, Subhani KN, Purakkat S (2021). Plasma-assisted electron beam evaporation of low stress aluminium films for MEMS applications. ISSS J Micro Smart Syst 10:33–39. https://doi.org/10.1007/s41683-021-00067-4
4. Schuhmann T, Protz J, Fields D, Yin H, Cross A, He W, Bowes D, Ronald K, Phelps A (2010) A MEMS fabrication approach for a 200GHz microklystron driven by a small-scaled pseudospark electron beam. Proc SPIE 7837:783705. https://doi.org/10.1117/12.864902
5. Zhou H, Chong BK, Stopford P, Mills G, Midha A, Donaldson L, Weaver JMR (2000) Nano-MEMS, Bio-Nano, molecular electronic and atomic beams-lithographically defined nano and micro sensors using float coating of resist and electron beam lithography. J Vac Sci Technol-Sect B 18(6):3594–3599. https://doi.org/10.1116/1.1321271
6. Robinson APG, Palmer RE, Tada T, Kanayama T, Shelley EJ, Preece JA (1998) Fullerene derivatives as novel resist materials for fabrication of MEMS devices by electron beam lithography. MRS Online Proc Libr Arch 546:219–224. https://doi.org/10.1557/PROC-546-219
7. Ichimura T, Ren Y, Kruit P (2014) A large current scanning electron microscope with MEMS-based multi-beam optics. Microelectron Eng 113:109–113. https://doi.org/10.1016/j.mee.2013.07.008
8. Lobato-Dauzier N, Denoual M, Sato T, Tachikawa S, Jalabert L, Fujita H (2019) Current driven magnetic actuation of a MEMS silicon beam in a transmission electron microscope. Ultramicroscopy 197:100–104. https://doi.org/10.1016/j.ultramic.2018.12.002
9. Kojima A, Ikegami N, Yoshida T, Miyaguchi H, Muroyama M, Nishino H, Yoshida S, Sugata M, Ohyi H, Koshida N, Esashi M (2014) Massively parallel electron beam direct writing (MPEBDW) system based on micro-electro-mechanical system (MEMS)/nanocrystalineSi emitter array. Proc SPIE 9049:904918. https://doi.org/10.1117/12.2046584
10. Kim H, Han C, Lee K, Seo H, Chun K (2002) The multi arrayed micro-scale electron optical system for electron beam lithography using MEMS technology. IEICE technical report 102(180):27–31
11. Grella L, Carroll A, Murray K, McCord MA, Tong WM, Brodie AD, Gubiotti T, Sun F, Kidwingira F, Kojima S, Petric P (2013) Digital pattern generator: an electron-optical MEMS for massively parallel reflective electron beam lithography. J Micro/Nanolithography, MEMS, MOEMS 12(3):031107. https://doi.org/10.1117/1.JMM.12.3.031107
12. Chase JG, Smith BW (2001) MEMS-based precision motion control approach to high-throughput-rate electron beam lithography. Proc SPIE 4592:450–461. https://doi.org/10.1117/12.449002
13. Kojima A, Ikegami N, Yoshida T, Miyaguchi H, Muroyama M, Yoshida S, Totsu K, Koshida N, Esashi M (2016) Development of a MEMS electrostatic condenser lens array for nc-Si surface electron emitters of the massive parallel electron beam direct-write system. Proc SPIE 9777:977712. https://doi.org/10.1117/12.2219338
14. Polasik SJ (2005) Accelerated assessment and representation of materials behavior via integrated electron-optical, focused ion beam and MEMS-based characterization methods. Doctoral dissertation, Ohio State University
15. Osterberg PM, Senturia SD (1997) M-TEST: a test chip for MEMS material property measurement using electrostatically actuated test structures. J Microelectromech Syst 6(2):107–118. https://doi.org/10.1109/84.585788
16. Bochobza-Degani O, Elata D, Nemirovsky Y (2002) An efficient DIPIE algorithm for CAD of electrostatically actuated MEMS devices. J Microelectromech Syst 11(5):612–620. https://doi.org/10.1109/JMEMS.2002.803280
17. Allen M, Raulli M, Maute K, Frangopol DM (2004) Reliability-based analysis and design optimization of electrostatically actuated MEMS. Comput Struct 82(13–14):1007–1020. https://doi.org/10.1016/j.compstruc.2004.03.009
18. Zhu G, Lévine J, Praly L, Peter YA (2006) Flatness-based control of electrostatically actuated MEMS with application to adaptive optics: a simulation study. J Microelectromech Syst 15(5):1165–1174. https://doi.org/10.1109/JMEMS.2006.880198

19. Zhang WM, Meng G (2007) Nonlinear dynamic analysis of electrostatically actuated resonant MEMS sensors under parametric excitation. IEEE Sens J 7(3):370–380. https://doi.org/10.1109/JSEN.2006.890158
20. Zhang WM, Meng G, Chen DI (2007) Stability, nonlinearity and reliability of electrostatically actuated MEMS devices. Sensors 7(5):760–796. https://doi.org/10.3390/s7050760
21. Guo ZJ, McGruer NE, Adams GG (2007) Modeling, simulation and measurement of the dynamic performance of an ohmic contact, electrostatically actuated RF MEMS switch. J Micromech Microeng 17(9):1899. https://doi.org/10.1088/0960-1317/17/9/019
22. De Pasquale G, Somà A (2010) Dynamic identification of electrostatically actuated MEMS in the frequency domain. Mech Syst Signal Process 24(6):1621–1633. https://doi.org/10.1016/j.ymssp.2010.01.010
23. Alsaleem FM, Younis MI, Ruzziconi L (2010) An experimental and theoretical investigation of dynamic pull-in in MEMS resonators actuated electrostatically. J Microelectromech Syst 19(4):794–806. https://doi.org/10.1109/JMEMS.2010.2047846
24. Wang B, Zhou S, Zhao J, Chen X (2011) Size-dependent pull-in instability of electrostatically actuated microbeam-based MEMS. J Micromech Microeng 21(2):027001. https://doi.org/10.1088/0960-1317/21/2/027001
25. Zhang WM, Tabata O, Tsuchiya T, Meng G (2011) Noise-induced chaos in the electrostatically actuated MEMS resonators. Phys Lett A 375(32):2903–2910. https://doi.org/10.1016/j.physleta.2011.06.020
26. Ouakad HM (2013) An electrostatically actuated MEMS arch band-pass filter. Shock Vib 20(4):809–819. https://doi.org/10.3233/SAV-130786
27. Caruntu DI, Martinez I (2014) Reduced order model of parametric resonance of electrostatically actuated MEMS cantilever resonators. Int J Non-Linear Mech 66:28–32. https://doi.org/10.1016/j.ijnonlinmec.2014.02.007
28. Ramini AH, Hennawi QM, Younis MI (2016) Theoretical and experimental investigation of the nonlinear behavior of an electrostatically actuated in-plane MEMS arch. J Microelectromech Syst 25(3):570–578. https://doi.org/10.1109/JMEMS.2016.2554659
29. Cass S (2001) MEMS in space. IEEE Spectr 38(7):56–61. https://doi.org/10.1109/6.931884
30. George T (2002) MEMS/NEMS development for space applications at NASA/JPL. Proc SPIE 4755:556–567. https://doi.org/10.1117/12.462856
31. Lafontan X, Pressecq F, Beaudoin F, Rigo S, Dardalhon M, Roux JL, Schmitt P, Kuchenbecker J, Baradat B, Lellouchi D, Le-Touze C (2003) The advent of MEMS in space. Microelectron Reliab 43(7):1061–1083. https://doi.org/10.1016/S0026-2714(03)00120-3
32. Farrar D, Schneider W, Osiander R, Champion JL, Darrin AG, Douglas D, Swanson TD (2003) Controlling micro electromechanical systems (MEMS) in space. AIP Conf Proc 654(1):180–186. https://doi.org/10.1063/1.1541293
33. George T (2003) Overview of MEMS/NEMS technology development for space applications at NASA/JPL. Proc SPIE 5116:136–148. https://doi.org/10.1117/12.497796
34. Rigaud S, Quadri G, Gilard O, Nicot JM (2006) Optical MEMS reliability in space environment. Proc SPIE 6186:61860P. https://doi.org/10.1117/12.661458
35. Shea HR (2006) Reliability of MEMS for space applications. Proc SPIE 6111:61110A. https://doi.org/10.1117/12.651008
36. Grönland TA, Rangsten P, Nese M, Lang M (2007) Miniaturization of components and systems for space using MEMS-technology. Acta Astronaut 61(1–6):228–233. https://doi.org/10.1016/j.actaastro.2007.01.029
37. Park IH (2004) MEMS based space telescope for extreme energy cosmic rays experiments. Nucl Phys B 134:196–201. https://doi.org/10.1016/j.nuclphysbps.2004.08.031
38. Kuhn JL, Barclay RB, Dutta SB, Freund MM, Greenhouse MA, Moseley SH (2000) Simulation of micro-electromechanical systems (MEMS) for the next generation space telescope. Am Astron Soc Meet Abstr 197:14.02
39. Ford HC, Clampin M, Illingworth GD, Kris JE, Olivier SS, Petro L, Sommagren GE (2003) Requirements for an optical 8-m space telescope with a MEMs deformable mirror to detect Earth-like planets around nearby stars. Proc SPIE 4854:554–557. https://doi.org/10.1117/12.460052

40. Li MJ, Adachi T, Allen C, Babu S, Bajikar S, Beamesderfer M, Bradley R, Denis K, Costen N, Ewin A, Franz D (2006) MEMS microshutter arrays for James webb space telescope. Proc SPIE 6415:64150B. https://doi.org/10.1117/12.706308
41. Park IH, Artikova S, Kim JE, Jeon JA, Jeong S, Jung A, Lee HY, Lee J, Na G, Nam S, Oh S (2007) MEMS space telescope obscura: pathfinder of future space missions. Bull Korean Astron Soc 32(2):21–21
42. Jeong SM, Nam SU, Park IH, Park JH, Lee J, Lee CH (2008) Telescope with using MEMS mirror for detecting Gamma Ray burst. Bull Korean Astron Soc 33(1):44–42
43. Jeon JA, Jin JY, Jung AR, Jeong SM, Kim JE, Kim M, Kim YK, Lee CH, Lee HY, Lee J, Na GW (2009) A Few Selected research topics of research center of MEMS space telescope. AAPPS Bull 19(4):33
44. Agrawal B, Kubby J (2011) Applications of MEMS in segmented mirror space telescopes. Proc SPIE 7931:793102. https://doi.org/10.1117/12.876503
45. Chen S, Sivanandam S, Moon DS (2016) Optical design of MEMS-based infrared multi-object spectrograph concept for the Gemini south telescope. Proc SPIE 9908:99087N. https://doi.org/10.1117/12.2230694
46. Jeon JA, Lee HY, Kim JE, Lee J, Park IH (2016) Performance of the second MEMS space telescope for observation of extreme lightning from space. J Korean Phys Soc 68(5):710–714. https://doi.org/10.3938/jkps.68.710
47. Miyamura N (2017) Adaptive optics using a MEMS deformable mirror for a segmented mirror telescope. Proc SPIE 10410:1041003. https://doi.org/10.1117/12.2273868
48. Konidaris NP, Kubby JA, Sheinis AI (2008) Small solutions to the large telescope problem: a massively replicated MEMS spectrograph. Proc SPIE 7018:70182I. https://doi.org/10.1117/12.787012
49. Park JH, Garipov G, Jeon JA, Jin JY, Jung A, Kim JE, Kim M, Kim YK, Klimov P, Khrenov B, Lee CH (2009) MEMS space telescope for the observation of extreme lightening (MTEL). J Korean Astron Soc 34(1):113–113
50. Gavel D (2009) MEMS devices in future astronomical telescope adaptive optics instruments. Proc SPIE 7209:72090E. https://doi.org/10.1117/12.812084
51. Scott C, Potsaid B, Wen JT (2010) Wide field scanning telescope using MEMS deformable mirrors. Int J Optomechatronics 4(3):285–305. https://doi.org/10.1080/15599612.2010.513720
52. Morzinski KM, Norton AP, Evans JW, Reza L, Severson SA, Dillon D, Reinig M, Gavel DT, Cornelissen S, Macintosh BA, Max CE (2012) MEMS practice: from the lab to the telescope. Proc SPIE 8253:825304. https://doi.org/10.1117/12.910964
53. Mészáros L, Jaskó A, Pál A, Csépány G (2014) Accurate telescope mount positioning with MEMS accelerometers. Publ Astron Soc Pac 126(942):769. https://doi.org/10.1086/677943
54. Haindl R, Trasischker W, Baumann B, Pircher M, Hitzenberger CK (2015) Three-beam doppler optical coherence tomography using a facet prism telescope and MEMS mirror for improved transversal resolution. J Mod Opt 62(21):1781–1788. https://doi.org/10.1080/09500340.2014.983569
55. Blase RC, Libardoni MJ, Miller GP, Miller KE, Phillips-Lander CM, Waite JH, Glein CR, Zhu H, Ghosh A, Venkatasubramanian A, Fan X (2020) Experimental coupling of a MEMS gas chromatograph and a mass spectrometer for organic analysis in space environments. ACS Earth Space Chem 4(10):1718–1729. https://doi.org/10.1021/acsearthspacechem.0c00131
56. Lu CJ, Steinecker WH, Tian WC, Oborny MC, Nichols JM, Agah M, Potkay JA, Chan HK, Driscoll J, Sacks RD, Wise KD (2005) First-generation hybrid MEMS gas chromatograph. Lab Chip 5(10):1123–1131. https://doi.org/10.1039/B508596A
57. Cook KA, Sastry AM (2005) An algorithm for selection and design of hybrid power supplies for MEMS with a case study of a micro-gas chromatograph system. J Power Sources 140(1):181–202. https://doi.org/10.1016/j.jpowsour.2004.06.071
58. Dexiang C (2007) Development of MEMS based micro gas chromatograph. Anal Instrum (Beijing) 3(153):9

59. Zampolli S, Elmi I, Mancarella F, Betti P, Dalcanale E, Cardinali GC, Severi M (2009) Real-time monitoring of sub-ppb concentrations of aromatic volatiles with a MEMS-enabled miniaturized gas-chromatograph. Sens Actuators, B Chem 141(1):322–328. https://doi.org/10.1016/j.snb.2009.06.021
60. Akao S, Iwaya T, Sakamoto T, Nagai H, Nakaso N, Tsuji T, Yamanaka K (2010) 3J-5 Development and evaluation of metal MEMS column for the ball SAW gas chromatograph. In: Proceedings of symposium on ultrasonic electronics, vol 31, pp 453–454. https://doi.org/10.24492/use.31.0_453
61. Iwaya T, Akao S, Sakamoto T, Tsuji T, Nakaso N, Yamanaka K (2011) 3Pa3–9 Development of high precision metal MEMS column for portable SAW gas chromatograph. In: Proceedings of symposium on ultrasonic electronics, vol 32, pp 367–368. https://doi.org/10.24492/use.32.0_367
62. Wright LK, Collin W, Serrano G, Zellers ET (2013) MEMS gas chromatograph for explosive marker compounds: temperature and flow rate effects on sensor-array detector responses and system performance. ECS Meet Abstr MA2013-02:2634–2634. https://doi.org/10.1149/MA2013-02/47/2634
63. Wang J, Nuñovero N, Lin Z, Nidetz R, Buggaveeti S, Zhan C, Kurabayashi K, Steinecker WH, Zellers ET (2016) A wearable MEMS gas chromatograph for multi-vapor determinations. Procedia Eng 168:1398–1401. https://doi.org/10.1016/j.proeng.2016.11.391
64. Blase R, Libardoni M, Miller G, Miller KE, Phillips-Lander CM, Waite H, Zhu H, Glein CR, Ghosh A, Venkatasubramanian A, Fan X (2020) Experimental coupling of a MEMS gas chromatograph and a mass spectrometer for organic analysis on Icy ocean world landers. ACS Earth Space Chem 4(10):1718–1729. https://doi.org/10.1021/acsearthspacechem.0c00131
65. Hauschild JP, Wapelhorst E, Müller J (2007) Mass spectra measured by a fully integrated MEMS mass spectrometer. Int J Mass Spectrom 264(1):53–60. https://doi.org/10.1016/j.ijms.2007.03.014
66. Wapelhorst E, Hauschild JP, Müller J (2007) Complex MEMS: a fully integrated TOF micro mass spectrometer. Sens Actuators, A 138(1):22–27. https://doi.org/10.1016/j.sna.2007.04.041
67. Roman PA, Getty SA, Herrero FA, Hu R, Jones HH, Kahle D, King TT, Mahaffy P (2008) A miniature MEMS and NEMS enabled time-of-flight mass spectrometer for investigations in planetary science. Proc SPIE 6959:69590G. https://doi.org/10.1117/12.777670
68. Wright S, Syms RR, Moseley R, Hong G, O'Prey S, Boxford WE, Dash N, Edwards P (2010) MEMS-based nanospray-ionization mass spectrometer. J Microelectromech Syst 19(6):1430–1443. https://doi.org/10.1109/JMEMS.2010.2082501
69. Russell Z, Russell PE, Glass J, Parker C, Stoner B, Gilchrist K, Piascik J (2009) Development of a micro mass spectrometer: analysis of particle behavior in MEMS Ion lens systems. Microsc Microanal 15(S2):242–243. https://doi.org/10.1017/S1431927609098870
70. Tassetti CM, Duraffourg L, Danel JS, Peyssonneaux O, Progent F, Machuron-Mandard X (2012) A MEMS electron impact ion source integrated in a micro-time-of-flight mass spectrometer. Procedia Eng 47:538–541. https://doi.org/10.1016/j.proeng.2012.09.203
71. Tassetti CM, Mahieu R, Danel JS, Peyssonneaux O, Progent F, Polizzi JP, Machuron-Mandard X, Duraffourg L (2013) A MEMS electron impact ion source integrated in a microtime-of-flight mass spectrometer. Sens Actuators, B Chem 189:173–178. https://doi.org/10.1016/j.snb.2013.02.087
72. Szyszka P, Grzebyk T, Górecka-Drzazga A, Dziuban JA (2016) MEMS ion source for mass spectrometer integrated on a chip. J Phys: Conf Ser 773:012099. https://doi.org/10.1088/1742-6596/773/1/012099
73. Vigne S, Alava T, Tassetti CM, Duraffour L, Progent F (2017) Optimization of an electron impact ion source on a MEMS time-of-flight mass spectrometer. Sens Actuators, B Chem 243:690–695. https://doi.org/10.1016/j.snb.2016.11.083
74. Vigne S, Alava T, Videlier H, Mahieu R, Tassetti CM, Duraffourg L, Progent F (2017) Gas analysis using a MEMS linear time-of-flight mass spectrometer. Int J Mass Spectrom 422:170–176. https://doi.org/10.1016/j.ijms.2017.03.011

75. Grzebyk T, Szyszka P, Górecka-Drzazga A, Dziuban JA (2019) Glow-discharge ion source for on-chip integrated miniature mems mass spectrometer. J Phys: Conf Ser 1407:012020. https://doi.org/10.1088/1742-6596/1407/1/012020
76. Grzebyk T, Szmajda T, Szyszka P, Gorecka-Drzazga A, Dziuban J (2020) Glow-discharge ion source for MEMS mass spectrometer. Vacuum 171:109008. https://doi.org/10.1016/j.vacuum.2019.109008
77. Chang TH, Struk D, Navaei M, Doroshenko VM, Laiko V, Moskovets E, Novoselov K, Dimandja JMD, Hesketh PJ (2020) Separation of volatile organic compounds using a MEMS separation column integrated with ion trap mass spectrometer. Sens Actuators, B Chem 307:127588. https://doi.org/10.1016/j.snb.2019.127588
78. Wright S, Syms RRA (2021) Shock-free ion transmission in a skimmer-based MEMS mass spectrometer vacuum interface. J Micromech Microeng 31(4):045010. https://doi.org/10.1088/1361-6439/abe247
79. Huang A, Yang EH (2009) MEMS thruster system for CubeSat orbital maneuver applications. Proc ASME 43857:931–936. https://doi.org/10.1115/IMECE2009-12675
80. Cahoy KL, Marinan AD, Novak B, Kerr C, Nguyen T, Webber M, Falkenburg G, Barg A, Berry K, Carlton A, Belikov R (2013) MEMS deformable mirror CubeSat testbed. Proc SPIE 8864:88640U. https://doi.org/10.1117/12.2024684
81. Marinan A, Cahoy K, Webber M, Belikov R, Bendek E (2014) Payload characterization for CubeSat demonstration of MEMS deformable mirrors. Proc SPIE 9148:91483Z
82. Jang SE, Han SH, Kim TG, Lee JK, Jang TS, Oh HU (2015) Structural design and verification of MEMS solid thruster for CubeSat application. J Korean Soc Aeronaut Space Sci 43(5):432–439
83. Douglas ES, Bendek E, Marinan A, Belikov R, Merck J, Cahoy KL (2017) The DeMi CubeSat: wavefront control with a MEMS deformable mirror in space. Am Astron Soc Meet Abstr 229:146.23
84. Rossi C (2002) Micropropulsion for space - a survey of MEMS-based micro thrusters and their solid propellant technology. Sens Updat 10(1):257–292. https://doi.org/10.1002/1616-8984. (200201)10:1%3C257::AID-SEUP257%3E3.0.CO;2-2
85. Chaalane A, Chemam R, Houabes M, Yahiaoui R, Metatla A, Ouari B, Metatla N, Mahi D, Dkhissi A, Esteve D (2015) A MEMS-based solid propellant microthruster array for space and military applications. J Phys: Conf Ser 660:012137. https://doi.org/10.1088/1742-6596/660/1/012137
86. Gill WA, Ali D, An BH, Syed WU, Saeed N, Al-shaibah M, Elfadel IM, Al Dahmani S, Choi DS (2020) MEMS multi-vibrating ring gyroscope for space applications. Microsyst Technol 26(8):2527–2533. https://doi.org/10.1007/s00542-020-04793-x
87. Keymeulen D, Ferguson MI, Breuer L, Peay C, Oks B, Cheng Y, Kim D, MacDonald E, Foor D, Terrile R, Arenki N (2007) Miniaturized self-adaptive tuning of MEMS gyroscope for space. Proc SPIE 6556:655613. https://doi.org/10.1117/12.723605
88. Liu J, Fu M, Meng C, Li J, Li K, Hu J, Chen X (2020) Consideration of thermo-vacuum stability of a MEMS gyroscope for space applications. Sensors 20(24):7172. https://doi.org/10.3390/s20247172
89. Lee J, Kim T (2014) Micro space power system using MEMS fuel cell for nano-satellites. Acta Astronaut 101:165–169. https://doi.org/10.1016/j.actaastro.2014.04.010
90. Marinaro DG, McMahon P, Wilson A (2008) Proton radiation effects on MEMS silicon strain gauges. IEEE Trans Nucl Sci 55(3):1714–1718. https://doi.org/10.1109/TNS.2008.921933
91. Gomes J, Shea HR (2011) Displacement damage effects in silicon MEMS at high proton doses. Proc SPIE 7928:79280G. https://doi.org/10.1117/12.873546
92. Patton ST, Frasca AJ, Talnagi JW, Hyman DJ, Phillips BS, Jones JG, Vaia RA, Voevodin AA (2013) Effect of space radiation on the leakage current of MEMS insulators. IEEE Trans Nucl Sci 60(4):3074–3083. https://doi.org/10.1109/TNS.2013.2263840
93. Bandi T, Polido-Gomes J, Neels A, Dommann A, Shea HR (2013) Making MEMS more suited for Space: Assessing the proton-radiation tolerance of structural materials for microsystems in orbit. Proc SPIE 8614:86140M. https://doi.org/10.1117/12.2004705

94. Sallese JM, Bouvet D (2007) Principles of space-charge based bi-stable MEMS: the junction-MEMS. Sens Actuators, A 133(1):173–179. https://doi.org/10.1016/j.sna.2006.03.024
95. Velásquez-García LF, Hill TF, Wilhite BA, Jensen KF, Epstein AH, Livermore C (2007) A MEMS singlet oxygen generator. Part I: device fabrication and proof of concept demonstration. J Microelectromechanical Syst 16(6):1482–1491. https://doi.org/10.1109/JMEMS.2007.902446
96. Hill TF, Velasquez-Garcia LF, Wilhite BA, Rawlins WT, Lee S, Davis SJ, Jensen KF, Epstein AH, Livermore C (2007) A MEMS singlet oxygen generator. Part II: experimental exploration of the performance space. J Microelectromechanical Syst 16(6):1492–1505. https://doi.org/10.1109/JMEMS.2007.907880
97. Livermore C, Hill TF, Velásquez-García L, Wilhite BA, Epstein AH, Jensen K, Rawlins WT, Lee S, Davis S (2007) Singlet oxygen generator on a chip for MEMS-based COIL. Proc SPIE 6454:64540G. https://doi.org/10.1117/12.707205
98. Mitsuishi I, Ezoe Y, Ishizu K, Moriyama T, Mita M, Yamasaki NY, Mitsuda K, Horade M, Sugiyama S, Riveros R, Yamaguchi H (2011) Novel ultra-lightweight and high-resolution MEMS X-ray optics for space astronomy. Sens Actuators A: Phys 188:411–416. In: 2011 16th international solid-state sensors, actuators and microsystems conference, pp 566–569. IEEE (2011)
99. Walko DA, Jung IW, Lopez D, Mukhopadhyay D, Schwartz CP, Shenoy GK, Wang J (2012) Temporal modulation of synchrotron x-rays using torsional MEMS mirrors. Proc SPIE 8502:85020H. https://doi.org/10.1117/12.930057
100. Chakraborty I, Tang WC, Bame DP, Tang TK (2000) MEMS micro-valve for space applications. Sens Actuators, A 83(1–3):188–193. https://doi.org/10.1016/S0924-4247(99)00382-9
101. Sawant A, Antonuk LE, El-Mohri Y, Li Y, Su Z, Wang Y, Yamamoto J, Zhao Q, Du H, Daniel J, Street R (2005) Segmented phosphors: MEMS-based high quantum efficiency detectors for megavoltage X-ray imaging. Med Phys 32(2):553–565. https://doi.org/10.1118/1.1854774
102. Mitsuishi I, Ezoe Y, Ishizu K, Moriyama T, Mita M, Yamasaki NY, Mitsuda K, Kanamori Y, Morishita K, Nakajima K (2012) Novel ultra-lightweight and high-resolution MEMS X-ray optics for space astronomy. Sens Actuators, A 188:411–416. https://doi.org/10.1016/j.sna.2012.02.042
103. Riveros RE, Yamaguchi H, Mitsuishi I, Takagi U, Ezoe Y, Kato F, Sugiyama S, Yamasaki N, Mitsuda K (2009) Magnetic field assisted finishing of ultra-lightweight and high-resolution MEMS X-ray micro-pore optics. Proc SPIE 7360:736013. https://doi.org/10.1117/12.823961
104. Ishikawa K, Ezoe Y, Numazawa M, Ogawa T, Sato M, Nakamura K, Ohashi T, Mitsuda K, Maeda R, Hiroshima H, Kurashima Y (2017) 12-inch X-ray optics based on MEMS process. Microsyst Technol 23(7):2815–2821. https://doi.org/10.1007/s00542-016-2980-6
105. Gilles JP, Megherbi S, Raynaud G, Parrain F, Mathias H, Leroux X, Bosseboeuf A (2008) Scanning electron microscopy for vacuum quality factor measurement of small-size MEMS resonators. Sens Actuators, A 145:187–193. https://doi.org/10.1016/j.sna.2007.11.026
106. Zhang C, Zhao H, Xu S, Chen N, Li K, Jiang X, Liu L, Liu Z, Wang L, Wong KK, Zou J (2020) Multiscale high-speed photoacoustic microscopy based on free-space light transmission and a MEMS scanning mirror. Opt Lett 45(15):4312–4315. https://doi.org/10.1364/ol.397733
107. Szentesi OI (1972) Stroboscopic electron mirror microscopy at frequencies up to 100 MHz. J Phys E: Sci Instrum 5(6):563. https://doi.org/10.1088/0022-3735/5/6/024
108. Qiu J, Ha G, Jing C, Baryshev SV, Reed BW, Lau JW, Zhu Y (2016) GHz laser-free time-resolved transmission electron microscopy: a stroboscopic high-duty-cycle method. Ultramicroscopy 161:130–136. https://doi.org/10.1016/j.ultramic.2015.11.006
109. Neudert A, McCord J, Schäfer R, Schultz L (2005) Subnanosecond vortex transformation in ferromagnetic film elements observed by stroboscopic wide-field Kerr microscopy. J Appl Phys 97(10):10E701. https://doi.org/10.1063/1.1850832
110. Jing C, Zhu Y, Liu A, Schliep K, Fu X, Zhao Y, Montgomery E, Rush W, Kanareykin A, Katz M, Lau J (2019) Tunable electron beam pulser for picoseconds stroboscopic microscopy in transmission electron microscopes. Ultramicroscopy 207:112829. https://doi.org/10.1016/j.ultramic.2019.112829

111. Bostanjoglo O, Rosin T (1977) Ultrasonically induced magnetic reversals observed by stroboscopic electron microscopy. Opt Acta: Int J Opt 24(6):657–664. https://doi.org/10.1080/713819603
112. Schliep KB, Katz MB, Lau JW (2018) A RF broadband biasing holder for ultrafast stroboscopic electron microscopy. Microsc Microanal 24(S1):1852–1853. https://doi.org/10.1017/S1431927618009741
113. Foerster M, Statuto N, Casals B, Hernández-Mínguez A, Finizio S, Mandziak A, Aballe L, Hernàndez Ferràs JM, Macia F (2019) Quantification of propagating and standing surface acoustic waves by stroboscopic X-ray photoemission electron microscopy. J Synchrotron Radiat 26(1):184–193. https://doi.org/10.1107/S1600577518015370
114. Whiteley S, Heremans FJ, Awschalom D, Holt M (2018) Nanoscale imaging of surface acoustic waves with stroboscopic scanning X-ray diffraction microscopy. Bull Am Phys Soc 63(1):E11-00009
115. Vinoy KJ, Varadan VK (2001) Design of reconfigurable fractal antennas and RF-MEMS for space-based systems. Smart Mater Struct 10(6):1211. https://doi.org/10.1088/0964-1726/10/6/310
116. Di Nardo S, Farinelli P, Kim T, Marcelli R, Margesin B, Paola E, Pochesci D, Vietzorreck L, Vitulli F (2013) Design of RF MEMS based switch matrix for space applications. Adv Radio Sci 11(C.7):143–152. https://doi.org/10.5194/ars-11-143-2013
117. Lucibello A, Marcelli R, Proietti E, Bartolucci G, Mulloni V, Margesin B (2015) Reliability of RF MEMS capacitive and ohmic switches for space redundancy configurations. Microsyst Technol 21(9):1903–1913. https://doi.org/10.1007/s00542-014-2124-9
118. Savin EA, Kirtaev RV, Zhukov AA, Didyk PI (2019) Design of a packaging-friendly double-topology RF MEMS switch for space applications. Microsyst Technol 25:51–55. https://doi.org/10.1007/s00542-018-3929-8
119. Singh H, Malhotra J (2017) RF-MEMS-based DPDT switch on silicon substrate for Ku-Band space-borne applications. Trans Electr Electron Mater 18(1):16–20. https://doi.org/10.4313/TEEM.2017.18.1.16
120. Ikuta T, Mase H, Shimizu R (1987) Simple line-sampling apparatus for line frequency stroboscopic observation of periodic phenomena using scanning electron microscopy. Rev Sci Instrum 58(10):1811–1816. https://doi.org/10.1063/1.1139662
121. Lynch ER, Boland FM (1988) Waveform parametrisation in stroboscopic scanning electron microscopy. J Phys E: Sci Instrum 21(2):197. https://doi.org/10.1088/0022-3735/21/2/015
122. Nakamae K, Fujioka H, Ura K (1990) Analysis of waveform distortion due to the transit time effect in stroboscopic scanning electron microscopy. Meas Sci Technol 1(9):894. https://doi.org/10.1088/0957-0233/1/9/011
123. Gradov OV, Jablokov AG, Skrynnik AA (2017) Periodimeter-regulometer for kinetic analysis of symbolic registerograms of correlation stroboscopic electron microscopy and microprobe chemical mapping data. Fundam Probl Radioelectron Instrum 17(3):769–772
124. Hafiane A, Petitgrand S, Gigan O, Bouchafa S, Bosseboeuf A (2003) Study of subpixel image processing algorithms for MEMS in-plane vibration measurements by stroboscopic microscopy. Proc SPIE 5145:169–179. https://doi.org/10.1117/12.500141
125. Blanche PA, LaComb L, Wang Y, Wu MC (2017) Diffraction-based optical switching with MEMS. Appl Sci 7(4):411. https://doi.org/10.3390/app7040411
126. Zamkotsian F, Timotijevic B, Lockhart R, Stanley RP, Lanzoni P, Luetzelschwab M, Canonica M, Noell W, Tormen M (2012) Optical characterization of fully programmable MEMS diffraction gratings. Opt Express 20(23):25267–25274. https://doi.org/10.1364/oe.20.025267
127. Suresh VG, DasGupta N, Bhattacharya S (2012) Tunable MEMS diffraction gratings. Proc SPIE 8549:854918. https://doi.org/10.1117/12.925113
128. Zhou G, Du Y, Cheo KK, Chau FS (2011) MEMS-driven diffraction gratings for rapid scanning of laser beams with very high optical resolution. Proc SPIE 8191:819105. https://doi.org/10.1117/12.901076
129. Han S, Ishfaque A, Phamduy P, Kim B (2020) Effect of squeeze air film damping on diffraction grating based bio-inspired MEMS directional microphone. Microsyst Technol 26(4):1203–1212. https://doi.org/10.1007/s00542-019-04650-6

130. Long H, Xi S, Liu D, Shi T, Xia Q, Liu S, Tang Z (2012) Tailoring diffraction-induced light distribution toward controllable fabrication of suspended C-MEMS. Opt Express 20(15):17126–17135. https://doi.org/10.1364/OE.20.017126
131. Gloor S, Bachmann AH, Epitaux M, von Niederhäusern T, Vorreau P, Matuschek N, Hsu K, Duelk M, Vélez C (2013) High-speed miniaturized swept sources based on resonant MEMS mirrors and diffraction gratings. Proc SPIE 8571:85712X. https://doi.org/10.1117/12.2007106
132. Ketchum RS, Blanche PA (2021) Diffraction efficiency characteristics for MEMS-based phase-only spatial light modulator with nonlinear phase distribution. Photonics 8(3):62. https://doi.org/10.3390/photonics8030062
133. Kim B, Schmittdiel MC, Degertekin FL, Kurfess TR (2005) Actively controlled diffraction grating interferometer MEMS devices. Proc SPIE 5721:151–158. https://doi.org/10.1117/12.592215
134. Bicen B, Garcia C, Hall NA, Okandan M, Cui W, Su QT, Miles RN, Degertekin L (2008) Diffraction based optical MEMS microphones and accelerometers with active electrostatic force feedback. J Acoust Soc Am 123(5):3230–3230. https://doi.org/10.1121/1.2933456
135. Suzuki Y, Tezuka SI (2019) Numerical simulation of 3D Fox–Li integral equation described by Rayleigh-Sommerfeld diffraction for MEMS-VCSEL. Opt Rev 26(5):430–435. https://doi.org/10.1007/s10043-019-00551-1
136. Jungwirth ME, Wick DV, Dereniak EL (2012) Theory and design of a MEMS-enabled diffraction limited adaptive optical zoom system. Proc SPIE 8373:83730T. https://doi.org/10.1117/12.917841
137. Grabarnik S, Emadi A, Wu H, De Graaf G, Vdovin G, Wolffenbuttel RF (2008) Fabrication of an imaging diffraction grating for use in a MEMS-based optical microspectrograph. J Micromech Microeng 18(6):064006. https://doi.org/10.1088/0960-1317/18/6/064006
138. Yu YT, Yuan WZ, Qiao DY, Yan B (2010) A simple numerical method to study the far-field diffraction of optical MEMS devices with a large array and complex element geometry. Sens Actuators, A 158(1):30–36. https://doi.org/10.1016/j.sna.2009.12.032
139. Day DR (2004) Multichannel gain equalizer MEMs surface topology effects on chromatic dispersion: a comparison of scalar diffraction modeling to actual product. Proc SPIE 5577:609–620. https://doi.org/10.1117/12.567485
140. van Zeijl HW, Simon K, Slabbekoorn JHCM, Buel WV, Gui CQ (2002) Waferstepper alignment for MEMS applications using diffraction gratings. MRS Online Proc Libr 729(1):581–586. https://doi.org/10.1557/PROC-729-U5.8
141. Shtyrkova K, Arguello L, Oesch D, Sanchez DJ, Kelly P, Tewksbury-Christle C, Smith J (2010) Experimental analysis of diffraction effects from a segmented MEMS deformable mirror for a closed loop adaptive optics system. Proc SPIE 7816:781608. https://doi.org/10.1117/12.861544
142. Wang C, Lu Q, Bai J, Yang G, Wang K, Liu D, Yang Y (2015) Highly sensitive lateral deformable optical MEMS displacement sensor: anomalous diffraction studied by rigorous coupled-wave analysis. Appl Opt 54(30):8935–8943. https://doi.org/10.1364/ao.54.008935
143. Li T-P, Li X-P, Yu YT, Yan B, Sun R-K, Qiao D-Y (2010) Experimental investigation of diffraction propertyl for a MEMS-based pitch-tunable grating. Acta Photonica Sin 39(4):618
144. Degertekin L, Jeelani K, Qureshi S, Hasler P, Bicen B, Cui W, Su QT, Miles RN (2008) Miniature diffraction-based optical MEMS microphones with integrated optoelectronics. J Acoust Soc Am 123(5):3229–3229. https://doi.org/10.1121/1.2933454
145. Reitterer J, Fidler F, Schmid G, Hambeck C, Saint J-W, Leeb W, Schmid U (2015) Beam-clipping-induced diffraction effects in MEMS laser scanners for autostereoscopic outdoor displays. Sens Actuators, A 233:582–589. https://doi.org/10.1016/j.sna.2015.06.024
146. Bui TD, Chen G (1997) Multiresolution moment-Fourier-wavelet descriptor for 2D pattern recognition. Proc, SPIE 3078:552–557. https://doi.org/10.1117/12.271746
147. Adcock B, Hansen AC, Kutyniok G, Ma J (2015) Linear stable sampling rate: optimality of 2D wavelet reconstructions from fourier measurements. SIAM J Math Anal 47(2):1196–1233. https://doi.org/10.1137/140959365

148. Pan SL, Zhou YH (2002) Plane type object recognition based on revised 2D Fourier-wavelet descriptor. Infrared Laser Eng 31(4):290–293
149. Walker JS (1997) Fourier analysis and wavelet analysis. Not AMS 44(6):658–670
150. Patton ST, Zabinski JS (2002) Failure mechanisms of a MEMS actuator in very high vacuum. Tribol Int 35(6):373–379. https://doi.org/10.1016/S0301-679X(02)00018-X
151. Dienel M, Naumann M, Sorger A, Tenholte D, Voigt S, Mehner J (2012) On the influence of vacuum on the design and characterization of MEMS. Vacuum 86(5):536–546. https://doi.org/10.1016/j.vacuum.2011.10.005
152. Fedeli P, Frangi A, Laghi G, Langfelder G, Gattere G (2017) Near vacuum gas damping in MEMS: simplified modeling. J Microelectromech Syst 26(3):632–642. https://doi.org/10.1109/JMEMS.2017.2686650
153. Grzebyk T, Turczyk K, Szyszka P, Górecka-Drzazga A, Dziuban J (2020) Pressure control system for vacuum MEMS. Vacuum 178:109452. https://doi.org/10.1016/j.vacuum.2020.109452
154. Eapen KC, Smallwood SA, Zabinski JS (2006) Lubrication of MEMS under vacuum. Surf Coat Technol 201(6):2960–2969. https://doi.org/10.1016/j.surfcoat.2006.06.011
155. Gan Z, Lin D, Wang X, Zhang H, Liu S (2007) Vacuum measurement on vacuum packaged MEMS devices. J Phys: Conf Ser 48(1):264. https://doi.org/10.1088/1742-6596/48/1/264
156. Ghionea S, Hull D, Williams K (2013) Characterization techniques for a MEMS electric-field sensor in vacuum. J Electrostat 71(6):1076–1082. https://doi.org/10.1016/j.elstat.2013.09.002
157. Chen SJ, Wu YC (2020) A new macro-model of gas flow and parameter extraction for a CMOS-MEMS vacuum sensor. Symmetry 12(10):1604. https://doi.org/10.3390/sym12101604
158. Tenholte D, Kurth S, Geßner T, Dötzel W (2008) A MEMS friction vacuum gauge suitable for high temperature environment. Sens Actuators, A 142(1):166–172. https://doi.org/10.1016/j.sna.2007.05.031
159. Grzebyk T, Knapkiewicz P, Szyszka P, Górecka-Drzazga A, Dziuban JA (2016) MEMS ion-sorption high vacuum pump. J Phys: Conf Ser 773:012047. https://doi.org/10.1088/1742-6596/773/1/012047
160. Vargo SE, Muntz EP (2001) Initial results from the first MEMS fabricated thermal transpiration-driven vacuum pump. AIP Conf Proc 585:502–509. https://doi.org/10.1063/1.1407602
161. Grzebyk T (2017) MEMS vacuum pumps. J Microelectromech Syst 26(4):705–717. https://doi.org/10.1109/JMEMS.2017.2676820
162. Górecka-Drzazga A (2009) Miniature and MEMS-type vacuum sensors and pumps. Vacuum 83(12):1419–1426. https://doi.org/10.1016/j.vacuum.2009.05.003
163. Grzebyk T, Górecka-Drzazga A, Dziuban JA, Maamari K, An S, Dankovic T, Feinerman A, Busta H (2015) Integration of a MEMS-type vacuum pump with a MEMS-type Pirani pressure gauge. J Vac Sci Technol B Nanotechnol Microelectron: Mater, Process, Meas, Phenom 33(3):03C103. https://doi.org/10.1116/1.4903448
164. Gannon AJ, Hobson GV, Shea MJ, Clay CS, Millsaps KT (2014) MEMS-scale turbomachinery based vacuum roughing pump. J Turbomach 136(10):101002. https://doi.org/10.1115/1.4027971
165. Gradov OV, Nasirov FA, Goncharova AA, Fischenko VK, Jablokow AG (2018) On-chip lens-less holographic trichoscopy/trichometry technologies: microinterferometric, Fourier spectral and correlographic techniques for clinical trichology. Morphologia 12(2):7–21. https://doi.org/10.26641/1997-9665.2018.2.7-21
166. Gradov OV, Aleksandrov PL, Gradova MA (2019) Study of mineral samples relevant for desert locations using software correlation spectral analysis of scanning electron microscopy registers: from 2D Fourier spectra to online analysis of statistics of integral spatial characteristics. Softw Syst Comput Methods 4:125–171. https://doi.org/10.7256/2454-0714.2019.4.31379
167. Skrynnik A, Oganessian VA, Jablokow AG, Gradov OV (2018) System for semiautomatic tissue classification based on optical diffractometer for biopolymer structure analysis. Morphologia 12(3):164–171. https://doi.org/10.26641/1997-9665.2018.3.164-171

168. Adamovich ED, Gradov OV, Nasirov FA (2018) Liquid metal microfluidics with morphometric monitoring: from liquid metal antenn towards the radiofrequency polarography. Fundam Probl Radioelectron Instrum 18(3):650–653. https://doi.org/10.13140/RG.2.2.33474.63682
169. Gradov OV, Nasirov PA, Jablokov AG (2018) Lensless on-chip-hemocytometry with secondary processing of cell images in the framework of an unconventional photometric model. Photonics Russ 12(7):716–729. https://doi.org/10.22184/1993-7296.2018.12.7.716.729
170. Sun X, Xu D, Xiong B, Wu G, Wang Y (2013) A wide measurement pressure range CMOS-MEMS based integrated thermopile vacuum gauge with an XeF2 dry-etching process. Sens Actuators, A 201:428–433. https://doi.org/10.1016/j.sna.2013.07.020
171. Zhang LM, Jiao BB, Yun SC, Kong YM, Ku CW, Chen DP (2017) A CMOS compatible MEMS pirani vacuum gauge with monocrystal silicon heaters and heat sinks. Chin Phys Lett 34(2):025101. https://doi.org/10.1088/0256-307X/34/2/025101
172. Taghvaei MA, Cicek PV, Allidina K, Nabki F, El-Gamal MN (2013) A 0.13-μm CMOS interface circuit for a MEMS resonator-based vacuum measurement system. IEEE Trans Circuits Syst I: Regul Pap 60(12):3136–3144. https://doi.org/10.1109/TCSI.2013.2265976
173. Gradov OV (2020) Lab-on-a-Chip with CMOS or CCD sensors as ESEM micro-subchamber and lens-less microscope for correlative light and electron microscopy (CLEM). In: Preprint of SDS-2020 conference [cancelled due to COVID-19]. https://doi.org/10.13140/RG.2.2.30685.97765
174. Fischenko VK, Goncharova AA, Dolgikh GI, Zimin PS, Subote AE, Klescheva NA, Golik AV (2021) Express image and video analysis technology QAVIS: application in system for video monitoring of Peter the Great Bay (Sea of Japan/East Sea). J Mar Sci Eng 9(10):1073. https://doi.org/10.3390/jmse9101073
175. Fischenko V, Mitnik L, Dolgikh G, Goncharova A, Zimin P, Klescheva N, Golik A (2022) QAVIS technology: measuring wave processes in coastal zones based on the analysis of internet video broadcast. In: IGARSS 2022 - IEEE international geoscience and remote sensing symposium. IEEE, pp 6801–6804. https://doi.org/10.1109/IGARSS46834.2022.9883487
176. Goncharova AA, Fischenko VK, Dubina VA (2012) Use the express-analysis program QAVIS for the satellite monitoring. Curr Probl Remote Sens Earth Space 9(3):293–298
177. Goncharova AA, Fischenko VK (2011) QAVIS – program for quick image and video analysis. In: Proceedings of 8th open german-russian workshop on pattern recognition and image understanding (OGRW-8), p 85
178. Gradov OV, Aleksandrov PL, Gradova MA (2019) Study of mineral samples relevant for desert locations using software correlation spectral analysis of scanning electron microscopy registers: from 2D Fourier spectra to online analysis of statistics of integral spatial characteristics. Geol Sci Repos (repository.geologyscience.ru), Article ID: 123456789
179. Gradov OV, Gradova MA, Maklakova IA, Kholuiskaya SN (2022) Towards electron-beam-driven soft/polymer fiber microrobotics for vacuum conditions. Mater Res Proc 21:370–383
180. Jiang B, Yang K, Wang J (2010) Design of the optical fiber MEMS infrasound sensor. In: Proceedings of the 2010 international conference on measuring technology and mechatronics automation. IEEE, pp 1049–1051
181. Si W, Fu C, Li D, Li H, Yuan P, Yu Y (2018) Directional sensitivity of a MEMS-based fiber-optic extrinsic fabry–perot ultrasonic sensor for partial discharge detection. Sensors 18(6):1975. https://doi.org/10.3390/s18061975
182. Si W, Fu C, Li H, Lv J, Xiong C, Yuan P, Yu Y (2019) Locating ultrasonic signals employing MEMS-On-Fiber sensors. Sensors 19(17):3696. https://doi.org/10.3390/s19173696
183. Xia J, Wang FY, Luo H, Hu YM, Xiong SD (2017) MEMS-based optic fiber fabry-perot sensor for underwater acoustic measurement with a wavelength-switched system. IOP Conf Ser: Mater Sci Eng 281(1):012002. https://doi.org/10.1088/1757-899X/281/1/012002
184. Liu B, Lin J, Wang J, Ye C, Jin P (2015) MEMS-based high-sensitivity Fabry-Perot acoustic sensor with a 45 angled fiber. IEEE Photonics Technol Lett 28(5):581–584. https://doi.org/10.1109/LPT.2015.2506480

185. Jiang BL, Yang KT, Wang JG, Wu YM (2011) Design of Fiber-optic MEMS acoustic sensor system for gunship acoustic detecting. Appl Mech Mater 55:719–722. https://doi.org/10.4028/www.scientific.net/AMM.55-57.719
186. Lu X, Wu Y, Gong Y, Rao Y (2018) A miniature fiber-optic microphone based on an annular corrugated MEMS diaphragm. J Lightwave Technol 36(22):5224–5229. https://doi.org/10.1109/JLT.2018.2868964
187. Karaca EM (2019) Design and development of fiber optic MEMS microphone measurement system. Electrical and electronics engineering, middle east technical university, Thesis, graduate school of natural and applied sciences
188. Dagang G, Po SNC, Hock FTE, Rongming L (2006) Design and optimization of dual optical fiber MEMS pressure sensor for biomedical applications. J Phys: Conf Ser 34(1):177. https://doi.org/10.1088/1742-6596/34/1/177
189. Ge Y, Wang M, Rong H, Chen X (2007) The design and fabrication of an optical fiber MEMS pressure sensor. Proc SPIE 6595:659541. https://doi.org/10.1117/12.726498
190. Zheng Z-X, Huang Y-Q (2012) Diaphragm type optical fiber MEMS pressure sensors based on FP cavity interference. Acta Photonica Sin 41(12):1488
191. Jiang X-F, Lin C, Xie H-H, Yan H-P, Zhang JC, Huang Y-Q (2014) Optic fiber MEMS pressure sensor based on white light interferometry. Acta Photonica Sin 43(10):1006003
192. Wang Y, Ni X, Wang M, Cui Y, Shi Q (2017) Demodulation of an optical fiber MEMS pressure sensor based on single bandpass microwave photonic filter. Opt Express 25(2):644–653. https://doi.org/10.1364/oe.25.000644
193. Qi X, Wang S, Jiang J, Liu K, Wang X, Yang Y, Liu T (2018) Fiber optic Fabry-Perot pressure sensor with embedded MEMS micro-cavity for ultra-high pressure detection. J Lightwave Technol 37(11):2719–2725. https://doi.org/10.1109/JLT.2018.2876717
194. Tian B, Li K, Liu J, Zhao N, Lin Q, Zhao Y, Jiang Z (2020) Eccentric reflective optical fiber MEMS micro-pressure sensor. J Micromech Microeng 30(8):085010. https://doi.org/10.1088/1361-6439/ab931a
195. Wang X, Wang S, Jiang J, Liu K, Xiao M, Chen X, Zhang D, Liu T (2018) An MEMS optical fiber pressure sensor fabricated by Au-Au thermal-compression bonding. Proc SPIE 10618:106180J. https://doi.org/10.1117/12.2294971
196. Liu J, Jia P, Feng F, An G, Liang T, Hong Y, Xiong J (2020) MgO single crystals MEMS-based fiber-optic Fabry-Perot pressure sensor for harsh monitoring. IEEE Sens J 21(4):4272–4279. https://doi.org/10.1109/JSEN.2020.3029152
197. Li XL, Wang JG, Wu YM (2013) Design of Optical fiber MEMS displacement sensor. Appl Mech Mater 312:766–770. https://doi.org/10.4028/www.scientific.net/AMM.312.766
198. Kim YS, Dagalakis NG, Choi YM (2018) Optical fiber Fabry-Pérot micro-displacement sensor for MEMS in-plane motion stage. Microelectron Eng 187:6–13. https://doi.org/10.1016/j.mee.2017.11.010
199. Orłowska K, Świątkowski M, Kunicki P, Kopiec D, Gotszalk T (2016) High-resolution and wide-bandwidth light intensity fiber optic displacement sensor for MEMS metrology. Appl Opt 55(22):5960–5966. https://doi.org/10.1364/ao.55.005960
200. Kim MG, Jo KW, Park YS, Jang WG, Lee JH (2009) Accuracy enhancement in the in-plane dynamic measurement of MEMS actuators using a laser Doppler vibrometer with a 45°-angled optical fiber. In: 2009 IEEE 22nd international conference on micro electro mechanical systems. IEEE, pp 963–966. https://doi.org/10.1109/MEMSYS.2009.4805545
201. Kim MG, Jo K, Park Y, Jang W, Lee JH (2007) Laser doppler vibrometer using a 45°-angled optical fiber for in-plane dynamic measurement of MEMS actuators. In: 2007 IEEE/LEOS international conference on optical MEMS and nanophotonics. IEEE, pp 63–64. https://doi.org/10.1109/OMEMS.2007.4373841
202. Nainita JJ (2014) MEMS based micro heaters in fiber tapering. IOSR J Electr Com Eng 9(4):64–67
203. Unamuno A, Yao J, Uttamchandani D (2005) Alignment and fixing of fiber optics based on electrothermal MEMS actuators. IEEE Photonics Technol Lett 17(4):816–818. https://doi.org/10.1109/LPT.2005.844006

204. Zhang X, Duan C, Liu L, Li X, Xie H (2015) A non-resonant fiber scanner based on an electrothermally-actuated MEMS stage. Sens Actuators, A 233:239–245. https://doi.org/10.1016/j.sna.2015.07.001
205. Seo Y-H, Jung K-H, Hwang K-M (2015) Two-dimensional electrothermal MEMS fiber scanner for forward-viewing endomicroscopic applications. Proc KMEMS 2015, Soc Micro-nano Syst
206. Seo YH, Park HC, Jeong KH (2016) Electrothermal MEMS fiber scanner with lissajous patterns for endomicroscopic applications. In: 2016 IEEE 29th international conference on micro electro mechanical systems (MEMS). IEEE, pp 367–370. https://doi.org/10.1109/MEMSYS.2016.7421637
207. Islam MS, Challa S, Yassin MH, Vankayala SS, Beharic J, Harnett CK (2022) MEMS bimorph fiber-gripping actuators. In: 2022 international conference on manipulation, automation and robotics at small scales (MARSS). IEEE, pp 1–6. https://doi.org/10.1109/MARSS55884.2022.9870247
208. Kochervinskii VV (2021) Application of ferroelectric polymers in engineering and medicine: preprint of a comprehensive bibliographic review, 2nd edn (ISBN: 6202395699). Palmarium Academic Publishing, Chişinău
209. Yin J, Liu T, Jiang J, Liu K, Wang S, Zhao B, Xue L, Mei Y, Pu Y, Yin J, Qin Z (2013) Investigation of temperature characteristic of MEMS-based optical fiber pressure sensor. Proc SPIE 9044:90440B. https://doi.org/10.1117/12.2036262
210. Miller MB, Meller SA, Wavering TA, Greene JA, Murphy KA (1998) Optical-fiber-interconnected MEMS sensors and actuators. Proc SPIE 3330:174–179. https://doi.org/10.1117/12.316971
211. Zickar M (2006) MEMS based optical cross connects for fiber optical communication. Dissertation, Université de Neuchâtel
212. Kim MG, Jo K, Kwon HS, Jang W, Park Y, Lee JH (2009) Fiber-optic laser doppler vibrometer to dynamically measure MEMS actuator with in-plane motion. J Microelectromech Syst 18(6):1365–1370. https://doi.org/10.1109/JMEMS.2009.2031698
213. Kim MS, Jo KW, Lee JH (2004) Dimensional controllability of the self-aligned microlens on the sidewall of optical fiber for compact SNOM optics. In: 3rd international conference on optical internet (Kanagawa, Japan). Institute of Electronics, Information and Communication Engineers, Kanagawa, Japan, pp 908–909. https://doi.org/10.3316/informit.200347411784753
214. Nobuyoshi T, Koga R, Toyota Y (2004) Standing wave ratio measurement by SNOM on the optical fiber interconnection. IEICE technical report 104(412):23–26
215. Kitahama Y, Itoh T, Aoyama JI, Nishikata K, Ozaki Y (2009) SERRS fiber probe: fabrication of silver nanoparticles at the aperture of an optical fiber used for SNOM. Chem Commun 43:6563–6565. https://doi.org/10.1039/b909603e
216. Smajic J, Hafner C (2011) Numerical analysis of a SNOM tip based on a partially cladded optical fiber. Opt Express 19(23):23140–23152. https://doi.org/10.1364/oe.19.023140
217. El Eter A, Hameed NM, Baida FI, Salut R, Filiatre C, Nedeljkovic D, Atie E, Bole S, Grosjean T (2014) Fiber-integrated optical nano-tweezer based on a bowtie-aperture nano-antenna at the apex of a SNOM tip. Opt Express 22(8):10072–10080. https://doi.org/10.1364/oe.22.010072
218. Wang C, Zaouk R, Malladi K, Taherabadi L, Madou M (2004) C-MEMS/NEMS: a novel technology for nanoscale material formation from graphite fiber to NI and SI nanowires. Proc ASME 41774:133–134. https://doi.org/10.1115/NANO2004-46043
219. Zheng BC, Xu F (2015) A compact fiber magnetic sensor based on graphene NEMS. In: Asia communications and photonics conference. Optical Society of America, p. AM4A-3
220. Basher S, Sun H, Hirai Y, Sakai K, Yoshida K, Furukawa H, Sukumaran SK, Yoshida T, Khosla A (2017) Ultrasonically assisted preparation of carbon fiber doped electriclly conductive micropatternable nanocomposite polymer for MEMS/Nems applications. ECS Meet Abstr 49(MA2017-02):2114. https://doi.org/10.1149/MA2017-02/49/2114
221. Liu ZY, Xu F (2017) Miniature sensor based on fiber-graphene-integrated NEMS. In: 2017 16th international conference on optical communications and networks (ICOCN). IEEE, pp 1–2. https://doi.org/10.1109/ICOCN.2017.8121538

222. Liu ZY, Yan SC, Xu F (2018) A fiber-optic magnetometer based on graphene NEMS using superparamagnetic nanoparticles. In: Conference on lasers and electro-optics/pacific Rim. Optical Society of America, p W3A-76
223. Damián MEL (2011) Vertically aligned carbon nanofiber synthesis on top of TiN films for NEMS devices. Master's thesis, Department of Microtechnology and Nanoscience, Chalmers University of Technology (Göteborg, Sweden)
224. Shi F, Guifu D, Yuchao W, Huiqing W, Hong W (2008) Fabrication of Cu-based carbon nanofiber composite film applied in MEMS contactor. In: 2008 3rd IEEE international conference on nano/micro engineered and molecular systems. IEEE, pp 27–30. https://doi.org/10.1109/ICOCN.2017.8121538
225. Imai S (2019) Thin-film diaphragms of cellulose nanofiber fabricated using high-concentration polar dispersion for application to MEMS actuators. Sens Actuators, A 289:134–143. https://doi.org/10.1016/j.sna.2019.01.007
226. Masouleh MM, Koohsorkhi J, Moghadam RA (2020) Direct writing of individual micro/nanofiber patterns suitable for flexible electronics using MEMS-based microneedle. Microelectron Eng 229:111345. https://doi.org/10.1016/j.mee.2020.111345
227. Koudehi MF, Pourmortazavi SM, Zibaseresht R, Mirsadeghi S (2021) MEMS-based PVA/PPy/MIP polymeric-nanofiber sensor fabricated by LIFT-OFF process for detection 2, 4-dinitrotoluene vapor. IEEE Sens J 21(7):9492–9499. https://doi.org/10.1109/JSEN.2021.3055797
228. Malygin GA (2012) Surface size effects under plastic deformation of microcrystals and nanocrystals. Phys Solid State 54:1606–1611. https://doi.org/10.1134/S1063783412080197
229. Trap D (2019) A proof of concept for a hair cell inspired carbon nanofiber based MEMS flow sensor. Dissertation, University of Groningen
230. Demir MM, Naseer M, Bechteler TF, Gurbuz Y, Menceloglu YZ (2003) Polyurethane nanofiber webs for sensor and actuator applications in microelectromechanical systems (MEMS). MRS Online Proc Libr Arch 782:529. https://doi.org/10.1557/PROC-782-A5.29
231. Deng M, Ding G, Wang Y, Wang Y, Wang H, Fu S (2009) MEMS-based carbon nanotube and carbon nanofiber Cu micro special electric contact. J Micromech Microeng 19(6):065001. https://doi.org/10.1088/0960-1317/19/6/065001
232. Wang YC, Ding GF, Wu HQ, Wang H, Yao JY, Yang CS (2007) MEMS-based carbon nanotube and carbon nanofiber copper micro special electric contact. Chin J Electron Devices 5:1542–1546
233. Fazio B (1988) Electric stimulation and neuromuscular diseases: effects on denervated and dystrophic muscle. Minerva Medica 79(3):205–208.
234. Pelizzari C, Mazzanti A, Raiser AG, Lopes ST, Graça DL, Ramos AT, Salbego FZ, Festugatto R, Beckmann DV, Souza LB, Cunha MG (2008) Neuromuscular electric stimulation in dogs with induced muscle atrophy. Arquivo Brasileiro de Medicina Veterinaria e Zootecnia 60:76–82. https://doi.org/10.1590/S0102-09352008000100012
235. Liu MC, Dai CL, Chan CH, Wu CC (2009) Manufacture of a polyaniline nanofiber ammonia sensor integrated with a readout circuit using the CMOS-MEMS technique. Sensors 9(2):869–880. https://doi.org/10.3390/s90200869
236. Tong S, Wang T, Zhang W (2008) Fault tolerant control for uncertain fuzzy systems with actuator failures. Int J Innov Comput, Inf Control 4(10):2461–2474
237. Lyu Z, Liu Z, Xie K, Chen CP, Zhang Y (2020) Adaptive tracking control for switched nonlinear systems with fuzzy actuator backlash. Fuzzy Sets Syst 385:60–80. https://doi.org/10.1016/j.fss.2019.02.008
238. Ruan Z, Yang Q (2020) Adaptive fuzzy output-constrained control of uncertain MISO nonlinear systems with actuator faults. IFAC 53(2):13739–13744. https://doi.org/10.1016/j.ifacol.2020.12.879
239. Boulouma S, Labiod S, Boubertakh H (2021) Adaptive fuzzy actuator failure compensation control for a class of MIMO nonlinear systems. Alger J Signals Syst 6(2):112–121. https://doi.org/10.51485/ajss.v6i2.128

240. Bounemeur A, Chemachema M (2021) Adaptive fuzzy fault-tolerant control using Nussbaum-type function with state-dependent actuator failures. Neural Comput Appl 33:191–208. https://doi.org/10.1007/s00521-020-04977-6
241. Evstifeev MI, Rozentsvein DV, Chelpanov IB (2010) Strength analysis of MEMS gyro elastic suspensions. Gyroscopy Navig 1(4):263–271. https://doi.org/10.1134/S207510871004005X
242. Dzhashitov VE, Pankratov VM (2011) Elastic suspension of a superminiature MEMS gyro. Gyroscopy Navig 2(4):239–246. https://doi.org/10.1134/S2075108711040079
243. Yuan T, Chen D, Chen J, Chen X, Wang X, Lu B (2014) A novel MEMS elastic-based dry electrode for electroencephalography measurement. Microsyst Technol 20(6):1125–1129. https://doi.org/10.1007/s00542-013-1799-7
244. Sun Y, Liu XY, Wang WH, Lansdorp BM (2006) An elastic MEMS device for cellular force measurement. In: Tech Dig 12th solid-state sensor, actuator and microsystems workshop. IEEE, pp 412–413. https://doi.org/10.31438/trf.hh2006.106
245. Bardaweel HK (2013) Tunable elastic fluidic resonant MEMS-type actuator. Eur Phys J Appl Phys 62(3):30901. https://doi.org/10.1051/epjap/2013120383
246. Truxal SC (2009) Elastic tunable grating powered by MEMS actuators for fluorescence spectroscopy applications. Dissertation, The University of Michigan
247. Yang Z, Ding G, Cai H, Zhao X (2008) A MEMS inertia switch with bridge-type elastic fixed electrode for long duration contact. IEEE Trans Electron Devices 55(9):2492–2497. https://doi.org/10.1109/TED.2008.927385
248. Li G, Shi G, Sui L, Yi F, Wang B (2015) The application of multilayer elastic beam in MEMS safe and arming system. AIP Adv 5(7):077151. https://doi.org/10.1063/1.4927495
249. Zega V, Langfelder G, Falorni LG, Comi C (2019) Hardening, softening, and linear behavior of elastic beams in MEMS: an analytical approach. J Microelectromech Syst 28(2):189–198. https://doi.org/10.1109/JMEMS.2019.2892296
250. Qin Y, Chen L, Hao Y, Liu S, Zou X (2020) A study on the elastic coefficients of setback micro-springs for a MEMS safety and arming device. Microsyst Technol 26(2):583–593. https://doi.org/10.1007/s00542-019-04558-1
251. Huang S, Zhang X (2007) Gradient residual stress induced elastic deformation of multilayer MEMS structures. Sens Actuators, A 134(1):177–185. https://doi.org/10.1016/j.sna.2006.05.026
252. Las HR, Hemker KJ, Witt R (1999) MEMS material microstructure and elastic property modeling. MRS Online Proc Libr 605(1):191–196. https://doi.org/10.1557/PROC-605-191
253. Ricote J, Algueró M, Chateigner D (2003) Tailoring of the elastic properties by texture control in ferroelectric thin films for MEMS. Mater Sci Forum 426(4):3433–3438. https://doi.org/10.4028/www.scientific.net/MSF.426-432.3433
254. Cho SW, Chasiotis I (2007) Elastic properties and representative volume element of polycrystalline silicon for MEMS. Exp Mech 47(1):37–49. https://doi.org/10.1007/s11340-006-0405-7
255. Ongwen NO, Ogam E, Otunga HO (2021) Ab initio study of elastic properties of orthorhombic cadmium stannate as a substrate for the manufacture of MEMS devices. Mater Today Commun 26:101822. https://doi.org/10.1016/j.mtcomm.2020.101822
256. Jing X, Chen D, Huang C, Chen X, Miao J, Liu J, Zhu J (2010) Elastic MEMS probe card based on the PDMS substrate. J Micromech Microeng 20(5):055038. https://doi.org/10.1088/0960-1317/20/5/055038
257. Ruhhammer J, Zens M, Goldschmidtboeing F, Seifert A, Woias P (2015) Highly elastic conductive polymeric MEMS. Sci Technol Adv Mater 16(1):015003. https://doi.org/10.1088/1468-6996/16/1/015003
258. Tao K, Lye SW, Miao J, Tang L, Hu X (2015) Out-of-plane electret-based MEMS energy harvester with the combined nonlinear effect from electrostatic force and a mechanical elastic stopper. J Micromech Microeng 25(10):104014. https://doi.org/10.1088/0960-1317/25/10/104014
259. Tas NR, Gui C, Elwenspoek M (2003) Static friction in elastic adhesion contacts in MEMS. J Adhes Sci Technol 17(4):547–561. https://doi.org/10.1163/15685610360554401

260. Braghin F, Leo E, Resta F (2008) Visco-elastic modelling of MEMS inertial sensors. Sens Lett 6(1):9–22. https://doi.org/10.1166/sl.2008.009
261. Ma Z, Zhao H, Du X, Zhou M, Ma X, Liu C, Ren L (2018) Evaluation of nanoindentation load-depth curve of MEMS bridge structures by calculating the critical elastic-plastic bending deflections. Appl Surf Sci 434:1–10. https://doi.org/10.1016/j.apsusc.2017.10.106
262. Somà A, Saleem MM, Margesin B, Armando M (2019) Pull-in tests of MEMS specimens for characterization of elastic–plastic behavior. Microsyst Technol 25(7):2525–2533. https://doi.org/10.1007/s00542-019-04396-1
263. Profunser DM, Vollmann J, Bryner J, Dual J (2002) Measurement and simulation of the laser-based thermo-elastic excitation and propagation of acoustic pulses for thin film and MEMS inspection. Proc SPIE 4703:21–30. https://doi.org/10.1117/12.469627
264. Barauskas R, Kausinis S, Tilmans HAC (2009) Investigation of thermo-elastic damping of vibrations of rectangular and ring-shaped MEMS resonators. J Vibroengineering 11(1):177–187
265. Jaakkola A, Prunnila M, Pensala T, Dekker J, Pekko P (2014) Determination of doping and temperature-dependent elastic constants of degenerately doped silicon from MEMS resonators. IEEE Trans Ultrason Ferroelectr Freq Control 61(7):1063–1074. https://doi.org/10.1109/TUFFC.2014.3007
266. Sorasen O, Vogl A, Chen X, Furuberg L, Johansen KM (2004) 3D modelling of MEMS components for coupled elastic-electrostatic characterization. In: 2004 international conference on MEMS, NANO and smart systems (ICMENS'04). IEEE, pp 509–514. https://doi.org/10.1109/ICMENS.2004.1509003
267. Fumagalli A, Quaranta G, Mantegazza P (2006) MEMS integrated electro-fluid-elastic modeling for aerospace applications. Proc ASME 42541:179–188. https://doi.org/10.1115/CANEUS2006-11037
268. Fumagalli A, Quaranta G, Mantegazza P (2007) A matrix-free solution to the electro-elastic problem for MEMS modelling and simulation. In: International conference on computational methods for coupled problems in science and engineering. CIMNE, pp 1–4
269. Lup AS, Ciuprina G, Ştefănescu A (2014) Extraction of effective elastic coefficients from a coupled structural electrostatic simulation of a MEMS switch. In: 2014 international symposium on fundamentals of electrical engineering (ISFEE). IEEE, pp 1–5. https://doi.org/10.1109/ISFEE.2014.7050539
270. Bedendo A, Reimbold M, Sausen A (2014) Evaluation of model ARX for elastic masses MEMS using the indexes RMSE, AIC, and BIC. J Control, Autom Electr Syst 25(2):195–205. https://doi.org/10.1007/s40313-013-0103-5
271. Reimbold MM, Chong W, Leandro GV, Ribas RP (2008) ARX model of elastic nuclei for rotors MEMS. ECS Trans 14(1):83. https://doi.org/10.1149/1.2956020
272. Cho S, Cárdenas-García JF, Chasiotis I (2005) Measurement of nanodisplacements and elastic properties of MEMS via the microscopic hole method. Sens Actuators, A 120(1):163–171. https://doi.org/10.1016/j.sna.2004.11.028
273. Sun C, Zhou ZF, Li WH, Huang QA (2013) In situ measurement of elastic modulus of individual layers for composite thin films by MEMS test structures. Appl Mech Mater 364:248–252. https://doi.org/10.4028/www.scientific.net/AMM.364.248
274. White CD, Xu R, Sun X, Komvopoulos K (2003) Dynamic MEMS devices for multiaxial fatigue and elastic modulus measurement. Proc SPIE 4980:63–74. https://doi.org/10.1117/12.476332
275. Sun C, Zhou ZF, Li WH, Huang QA (2012) Measurement of elastic modulus and residual stress of individual layers for composite films by resonant frequency of MEMS structures. In: Proceedings of the sensors. IEEE, pp 1–4. https://doi.org/10.1109/ICSENS.2012.6411372
276. Alves GMA, Mansano RD (2016) Use of protek-PSB as structural material for MEMS: measurement of elastic modulus. In: 2016 31st symposium on microelectronics technology and devices (SBMicro). IEEE, pp 1–4. https://doi.org/10.1109/SBMicro.2016.7731359
277. Lin JT, Shuvra PD, Liao W, McNamara S, Walsh KM, Arutt CN, Gong H, Davison JL, Alles ML, Alphenaar BW (2017) Surface carrier concentration effect on elastic modulus

of piezoelectric MEMS silicon cantilevers. In: Proceedings of the 2017 19th international conference on solid-state sensors, actuators and microsystems (TRANSDUCERS). IEEE, pp 1175–1178. https://doi.org/10.1109/TRANSDUCERS.2017.7994263
278. Beni YT, Koochi A, Abadyan M (2011) Theoretical study of the effect of Casimir force, elastic boundary conditions and size dependency on the pull-in instability of beam-type NEMS. Phys E 43(4):979–988. https://doi.org/10.1016/j.physe.2010.11.033
279. Abadyan MR, Beni YT, Noghrehabadi A (2011) Investigation of elastic boundary condition on the pull-in instability of beam-type NEMS under van der Waals attraction. Procedia Eng 10:1724–1729. https://doi.org/10.1016/j.proeng.2011.04.287
280. Magrab EB (2012) Vibrations of elastic systems: with applications to MEMS and NEMS. Solid Mech Its Appl (SMIA) 184:1–489. https://doi.org/10.1007/978-94-007-2672-7
281. Langari A, Yekrangi A, Soroush R, Fakhraei M, Hafshejani MK (2014) Numerical study of size-dependent instability of nems considering molecular force and elastic support conditions. ARPN J Eng Appl Sci 9(3):382–387
282. Fei J, Zhou J (2012) Robust adaptive control of MEMS triaxial gyroscope using fuzzy compensator. IEEE Trans Syst, Man, Cybern, Part B 42(6):1599–1607. https://doi.org/10.1109/TSMCB.2012.2196039
283. Fazlyab M, Pedram MZ, Salarieh H, Alasty A (2013) Parameter estimation and interval type-2 fuzzy sliding mode control of a z-axis MEMS gyroscope. ISA Trans 52(6):900–911. https://doi.org/10.1016/j.isatra.2013.05.004
284. Fei J, Yan W (2014) Adaptive control of MEMS gyroscope using global fast terminal sliding mode control and fuzzy-neural-network. Nonlinear Dyn 78:103–116. https://doi.org/10.1007/s11071-014-1424-z
285. Asad YP, Shamsi A, Tavoosi J (2017) Backstepping-based recurrent type-2 fuzzy sliding mode control for MIMO systems (MEMS triaxial gyroscope case study). Internat J Uncertain Fuzziness Knowl-Based Syst 25:213–233. https://doi.org/10.1142/S0218488517500088
286. Vu HS, Nguyen QD, Huang SC (2021) Robust observer based on fixed-time sliding mode control of position/velocity for a TS fuzzy MEMS gyroscope. IEEE Access 9:96390–96403. https://doi.org/10.1109/ACCESS.2021.3095465
287. Vafaie RH, Mohammadzadeh A, Piran MJ (2021) A new type-3 fuzzy predictive controller for MEMS gyroscopes. Nonlinear Dyn 106(1):381–403. https://doi.org/10.1007/s11071-021-06830-4
288. Shao X, Si H, Zhang W (2021) Fuzzy wavelet neural control with improved prescribed performance for MEMS gyroscope subject to input quantization. Fuzzy Sets Syst 411:136–154. https://doi.org/10.1016/j.fss.2020.08.005
289. Wang Z, Fei J (2021) Fractional-order terminal sliding-mode control using self-evolving recurrent Chebyshev fuzzy neural network for MEMS gyroscope. IEEE Trans Fuzzy Syst 30(7):2747–2758. https://doi.org/10.1109/TFUZZ.2021.3094717
290. Wang X, Wu S, Yin J, Moradi Z, Safa M, Khadimallah MA (2023) On the electromechanical energy absorption of the reinforced composites piezoelectric MEMS via adaptive neuro-fuzzy inference system and MCS theory. Compos Struct 303:116246. https://doi.org/10.1016/j.compstruct.2022.116246
291. Zhang L, Liu J, Lai J, Xiong Z (2014) Performance analysis of adaptive neuro fuzzy inference system control for MEMS navigation system. Math Probl Eng 2014:961067. https://doi.org/10.1155/2014/961067
292. El-Sheimy N, Abdel-Hamid W, Lachapelle G (2004) An adaptive neuro-fuzzy model for bridging GPS outages in MEMS-IMU/GPS land vehicle navigation. In: Proceedings of the 17th international technical meeting of the satellite division of the institute of navigation. Institute of Navigation, pp 1088–1095
293. Abdolkarimi ES, Mosavi MR (2023) A modified neuro-fuzzy system for accuracy improvement of low-cost MEMS-based INS/GPS navigation system. Wireless Pers Commun 129(2):1369–1392. https://doi.org/10.1007/s11277-023-10194-w
294. El-Shafie A, Noureldin A, Bayoumi M (2007) An augmented wavelet-neuro-fuzzy module for enhancing MEMS based navigation systems. In: 2007 IEEE international conference on

signal processing and communications 2007. IEEE, pp 1339–1342. https://doi.org/10.1109/ICSPC.2007.4728575
295. Varadan VK, Varadan VV, Motojima S (1996) Three-dimensional polymeric and ceramic MEMS and their applications. Proc SPIE 2722:156–164. https://doi.org/10.1117/12.240440
296. Hossfeld J, Paatzsch T, Schulze J, Neumeier M, Weber L, Bauer HD, Ehrfeld W (1999) Polymeric optical MEMS. Proc SPIE 3680:637–645. https://doi.org/10.1117/12.341256
297. Ruben KA, Flaim TD, Li C (2003) Polymeric protective coatings for MEMS wet-etch processes. Proc SPIE 5342:212–220. https://doi.org/10.1117/12.523965
298. Su YC, Lin L (2005) Localized bonding processes for assembly and packaging of polymeric MEMS. IEEE Trans Adv Packag 28(4):635–642. https://doi.org/10.1109/TADVP.2005.858333
299. Bachmann D, Kuhne S, Hierold C (2005) Fabrication process of polymeric springs for MEMS applications. In: The 13th international conference on solid-state sensors, actuators and microsystems, 2005 (TRANSDUCERS'05). IEEE, pp 1416–1419. https://doi.org/10.1109/SENSOR.2005.1497347
300. Zhong ZW, Wang ZF, Tan YH (2006) Chemical mechanical polishing of polymeric materials for MEMS applications. Microelectron J 37(4):295–301. https://doi.org/10.1016/j.mejo.2005.05.016
301. Lang U, Rust P, Dual J (2008) Towards fully polymeric MEMS: fabrication and testing of PEDOT/PSS strain gauges. Microelectron Eng 85(5–6):1050–1053. https://doi.org/10.1016/j.mee.2008.01.051
302. Takahashi Y, Teh KS, Lu YW (2008) Fabrication and characterization of nano-structuring polymeric surfaces for MEMS applications. Proc SPIE 48746:209–213. https://doi.org/10.1115/IMECE2008-68893
303. Ozaydin-Ince G, Coclite AM, Gleason KK (2011) CVD of polymeric thin films: applications in sensors, biotechnology, microelectronics/organic electronics, microfluidics, MEMS, composites and membranes. Rep Prog Phys 75(1):016501. https://doi.org/10.1088/0034-4885/75/1/016501
304. Basu AK, Basu A (2020) Greyscale lithography-based fabrication of thin polymeric cantilevers for MEMS/NEMS applications. CSVTU Int J Biotechnol, Bioinform Biomed 5(2):40–44
305. Li KH, Liu LH, Wang PC (2013) Fabrication of robust polyaniline thin film electrodes on the surface of elastomeric poly (dimethylsiloxane) for polymeric MEMS applications. In: 2013 8th international microsystems, packaging, assembly and circuits technology conference (IMPACT). IEEE, pp 330–333. https://doi.org/10.1109/IMPACT.2013.6706680
306. Pelegrini MV, Pereyra I (2009) Study of a-Si1-XCX: H thin films obtained by PECVD in temperatures lower than 250 ºC aiming applications in optics, thin films devices on polymeric substrate and MEMS. ECS Trans 23(1):127. https://doi.org/10.1149/1.3183710
307. Elias AL, Harris KD, Bastiaansen CW, Broer DJ, Brett MJ (2005) Polymeric helices with submicron dimensions for MEMS devices. Proc SPIE 5836:41–55. https://doi.org/10.1117/12.608584
308. Diemeer MBJ, Dekker R (2002) MEMS VOA with polymeric thermal microactuators. In: 28 European conference on optical communication. IEEE, pp 1–2
309. Lago H, Zakaria Z, Jamlos MF, Soh PJ (2019) A wideband reconfigurable folded planar dipole using MEMS and hybrid polymeric substrates. AEU-Int J Electron Commun 99:347–353. https://doi.org/10.1016/j.aeue.2018.12.011
310. Rathna K, Haji-Saeed B, Pyburn D, Leon R, Sengupta SK, Testorf M, Goodhue W, Khoury J, Drehman A, Woods CL, Kierstead J (2003) Photoconductive optically driven MEMS spatial light modulator utilizing a GaAs substrate and polymeric membrane mirror. In: Frontiers in optics. Optical Society of America, p TuV5
311. Bachmann D, Kuhne S, Hierold C (2007) MEMS scanning mirror supported by soft polymeric springs and actuated by electrostatic charge separation. In: 2007 IEEE 20th international conference on micro electro mechanical systems (MEMS). IEEE, pp 723–726. https://doi.org/10.1109/MEMSYS.2007.4433088

312. Gradov OV, Blinov ND, Shibryaeva LS (2022) PHB-based MEMS and biomimetic MEMS based on biodegradable polymer and composite fibers with λ-selective photoresponces. Preprint IBCP RAS
313. Colinjivadi KS, Cui Y, Ellis M, Skidmore G, Lee JB (2008) De-tethering of high aspect ratio metallic and polymeric MEMS/NEMS parts for the direct pick-and-place assembly of 3D microsystem. Microsyst Technol 14(9):1621–1626. https://doi.org/10.1007/s00542-008-0579-2
314. Colinjivadi KS, Cui Y, Ellis M, Lee JB, Skidmore G (2006) De-tethering of metallic and polymeric MEMS/NEMS parts for the direct pick-and-place assembly of 3D microsystem. Proc ASME 47756:193–198. https://doi.org/10.1115/IMECE2006-14818
315. Li D, Tumkor S, Manoochehri S, Pochiraju K (2009) Fabrication of polymeric and polymeric nano-composite MEMS structures. Proc ASME 43857:309–313. https://doi.org/10.1115/IMECE2009-12460
316. Birkelund K, Nørgaard L, Thomsen EV (2011) Enhanced polymeric encapsulation for MEMS based multi sensors for fisheries research. Sens Actuators, A 170(1–2):196–201. https://doi.org/10.1016/j.sna.2011.06.008
317. Kim K, Liu X, Zhang Y, Cheng J, Wu XY, Sun Y (2008) Mechanical characterization of polymeric microcapsules using a force-feedback MEMS microgripper. In: 2008 30th annual international conference of the IEEE engineering in medicine and biology society. IEEE, pp 1845–1848. https://doi.org/10.1109/IEMBS.2008.4649539
318. Kuroiwa T, Ishikawa N, Obara D, Vinet F, Ang ES, Guelbi A, Soucemarianadin A (2003) Dispensing of polymeric fluids for BIO-MEMS applications. NIP & DFC (Soc Imaging Sci Technol) 19(1):884–890. https://doi.org/10.2352/ISSN.2169-4451.2003.19.1.art00104_2
319. Almog RO, Sverdlov Y, Fishelson N, Shmilovich T, Rabinovich E, Shacham-Diamand YY (2010) Electrochemical micro technologies for polymeric MEMS and biochip applications. ECS Trans 25(16):17–22. https://doi.org/10.1149/1.3300058
320. Bakke T, Sullivan CT (2003) Integration of polymeric optical waveguides with SOI MEMS structures. Proc SPIE 4997:109–122. https://doi.org/10.1117/12.479462
321. Zhang B, Kahn MTE (2006) Overview and improving fiber optic gyroscope based on MEMS/NEMS fabrication. J Phys: Conf Ser 34:025. https://doi.org/10.1088/1742-6596/34/1/025
322. Goodwin S, Carlson J, Rogers S, Kim J, Kim C, Brady D, Stoner BR (2007) Dynamic aperture optical arrays based on polymeric MEMS actuators for large scale coding elements with application in visible to MWIR. Proc SPIE 6714:67140C. https://doi.org/10.1117/12.736074
323. Jang C, Goswami A, Han B, Ham SJ (2009) In situ measurement of gas diffusion properties of polymeric seals used in MEMS packages by optical gas leak testing. J Micro/Nanolithogr MEMS MOEMS 8(4):043025. https://doi.org/10.1117/1.3227904
324. Kullberg RC, Phillip BL (2010) Characterization of polymeric getter materials for MEMS/MOEMS and other microelectronic package service. Proc SPIE 7592:759206. https://doi.org/10.1117/12.843859
325. Chasiotis I, Sottos N (2007) A novel MEMS platform for mechanical testing of polymeric nanofibers (Final report: 1 Aug 2006–30 Apr 2007; NTIS Issue Number: 200822; Contract Number: W911NF-06-1-0356). Illinois University at Urbana-Champaign. Beckman Institute for Advanced Sciences and Technology. https://ntrl.ntis.gov/NTRL/dashboard/searchResults/titleDetail/ADA482977.xhtml
326. Singh RA, Yoon ES, Kim HJ, Kong H, Jeong HE, Suh KY (2005) Submicron-scale polymeric patterns for tribological application in MEMS/NEMS. KTS Int J 6(2):33–38
327. Gradov OV, Gradova MA (2016) Methods of electron microscopy of biological and abiogenic structures in artificial gas atmospheres. Surf Eng Appl Electrochem 52(1):117–125. https://doi.org/10.3103/S1068375516010063
328. Gradov OV, Gradova MA (2014) Cryoelectron microscopy as a functional instrument for system biology, structural analysis and experimental manipulations with living cells. Probl Cryobiol Cryomedicine 24(3):193–210. https://doi.org/10.15407/cryo24.03.193

329. Gradov OV (2023) Cryoconveyor protocols in correlative light and electron microscopy: from multilevel imaging to modeling of biophysical effects and "cryotheranostics". Vest MSU Seriya 16. Biologiya 78(3S):57–62. (In Russia) https://doi.org/10.55959/MSU0137-0952-16-78-3S-10

Chapter 15
Time-Resolved Correlation MicroPorometry (TRCMP) and Melting Point Correlation-Spectral MicroPorometry (MPCSMP) of Gas-Filled Polymer Composites Formed Under the Action of Various Pore-Forming Agents

Elena Grigorieva, Oleg Gradov, Irina Maklakova, Anatoly Popov, and Anna Ratnovskaya

Abstract This review material discusses the most modern approaches to (micro) porometry of gas-filled composites for civil engineering, ecology, and hydraulics. It is proposed to consider some of these materials as smart materials. Trends in technology development are compared, as a result of which a conclusion is made about the transition from additive or average porosity analysis to the measurement of individual pores with their further morphometric processing and accumulation of their statistics to characterize the material. A new approach to microporometric analysis in real time is proposed, based on laser diffraction or diffraction emulation using a two-dimensional Fourier transform and some other mathematical operations in real time. We present images of the porometry of individual pores during the melting of the material, as well as their correlograms, which clearly change during the collapse of the pores, as with thermal extrusion and the use of porogens. It talks about the influence of a number of factors on the geometry and topology, and not just on the size of the pores formed with their use.

Keywords Porometry · Porosimetry · Gas-filled polymer composites · Porogenes · Laser porometry · Lens-less porometry · Dynamic correlation porometry · 2D FFT · Thermal heating stage · Melting point

E. Grigorieva · A. Popov · A. Ratnovskaya
Emanuel Institute of Biochemical Physics, Kosygina 4, Moscow 119334, Russia

O. Gradov (✉) · I. Maklakova
N.N. Semenov Institute of Chemical Physics, Kosygina 4, Moscow 119991, Russia
e-mail: o.v.gradov@gmail.com

H. Altenbach and V. A. Eremeyev (eds.), *Dynamics of Discrete and Continuum Structures and Media*, Advanced Structured Materials 221,
https://doi.org/10.1007/978-3-031-75626-9_15

15.1 Introduction: From Additive Porosimetry to MicroPorometry

The main problem of the research and development of gas-filled polymer composites for civil engineering and hydraulic engineering is to study the shape and size of pores formed under the action of various pore-forming agents (thresholds), since the main physical properties, that are the subject of interest for application of such gas-filled systems, are determined by the shape and size of pores. It should be noted that classification of such porous materials is primarily based on the nature of pores: in the most primitive classification, foams or foam plastics (closed-pore systems), and protoplasts (open-pore systems) are known, while in the more progressive classifications [1] there are «closed-pore», «open-pore», integral structure and syntactic foams (in early versions considered as «closed-pore» structures, but later separated into a specific type of structures, since the closed nature of their pores is determined by the introduction of hollow spheres into a low-viscosity substance during the molding process). The properties of porous polymers/composites determined by their place in such classifications include thermal conductivity (and, consequently, heat insulation and fire-preventing properties of the material), damping (vibration and sound isolation), and other parameters related to the reactivity (or resistance) of the polymer to the certain environmental factors. Therefore, the control of not only the size but also of the morphology of pores in such polymer systems is inevitably an obligatory condition for their successful application, if we consider the prospects for the development of smart materials based on them [2, 3].

However, the problem of an objective study of the size of complex pores for such systems has not yet been completely solved. Conventional porosimetry (or porometry) methods based on filling the pores with a substance (mercury and other liquid metals) provide only additive characteristics of porosity, but do not answer the question about the geometry and orientation of the pores. Any calculations are carried out for the pores of a single averaged geometry, which can be true only for syntactic foams, in which the shape of the pores is maintained by the hollow shells embedded into them. However, porosimetry of syntactic foams is not feasible within the framework of the standard approaches, due to the absence of through pores in these spheroplastics, which determines the inapplicability of gas-dynamic capillary flow porosimetry and mercury porometry to these materials, as well as makes no sense the application of the principles of liquid and gas pycnometry in the indirect estimation of the pore size by the effective density. Reducibility of the initial data of most of the above porosimetry principles to percolation has led to the formation of the term «pOrcolation» («percolation through pores»), which determines the applicability of invasive-percolation models in liquid porosimetry [4]. However, this abstraction does not solve the problem of objective pore size estimation due to the diversity of their physical-geometric models.

15.2 Model-Based Approach: From ad hoc Estimation of the Pore Geometry to Individual Pore Morphometric Measurements

The main source of problems related to the bias of simple additive porosimetry methods is the indirect nature and difference of physical models required for processing/interpreting the initial experimental data. The «nominal» models (Horvath-Kawazoe [5], Katz-Thompson [6, 7], van Genuchten [8], etc.) are not equivalent in geometry, physics, and algorithmics of calculations, and some of them (for example, the well-known Derjaguin–Broekhoff–De Boer model [9]) can be effectively applied only in the case of solving the inverse problem of a specific measurement method (in the last cited case—ellipsometric porometry, in which the target optical parameters of the sample surface are determined on the basis of some pre-selected model used in solving the inverse ellipsometry problem). The physical geometry of the model, which determines the correctness of calculations and their applicability to the specific pore morphology or to a specific class of samples, may be different. An example of models that are not sufficiently reducible to each other, is the lattice model [10], a set of hierarchical approaches [11] and a model solved on the basis of the method of Boltzmann lattice equations (including taking into account collisions using the Batnagar–Gross–Krook technique). When moving to the real morphology of interconnected pores, LBM-type models that do not provide solutions to the Navier–Stokes equations are no longer exhaustive in describing capillary pressure and saturation [12], which makes it necessary to take into account (and approximate) the real morphology of pores that are unequal in diameter in different sections. Thus, in mercury porosimetry, since the 1980s, methods have been developed for taking into account the actual morphology of pores in networks, including necks of pores [13], as well as the other bodies of revolution (approximated by the analytical surfaces, which can be found in the literature for effective approximation of the pore morphology [14]). In this regard, for correct porosimetry with effective approximation of the pore surface/correct determination of their volume delimited by one or another analytical surface, first of all, it is necessary to study the morphology of pores by morphometry using mathematical software that provides measurements of not only simple (length, width, perimeter) but also more complex calculated characteristics (ellipticity coefficients, roundness, and elongation; shell areas; maximum and minimum Feret diameters; tortuosity and convexity coefficients; compactness and filling coefficients; moments of inertia, etc.). There are publications of the last 10–15 years, corresponding to this trend, although incomplete in the formal data interpretation, in which the methods of porometry, porosimetry, modeling, and image analysis of single pores and their networks are integrated [15]. However, despite the obvious nature of the errors of the single additive methods, the introduction of the above-mentioned multiparametric approaches is very slow, and many newly emerging porometry methods, positioned as qualitatively new, are in fact based on the same principles as previously known methods (with the introduction of liquid into a medium porosity of which is analyzed [16]) [17].

15.3 From Simple Microscopy and Morphometric Measurements to Laser-Assisted Techniques Including 2D Diffraction-Based Ones: 2D FFT and Multimodal Correlation-Spectral Analysis

Speaking about the necessity for an individual analysis of the pore morphology during morphometry, we inevitably encounter the use of optical methods, as a rule, optimal in resolution for the analysis of small/medium pores—from submicron to decamicron sizes. An effective tool for obtaining a strictly coherent beam of a given wavelength in order to unambiguously compare it with the size of micro- and ultramicropores in the optical range is a laser, which is becoming publicly available due to the introduction of cheap technologies for manufacturing diode-pumped semiconductor lasers (DPSSL) [18]. The trend towards a combination of gas-adsorption spectroscopy (using wavelength-tunable diode lasers—TDLS) and optical or diffusion porosimetry [19, 20] has begun the convergence of optical spectral and optical «structurometric» methods. However, according to physical criteria, most of these methods are also additive, like turbidimetry of inhomogeneous media with the only exception that, instead of inhomogeneous liquid dispersion, there is scattering on pores in a solid (informatively displayed on two-dimensional Fourier spectra in methods such as laser diffractometry) [21]. It is advisable to develop this approach towards quantifying/parametrizing the pore anisotropy, statistical analysis of the distribution of their sizes by integral frequency characteristics, establishing clear correlational (in the above cases, correlation-spectral) relationships between pores of different localization, different genesis, variously approximated by the certain analytical surfaces. This would give a qualitatively new level of understanding of the structure of materials for which porosity is one of the target properties. Meanwhile, the use of laser technology in porosimetry usually does not reach diffractional structurography and clustering of pores according to a complex of their geometric or topological properties in multidimensional space, but is an «appendage» to the additive physical methods of the pore measuring, such as porosimetry using ellipsometric criteria (and solving the inverse problem) [22] or capillary flow porometry [23]. In fact, most of the porosimetric works carried out nowadays using lasers (including TDL) differs little in technique and design of the experiment from the works of the 1980s, excluding the level of automation. In some cases, we are talking about the same processes, but viewed from the other side—it is possible to analyze spectroscopically the medium passing through the pores, or it is possible to analyze its sorption on the walls of the pores or the degree of leakage. Thus, the «gas diffusion» aspect of the process of passing through the pores was studied (using resonant laser beams) back in the early 1980s [24], and the «collision» (that is, associated with the collisions with walls), or the «absorption» aspect of it is being studied nowadays [25, 26]. In other words, until now there has been no transition to hybrid (laser spectroscopic and laser-structurographic) techniques and ideology of experiment planning. This leads a number of branches of materials science and the study of natural media to stagnation, since the porosity parameters remain a «thing-in-itself»—there is no internal correlation between the pool of laser diffraction (micro-/mesostructural) information and a pool of laser spectroscopic (concentration, ratiometric) information for individual classes of pores (both dimensional and geometric).

The authors of this paper do not claim that the existing state of the technique is insufficient/inadequate to the requirements of multiparametric porosimetric tasks, but they want to emphasize the divergent nature of the development of laser technologies in porometry, which prevents their fusion. While spectral laser measurements have reached such a high level that they can be used for fairly accurate analysis of the content of individual stable isotopes in a liquid within the pores (which is used in geochemistry and hydrology [27–30]), and Raman laser spectroscopy allows, in parallel with the implementation of its conventional functions, also to determine pH of the liquid (water) in the pores [31, 32], the structural correspondence of these data to the size of the pores and their geometry has not yet been achieved and, unfortunately, most likely, is not set as a promising task, despite the presence of confocal Raman microscopes and mass spectrometric imaging systems (which can control isotopy) with MALDI (laser matrix-activated desorption/ionization). Meanwhile, the geometry of pores is actively studied by CLSM (confocal laser scanning microscopy) [33–37], but without determining the chemical composition of adsorption/chemisorption products on the pore walls, not to mention such fine analyses, which are carried out with isotopic accuracy for «non-imaging», «non-mapping» methods of analyzing substances in pores, annotated above. The exceptions to the described state are the few precedents of work aimed at determining wettability of the pore surface and the water-repellent properties of the pore wall material using laser confocal microscopes with fixed laser wavelengths that do not involve chemical and isotopic analysis [38, 39], excluding the so-called «ratiometric fluorescence imaging» using stoichiometrically bound labels or dyes that are one-to-one corresponded to the specific proteins that cannot be visualized without the use of labels for confocal fluorescent laser scanning microscopy (therefore, it is obvious that the only substances whose adsorption in pores can be evaluated by confocal laser microscopy are proteins [40, 41]). As a rule, laser methods of the contact angle measuring and tensiometry in porometry, in principle, do not occur, so their detailed consideration in this paper is not necessary. Therefore, defining the achievements of CLSM and laser spectroscopic (including Raman and isotope) analysis as the highest form of meeting analytical requirements for chemical porosimetry, we can't help but note that they are ideologically disconnected at the moment, and without analysis and analytical approximation of the pore surfaces, being applied separately, they are often meaningless. An integrative approach would have to include measurements of pore shapes in the separate spectral ranges of registration at the corresponding wavelengths of laser sources when calibrating the detector and the control driver program of the visualizing device by colorimetric/spectrophotometric temperature adequate to the wavelength range at which differences in the pore morphology are contrastingly recorded, and scanning by depth (as in CLSM) or by angle (as in tomography [42, 43]) is performed with parallel (either multilayer or multi-angle) 2D-Fourier spectroscopy (or laser diffractometry) of the visualized structures/pores for parametrization of their anisotropy and frequency characteristics, that is, for advanced analysis of the porous structures.

All this can be realized with an optical unit integrated with a single microscope (and in the case of multi-angle DPSSL imaging—not even a confocal one).

15.4 Our Instruments and Some Preliminary Results

To dynamically analyze the size and shape of pores during melting, we used a thermoanalytical complex based on a two-channel thermohygrometer CENTER-314 with RS-232-to-USB output (Fig. 15.1b), installed together with a digitized time-lapse camera on a microscope for melting point analysis NAGEMA PHMK 05 (see

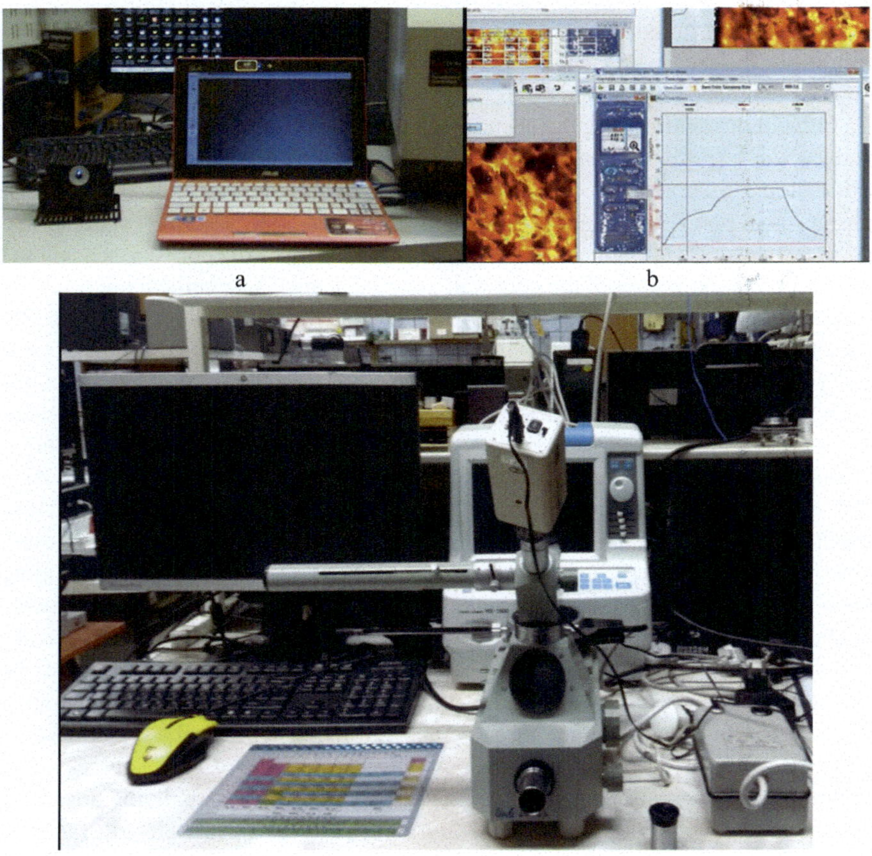

Fig. 15.1 Compact lens-less time-resolved laser diffraction-based correlation porometer developed by O. Gradov group in 2017–2018 (**a**) and GUI of the system for dynamic porometry (**b**) in the programmable/controllable melting process on the NAGEMA heating stage for local thermal analysis (**c**)

(a)

Fig. 15.2 **a** Heating from room temperature to 80 °C—long before the melting point. **b** Heating from 81 °C to 104 °C. **c** Heating from 105 °C to 120 °C. The optical clearing point corresponds to the melting point and the beginning of polymer flow (frame 61) at 114 °C. Already at 120 °C, the formation of an open pore or pit in the place of a porogene particle becomes obvious. **d** Heating from 121 °C to 131 °C only emphasizes the edges of the pores, which retain their shape upon further cooling

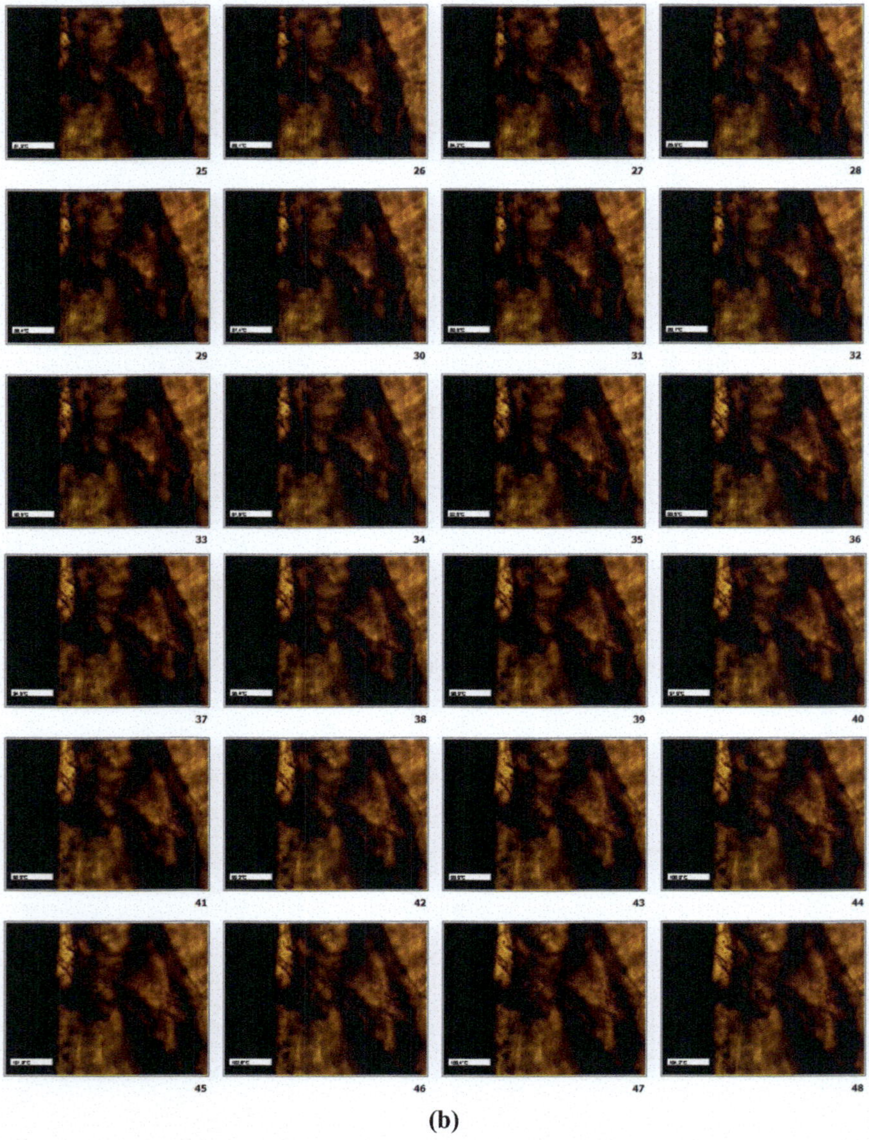

Fig. 15.2 (continued)

9

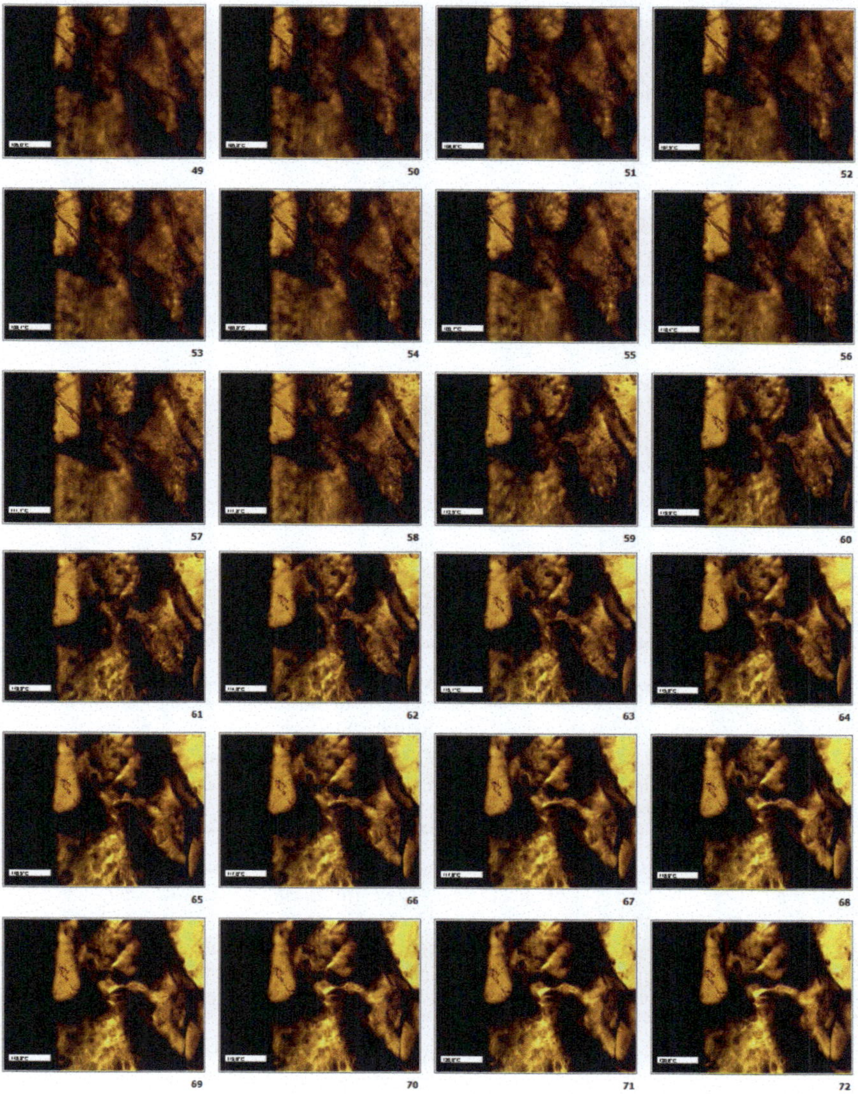

(c)

Fig. 15.2 (continued)

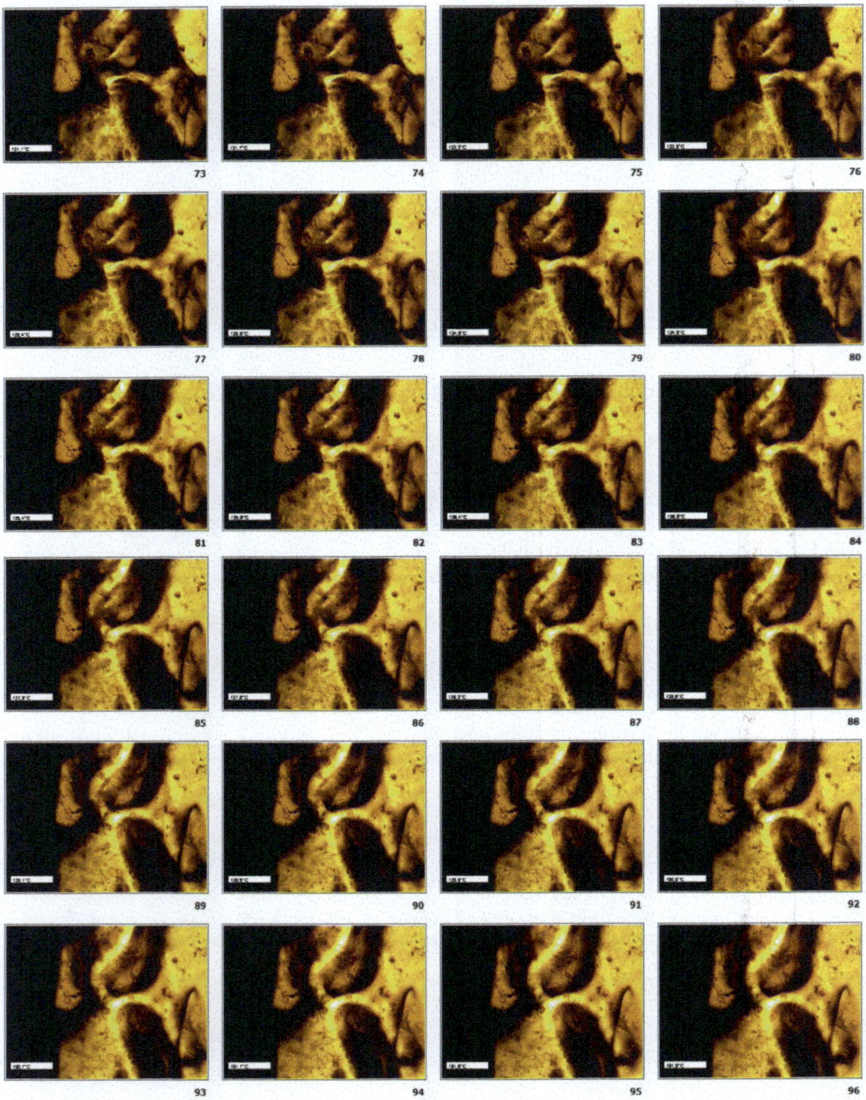

(d)

Fig. 15.2 (continued)

Fig. 15.1c) with software based on the FFTW QAVIS library (developed by POI FEB RAS). As a result, it became possible not only to conduct a sample from room temperature to the melting point and further with every second (or more high-speed) recording, observing the formation of pores during the melting of a raw material with different porogens (see an example of a series of such images in Fig. 15.2; frames 1–96), but also to obtain an in situ—real-time analysis of the evolution of the pore shape relative to other pores or the same pore at an early stage using correlogram methods or correlation-spectral techniques (see Fig. 15.3).

Specifically, Fig. 15.3 shows examples of time-resolved correlation-spectral porometry results on images and correlograms, precisely static pores of filled polymer composites in 47.3 °C (upper images) and pores of the polymer composite at the melting point 113.3 °C (lower images). It can be noted that frame correlograms during melting process and pore collapse (as well as when they appear due to porogenes) are qualitatively different from interframe correlograms before reaching the melting point of the sample. The limitation on the size of the article for the article does not allow us to dwell in detail on the methods proposed by our group and provide many illustrations of this technique; therefore, we provide links to some of the team's works on this topic [44–50].

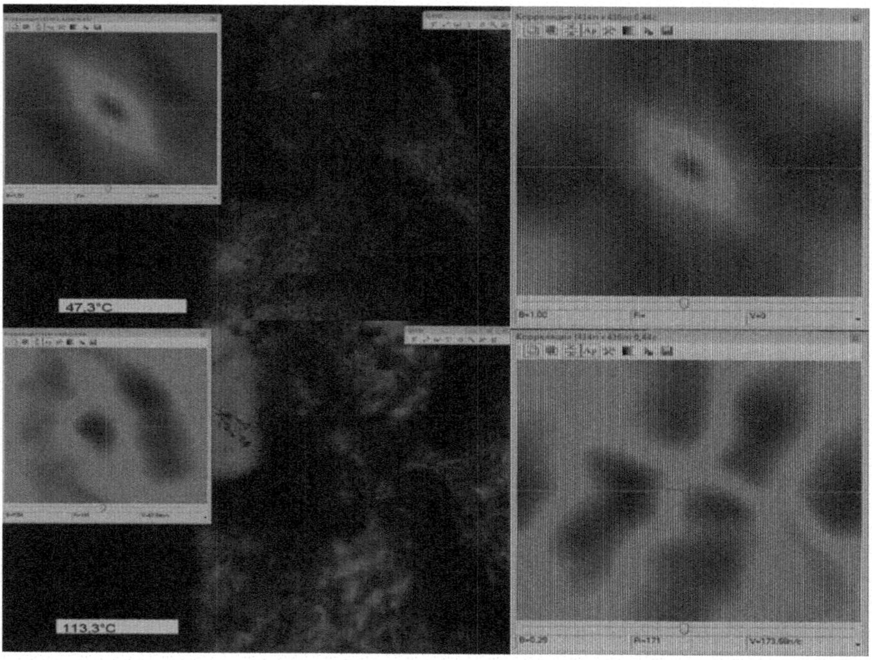

Fig. 15.3 Examples of time-resolved correlation-spectral porometry results on images and correlograms: static pores of filled polymer composites in 47.3 °C (upper images) and movable pores of the polymer composite at the melting point 113.3 °C (lower images). It can be noted that frame correlograms during melting process and pore collapse (as well as when they appear due to porogenes) are qualitatively different from interframe correlograms before reaching the melting point

15.5 Conclusions

The developed approaches to optical 2D/3D representation and time-resolved porometry of individual pores and complex pore topologies on lens-less microscopes, microthermographs with heating stages, and correlative porometry techniques provide statistics not only for the average pore sizes but also for their shapes or topologies. This makes it possible to more accurately study the effects of pores in mechanical, sorption, substance-releasing, or substance-accumulating applications. The ability to trace the formation of single pores under conditions of heating to the melting point and further cooling of the polymer gives new impetus to the development of thermomechanical studies of structured porous materials.

References

1. Bobovich BB (2009) Non-metallic structural materials. MGIU, Moscow, 384 p
2. Maruszewski BM, Starosta R, Restuccia L (1995) Thermodynamic model of an elasto-visco-elastoporous smart material architecture. Proc SPIE 2442:524–531. https://doi.org/10.1117/12.208870
3. Ponomarenko AT, Figovsky O, Shevchenko VG (2008) Multifunctional polymer composites for “intellectual” structures: present state, problems. Futur Adv Mater Res 47:81–84. https://doi.org/10.4028/www.scientific.net/AMR.47-50.81
4. Bak BD, Kalmár-Nagy T (2017) Porcolation: an invasion percolation model for mercury porosimetry. Fluct Noise Lett 16(1):1750008. https://doi.org/10.1142/S0219477517500080
5. Saito A, Foley HC (1995) Argon porosimetry of selected molecular sieves: experiments and examination of the adapted Horvath-Kawazoe model. Microporous Mater 3(4–5):531–542. https://doi.org/10.1016/0927-6513(94)00063-2
6. El-Dieb AS, Hooton RD (1994) Evaluation of the Katz-Thompson model for estimating the water permeability of cement-based materials from mercury intrusion porosimetry data. Cem Concr Res 24(3):443–455. https://doi.org/10.1016/0008-8846(94)90131-7
7. Nishiyama N, Yokoyama T (2014) Estimation of permeability of sedimentary rocks by applying water-expulsion porosimetry to Katz and Thompson model. Eng Geol 177:75–82. https://doi.org/10.1016/j.enggeo.2014.05.016
8. Leech C, Lockington D, Hooton RD (2006) Estimation of water retention curve from mercury intrusion porosimetry and van Genuchten model. ACI Struct J 103(2):291. https://www.webofscience.com/wos/WOSCC/full-record/000237815500015
9. Rebrov EV, Creatore M, Protasova LN, Muraza O, van de Sanden MCM, Schouten JC (2008) Characterization of the porous structure of mesoporous thin films by ellipsometric porosimetry using the modified Deriaguin-Broekhoff-De Boer model. In: Mesostructured materials: design and opportunities: proceedings of the 6th international mesostructured materials symposium (IMMS 2008), September 9–11 2008, Namur, Belgium, P-054
10. Porcheron F, Monson PA (2005) Dynamic aspects of mercury porosimetry: a lattice model study. Langmuir 21(7):3179–3186. https://doi.org/10.1021/la047596e
11. Rigby SP (2000) A hierarchical structural model for the interpretation of mercury porosimetry and nitrogen sorption. J Colloid Interface Sci 224(2):382–396. https://doi.org/10.1006/jcis.1999.6707
12. Vogel HJ, Tölke J, Schulz VP, Krafczyk M, Roth K (2005) Comparison of a lattice-Boltzmann model, a full-morphology model, and a pore network model for determining capillary pressure–saturation relationships. Vadose Zone J 4(2):380–388. https://doi.org/10.2136/vzj2004.0114

13. Lapidus GR, Lane AM, Ng KM, Conner WC (1985) Interpretation of mercury porosimetry data using a pore-throat network model. Chem Eng Commun 38(1–2):33–56. https://doi.org/10.1080/00986448508911294
14. Krivoshapko SN, Ivanov VN (2019) Encyclopedia of analytical surfaces. URSS, Moscow, 560 p
15. Gribble CM, Matthews GP, Laudone GM, Turner A, Ridgway CJ, Schoelkopf J, Gane PA (2011) Porometry, porosimetry, image analysis and void network modelling in the study of the pore-level properties of filters. Chem Eng Sci 66(16):3701–3709. https://doi.org/10.1016/j.ces.2011.05.013
16. de Castro AR, Omari A, Ahmadi-Sénichault A, Bruneau D (2014) Toward a new method of porosimetry: principles and experiments. Transp Porous Media 101(3):349–364. https://doi.org/10.1007/s11242-013-0248-5
17. Volfkovich YM, Bagotzky VS (1994) The method of standard porosimetry: 1. Principles and possibilities. J Power Sources 48(3):327–338. https://doi.org/10.1016/0378-7753(94)80029-4
18. Lichtenstein N, Schmidt BE, Fily A, Weiß S, Arlt S, Pawlik S, Sverdlov B, Muller J, Harder CS (2004) DPSSL and FL pumps based on 980-nm telecom pump laser technology: changing the industry. Proc SPIE 5336:77–83. https://doi.org/10.1117/12.558737
19. Kuu WY, O'Bryan KR, Hardwick LM, Paul TW (2011) Product mass transfer resistance directly determined during freeze-drying cycle runs using tunable diode laser absorption spectroscopy (TDLAS) and pore diffusion model. Pharm Dev Technol 16(4):343–357. https://doi.org/10.3109/10837451003739263
20. Mei L, Svanberg S, Somesfalean G (2012) Combined optical porosimetry and gas absorption spectroscopy in gas-filled porous media using diode-laser-based frequency domain photon migration. Opt Express 20(15):16942–16954. https://doi.org/10.1364/OE.20.016942
21. Svensson T, Andersson-Engels S, Johansson J, Folestad S (2008) Laser spectroscopy for assessing structural properties of turbid solids: towards optical porosimetry. In: Laser applications to chemical, security and environmental analysis (Optical Society of America, 2008). ID: LTuC4. https://doi.org/10.1364/LACSEA.2008.LTuC4
22. Urbanowicz AM, Meshman B, Schneider D, Baklanov MR (2008) Stiffening and hydrophilisation of SOG lowk material studied by ellipsometric porosimetry, UV ellipsometry and laserinduced surface acoustic waves. Phys Status Solidi (A) 205(4):829–832. https://doi.org/10.1002/pssa.200777749
23. Vergauwen TEM, Michiels R, Torbeyns D, Meire M, De Bruyne M, De Moor RJG (2010) Investigation of coronal leakage of root fillings after smear layer removal with EDTA or Er, Cr: YSGG laser through capillary flow porometry. Photomed Laser Surg 28(S2):S-43. https://doi.org/10.1089/pho.2009.2575
24. Karlov NV, Kirillov AA, Kravchenko VA, Orlov AN, Petrov IN (1983) Altering of gas diffusion through a fine-pore membrane by a resonant laser beam. Tech Phys Lett 9:954–958
25. Svensson T, Adolfsson E, Burresi M, Savo R, Xu CT, Wiersma DS, Svanberg S (2013) Pore size assessment based on wall collision broadening of spectral lines of confined gas: experiments on strongly scattering nanoporous ceramics with fine-tuned pore sizes. Appl Phys B 110:147–154. https://doi.org/10.1007/s00340-012-5011-z
26. Al-Saudi A, Al-Basheer W, Aljalal A, Gasmi K, Qari S Sr (2019) Estimation of pore sizes using laser absorption in molecularoxygen gas enclosed in mesoporous alumina. Proc SPIE 11028:1102818. https://doi.org/10.1117/12.2519604
27. Hendry MJ, Schmeling E, Wassenaar LI, Barbour SL, Pratt D (2015) Determining the stable isotope composition of pore water from saturated and unsaturated zone core: improvements to the direct vapour equilibration laser spectrometry method. Hydrol Earth Syst Sci 19(11):4427–4440. https://doi.org/10.5194/hess-19-4427-2015
28. Hendry MJ, Richman B, Wassenaar LI (2011) Correcting for methane interferences on δ2H and δ18O measurements in pore water using H2O (liquid)–H2O (vapor) equilibration laser spectroscopy. Anal Chem 83(14):5789–5796. https://doi.org/10.1021/ac201341p
29. Oerter EJ, Perelet A, Pardyjak E, Bowen G (2017) Membrane inlet laser spectroscopy to measure H and O stable isotope compositions of soil and sediment pore water with high

sample throughput. Rapid Commun Mass Spectrom 31(1):75–84. https://doi.org/10.1002/rcm.7768
30. Widerlund A, Nowell GM, Davison W, Pearson DG (2012) High-resolution measurements of sulphur isotope variations in sediment pore-waters by laser ablation multicollector inductively coupled plasma mass spectrometry. Chem Geol 291:278–285. https://doi.org/10.1016/j.chemgeo.2011.10.018
31. Peltzer ET, Luna M, Walz PM, Zhang X, Brewer PG (2010) In situ measurement of pore-water pH in Anoxic sediments using laser Raman spectrometry. In: AGU fall meeting abstracts
32. Peltzer ET, Walz PM, Luna M, Zhang X, Brewer PG (2015) In situ determination of pore-water pH in reducing sediments near methane seeps and vents by laser Raman spectroscopy. In: AGU fall meeting abstracts
33. Petford N, Davidson G, Miller JA (1999) Pore structure determination using confocal scanning laser microscopy. Phys Chem Earth Part A 24(7):563–567. https://doi.org/10.1016/S1464-1895(99)00080-0
34. Menéndez B, David C, Nistal AM (2001) Confocal scanning laser microscopy applied to the study of pore and crack networks in rocks. Comput Geosci 27(9):1101–1109. https://doi.org/10.1016/S0098-3004(00)00151-5
35. Ying FX, Yang SS, Zhang M, Li YX, Zhou HY (2002) Application of laser scanning confocal microscope to the measurement of pore texture in reservoirs. Acta Sedimentol Sin 20(1):75–79. http://119.78.100.143/article/id/2326
36. Saito Y, Tanaka S, Uchida N, Uematsu K (2002) New characterization method for pore and packing structure in powder compacts using confocal laser scanning microscope. J Electron Microsc 51(4):215–223. https://doi.org/10.1093/jmicro/51.4.215
37. Bagherzadeh R, Latifi M, Najar SS, Tehran MA, Kong L (2013) Threedimensional pore structure analysis of Nano/Microfibrous scaffolds using confocal laser scanning microscopy. J Biomed Mater Res, Part A 101(3):765–774. https://doi.org/10.1002/jbm.a.34379
38. Müehl GJH, Rüehlmann J, Goebel MO, Bachmann J (2012) Application of confocal laser scanning microscopy (CLSM) to visualize the effect of porous media wettability on unsaturated pore water configuration. J Soils Sediments 12(1):75–85. https://doi.org/10.1007/s11368-011-0395-7
39. Zoghlami K, GómezGras D, Corbella M, Darragi F (2008) Laser scanning confocal microscopy characterization of water repellent distribution in a sandstone pore network. Microsc Res Tech 71(11):816–821. https://doi.org/10.1002/jemt.20624
40. Suh CW, Kim MY, Choo JB, Kim JK, Kim HK, Lee EK (2004) Analysis of protein adsorption characteristics to nano-pore silica particles by using confocal laser scanning microscopy. J Biotechnol 112(3):267–277. https://doi.org/10.1016/j.jbiotec.2004.05.005
41. Kim M, Seo C, Lee B, Kim H, Lee E (2007) Imaging analysis of diffusive adsorption of a protein to nano-pore silica particles by using confocal laser scanning microscopy. Theory Appl Chem Eng 10(1):428–428
42. Shah SM, Yang J, Crawshaw JP, Gharbi O, Boek ES (2013) Predicting porosity and permeability of carbonate rocks from core-scale to pore-scale using medical CT, confocal laser scanning microscopy and micro CT. In: SPE annual technical conference and exhibition - society of petroleum engineers meeting. https://doi.org/10.2118/166252-MS
43. Seyfarth M, Holldorf J, Pagenkemper SK (2012) Investigation of shrinkage induced changes in soil volume with laser scanning technique and automated soil volume determination—a new approach/method to analyze pore rigidity limits. Soil Tillage Res 125:105–108. https://doi.org/10.1016/j.still.2012.06.012
44. Gradov OV (2019) Multiangle laser porosimetry and electron beam porosimetry for scaffolds, decellularized extracellular matrix samples and tissue-like models, including ESEM-based & CLEM-based realizations. Genes Cells 14(S1):69–70. https://doi.org/10.23868/gc122421
45. Gorshenev VN, Gradov OV, Gradova MA (2019) Differential estimation of structural properties of biomimetic materials for tissue engineering using real time correlation spectral analysis and structure-preserving maps (morphisms) in category theory framework. Genes Cells 14(S1):68–69. https://doi.org/10.23868/gc122415

46. Gorshenev VN, Olkhov AA, Gradov OV, Gradova MA, Aleksandrov PL (2019) Emergent topological approach for integration of physical properties of solid and soft matter media in porous scaffolds and tissue-engineered constructs: case of PHB and HAp with multi-angle CLSEM visualization. Genes Cells 14(S1):69. https://doi.org/10.23868/gc122418
47. Maklakova IA, Gradov OV, Gradova MA, Aleksandrov PL (2021) Comparison of SEM-assisted nanoporometric and microporometric morphometric techniques applied for the ultramicroporous polymer films. Key Eng Mater 899:660–674. https://doi.org/10.4028/www.scientific.net/KEM.899.660
48. Grigorieva EA, Olkhov AA, Gradov OV, Gradova MA (2021) Thermal behavior of the porous polymer composites based on LDPE and natural fillers studied by real time thermal microscopy. Key Eng Mater 899:644–659. https://doi.org/10.4028/www.scientific.net/KEM.899.644
49. Grigorieva E, Gradov O, Gradova M, Maklakova I (2022) Towards multi-angle multispectral optical 3D porometry and lens-less porometry of civil engineering composites and geocomposites including biodegradable ones. Adv Transdiscipl Eng 31:101–113. https://doi.org/10.3233/ATDE220857
50. Grigorieva EA, Gradov OV, Gradova MA, Maklakova IA (2023) Towards multi-angle multichannel optical porometry and scanning electron microscopic porometry of LDPE composites including geotechnical biodegradable ones. Adv Struct Mater 195:105–118. https://doi.org/10.1007/978-3-031-28744-2_6

Chapter 16
Finite Element Modeling of Uniaxial Compression of Scaffolds

Roman V. Karotkiyan, Evgeniy V. Sadyrin, Andrey L. Nikolaev, Sergei V. Chapek, and Andrey S. Vasiliev

Abstract In the present work, finite element modeling of the uniaxial compression process of polymer scaffolds with increasing wall thickness values was carried out. The method for using X-ray computed microtomography to construct the finite element model was proposed, and evolution of the deformation process of internal structure of the scaffolds was studied. The mechanical characteristics of the polymer material used in printing of the scaffolds were evaluated using nanoindentation. The results obtained may be useful for optimizing the design of scaffolds and increasing their efficacy in regenerative medicine applications.

16.1 Introduction

A scaffold is a temporary mechanical structure that mimics the extracellular matrix of bone tissue and creates an optimal environment for the restoration of damaged bone. A high level of binding between the scaffold and organic tissues is necessary for successful completion of all stages of bone restoration process. After the implantation stage, the process of bone tissue growth begins on the surfaces of the scaffold. The speed of this process depends on the following factors: the ratio of scaffold porosity to bone porosity, the roughness of the outer surface of the scaffold, the compatibility of the implant material with bone tissue, and the scaffold chemical composition. In addition, mechanical properties of the scaffolds should be as close as possible to

R. V. Karotkiyan · E. V. Sadyrin (✉) · A. L. Nikolaev · A. S. Vasiliev
Laboratory for Mechanics of Biomaterials, Don State Technical University, 1 Gagarin Square, 344003 Rostov-on-Don, Russian Federation
e-mail: e.sadyrin@sci.donstu.ru

A. L. Nikolaev
e-mail: andreynicolaev@eurosites.ru

S. V. Chapek
The Smart Materials Research Institute, Southern Federal University, 178/24 Sladkova Street, 344090 Rostov-on-Don, Russian Federation
e-mail: chapek@sfedu.ru

H. Altenbach and V. A. Eremeyev (eds.), *Dynamics of Discrete and Continuum Structures and Media*, Advanced Structured Materials 221,
https://doi.org/10.1007/978-3-031-75626-9_16

the biological material they substitute in order to conduct optimum maintenance of biodegradation properties [1].

One of the main characteristics of a scaffold is its porosity. In addition to the size of the pores, it is important that the pores are interconnected, forming various cavities. These cavities serve as a place for the germination of blood vessels and the movement of nutrients. Organic substances, including metabolic waste from the body, are also deposited in the pores. To reduce residual components from the implant and ensure successful healing, it is important to ensure the maximum degree of degradability of the composition, as well as pay attention to the roughness and internal geometry of the scaffold, which affect the connection and fusion of the implant with the bone and the proliferation ability of cells in the pores of the scaffold. Since porosity is a key characteristic influencing the interaction of body cells with the scaffold, the main challenge is the selection of optimal materials with a controlled microstructure. Some synthetic materials, such as polyester resins, are excellent for creating complex porous structures because they can be easily subjected to various pore formation techniques, such as phase separation and gas foaming ones [2], and have the desired mechanical properties.

On the other hand, a variety of methods was developed for creating porous scaffolds from materials of natural origin, such as chitosan, starch, collagen, gelatin, and other kinds of substances. In such a case, good biocompatibility, renewability, low cost, and low density can be achieved. However, the drawbacks of natural materials in such a case are the high degree of moisture absorption, poor wettability with non-polar polymers, low-adhesion of untreated fibers with polymer–matrix facilitates the retardation of significant bonding during aging [3].

Although high porosity enhances the biological performance of porous implants, it lowers their mechanical properties [2, 3]. However, at these porosity values, strength limitations may occur, especially if the damaged bone is loaded and cannot be completely isolated from external loads. Required porosity values also differ for different types of bone tissues. The number of connecting channels between the pores in the scaffold plays an important role in the regeneration process, as it affects the rate and efficiency of its decomposition after the healing process. Interpore flow, which characterizes the size and number of channels, is directly related to this process. Architectural parameters of the matrix, such as porosity, pore size, ability to interact between cells, as well as the biological, and mechanical properties of the materials from which they are made, are crucial for the osteogenic signaling expression and differentiation during cell colonization and bone tissue formation in vitro and in vivo [4–6].

These properties of the matrix architecture also significantly influence the quantitative parameters of bone tissue formation *in vivo* [7–10]. Optimal matrix structure should induce osteogenic differentiation and signal expression of the recruited cell population. Together with biological and chemical factors such as growth proteins, optimized design and matrix architecture parameters are integral to achieving optimal bone tissue engineering. One of the methods that can accurately reproduce these parameters using matrix microsizes is the stereophotolithographic method [11, 12]. This method, based on photopolarization and the use of computer graphics, allows the rapid creation of 3D cell matrices with specific designs, including porosity, pore sizes, internal communications and mechanical properties [13–15].

Modern scientific approaches made it possible to develop a wide range of composite materials with osteoconductive and osteoinductive properties. However, the porosity and structure of these materials are highly dependent on the scaffold fabrication process. The review article [16] discusses various methods for manufacturing scaffolds, and their advantages and disadvantages and concludes that 3D printing is a more advantageous approach compared to traditional manufacturing technologies due to the possibility of creating customized designs, computer-controlled process control, anisotropic scaffold formation, and better processing conditions [16–24]. However, the challenge remains to find the optimal balance between mechanical properties, porosity, structure, and pore size. Micro-CT in situ testing was successfully utilized for understanding the efficacy of different geometries to withstand the mechanical loads [25, 26] and for evaluation of the 3D printing quality [27]. Computer simulation is an effective tool for predicting mechanical performance as a function of various biological and biocompatible material [28–34] design parameters. Mathematical modeling is capable to save time during scaffold design process and help researchers to better understand their mechanical behavior. This paper examines the modeling of the behavior of polymer scaffolds, based on reconstructed geometric models obtained by micro-CT under load, namely the occurrence of deformations.

16.2 Materials and Methods

A set of scaffolds with dimensions of 12.5 x 12.5 x 7.5 mm and wall thicknesses from 0.5 to 1.5 mm, with a pitch of 0.2 mm were manufactured using digital light processing (DLP) technique. The material used was transparent photopolymer resin FunToDo Nano Clear (FunToDo Nano Clear, Alkmaar, The Netherlands) with ultrahigh resolution. The work used a Deben CT5000TEC uniaxial compression device, which allows loading the material with a maximum force of up to 5 kN at a constant loading rate in the range from 0.03 to 3.00 mm/min. During the process, the installation software records stress and strain data on the material being tested. The device is housed in a Xradia 520 Versa computed X-ray microtomography (micro-CT) unit (Carl Zeiss X-ray Microscopy, Inc., Pleasanton, USA), which includes an X-ray source, a rotary table, and a detector. The turntable on which the scaffold sample is placed can be rotated 360° to fully capture internal cracks in the scaffold. X-rays emitted by the source pass through the sample on a turntable and are detected by a detector. The detector converts the attenuated X-ray signal into a high-resolution digital image that reflects the internal information of the sample and serves as the basis for subsequent 3D modeling thus providing high-resolution images of samples even when they are located at a considerable distance from the X-ray source. This approach makes it possible to observe the continuous development of a 3D network of cracks in samples under axial compression in laboratory conditions.

Micro-CT scanning of the samples was conducted using an Xradia Versa 520 unit (Carl Zeiss X-ray Microscopy, Inc., Pleasanton, USA). The micro-CT scanning parameters for all samples for construction of the finite element modelling (FEM) were as follows: X-ray tube voltage 30 kV, power 1.5 V, pixel size 28.8 μm, sample rotation by 360o, frames per image 1, and exposure time on each frame 1 s. No X-ray tube filter (air) was used (selected based on the observed transmittance values according to the recommendations of the Xradia Versa 520 User's Guide A003030 Rev. B.). In the process of scanning, 1601 projections of each sample were obtained. For the in situ scanning in the course of the compression test, the following parameters were utilized: X-ray tube voltage 40 kV, current 2.5 W, voxel size 28.0 μm, exposure time 5 s, and X-ray tube filter LE1, and 1301 projections were collected for each scan. An additional set of 10 projections was also collected with identical parameters, but with a filter on the LE3 X-ray tube for subsequent correction of ring artifacts. Reconstructions of 3D images were carried out with automatically adjusted values of center shift, Gaussian blur filter 0.5. The beam hardening parameter was used to correct the artifacts in reconstructed images in the form of depressions and shaded areas between dense objects, caused by a stronger absorption of X-ray photons in the middle of an object, compared to the edges (associated with an increase in beam hardening). For each scan, the sample was placed at the closest possible distance to the X-ray source. The flat panel detector was used to acquire the images of 1536 x 972 pixels. The acquisition was performed with camera binning factor = 2. The exposure time was selected to maintain count (intensity) values 5000 with the selected source parameters. During each tomography procedure, 10 reference (air) X-ray images were acquired with equal time intervals between them. For the reconstruction of the set of virtual slices Scout-and-Scan Reconstructor 12.0.8086.19558 (Carl Zeiss X-ray Microscopy, Inc., USA) software was used.

VGstudioMAX 3.5 software (Volume Graphics, Heidelberg, Germany) was used to process the resulting images, and InVesalius 3 software (Centro de Tecnologia da Informação Renato Arche, Campinas Brazil, Fig. 16.1) was used to convert the images into STL format for further work. This open source software is used to reconstruct computed tomography and magnetic resonance images. Utilizing this software, DICOM segmentation of the resulting reconstructions was carried out. This segmentation method consists of a combination of automatic clipping of pixels with certain X-ray density values based on threshold binarization and subsequent manual adjustment of images. Thus, the scaffold was segmented using the Phong method [35], after which volumetric rendering of the segmented areas was performed and a "monolithic" digital 3D model of the scaffold was built, which was saved in STL format.

As can be seen in Fig. 16.1, after processing the model a lot of image artifacts remain. In the present research, the SpaceClaim module included in the ANSYS Mechanical Premium (ANSYS, Inc., Canonsburg, USA) software package was used to remove such image artifacts. Initially, files with the STL extension are models used at the initial stage of modeling, namely representing geometry without physical parameters. SpaceClaim treats the STL model as a facet body. For calculations in the "Static structural" module, the model was converted into a solid model, reduced and trimmed in accordance with the real sample. Figure 16.2 shows the scaffold model after editing.

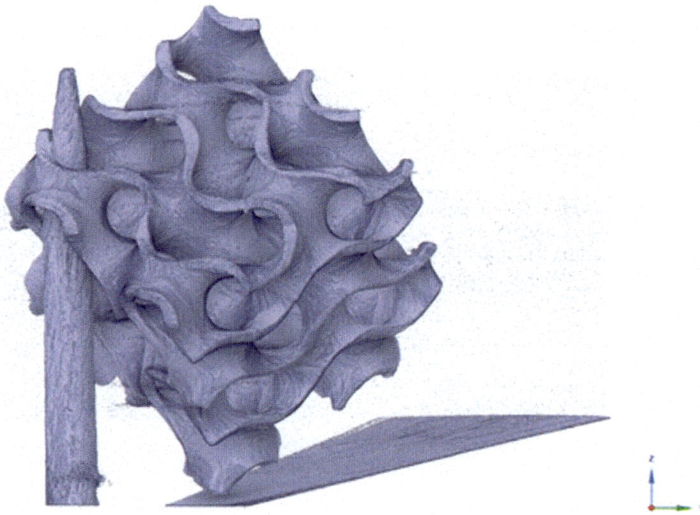

Fig. 16.1 Image of the scaffold obtained after processing in InVesalius 3

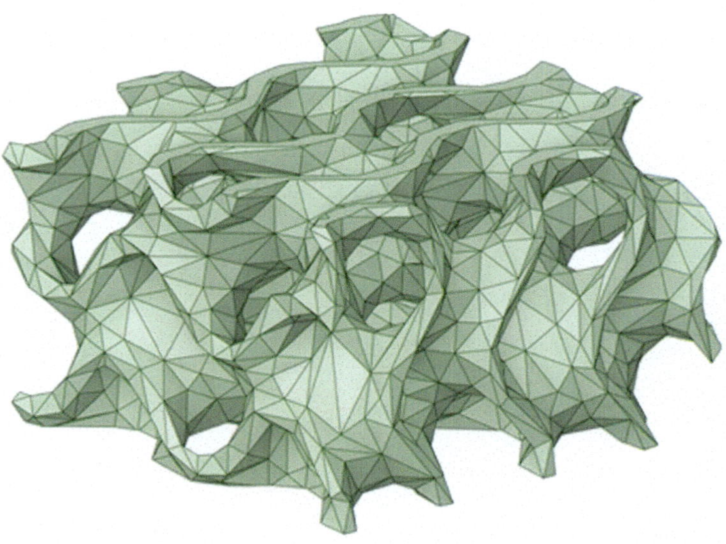

Fig. 16.2 Solid model of the scaffold after processing in SpaceClaim

The following material characteristics were used in the modeling, given in Table 16.1.

The value of Young's modulus was obtained during preliminary experiments on a Nanotest 600 Platform 3 nanoindentation test unit (Micro Materials, Wrexham, UK). The experiments were carried out with a constant temperature control

Table 16.1 The mechanical characteristics utilized for mathematical modeling

Characteristic name	Value
Young's modulus E, MPa	1570
Poisson's ratio ν	0.3 [36, 37]

of 27.0 ± 0.1 °C in a closed instrument room. A Berkovich diamond pyramid was used as an indenter, loaded using the pendulum method according to the following profile: the load increased linearly for 30 s, remained constant for 30 s at the maximum penetration force (to minimize the creep reaction), then decreased linearly for 30 s. Correction of temperature drift during the experiments was carried out using nanoindentometer software. Three series of experiments were carried out, and the value of the maximum load P_{max} for these series was 2.5, 5.0, 7.5 mN, respectively. Within each series, 12 indentations were performed. The results were then analyzed using the Oliver-Pharr method [38, 39] and averaged.

When modeling deformation, displacement was applied to the scaffolds on the upper faces (as indicated in Fig. 16.3), and fastening was applied to the lower faces from the outside. When constructing the finite element mesh of the scaffold, elements with a size of 0.1 mm were used.

The movement per scaffold was represented as a displacement of 1 mm along the Y axis. During the simulation, it was possible to record the starting points of deformation and the behavior of the scaffold, up to complete fracture.

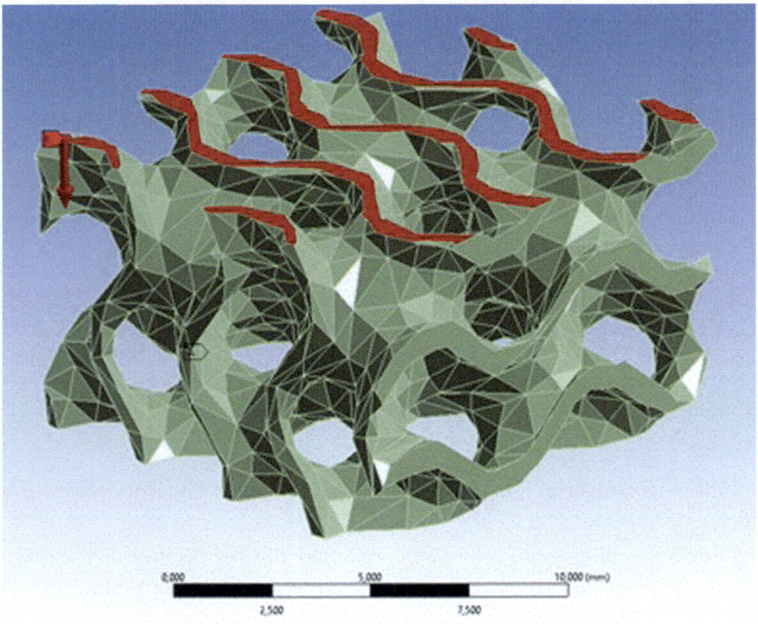

Fig. 16.3 The model of the scaffold with the load applied

It is theoretically possible to simulate the complete compression process of the scaffold, including its residence time in micro-CT, which allows to consider in more detail the possible fracture of the material during intermediate loads.

16.3 Results and Discussion

Modeling of the process of uniaxial compression of the scaffolds was carried out in ANSYS program taking into account full displacement, and the results are shown in Fig. 16.4.

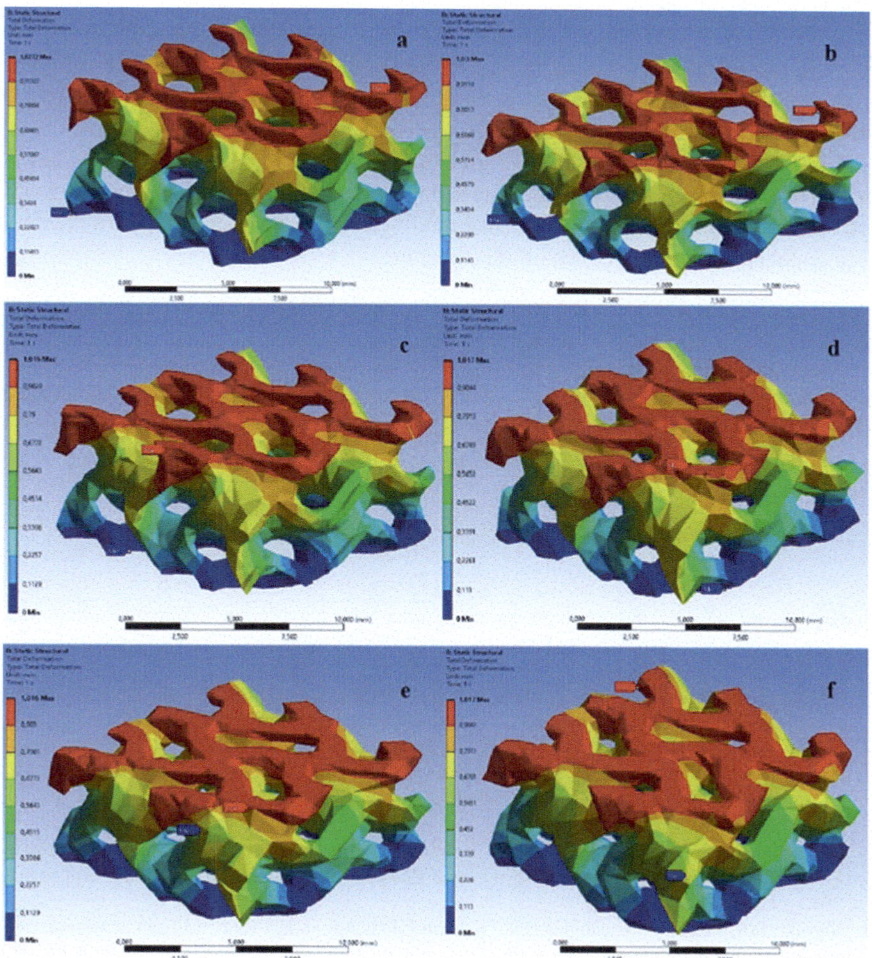

Fig. 16.4 Total displacement results for the scaffolds with the wall thickness: **a** 0.5 mm, **b** 0.7 mm, **c** 0.9 mm, **d** 1.1 mm, **e** 1.3 mm, and **f** 1.5 mm

From the obtained results of total displacement it follows that with increasing scaffold thickness, the load exerted on the object is more distributed, and maximum points and areas with a high rate of displacement are located in areas located at a distance from the center of the object, which corresponds to the general behavior of the scaffold compressed *in situ* in the micro-CT unit. In the future research, it is planned to construct diagrams of displacement versus load based on the proposed scaffold models and to use well-established analytical mathematical models [40–43] for the verification of the results.

16.4 Conclusions

To study the deformation behavior of polymer DLP-printed scaffolds with different values of wall thickness and the evolution of the internal failure structure under different loading conditions, FEM was carried out using ANSYS software. Total displacement results were obtained for each scaffold type. Based on the simulation results, it is also possible to consider the behavior of the scaffold structure under load. Areas of general movement were distributed over the surface, and areas of stresses were concentrated on the edges of the structure.

Acknowledgements The work was supported by the grant of the Russian Science Foundation, grant number 22-49-08014. All micro-CT experiments were conducted in the Resource Center of Collective Usage, Don State Technical University (http://nano.donstu.ru). The authors express their gratitude to the head of the department of "Theoretical and Applied Mechanics" (Don State Technical University, Rostov-on-Don), Dr. Panfilov I. A. for his help in formulating the problem.

Competing Interests The authors have no conflicts of interest to declare that are relevant to the content of this chapter.

References

1. Gómez S, Vlad MD, López J, Fernández E (2016) Design and properties of 3D scaffolds for bone tissue engineering. Acta Biomater. https://doi.org/10.1016/j.actbio.2016.06.032
2. Śmiga-Matuszowicz M, Janicki B, Jaszcz K, Łukaszczyk J, Kaczmarek M, Lesiak M, Sieron AL, Simka W, Mierzwinski W, Kusz D (2014) Novel bioactive polyester scaffolds prepared from unsaturated resins based on isosorbide and succinic acid. Mater Sci Eng: C. https://doi.org/10.1016/j.msec.2014.08.069
3. Chowdhury MNK, Beg MDH, Khan MR (2013) Biodegradability of nanoparticle modified fiber reinforced polyester resin nanocomposite. Procedia Eng. https://doi.org/10.1016/j.proeng.2013.12.203
4. Shimko DA, Shimko VF, Sander EA, Dickson KF, Nauman EA (2005) Effect of porosity on the fluid flow characteristics and mechanical properties of tantalum scaffolds. J Biomed Mater Res Part B: Appl Biomater. https://doi.org/10.1002/jbm.b.30229
5. Khatiwala CB, Peyton SR, Putnam AJ (2006) Intrinsic mechanical properties of the extracellular matrix affect the behavior of pre-osteoblastic MC3T3-E1 cells. Am J Physiol Cell Physiol. https://doi.org/10.1152/ajpcell.00455.2005

6. Roy TD, Simon JL, Ricci JL et al (2003) Performance of degradable composite bone repair products made via three-dimensional fabrication techniques. J Biomed Mater Res A. https://doi.org/10.1002/jbm.a.10582
7. Chu TM, Orton DG, Hollister SJ et al (2002) Mechanical and in vivo performance of hydroxyapatite implants with controlled architectures. Biomaterials. https://doi.org/10.1016/S0142-9612(01)00243-5
8. Gauthier O, Bouler JM, Aguado E et al (1998) Macroporous biphasic calcium phosphate ceramics: influence of macropore diameter and macroporosity percentage on bone ingrowth. Biomaterials. https://doi.org/10.1016/S0142-9612(97)00180-4
9. Oh SH, Park IK, Kim JM et al (2007) In vitro and in vivo characteristics of PCL scaffolds with pore size gradient fabricated by a centrifugation method. Biomaterials. https://doi.org/10.1016/j.biomaterials.2006.11.024
10. Roy TD, Simon JL, Ricci JL et al (2003) Performance of degradable composite bone repair products made via three-dimensional fabrication techniques. J Biomed Mater Res A. https://doi.org/10.1002/jbm.a.10582
11. Hollister SJ (2005) Porous scaffold design for tissue engineering. Nat Mater. https://doi.org/10.1038/nmat1421
12. Leong KF, Cheah CM, Chua CK (2003) Solid freeform fabrication of three-dimensional scaffolds for engineering replacement tissues and organs. Biomaterials. https://doi.org/10.1016/S0142-9612(03)00030-9
13. Bartolo PJ, Almeida HA, Rezende RA, et al (2008) Advanced processes to fabricate scaffolds for tissue engineering. In: Bidanda B, Bartolo PJ (eds) Virtual prototyping & bio manufacturing in medical applications. https://doi.org/10.1007/978-0-387-68831-2_8
14. Hutmacher DW, Sittinger M, Risbud MV (2004) Scaffold-based tissue engineering: rationale for computer-aided design and solid free-form fabrication systems. Trends Biotechnol. https://doi.org/10.1016/j.tibtech.2004.05.005
15. Pham DT, Gault RS (1998) A comparison of rapid prototyping technologies. Int J Mach Tools Manuf. https://doi.org/10.1016/S0890-6955(97)00137-5
16. Bagde AD et al (2019) State of the art technology for bone tissue engineering and drug delivery. Irbm. https://doi.org/10.1016/j.irbm.2019.03.001
17. Thavornyutikarn B, Chantarapanich N, Sitthiseripratip K, Thouas GA, Chen Q (2014) Bone tissue engineering scaffolding: computer-aided scaffolding techniques. Prog Biomater. https://doi.org/10.1007/s40204-014-0026-7
18. Wen Y et al (2017) 3D printed porous ceramic scaffolds for bone tissue engineering: a review. Biomater Sci. https://doi.org/10.1039/C7BM00315C
19. Zhu W, Ma X, Gou M, Mei D, Zhang K, Chen S (2016) 3D printing of functional biomaterials for tissue engineering. Curr Opin Biotechnol. https://doi.org/10.1016/j.copbio.2016.03.01
20. Mandrycky C, Wang Z, Kim K, Kim DH (2016) 3D bioprinting for engineering complex tissues. Biotechnol Adv. https://doi.org/10.1016/j.biotechadv.2015.12.011
21. Gao B, Yang Q, Zhao X, Jin G, Ma Y, Xu F (2016) 4D Bioprinting for Biomedical Applications. Trends Biotechnol. https://doi.org/10.1016/j.tibtech.2016.03.004
22. Travitzky N et al (2014) Additive manufacturing of ceramic-based materials. Adv Eng Mater. https://doi.org/10.1002/adem.201400097
23. Inzana JA et al (2014) 3D printing of composite calcium phosphate and collagen scaffolds for bone regeneration. Biomaterials. https://doi.org/10.1016/j.biomaterials.2014.01.064
24. Bergmann C et al (2010) 3D printing of bone substitute implants using calcium phosphate and bioactive glasses. J Eur Ceram Soc. https://doi.org/10.1016/j.jeurceramsoc.2010.04.037
25. Mitrin BI, Chapek SV, Sadyrin EV, Swain MV (2021) Mechanical properties and failure mechanisms of 3D-printed PLA scaffolds: A preliminary study. IOP Conf Ser Mater Sci Eng. https://doi.org/10.1088/1757-899X/1029/1/012074
26. Ribeiro JF, Oliveira SM, Alves JL, Pedro AJ, Reis RL, Fernandes EM, Mano JF (2017) Structural monitoring and modeling of the mechanical deformation of three-dimensional printed poly (*epsilon*-caprolactone) scaffolds. Biofabrication. https://doi.org/10.1088/1758-5090/aa698e

27. Sadyrin EV, Nikolaev AL, Chapek SV, Nazarenko DV, Aizikovich SM, Wang YC (2023) Manufacturing quality evaluation of photopolymer resin 3d-printed scaffolds using microtomography. In: Sixty shades of generalized continua: dedicated to the 60th birthday of prof. Victor A. Eremeyev. https://doi.org/10.1007/978-3-031-26186-2_38
28. Zelentsov VB, Sadyrin EV, Mitrin BI, Swain MV (2023) Mathematical tools for recovery of the load on the fissure according to the micro-CT results. J Mech Behav Biomed. https://doi.org/10.1016/j.jmbbm.2022.105625
29. Farina E, Gastaldi D, Baino F, Vernè E, Massera J, Orlygsson G, Vena P (2021) Micro computed tomography based finite element models for elastic and strength properties of 3D printed glass scaffolds. Acta Mech Sinica. https://doi.org/10.1007/s10409-021-01065-3
30. Sadyrin EV, Mitrin BI, Yogina DV, Swain MV (2021) Preliminary study of distribution of mechanical properties and mineral density by depth of liquid saturated carious dentine. IOP Conf Ser Mater Sci Eng. https://doi.org/10.1088/1757-899X/1029/1/012056
31. Sadyrin EV, Yogina DV, Swain MV, Maksyukov SY, Vasiliev AS (2021) Efficacy of dental materials in terms of apparent mineral density restoration: composite resin, glass ionomer cement and infiltrant. Open Access Compos Part C. https://doi.org/10.1016/j.jcomc.2021.100192
32. Sadyrin EV, Kislyakov EA, Karotkiyan RV, Yogina DV, Drogan EG, Swain MV, Maksyukov SY, Nikolaev AL, Aizikovich SM (2020) Influence of citric acid concentration and etching time on enamel surface roughness of prepared human tooth: in vitro study. In: Plasticity, damage and fracture in advanced materials. https://doi.org/10.1007/978-3-030-34851-9_8
33. Kislyakov EA, Karotkiyan RV, Sadyrin EV, Mitrin BI, Yogina DV, Kheygetyan AV, Maksyukov SY (2020) Nanoindentation derived mechanical properties of human enamel and dentine subjected to etching with different concentrations of citric acid. In: Modeling, synthesis and fracture of advanced materials for industrial and medical applications. https://doi.org/10.1007/978-3-030-48161-2_5
34. Melnikova G, Kuznetsova T, Lapitskaya V, Petrovskaya A, Chizhik S, Zykova A, Safonov V, Aizikovich S, Sadyrin E, Sun W, Yakovin S (2021) Nanomechanical and nanotribological properties of nanostructured coatings of tantalum and its compounds on steel substrates. Nanomaterials. https://doi.org/10.3390/nano11092407
35. Jeong T, Shin HJ (2014) An approximation technique for real-time rendering of phong reflection model with image-based lighting. J Korea Comput Graph Soc 20(1):13–19 (2014)
36. Lovo JFP, Camargo ILD, Erbereli R, Morais MM, Fortulan CA (2020) Vat photopolymerization additive manufacturing resins: analysis and case study. Mater Res. https://doi.org/10.1590/1980-5373-MR-2020-0010
37. Mori KI (2001) Simulation of material processing: theory, methods and application: proceedings of the 7th international conference NUMIFORM 2001
38. Oliver WC, Pharr GM (1992) An improved technique for determining hardness and elastic modulus using load and displacement sensing indentation experiments. J Mater Res. https://doi.org/10.1557/JMR.1992.1564
39. Oliver WC, Pharr GM (2004) Measurement of hardness and elastic modulus by instrumented indentation: advances in understanding and refinements to methodology. J Mater Res. https://doi.org/10.1557/jmr.2004.19.1.3
40. Argatov I, Mishuris G (2016) Articular contact mechanics from an asymptotic modeling perspective: a review. Front Bioeng Biotechnol. https://doi.org/10.3389/fbioe.2016.00083
41. Vasiliev AS, Volkov SS, Sadyrin EV, Aizikovich SM (2020) Simplified analytical solution of the contact problem on indentation of a coated half-space by a conical punch. Mathematics. https://doi.org/10.3390/math8060983
42. Panfilov I, Beskopylny AN, Meskhi B (2023) Numerical simulation of heat transfer and spread of virus particles in the car interior. Mathematics. https://doi.org/10.3390/math11030784
43. Beskopylny AN, Panfilov I, Meskhi B (2022) Modeling of flow heat transfer processes and aerodynamics in the cabins of vehicles. Fluids. https://doi.org/10.3390/fluids7070226

Chapter 17
Coupled Traveling Thermo-Chemo-Mechanical Waves in Elastic Media

Anna G. Knyazeva

Abstract The Burgers equation also describes a wide class of nonlinear effects arising in wave propagation theory, plasma physics, and acoustics. The Burgers type equation follows from the equations of the coupled nonlinear theory of thermoelasticity. In this paper it is shown that a generalization of the Burgers equation can be approached by constructing traveling wave type solutions in coupled models of solid-phase combustion. Examples of partial problems in which the automodel solution is a thermo-chemo-mechanical wave running with a uniform velocity are presented. They include a model considering only thermal expansion, a model considering concentration expansion; a generalization of the model for finite strains; and a model considering the dependence of the reaction velocity on the stress work. The wave velocity is determined by the method of jointed asymptotic expansions. A brief review of the author's works with the development of coupled models for practical applications to the synthesis of new materials and coatings; to different variants of surface treatment is presented.

Keywords Thermoelasticity · Coupling model · Traveling wave · Asymptotic solution · Practical examples

Abbreviations

Basic designations

T, θ	Temperature
t, τ	Time
$x, \xi, X, \xi_1 = \xi V$	Spatial coordinate in the direction of front propagation
σ_{ij}, s	Stresses

A. G. Knyazeva (✉)
Institute of Strength Physics and Materials Science SB RAS, Tomsk, Russia
e-mail: anna@ispms.ru

H. Altenbach and V. A. Eremeyev (eds.), *Dynamics of Discrete and Continuum Structures and Media*, Advanced Structured Materials 221,
https://doi.org/10.1007/978-3-031-75626-9_17

$\varepsilon_{ij}, \varepsilon, e$	Strains (deformation)
$\mathbf{u}$	Displacement vector
ρ, ρ_0	Density
c_ε	Heat capacity at constant strains
λ_T	Thermal conductivity coefficient
α_T	Thermal expansion coefficient
α_k	Concentration expansion coefficients
K	Bulk module
λ, μ	Lame coefficients
δ_{ij}	Kronecker delta
ω, ω_0	Coupling coefficient
T_*, t_*, x_*	Are scales for temperature, time and spatial coordinate

$$x_* = \sqrt{\kappa_T t_*},\ T_* = T_0 + \frac{Q_{ch}}{c_\varepsilon \rho_0};\ \kappa_T = \frac{\lambda_T}{\rho c_\varepsilon}$$

Q_{ch}	Chemical heat release
k_0	Pre-exponential factor
E_a	Activation energy for chemical reaction

$$\mathrm{v}_n = V\sqrt{\kappa_T / t_*}$$

Dimensionless parameters of the models:

$\sigma = \frac{T_0}{T_* - T_0}$, $\beta = \frac{RT_*}{E}$, $\beta = \frac{\gamma}{1+\sigma}$, $\gamma = \frac{RT_*^2}{E(T_* - T_0)}$;

$\alpha^2 = \frac{\rho_0}{\lambda+2\mu}\frac{\kappa_T}{t_*}$; $\sigma = \frac{T_0}{(T_* - T_0)}$; $g = \frac{\alpha_B - \alpha_A}{\alpha_T (T_* - T_0)}$;

$\alpha_1 = (V\alpha)^2$ (in models with constant density);

$\alpha_1 = \frac{M^2}{\rho_0(\lambda+2\mu)} = \frac{\rho^2(\mathrm{v}_n - \mathrm{v})^2}{\rho_0^2 \frac{(\lambda+2\mu)}{\rho_0}} = (V\alpha)^2$ (in the model with varying density);

$V = \frac{\mathrm{v}_n}{\sqrt{\kappa_T / t_*}}$; $m = \frac{V}{\gamma}$ or $m = \frac{\overline{m}}{\gamma}$ (depending on the problem);

$\overline{m} = \frac{M}{\rho_0 \sqrt{\kappa_T / t_*}} = \frac{\rho(\mathrm{v}_n - \mathrm{v})}{\rho_0 \sqrt{\kappa_T / t_*}}$;

$\omega = \frac{(3\alpha_T K)^2 (T_* - T_0)}{c_\varepsilon \rho_0 (\lambda+2\mu)}$;

$\delta = \varepsilon_* = \frac{3K\alpha_T}{\lambda+2\mu}(T_* - T_0) << 1$

Some functions for chemical heat release:

$\phi_2(\theta) = \exp\left(-\frac{1+\sigma}{\theta+\sigma} \cdot \frac{1-\theta}{\gamma}\right)$; $\phi_1(\eta) = 1 - \eta$

17.1 Introduction

The Burgers equation [1] attracts considerable attention due to its applicability in shock wave theory and mathematical modeling of turbulence, in heat conduction and diffusion problems; in the study of fluid flow through porous media, in the description of the behavior of elastic and viscoelastic media, etc. The Burgers equation also describes a wide class of nonlinear effects arising in wave propagation theory, plasma physics, and acoustics. In the classical version, the one-dimensional Burgers equation has the following form:

$$\frac{\partial u}{\partial \tau} + u\frac{\partial u}{\partial x} = \nu\frac{\partial^2 u}{\partial x^2}.$$

In most studies, the solution to this equation is sought on a finite interval,

$$x \in (a, b)$$

with given initial and boundary conditions of the form

$$u(x, t_0) = h(x);$$

$$B_1 u(a, t) = g_1(t);\ \ B_2 u(b, t) = g_2(t)$$

or in the form of a traveling wave.

This equation is remarkable in that it has an exact analytic solution.

The simplest generalizations of this equation are

$$\frac{\partial u}{\partial \tau} + a(x, t)u\frac{\partial u}{\partial x} = b(x, t)\frac{\partial^2 u}{\partial x^2} + F(x, t),$$

or

$$\frac{\partial u}{\partial \tau} + au\frac{\partial u}{\partial x} = b\frac{\partial^2 u}{\partial x^2} + cu(1 - u)$$

and their two- and three-dimensional analogs attract many researchers. The first equation has an application to the theory of turbulence [2]. The second one is the Burgers-Fisher equation [3]. The coupled systems of Burgers equations are popular [4–6].

New approximate and exact analytical solutions appear, and efficient numerical methods are developed. For example, in [7], the diagonal Padé approximation is applied in the development of a numerical algorithm. In [8], a finite-difference scheme of second-order approximation for the numerical solution of the one-dimensional Burgers equation is presented. The authors of [9] reduce the Burgers equation to the heat conduction equation using the Cole-Hopf transform and then solve numerically on a finite segment. In [10] work, the application of straight line method is illustrated to solve the Burgers equation. In [11], radial basis functions are used in the numerical solution of the Burgers equation. The double power series method is applied in [12] to construct an approximate solution of the nonlinear Burgers equation. The authors of [13] analyze solutions of the Burgers equation with variable coefficients and with linear damping. Exact solutions of the one-dimensional Burgers equation are sought in [14] using similarity transformations method. Etc.

In the present work, we will show that a generalization of the Burgers equation can be reached by constructing traveling wave solutions in coupled models of solid-phase combustion. Stresses and deformations are a consequence of transformations and can influence them through different channels. As a result, a single thermo-chemo-mechanical wave running at a constant velocity appears.

17.2 Burgers Equation in the Theory of Thermoelasticity

The dynamic coupled theory of thermoelasticity [15] includes the heat conduction equation

$$\rho c_\varepsilon \frac{\partial T}{\partial t} + 3K\alpha_T T \frac{\partial \varepsilon_{kk}}{\partial t} = \nabla \cdot (\lambda_T \nabla T), \tag{17.1}$$

motion equation

$$\rho \frac{\partial^2 \mathbf{u}}{\partial t^2} = \nabla \cdot \boldsymbol{\sigma}, \tag{17.2}$$

and defining relationships

$$\sigma_{ij} = 2\mu\varepsilon_{ij} + \delta_{ij}[\lambda\varepsilon_{kk} - 3K\alpha_T(T - T_0)], \tag{17.3}$$

$$\varepsilon_{ij} = \frac{1}{2}\left(\frac{\partial u_i}{\partial x_j} + \frac{\partial u_j}{\partial x_i}\right), \tag{17.4}$$

where T—is the temperature; t- is the time; ρ—is the density; c_ε—is the heat capacity at constant strains; $K = \lambda + 2\mu/3$—is isothermal bulk module; λ and μ – Lame coefficients; α_T—is the линейный коэффициент теплового расширения; λ_T—is the thermal conductivity coefficient; $\nabla \ \ldots = div \ldots$; $\nabla\ldots = grad \ldots$; $\varepsilon_{kk} = \varepsilon_{11} + \varepsilon_{22} + \varepsilon_{33} = \nabla \cdot \mathbf{u}$; $\mathbf{u}$—is the displacement vector with components u_i; σ—is stress тензор with components σ_{ij}; ε_{ij}—components of strain tensor; δ_{ij}—Kronecker delta; T_0—is temperature for non-deformed state.

The simplest classical problem of the theory of temperature stresses is formulated as a problem of thermal shock on the surface of a half-space free from the action of mass and external mechanical forces. In the first boundary value problem it is considered that at a moment of time $t = 0$ the temperature of the environment (or the surface of a solid body) suddenly changes from the value T_0 to the value T_s. Consequently, in this case, the deformation of the body can be caused only by time-varying heating or cooling of its surface; only one component of the displacement vector is different from zero $u = u_1$ (component of the displacement vector in the direction of the axis Ox) and one component of the strain tensor $\varepsilon = \varepsilon_{11}$. Then $\varepsilon_{kk} = \varepsilon_{11} = \varepsilon = \partial u / \partial x$ and from (17.3) we have

$$\sigma_{11} = (\lambda + 2\mu)\varepsilon - 3K\alpha_T(T - T_0);\ \sigma_{22} = \sigma_{33} = \lambda\varepsilon - 3K\alpha_T(T - T_0). \quad (17.5)$$

The other components of the stress tensor are equal to zero.

The solution of the simplest dynamic problem is reduced to solving the coupled equations

$$\rho c_\varepsilon \frac{\partial T}{\partial t} = \lambda_T \frac{\partial^2 T}{\partial x^2} - 3K\alpha_T T \frac{\partial \varepsilon}{\partial t}; \quad (17.6)$$

$$\rho \frac{\partial^2 u}{\partial t^2} = \frac{\partial \sigma_{11}}{\partial x} \quad (17.7)$$

with boundary conditions

$$x = 0 :\ T = T_s,\ u = 0 \text{ (or } \sigma_{11} = 0), \quad (17.8)$$

$$x \to \infty :\ \lambda_T \frac{\partial T}{\partial x} = 0 \quad (17.9)$$

and the initial condition

$$t = 0 :\ \frac{\partial u}{\partial t} = 0,\ \frac{\partial u}{\partial t} = 0 \quad (17.10)$$

Using the Cauchy relations (17.4) and Duhamel-Neumann relations (17.5), the problem can be reformulated in stresses or in displacements. In the latter case instead of (17.6), (17.7) we have:

$$\rho c_\varepsilon \frac{\partial T}{\partial t} = \lambda_T \frac{\partial^2 T}{\partial x^2} - 3K\alpha_T T \frac{\partial^2 u}{\partial t \partial x};$$

$$\rho \frac{\partial^2 u}{\partial t^2} = (\lambda + 2\mu) \frac{\partial^2 u}{\partial x^2} - 3K\alpha_T \frac{\partial T}{\partial x}.$$

It is convenient to further write Eq. (17.7) in the form

$$\rho \frac{\partial^2 \varepsilon}{\partial t^2} = (\lambda + 2\mu) \frac{\partial^2 \varepsilon}{\partial x^2} - 3K\alpha_T \frac{\partial^2 T}{\partial x^2}. \quad (17.11)$$

In the classical theory of thermoelasticity [16] taking into account the smallness of the coupling coefficient

$$\omega_0 = \frac{(3K\alpha_T)^2}{\lambda + 2\mu} \frac{T_0}{c_\varepsilon \rho} \tag{17.12}$$

the heat conduction equation is linearized at the temperature of the undeformed state T_0. In this case, however, we must assume that $\partial \varepsilon_{kk} / \partial t$—is also a small value, which is not always obvious [17]. In most of the works, in which it is required to estimate the temperature stresses, the coupling effect is neglected. The solutions of linear theory problems are well investigated. These solutions are waves that decay rapidly as they move away from the heated surface. Solutions of the traveling wave type do not exist in the linear theory of thermoelasticity. Such solutions appear in coupled nonlinear theory [18, 19].

For definiteness, let us assume that $T_s > T_0$. Then in the variables

$$\theta = \frac{T - T_0}{T_s - T_0}, \; X = \frac{x}{x_*}, \; \tau = \frac{t}{t_*}, \; e = \frac{\varepsilon}{\varepsilon_*}, \tag{17.13}$$

where

$$x_* = \sqrt{\kappa_T t_*}, \; \varepsilon_* = \frac{3K\alpha_T}{\lambda + 2\mu}(T_s - T_0), \; \kappa_T = \frac{\lambda_T}{\rho c_\varepsilon}, \tag{17.14}$$

and the time scale t_* does not have principal importance, Eqs. (17.6) and (17.11) take the form

$$\frac{\partial \theta}{\partial \tau} = \frac{\partial^2 \theta}{\partial X^2} - \omega(\theta + \sigma)\frac{\partial e}{\partial \tau}; \tag{17.15}$$

$$\frac{\partial^2 e}{\partial X^2} - \frac{\partial^2 \theta}{\partial X^2} = \alpha^2 \frac{\partial^2 e}{\partial \tau^2}, \tag{17.16}$$

where

$$\omega = \frac{(3K\alpha_T)^2}{\lambda + 2\mu} \frac{T_s - T_0}{c_\varepsilon \rho}, \; \sigma = \frac{T_0}{T_s - T_0}, \; \alpha^2 = \frac{\rho}{\lambda + 2\mu} \frac{\kappa_T}{t_*} \tag{17.17}$$

Note that in this simple problem the time scale, in principle, can be determined by assuming that

$$\alpha^2 = \frac{\rho}{\lambda + 2\mu} \frac{\kappa_T}{t_*} = 1.$$

However, further on the time scale will be defined differently, and α remains a parameter.

Let us consider the solutions of a nonlinear system of equations representing waves with constant profile traveling with velocity V. Let's go to the coordinate $\xi = X - V\tau$, assuming that the wave moves to the right. Then the solution must satisfy the system of equations

$$-V\frac{d\theta}{d\xi} = \frac{d^2\theta}{d\xi^2} + \omega(\theta + \sigma)V\frac{de}{d\xi}; \tag{17.18}$$

$$\frac{d^2 e}{d\xi^2} - \frac{d^2\theta}{d\xi^2} = (\alpha V)^2\frac{d^2 e}{d\xi^2} \tag{17.19}$$

and conditions

$$\xi \to -\infty : \cdots \theta = \theta_1 = 1 \text{ and } \xi \to +\infty \cdots \theta = \theta_2 = 0\,. \tag{17.20}$$

Equations (17.18), (17.19) are easily integrable. Taking into account the condition of absence of perturbations at infinity ($\xi \to +\infty$) we have

$$\frac{de}{d\xi} = \frac{1}{1-(\alpha V)^2}\frac{d\theta}{d\xi},\; e = \frac{1}{1-(\alpha V)^2}\theta. \tag{17.21}$$

Hence, the heat conduction equation takes the form

$$-V\left(1 + \frac{\omega}{1-(\alpha V)^2}(\theta + \sigma)\right)\frac{d\theta}{d\xi} = \frac{d^2\theta}{d\xi^2}. \tag{17.22}$$

Under condition $(\alpha V)^2 > 1$ or $V^2 > \frac{1}{\alpha^2}$, Eq. (17.22) coincides in form with the Burgers equation (see [1,18.] et al.) written in automodel variables. Indeed, using the notations

$$U = \omega^{-1}\left((\alpha V)^2 - 1 - \omega\sigma\right),\; \nu = \frac{(\alpha V)^2 - 1}{V\omega}, \tag{17.23}$$

we obtain the equation

$$-U\frac{d\theta}{d\xi} + \theta\frac{d\theta}{d\xi} = \nu\frac{d^2\theta}{d\xi^2}, \tag{17.24}$$

The "normal" (non-automodel) form of this equation is

$$\frac{\partial\theta}{\partial\tau} + \theta\frac{\partial\theta}{\partial X} = \nu\frac{\partial^2\theta}{\partial X^2}. \tag{17.25}$$

Solutions of the Eq. (17.25) in the form of a wave of constant profile (17.24) exist at $V^2 > (1 + \omega\sigma)/\alpha^2$ or $v_n^2 > C_L^2(1 + \omega_0)$, where $v_n = V\sqrt{\kappa_T/t_*}$—is dimensional wave velocity; $C_L = \left[(\lambda + 2\mu)/\rho\right]^{1/2}$—sound velocity, $\omega_0 \equiv \omega\sigma$.

The exact solution of the nonlinear Eq. (17.24) satisfying the conditions (17.20) can be represented as

$$\theta = 1 - \frac{1}{1 + \exp(-\xi/(2\nu))}, \; U = \frac{1}{2},$$

from where it is easy to find the velocity V for given values of temperature in front of and behind the wave front. This solution represents a weak shock wave. Since strains and stresses are related to temperature by linear relations (17.5), the strains and stresses in the wave front also satisfy an equation of the form (17.24). This makes it possible to study the evolution of an initial perturbation of a given shape into a stationary profile. In particular, it was demonstrated in [1] that in a thermally conductive medium the temperature remains continuous during the transition through a shock wave. At $\nu \to 0$ solutions of Eq. (17.20) converge to the shock-wave discontinuous solutions of equation.

$$\frac{\partial\theta}{\partial\tau} + \theta\frac{\partial\theta}{\partial X} = 0, \tag{17.26}$$

satisfying the conditions

$$U = \frac{1}{2}(\theta_1 + \theta_2), \theta_1 > U > \theta_2.$$

A similar conclusion was made in [20] for a viscoelastic medium.

17.3 Solid-Phase Combustion Models for Elastic Media

17.3.1 Model of Solid-Phase Combustion with Thermal Expansion

Let there is an internal heat source in the medium due to an exothermic chemical reaction. In the case when only thermal stresses and deformations are significant, and the chemical transformation can be described by the total (or general) scheme "solid reactant—solid product", the equation of motion remains the same (see (17.7) or (17.11)), and the one-dimensional heat conduction equation takes the following form

$$\rho c_\varepsilon \frac{\partial T}{\partial t} = \lambda_T \frac{\partial^2 T}{\partial x^2} - 3K\alpha_T T \frac{\partial \varepsilon}{\partial t} + Q_{ch} k_0 \phi_1(\eta)\phi_2(T). \tag{17.27}$$

The transformation level (reaction product fraction) η satisfies the equation

$$\frac{d\eta}{dt} = k_0 \phi_1(\eta)\phi_2(T) = \Phi(\eta, T) \tag{17.28}$$

where it is assumed that the summary reaction rate satisfies the Arrhenius law

$$\phi_2(T) = \exp\left(-\frac{E_a}{RT}\right).$$

Here Q_{ch}—is chemical heat release; $\phi_1(\eta)$—is kinetic function; k_0—pre-exponential factor; E_a—is activation energy of a chemical reaction; R—universal gas constant.

In dimensionless variables (17.13) (with replacement T_s by T_*, which is not yet defined) Eqs. (17.27) and (17.28) take the form

$$\frac{\partial \theta}{\partial \tau} = \frac{\partial^2 \theta}{\partial X^2} - \omega(\theta + \sigma)\frac{\partial e}{\partial \tau} + \overline{Q}_{ch}\phi_1(\eta)\phi_2(\theta); \tag{17.29}$$

$$\frac{d\eta}{d\tau} = \phi_1(\eta)\phi_2(\theta), \tag{17.30}$$

where

$$\phi_2(\theta) = \exp\left(-\frac{1+\sigma}{\theta+\sigma}\cdot\frac{1-\theta}{\gamma}\right); \sigma = \frac{T_0}{T_* - T_0},$$

$$\gamma = \frac{RT_*^2}{E(T_* - T_0)} << 1, \beta = \frac{RT_*}{E} = \frac{\gamma}{1+\sigma}, \overline{Q}_{ch} = \frac{Q_{ch}}{c_\varepsilon \rho (T_* - T_0)}$$

Some transformations of the variables are shown in the Appendix.

Defining the scale temperature as the temperature of reaction products in the thermal theory of combustion [21]:

$$\overline{Q}_{ch} = 1 : T_* = T_0 + Q/c_\varepsilon \rho = T_{b0} \tag{17.31}$$

we reduce the number of parameters. In this case, it is convenient to define the time scale as the characteristic time of the chemical reaction:

$$t_* = k^{-1}\exp\left(\frac{E}{RT_*}\right). \tag{17.32}$$

The solution (17.29), (17.30) of the traveling wave type (wave with constant profile) satisfies the equation system, which is obtained after the transition from

(17.29), (17.30) to the moving coordinate system $\xi = X - V\tau$ and rejection of time derivatives,

$$-V\frac{d\theta}{d\xi} = \frac{d^2\theta}{d\xi^2} + \omega(\theta+\sigma)V\frac{de}{d\xi} + \phi_1(\eta)\phi_2(\theta); \tag{17.33}$$

$$-V\frac{d\eta}{d\xi} = \phi_1(\eta)\phi_2(\theta) \tag{17.34}$$

and (17.19). This system can be partially integrated. Using (17.21) and (17.34), we find from (33)

$$-V\left(1+\frac{\omega}{1-(\alpha V)^2}(\theta+\sigma)\right)\frac{d\theta}{d\xi} = \frac{d^2\theta}{d\xi^2} - V\frac{d\eta}{d\xi}. \tag{17.35}$$

Integrating Eq. (17.35) taking into account that in undisturbed matter $\eta = 0$, we obtain

$$-V\left(1+\frac{\omega\sigma}{1-(\alpha V)^2}\right)\theta - \frac{\omega V}{1-(\alpha V)^2}\frac{\theta^2}{2} = \frac{d\theta}{d\xi} - V\eta. \tag{17.36}$$

Extending ξ to the region of reaction products and assuming that there $\eta = 1$, we come to the quadratic equation for the temperature of products

$$B\theta_b^2 + (2B\sigma - 1)\theta_b + 1 = 0, \tag{17.37}$$

where

$$B = \frac{\omega}{2\left((\alpha V)^2 - 1\right)}.$$

The number of solutions of this equation obviously depends on the parameters $\omega,\ V,\ \alpha,\ \sigma$ [22].

At $B \le 0$, which is possible if $V < \alpha^{-1}$ or $V_n^2 < (\lambda + 2\mu)/\rho$, Eq. (17.37) has a single real solution.

Each value $B < 0$ corresponds to only one temperature $\theta_b = \theta_1 < 1$. When $B \to 0$ ($\omega \to 0$) we have $\theta_b \to 1$.

At $B > 0$ quadratic Eq. (17.37) has two valid solutions: $\theta_b = \theta_A$ and $\theta_b = \theta_B$. If α, ω, σ are fixed, then each temperature value corresponds to a different velocity values V, satisfying the condition $V > \alpha^{-1}$ or $v_n^2 > (\lambda + 2\mu)/\rho$, T, i.e., we have an infinite set of solutions. To two values of temperature θ_A and θ_B correspond two stationary points of the heat conduction Eq. (17.33) at $X \to -\infty$ (in the region of reaction products), similar to the stationary points of the Burgers equation [1], the point A being a stable special point.

An exact solution to this problem cannot be obtained until the end.

17.3.2 The Simplest Asymptotics

Approximate solution of the problem (17.34), (17.35) with conditions

$$\begin{aligned} &\xi \to -\infty : \frac{d\theta}{d\xi} = 0\,(\theta \to \theta_B);\ \eta = 1 \text{ (it is area of reaction products).} \\ &\xi \to +\infty : \theta = 0;\ \eta = 0 \text{ (it is area of reagents).} \end{aligned} \tag{17.38}$$

we perform by the method of joint asymptotic expansions [23].

We pass in the stationary problem (17.34), (17.35), (17.38) to a new variable $\xi_1 = \xi V$, which is equivalent to the introduction of a new spatial scale, i.e.,

$$\xi_1 = \frac{x}{\kappa_T / \mathrm{v}_n}. \tag{17.39}$$

In this case, the definition of velocity remains unchanged:

$$V = \frac{\mathrm{v}_n}{V_*} = \frac{\mathrm{v}_n}{\sqrt{\kappa_T / t_*}}$$

As a result, we come to the equations

$$\begin{aligned} &-\left[1 + \frac{\omega(\theta+\sigma)}{1-\alpha_1}\right]\frac{d\theta}{d\xi_1} = \frac{d^2\theta}{d\xi_1^2} - \frac{d\eta}{d\xi_1}; \\ &-m\gamma\frac{d\eta}{d\xi_1} = \phi_1(\eta)\phi_2(\theta), \end{aligned} \tag{17.40}$$

where $m = V^2\gamma^{-1}$; $\alpha_1 = \alpha^2 V^2$, with the same boundary conditions (17.38).

Let's introduce a variable $p = d\theta / d\xi_1$, which lowers the order of the system:

$$\begin{aligned} &-\left[1 + \frac{\omega(\theta+\sigma)}{1-\alpha_1}\right] = \frac{dp}{d\theta} - \frac{d\eta}{d\theta}; \\ &-m\gamma\frac{d\eta}{d\theta}p = \phi_1(\eta)\phi_2(\theta). \end{aligned} \tag{17.41}$$

The boundary conditions to (17.41) are:

$$\begin{aligned} &\theta \to 0 : p = 0, \eta = 0; \\ &\theta \to \theta_b : p = 0, \eta = 1. \end{aligned} \tag{17.42}$$

The problem in the outer region (inert problem) is obtained after neglecting in the heating region the heat release from the chemical reaction:

$$-\left[1+\frac{\omega(\theta+\sigma)}{1-\alpha_1}\right]=\frac{dp}{d\theta};\ \theta\to 0: p\to 0;\ \eta=0. \tag{17.43}$$

The first integral has the form

$$p=-\left(1+\frac{\omega\sigma}{1-\alpha_1}\right)\theta-\frac{\omega}{1-\alpha_1}\frac{\theta^2}{2}+C;\ \eta=0. \tag{17.44}$$

From the condition in reagents $C=0$. Hence,

$$\frac{d\theta}{d\xi_1}=-\left(1+\frac{\omega\sigma}{1-\alpha_1}\right)\theta-\frac{\omega}{1-\alpha_1}\frac{\theta^2}{2}.$$

Subsequent integration gives:

$$\frac{\theta}{1+\left(\theta+\frac{\sigma}{2}\right)\frac{\omega}{1-\alpha_1}}=C_1\exp\left(-\left(1+\frac{\omega\sigma}{1-\alpha_1}\right)\xi\right). \tag{17.45}$$

The integration constant C_1 can be found from the joining condition with the inner solution.

In the special case $\omega=0$ the solution of the external problem has the form

$$p=-\theta \ \text{ and } \ \theta=C_1\exp(-\xi_1)\,, \tag{17.46}$$

which, obviously, completely coincides with the solution of the classical problem of combustion theory.

To solve the internal problem we pass to the internal variable

$$t=\frac{1-\theta}{\gamma},\, d\theta=-\gamma dt. \tag{17.47}$$

Then the problem in the inner region (in the zone of chemical reactions) takes the form

$$\begin{aligned}
&-\gamma\left[1+\frac{\omega(1-\gamma t+\sigma)}{1-\alpha_1}\right]=-\frac{dp}{dt}+\frac{d\eta}{dt};\\
&m\frac{d\eta}{dt}p=\phi_1(\eta)\phi_2(t);\\
&t\to t_b=\frac{1-\theta_b}{\gamma}: p\to 0;\ y=1,
\end{aligned} \tag{17.48}$$

where

$$\phi_2(t)=\exp\left[-\frac{(1+\sigma)t}{(1+\sigma-\gamma t)}\right].$$

The representation of this function in the form of a Taylor series in powers of a small parameter with accuracy up to the second order of magnitude summands will have the following form

$$\phi_2(t,\gamma) \approx \phi_2(t,0)\left[1 - \frac{t^2}{1+\sigma} + \cdots\right] = \Phi_0 + \Phi_1 + \cdots. \tag{17.49}$$

We find the solution of the inner problem (17.48) in the form

$$\begin{aligned} p &= p_0 + \gamma p_1 + \gamma^2 p_2 + \cdots; \\ \eta &= \eta_0 + \gamma \eta_1 + \gamma^2 \eta_2 + \cdots; \\ m &= m_0 + \gamma m_1 + \gamma^2 m_2 + \cdots \end{aligned} \tag{17.50}$$

Substituting (17.50) into (17.48), we find for the zero approximation

$$\begin{aligned} \frac{dp_0}{dt} &= \frac{d\eta_0}{dt}; \\ m_0 p_0 \frac{d\eta_0}{dt} &= (1-\eta_0)\Phi_0, \end{aligned}$$

for the first

$$\begin{aligned} -\left[1 + \frac{\omega(\sigma+1)}{1-\alpha_1}\right] &= -\frac{dp_1}{dt} + \frac{d\eta_1}{dt}; \\ m_0 p_0 \frac{d\eta_1}{dt} + (p_1 m_0 + p_0 m_1)\frac{d\eta_0}{dt} &= (1-\eta_0)\Phi_1 - \eta_1 \Phi_0, \end{aligned}$$

for the second

$$\begin{aligned} &\frac{\omega t}{1-\alpha_1} = -\frac{dp_2}{dt} + \frac{d\eta_2}{dt}; \\ &m_0 p_0 \frac{d\eta_2}{dt} + (p_1 m_0 + p_0 m_1)\frac{d\eta_1}{dt} + (p_2 m_0 + p_1 m_1 + p_0 m_2)\frac{d\eta_0}{dt} = \\ &= (1-\eta_0)\Phi_2 - \eta_1 \Phi_1 - \eta_2 \Phi_0 \end{aligned}$$

and for the following

$$-\frac{dp_k}{dt}+\frac{d\eta_k}{dt}=0;$$

$$m_0 p_0\frac{d\eta_k}{dt}+(p_1m_0+p_0m_1)\frac{d\eta_{k-1}}{dt}+(p_2m_0+p_1m_1+p_0m_2)\frac{d\eta_{k-2}}{dt}+$$
$$+(p_3m_0+p_2m_1+p_0m_3+m_2p_1)\frac{d\eta_{k-3}}{dt}+\cdots$$
$$+(m_0p_k+m_1p_{k-1}+m_2p_{k-2}+\cdots+m_kp_0)\frac{d\eta_0}{dt}=$$
$$=(1-y_0)\Phi_k-\eta_1\Phi_{k-1}-\eta_2\Phi_{k-2}-\cdots-\eta_k\Phi_0,$$

where

$$\Phi_0=\phi_2(t,\gamma=0);\ \Phi_k=\left.\frac{\partial^k\phi_2(t,\gamma)}{\partial\gamma^k}\right|_{\gamma=0}.$$

Solution of the inner problem in the zero approximation taking into account the condition

$$t\to t_b=\frac{1-\theta_b}{\gamma}:\ p_0\to 0;\ \eta_0\to 1$$

looks like

$$p_0=\eta_0-1;\ m_0(1-\eta_0)=e^{-t_b}-e^{-t}. \tag{17.51}$$

Passing in (17.51) to the outer limit $t\to\frac{1}{\gamma}$ and then equating the inner solution to the outer solution written for $t\to t_b$, we arrive to the system of equalities

$$\left(1+\frac{\omega\sigma}{1-\alpha_1}\right)\theta_b+\frac{1}{2}\frac{\omega}{1-\alpha_1}\theta_b^2-1=0, \tag{17.52}$$

$$m_0=\exp\left(-\frac{1-\theta_b}{\gamma}\right)-\exp\left(-\frac{1}{\gamma}\right)\approx\exp\left(-\frac{1-\theta_b}{\gamma}\right). \tag{17.53}$$

Equation (17.52) coincides with Eq. (17.37), so that the subsequent approximations for the temperature of the products are zero.

As stated above, depending on the model parameters, Eq. (17.52) may have one positive solution, two positive solutions, or no solutions.

In the case $\alpha<1$, the only positive root of Eq. (17.52) $\theta_b<1$ and the front velocity $m_0<1$ decrease with increasing of the coupling coefficient ω and with increasing σ (Fig. 17.1). Both temperature and front velocity are smaller than the corresponding values for the classical theory.

In the case of high-velocity regime $\alpha>1$ from (17.52) we have

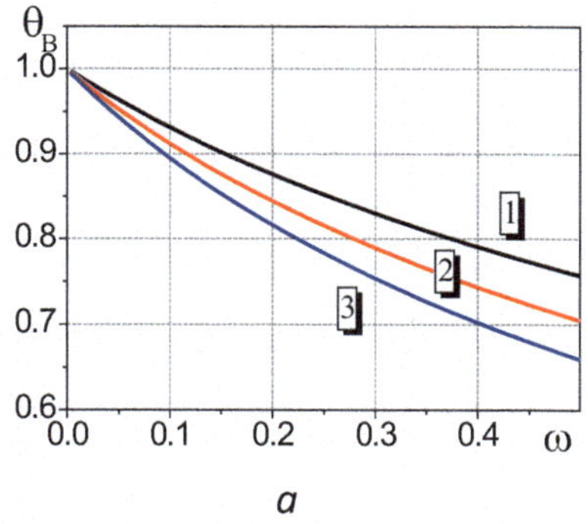

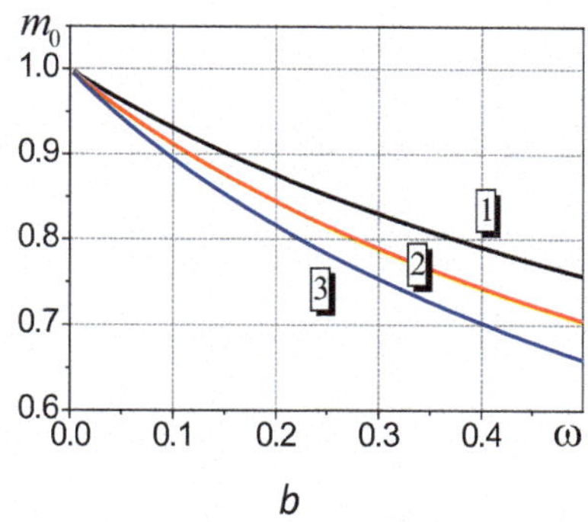

Fig. 17.1 Dependence of product temperature and front velocity for the low-velocity regime on parameters in the model with small deformations. $\gamma = 0.1$: **a** $\alpha_1 = 0, 1$; $\sigma = 1. - 0, 2$; $2. - 0, 4$; $3. - 0, 6$. **b** $\sigma = 0, 4$; $\alpha_1 = 1. - 0, 025$; $2. - 0, 15$; $3. - 0, 3$

$$\theta_{b,\pm} = \frac{1 - \omega_1\sigma \pm \sqrt{(1 - \omega_1\sigma)^2 - 2\omega_1}}{\omega_1},$$

where $\omega_1 = \frac{\omega}{\alpha_1 - 1}$.

Both temperature values are real if the inequality is satisfied

$$\omega_1^2\sigma^2 - 2\omega_1(\sigma + 1) + 1 = \sigma_1^2(\omega_1 - \omega_{1,+})(\omega_1 - \omega_{1,-}) > 0,$$

where

$$\omega_{1,\pm} = \frac{\sigma + 1}{\sigma^2}\left[1 \pm \sqrt{1 - \frac{\sigma^2}{(\sigma + 1)^2}}\right].$$

Since $\frac{\sigma^2}{(\sigma+1)^2} < 1$ (always), then $\omega_{1-} > 0$ and $\omega_{1+} > 0$.

At least one root of Eq. (17.52) in this case is positive if $\omega_1 < \omega_* = 1/\sigma$. Consequently, high-velocity modes of transformation exist only in the parameter region below the curve $\omega_{1,-}(\sigma)$, i.e., at small ω (which corresponds to the real area of variation of this parameter) and high values α_1:

$$\omega < (\alpha_1 - 1)\frac{\sigma + 1}{\sigma^2}\left[1 - \sqrt{1 - \frac{\sigma^2}{(\sigma + 1)^2}}\right].$$

It is obvious (Fig. 17.2) that the higher α_1, the higher values of the coupling coefficient, the higher values of the coupling coefficient can exist the high-velocity mode.

From the two values of the product temperature θ_{b2} and θ_{b1} at $\alpha > 1$ the below curve $\omega_{1,-}$, we choose the lower value to calculate the velocity. We assume that

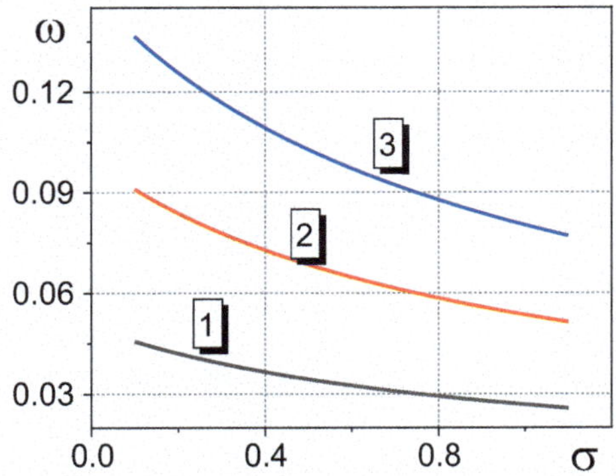

Fig. 17.2 Values of the coupling coefficient, below which there is a high-speed two-temperature regime, depending on the σ. $\alpha_1 = (1)$ 1.1; (2) 1.2; (3) 1.3

the lower temperature corresponds to the reaction zone and the higher temperature corresponds to the products.

Both temperature and combustion rate depend on the parameters of the problem, as shown in Fig. 17.3. Obviously, at $\omega = 0$ we have $\theta_b = 1$ and which coincides with the known solutions of the theory of solid-phase combustion. As σ increases, the smaller temperature increases and the larger temperature decreases. The smaller the coupling coefficient ω, the higher the larger temperature (product temperature). Both the temperature in the reaction zone and the fast wave velocity increase rapidly with increasing ω.

As the numerical study of the problem [22] shows, both temperatures $\theta_{b2} > \theta_{b1} > 1$ (for $\alpha > 1$) have a physical meaning, which actually indicates a two-temperature structure of the combustion front. A similar result is obtained in more complicated cases [24–26].

Thus, the asymptotic analysis of the problem with conditions typical for combustion theory, carried out above, gives two types of traveling wave solutions.

The first continuous solution $\left(V < \alpha^{-1}\right)$ at $\omega \to 0$ converges to the solution of the simplest classical problem of the theory of solid-phase combustion, which includes the equations

$$\frac{\partial \theta}{\partial \tau} = \frac{\partial^2 \theta}{\partial X^2} + \theta_0^{-1} \phi_1(\eta)\phi_2(\theta);$$
$$\frac{d\eta}{d\tau} = \phi_1(\eta)\phi_2(\theta)$$

or in automodel variables

$$-V\frac{d\theta}{d\xi} = \frac{d^2\theta}{d\xi^2} + \phi_1(\eta)\phi_2(\theta);$$
$$-V\frac{d\eta}{d\xi} = \phi_1(\eta)\phi_2(\theta)$$

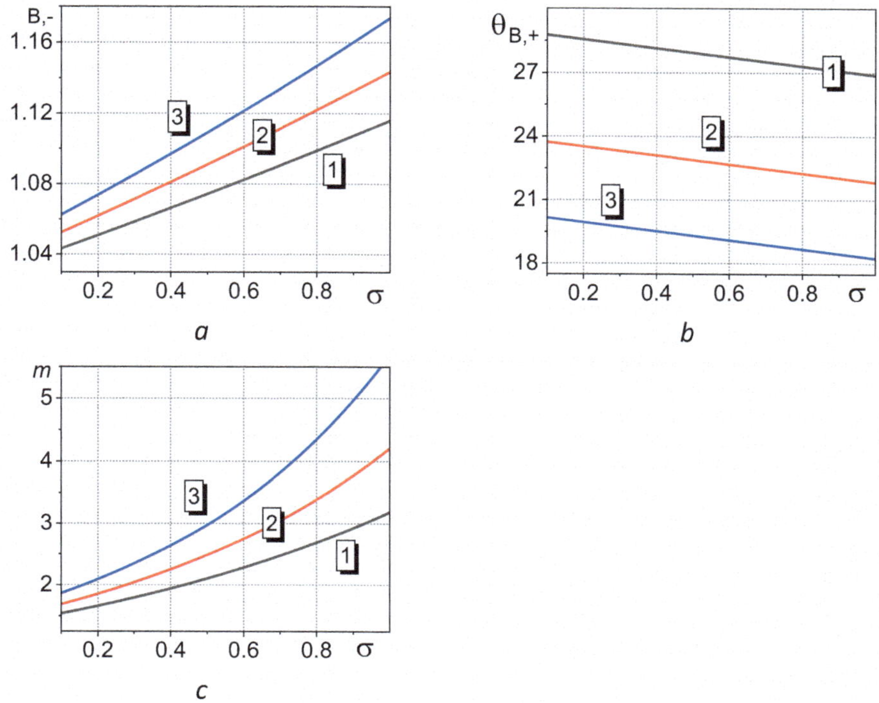

Fig. 17.3 Dependence of reaction zone temperature (**a**), product temperature (**b**) and mass rate of combustion (c) on σ. Coupling coefficient:$\omega = 1. - 0, 1;\ 2. - 0, 12;\ 3. - 0, 14; \alpha_1 = 2.5; \gamma = 0.1$

with boundary conditions (17.38).

The latter equations together with (17.19), (17.20) form a system of equations of the uncoupled theory of thermoelasticity written in automodel variables. According to the theory of combustion [21], this problem has a single solution. For any value $\omega \geq 0$, the solution is also singular; a wave of constant profile propagates at a speed less than the speed of sound.

The solutions of the second type $\left(V > \alpha^{-1}\right)$ are also continuous and are shock-wave solutions. These solutions describe shock waves in a heat conducting medium with an exothermic chemical reaction or solid-phase detonation.

Using the designations U, ν (17.23), we present Eq. (17.40) in the form

$$-U\frac{d\theta}{d\xi} + \theta\frac{d\theta}{d\xi} = \nu\frac{d^2\theta}{d\xi^2} + \nu\phi_1(\eta)\phi_1(\theta). \tag{17.54}$$

At $\nu \to 0$, which corresponds to $V \to \alpha^{-1}$, the continuous solutions of Eq. (17.54) converge to the shock-wave discontinuous solutions of Eq. (17.26).

The stresses and strains in the automodel problem are completely determined by the temperature and conversion degree, which is easy to show using the integrals of the equation of motion (17.21) and the relations between stresses and strains written in dimensionless variables. For the components of the stress tensor we have

$$s_{11} = s = \frac{(\alpha V)^2 \theta}{1 - (\alpha V)^2}, \; s_{22} = s_{33} = \frac{\left[1 - (\alpha V)^2 + h\right]}{1 - (\alpha V)^2} \theta, \quad (17.55)$$

where

$$s_{ij} = \sigma_{ij} \big/ \sigma_*, ; \sigma_* = 3K\alpha_T (T_* - T_0); \; h = \frac{\lambda}{\lambda + 2\mu}.$$

Thus, in the noncoupled problem ($\omega = 0$), a stress (strain) wave of constant profile always runs with the velocity of the chemical conversion wave, less than the velocity of sound, if it is not amplified by an external mechanical action. In the coupled problem ($\omega \neq 0$) there are two types of constant profile waves. The first, subsonic, is a consequence of the nonlinear dependence of the chemical heat release rate on temperature (nonlinear interaction of heat conduction and chemical conversion). The second wave exists due to energy transfer by the wave of mechanical disturbances ("hydrodynamic" energy transfer) or due to nonlinear interaction of heat transfer and mechanical disturbance propagation. The linearized system of equations has no automodel solutions of this type.

Note that in the case of $\phi_1(\eta) = 1$ Eq. (17.35) is a special case of the Liénard equation, whose solution at fixed parameters, including a given value of V, exists and is unique [27]:

$$\frac{d^2\theta}{d\xi^2} + h(\theta)\frac{d\theta}{d\xi} + \phi_2(\theta) = 0.$$

The type of solution essentially depends on the function standing at the first derivative, including the values of the parameters included in it.

Similar results occur in the other cases described below.

17.3.3 *Model of Solid-Phase Combustion with Concentration Expansion*

In the model of solid-phase transformation discussed above, it was assumed that the consequence of the chemical reaction is heat release during the reaction, leading to the appearance of a temperature gradient and thermal stresses. However, a chemical reaction is always accompanied by a change in properties or a change in volume.

Within the framework of the small strain, displacement and velocity approximation used above, the change in volume during the reaction can be accounted for in the Duhamel-Neumann equations by augmenting (17.3) with a summand to account for the change in composition:

$$\sigma_{ij} = 2\mu\varepsilon_{ij} + \delta_{ij}[\lambda\varepsilon_{kk} - Kw], \tag{17.56}$$

where

$$w = 3\left[\alpha_T(T - T_0) + \sum_{k=1}^{n} \alpha_k(C_k - C_{k0})\right],$$

C_k—are concentrations of the components involved in the reactions; α_k—are concentration expansion coefficients; n- number of components.

Assuming further that the chemical reaction can be described by the simplest summary scheme $A \to B$, for the function w we obtain

$$w = 3[\alpha_\text{T}(T - T_0) + (\alpha_B - \alpha_A)\,\eta]. \tag{17.57}$$

Here α_B and α_A represent the concentration expansion coefficients for the product and reactant, which in thermodynamics are defined in the same way as the coefficient of thermal expansion and are directly linked to the partial specific volumes of the substances involved in the reaction [28]. In fact, the difference of the coefficients reflects the fact of volume change in the reaction. (This, however, does not exclude the dependence of physicochemical properties on temperature and conversion degree). In the case of small deformations, accelerations, and displacements, the equations of motion remain in the form (17.7), and the continuity equation is not required. Considering (17.57), we find:

$$\rho\frac{\partial^2\varepsilon}{\partial t^2} = (\lambda + 2\mu)\frac{\partial^2\varepsilon}{\partial x^2} - 3K\left(\alpha_T\frac{\partial^2 T}{\partial x^2} + (\alpha_B - \alpha_A)\frac{\partial^2\eta}{\partial x^2}\right) \tag{17.58}$$

In variables (17.13) we have Eqs. (17.15), (17.30) and

$$\frac{\partial^2 e}{\partial X^2} - \left(\frac{\partial^2\theta}{\partial X^2} + g\frac{\partial^2\eta}{\partial X^2}\right) = \alpha^2\frac{\partial^2 e}{\partial\tau^2}, \tag{17.59}$$

where $e = \frac{\varepsilon}{\varepsilon_*}$, $\varepsilon_* = \frac{3K\alpha_T}{\lambda+2\mu}(T_* - T_0)$; $g = (\alpha_B - \alpha_A)\big/(\alpha_T(T_* - T_0))$—is a dimensionless parameter characterizing the reaction occurring in the volume of the substance. At $g > 0$ reaction proceeds with increasing volume, at $g < 0$—reduced in volume, at $g = 0$—specific volume does not change during the reaction.

In a stationary wave traveling to the right with velocity V, from (17.59) we have:

$$\frac{d^2 e}{d\xi^2} - \left(\frac{d^2\theta}{d\xi^2} + g\frac{d^2\eta}{d\xi^2}\right) = (\alpha V)^2\frac{d^2 e}{d\xi^2}.$$

Consequently, in a combustion wave, deformations are related to temperature and degree of transformation by the relation

$$e = \frac{1}{1-(\alpha V)^2}\left[\theta + g\,\eta\right]. \tag{17.60}$$

Then, similarly to (17.55) we have

$$s = \frac{(\alpha V)^2}{1-(\alpha V)^2}\left[\theta + g\,\eta\right].$$

These relations indicate that the traveling wave of chemical reaction (autowave) is accompanied by stress and strain waves that completely repeat the structure of the reaction wave.

In this case the stationary heat conduction equation takes the form

$$-V\left(1+\frac{\omega\,(\theta+\sigma)}{1-(\alpha V)^2}\right)\frac{d\theta}{d\xi} = \frac{d^2\theta}{d\xi^2} + \phi_1(\eta)\,\phi_2(\theta)\left(1-\frac{\omega\,(\theta+\sigma)}{1-(\alpha V)^2}\,g\right). \tag{17.61}$$

Using the notations U, ν (17.23), we represent the Eq. (17.61) in the form

$$-U\frac{d\theta}{d\xi} + \theta\frac{d\theta}{d\xi} = \nu\frac{d^2\theta}{d\xi^2} + \nu\phi_1(\eta)\phi_1(\theta)\left(1-\frac{\omega\,(\theta+\sigma)}{1-(\alpha V)^2}\,g\right) \tag{17.62}$$

And this is the automodel form of the equation

$$\frac{\partial\theta}{\partial\xi} + \theta\frac{d\theta}{d\xi} = \nu\frac{d^2\theta}{d\xi^2} + \nu\phi_1(\eta)\phi_1(\theta)\left(1-\frac{\omega\,(\theta+\sigma)}{1-(\alpha V)^2}\,g\right). \tag{17.63}$$

Thus, we again obtain that at $\nu \to 0$ the solutions of Eq. (17.63) converge to the shock-wave discontinuous solutions of Eq. (17.26).

The approximate solution of the problem of stationary front propagation, as above, can be found by the method of joint asymptotic expansions.

In variables θ and $p = d\theta/d\xi_1$, where ξ_1 is determined by the equality (17.39), the stationary problem will include the equations

$$-\left(1+\frac{\omega\,(\theta+\sigma)}{1-\alpha_1}\right) = \frac{dp}{d\theta} - \frac{d\eta}{d\theta}\left(1-\frac{\omega\,(\theta+\sigma)}{1-\alpha_1}\,g\right);$$

$$-m\gamma\frac{d\eta}{d\theta}p = \phi_1(\eta)\phi_2(\theta)$$

and boundary conditions (17.42).

The outer problem does not differ from the previous one. Its solution has the form (17.45).

Turning to the internal variable, we obtain the problem

$$-\gamma\left[1+\frac{\omega(1+\sigma-\gamma t)}{1-\alpha_1}\right]=-\frac{dp}{dt}+\frac{d\eta}{dt}\left[1-\frac{\omega(1+\sigma-\gamma t)}{1-\alpha_1}g\right];$$
$$m\frac{d\eta}{dt}p=\phi_1(\eta)\phi_2(t); \tag{17.64}$$
$$t\to t_b=\frac{1-\theta_b}{\gamma}:p\to 0;\ \eta=1.$$

Its solution is found in the form (17.40), similarly to the previous one. We obtain a series of problems:

For zero approximation we have:

$$\frac{dp_0}{dt}=\frac{d\eta_0}{dt}\left[1-g\frac{\omega(1+\sigma)}{1-\alpha_1}\right];$$
$$m_0p_0\frac{d\eta_0}{dt}=(1-\eta_0)\Phi_0.$$

For the first approximation -

$$-\left[1+\frac{\omega(\sigma+1)}{1-\alpha_1}\right]=-\frac{dp_1}{dt}+\frac{d\eta_1}{dt}\left(1-\frac{g\omega}{1-\alpha_1}(1+\sigma)\right)+\frac{d\eta_0}{dt}\frac{g\omega}{1-\alpha_1}t;$$
$$m_0p_0\frac{d\eta_1}{dt}+(p_1m_0+p_0m_1)\frac{d\eta_0}{dt}=(1-\eta_0)\Phi_1-\eta_1\Phi_0.$$

The solution for the zero approximation, taking into account the condition in the reaction products, is:

$$p_0=(\eta_0-1)\left[1-g\frac{\omega(1+\sigma)}{1-\alpha_1}\right]; \tag{17.65}$$

$$m_0(1-\eta_0)\left[1-g\frac{\omega(1+\sigma)}{1-\alpha_1}\right]=e^{-\frac{1-\theta_b}{\gamma}}-e^{-t}. \tag{17.66}$$

Equating the outer solution at $\theta\to\theta_b$ with the outer limit of the inner solution (17.65) we find the equation for the temperature of products:

$$\frac{\omega}{1-\alpha_1}\frac{\theta_b^2}{2}+\left(1+\frac{\omega\sigma}{1-\alpha_1}\right)\theta_b-1+g\frac{\omega(1+\sigma)}{1-\alpha_1}=0. \tag{17.67}$$

The outer limit for (17.66) gives the equation for the front velocity:

$$m_0\left[1-g\frac{\omega(1+\sigma)}{1-\alpha_1}\right]=\exp\left(-\frac{1-\theta_b}{\gamma}\right)-\exp\left(-\frac{1}{\gamma}\right). \tag{17.68}$$

The parameter g affects the characteristics of the co-propagating waves ambiguously. Thus, at $\alpha_1<1$, the product temperature and wave velocity are always less than unity (Fig. 17.4a). Moreover, when concentration expansion is taken into account, the wave velocity tends to zero as the coherence coefficient increases. With g increase, the positive roots of Eq. (17.67) disappear and the wave does not exist. This occurs at $g > \frac{1-\alpha_1}{\omega(1+\sigma)}$.

If the reaction proceeds with volume reduction, the positive solution (17.67) always exists. In contrast to the previous variant, the temperature of the products can be both greater and slightly less than unity (Fig. 17.4b, left). The wave velocity can both decrease with the increase of the binding coefficient and increase (Fig. 17.4b, right).

For the high-velocity regime, if the reaction proceeds with volume expansion, the region of existence of two-temperature regimes changes insignificantly. The real roots of Eq. (17.67) exist if the inequality is satisfied

$$\omega_1^2\left[\sigma^2 - 2g(1+\sigma)\right] - 2\omega(1+\sigma) + 1 > 0,$$

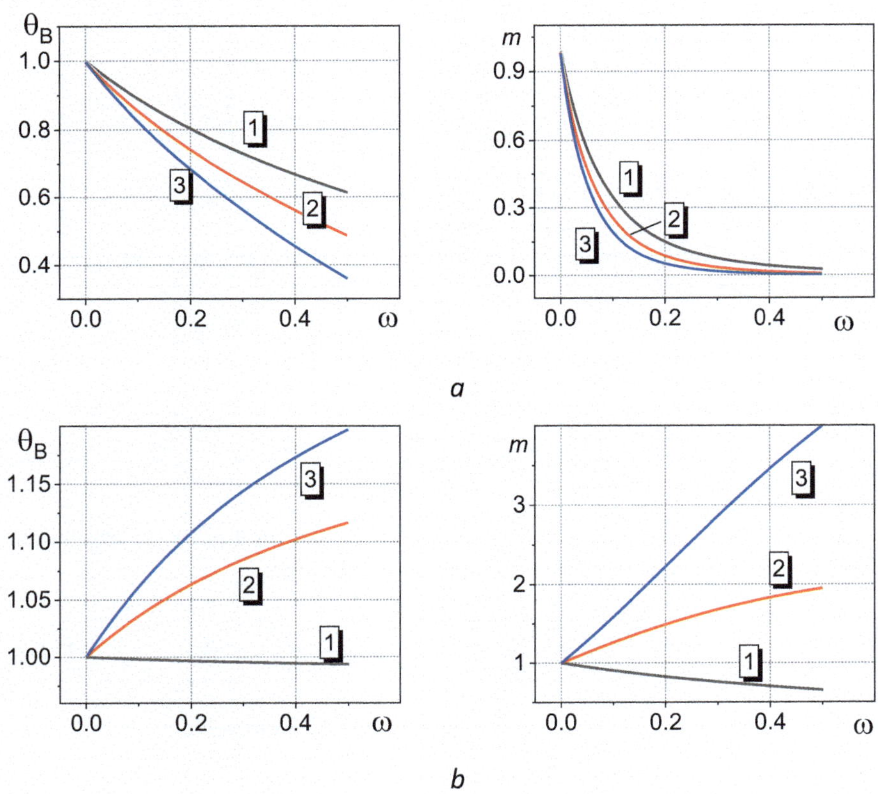

Fig. 17.4 Dependence of product temperature (left) and front velocity (right) for a slow wave. $\alpha = 0.1$; $\gamma = 0.1$; $\sigma = 0.3$. **a** The reaction proceeds with increasing volume: $g =$ (1) 0.25; (2) 0.5; (3) 0.75. **b** The reaction proceeds with volume decrease: $g =$(1) -0.6; (2) 0.9; (3) 1.1

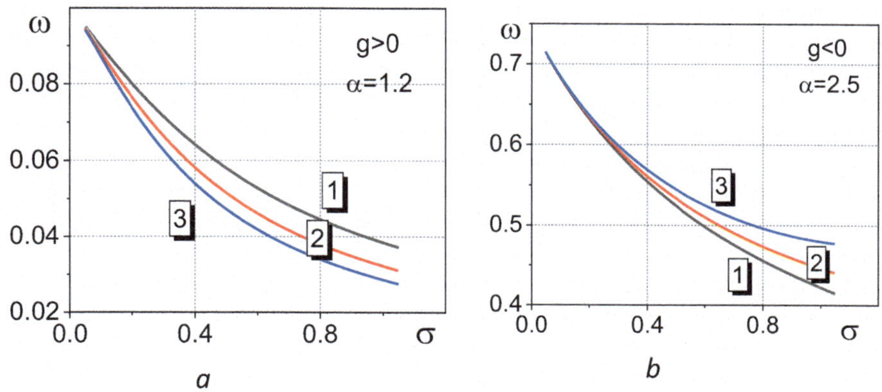

Fig. 17.5 Dependence on σ of the value of the coupling coefficient, below which, there exists a two-temperature solution of the problem. **a** The reaction proceeds with increasing volume: g = 1. 0.25; 2. 0.5; 3. 0.75; **b** The reaction proceeds with volume decrease: g = 1. −0.02; 2. −0.04; 3. −0.06

In this case, both roots are positive under the same condition as above. An example of the dependence of the value of the coupling coefficient, below which this regime exists, on the parameter σ at different values of g is shown in Fig. 17.5. The region of negative values of g, where this mode can be detected, depends on α_1.

In a reaction with increasing volume, the reaction zone temperature and wave velocity increase with increasing σ, g (Fig. 17.6a), but the second temperature decreases (not shown in the figures). If the reaction proceeds with decreasing volume, the influence of |g| is opposite (Fig. 17.6b).

17.3.4 Generalization of the Model to the Case of Finite Deformations

The above-mentioned coupled models of solid-phase combustion take place if the change in the density of the medium can be neglected and the deformations are small. If these conditions are not fulfilled, then in the system of Eqs. (17.6), (17.23), (17.33) the partial derivatives in time should be replaced by full derivatives ($\partial .../\partial t \rightarrow d.../dt = \partial ../\partial t + \mathbf{v}\nabla ...$), and also to supplement these equations by the continuity equation.

$$\frac{d\rho}{dt} + \rho\nabla\cdot\mathbf{v} = 0.$$

Linear relations between the components of stress and strain tensors are written in increments:

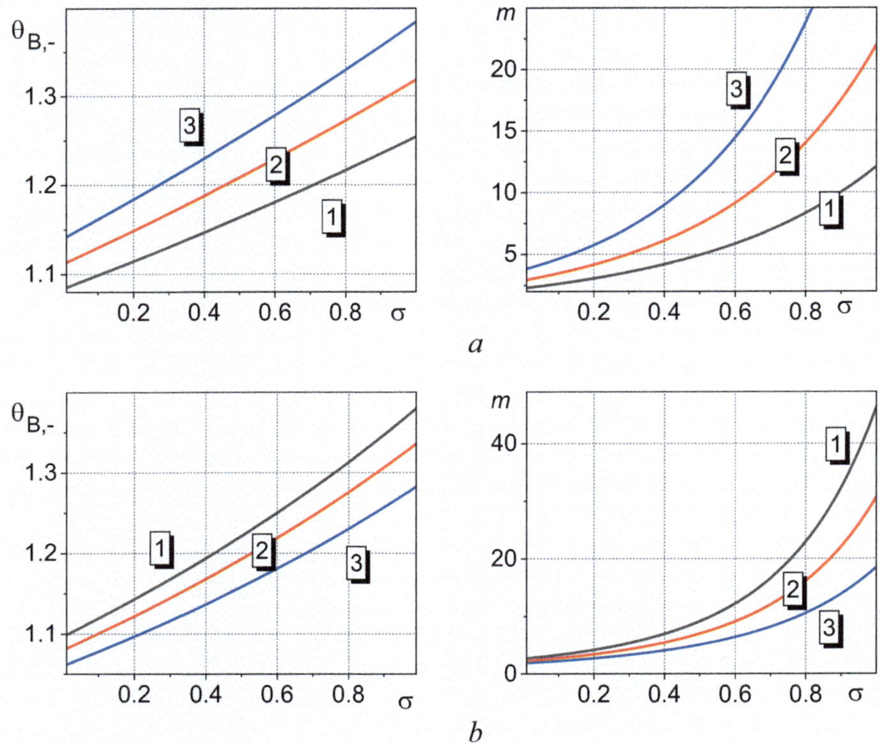

Fig. 17.6 Reaction zone temperature (left) and wave speed (right) for high-velocity mode. **a** The reaction proceeds with volume expansion: ω = 0.02; α = 1.2; g = 1. 0.25; 2. 0.5;3. 0.75; **b** The reaction proceeds with volume decrease: ω = 0.25; α = 0.25; g = 1. −0.02; 2. −0.1; 3. −0.2

$$d\sigma_{ij} = 2\mu d\varepsilon_{ij} + \delta_{ij}(\lambda d\varepsilon_{kk} - K d\mathrm{w}).$$

Then the one-dimensional problem of reaction front propagation with the total scheme "reagent-product" takes the following form

$$\rho c_\varepsilon \frac{dT}{dt} = \lambda_T \frac{\partial^2 T}{\partial x^2} - 3\rho\alpha_T T \frac{d}{dt}\left(\frac{K}{\rho}\varepsilon\right) + Q_{ch} k_0 \phi_1(\eta)\phi_2(T);$$
$$\frac{d\eta}{dt} = k_0 \phi_1(\eta)\phi_2(T);$$
$$\rho \frac{d\mathrm{v}}{dt} = \frac{\partial \sigma_{11}}{\partial x};$$
$$\frac{d\rho}{dt} + \rho \frac{d\mathrm{v}}{dx} = 0;$$
$$\frac{d\sigma_{11}}{dt} = (\lambda + 2\mu)\frac{d\varepsilon}{dt} - K\frac{d\mathrm{w}}{dt},$$

where, as above, in one-dimensional approximation $\varepsilon_{kk} \approx \varepsilon_{11} = \varepsilon$. Consequently, in the coordinate system associated with the reaction front moving to the right with velocity v_n, we have the equation system

$$
\begin{aligned}
&-M\, c_\varepsilon \frac{dT}{dx} = \lambda_T \frac{d^2T}{dx^2} + 3\alpha_T TM \frac{d}{dx}\left(\frac{K}{\rho}\varepsilon\right) + Q_{ch} k_0 \phi_1(\eta)\phi_2(T);\\
&-M \frac{d\eta}{dx} = \rho k_0 \phi_1(\eta)\phi_2(T);\\
&-M \frac{d\mathrm{v}}{dx} = \frac{d\sigma_{11}}{dx};\\
&-M \frac{d\rho}{dx} + \rho^2 \frac{d\mathrm{v}}{dx} = 0;\\
&\frac{d\sigma_{11}}{dx} = (\lambda + 2\mu)\frac{d\varepsilon}{dx} - K\frac{d\mathrm{w}}{dx},
\end{aligned}
\tag{17.69}
$$

where $M = \rho(\mathrm{v_n} - \mathrm{v})$—mass rate of "combustion", and w follows from (17.57). These five equations are not enough to find the six unknowns T,η,ρ,ε,v,σ_{11}

Different approximations can be used to close this system. Let us use an approximation similar to [19].

Let us assume, for simplification of the problem $\varepsilon_{kk} \approx \varepsilon \approx \rho_0 / \rho - 1$ and we will restrict ourselves to thermal deformations, $\mathrm{w} = 3\alpha_T(T - T_0)$.

Hence,

$$
\frac{d\varepsilon}{dx} = -\frac{\rho_0}{\rho^2}\frac{d\rho}{dx}. \tag{17.70}
$$

The third and fifth equations of the system (17.69) give

$$
-M \frac{d\mathrm{v}}{dx} = (\lambda + 2\mu)\frac{d\varepsilon}{dx} - 3K\alpha_T \frac{dT}{dx}
$$

or taking into account (17.70)

$$
M \frac{d\mathrm{v}}{dx} = (\lambda + 2\mu)\frac{\rho_0}{\rho^2}\frac{d\rho}{dx} + 3K\alpha_T \frac{dT}{dx}.
$$

From the fourth equation of the system (17.69) the equation

$$
\frac{d\rho}{dx} = \frac{\rho^2}{M}\frac{d\mathrm{v}}{dx}. \tag{17.71}
$$

follows; then we find

$$\frac{d\varepsilon}{dx} = -\frac{\rho_0}{M}\frac{d\mathrm{v}}{dx};$$
$$M\frac{d\mathrm{v}}{dx} = (\lambda + 2\mu)\frac{\rho_0}{M}\frac{d\mathrm{v}}{dx} + 3K\alpha_T\frac{dT}{dx}.$$

The third Eq. (17.69) gives the stresses.

As a result, we find the relationship of velocity, density, strains and stresses with temperature:

$$\frac{d\mathrm{v}}{dx} = \frac{3\alpha_T MK}{\left[M^2 - \rho_0(\lambda + 2\mu)\right]}\frac{dT}{dx};$$
$$\frac{1}{\rho^2}\frac{d\rho}{dx} = \frac{1}{M}\frac{d\mathrm{v}}{dx} = \frac{3\alpha_T K}{\left[M^2 - \rho_0(\lambda + 2\mu)\right]}\frac{dT}{dx};$$
$$\frac{d\varepsilon}{dx} = -\frac{\rho_0}{M}\frac{d\mathrm{v}}{dx} = -\frac{\rho_0 3\alpha_T K}{\left[M^2 - \rho_0(\lambda + 2\mu)\right]}\frac{dT}{dx};$$
$$\frac{\partial\sigma_{11}}{\partial x} = -M\frac{d\mathrm{v}}{dx} = -\frac{M^2 3\alpha_T K}{\left[M^2 - \rho_0(\lambda + 2\mu)\right]}\frac{dT}{dx}$$

and taking into account conditions in unperturbed matter we obtain:

$$\mathrm{v} = \frac{3MK\alpha_T(T - T_0)}{\left[M^2 - \rho_0(\lambda + 2\mu)\right]};\ \rho = \frac{\rho_0}{1 - \frac{3K\alpha_T(T - T_0)}{M^2 - \rho_0(\lambda+2\mu)}\rho_0};$$
$$\varepsilon = -\frac{3\rho_0 K\alpha_T(T - T_0)}{\left[M^2 - \rho_0(\lambda + 2\mu)\right]};\ \sigma_{11} = -\frac{3M^2K\alpha_T(T - T_0)}{\left[M^2 - \rho_0(\lambda + 2\mu)\right]}. \tag{17.72}$$

The first equation of the system (17.69) taking into account the second equation takes the form:

$$-M\,c_\varepsilon\frac{dT}{dx} = \lambda_T\frac{d^2T}{dx^2} + 3\alpha_T TM\frac{d}{dx}\left(\frac{K}{\rho}\varepsilon\right) - Q_{ch}\frac{M}{\rho}\frac{d\eta}{dx}.$$

Using the transformations

$$Z = \frac{d}{dx}\left(\frac{K}{\rho}\varepsilon\right) = \frac{K}{\rho}\frac{d\varepsilon}{dx} - \frac{K\varepsilon}{\rho^2}\frac{d\rho}{dx} =$$
$$= -\frac{K}{\rho}\frac{\rho_0 3\alpha_T K}{\left[M^2 - \rho_0(\lambda + 2\mu)\right]}\frac{dT}{dx} - K\varepsilon\frac{3\alpha_T K}{\left[M^2 - \rho_0(\lambda + 2\mu)\right]}\frac{dT}{dx}$$

or next -

$$Z = -\frac{3\alpha_T K^2}{\left[M^2 - \rho_0(\lambda + 2\mu)\right]}\left(\varepsilon + \frac{\rho_0}{\rho}\right)\frac{dT}{dx} = -\frac{3\alpha_T K^2}{\left[M^2 - \rho_0(\lambda + 2\mu)\right]}\left(2\frac{\rho_0}{\rho} - 1\right)\frac{dT}{dx} =$$
$$= -\frac{3\alpha_T K^2}{\left[M^2 - \rho_0(\lambda + 2\mu)\right]}\left(1 - \frac{6K\alpha_T(T - T_0)}{M^2 - \rho_0(\lambda + 2\mu)}\rho_0\right)\frac{dT}{dx},$$

we get

$$-M\, c_\varepsilon \frac{dT}{dx} = \lambda_T \frac{d^2T}{dx^2} -$$
$$-T\frac{(3\alpha_T K)^2 M}{\left[M^2 - \rho_0(\lambda + 2\mu)\right]}\left(1 - \frac{6K\alpha_T(T - T_0)}{M^2 - \rho_0(\lambda + 2\mu)}\rho_0\right)\frac{dT}{dx} - Q_{ch}\frac{M}{\rho}\frac{d\eta}{dx}. \tag{17.73}$$

Let's pass in Eq. (17.73) and in the second equation of the system (17.69), and also in relations (17.72) to the variables

$$\theta,\ \overline{\rho} = \frac{\rho}{\rho_0},\ \overline{\mathrm{v}} = \frac{\mathrm{v}}{\sqrt{\kappa_T / t_*}},\ \xi, e = \frac{\varepsilon}{\varepsilon_*},\ s = \frac{\sigma_{11}}{\sigma_*},$$

where $\sigma_* = 3K\alpha_T(T_* - T_0)$, and the scale for deformations ε_* becomes a new small parameter:

$$\delta = \varepsilon_* = \frac{3K\alpha_T}{\lambda + 2\mu}(T_* - T_0) << 1.$$

The equations of heat conduction and kinetics will take the form

$$-\overline{m}\frac{d\theta}{d\xi} = \frac{d^2\theta}{d\xi^2} + (\theta + \sigma)\frac{\overline{m}\omega}{1 - \alpha_1}\left(1 + \frac{2\delta\theta}{1 - \alpha_1}\right)\frac{d\theta}{d\xi} - \frac{\overline{m}}{\overline{\rho}}\frac{d\eta}{d\xi}; \tag{17.74}$$

$$-\overline{m}\frac{d\eta}{d\xi} = \overline{\rho}\phi_1(\eta)\phi_2(\theta). \tag{17.75}$$

The solution of the stationary problem with conditions (17.38) allows us to determine all other quantities in the thermomechanical combustion wave:

$$\overline{\mathrm{v}} = -\frac{\overline{m}\cdot\delta\theta}{1 - \alpha_1};\ \overline{\rho} = \frac{1}{1 + \frac{\delta\theta}{1 - \alpha_1}};\ e = \frac{\theta}{1 - \alpha_1};\ s = \frac{\alpha_1\theta}{1 - \alpha_1}. \tag{17.76}$$

Here

$$\alpha_1 = \frac{M^2}{\rho_0(\lambda + 2\mu)} = \frac{\rho^2(\mathrm{v_n} - \mathrm{v})^2}{\rho_0^2\frac{(\lambda + 2\mu)}{\rho_0}}.$$

First, however, let us show that the obtained heat conduction equation at $\alpha_1 > 1$ is a limiting version of the generalized shock-wave equation. By means of simple transformations and taking into account (17.75), Eq. (17.74) is reduced to the form

$$-U\frac{d\theta}{d\xi} + (\theta)\frac{d\theta}{d\xi} = \nu\frac{d^2\theta}{d\xi^2} + \nu\phi_1(\eta)\phi_2(\theta),$$

where

$$U = \omega^{-1}(\alpha_1 - 1 - \omega\sigma);\ \nu = \frac{\alpha_1 - 1}{\overline{m}\omega};\ (\theta) = \theta\left(1 + \frac{2\delta\sigma}{1-\alpha_1}\right) + \frac{2\delta\theta^2}{1-\alpha_1}.$$

At $\nu \to 0$ we come to the equation

$$-U\frac{d\theta}{d\xi} + (\theta)\frac{d\theta}{d\xi} = 0,$$

and this is the automodel form of the equation

$$\frac{\partial\theta}{\partial\tau} + (\theta)\frac{d\theta}{d\xi} = 0.$$

Since there is a linear relationship between s, e and temperature, these quantities also satisfy a similar nonlinear equation.

To find the temperature of the products and the wave velocity similarly to the previous one, we pass to a new variable $\xi_1 = \xi\overline{m}$ and we emphasize the obviously small parameter γ:

$$-\left(1 + \frac{\omega(\theta+\sigma)}{1-\alpha_1}\left(1 + \frac{2\delta\theta}{1-\alpha_1}\right)\right)\frac{d\theta}{d\xi_1} = \frac{d^2\theta}{d\xi_1{}^2} - \frac{1}{\bar{\rho}}\frac{d\eta}{d\xi_1};$$

$$-m\gamma\frac{d\eta}{d\xi_1} = \bar{\rho}\phi_1(\eta)\phi_2(\theta),$$

where $m = \frac{\overline{m}}{\gamma}$.

In the heating zone we have the equation:

$$-\left(1 + \frac{\omega(\theta+\sigma)}{1-\alpha_1}\left(1 + \frac{2\delta\theta}{1-\alpha_1}\right)\right)\frac{d\theta}{d\xi_1} = \frac{d^2\theta}{d\xi_1^2},$$

or:

$$-\left(1 + \frac{\omega(\theta+\sigma)}{1-\alpha_1}\left(1 + \frac{2\delta\theta}{1-\alpha_1}\right)\right) = \frac{dp}{d\theta}.$$

The solution of this equation has the form

$$p = -\left(C\theta + B\theta^2 + A\theta^3\right), \tag{17.77}$$

where

$$C = 1 + \frac{\omega\sigma}{1-\alpha_1};\ B = \frac{1}{2}\frac{\omega}{1-\alpha_1}\left(1 + \frac{2\delta\sigma}{1-\alpha_1}\right);\ A = \frac{1}{3}\frac{2\delta\omega}{(1-\alpha_1)^2}.$$

The internal problem (the problem for the reaction zone) has the form:

$$\begin{aligned}
&-\gamma\left(1 + \frac{\omega}{1-\alpha_1}(1+\sigma-\gamma t)\left(1 + \frac{2\delta}{1-\alpha_1}(1-\gamma t)\right)\right) = -\frac{dp}{dt} + \frac{d\eta}{dt};\\
&m\left(1 + \frac{\delta}{1-\alpha_1}(1-\gamma t)\right)\frac{d\eta}{dt}p = \phi_1(\eta)\phi_2(\theta);\\
&t \to t_b = \frac{1-\theta_b}{\gamma} : p \to 0;\ \eta = 1.
\end{aligned} \tag{17.78}$$

This problem differs from [23], where a coarser approximation is used. In the zero approximation the solution of the inner problem differs little from (17.51)

$$p_0 = \eta_0 - 1;\ m_0\left(1 + \frac{\delta}{1-\alpha_1}\right)(1-\eta_0) = e^{-t_b} - e^{-t}.$$

Then, using the condition of splicing of internal and external solutions, we come to the equations for the temperature of reaction products and front velocity:

$$C\theta_b + B\theta_b^2 + A\theta_b^3 - 1 = 0. \tag{17.79}$$

$$m_0\left(1 + \frac{\delta}{1-\alpha_1}\right) = \exp\left(-\frac{1-\theta_b}{\gamma}\right) - \exp\left(-\frac{1}{\gamma}\right). \tag{17.80}$$

Depending on the value and sign of the discriminant of Eq. (17.79), it has one or three valid solutions. The analysis shows that in the real area of parameter variation this equation has either one positive root, or two positive roots, or three positive roots.

The analysis of a similar problem [23], in which slightly different coefficients are obtained for a coarser approximation (near the parameter δ in B and A no 2, and the bracket on the left is missing in the equation for velocity), showed the following.

If $\alpha_1 < 1$, then the values $\theta_b < 1$ give the mass combustion velocity for the subsonic regime, and both the temperature and velocity values depend weakly on the parameter δ (Fig. 17.4). A different situation is observed in the supersonic region. If α_1 little exceeds unity, there is a single positive root of the equation for temperature (17.79), which can be either greater or less than unity. This combustion regime is one-temperature (the temperature θ_b can be either greater or less than unity). For $\theta_b < 1$, we have a velocity m slightly greater than unity. Otherwise—$m >> 1$. At α_1, exceeding some value of α_{1*}, and δ smaller than some value of δ_* (their specific values depend on other parameters of the problem), Eq. (17.79) has three positive roots, i.e., we obtain a "three-temperature" combustion mode.

17.3.5 Dependence of Reaction Rate on Stress Work

The possible dependence of the rate of transformations in the solid phase on stresses has been emphasized by many authors, including theoretical works. Without an explicit external mechanical influence, stresses and deformations may be a consequence of the fact of chemical transformation itself and the accompanying physical processes of different nature. Based on the analysis of literature in [29], the reaction rate is presented in the form of

$$\Phi(\eta, T, \Pi) = k_0\phi_1(\eta)\exp\left(-\frac{E_a}{RT}\right)\exp\left(\frac{k_\sigma\Pi}{RT}\right), \tag{17.81}$$

which reflects the fact that a chemical reaction can be influenced in two ways—by changing the internal energy (i.e., temperature) and by doing work Π. The work in this case is done by mechanical stresses, which are a consequence of the change in temperature and composition in the reaction. I.e., the processes are, in fact, related. Let us define work in accordance with the formula:

$$\Pi = -\sum_{i,j}\sigma_{ij}\varepsilon_{ij}.$$

In the above models only one of the components of the strain tensor is different from zero, therefore

$$\Pi = -\sigma_{11}\varepsilon_{11}.$$

In dimensionless variables -

$$\frac{\Pi}{\sigma_*\varepsilon_*} = -se.$$

Then in the case of thermal stresses only we find:

$$\frac{\Pi}{\sigma_*\varepsilon_*} = -\frac{\alpha_1}{[1-\alpha_1]^2}\theta^2.$$

Consequently, the chemical reaction rate function in this model will take the form

$$\phi_1(\eta)\phi_2(\theta) = \phi_1(\eta)\exp\left[-\frac{(1-\theta)(1+\sigma)+\omega\gamma s_0\theta^2}{\gamma(\theta+\sigma)}\right], \tag{17.82}$$

where

$$s_0 = \frac{k_\sigma \rho c_\varepsilon}{R} \frac{\alpha_1}{[1-\alpha_1]^2} > 0.$$

When constructing the asymptotic solution, only the problem for the inner region will partially change. Representation of the function

$$\phi_2(t) = \exp\left[-\frac{t(1+\sigma) + \omega s_0 (1-t\gamma)^2}{(1-t\gamma+\sigma)}\right]$$

in the form of a Taylor series in powers of a small parameter γ with accuracy up to the second order of smallness instead of (17.49) will have the following form

$$\phi_2(t) \approx \exp\left(-t + \frac{\omega s_0}{1+\sigma}\right)\left[1 - \gamma\left(t^2 + t\frac{\omega s_0}{(1+\sigma)^2}(1+2\sigma) - \frac{\omega s_0}{1+\sigma}\right) + \cdots\right].$$

Hence, instead of (17.53) in the zero approximation we obtain:

$$m_0 \approx \exp\left(\frac{\omega s_0}{1+\sigma}\right)\left[\exp\left(-\frac{1-\theta_b}{\gamma}\right) - \exp\left(-\frac{1}{\gamma}\right)\right],$$

which obviously indicates that if the chemical reaction rate increases under the influence of stresses ($s_0 > 0$), the wave velocity also increases. The situation is also possible when $s_0 < 0$.

17.4 Discussion (Instead of Conclusion)

The simple coupled models and their traveling wave solutions presented in the paper have quite a practical application, for example, to the synthesis of new materials and coatings under different technological conditions [25, 30–32] and are generalized for more complex situations.

For example, in recent years, models of the theory of heat conduction and thermoelasticity, in which the finiteness of the relaxation time of the heat flux is taken into account, have become popular [33, 34]. Taking this feature into account leads to the hyperbolic type heat conduction equation, which, taking into account the chemical source of heat release, takes the following form

$$\rho c_\varepsilon \left[t_T \frac{\partial^2 T}{\partial t^2} + \frac{\partial T}{\partial t}\right] + 3K\alpha_T T \frac{\partial \varepsilon_{kk}}{\partial t} + t_T \frac{\partial}{\partial t}\left(3K\alpha_T T \frac{\partial \varepsilon}{\partial t}\right) =$$
$$= \nabla \cdot (\lambda_T \nabla T) + Q_{ch}\left[\Phi(\eta, T) + t_T \frac{\partial}{\partial t}\Phi(\eta, T)\right].$$

The traveling wave solution even of the linearized problem [24] leads to velocity changes in both fast and slow modes and to the appearance of additional modes of co-propagation of thermal mechanical waves. The nonlinear problem in this case is also partially integrated. In the moving coordinate system, the stationary heat conduction equation, taking into account the integrals (17.21) of the equation of motion in dimensionless variables, takes the following form:

$$-V\frac{d\theta}{d\xi}+V^2\tau_T\frac{d^2\theta}{d\xi^2}=\frac{d^2\theta}{d\xi^2}+\omega\left(\theta+\sigma\right)V\frac{1}{1-(\alpha V)^2}\frac{d\theta}{d\xi}-V\frac{d\eta}{d\xi}-$$
$$-\frac{V^2\tau_T\omega}{1-(\alpha V)^2}\frac{d}{d\xi}\left[(\theta+\sigma)\frac{d\theta}{d\xi}\right]+V^2\tau_T\frac{d^2\eta}{d\xi^2}.$$

where $\tau_T=t_T\big/t_*$, and all other notations are similar to the previous one. The second equation will be Eq. (17.34). The boundary conditions (17.38) in the stationary problem do not change.

Another example is related to the consideration of heat losses during the synthesis of a coating on a substrate [35–37]. Heat losses in this case are possible by convection and radiation, as well as by heat conduction into the substrate. If the layer of material in which the synthesis takes place is thin, the temperature distribution along the thickness can be neglected in it. Then a summand reflecting effective convective heat losses will appear in the heat conduction equation, including in the coupled formulation [31], but it is possible to take into account the substrate properties explicitly, which will lead to the appearance of effective properties and to the appearance of an effective coupling coefficient that takes into account the material properties [38]. Under such conditions, independent modes of reaction propagation are also possible. To find the conditions for their existence, it is enough to pass to "new variables", by means of which the problem will be reduced to the already solved one (Sects. 3.1 and 3.2).

The rheological properties of the medium affect not only the dynamics of wave propagation [39, 40], but also the chemical transformations associated with mechanical processes [41]. In models for viscoelastic media, traveling wave type solutions are also possible, when the chemical reaction front is the union of thermal, mechanical, and chemical waves. Coupled models of this type have been studied in [26, 42]. Thus, in the Maxwell-type medium model, the relations between the components of stress and strain tensors can be represented as [43]

$$\frac{\partial\sigma_{ij}}{\partial t}+\frac{\mu}{\kappa}\sigma_{ij}=2\mu\frac{\partial\varepsilon_{ij}}{\partial t}+\delta_{ij}\left[\frac{\partial}{\partial t}(\lambda\varepsilon_{kk}-Kw)+\frac{\mu}{\kappa}K(\varepsilon_{kk}-w)\right],$$

where, as in problem 3.1, $w=3\alpha_T(T-T_0)$.

In this case, a stationary problem similar to the previous one in the same dimensionless variables will include Eqs. (17.33), (17.34) and

$$(\alpha V)^2\frac{d^2e}{d\xi^2}=\frac{d^2S}{d\xi^2}\text{ or }S=(\alpha V)^2e\text{ and }-V\frac{dS}{d\xi}+\frac{S}{\tau_v}=-V\frac{de}{d\xi}+V\frac{d\theta}{d\xi}+\frac{he-\theta}{\tau_v}.$$

New parameters appear in the model: $h = K\big/(\lambda + 2\mu)$—ratio of volume and longitudinal velocities of sound and $\tau_v = \left(\kappa \upsilon_n^2\right)\big/(\mu \kappa_T)$—time (or rate) of viscous stress relaxation. The asymptotic solution in an acceptable form can be constructed for two limiting cases: $\tau_v << 1$ and $\tau_v >> 1$. It was shown in [26] that the rheological features of the medium have a significant influence both on the number of possible transformation modes and on their characteristics (front velocity, its structure, and product temperature). Supplementing the model with melting and viscosity dependence on the phase state and reaction product fraction significantly complicates the problem, which is also relevant to practical problems [44, 45]. However, the numerical study shows the possibility of propagation of both forced (controlled by external influence) and independent thermo-chemo-mechanical waves.

Finally, we note that the study of the thermomechanical stability of such coupled waves is also possible using classical approaches [46–50].

Acknowledgements The work was performed according to the Government research assignment for ISPMS SB RAS, project FWRW-2022-0003

Appendix

Some transformations of variables and function:

$$\beta = \frac{RT_*}{E} = \frac{RT_*^2}{E(T_* - T_0)} \frac{(T_* - T_0)}{T_*} = \gamma \frac{(T_* - T_0)}{T_* + T_0 - T_0} = \frac{\gamma}{1+\sigma}; \ \sigma = \frac{T_0}{T_* - T_0}$$

$$\phi_2(\theta) = \exp\left[-\frac{E}{RT} + \frac{E}{RT_*}\right] = \exp\left[\frac{E}{R}\frac{T - T_*}{TT_*}\right] = \exp\left[\frac{E}{R}\frac{(T_* - T_0)\theta + T_0 - T_*}{T_*((T_* - T_0)\theta + T_0)}\right] =$$

$$= \exp\left[\frac{1}{\beta}\frac{\theta - 1}{(\theta + \sigma)}\right] = \exp\left[\frac{1+\sigma}{\gamma}\frac{\theta - 1}{(\theta + \sigma)}\right] \equiv \exp\left[-\frac{1+\sigma}{\gamma}\frac{1 - \theta}{(\theta + \sigma)}\right]$$

With stresses.

Consider the function:

$$\phi_2(\theta) = \exp\left[-\frac{E}{RT} + \frac{E}{RT_*} \mp k_\sigma \sigma_* \varepsilon_* \frac{\overline{\Pi}}{RT}\right].$$

$$-\frac{E}{RT} + \frac{E}{RT_*} \mp k_\sigma \sigma_* \varepsilon_* \frac{\overline{\Pi}}{RT} = \frac{E}{R}\left(\frac{T - T_* \mp \frac{k_\sigma \sigma_* \varepsilon_*}{E} T_* \overline{\Pi}}{TT_*}\right) =$$

$$\frac{E}{R}\frac{(T_* - T_0)\theta + T_0 - T_* \mp \frac{k_\sigma \sigma_* \varepsilon_*}{E} T_* \overline{\Pi}}{((T_* - T_0)\theta + T_0)T_*} = \frac{E}{RT_*}\frac{\theta - 1 \mp \frac{k_\sigma \sigma_* \varepsilon_*}{E(T_* - T_0)} T_* \overline{\Pi}}{(\theta + \sigma)} =$$

$$= \frac{1}{\beta}\frac{\theta - 1 \mp \frac{k_\sigma \sigma_* \varepsilon_*}{E(T_* - T_0)} T_* \overline{\Pi}}{(\theta + \sigma)} = \frac{\theta - 1 \mp \frac{k_\sigma \sigma_* \varepsilon_*}{E(T_* - T_0)} T_* \overline{\Pi}}{\gamma(\theta + \sigma)}(1 + \sigma)$$

Let us consider separately the last summand in the numerator:

$$\frac{k_\sigma \sigma_* \varepsilon_*}{E(T_* - T_0)} T_* \overline{\Pi} = \frac{(3K\alpha_T)^2 (T_* - T_0)}{\lambda + 2\mu} \frac{k_\sigma}{E} T_* \overline{\Pi} =$$
$$= \frac{(3K\alpha_T)^2 (T_* - T_0)}{(\lambda + 2\mu)\rho c_\varepsilon} \frac{k_\sigma \rho c_\varepsilon}{E} T_* \overline{\Pi} = \omega \frac{k_\sigma \rho c_\varepsilon}{ER} T_* \overline{\Pi} R =$$
$$= \omega\beta \frac{k_\sigma \rho c_\varepsilon}{R} \overline{\Pi} \equiv \omega \frac{\gamma}{1+\sigma} \frac{k_\sigma \rho c_\varepsilon}{R} \overline{\Pi} = -\omega \frac{\gamma}{1+\sigma} \frac{k_\sigma \rho c_\varepsilon}{R} \frac{(\alpha V)^2}{\left[1 - (\alpha V)^2\right]^2} \theta^2$$

Hence,

$$\exp\left[-\frac{E}{RT} + \frac{E}{RT_*} \mp k_\sigma \sigma_* \varepsilon_* \frac{\overline{\Pi}}{RT} \right] = \exp\left[\frac{\theta - 1 \mp \frac{k_\sigma \sigma_* \varepsilon_*}{E(T_* - T_0)} T_* \overline{\Pi}}{\gamma(\theta + \sigma)} (1 + \sigma) \right] =$$
$$= \exp\left[\frac{(\theta - 1)(1 + \sigma) \pm \omega \gamma s_0 \theta^2}{\gamma(\theta + \sigma)} \right],$$

where

$$s_0 = \frac{k_\sigma \rho c_\varepsilon}{R} \frac{(\alpha V)^2}{\left[1 - (\alpha V)^2\right]^2} > 0.$$

References

1. Whitham GB (1974) Linear and nonlinear waves. Wiley, New York
2. Woyczynski WA (1998) Burgers–KPZ turbulence: gottingen lectures. Springer, Berlin, 318 p. https://doi.org/10.1007/BFb0093107
3. Liu W, Han M (2024) Bifurcations of traveling wave solutions of a generalized Burgers-Fisher equation. J Math Anal Appl 533:128012. https://doi.org/10.1016/j.jmaa.2023.128012
4. Ahmad H, Khan TA, Cesarano C (2019) Numerical solutions of coupled burgers' equations. Axioms 8:119. https://doi.org/10.3390/axioms8040119
5. Babu A, Han B (2021) Noufal Asharaf numerical solution of the viscous Burgers' equation using localized differential quadrature method. Sci Notes Leningr State Univ 4:100044. http://creativecommons.org/licenses/by-nc-nd/4.0/
6. Kapoor M (2020) Exact solution of coupled 1D non-linear Burgers' equation by using Homotopy Perturbation Method (HPM): a review. J Phys Commun 4:095017. https://doi.org/10.1088/2399-6528/abb218
7. Altıparmak K, Ozis T (2011) Numerical solution of Burgers' equation with factorized diagonal Pade´ approximation. Int J Numer Methods Heat Fluid Flow 21(3):310–319. https://doi.org/10.1108/09615531111108486
8. Markov VV, Utesinov VN (2020) Difference scheme for the numerical solution of the burgers equation. Comput Math Math Phys 60(12):1985–1989
9. Salkuyeh D, Sharafeh FS (2009) On the numerical solution of the Burgers's equation. Int J Comput Math 86(8):1334–1344. https://doi.org/10.1080/00207160701864434

10. Biazar J, Ayati Z, Shahbazi S (2014) Solution of the burgers equation by the method of lines. Am J Numer Anal 2(1):1–3. https://doi.org/10.12691/ajna-2-1-1
11. Almansor AJS, Tavakoli A (2024) A novel approach based on minimization for numerical solutions of Burgers' equation. PartL Differ Equ Appl Math 9:100598. https://doi.org/10.1016/j.padiff.2023.100598
12. Makwana PR, Chauhan JP, Chauhan RB, Parikh AK (2024) Solution of nonlinear Burger's equation arising in longitudinal dispersion phenomena. Results Control Optim 14:100370. https://doi.org/10.1016/j.rico.2023.100370
13. De la cruz Guerrero RA, Lu Y, Wang XT (2024) Riemann problem for a general variable coefficient Burgers equation with time-dependent damping. Int J Non–Linear Mech 162:104703. https://doi.org/10.1016/j.ijnonlinmec.2024.104703
14. Kumar M, Kumar R (2015) Some more solutions of burgers' equation. J Phys: Conf Ser 574:012038. https://doi.org/10.1088/1742-6596/574/1/012038
15. Nowacki W (1975) Dynamic problems of thermoelasticity. Springer Science & Business Media, 436 p
16. Kovalenko AD (1975) Thermoelasticity (in Russian). Vica Shcola, Kiev, 216 p
17. Nenakhov EV, Kartashov EM (2022) Estimates of temperature stresses in dynamic thermoelasticity models (in Russian). [Vestnik MGTU imeni N.E. Baumana, Ser. Estestvennye nauki], 100(1):88–106. https://doi.org/10.18698/1812-3368-2022-1-88-106
18. Bland D (1969) Nonlinear dynamic elasticity. Blaisdell, Waltham
19. Knyazeva AG (2003) Solution of the thermoelasticity problem in the form of a traveling wave and its application to analysis of possible regimes of solid-phase transformations. J Appl Mech Tech Phys 44(2):164–173. https://doi.org/10.1023/A:1022580023603
20. Varley E, Rogers TG (1967) The propagation of high frequency, finite acceleration pulses and shocks in viscoelastic materials. Proc Roy Soc, Ser A 296(1447):498–518. https://doi.org/10.1098/rspa.1967.0041
21. Zeldovich YB, Barenblatt GI, Librovich VB, Makhviladze GM (1980) Mathematical theory of combustion and explosion[(in Russian) [Matematicheskaia teoriia goreniia I vzryva]. Nauka, Moscow
22. Timokhin AM, Knyazeva AG (1996) Regimes of propagation of the solid-phase reaction front in the coupled thermomechanical model of solid-phase combustion (in Russian). Khimicheskaya Fizika, 15(10):1497–1514
23. Knyazeva AG (2004) Wave velocity of solid-phase combustion. Asymptotic analysis (in Rusian). Physical mesomechanics (Fizicheskaya mezomekhanika) 7(3):63–70
24. Knyazeva AG, Dyukarev EA (1995) Stationary wave of a chemical reaction in a deformable medium with finite relaxation time of the heat flux. Combust Expl Shock Waves 31(3):304–312. https://doi.org/10.1007/BF00742675
25. Knyazeva AG, Dyukarev EA (1999) Model of detonation of lead aside (PbN3) with regard to fracture. Int J Fract 100(2):197–205. https://doi.org/10.1023/A:1018788304044
26. Knyazeva AG, Sorokova SN (2006) Steady regimes of conversion in a viscoelastic medium. Combust, Explos, Shock Waves 42(5):549–558. https://doi.org/10.1007/s10573-006-0087-6
27. Reissig R, Sansone G, Conti R (1974) Nonlinear differntial equations of higher order. Noordhoff International Publication, Leyden
28. Knyazeva AG (1996) Introduction into locally equilibrium thermodynamics of physicochemical transformations in deformable media. Izd. Tomsk University, Tomsk (in Russian)
29. Knyazeva AG (1993) Hot spot thermal explosion in deformed solid. Combust Explos Shock Waves 29:419–428. https://doi.org/10.1007/BF00782966
30. Knyazeva AG, Dyukarev EA (1998) Model of autowave propagation of solid-state low-temperature chlorination of butyl chloride. Combust Explos Shock Waves 34:556–565. https://doi.org/10.1007/BF02672679
31. Knyazeva AG, Kryukova ON (2022) A coupled model of controlled synthesis, of a composite on a substrate. Lobachevskii J Math 43(7):1878–1893. https://doi.org/10.1134/S1995080222100183

32. Knyazeva AG, Bukrina NV (2021) A coupled model of composite synthesis in combustion regime. Combust Theory Model 26(31):152–178. https://doi.org/10.1080/13647830.2021.1996634
33. Vitokhin EY, Babenkov MB (2017) Influence of boundary conditions on the solution of a hyperbolic thermoelasticity problem. Continuum Mech Thermodyn 29:457–475. https://doi.org/10.1007/s00161-016-0540-z
34. Vitokhin EY, Ivanova EA (2017) Dispersion relations for the hyperbolic thermal conductivity, thermoelasticity and thermoviscoelasticity. Continuum Mech Thermodyn 29:1219–1240. https://doi.org/10.1007/s00161-017-0574-x
35. Knyazeva AG, Kryukova ON (2021) The synthesis of composites with reinforcing particles on a thin substrate. Mater Phys Mech 47(2):254–265. https://doi.org/10.18149/MPM.4722021_8
36. Bakinovskii AA, Knyazeva AG, Krinitcyn MG, Kryukova ON, Pobol IL, Fedorov VV, Rajczyk J (2019) Electron beam assisted deposition of Ni–Al coatings onto steel substrate. Int J Self Propag High Temp Synth 28(4):245–255. https://doi.org/10.3103/S1061386219040034
37. Sorokova SN, Knyazeva AG (2010) Numerical study of the influence of the technological parameters on the composition and stressed-deformed state of a coating synthesized under electron-beam heating. Theor Found Chem Eng 44:172–185. https://doi.org/10.1134/S0040579510020089
38. Knyazeva AG (2024) Thermomechanical coupled model of coating synthesis on a substrate (in Russian). PRNPU Mech Bull 1. https://doi.org/10.15593/perm.mech/2024.1.05
39. Liapidevskii VY, Pukhnachev VV, Tani A (2011) Nonlinear waves in incompressible viscoelastic maxwell medium. Wave Motion 48:727–737. https://doi.org/10.1016/j.wavemoti.2011.04.002
40. Berjamin H, Destrade M, Saccomandi G (2024) Singular travelling waves in soft viscoelastic solids of rate type. Eur J Mechanics: A Solids 103:105144. https://doi.org/10.1016/j.euromechsol.2023.105144
41. Petrenko S, Freidin A Eric Charkaluk chemical reaction planar fronts with a viscoelastic reaction product. https://doi.org/10.48550/arXiv.2104.01074
42. Knyazeva AG, Dyukarev EA (2000) Model for the propagation of a stationary reaction front in a viscoelastic medium. Combust Explos Shock Waves 36:452–461. https://doi.org/10.1007/BF02699475
43. Boley B, Weiner J (1960) Theory of thermal stresses. Wiley, New York
44. Knyazeva AG (2023) Influence of viscosity change in the vicinity of melting temperature on the dynamics of coating synthesis (in Russian). Izvestiya Vuzov, Fizika. 66(9)(790):47–58. https://doi.org/10.17223/00213411/66/9/5
45. Knyazeva AG (2024) Thermo-viscous-elastic model of surface layer modification with changing viscosity. J Appl Mech Tech Phys
46. Knyazeva AG (1993) Combustion wave propagation through deformed solids. Combust Explos Shock Waves 29:299–303. https://doi.org/10.1007/BF00797645
47. Knyazeva AG, Sorokova SN (2006) Stability of the combustion wave in a viscoelastic medium to small one-dimensional perturbations. Combust Explos Shock Waves 42:411–420. https://doi.org/10.1007/s10573-006-0070-2
48. Knyazeva AG (2010) Solid-phase combustion in a plane stress state. 1. Stationary combustion wave. J Appl Mech Tech Phys 51:164–173. https://doi.org/10.1007/s10808-010-0025-y
49. Knyazeva AG (2010) Solid-phase combustion in a plane stress state. 2. Stability to small perturbations. J Appl Mech Tech Phys 51:317–323. https://doi.org/10.1007/s10808-010-0043-9
50. Knyazeva AG (2011) Thermomechanical stability of solid phase conversion front to two dimensoinal perturbations (in Russian). Vesnt Perm Gos Tekh Univ Mekhanika (4):88–123

Chapter 18
Painleve's Paradox When Considering Heridicatory Friction

Vladimir S. Metrikin and Leonid A. Igumnov

Abstract The work is devoted to the study of the nonlinear dynamics of frictional vibration systems, which take into account, when developing mathematical models of vibration systems, an essentially nonlinear function that describes the work of friction forces with memory. The problem considered in the work combines two: the well-known problem of the Painlevé paradox and the problem of studying dynamic systems taking into account hereditary friction forces. In both cases, we are talking about the idealization of real elastic bodies and flexible bonds with ideal absolutely hard and rigid ones. Using the well-known experimentally proven hypothesis of A. Yu. Ishlinsky and Kragelsky about the hereditary dependence of the coefficient of friction of relative rest at low speeds of movement of bodies, the dynamic system under study is transformed from autonomous to non-autonomous, in which new types of movements appear that were not previously observed in the study of vibration systems. In particular, taking into account the Painlevé paradox entails the previously unexpected possibility of the occurrence of instability and self-oscillations, and taking into account the heredity of the friction coefficient leads to the appearance of chaotic movements of bodies. To study these combined problems, a numerical and analytical method has been developed to study the behavior of the dynamic characteristics of friction vibration systems depending on the chosen model of the relative static friction coefficient. The new bifurcation diagrams presented in this work, based on the main parameters, demonstrate the presence of arbitrarily complex periodic regimes of body motion, including chaotic ones. It has been established that the transition to chaotic modes of motion occurs according to the well-known period doubling scenario. Analytical relations are presented for point mappings of Poincaré surfaces, which made it possible to obtain succession functions illustrating arbitrarily complex modes of body motion, including chaotic ones. It is noted that such body movements were not found when studying vibration systems without taking into account hereditary friction.

V. S. Metrikin · L. A. Igumnov (✉)
National Research Lobachevsky State University of Nizhny Novgorod, Nizhny Novgorod, Russia
e-mail: igumnov@mech.unn.ru

H. Altenbach and V. A. Eremeyev (eds.), *Dynamics of Discrete and Continuum Structures and Media*, Advanced Structured Materials 221,
https://doi.org/10.1007/978-3-031-75626-9_18

Keywords Modeling · Friction systems · Relative static friction coefficient · Painlevé paradox · Friction with memory · Point mapping · Self-oscillations · Stability · Chaos

18.1 Introduction

It is known [1–6] that friction causes instability and self-oscillations, which is something paradoxical, since we are accustomed to the fact that friction slows down and stops movement. It turns out that this is not always the case, and the paradox is resolved by the fact that friction acts as a distributor of energy for some other source of energy not related to friction. By itself, it can really only slow down movement. It is clear that self-excited self-oscillations are undesirable for many real structures such as drilling rigs, brake devices, plain bearings, friction clutch mechanisms, and many others.

The simplified model with which the process of studying systems with the paradoxical role of dry friction first began was studied by Van der Pol [1], and consisted of a mass that is pulled through a spring at a constant speed along a horizontal rough plane. The massive body moves slowly under the influence of the force of the stretched spring. This model can also be represented as a rough tape moving at a constant speed, and the end of the spring attached to the body is motionless. We will consider in the future the latest version of this model. Real prototypes of the described schematic model can also serve as brake pads of a car, tram, railway car, and other various mechanical braking devices. Subsequently, many variants of this model were considered in works [7–18] and others. In these works, it was assumed that the force of dry friction does not depend on the relative speed of the rubbing surfaces, and its action is directed against the speed and is proportional in magnitude to normal pressure. And when the mass moves along with the tape, as if attached to it, the static friction force can be any within certain limits. However, it is experimentally known [2] that the friction coefficient of relative rest is not a constant value, but is hereditary from the time of relative rest of interacting bodies. The corresponding mathematical model in dimensionless time $\tau = \omega t$, coordinate $\xi = kx/(f_* P)$ and $\varepsilon(\tau_k) = (f(\tau_k) - f_*)/f_*$, characterizing the functional dependence of the coefficient of relative static friction (RTF), can be written in the form

$$\ddot{\xi} = -\xi - \text{sgn}\,(\dot{\xi} - \theta) + \alpha(\dot{\xi} - \theta), \quad \dot{\xi} \neq \theta, \tag{18.1}$$

$$|\,\xi\,| \leq 1 + \varepsilon(\tau_k), \quad \dot{\xi} = \theta. \tag{18.2}$$

Here $f(\tau)$ is a piecewise continuous monotonically non-decreasing function that describes the functional dependence of the RTF on the time τ of relative rest of the bodies rubbing during joint motion, τ_k—the time interval of the relative rest of the bodies, θ- the dimensionless speed of the rough tape, α—the friction coefficient of the relative motion of the bodies, f_*—the friction coefficient slip.

From the structure of the differential equations of system (18.1), (18.2) it follows that its three-dimensional phase space $F(\xi, \dot{\xi}, \tau)$ is divided by a plane $\dot{\xi} = \theta$ into three subspaces $F_+(\xi, \dot{\xi} > \theta, \tau), F_-(\xi, \dot{\xi} < \theta, \tau), F_0(\xi, \dot{\xi} = \theta, \tau)$. In subspaces $F_+(\xi, \dot{\xi} > \theta, \tau), F_-(\xi, \dot{\xi} < \theta, \tau)$, , the motion of phase trajectories obeys, according to (18.1) and (18.2), equations, each of which is a harmonic oscillator with its own equilibrium state, respectively, at a point $\xi = 1 - \alpha\theta, \dot{\xi} = 0$ or at a point $\xi = -1 - \alpha\theta, \dot{\xi} = 0$. The movements of the bodies in them correspond to instantaneous stops relative to the belt, caused by a change in the direction of sliding performed at speed $\dot{\xi} - \theta$. In subspace $F_0(\xi, \dot{\xi} = \theta, \tau)$ there is no slipping, the mass moves along with the belt at a speed θ due to the fact that the force of the spring does not exceed the magnitude of the force of static friction. In phase space, this corresponds to the presence of a segment of sliding movements $L(|\xi| \le 1, \dot{\xi} = \theta)$ (a segment of the junction of phase trajectories) [19, 20]. The state of equilibrium is stable $\xi = 1 - \alpha_1\theta, \dot{\xi} = 0$, like a center. Near it, an infinite number of harmonic oscillations are possible up to a certain limiting amplitude. These harmonic vibrations do not die out, despite the presence of friction. The friction seems to disappear. All other movements, after several oscillations with a decreasing range, lead to temporary movement of the body along with the moving belt. The imaging point moves according to Eq. (18.2). Therefore, it is obvious that the study of nonlinear dynamics can be carried out using a point display [19–21] of subsequent times of relative rest of the body and the tape. This dependence (following function), according to (18.1) and (18.2), can be represented in the form $\psi(\tau_{k+1}) = \varphi(\tau_k)$, where

$$\psi(\tau) = \theta(\tau) - \varepsilon(\tau),$$
$$\varphi(\tau) = 1 - (-1)^j[\varepsilon(\tau) - 2j + 1], 2(j - 1) < \varepsilon(\tau) < 2j, j = 1, 2, ...$$

Thus, Fig. 18.1 shows the functions of succession $\tau_{k+1} = g(\tau_k)$ and Lamerey's ladder for $\theta = 1.7$ and various values of the parameter $\varepsilon^* = (f^* - f_*)/f_*$, where f^* is the maximum value of RTF with piecewise linear approximation of the functional dependence of RTF, and f_* is the minimum value of RTF. From Fig. 18.1 it is clear that at $\varepsilon^* = 2.3$ there is a double unstable fixed point (the point of intersection of the bisector and the succession function), corresponding to the periodic motion of the body with two long stops (with two relative rests of the body with the tape). With an increase ε^* ($\varepsilon^* = 2.9$), bifurcation occurs and four stable fixed points are born, corresponding to the periodic movement of the body with four long stops (with four relative rests of the body with the tape). A further increase ε^* ($\varepsilon^* = 3; 6$) brings the system into a stochastic mode (through doubling number of fixed points). From this figure, it also follows that in the space of system parameters, these last two fixed points belong to the region of chaotic movements of the body [21].

Figure 18.2 shows bifurcation diagrams [19–21] by parameter ε^* (abscissa axis) for dimensionless belt speeds $\theta = 1.7$ (Fig. 18.2a) and $\theta = 1.2$ (Fig. 18.2b) and by parameter θ (abscissa axis) (Fig. 18.2c) for $\varepsilon^* = 3.$. On the ordinate axes in all

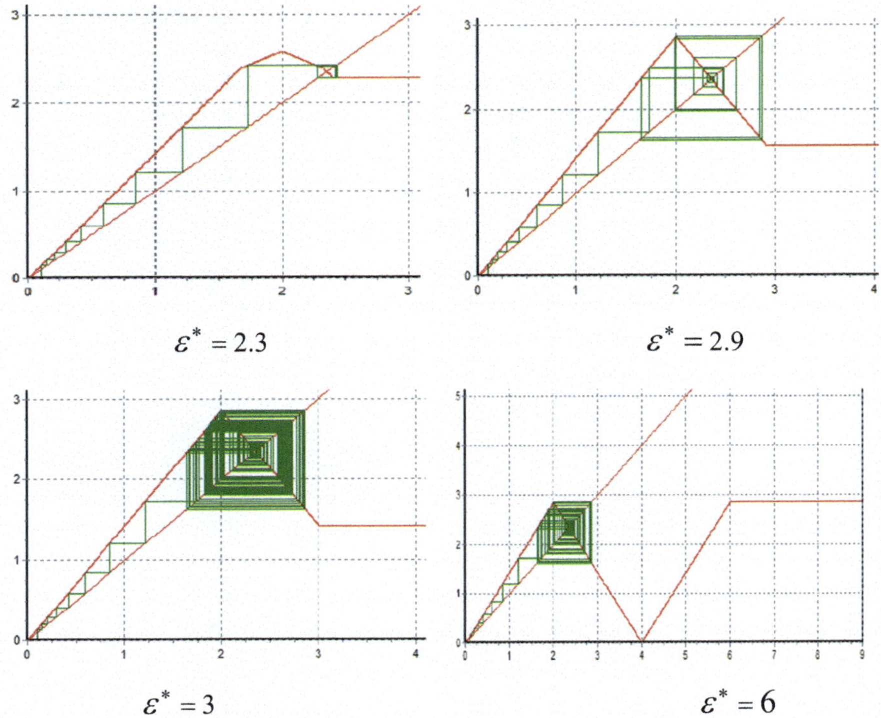

Fig. 18.1 Type of succession function at $\theta = 1.7$

Figs. 18.2a, b, c) the successive values of the times of the relative rest of the body and the ribbon are plotted τ_k, $k = 1, 2, 3....$

The results of the calculations presented in Fig. 18.2 prove once again that the transition to chaotic movements occurs according to the scenario of doubling the number of fixed points of the point mapping of the Pu-ankare surface (doubling the number of long stops of the body with the tape) [21]. It should be noted that studies of the dynamic system (18.1), (18.2) under the assumption of constancy of the RTF without taking into account the heredity of the friction coefficient of relative rest showed that no matter how complex and chaotic movements were observed. From Fig. 18.2a and b it is clear that with increasing periodic movements with one long stop are first observed, and with increasing ε^* time the long stop also increases, then the time of the long stop remains constant until a certain bifurcation value ε^*, after which periodic movement with two, four, and so on ad infinitum long stops. Then periodic movements turn into chaotic ones.

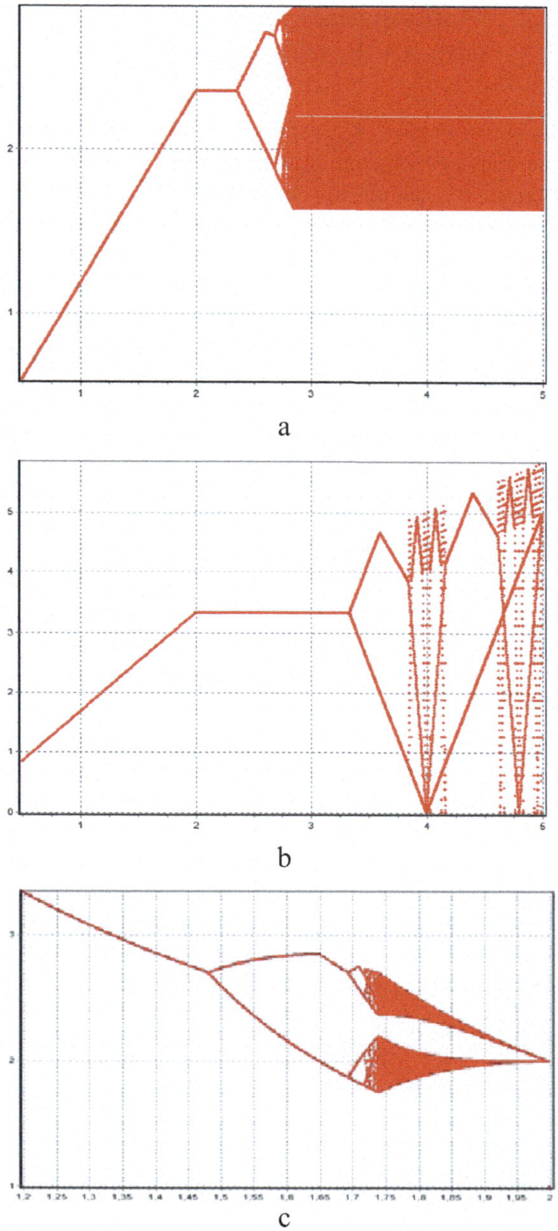

Fig. 18.2 Bifurcation diagrams by parameter ε^* for $\theta = 1.7$ (Fig. 18.2a) and $\theta = 1.7$ (Fig. 18.2b) and by parameter θ for $\varepsilon^* = 3$ (Fig. 18.2c)

18.2 Mathematical Model of a Friction System with a Vibration Limiter

The above studies of a dynamic system with a constant belt speed indicate that for a number of technical structures, the constancy of the belt speed and the absence of constantly acting external influences on the body significantly reduces the entire range of possible working movements suitable for preliminary research using mathematical models. There are quite a lot of such systems, for example: the Painlevé-Klein system about the movement under the action of a constant force of a rod between parallel guides, one of which is smooth and the other rough, a brake pad pressed against a wheel rotated by a moment, and many others. Therefore, we will include additional terms in the mathematical model (18.1) and (18.2), which will provide the opportunity to study the movement of the belt with a speed $V(t)$ changing over time and external influence on the body $f(t)$. This inclusion is especially important when studying the dynamics of deep-water wells for oil and gas production, as well as driving piles of large diameter and long length, when building special-purpose bridges, offshore wind turbines, and so on. We will also assume that the movement of the body in the direction of movement of the belt with speed $V(t)$ is limited by a wall located at a distance "a" from the equilibrium position of the body with the belt stationary.

The mathematical model in this case can be written in dimensionless form as follows:

$$\ddot{\xi} = -\xi - \mathrm{sgn}(\dot{\xi} - \theta) + \alpha(\dot{\xi} - \theta) + F(\tau), \dot{\xi} \neq \theta, \xi < b \tag{18.3}$$

$$|\xi - F(\tau) + \dot{\theta}| \leq 1 + \varepsilon_k, \dot{\xi} = \theta, \xi < b \tag{18.4}$$

$$\dot{\xi}^+ = -R\dot{\xi}^-, \xi = b, \dot{\xi}^- > 0 \tag{18.5}$$

The meaning of the parameters in system (18.3)–(18.5) is obvious: $F(\tau)$—dimensionless external force acting on the body, b—dimensionless initial distance of the body from the vibration limiter. The phase space is therefore truncated along the phase coordinate $\xi \leq b$. In subspaces $F_+(\xi, \dot{\xi} > \theta, \tau)$, $F_-(\xi, \dot{\xi} < \theta, \tau)$, the movement of phase points obeys, according to (3)-(5), equations $\ddot{\xi} = -\xi \pm 1 + \alpha(\dot{\xi} - \theta) + F(\tau)$, $\dot{\xi} \neq \theta, \xi < b$, each of which is a harmonic oscillator, which is affected by an external force.

18.3 Dynamics of the System Taking into Account the Wall

Example 1: $F(\tau) = A\sin(\Omega\tau), \dot{\theta} = const.$

The mathematical model, using (18.3)–(18.5), will be as follows

$$\ddot{\xi} = -\xi - \text{sgn}\,(\dot{\xi} - \theta) + \alpha(\dot{\xi} - \theta) + A\sin(\Omega\tau), \quad \dot{\xi} \neq \theta, \xi < b \tag{18.6}$$

$$|\xi - A\sin(\Omega\tau)| \leq 1 + \varepsilon_k, \dot{\xi} = \theta, \xi < b. \tag{18.7}$$

$$\dot{\xi}^{+} = -R\xi^{-}, \xi = b, \dot{\xi}^{-} > 0$$

Phase trajectories in subspaces $F_{+}(\xi, \dot{\xi} > \theta, \tau)$ and $F_{-}(\xi, \dot{\xi} < \theta, \tau)$ are described by the equations

$$\ddot{\xi} = -\xi \pm 1 + \alpha(\dot{\xi} - \theta) + A\sin(\Omega\tau), \dot{\xi} \neq \theta, \xi < b, \tag{18.8}$$

and in subspace $F_0(\xi, \dot{\xi} = \theta, \tau)$—by equations

$$\dot{\xi} = \theta, |\xi - A\sin(\Omega\tau)| \leq 1 + \varepsilon_k, \xi < b \tag{18.9}$$

18.3.1 Results of Numerical Calculations

Figure 18.3 shows phase portraits for different values of the velocity restitution coefficient$R$. In Fig. 18.3a $R = 0.5$, and in Fig. 18.3b $R = 0.7$. Rice. 3a correspond to the following parameter values: the parameter ε^{*} is 3, the belt speed is constant and equal to 1, the external force is variable F(t) = 0.25cos(2t), the wall coordinate is 2, the coefficient $\mathscr{R} = 0.5$. Rice. 3c differs from Fig. 18.3a only in the value of the coefficient R, which is equal to 0.7. From these figures it is clear that when the coefficient R = 0.5, the system has a stable limit cycle that describes the frictional oscillations of the body with three long stops of the form: the first stop is followed by an impact on the wall and then a section in the half-space $\xi < \theta$, after followed by 2 more similar turns, and then the cycle repeats. When the coefficient R = 0.7, the behavior of the system is chaotic.

In Figs. 18.4 and 18.5, where the sequence of times of relative rest of the body with the tape τ_k is plotted along the ordinate axes, and the distance of installation of the wall (Fig. 18.4) and the values of the belt speed (Fig. 18.5) are plotted along the abscissa axes.

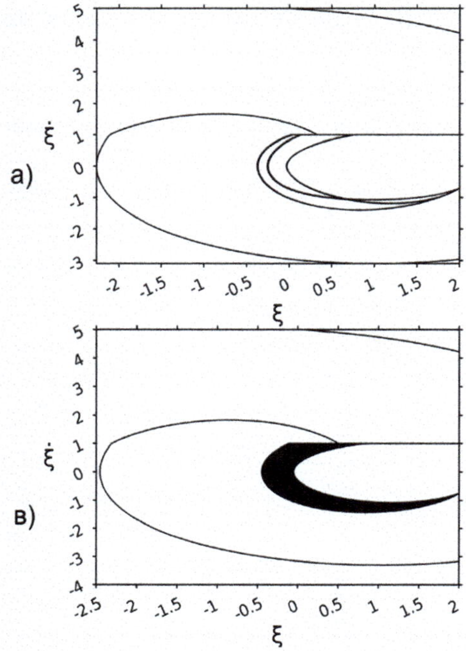

Fig. 18.3 Phase portraits for $R = 0.5$ (Fig. 18.3a) and $R = 0.7$ (Fig. 18.3c)

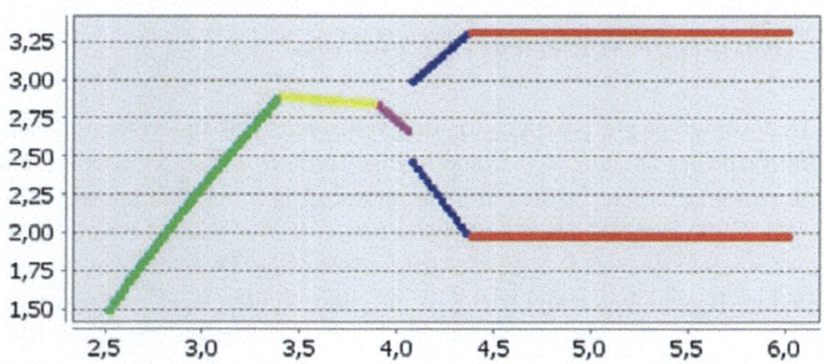

Fig. 18.4 Bifurcation diagram according to the parameter "b"—wall—limiter

Figure 18.6 shows phase portraits in the plane (τ, ξ) at different values of amplitude A. From Fig. 18.6 it is clear that with increasing amplitude, chaotic movements (wide red stripes) become more intense and are located in a larger space of the phase plane. Moreover, in the first case, modes of body movement are observed with a body speed less than or equal to the belt speed, when, as in the second case, body movements occur both with and without advancing the belt speed. The phase portraits shown in Figs. 18.3 and 18.6 were not observed when studying frictional

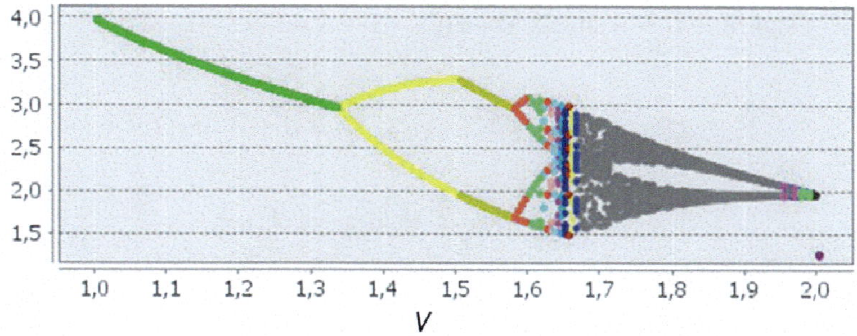

Fig. 18.5 Bifurcation diagram for the parameter V—belt speed

dynamic systems with friction without taking into account the heredity of the friction coefficient of relative rest.

Example 2: $\theta(\tau) = B + A\sin(\Omega\tau)$, $F(\tau) = 0$.

The mathematical model, using (18.3)–(18.5), is written as

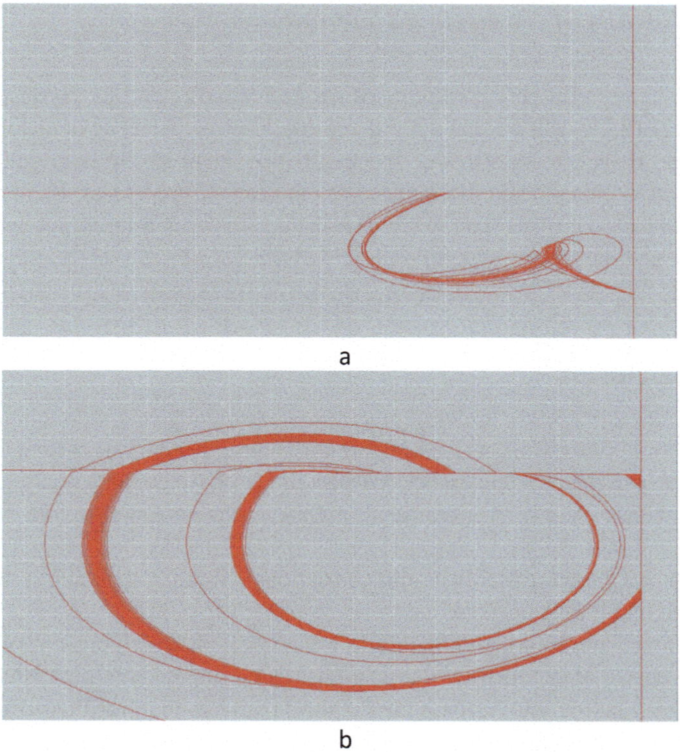

Fig. 18.6 Phase portraits in the plane (τ, ξ) for A = 1 (Fig. 18.6a) and A = 2 (Fig. 18.6b)

$$\ddot{\xi} + \xi = -\text{sign}(\dot{\xi} - B - A\sin(\Omega\tau)) + \alpha(\dot{\xi} - B - A\sin(\Omega\tau)), \dot{\xi} \neq B + A\sin(\Omega\tau) \tag{18.10}$$

$$|\xi + A\Omega\cos(\Omega\tau)| \leq 1 + \varepsilon(\tau_k),\ \dot{\xi} = B + A\sin(\Omega\tau) \tag{18.11}$$

$$\dot{\xi}^{+} = -R\xi^{-}, \xi = b, \dot{\xi}^{-} > 0$$

In three-dimensional phase space in subspaces $F_{+}(\xi, \dot{\xi} > \theta, \tau)$ and $F_{-}(\xi, \dot{\xi} < \theta, \tau)$ are described by the equations

$$\ddot{\xi} + \xi = A sin(\Omega\tau) \mp 1 + \alpha(\dot{\xi} - B - A\sin(\Omega\tau)), \dot{\xi} \neq \theta, \xi < b, \tag{18.12}$$

and in subspace $F_0(\xi, \dot{\xi} = \theta, \tau)$—by equations

$$\dot{\xi} = \theta, |\xi + A\Omega\cos(\Omega\tau)| \leq 1 + \varepsilon_k, \xi < b$$

18.3.2 *Results of Numerical Calculations*

Figure 18.7 shows bifurcation diagrams θ for $A = 0$ and $\varepsilon_* = 3.2;\ 3.7$ respectively.

From Fig. 18.7 it follows that with increasing ε_* the occurrence of chaotic movements of the body occurs at lower speeds of movement of the belt, and the area of chaotic movements of the body increases in area. A process of doubling of the Feigenbaum period is observed (an increase in the number of long-term relative stops of the body and the tape).

18.4 Conclusion

For the first time, a study of a non-autonomous dynamic system was carried out taking into account both the problem associated with the Painlevé paradox and the problem associated with the study of vibrational, highly nonlinear systems with hereditary friction.

A method for studying highly nonlinear frictional vibration dynamic systems with friction with memory is presented.

For the first time, bifurcation diagrams are presented, which make it possible to show the scenario of the birth of chaotic modes of motion in dynamic systems while simultaneously taking into account the Painlevé paradox and hereditary friction.

The bifurcation diagrams presented (for the first time) made it possible to prove that the scenario of chaos in the considered friction systems occurs as a result of period doubling.

Fig. 18.7 Bifurcation diagrams by parameter θ at $\varepsilon_* = 3.2$ (**a**), 3.7 (**b**)

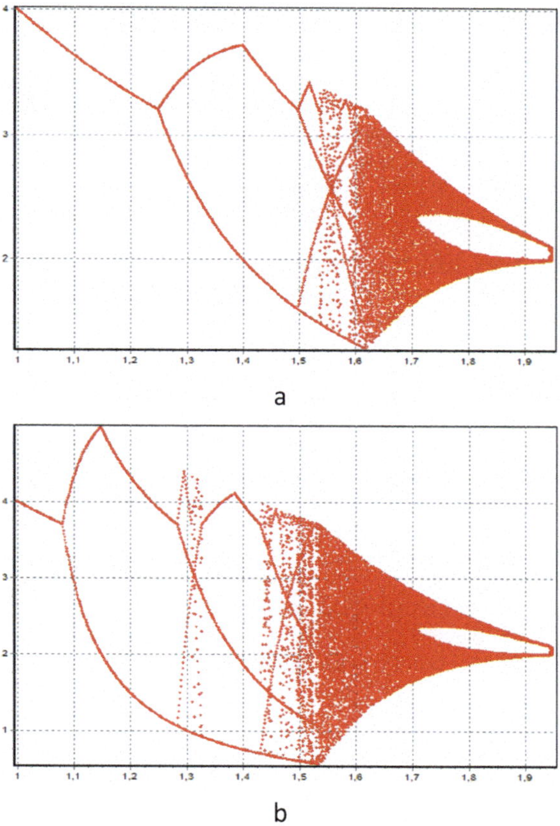

Acknowledgements The work was carried out with the financial support of the Ministry of Science and Higher Education of the Russian Federation (Project No. FSWR-2023-0036).

References

1. Van-der-Pol B (1935) Nelineynaya teoriya elektritheskix kolebaniy. M.: Svas-izdat. 91s
2. Ishlinskiy AY, Kragelskiy IV (1944) Gurn tehn fiziki, vol 4/5, pp 276–282
3. Vetukov MM, Dobrovolskiy SV, Nagaev RF (1990) Izv. AN SSSR MTT 1:23–28
4. Metrikin VS, Nagaev RF, Stepanova VV (1996) PMM T. 60 vol 5, pp 859–864
5. Zaiyzev MV, Metrikin VS (2012) Vestnik Nizgegorodskogo universiteta im. N.I. Lobathevskogo 3(1):141–146
6. Vetukov MM, Platovskih MY (2015) Trudi Vserossiyskoy nauthno-metodithzeskoy konferenii "Sovremennii problemmi mehaniki i ee prepodavanie v vuze.Sankt-Peterburg, Voenno-kosmitheskaya akademiya imeni A.F. Mozhayskogo. Tom 1, pp 58–63
7. Wang C, Nilsson H, Yang J, Petit O (2016) 1D–3D coupling for hydraulic system transient simulations. J Comput Phys Commun 135–149

8. Gómez SS, Geurts CPW, Metrikine A (2018) On the importance of soil damping for tall buildings loaded by wind. J Eng Struct 163:426–435
9. Margielewicz J, Gaska D, Litak G (2019) Evolution of the geometric structure of strange attractors of a quasi-zero stiffness vibration isolator. J Chaos, Solitons Fractals 118:47–57
10. Liu Y, Pavlovskaia E, Wiercigroch M, Peng Z (2014) J Non-Linear Mech 30–46
11. Luo GW (1988) и др. Л.: Mashinostroeniye, 379 p
12. Chávez JP, Liu Y, Pavlovskaia E, Wiercigroch M (2016) Path-following analysis of the dynamic alresponseo fapiecewise-linear capsule system. Commun Nonlinear Sci Numer Simulat 102–116
13. Liu Y, Jiang H, Pavlovskaia E, Wiercigroch M (2017) Experimental investigation of the vibro-impact capsule system. Procedia IUTAM 22:237–247
14. Savi MA, Pereira-Pinto FHI, Viola FM, de Paula AS, Bernardini D, Litak G, Rega G (2017) Using 0–1 test to diagnose chaos on shape memory alloy, dynamical systems. J Chaos, Solitons Fractals 307–324
15. Kragelskiy I.V. Treniye i iznos // izdaniye 2e, pererabotannoye i dopolnennoye – Izdatelstvo «Mashinostroeniye», M: 1968 – 480 s.
16. Howe PG, Benton DP, Puddington IE (1955) Canadian. J Chemistry 33:1375–1380
17. Chávez JP, Brzeski P, Perlikowski P (2017) Int J Non-Linear Mech. https://doi.org/10.1016/j.ijnonlinmec.2017.02.018
18. Defant F, Albertelli P (2020) J Sound Vib 1–18
19. Feygin MI (1994) Vinuzhgdenniye kolebaniya sistem s rezrivnivi nelineynostiami. M.: Nauka, 285 p
20. Neymark YI (1972) Metod tothesnih otobrazheniy v theorii nelineynih kolebanii. M.: Nauka, 471 p
21. SHuster G (1988) Determinirovanniy haus. M.:Mir, 237 p

Chapter 19
On the Dynamic Green's Function of the Three-Dimensional Infinite Anisotropic Medium

Thomas M. Michelitsch

Abstract The propagation of an acoustic wave generated by a δ-pulse unit point force in a three-dimensional anisotropic medium is considered. A general expression for the retarded 3×3 *elastodynamic* displacement Green's function is derived for arbitrary anisotropic symmetry. The initial integration problem of four integrations is reduced to only one φ-integration problem combined with the problem of solving three algebraic equations. The evaluation of expressions of Green's function for the behavior at infinity yields the radiation condition and describes the acoustic wave propagation and gives the possible distinguished propagation directions from the source point for outgoing acoustic waves (high symmetry directions of the crystal beside other distinguished directions). For high symmetry directions of the crystal this condition yields three outgoing singular acoustic waves, one with longitudinal and two with transverse polarization. For elastic isotropy of the medium we obtain the result known from literature [17] which was derived there by using other techniques. For hexagonal symmetry of the medium we derive closed-form expressions of the farfield waves for the high symmetry directions (i.e. for propagation directions parallel and in the plane perpendicular to the hexagonal c-axis).

Keywords Acoustic problem · Elastodynamic Green's function · Dynamic Green's function · Retarded Green's function · Causality · Elastic medium · Elastic anisotropy · Wave propagation

19.1 Introduction

In recent years the modeling of dynamic processes in composite structures has become of great interest due to widespread applications in ultrasonics technology. In the description of the effective characteristics, Green's function methods are powerful tools. Especially useful are closed-form representations.

T. M. Michelitsch (✉)
Sorbonne Université, CNRS, Institut Jean Le Rond d'Alembert, F-75005 Paris, France
e-mail: thomas.michelitsch@sorbonne-universite.fr

H. Altenbach and V. A. Eremeyev (eds.), *Dynamics of Discrete and Continuum Structures and Media*, Advanced Structured Materials 221,
https://doi.org/10.1007/978-3-031-75626-9_19

In numerous papers Green's function of elastostatics have been derived. In the isotropic elasticity, Green's function of statics was derived by Kelvin (Sir Thomason) [4]. For the infinite medium of hexagonal symmetry, closed-form representations of Green's functions for displacements have been derived by Lifshitz and Rosenzweig [6] and more compactly by Kröner [7]. For the electroelastic infinite medium of hexagonal symmetry, the explicit static Green's function has been derived by Michelitsch [9] by employing an integral transformation technique which is outlined in [10]. This closed-form formulation of the electroelastic Green's function has turned out to be a useful tool for the explicit evaluation of the electroelastic analogue of the Eshelby tensor [11] which represents a key quantity in the treatment of inclusion problems. For the semi-infinite piezoelectric medium of hexagonal symmetry, closed-form expressions for the static Green's function are derived by Karapetian et al. [12] only recently by utilizing an advanced method in potential theory that was discovered by Fabrikant [8].
A treatment of the space-frequency representation of the dynamic Green's function which represents the solution for the dynamic response of a time harmonic point force was given by Norris [13]. In that paper, time-harmonic Green's functions for anisotropic piezoelectric, thermoelastic, and poroelastic solids are constructed where the similarity in the theories is outlined. Another representation of the frequency-space-representation of the dynamic Green's function was given by Willis [20] among other authors. The wave propagation in elastic solids was analyzed by Achenbach in [18]. Only for the isotropic elasticity, a full closed-form solution for the frequency-space-representation of the dynamic Green's function is known in the literature [17]. In second-order gradient electrodynamics retarded Green's functions and related quantities were derived by Lazar [19]. The time-domain dynamic response function which represents the dynamic displacement response to a unit point force having Heaviside step function time dependence in elastic anisotropic media was treated by Every and Kim [15]. In their paper, special emphasis is given on singularities and discontinuities that appear in dynamic response functions. A treatment of the dynamic Green's function of the space-frequency representation of the infinite transversely isotropic medium was given by Tverdokhlebov and Rose [16]. In that paper, a formula is developed for the case of weak anisotropy that allows especially a convenient treatment of scattering fields in the farfield approximation. An advanced analysis of the dynamic response of transient electroacoustic waves in piezoelectric solids was given by Sosa and Khutoryansky (1999, 2001) [1, 2]. Utilizing this approach a general expression for the Green's function was also derived by Daros and Antes [3].

In the present paper, our goal is to develop a general method for the construction of the space-time representation of the causal (retarded) Green's function. Our approach complements the existing literature and is closely related to that introduced by Sosa and Khutoryansky (1999, 2001) for piezoelectric solids, but with focus on farfield expressions for purely elastic solids of arbitrary anisotropic symmetry. We show that this approach yields useful representations especially for the farfield approximation of the dynamic Green's function. The constructed Green's function represents the dynamic displacement-response of a δ-pulse-point-force. We formulate our method for arbitrary symmetry and anisotropy of the infinitely extended medium. The presented formulation of the dynamic Green's function in the space-time domain may be especially useful for a treatment of scattering problems, for instance the scattering of acoustic waves at impurities or inclusions.

19.2 Equations of Motion

We start with the field equations for the stress tensor $\boldsymbol{\sigma}$ and the elastic displacements $\boldsymbol{u}$. The governing equilibrium conditions are given by

$$\rho_M \frac{d^2}{dt^2} u_i = \partial_j \sigma_{ij} + f_i \,, \quad \sigma_{ij} = \sigma_{ji} \tag{19.1}$$

∂_j indicates the spatial derivatives with respect to the Cartesian space coordinates $x_j = (x, y, z)$, $\frac{d}{dt}$ is the time derivative, $\boldsymbol{f}$ the density of body forces and ρ_M the mass density. As constitutive equations we assume the anisotropic Hooke's law:

$$\sigma_{ij} = C_{ijkl} \epsilon_{kl} \tag{19.2}$$

where C_{ijkl} denote the elastic moduli. They have the symmetry properties $C_{ijkl} = C_{klij} = C_{jikl} = C_{ijlk}$. The elastic displacement field $\boldsymbol{u}$ is related to the tensor $\boldsymbol{\epsilon}$ of elastic strains by the ansatz

$$\epsilon_{ij} = \frac{1}{2} \left(\partial_i u_j + \partial_j u_i \right) \tag{19.3}$$

Putting ansatz (19.3) using (19.2) into the field Eq. (19.1) we obtain a 3×3 second-order differential equation (wave equation) for the displacement field $\boldsymbol{u}$ of the form:

$$[\boldsymbol{T}(\nabla) - \rho_M \frac{d^2}{dt^2} \mathbf{1}] \boldsymbol{u}(\boldsymbol{r}, t) + \boldsymbol{f}(\boldsymbol{r}, t) = 0 \tag{19.4}$$

∇ indicates the spatial gradient operator, $\mathbf{1}$ denotes the 3×3 unit tensor and $\boldsymbol{r} = (x, y, z)$ the space vector and t the time coordinate. $\boldsymbol{T}(\nabla)$ indicates the fundamental 3×3 second-order tensor differential operator of elasticity and is given by

$$T_{ij}(\nabla) = C_{ipjq} \partial_p \partial_q \tag{19.5}$$

The displacement field $\boldsymbol{u}$ can then be represented by the dynamic 3×3 Green's function $\boldsymbol{G}$ according to

$$\boldsymbol{u}(\boldsymbol{r}, t) = \int \int \boldsymbol{G}(\boldsymbol{r} - \boldsymbol{r}^{'}, t - t^{'}) \boldsymbol{f}(\boldsymbol{r}^{'}, t^{'}) \mathrm{d}^3 \boldsymbol{r}^{'} \mathrm{d}t^{'} \tag{19.6}$$

The requirement that $\boldsymbol{u}$ must be a solution of (19.4) yields as defining wave equation for the dynamic Green's function

$$[\boldsymbol{T}(\nabla) - \rho_M \frac{d^2}{dt^2}\mathbf{1}]\boldsymbol{G}(\boldsymbol{r}, t) + \delta^3(\boldsymbol{r})\,\delta(t)\mathbf{1} = 0 \tag{19.7}$$

where $\delta^3(\boldsymbol{r})$ and $\delta(t)$ represent the spatial δ-function and the time-dependent δ-function, respectively. The above-defined dynamic Green's function has the following physical interpretation:
$G_{ij}(\boldsymbol{r}, t)$ gives the elastic displacement in x_i-direction at space-point $\boldsymbol{r}$ and time t caused by a unit δ-pulse-point-force at $\boldsymbol{r}' = 0$ and $t' = 0$ in x_j-direction. We furthermore observe the symmetry property $G_{ij} = G_{ji}$.

Here we note an important observation: on the one hand, since Eq. (19.7) is a partial second-order differential equation, its solution $\boldsymbol{G}$ has the property of time inversion symmetry being invariant against the replacement $t \to -t$. On the other hand, we have the physical requirement of causality for the Green's function to guarantee the fact that the displacement response at space-point $\boldsymbol{r}$ and time t can only take place *after* being excited by a source at space-point $\boldsymbol{r}'$ and time $t' < t$. As a consequence thereof the causal (retarded) Green's function $\boldsymbol{G}$ must fulfill the property $\boldsymbol{G} \neq 0$ for $t > 0$ and $\boldsymbol{G} = 0$ for $t < 0$. This is an obvious contradiction to the time inversion-symmetry-property of Eq. (19.7) that can only be eliminated by introducing an infinitesimal first-order damping term $\epsilon \frac{d}{dt}$ ($\epsilon \to 0+$) into (19.7) and (19.4), breaking the time inversion symmetry and guaranteeing causality as we shall show in the subsequent section. Simultaneously this infinitesimal damping term is necessary to regularize our problem as we depict in the next section. The (causal) dynamic Green's function (often denoted as causal propagator) has the form

$$\boldsymbol{G}(\boldsymbol{r}, t) = \Theta(t)\boldsymbol{G}(\boldsymbol{r}, t) \tag{19.8}$$

where we introduced the Heaviside Θ-function (step function) which is defined by $\Theta(t) = 1$ if $t > 0$ and $\Theta(t) = 0$ if $t < 0$.

19.3 Outline of the Method

In this section, our goal is to derive a general method that allows us to construct the causal dynamic 3×3 Green's function defined by Eq. (19.7) for the infinite medium with arbitrary symmetry. To that end we represent Green's function $\boldsymbol{G}$ by Fourier transformation according to

$$\boldsymbol{G}(\boldsymbol{r}, t) = \frac{1}{(2\pi)^4} \int\int \exp(i\omega t) \exp(i\boldsymbol{k} \cdot \boldsymbol{r}) \tilde{\boldsymbol{G}}(\boldsymbol{k}, \omega) \mathrm{d}^3\boldsymbol{k}\mathrm{d}\omega \tag{19.9}$$

The 3×3 Fourier transform $\tilde{\boldsymbol{G}}(\boldsymbol{k}, \omega)$ of Green's function can then be obtained by using Eq. (19.7)

$$\tilde{\boldsymbol{G}}(\boldsymbol{k}, \omega) = [\boldsymbol{T}(\boldsymbol{k}) - \rho_M \omega^2 \mathbf{1}]^{-1} \tag{19.10}$$

In order to obtain a convenient formulation, we now introduce the (orthonormal) eigenvectors $\boldsymbol{v}_\lambda(\boldsymbol{k})$ with the eigenvalues $\rho_M \mu_\lambda^2(\boldsymbol{k})$ $(\lambda = 1, 2, 3)$ of $T_{ij}(\boldsymbol{k}) = C_{ipjq} k_p k_q$ which are defined by

$$\boldsymbol{T}(\boldsymbol{k}) \boldsymbol{v}_\lambda(\boldsymbol{k}) = \rho_M \mu_\lambda^2(\boldsymbol{k}) \boldsymbol{v}_\lambda(\boldsymbol{k}) \quad , \lambda = 1, 2, 3 \tag{19.11}$$

Because of elastic stability the eigenvalues fulfill the property $\mu_\lambda^2(\boldsymbol{k}) > 0$ for $\boldsymbol{k} \neq 0$. The $\mu_\lambda(\boldsymbol{k})$ can be physically interpreted as eigenfrequencies of three harmonic oscillators corresponding to three elastic waves with wave vector $\boldsymbol{k}$ and polarization vector $\boldsymbol{v}_\lambda(\boldsymbol{k})$. We observe that these three waves having wave vector $\boldsymbol{k}$ represent homogeneous solutions of wave Eq. (19.4). Obviously there are three distinguished frequency branches $\mu_\lambda(\boldsymbol{k})$. As a consequence of the symmetry property $T_{ij} = T_{ji}$, the normalized eigenvectors $\boldsymbol{v}_\lambda$ represent an orthonormal basis and fulfill

$$\boldsymbol{v}_\lambda(\boldsymbol{k}) \cdot \boldsymbol{v}_\mu(\boldsymbol{k}) = \delta_{\lambda\mu} \tag{19.12}$$

together with the completeness property

$$\sum_\lambda^3 \boldsymbol{v}_\lambda(\boldsymbol{k}) \otimes \boldsymbol{v}_\lambda(\boldsymbol{k}) = \mathbf{1} \tag{19.13}$$

where $\otimes$ stands for dyadic multiplication. Using the set $\{\boldsymbol{v}_\lambda \otimes \boldsymbol{v}_\lambda\}$ as tensor basis, we can represent the $\boldsymbol{T}$-tensor in its spectral representation

$$\boldsymbol{T}(\boldsymbol{k}) = \sum_\lambda^3 \rho_M \mu_\lambda^2(\boldsymbol{k}) \boldsymbol{v}_\lambda(\boldsymbol{k}) \otimes \boldsymbol{v}_\lambda(\boldsymbol{k}) \tag{19.14}$$

By employing these properties, the Fourier transform (19.10) can be written as

$$\tilde{\boldsymbol{G}}(\boldsymbol{k}, \omega) = \frac{1}{\rho_M} \sum_\lambda^3 \boldsymbol{v}_\lambda(\boldsymbol{k}) \otimes \boldsymbol{v}_\lambda(\boldsymbol{k}) g_\lambda(\boldsymbol{k}, \omega) \tag{19.15}$$

where we have introduced the abbreviation

$$g_\lambda(\boldsymbol{k}, \omega) = \frac{1}{\mu_\lambda^2(\boldsymbol{k}) - \omega^2} \tag{19.16}$$

First we start with the calculation of $\boldsymbol{G}(\boldsymbol{k}, t)$ which is given by

$$G(\boldsymbol{k}, t) = \frac{1}{2\pi} \int_{-\infty}^{\infty} \tilde{G}(\boldsymbol{k}, \omega) \exp(i\omega t) \mathrm{d}\omega \tag{19.17}$$

From Eq. (19.9), we observe that $G(\boldsymbol{r}, t)$ and $G(\boldsymbol{k}, t)$ are related by

$$G(\boldsymbol{r}, t) = \frac{1}{(2\pi)^3} \int \exp(i\boldsymbol{k} \cdot \boldsymbol{r}) G(\boldsymbol{k}, t) \mathrm{d}^3\boldsymbol{k} \tag{19.18}$$

We can rewrite Eq. (19.17) as

$$G(\boldsymbol{k}, t) = \frac{1}{\rho_M} \sum_{\lambda}^{3} \boldsymbol{v}_\lambda(\boldsymbol{k}) \otimes \boldsymbol{v}_\lambda(\boldsymbol{k}) g_\lambda(\boldsymbol{k}, t) \tag{19.19}$$

where we denote

$$g_\lambda(\boldsymbol{k}, t) = \frac{1}{2\pi} \int_{-\infty}^{\infty} g_\lambda(\boldsymbol{k}, \omega) \exp(i\omega t) \mathrm{d}\omega \tag{19.20}$$

Obviously integral (19.20) is not well defined because of the singularities of $g_\lambda(\boldsymbol{k}, \omega)$ at $\omega_{1,2} = \pm\mu_\lambda(\boldsymbol{k})$ which are located on the real ω-axis (Eq. (19.16)). In order to obtain a regularization of (19.20), we introduce an infinitesimal damping $\epsilon \to 0+$ ($\epsilon > 0$). The amplitude $g_\lambda(\boldsymbol{k}, t)$ fulfills then the equation of motion of an infinitesimally damped harmonic oscillator with eigenfrequency $\mu_\lambda > 0$ excited by a δ-pulse

$$\left\{ \frac{d^2}{dt^2} + 2\epsilon \frac{d}{dt} + \mu_\lambda^2(\boldsymbol{k}) \right\} g_\lambda(\boldsymbol{k}, t) = \delta(t) \tag{19.21}$$

For the regularized $g_\lambda(\boldsymbol{k}, \omega)$ we now obtain

$$g_\lambda(\boldsymbol{k}, \omega) = \frac{1}{\mu_\lambda^2(\boldsymbol{k}) - \omega^2 + 2i\epsilon\omega} \tag{19.22}$$

Expression (19.20) becomes with (19.22) a regular well-defined integral. The integrand has now singularities at $\omega_{1,2} = i\epsilon \pm \mu_\lambda$ which are infinitesimally displaced from the real ω-axis into the upper complex ω-plane. In order to evaluate integral (19.20) via residue theorem, we have to find a closed integration path in the complex ω-plane which contains on the one hand the real ω-axis and yields on the other hand in a limiting process as only non-zero contribution integral (19.20). At this point, we apply a very common method of integration that is described in many introductory textbooks of theoretical physics (see, e.g. [21]). We therefore do not repeat its detailed derivation here. Its brief description is as follows: For $t > 0$ we can calculate (19.20) using residue theorem when we close the integration path by a semicircle around the upper complex ω-plane ($Im\,\omega > 0$) with radius $|\omega| \to \infty$. For $t < 0$ we have to close the integration path by a semicircle around the lower ω-plane ($Im\,\omega < 0$). In the limiting case when the radii of the semicircles tend to infinity ($|\omega| \to \infty$), the contributions of the integrals over the semicircles to (19.20) are tending to zero faster than $1/|\omega|$. Since the singularities $\omega_{1,2}$ are located in the upper complex ω-plane, there are non-zero contributions to (19.20) only for $t > 0$. For $t < 0$ (19.20) is vanishing. We thus can write

$$\begin{aligned} g_\lambda(\boldsymbol{k}, t) &= \frac{1}{2\pi} \oint g_\lambda(\boldsymbol{k}, \omega) \exp(i\omega t) \mathrm{d}\omega \\ &= 2\pi i \sum Res[g_\lambda(\boldsymbol{k}, \omega) \exp(i\omega t)] \end{aligned} \tag{19.23}$$

where $Res[g_\lambda(\boldsymbol{k}, \omega) \exp(i\omega t)]$ denote the residues of function $g_\lambda(\boldsymbol{k}, \omega) \exp(i\omega t)$. By evaluating this expression we obtain

$$g_\lambda(\boldsymbol{k}, t) = \Theta(t) \exp(-\epsilon t) \frac{\sin \mu_\lambda t}{\mu_\lambda} \tag{19.24}$$

This result is well known as *causal* (retarded) Green's function of the damped harmonic oscillator (e.g. [21]). The exponential factor $\exp(-\epsilon t)$ can be omitted since we assume that ϵ is chosen small enough that the limiting case $t\epsilon \to 0+$ is fulfilled for finite t. The Heaviside function (Θ-function) in (19.24) describes the causality of Green's function. The causality is automatically fulfilled by introducing the infinitesimal damping into Eq. (19.21). Putting (19.24) into (19.19) then gives the causal $\boldsymbol{G}(\boldsymbol{k}, t)$. The remaining task is now to perform the $\boldsymbol{k}$-integration of (19.19) by evaluating integral (19.18) which takes the form

$$\begin{aligned} \boldsymbol{G}(\boldsymbol{r}, t) = & \frac{\Theta(t)}{(2\pi)^3 \rho_M} \sum_\lambda^3 \int_0^{2\pi} \mathrm{d}\varphi \\ & \times \int_0^\pi \mathrm{d}\vartheta \sin\vartheta \frac{1}{\mu_\lambda(\hat{\boldsymbol{k}})} \boldsymbol{v}_\lambda(\hat{\boldsymbol{k}}) \otimes \boldsymbol{v}_\lambda(\hat{\boldsymbol{k}}) \\ & \times \int_0^\infty \mathrm{d}k\, k \cos(kr\cos(\vartheta)) \sin(tk\mu_\lambda(\hat{\boldsymbol{k}})) \end{aligned} \tag{19.25}$$

Here we have introduced $\boldsymbol{k} = k\hat{\boldsymbol{k}}$ where $\hat{\boldsymbol{k}}$ denotes the unit vector in $\boldsymbol{k}$-direction. We furthermore make use of the homogeneity property

$$\mu_\lambda(k\hat{\boldsymbol{k}}) = k\mu_\lambda(\hat{\boldsymbol{k}}) \tag{19.26}$$

The unit vector $\hat{\boldsymbol{k}}$ can be represented as

$$\hat{\boldsymbol{k}}(u, \varphi) = \boldsymbol{\xi}(\varphi)\sqrt{(1-u^2)} + \boldsymbol{e}_3 u \tag{19.27}$$

with ϑ denoting the angle between space vector $\boldsymbol{r}$ and wave vector $\mathbf{k}$ with $u = \cos\vartheta$ and $+\sqrt{(1-u^2)} = \sin\vartheta$ $(0 \le \vartheta \le \pi)$. $\boldsymbol{e}_3 = \boldsymbol{r}/r$ denotes the unit vector in the $\boldsymbol{r}$-direction. The unit vector $\boldsymbol{\xi}$ is perpendicular to $\boldsymbol{r}$ and has the form [9, 10]

$$\boldsymbol{\xi}\,(\varphi) = \boldsymbol{e}_1 \cos\varphi + \boldsymbol{e}_2 \sin\varphi \tag{19.28}$$

The vectors $\boldsymbol{e}_i$ form a useful orthonormal basis

$$\boldsymbol{e}_1 = \frac{1}{\rho}\begin{pmatrix} -y \\ x \\ 0 \end{pmatrix}, \ \boldsymbol{e}_2 = \frac{1}{\rho r}\begin{pmatrix} -zx \\ -zy \\ \rho^2 \end{pmatrix}, \ \boldsymbol{e}_3 = \frac{1}{r}\begin{pmatrix} x \\ y \\ z \end{pmatrix}, \tag{19.29}$$

where $\rho^2 = x^2 + y^2$, $r^2 = \rho^2 + z^2$. The basis vectors $\boldsymbol{e}_i$ fulfill $\boldsymbol{e}_i = \frac{1}{2}\epsilon_{ijk}\boldsymbol{e}_j \times \boldsymbol{e}_k$. Here ϵ_{ijk} denotes the anti-symmetric permutation tensor. In order to evaluate (19.25) it is convenient to use the relation

$$\sin(tk\mu_\lambda(\hat{\boldsymbol{k}})) = \frac{-1}{k\mu_\lambda(\hat{\boldsymbol{k}})}\frac{d}{dt}\cos(tk\mu_\lambda(\hat{\boldsymbol{k}})) \tag{19.30}$$

Then Eq. (19.25) takes the form

$$\begin{aligned} &\boldsymbol{G}(\boldsymbol{r},t) = \\ &-\frac{\Theta(t)}{(2\pi)^3\rho_M}\frac{d}{dt}\sum_{\lambda}^{3}\int_0^{2\pi} \mathrm{d}\varphi \int_{-1}^{+1} \mathrm{d}u \frac{1}{\mu_\lambda^2(\hat{\boldsymbol{k}})}\boldsymbol{v}_\lambda(\hat{\boldsymbol{k}}) \otimes \boldsymbol{v}_\lambda(\hat{\boldsymbol{k}}) \\ &\times \int_0^\infty \mathrm{d}k \cos(kru)\cos(tk\mu_\lambda(\hat{\boldsymbol{k}})) \end{aligned} \tag{19.31}$$

where $\hat{\boldsymbol{k}} = \hat{\boldsymbol{k}}(u,\varphi)$ (Eq. (19.27)). To simplify the above equation, a consideration of symmetry properties is useful: the operator $T_{ij}(\boldsymbol{k}) = C_{irjs}k_r k_s$ is a bilinear function of k_i depending only on the products $\{k_r k_s\}$ which obviously are invariant under inversion $\boldsymbol{k} \to -\boldsymbol{k}$. Consequently $\mu_\lambda^2(\boldsymbol{k})$ and $\mu_\lambda(\boldsymbol{k}) = +\sqrt{\mu_\lambda^2(\boldsymbol{k})}$ are even functions under inversion

$$\mu_\lambda(-\boldsymbol{k}) = +\mu_\lambda(\boldsymbol{k}) \tag{19.32}$$

since they are always functions of the set of products $\{k_r k_s\}$ and not affected by the signs of the k_i. As we show below this symmetry property is crucial to obtain a very compact formulation of Green's function. To that end we note that the eigenvectors fulfill the property

$$\boldsymbol{v}_\lambda(-\boldsymbol{k}) = p\boldsymbol{v}_\lambda(\boldsymbol{k}) \tag{19.33}$$

with $p^2 = 1$ thus $p = \pm 1$. This follows directly from the orthonormality relation (19.12). From these observations, we can conclude that obviously $\boldsymbol{v}_\lambda(\hat{\boldsymbol{k}}) \otimes \boldsymbol{v}_\lambda(\hat{\boldsymbol{k}})/\mu_\lambda^2(\hat{\boldsymbol{k}})$ is an even function with respect to inversion. We then observe that the surface-integral

$$0 = \int_0^{2\pi} \mathrm{d}\varphi \int_{-1}^{+1} \mathrm{d}u \left[\frac{1}{\mu_\lambda^2(\hat{\boldsymbol{k}})} \boldsymbol{v}_\lambda(\hat{\boldsymbol{k}}) \otimes \boldsymbol{v}_\lambda(\hat{\boldsymbol{k}}) \right. \\ \left. \times \sin(kru) \sin(tk\mu_\lambda(\hat{\boldsymbol{k}})) \right] \tag{19.34}$$

is vanishing since its integrand is an odd function under inversion. Using this property we arrive at[1]

$$\boldsymbol{G}(\boldsymbol{r}, t) = -\frac{\Theta(t)}{(2\pi)^3 \rho_M} \frac{d}{dt} \sum_\lambda^3 \int_0^{2\pi} \mathrm{d}\varphi \int_{-1}^{+1} \mathrm{d}u \frac{1}{\mu_\lambda^2(\hat{\boldsymbol{k}})} \\ \times \boldsymbol{v}_\lambda(\hat{\boldsymbol{k}}) \otimes \boldsymbol{v}_\lambda(\hat{\boldsymbol{k}}) \int_0^\infty \mathrm{d}k \cos(k[ru - t\mu_\lambda(\hat{\boldsymbol{k}})]) \tag{19.35}$$

Taking into account that the k-integration yields a δ-function, we obtain

$$\boldsymbol{G}(\boldsymbol{r}, t) = -\frac{\Theta(t)}{8\pi^2 \rho_M r} \frac{d}{dt} \sum_\lambda^3 \int_0^{2\pi} \mathrm{d}\varphi \int_{-1}^{+1} \mathrm{d}u \\ \times \left[\frac{1}{\mu_\lambda^2(\hat{\boldsymbol{k}})} \boldsymbol{v}_\lambda(\hat{\boldsymbol{k}}) \otimes \boldsymbol{v}_\lambda(\hat{\boldsymbol{k}}) \right. \\ \left. \times \delta\left(u - \frac{t\mu_\lambda(u, \varphi)}{r} \right) \right] \tag{19.36}$$

The (u, φ)-dependencies of μ_λ and $\boldsymbol{v}_\lambda$ depend on the orientation of $\boldsymbol{r}$ (compare Eq. (19.27)). We note here that the eigenvalues $\mu_\lambda(\hat{\boldsymbol{k}}) = \mu_\lambda(u, \varphi)$ are functions defined on the unit sphere $|\hat{\boldsymbol{k}}| = 1$. By choosing a coordinate free representation, we can write (19.36) alternatively as integral over the unit sphere $|\hat{\boldsymbol{k}}| = 1$[2]:

[1] We observe from Eq. (19.34) that the term $\cos(k[ru - t\mu_\lambda(\hat{\boldsymbol{k}})])$ could also be replaced by $\cos(k[ru + t\mu_\lambda(\hat{\boldsymbol{k}})])$. Because of the inversion symmetry $\mu_\lambda(-\hat{\boldsymbol{k}}) = \mu_\lambda(\hat{\boldsymbol{k}})$, both variants lead to the same result for Green's function (Eq. (19.43) below).

[2] Because of the inversion symmetry of the integrand of (19.37), we observe that both signs in $\delta(\hat{\boldsymbol{k}} \cdot \boldsymbol{r} \pm t\mu_\lambda(\hat{\boldsymbol{k}}))$ lead to the same result.

$$
\begin{aligned}
\boldsymbol{G}(\boldsymbol{r},t) = & -\frac{\Theta(t)}{8\pi^2\rho_M}\frac{d}{dt}\sum_{\lambda}^{3}\int_{|\hat{\boldsymbol{k}}|=1} \mathrm{d}\Omega(\hat{\boldsymbol{k}}) \\
& \times\left[\frac{1}{\mu_\lambda^2(\hat{\boldsymbol{k}})}\boldsymbol{v}_\lambda(\hat{\boldsymbol{k}})\otimes\boldsymbol{v}_\lambda(\hat{\boldsymbol{k}})\delta(\hat{\boldsymbol{k}}\cdot\boldsymbol{r}-t\mu_\lambda(\hat{\boldsymbol{k}}))\right]
\end{aligned}
\tag{19.37}
$$

The contributions to (19.37) are determined by the zeros of the equations

$$
\hat{\boldsymbol{k}}\cdot\boldsymbol{r}-t\mu_\lambda(\hat{\boldsymbol{k}})=0 \ , \lambda=1,2,3 \tag{19.38}
$$

The solutions $\hat{\boldsymbol{k}}_\lambda(\varphi,t/r)=\hat{\boldsymbol{k}}(u_\lambda(\varphi,t/r),\varphi)$ of (19.38) represent curves on the unit sphere $|\hat{\boldsymbol{k}}|=1$ which can be parameterized as follows:

$$
\begin{aligned}
\hat{\boldsymbol{k}}_\lambda\left\{\varphi,\frac{t}{r}\right\} &= \hat{\boldsymbol{k}}\left\{u_\lambda(\varphi,\frac{t}{r}),\varphi\right\} \\
\hat{\boldsymbol{k}}_\lambda\left\{\varphi,\frac{t}{r}\right\} &= \sqrt{1-u_\lambda^2(\varphi,\frac{t}{r})}\boldsymbol{\xi}(\varphi)+\boldsymbol{e}_3u_\lambda(\varphi,\frac{t}{r})
\end{aligned}
\tag{19.39}
$$

where the $u_\lambda(\varphi,t/r)$ denote only *those* zeros of the equations

$$
u-\frac{t\mu_\lambda(u,\varphi)}{r}=0, \quad \lambda=1,2,3 \tag{19.40}
$$

which fulfill additionally the condition

$$
1-u^2\geq 0 \tag{19.41}
$$

The solutions u_λ of (19.40) which fulfill additionally (19.41) represent the zeros of the arguments in the δ-functions in Eq. (19.36). Condition (19.41) takes into account that the integration limits are $u=\pm 1$. For further evaluation of (19.36) we have to determine these solutions.

To discuss the solutions of (19.40), it is convenient to introduce the minimum and maximum values μ_λ^{Min} and μ_λ^{Max} with $0<\mu_\lambda^{Min}\leq\mu_\lambda(\hat{\boldsymbol{k}})\leq\mu_\lambda^{Max}$. We observe, because of the property $\mu_\lambda(u,\varphi)t/r>0$, that the solutions of (19.40) always fulfill $u_\lambda(\varphi,t/r)>0$. However, only those solutions of (19.40) affect (19.36) which fulfill additionally (19.41). In order to determine them, three distinguished t/r-ranges have to be considered for each fixed λ:

Range I:
$t\mu_\lambda^{Min}/r\leq t\mu_\lambda/r\leq t\mu_\lambda^{Max}/r\leq 1$; $u_\lambda(t/r,\varphi)<1$ for all φ.

Range II:
$\mu_\lambda^{Min}t/r\leq 1\leq\mu_\lambda^{Max}t/r$; $u_\lambda(t/r,\varphi)<1$ is fulfilled for distinguished φ-ranges only.

Range III:
$1 < \mu_\lambda^{Min} t/r \le \mu_\lambda t/r \le \mu_\lambda^{Max} t/r;\ u_\lambda(t/r, \varphi) > 1$ thus (19.40) has no solutions which fulfill (19.41).

In the general case of arbitrary symmetry of the medium, the solutions of (19.40) which fulfill (19.41) for fixed λ's consist in several steady solution segments $u_\lambda^{(s_\lambda)}(\varphi, t/r)$ $(s_\lambda = 1, ..., m_\lambda)$ where $m_\lambda = m_\lambda(t/r)$ denotes the number of steady solution segments. When the entire φ interval $[0, 2\pi]$ is passed, the solutions pass all m_λ segments $u_\lambda^{(s_\lambda)}(\varphi, t/r)$ sequentially. Each segment $u_\lambda^{(s_\lambda)}$ represents the unique solution of (19.40) fulfilling (19.41) for a certain φ-interval. The introduction of steady segments takes into account that the solutions of (19.40) fulfilling (19.41) may behave discontinuously, jumping from one segment $u_\lambda^{(p)}(\varphi, t/r)$ to the next $u_\lambda^{(p+1)}(\varphi, t/r)$ when φ is continuously passed. In our subsequent considerations segment index s_λ is omitted whenever possible.

In order to derive a compact formulation for Green's function from (19.36), it is convenient to introduce characteristic functions $\Theta_{s_\lambda}(\varphi, t/r)$ with $\Theta_{s_\lambda} = 1$ only in that very φ-interval where the solution assumes segment $u_\lambda^{(s_\lambda)}$ and $\Theta_{s_\lambda} = 0$ outside that φ-interval. The above-introduced characteristic functions then fulfill the condition

$$\sum_{s_\lambda}^{m_\lambda} \Theta_{s_\lambda}(\varphi, \frac{t}{r}) = \Theta_\lambda(t/r) \tag{19.42}$$

with $\Theta_\lambda(t/r) = 1$ for t/r in the above-defined range I, $0 < \Theta_\lambda(t/r) < 1$ for t/r in range II, and $\Theta_\lambda(t/r) = 0$ for t/r in range III.

Thus we finally obtain for Green's function (19.36) the representation

$$\begin{aligned} \boldsymbol{G}(\boldsymbol{r}, t) = &-\frac{\Theta(t)}{8\pi^2 \rho_M r} \frac{d}{dt} \sum_\lambda^3 \sum_{s_\lambda}^{m_\lambda} \int_0^{2\pi} d\varphi\, \Theta_{s_\lambda}(\varphi, \frac{t}{r}) \\ &\times \left[\frac{\Theta\{1 - u_\lambda^{(s_\lambda)}(\varphi, \frac{t}{r})\}}{B_\lambda^{(s_\lambda)}[\hat{\boldsymbol{k}}_\lambda^{(s_\lambda)}(\varphi, \frac{t}{r})] \mu_\lambda^2 [\hat{\boldsymbol{k}}_\lambda^{(s_\lambda)}(\varphi, \frac{t}{r})]} \right. \\ &\left. \times \boldsymbol{v}_\lambda[\hat{\boldsymbol{k}}_\lambda^{(s_\lambda)}(\varphi, \frac{t}{r})] \otimes \boldsymbol{v}_\lambda[\hat{\boldsymbol{k}}_\lambda^{(s_\lambda)}(\varphi, \frac{t}{r})] \right] \end{aligned} \tag{19.43}$$

The $B_\lambda^{(s_\lambda)}$ are given by

$$\begin{aligned} B_\lambda^{(s_\lambda)}[\hat{\boldsymbol{k}}_\lambda^{(s_\lambda)}(\varphi, \tfrac{t}{r})] &= B_\lambda^{(s_\lambda)}(u_\lambda^{(s_\lambda)}, \varphi) \\ &= |1 - \frac{t}{r} \frac{\partial \mu_\lambda}{\partial u} [\hat{\boldsymbol{k}}(u = u_\lambda^{(s_\lambda)}, \varphi)]| \end{aligned} \tag{19.44}$$

The above representation (19.43) of Green's function reduces our initial integration problem (19.9) considerably, namely to only one single φ-integration-problem combined with the algebraic problem which consists in finding the zeros u_λ of the three equations (19.40) fulfilling (19.41). The $\Theta(1 - u_\lambda)$-functions in (19.43) guarantee that only those φ-ranges with $u_\lambda(\varphi, \frac{t}{r}) < 1$ contribute to (19.43).

Expression (19.43) generally holds for infinite media with arbitrary elastic anisotropy and symmetry. In view of (19.40) follows that for fixed r and $t \to \infty$ the step functions in (19.43) vanish in the entire φ-range $[0, 2\pi]$. As a consequence thereof, Green's function is vanishing in the limiting case of $t \to \infty$ for each fixed space-point $\boldsymbol{r}$. This can physically be interpreted by runtime effects.

Subsequently we show that the above representation (19.43) of the dynamic Green's function is highly convenient in order to obtain in a direct way the full closed-form solutions for the isotropic medium (Sect. 19.5) and for the farfield solution in the case of hexagonal symmetry of the medium (Sect. 19.6). In the next section, we derive general asymptotic expressions for Green's function from Eq. (19.43) at infinity $r \to \infty$ holding for arbitrary symmetry of the medium.

19.4 Asymptotic Expressions for the Green's Function

In this section, we derive asymptotic expressions for Green's function (19.43) far away from the source point $\boldsymbol{r}' = 0$ for $r \to \infty$ and $t > 0$. To that end we have to consider the $1/r$-expansion of (19.43) which is obtained by evaluating the expression

$$S_\lambda = \frac{d}{dt}\left\{\frac{\boldsymbol{v}_\lambda(\hat{\boldsymbol{k}}_\lambda) \otimes \boldsymbol{v}_\lambda(\hat{\boldsymbol{k}}_\lambda)\Theta(1 - u_\lambda)}{\mu_\lambda^2(\hat{\boldsymbol{k}}_\lambda)[B_\lambda(\hat{\boldsymbol{k}}_\lambda)]}\right\} \tag{19.45}$$

to arrive at

$$S_\lambda = -\frac{du_\lambda}{dt}\left\{\frac{\boldsymbol{v}_\lambda(\hat{\boldsymbol{k}}_\lambda) \otimes \boldsymbol{v}_\lambda(\hat{\boldsymbol{k}}_\lambda)\delta(1 - u_\lambda)}{\mu_\lambda^2(\hat{\boldsymbol{k}}_\lambda)[B_\lambda(\hat{\boldsymbol{k}}_\lambda)]}\right\} + O(\frac{1}{r}) \tag{19.46}$$

where the δ-function comes into play from the time derivative of the Θ-function and $\boldsymbol{O}(1/r)$ denotes remaining terms of at least first-order in $1/r$ corresponding to a $1/r^2$-order term of (19.43). This is a consequence of the property

$$\frac{du_\lambda}{dt} = \frac{\mu_\lambda}{r} + \frac{t}{r}\frac{\partial \mu_\lambda}{\partial u}\frac{du_\lambda}{dt} \tag{19.47}$$

Collecting terms $\frac{du_\lambda}{dt}$ we obtain an expression of order $1/r$:

$$\frac{du_\lambda}{dt} = \frac{\mu_\lambda}{r\left(1 - \frac{t}{r}\frac{\partial \mu_\lambda}{\partial u}\right)} = \frac{\mu_\lambda}{r}\frac{p_\lambda}{B_\lambda} \tag{19.48}$$

Here we have introduced $p_\lambda = sign(1 - \frac{t}{r}\frac{\partial \mu_\lambda}{\partial u})$ that can take the values ± 1. Equation (19.46) can now be evaluated by using (19.48). The factor μ_λ / r can be absorbed into the δ-function:

$$\frac{d}{dt}\Theta(1 - u_\lambda) = -\delta(t - \frac{r}{\mu_\lambda})\frac{p_\lambda}{B_\lambda} \tag{19.49}$$

To obtain the farfield approximation in the evaluation of (19.43), only the δ-term of (19.46) has to be taken into account as dominant term for $r \to \infty$. Thus (19.43) yields then an expression of the form

$$\boldsymbol{G}_\infty(\boldsymbol{r}, t) = \sum_{\lambda=1}^{3} \boldsymbol{G}_\infty^{(\lambda)}(\boldsymbol{r}, t) \tag{19.50}$$

where

$$\boldsymbol{G}_\infty^{(\lambda)}(\boldsymbol{r}, t) = \frac{p_\lambda}{8\pi^2 \rho_M r}\frac{\delta(t - \frac{r}{\mu_\lambda})}{\mu_\lambda^2}\int_0^{2\pi} \boldsymbol{v}_\lambda \otimes \boldsymbol{v}_\lambda \frac{d\varphi}{B_\lambda^2} \tag{19.51}$$

The quantities $B_\lambda = B_\lambda(u = 1, \varphi)$ in general depend on φ, whereas $\mu_\lambda(u = 1) = \mu_\lambda(\boldsymbol{e}_3)$ in (19.51) do not. The eigenvectors $\boldsymbol{v}_\lambda$ depend on φ under certain circumstances as will be discussed. To evaluate (19.51), we have to consider the expressions $\frac{\partial \mu_\lambda}{\partial u}$ for $u \to 1$ since the δ-functions are contributing when $u_\lambda = \mu_\lambda t / r = 1$. To that end we write

$$\hat{\boldsymbol{k}} = \boldsymbol{e}_3 + \delta\hat{\boldsymbol{k}} \tag{19.52}$$

where $\delta\hat{\boldsymbol{k}}$ is a small deviation given by (compare (19.27))

$$\delta\hat{\boldsymbol{k}} = \boldsymbol{e}_3(u - 1) + \boldsymbol{\xi}(\varphi)\sqrt{1 - u^2} \tag{19.53}$$

where $\boldsymbol{e}_3 = \boldsymbol{r}/r$. We now assume that μ_λ can be expanded around $\boldsymbol{e}_3$ according to

$$\mu_\lambda(\boldsymbol{e}_3 + \delta\hat{\boldsymbol{k}}) = \mu_\lambda(\boldsymbol{e}_3) + \boldsymbol{\alpha} \cdot \delta\hat{\boldsymbol{k}} + \frac{1}{2}\delta\hat{\boldsymbol{k}}^T \boldsymbol{\gamma} \delta\hat{\boldsymbol{k}} + \cdots \tag{19.54}$$

Here we denote

$$\begin{aligned} \boldsymbol{\alpha} &= \nabla_k \mu_\lambda(\boldsymbol{k} = \boldsymbol{e}_3) \\ \boldsymbol{\gamma} &= \nabla_k \otimes \nabla_k \mu_\lambda(\boldsymbol{k} = \boldsymbol{e}_3) \end{aligned} \tag{19.55}$$

where ∇_k indicates the gradient operator with respect to $\boldsymbol{k}$. By taking into account the u-dependence of $\delta\hat{\boldsymbol{k}}$ of (19.53), we obtain the representation

$$\begin{aligned} \mu_\lambda(\boldsymbol{e}_3 + \delta\hat{\boldsymbol{k}}) = \mu_\lambda(\boldsymbol{e}_3) + (u-1)\boldsymbol{\alpha}\cdot\boldsymbol{e}_3 + \sqrt{1-u^2}\boldsymbol{\alpha}\cdot\boldsymbol{\xi} \\ + \tfrac{1}{2}\boldsymbol{e}_3\boldsymbol{\gamma}\boldsymbol{e}_3(u-1)^2 \\ + \tfrac{1}{2}(\boldsymbol{\xi}\boldsymbol{\gamma}\boldsymbol{e}_3 + \boldsymbol{e}_3\boldsymbol{\gamma}\boldsymbol{\xi})\sqrt{1-u^2}(u-1) \\ + \tfrac{1}{2}\boldsymbol{\xi}\boldsymbol{\gamma}\boldsymbol{\xi}(1-u^2) + O(1-u)^{3/2} \end{aligned} \tag{19.56}$$

where $O(1-u)^{3/2}$ denotes higher order terms containing the common factor $(1-u)^{3/2}$. The partial derivative with respect to u of (19.56) yields[3]

$$\begin{aligned} \frac{\partial\mu_\lambda}{\partial u}(\boldsymbol{e}_3 + \delta\hat{\boldsymbol{k}}) = \boldsymbol{\alpha}\cdot\boldsymbol{e}_3 - \tfrac{u}{\sqrt{1-u^2}}\boldsymbol{\alpha}\cdot\boldsymbol{\xi} \\ + \boldsymbol{e}_3\boldsymbol{\gamma}\boldsymbol{e}_3(u-1) \\ + (\boldsymbol{\xi}\boldsymbol{\gamma}\boldsymbol{e}_3 + \boldsymbol{e}_3\boldsymbol{\gamma}\boldsymbol{\xi})\tfrac{1}{2}\sqrt{\tfrac{1-u}{1+u}}(1+2u) \\ - \boldsymbol{\xi}\boldsymbol{\gamma}\boldsymbol{\xi}u + O(1-u)^{1/2} \end{aligned} \tag{19.57}$$

For $u \to 1$ the term $\frac{u}{\sqrt{(1-u^2)}}\boldsymbol{\alpha}\cdot\boldsymbol{\xi}$ tends to infinity if $\boldsymbol{\alpha}\cdot\boldsymbol{\xi} \neq 0$. For arbitrary elastic anisotropy, there are only few distinguished $\boldsymbol{r}$-directions where $\boldsymbol{\alpha}\cdot\boldsymbol{\xi} = 0$. Only for these distinguished directions (19.57) yields finite values. As a consequence thereof, in view of (19.51), propagation of farfield waves is possible only in these directions. For all other $\boldsymbol{r}$-directions where $\boldsymbol{\alpha}\cdot\boldsymbol{\xi} \neq 0$, the integral (19.51) is vanishing due the singular behavior of (19.44). Then the propagation of farfield waves (and the corresponding energy flux) is suppressed. This is an ultimate consequence of the elastic anisotropy which causes the u-dependence of μ_λ.

Before considering the possible distinguished propagation directions $\boldsymbol{r}/r = \boldsymbol{e}_S$ fulfilling $\boldsymbol{\alpha}\cdot\boldsymbol{\xi} = 0$, we discuss the physical meaning of the term $\boldsymbol{\alpha}\cdot\boldsymbol{\xi}$: This term expresses (since $\boldsymbol{\xi}$ is perpendicular to $\boldsymbol{r}$) the derivative in the $\boldsymbol{\xi}$-direction perpendicular to $\boldsymbol{r}$. To clarify this we consider the gradient

$$\nabla_k \mu_\lambda(\boldsymbol{k}) = \hat{\boldsymbol{k}}\mu_\lambda(\hat{\boldsymbol{k}}) + \tilde{\nabla}_k \mu_\lambda(\hat{\boldsymbol{k}}) \tag{19.58}$$

[3] Here superscripts λ of $\boldsymbol{\alpha}$ and $\boldsymbol{\gamma}$ are omitted.

which has the physical meaning of *group velocity*. This representation is obtained by using $\nabla_{\boldsymbol{k}} = \hat{\boldsymbol{k}}\frac{d}{dk} + \frac{1}{k}\tilde{\nabla}_{\boldsymbol{k}}$ together with the homogeneity property $\mu_\lambda(k\hat{\boldsymbol{k}}) = k\mu_\lambda(\hat{\boldsymbol{k}})$. Note furthermore that the first term in the above equation $\hat{\boldsymbol{k}}\mu_\lambda(\hat{\boldsymbol{k}})$ has the physical meaning of *phase velocity*. When we now put $\hat{\boldsymbol{k}} = \boldsymbol{e}_3 = \boldsymbol{r}/r$, we see that the term $\boldsymbol{\alpha} \cdot \boldsymbol{e}_3 = \mu_\lambda(\boldsymbol{e}_3)$ $(u = 1)$ coincides with the phase velocity and $\boldsymbol{\alpha} \cdot \boldsymbol{\xi}$ measures the deviation between phase- and group velocity. Thus we can state (where we assume the source point being located in the origin): *In* $\boldsymbol{r}$*-directions, where group- and phase velocities do not coincide, no radiation, i.e. wave propagation occurs.* This is caused by the singularity of B_λ at $u = 1$ for these points as a consequence of $\boldsymbol{\alpha} \cdot \boldsymbol{\xi} \neq 0$ (compare Eq. (19.57)). Physically this can be interpreted as follows: if group and phase velocity does not coincide *dispersion* occurs, that is, a wave package cannot propagate stably and "survive" until infinity ($r \to \infty$). Hence we can formulate the *radiation condition* in the anisotropic elasticity as follows: the possible (distinguished) $\boldsymbol{r}$-directions $\boldsymbol{e}_3$ from the source point in which propagation of farfield waves is possible are determined by the condition

$$\boldsymbol{\alpha} \cdot \boldsymbol{\xi} = \boldsymbol{\xi} \cdot \nabla_{\boldsymbol{k}}\mu_\lambda(\boldsymbol{k} = \boldsymbol{e}_3) = 0 \tag{19.59}$$

For these directions $\hat{\boldsymbol{k}} = \boldsymbol{r}/r$ then group- and phase velocity coincide and, as a consequence, a wave package can stably propagate without dispersion in that direction without decay. The notation $\boldsymbol{e}_3 = \hat{\boldsymbol{e}}_S$ indicates that there are only special distinguished directions which fulfill (19.59), namely high symmetry directions beside a few other distinguished directions. For such directions (19.57) yields for $u = 1$ (corresponding to $\hat{\boldsymbol{k}} = \boldsymbol{e}_3 = \hat{\boldsymbol{e}}_S$) finite expressions of the form

$$\frac{\partial \mu_\lambda}{\partial u}(\hat{\boldsymbol{e}}_S) = \mu_\lambda(\hat{\boldsymbol{e}}_S) - \boldsymbol{\xi}\boldsymbol{\gamma}^{(\lambda)}\boldsymbol{\xi} \tag{19.60}$$

In the following we consider only *high symmetry directions* $\hat{\boldsymbol{e}}_S$ of the crystal. In that case all three μ_λ $(\lambda = 1, 2, 3)$ *simultaneously* fulfill (19.59). Thus we obtain for high symmetry propagation directions from the source point *three* corresponding outgoing farfield waves.[4] Relation (19.60) can also be interpreted as follows: the surfaces $\mu_\lambda(\boldsymbol{k}) = const$ have the property of being perpendicular to $\boldsymbol{r}$ at stationary points (extremal values) in directions $\hat{\boldsymbol{e}}_S$. Without loss of generality, we can choose the $\boldsymbol{e}_i$ such ($\boldsymbol{\xi}(\varphi) = \boldsymbol{e}_1 \cos\varphi + \boldsymbol{e}_2 \sin\varphi$ thus $\boldsymbol{e}_1 \times \boldsymbol{e}_2 = \hat{\boldsymbol{e}}_S$) that the $\boldsymbol{e}_i$ are eigenvectors of matrix $\boldsymbol{\gamma}^{(\lambda)}$ with corresponding eigenvalues $\gamma_i^{(\lambda)}$. Taking this into account we can write for (19.60):

$$\frac{\partial \mu_\lambda}{\partial u}(\boldsymbol{e}_S) = \mu_\lambda(\boldsymbol{e}_S) - (\gamma_1^{(\lambda)} \cos^2\varphi + \gamma_2^{(\lambda)} \sin^2\varphi) \tag{19.61}$$

and correspondingly, we obtain for B_λ by making use of the fact $u = \mu_\lambda(\boldsymbol{e}_S)t/r = 1$ thus $t/r = 1/\mu_\lambda(\boldsymbol{e}_S)$ and using Eq. (19.44)

[4] Additionally there are distinguished *non-high symmetry* directions for which one single μ_λ may fulfill (19.59), corresponding to only one single outgoing wave.

$$B_\lambda(\boldsymbol{e}_S, \varphi) = \frac{1}{\mu_\lambda(\boldsymbol{e}_S)} |(\gamma_1^{(\lambda)} \cos^2\varphi + \gamma_2^{(\lambda)} \sin^2\varphi)| \tag{19.62}$$

Here we have to note that there are two types of points (directions) $\boldsymbol{e}_S$. There are directions where $\gamma_i^{(\lambda)}$ are both of the same sign (elliptic points on the surface $\mu_\lambda(\hat{\boldsymbol{k}}) = const$) and other directions with opposite signs of $\gamma_i^{(\lambda)}$ (hyperbolic points of $\mu_\lambda = const$). Equation (19.51) takes the form

$$\boldsymbol{G}_\infty(\boldsymbol{r}, t) = \frac{1}{8\pi^2 \rho_M r} \sum_{\lambda=1}^{3} p_\lambda \delta(t - \frac{r}{\mu_\lambda}) \boldsymbol{I}_\lambda \tag{19.63}$$

where it is denoted as

$$\begin{aligned} \boldsymbol{I}_\lambda &= \int_0^{2\pi} \frac{\boldsymbol{v}_\lambda \otimes \boldsymbol{v}_\lambda}{B_\lambda^2 \mu_\lambda^2} \mathrm{d}\varphi \\ &= \int_0^{2\pi} \frac{\boldsymbol{v}_\lambda \otimes \boldsymbol{v}_\lambda}{|\gamma_1^{(\lambda)} \cos^2\varphi + \gamma_2^{(\lambda)} \sin^2\varphi|^2} \mathrm{d}\varphi \end{aligned} \tag{19.64}$$

The above integral is well defined only if the $\gamma_i^{(\lambda)}$ have the same signs (elliptic case). In the case that the $\gamma_i^{(\lambda)}$ have opposite signs (hyperbolic case), integral (19.64) is not well defined since there appear zeros in the denominator of the integrand of (19.64) for certain φ. This case requires a regularization technique by introducing an infinitesimal imaginary part $i\epsilon$ ($\epsilon \to 0+$) into the denominator of the integrand of (19.64). Such "catastrophic" singular behavior of the corresponding farfield amplitude can be classified by catastrophe theory [14] and lies beyond the scope of this paper. Corresponding singularities and "sharp" features occurring also in the dynamic response function[5] which represents the displacements caused by a point force with Heaviside step function time dependence are treated elsewhere [15].

In this paper we confine ourselves on *elliptic* points, i.e. we assume $sign(\gamma_1^{(\lambda)}) = sign(\gamma_2^{(\lambda)}) = p_\lambda$. As already mentioned there may be two distinguished cases in the evaluation of integral (19.64):

Case I: The eigenvectors $\boldsymbol{v}_\lambda$ in (19.64) are independent on angle φ. This is caused by Eq. (19.27) which shows for $u = 1$ no φ-dependence of the eigenvectors $\boldsymbol{v}_\lambda(\hat{\boldsymbol{k}})$.

Case II: (degenerate case): The sound velocities $\mu_{1,2} = \mu_t$ of the transverse polarized waves are degenerate since $\boldsymbol{e}_S = \boldsymbol{v}_3$ is an axis of rotational symmetry. As a consequence the curvatures coincide, i.e. $\gamma_1^\lambda = \gamma_2^\lambda = H_\lambda$. Therefore the transverse polarization vectors $\boldsymbol{v}_{1,2}$ are only determined by spanning the degenerate transverse subspace according to

$$\boldsymbol{v}_1 \otimes \boldsymbol{v}_1 + \boldsymbol{v}_2 \otimes \boldsymbol{v}_2 = \mathbf{1} - \hat{\boldsymbol{e}}_S \otimes \hat{\boldsymbol{e}}_S \tag{19.65}$$

[5] The dynamic Green's function represents the time derivative of the dynamic response function.

For $u = 1$ the $\boldsymbol{v}_{1,2}$ actually have indefinite φ-dependence. But in order to define integral (19.64) *uniquely* for this case, it is required to continue the vector functions $\boldsymbol{v}_\lambda(\hat{\boldsymbol{k}}_\lambda)$ $(\lambda = 1, 2)$ analytically for $u_\lambda \to 1$ in such a way that integral (19.64) is a steady and unique function of u_λ. This assumption leads to a (non-unique) φ-dependence of the transverse polarization vectors. Without loss of generality they can be chosen as $\boldsymbol{v}_1 = \boldsymbol{\xi}(\varphi)$ and $\boldsymbol{v}_2 = \frac{d}{d\varphi}\boldsymbol{\xi}(\varphi)$.

Case I. We first discuss the case where the $\boldsymbol{v}_\lambda$ do not depend on φ: then we can write $\boldsymbol{I}_\lambda = \boldsymbol{v}_\lambda \otimes \boldsymbol{v}_\lambda \mathcal{A}_\lambda$ with $\mathcal{A}_\lambda$ given by

$$\begin{aligned}\mathcal{A}_\lambda &= \int_0^{2\pi} \frac{1}{(\alpha_\lambda + \beta_\lambda \cos 2\varphi)^2} d\varphi \\ &= -\frac{d}{d\alpha_\lambda} \int_0^{2\pi} \frac{1}{(\alpha_\lambda + \beta_\lambda \cos 2\varphi)} d\varphi\end{aligned} \tag{19.66}$$

where it is denoted as $\alpha_\lambda = \frac{1}{2}(\gamma_1^{(\lambda)} + \gamma_2^{(\lambda)})$ and $\beta_\lambda = \frac{1}{2}(\gamma_1^{(\lambda)} - \gamma_2^{(\lambda)})$. Only for $|\beta_\lambda| < |\alpha_\lambda|$ the integral is regular, which corresponds to our assumption of elliptic points. Evaluation of $\mathcal{A}_\lambda$ then yields

$$\begin{aligned}\mathcal{A}_\lambda &= 2\pi \frac{|\alpha_\lambda|}{(\alpha_\lambda^2 - \beta_\lambda^2)^{\frac{3}{2}}} \\ &= \pi \frac{|\gamma_1^{(\lambda)} + \gamma_2^{(\lambda)}|}{(\gamma_1^{(\lambda)}\gamma_2^{(\lambda)})^{\frac{3}{2}}} \\ &= 2\pi \frac{|H_\lambda|}{K_\lambda^{\frac{3}{2}}}\end{aligned} \tag{19.67}$$

In the last equation we have introduced the average curvature $H_\lambda = (\gamma_1^{(\lambda)} + \gamma_2^{(\lambda)})/2$ and the Gaussian curvature $K_\lambda = \gamma_1^{(\lambda)}\gamma_2^{(\lambda)}$. With this result we finally obtain by using Eq. (19.63), when we assume that $\hat{\boldsymbol{e}}_S$ corresponds *for all* $\lambda = 1, 2, 3$ to *elliptic* points (directions of propagation) the farfield approximation:

$$\begin{aligned}\boldsymbol{G}_\infty(r\hat{\boldsymbol{e}}_S, t) &= \boldsymbol{G}_\infty^t(r\hat{\boldsymbol{e}}_S, t) + \boldsymbol{G}_\infty^l(r\hat{\boldsymbol{e}}_S, t) \\ &= \frac{1}{4\pi\rho_M r} \sum_{\lambda=1}^{3} \frac{H_\lambda}{K_\lambda^{\frac{3}{2}}} \delta(t - \frac{r}{\mu_\lambda}) \hat{\boldsymbol{e}}_\lambda \otimes \hat{\boldsymbol{e}}_\lambda\end{aligned} \tag{19.68}$$

where $\hat{\boldsymbol{e}}_S$ indicates a high symmetry direction of the crystal from the source point. Here we have put $\hat{\boldsymbol{e}}_\lambda = \boldsymbol{v}_\lambda$. Without loss of generality we have $\boldsymbol{v}_3 = \hat{\boldsymbol{e}}_S$ denoting the longitudinal polarization vector corresponding to the wave propagating with sound velocity μ_3. Equation (19.68) describes three outgoing waves, one ($\boldsymbol{G}^l_\infty$) being longitudinally polarized ($\lambda = 3$) and two ($\boldsymbol{G}^t_\infty$) with transverse polarization ($\lambda = 1, 2$).

Case II. We now consider the case of $\boldsymbol{v}_\lambda$ being φ-dependent. Integral (19.64) then yields an expression of the form

$$\boldsymbol{G}_\infty(r\hat{\boldsymbol{e}}_S, t) = \boldsymbol{G}^t_\infty(r\hat{\boldsymbol{e}}_S, t) + \boldsymbol{G}^l_\infty(r\hat{\boldsymbol{e}}_S, t) \tag{19.69}$$

with

$$\begin{aligned} \boldsymbol{G}^t_\infty(r\hat{\boldsymbol{e}}_S, t) &= \frac{1}{8\pi\rho_M r}\left(\frac{1}{H_1^2} + \frac{1}{H_2^2}\right)[\mathbf{1} - \boldsymbol{e}_S \otimes \boldsymbol{e}_S]\delta\left(t - \frac{r}{\mu_t}\right) \\ \boldsymbol{G}^l_\infty(r\hat{\boldsymbol{e}}_S, t) &= \frac{1}{4\pi\rho_M r}\frac{1}{H_3^2}\boldsymbol{e}_S \otimes \boldsymbol{e}_S\delta\left(t - \frac{r}{\mu_l}\right) \end{aligned} \tag{19.70}$$

where we have again assumed that $\hat{\boldsymbol{e}}_S$ corresponds to elliptic points for all $\lambda = 1, 2, 3$. Expression $\boldsymbol{G}^t_\infty(r\hat{\boldsymbol{e}}_S, t)$ represents two transverse waves propagating with twofold degenerate sound velocity $\mu_t = \mu_1 = \mu_2$ and $\boldsymbol{G}^l_\infty(r\hat{\boldsymbol{e}}_S, t)$ describes one longitudinal polarized wave propagating with sound velocity $\mu_l = \mu_3$.

In both Cases I and II the expressions (19.68) and (19.70) describe three singular outgoing waves (one longitudinal and two transverse) propagating with velocities $\mu_\lambda(\hat{\boldsymbol{e}}_S)$ and polarization-directions $\boldsymbol{v}_\lambda(\hat{\boldsymbol{e}}_S)$. The above results (19.68) and (19.70) represent the *radiation* of outgoing waves for the general anisotropic elasticity.

In the case of elastic isotropy expression, (19.68) simplifies considerably since the velocities μ_λ then are constant and independent on the direction of propagation $\boldsymbol{e}_3$. Consequently then is $B_\lambda = 1$. When we perform the limiting transition to elastic isotropy, the above expressions (19.68) and (19.69) coincide with the corresponding terms of δ-function type in the expression for the isotropic Green's function (Eq. (19.75) below).

From the above expressions (19.68) and (19.70) of the asymptotic elastodynamic Green's function in the space-time-domain, we obtain immediately its representation in the space-frequency-domain

$$\boldsymbol{G}_\infty(\boldsymbol{r}, \omega) = \int_{-\infty}^{+\infty} \exp(-i\omega t)\boldsymbol{G}_\infty(\boldsymbol{r}, t)\mathrm{d}t \tag{19.71}$$

Thus we obtain for $\boldsymbol{G}_\infty(\boldsymbol{r}, \omega)$ corresponding to (19.68) ff.

$$\boldsymbol{G}_\infty(r\boldsymbol{e}_S, \omega) = \frac{1}{4\pi\rho_M r}\sum_{\lambda=1}^{3}\frac{H_\lambda}{K_\lambda^{\frac{3}{2}}}\exp\left(-\frac{i\omega r}{\mu_\lambda[\boldsymbol{e}_s]}\right)\hat{\boldsymbol{e}}_\lambda \otimes \hat{\boldsymbol{e}}_\lambda \tag{19.72}$$

This expression represents the leading order in $1/r$ (farfield-approximation) of Green's function in the space-frequency domain, and the analogue result is obtained from Eq. (19.70).

19.5 Application to the Elastic Isotropic Medium

In this section we evaluate (19.43) for the case of an elastic isotropic medium. It is the only case that can be evaluated in full closed form.

In order to determine the eigenvalues and eigenvectors of the tensor $T_{ij}(\boldsymbol{k}) = C_{ipjq}k_pk_q$ we obtain for (19.14)

$$\frac{1}{\rho_M}\boldsymbol{T}(\boldsymbol{k}) = k^2\left[\frac{\mu}{\rho_M}(\mathbf{1} - \hat{\boldsymbol{k}} \otimes \hat{\boldsymbol{k}}) + \frac{(\lambda + 2\mu)}{\rho_M}\hat{\boldsymbol{k}} \otimes \hat{\boldsymbol{k}}\right] \tag{19.73}$$

where λ, μ denote the Lame-constants of the medium. From Eq. (19.73) we observe one longitudinal eigenvector $\boldsymbol{v}_3(\boldsymbol{k}) = \hat{\boldsymbol{k}}$ with eigenvalue $\mu_3^2(\hat{\boldsymbol{k}}) = (\lambda + 2\mu)/\rho_M$ and two transverse eigenvectors $\boldsymbol{v}_{1,2}(\boldsymbol{k})$ with the twofold degenerate eigenvalue $\mu_1^2(\hat{\boldsymbol{k}}) = \mu_2^3(\hat{\boldsymbol{k}}) = \mu/\rho_M$. As a consequence of (19.13) the transverse eigenvectors $\boldsymbol{v}_{1,2}(\boldsymbol{k})$ fulfill the property

$$\boldsymbol{v}_1(\boldsymbol{k}) \otimes \boldsymbol{v}_1(\boldsymbol{k}) + \boldsymbol{v}_2(\boldsymbol{k}) \otimes \boldsymbol{v}_2(\boldsymbol{k}) = \mathbf{1} - \hat{\boldsymbol{k}} \otimes \hat{\boldsymbol{k}} \tag{19.74}$$

where the right-hand side of this equation denotes the projection operator into the two-dimensional transverse subspace. At this point we state that one longitudinal and two transverse eigenvectors of (19.73) correspond to the polarization directions of acoustic waves: The longitudinal polarized wave is propagating with phase-velocity $v_L = \sqrt{(\lambda + 2\mu)/\rho_M}$ and its polarization-direction $\boldsymbol{v}_3 = \hat{\boldsymbol{k}}$ is parallel to its propagation-direction $\boldsymbol{k}$. Both transverse polarized acoustic waves propagate with phase velocity $v_T = \sqrt{\mu/\rho_M}$ and have polarization vectors $\boldsymbol{v}_{1,2}$ in the plane perpendicular to the direction $\hat{\boldsymbol{k}}$ of propagation. In other words, for each wave vector $\boldsymbol{k}$ exist one longitudinal polarized and two transverse polarized acoustic waves. We emphasize an important property of the isotropic medium: the eigenvalues (wave propagation velocities) are independent on φ and u thus $\mu_\lambda(\hat{\boldsymbol{k}}) = const$. Thus phase- and group velocities coincide for any $\boldsymbol{k}$-direction. Consequently no dispersion occurs as in the case of elastic anisotropy. This is the crucial point that allows us to perform the φ-integration of (19.43) to obtain Green's function in full closed form:

$$\mathbf{G}(\mathbf{r},t) = -\frac{\Theta(t)}{4\pi\rho_M}\frac{t}{r^3}\left[1-3\frac{1}{r^2}\mathbf{r}\otimes\mathbf{r}\right] \times\left\{\Theta(1-\frac{v_T t}{r})-\Theta(1-\frac{v_L t}{r})\right\} + \frac{\Theta(t)}{4\pi\rho_M r}\left[\frac{\delta(t-\frac{r}{v_T})}{v_T^2}[\mathbf{1}-\frac{1}{r^2}\mathbf{r}\otimes\mathbf{r}] + \frac{\delta(t-\frac{r}{v_L})}{v_L^2}\frac{1}{r^2}\mathbf{r}\otimes\mathbf{r}\right] \tag{19.75}$$

Here $v_L = \sqrt{(\lambda+2\mu)/\rho_M}$ and $v_T = \sqrt{\mu/\rho_M}$ indicate the longitudinal and transverse sound velocities, respectively. The Θ-term only contributes for $r/v_L < t < r/v_T$. The right-hand side term proportional to $\delta\,(r - v_T t)$ describes a singular wave surface having its displacement perpendicular to the direction of propagation (transverse polarized wave). The term proportional to $\delta\,(r - v_L t)$ describes a singular wave surface having its displacement parallel to the direction of propagation (longitudinal polarized wave). These two δ-terms correspond to the asymptotic expression of Eq. (19.69) as $r \to \infty$ indicating the radiation of three waves. The Θ-term describes the contributions for space-points $\boldsymbol{r}$ on a sphere r during the time period $r/v_L < t < r/v_T$ between the faster wave having longitudinal polarization and the slower waves with transverse polarization.

Expression (19.75) coincides with the known expression in the literature [17] which was derived there by employing other techniques.

19.6 Application to a Medium with Hexagonal Symmetry

In this section, we apply Eq. (19.43) to hexagonal symmetry (transversely isotropy) of the medium with special emphasis to the farfield solutions (Eqs. (19.68) ff.). To that end we consider the wave propagation along the hexagonal high symmetry directions, namely along the c-axis and in the plane perpendicular to the c-axis. The wave propagation in the plane perpendicular to the c-axis is described by Eq. (19.68) whereas for wave propagation along the c-axis hold Eq. (19.69) ff. For hexagonal symmetry the tensor operator $T_{ij}(\boldsymbol{k}) = C_{irjs}k_r k_s$ can conveniently be written as [9]

$$\begin{aligned} \boldsymbol{T}(\boldsymbol{k}) = {} & T_{b\perp}\boldsymbol{e}_{b\perp}\otimes\boldsymbol{e}_{b\perp} + T_b\boldsymbol{e}_b\otimes\boldsymbol{e}_b \\ & + T_{bc}(\boldsymbol{e}_b\otimes\boldsymbol{e}_c + \boldsymbol{e}_c\otimes\boldsymbol{e}_b) + T_c\boldsymbol{e}_c\otimes\boldsymbol{e}_c \end{aligned} \tag{19.76}$$

with

$$\boldsymbol{k} = k_b \boldsymbol{e}_b + k_c \boldsymbol{e}_c \tag{19.77}$$

The components $k_b = \sqrt{k_x^2 + k_y^2}$ and $k_c = k_z$ of $\boldsymbol{k}$ denote the $\boldsymbol{k}$-components perpendicular and parallel to the hexagonal c-axis, respectively. The unit vectors $\boldsymbol{e}_b$, $\boldsymbol{e}_{b\perp}$ are perpendicular and $\boldsymbol{e}_c$ is parallel to the c-direction and are determined by the orientation of the $\boldsymbol{k}$-vector with respect to the c-axis:

$$\begin{aligned} \boldsymbol{e}_b = \frac{1}{k_b}\begin{pmatrix} k_x \\ k_y \\ 0 \end{pmatrix} \quad \boldsymbol{e}_{b\perp} &= \frac{1}{k_b}\begin{pmatrix} -k_y \\ k_x \\ 0 \end{pmatrix} \\ \boldsymbol{e}_c &= \begin{pmatrix} 0 \\ 0 \\ 1 \end{pmatrix} \end{aligned} \tag{19.78}$$

In (19.76) the quantities

$$T_{b\perp} = C_{66} k_b^2 + C_{44} k_c^2 \tag{19.79}$$

$$T_b = C_{11} k_b^2 + C_{44} k_c^2 \tag{19.80}$$

$$T_{bc} = (C_{13} + C_{44})\, k_b k_c \tag{19.81}$$

$$T_c = C_{44} k_b^2 + C_{33} k_c^2 \tag{19.82}$$

have been introduced. $C_{11}, C_{44}, C_{66}, C_{13}, C_{33}$ denote the 5 independent hexagonal elastic constants in Voigt's notation.

The evaluation of (19.43) requires the determination of the eigenvalues and eigenvectors of $\boldsymbol{T}(\boldsymbol{k})$. From (19.76) we observe that obviously one eigenvalue $\mu_1^2(\boldsymbol{k})$ is given by

$$\mu_1^2(\boldsymbol{k}) = \frac{T_{b\perp}}{\rho_M} \tag{19.83}$$

with the corresponding eigenvector

$$\boldsymbol{v}_1(\boldsymbol{k}) = \boldsymbol{e}_{b\perp} \tag{19.84}$$

The remaining problem consists in determining the eigenvalues and eigenvectors of a 2×2 problem. The eigenvalues of this problem are given by the roots of the characteristic equation

$$\mu^4 - \mu^2 \frac{(T_b + T_c)}{\rho_M} + \frac{T_b T_c - T_{bc}^2}{\rho_M^2} = 0 \tag{19.85}$$

and we obtain the eigenvalues $\mu_{2,3}^2(\boldsymbol{k})$

$$\begin{aligned} \mu_2^2 &= \frac{T_b + T_c - \alpha}{2\rho_M} \\ \mu_3^2 &= \frac{T_b + T_c + \alpha}{2\rho_M} \end{aligned} \tag{19.86}$$

Here we have put

$$\alpha^2 = (T_c - T_b)^2 + 4T_{bc}^2 \tag{19.87}$$

The eigenvectors $\boldsymbol{v}_{2,3}$ can then be written as

$$\begin{aligned} \boldsymbol{v}_2 &= \frac{1}{\sqrt{2\alpha}}\{-\sqrt{\alpha - \beta}\boldsymbol{e}_b + \sqrt{\alpha + \beta}\boldsymbol{e}_c\} \\ \boldsymbol{v}_3 &= \frac{1}{\sqrt{2\alpha}}\{\sqrt{\alpha + \beta}\boldsymbol{e}_b + \sqrt{(\alpha - \beta)}\boldsymbol{e}_c\} \end{aligned} \tag{19.88}$$

where

$$\beta = (T_b - T_c) \tag{19.89}$$

Let us consider (19.43) for the case where the direction of the spatial vector $\boldsymbol{r}$ is *parallel* to the hexagonal c-axis:

$$\boldsymbol{r} = r\boldsymbol{e}_c \tag{19.90}$$

where $r = |z|$ and $\boldsymbol{e}_c = \boldsymbol{e}_z$. Note that in this case the quantities (19.79)-(19.82) are only functions of $u = \cos\vartheta$ but due to the hexagonal symmetry independent on angle φ in (19.64). We then have $\alpha + \beta = 0$ and $\alpha = T_c - T_b$. The sound velocities $\mu_1 = \mu_2$ coincide and correspond to the degenerate eigenvectors $\boldsymbol{v}_{1,2} = (\boldsymbol{e}_{b\perp}, -\boldsymbol{e}_b)$ which indicate transverse polarization. The longitudinal polarization direction $\boldsymbol{e}_c$ then corresponds to the sound velocity μ_3. This allows us to rewrite (19.43):

$$\boldsymbol{G}(\boldsymbol{r}, t) = -\Theta(t)\frac{1}{8\pi^2\rho_M r}\frac{d}{dt}\sum_{\lambda}^{3}\frac{\boldsymbol{I}_\lambda(u_\lambda)\Theta[1 - u_\lambda]}{[B_\lambda(u_\lambda)\mu_\lambda^2(u_\lambda)]} \tag{19.91}$$

where B_λ is given by (19.44). The basis vectors (19.78) assume the form $\boldsymbol{e}_b = (\cos\varphi, \sin\varphi, 0)$, $\boldsymbol{e}_{b\perp} = (-\sin\varphi, \cos\varphi, 0)$ and $\boldsymbol{e}_c = (0, 0, 1)$. To evaluate (19.91) we have to determine the tensors $\boldsymbol{I}_\lambda$ $(\lambda = 1, 2, 3)$ that are given by the integrals

$$\boldsymbol{I}_\lambda = \int_0^{2\pi} \mathrm{d}\varphi \boldsymbol{v}_\lambda(u_\lambda, \varphi) \otimes \boldsymbol{v}_\lambda(u_\lambda, \varphi) \tag{19.92}$$

where the eigenvectors $\boldsymbol{v}_\lambda$ are given by (19.84) and (19.88), respectively. To obtain (19.91) we have used the following properties which are fulfilled for hexagonal symmetry only if $\boldsymbol{r}$ is parallel to the c-axis (i.e. $r = |z|$): only the eigenvectors $\boldsymbol{v}_\lambda$ for $\lambda = 1, 2$ depend on φ. The zeros of Eq. (19.40) $u_\lambda = u_\lambda(t/r)$ are only functions of t/r and the hexagonal material characteristics $\{C_{ab}/\rho_M\}$ but are independent on φ. A derivation of a closed-form expression then is feasible. In that case the existing φ-integration problem consists only in evaluating the integrals (19.92) to arrive at

$$\begin{aligned}
\boldsymbol{I}_1 &= \pi[\mathbf{1} - \boldsymbol{e}_c \otimes \boldsymbol{e}_c] \\
\boldsymbol{I}_2 &= \frac{\pi}{2\alpha(u_\lambda)}[(\alpha(u_\lambda) - \beta(u_\lambda))\mathbf{1} + (\alpha(u_\lambda) + 3\beta(u_\lambda))\boldsymbol{e}_c \otimes \boldsymbol{e}_c] \\
\boldsymbol{I}_3 &= \frac{\pi}{2\alpha(u_\lambda)}[(\alpha(u_\lambda) + \beta(u_\lambda))\mathbf{1} + (\alpha(u_\lambda) - 3\beta(u_\lambda))\boldsymbol{e}_c \otimes \boldsymbol{e}_c]
\end{aligned} \tag{19.93}$$

Farfield Solutions. Our next interest is focused on the wave propagation at infinity ($r \to \infty$) along high symmetry directions of the hexagonal medium, i.e. the c-axis and the plane perpendicular to the c-axis.
First we treat the propagation along the hexagonal c-axis ($\boldsymbol{r} = r\boldsymbol{e}_c$). Expression (19.69) writes

$$\boldsymbol{G}_\infty(r\boldsymbol{e}_c, t) = \boldsymbol{G}^t_\infty(r\boldsymbol{e}_c, t) + \boldsymbol{G}^l_\infty(r\boldsymbol{e}_c, t) \tag{19.94}$$

where the terms $\boldsymbol{G}^{t,l}_\infty(r\boldsymbol{e}_c, t)$ stand for transverse and longitudinal polarization directions, respectively. They are obtained from Eq. (19.70)

$$\begin{aligned}
\boldsymbol{G}^t_\infty(r\boldsymbol{e}_c, t) &= \frac{1}{8\pi\rho_M r}\left(\frac{1}{H_1^2} + \frac{1}{H_2^2}\right)[\mathbf{1} - \boldsymbol{e}_c \otimes \boldsymbol{e}_c]\delta\left(t - \frac{r}{\mu_t}\right) \\
\boldsymbol{G}^l_\infty(r\boldsymbol{e}_c, t) &= \frac{1}{4\pi\rho_M r}\frac{1}{H_3^2}\boldsymbol{e}_c \otimes \boldsymbol{e}_c\delta\left(t - \frac{r}{\mu_l}\right)
\end{aligned} \tag{19.95}$$

where the H_λ denote the curvatures of the sound velocities μ_λ given in (19.67). The sound velocities μ_λ (where $\mu_t = \mu_1 = \mu_2$ and $\mu_l = \mu_3$) are obtained as

$$\mu_t = \sqrt{\frac{C_{44}}{\rho_M}}$$
$$\mu_l = \sqrt{\frac{C_{33}}{\rho_M}} \tag{19.96}$$

To determine H_λ we put

$$\boldsymbol{\xi} = \boldsymbol{e}_b \tag{19.97}$$

By using the representation of the gradient operator

$$\nabla_{\boldsymbol{k}} = \boldsymbol{e}_b \frac{\partial}{\partial k_b} + \frac{1}{k_b} \boldsymbol{e}_{b\perp} \frac{\partial}{\partial \varphi} + \boldsymbol{e}_c \frac{\partial}{\partial k_c} \tag{19.98}$$

we obtain (compare Eqs. (19.60) and (19.27) and (19.28)) $\gamma_1^{(\lambda)} = \gamma_2^{(\lambda)} = H_\lambda$ which are defined by the expression $\boldsymbol{\xi}\boldsymbol{\gamma}^{(\lambda)}\boldsymbol{\xi}$. The average curvatures H_λ are obtained by performing the second-order derivatives

$$H_\lambda = \frac{\partial^2}{\partial k_b^2} \mu_\lambda (k_b = 0, k_c = 1) \tag{19.99}$$

to arrive at

$$H_1 = \frac{C_{66}}{\sqrt{\rho_M C_{44}}}$$
$$H_2 = \frac{1}{\sqrt{\rho_M C_{44}}} (C_{11} - \frac{(C_{13} + C_{44})^2}{C_{33} - C_{44}}) \tag{19.100}$$
$$H_3 = \frac{1}{\sqrt{\rho_M C_{33}}} (C_{44} + \frac{(C_{13} + C_{44})^2}{C_{33} - C_{44}})$$

Expression (19.95) describes outgoing waves propagating with sound velocities μ_i and being polarized perpendicular (for $i = 1, 2$) and parallel (for $i = 3$) to the hexagonal c-axis (propagation direction) corresponding to the transverse (for $i = 1, 2$) and longitudinal (for $i = 3$) polarization directions.
When performing the limiting transition to the isotropic medium $C_{66}, C_{44} \to \mu$, $C_{11}, C_{33} \to \lambda + 2\mu$ and $C_{13} \to \lambda$ (λ, μ indicate Lame's constants), we see that expression (19.94) yields the δ-terms of the isotropic Green's function (19.75).

We now consider the wave propagation in the plane perpendicular to the hexagonal c-axis for $r \to \infty$. The space vector then is $\boldsymbol{r} = r\boldsymbol{e}_b$ where $\boldsymbol{e}_b$ is a unit vector in that plane defined by (19.78). In this case we put $\alpha - \beta = 0$ and $\beta = T_b - T_c$. The farfield expression now can be written as (Eq. (19.68))

$$\boldsymbol{G}_{\infty}(r\boldsymbol{e}_b, t) = \boldsymbol{G}^t_{\infty}(r\boldsymbol{e}_b, t) + \boldsymbol{G}^l_{\infty}(r\boldsymbol{e}_b, t) \tag{19.101}$$

with $= \boldsymbol{G}^{t,l}_{\infty}(r\boldsymbol{e}_b, t)$ denoting the parts of transverse and longitudinal polarizations, respectively. Equation (19.101) now indicates the representations

$$\begin{aligned}\boldsymbol{G}^t_{\infty}(r\boldsymbol{e}_b, t) = \\ \frac{1}{4\pi\rho_M r}\Bigg(\frac{H_1}{K_1^{\frac{3}{2}}}\delta(t - \frac{r}{\mu_1})\boldsymbol{e}_{b\perp} \otimes \boldsymbol{e}_{b\perp} \\ + \frac{H_2}{K_2^{\frac{3}{2}}}\delta(t - \frac{r}{\mu_2})\boldsymbol{e}_c \otimes \boldsymbol{e}_c\Bigg)\end{aligned} \tag{19.102}$$

and

$$\boldsymbol{G}^l_{\infty}(r\boldsymbol{e}_b, t) = \frac{1}{4\pi\rho_M r}\frac{H_3}{K_3^{\frac{3}{2}}}\delta(t - \frac{r}{\mu_3})\boldsymbol{e}_b \otimes \boldsymbol{e}_b \tag{19.103}$$

where it is denoted as (compare Eq. (19.67))

$$\begin{aligned} H_\lambda &= \tfrac{1}{2}(\gamma_1^{(\lambda)} + \gamma_2^{(\lambda)}) \\ K_\lambda &= \gamma_1^{(\lambda)}\gamma_2^{(\lambda)} \end{aligned} \tag{19.104}$$

Now we only have to determine the quantities $\gamma_1^{(\lambda)}$ for $\boldsymbol{r} = r\boldsymbol{e}_b$. To that end we put (compare Eqs. (19.60) and (19.27) and (19.28))

$$\boldsymbol{\xi} = \xi_1\boldsymbol{e}_{b\perp} + \xi_2\boldsymbol{e}_c \tag{19.105}$$

denoting the unit vectors perpendicular to $\boldsymbol{r}$ ($\xi_1^2 + \xi_2^2 = 1$). By using (19.98) we obtain for the γ_i (for $\lambda = 1, 2, 3$)

$$\begin{aligned} \gamma_1^{(\lambda)} &= \frac{1}{k_b}\frac{\partial}{\partial k_b}\mu_\lambda(k_b = 1, k_c = 0) \\ \gamma_2^{(\lambda)} &= \frac{\partial^2}{\partial k_c^2}\mu_\lambda(k_b = 1, k_c = 0) \end{aligned} \tag{19.106}$$

to arrive at

$$\gamma_1^{(1)} = \sqrt{\frac{C_{66}}{\rho_M}}$$
$$\gamma_2^{(1)} = \frac{C_{44}}{\sqrt{C_{66}\rho_M}} \tag{19.107}$$

$$\gamma_1^{(2)} = \sqrt{\frac{C_{44}}{\rho_M}}$$
$$\gamma_2^{(2)} = \frac{1}{\sqrt{\rho_M C_{44}}}(C_{33} - \frac{(C_{13}+C_{44})^2}{C_{11}-C_{44}}) \tag{19.108}$$

$$\gamma_1^{(3)} = \sqrt{\frac{C_{11}}{\rho_M}}$$
$$\gamma_2^{(3)} = \frac{1}{\sqrt{\rho_M C_{11}}}(C_{44} + \frac{(C_{13}+C_{44})^2}{C_{11}-C_{44}}) \tag{19.109}$$

The quantities $\gamma_i^{(1,2)}$ correspond to two transverse polarized waves which propagate with sound velocities $\mu_1 = \sqrt{C_{66}/\rho_M}$ and $\mu_2 = \sqrt{C_{44}/\rho_M}$ having polarization vectors $\boldsymbol{e}_{b\perp}$ and $\boldsymbol{e}_c$, respectively. $\gamma_i^{(3)}$ corresponds to one longitudinal polarized wave propagating with sound velocity $\mu_3 = \sqrt{C_{11}/\rho_M}$ with polarization vector $\boldsymbol{e}_b$.

We observe again in the limiting transition to isotropy that expression (19.101) coincides with the δ-terms of the isotropic Green's function (19.75).

19.7 Conclusions

A general method for the construction of the dynamic 3×3 Green's function for the anisotropic elasticity for an infinitely extended medium has been derived and applied to the cases of elastic isotropy and hexagonal symmetry of the medium. The resulting representation of the elastodynamic Green's function of the anistropic infinite body (Eq. (19.43)) reduces the initial integration problem of four integrations (compare Eq. (19.9)) to *only one φ-integration* problem (Eq. (19.43)) combined with the solution problem of three algebraic Eqs. ((19.40) or (19.38)). As a consequence thereof the mathematical effort is considerably reduced and makes the method useful for many applications. The method leads in a direct and immediate way to closed-form farfield solutions for the transversely isotropic medium, as well as to the well-known exact solution of the elastic isotropic medium. The obtained expression (19.43) is highly convenient for applications to various acoustic problems, e.g. acoustic wave propagation or scattering of acoustic waves at inhomogeneities. Especially useful for many applications are the derived expressions (19.68) ff. for the asymptotic behavior at infinity ($r \to \infty$) valid for any symmetry of the medium.

Appendix

Here we consider the important relation between the elastodynamic Green's function $\boldsymbol{G}(\boldsymbol{r}, t)$ which represents the displacements due to a unit point force with δ-time dependence and the response function (susceptibility function) $\boldsymbol{H}(\boldsymbol{r}, t)$ which represents the displacement response caused by a unit point force with Θ-time dependence. We furthermore consider the relation between the dynamic response function and the *elastostatic* Green's function. The dynamic response function $\boldsymbol{H}$ is defined by

$$[\boldsymbol{T}(\nabla) - \rho_M \frac{d^2}{dt^2}\mathbf{1}]\boldsymbol{H}(\boldsymbol{r}, t) + \delta^3(\boldsymbol{r})\Theta(t)\mathbf{1} = 0 \tag{A.1}$$

We observe in view of Eq. (19.7) when we perform the time derivative of the above equation the relation

$$\boldsymbol{G}(\boldsymbol{r}, t) = \frac{d}{dt}\boldsymbol{H}(\boldsymbol{r}, t) \tag{A.2}$$

To work out the relation between $\boldsymbol{H}$ and the elastostatic Green's function $\boldsymbol{G}^{static}(\boldsymbol{r})$ we remember Eq. (19.24)

$$g_\lambda(\boldsymbol{k}, t) = \Theta(t)\exp(-\epsilon t)\frac{\sin \mu_\lambda t}{\mu_\lambda} \tag{A.3}$$

which fulfills equation (21)

$$\left\{\frac{d^2}{dt^2} + 2\epsilon\frac{d}{dt} + \mu_\lambda^2(\boldsymbol{k})\right\} g_\lambda(\boldsymbol{k}, t) = \delta(t) \tag{A.4}$$

Integration of this equation with respect to time yields

$$\left\{\frac{d^2}{dt^2} + 2\epsilon\frac{d}{dt} + \mu_\lambda^2(\boldsymbol{k})\right\} h_\lambda(\boldsymbol{k}, t) = \Theta(t) \tag{A.5}$$

where $h_\lambda(\boldsymbol{k}, t)$ then is given by

$$h_\lambda(\boldsymbol{k}, t) = \int_{-\infty}^{t} g_\lambda(\boldsymbol{k}, \tau)d\tau \tag{A.6}$$

and yields

$$h_\lambda(\boldsymbol{k}, t) = \Theta(t) \frac{(1 - \cos(\mu_\lambda t) \exp(-\epsilon t))}{\mu_\lambda^2} \tag{A.7}$$

For finite times t the exponential factor $\exp(-\epsilon t)$ can be put equal to 1. This factor becomes important only when we perform the limiting transition $t \to \infty$. Then the infinitesimal small but constant finite damping ϵ becomes important and causes the physically required relaxation for $t >> 1/\epsilon$. The functions h_λ fulfill relation (A.5) and represent dynamic response functions (susceptibilities) of harmonic oscillators with eigenfrequencies μ_λ. Using Eq. (A.7) we can construct the complete dynamic response function by time integration of (19.19) to arrive at

$$\boldsymbol{H}(\boldsymbol{k}, t) = \frac{1}{\rho_M} \sum_{\lambda}^{3} \boldsymbol{v}_\lambda(\boldsymbol{k}) \otimes \boldsymbol{v}_\lambda(\boldsymbol{k}) h_\lambda(\boldsymbol{k}, t) \tag{A.8}$$

with

$$\boldsymbol{H}(\boldsymbol{r}, t) = \frac{1}{(2\pi)^3} \int \exp(i\boldsymbol{k} \cdot \boldsymbol{r}) \boldsymbol{H}(\boldsymbol{k}, t) \mathrm{d}^3\boldsymbol{k} \tag{A.9}$$

The dynamic response function $\boldsymbol{H}$ fulfills (A.1) what can be seen also by time integration of Eq. (19.7). Let us consider now the relation between the dynamic response function $\boldsymbol{H}$ and the static Green's function. We have the postulation that the relation (see also Norris [13])

$$\lim_{t \to \infty} \boldsymbol{H}(\boldsymbol{r}, t) = \boldsymbol{G}^{static}(\boldsymbol{r}) \tag{A.10}$$

is fulfilled due to physical relaxation. We show now by using (A.7) that (A.10) is indeed fulfilled: From equation (A.7) follows immediately (the exponential factor has now to be taken into account since it is assumed that the damping ϵ is a infinitesimal small but finite constant)

$$\lim_{t \to \infty} h_\lambda(\boldsymbol{r}, t) = h_{\lambda\infty} = \frac{1}{\mu_\lambda^2} \tag{A.11}$$

This result also corresponds to the static solution of Eq. (A.5) and represents the static solution of the harmonic oscillator with eigenfrequency μ_λ which we obtain when we put all time derivatives in that equation equal to zero. Correspondingly the limiting case $t \to \infty$ of (A.8) becomes

$$\boldsymbol{H}(\boldsymbol{k}, \infty) = \frac{1}{\rho_M} \sum_{\lambda}^{3} \boldsymbol{v}_\lambda(\boldsymbol{k}) \otimes \boldsymbol{v}_\lambda(\boldsymbol{k}) \frac{1}{\mu_\lambda^2} = \boldsymbol{T}^{-1}(\boldsymbol{k}) \tag{A.12}$$

This expression coincides with the Fourier transform of the *static* Green's function. Hence we observe the demanded property (A.10) that $\boldsymbol{H}(\boldsymbol{r},\infty) = \boldsymbol{G}^{static}(\boldsymbol{r})$ by performing (A.9). Obviously $\boldsymbol{H}(\boldsymbol{r},\infty)$ fulfills the static equation

$$\boldsymbol{T}(\nabla)\,\boldsymbol{H}(\boldsymbol{r},\infty) + \delta^3(\boldsymbol{r})\,\mathbf{1} = 0 \tag{A.13}$$

In view of the limiting case (A.12) we decompose the response function $\boldsymbol{H}$ into a dynamic and a static contribution

$$\boldsymbol{H}(\boldsymbol{r},t) = \boldsymbol{G}^{static}(\boldsymbol{r}) + \tilde{\boldsymbol{H}}(\boldsymbol{r},t) \tag{A.14}$$

with vanishing dynamic part $\tilde{\boldsymbol{H}}(\boldsymbol{r},t) \to 0$ for $t \to \infty$. This representation corresponds to those obtained by Norris [13], (Eq. (3.23) therein), with

$$\tilde{\boldsymbol{H}}(\boldsymbol{r},t) = \frac{1}{(2\pi)^3}\int \exp(i\boldsymbol{k}\cdot\boldsymbol{r})\tilde{\boldsymbol{H}}(\boldsymbol{k},t)\mathrm{d}^3\boldsymbol{k} \tag{A.15}$$

and in view of (A.7) and (A.8) with

$$\tilde{\boldsymbol{H}}(\boldsymbol{k},t) = -\Theta(t)\exp(-\epsilon t)\frac{1}{\rho_M}\sum_{\lambda}^{3}\boldsymbol{v}_\lambda(\boldsymbol{k})\otimes\boldsymbol{v}_\lambda(\boldsymbol{k})\frac{\cos(\mu_\lambda t)}{\mu_\lambda^2} \tag{A.16}$$

The contribution $\tilde{\boldsymbol{H}}(\boldsymbol{r},t)$ yields (compare Eq. (19.37))

$$\begin{aligned}\tilde{\boldsymbol{H}}(\boldsymbol{r},t) = -\frac{\Theta(t)}{8\pi^2\rho_M}\sum_{\lambda}^{3}\int_{|\hat{k}|=1}\mathrm{d}\Omega(\hat{\boldsymbol{k}})\Big[\frac{1}{\mu_\lambda^2(\hat{\boldsymbol{k}})}\boldsymbol{v}_\lambda(\hat{\boldsymbol{k}})\otimes\boldsymbol{v}_\lambda(\hat{\boldsymbol{k}})\\ \times\delta(\hat{\boldsymbol{k}}\cdot\boldsymbol{r} - t\mu_\lambda(\hat{\boldsymbol{k}}))\Big]\end{aligned} \tag{A.17}$$

where we have now omitted the exponential damping factor. This leads to the final expression (compare (19.43)) for the dynamic part of the response function

$$\begin{aligned}\tilde{\boldsymbol{H}}(\boldsymbol{r},t) = -\frac{\Theta(t)}{8\pi^2\rho_M r}\sum_{\lambda}^{3}\sum_{s_\lambda}^{m_\lambda}\int_0^{2\pi}\mathrm{d}\varphi\,\Theta_{s_\lambda}(\varphi)\\ \times\Bigg[\frac{\Theta\{1-u_\lambda^{(s_\lambda)}(\varphi,\frac{t}{r})\}}{B_\lambda^{(s_\lambda)}[\hat{\boldsymbol{k}}_\lambda^{(s_\lambda)}(\varphi,\frac{t}{r})]\mu_\lambda^2[\hat{\boldsymbol{k}}_\lambda^{(s_\lambda)}(\varphi,\frac{t}{r})]}\\ \times\boldsymbol{v}_\lambda[\hat{\boldsymbol{k}}_\lambda^{(s_\lambda)}(\varphi,\frac{t}{r})]\otimes\boldsymbol{v}_\lambda[\hat{\boldsymbol{k}}_\lambda^{(s_\lambda)}(\varphi,\frac{t}{r})]\Bigg]\end{aligned} \tag{A.18}$$

where the quantities herein are defined next to Eq. (19.43). We note that the above expression is related to Eq. (19.43) by time integration.

References

1. Sosa H, Khutoryansky N (1999) Transient dynamic response of piezoelectric bodies subjected to internal electric impulses. Int J Solids Struct 36:5467–5484
2. Sosa H, Khutoryansky N (2001) Further analysis of the transient dynamic response of piezoelectric bodies subjected to electric impulses. Int J Solids Struct 38:2101–2114
3. Daros CH, Antes H (2000) Dynamic Fundamental Solutions for transversely isotropic piezoelectric materials of crystal class 6mm. Int J Solids Struct 37:1639–1658
4. Thompson SW, Kelvin L (1882) Mathematical and physical systems. Cambridge University Press
5. Fredholm I (1900) Sur les équations de l'équilibre d'un corps solide élastique. *Acta math*, Stockh. 23, 1
6. Lifshitz IM, Rosenzweig LN (1947) Zh eksper teor Fiz 17:783
7. Kröner E (1953) Das Fundamentalintegral der anisotropen elastischen Differentialgleichungen. Z Physik 136:402
8. Fabrikant IV (1989) Application of potential theory in mechanics. In: Mathematics and its applications. Kluwer Academic Publishers, Dordrecht
9. Michelitsch T (1997) Calculation of the electroelastic Green's function of the hexagonal infinite medium. Z Physik B 104:497
10. Michelitsch T, Wunderlin A (1996) Fundamental solution of the incompatibility problem in three-dimensional infinite anisotropic elasticity theory. Acta Mechanica 119:25
11. Michelitsch T, Levin VM (2000) Inclusions and inhomogeneities in electroelastic media with hexagonal symmetry. Eur Phys J B 14:527
12. Karapetian E, Sevostianov I, Kachanov M (2000) Point force and point electric charge in infinite and semi-infinite transversely isotropic piezoelectric solids. Philos Mag B 80(3):331
13. Norris AN (1994) Dynamic Green's functions in anisotropic piezoelectric, thermoelastic and poroelastic solids. Proc R Soc Lond A 447:175–188
14. Thom R The two-fold way of catastrophe theory. In: Hilton P (ed) Structural stability, the theory of catastrophes, and applications in the sciences. Lecture notes in mathematics, vol 525. Springer, Berlin, Heidelberg. https://doi.org/10.1007/BFb0077851
15. Every AG, Kim KY (1994) Time domain dynamic response functions of elastically anisotropic solids. J Acoust Soc Am 95(5):2505
16. Tverdokhlebov A, Rose J (1988) On Green's functions for elastic waves in anisotropic media. J Acoust Soc Am 83(1):118
17. Shermergor TD (1977) The theory of elasticity of microinhomogeneous media. Nauka, Moscow (in Russian)
18. Achenbach JD (1984) Wave propagation in elastic solids. Elsevier, New York
19. Lazar M (2020) Second gradient electrodynamics: Green functions, wave propagation, regularization and self-force. Wave Motion 95:102531
20. Willis JR (1980) Polarization approach to the scattering of elastic waves-I. Scattering by a single inclusion. J Mech Phys Solids 28:287
21. Haake F (1983) Einführung in die theoretische Physik. Physik-Verlag Weinheim

Chapter 20
A Contact Problem with Large Elastic–plastic Strain, Friction, and Close to Exact Analytical Solution

Pavel G. Morrev and Alexey A. Markin

Abstract A multi-parametrical contact problem on compression an axisymmetric elastic–plastic body involving large strain and friction is considered. A set of parameters consists of 3 geometrical characteristics and 4 material constants. Based on several simplifying hypotheses, an inexact analytical solution is deduced. Comparative analysis with a FEM solution demonstrates high accuracy of the analytical one and this accuracy can be regulated by means of the geometric parameters. A-priory error estimation is proposed. The effect of stabilization parameter on the accuracy for a nodal integrated stabilized finite element is studied. The analytical solution obtained can be used as a benchmark for program codes in 3D contact mechanics—ANSYS, NASTRAN, DEFORM, ABACUS, Super Forge, etc.

Keywords Contact problem · Elastic–plastic deformation · Large strain · Error estimation · Benchmark · Contact algorithm

20.1 Introduction

Contact problems are the most complicated boundary value ones in solid mechanics. Here, we encounter the following difficulties: a definitional domain of tensor fields changes in time and is unknown in advance; the same about a contact surface and medium properties; boundary values of tensor fields are specified in some implicit form and cannot be directly entered the right hand side of the final system of equations; strong nonlinearity of both physical and geometrical types. By this reason, for large elastic–plastic deformation with friction, an exact analytical solution of any contact problem is unknown so far. Note that frictionless upsetting a uniform

P. G. Morrev (✉)
Oryol State University, Oryol, Russia
e-mail: paulorel@mail.ru

A. A. Markin
Tula State University, Tula, Russia

H. Altenbach and V. A. Eremeyev (eds.), *Dynamics of Discrete and Continuum Structures and Media*, Advanced Structured Materials 221,
https://doi.org/10.1007/978-3-031-75626-9_20

cylinder yields a trivial exact solution (here, we deal with uniform deformation) but this example is poorly useful. Note also that in case of small elastic deformation we can again arrive to an exact solution in the Hertz problem [1]. So, contact problems of general type which are often encountered in practice applications need numerical methods. These methods are based on variational principles of functional analysis and appear to be complicated in their implementation. Many special methods and computer programs have been elaborated up to date—ANSYS, NASTRAN, ABAQUS, DEFORM, etc.—and new methods and software continue to appear. In this connection, a natural question arises: how to determine the accuracy of new approaches elaborated? Exact analytical solutions would be the best choice but, as it was aforementioned, we do not know how to construct them. Hence we are forced to address the "reliable" numerical solutions. For example, in [2] the cylindrical billet upsetting is considered; a contact surface is of type "stone wall" (i.e., with infinite frictional coefficient); a hardening law is of the power/linear form. The solution is represented as a set of color pictures of the stress tensor components and plastic strain. This solution is used as a benchmark in [3] where the results are adduced in the form of isoline patterns. One can see the qualitative similarity between [3] and [2] but one cannot assert anything about the numerical difference between them. In addition, the finite element mesh is too coarse here and we know nothing about difference between the exact solution and the obtained numerical one. The results [2] were also used in [4] where the billet upsetting is again studied by another method that allows refining a finite element mesh and approaching the exact solution.

One can propose another approach when the dependence "force–displacement" or "pressure–displacement" is studied by refinement of a finite element mesh. In ideal, this dependence tends to an exact solution. In practice, we can observe different limits for different methods in the same problem. For example, in [5] and [6] a common benchmark is considered: an axisymmetric compression of a quasi-incompressible hyper-elastic block is performed. A continues stress tetrahedron is adopted in [5] whereas a finite strain Raviart-Thomson tetrahedron is proposed in [6]. One can see that the results of the works (Fig. 10 [6] and Fig. 8 [5]) differ by mesh refinement. May be they will converge to a common value at the limit but we cannot assert this.

If any parametrical method is employed—penalty, stabilization, or regularization—with a free parameter p (as a rule, p is distance or inverse stiffness) then it is usual to consider that unlimited refinement of the space–time discretisation at the limit $p \to 0$ results in exact solution. Unfortunately, this may be impossible in practice. Let us return to the billet upsetting test [2]. Here, convergence failed with the linear hardening model but this model run to completion by increasing the stabilization parameter p relatively the recommended "optimal" value $p = 0.05$. That is, the limit transition $p \to 0$ appears to be impossible in practice and one can say nothing about the exact solution in that test.

In the overwhelming majority of works, authors confine themselves to numerical examples with some fixed discretization. At best, a comparison between two or more methods is performed in order to obtain similar results as it was done in [7] where the penalty method and the augmented Lagrangian one were compared.

In this work, a compromise approach is proposed. Based on several simplifying hypotheses, an inexact analytical solution is deduced. Its accuracy is known as a-priory and can be regulated by geometric parameters. Comparative analysis with several FEM solutions is performed to evaluate the real accuracy of the analytical solution proposed.

20.2 Problem Statement

Let us consider an axisymmetric body with two plane butt-ends of a shape defined below. Let X and R be the axial and radial Lagrange coordinates as it is shown in Fig. 20.1 and R0, R1, S_0, S_1—radii and areas of the butt-ends where $R_0 < R_1$ and $S_0 < S_1$; S_X—cross section area corresponding to coordinate X. The body is compressed by two parallel plates and one of them is motionless while another is displaced under pressing force F—see Fig. 20.1. Let s_x be cross section of the deformed body corresponding to the Euler coordinate $x = x(X)$ (we adopt the plane section hypothesis—see further). For the sake of simplicity, cross section as a set of material points and cross section as area are denoted as s_x or S_x and this will not cause any misunderstanding. A friction law at the butt-ends will be specified later.

Let us analyze the stress–strain state of the deformed body at a point (x, r) where r—radial Euler coordinate. The mechanical balance equation reads as

$$\nabla \cdot \boldsymbol{\sigma} = \mathbf{0}$$

or, in cylindrical coordinates (x, r, φ), as [8]:

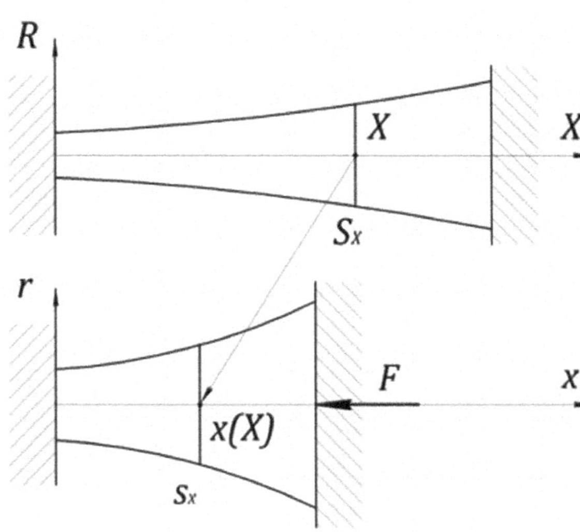

Fig. 20.1 Coordinate systems

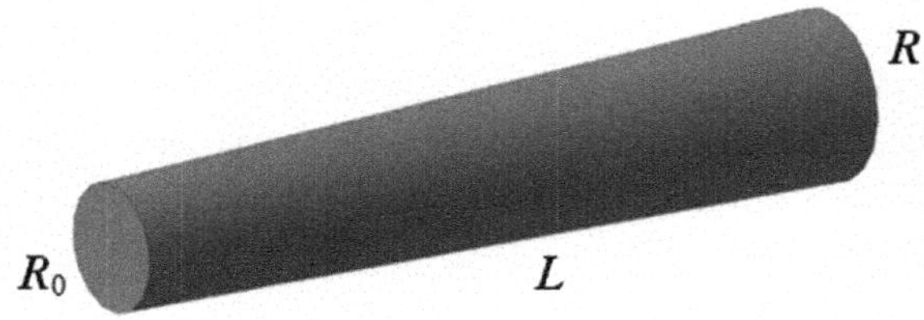

Fig. 20.2 A spatial view of a body deformed where $R_0 < R_1 < < L$

$$(r\sigma^{rx})_{,r} + r\sigma^{xx}_{,x} = 0 \tag{20.1a}$$

$$\sigma^{rr}_{,r} + \sigma^{xr}_{,x} + r^{-1}(\sigma^{rr} - \sigma^{\phi\phi}) = 0. \tag{20.1b}$$

Boundary values will be specified below. This statement is too complicated therefore re-formulate it such that the axial component σ^{xx} along with its axial derivatives appear to be constant over any cross section. In order to meet this requirement, the body is considered an oblong truncated cone—see Fig. 20.2. Suppose also that $\sigma^{rr} = \sigma^{\varphi\varphi}$. Then, Eqs. (20.1a and 20.1b) are easily integrated. Equation (20.1a) results in

$$\sigma^{rx} = -\sigma^{xx}_{,x} r/2 + C_x/r$$

where C_x—parametrical integration constant which is equal to zero in order to avoid the singularity at $r = 0$ therefore

$$\sigma^{rx} = -\sigma^{xx}_{,x} r/2 \tag{20.2}$$

while (20.1b) along with (20.2) result in

$$\sigma^{rr}_{,r} = \sigma^{xx}_{,xx} r/2$$

and integration in r yields

$$\sigma^{rr} = \sigma^{xx}_{,xx} r^2/4 + B_x \tag{20.3}$$

where B_x – another parametrical integration constant which is determined from the boundary conditions at the free surface:

$$\mathbf{n} \cdot \boldsymbol{\sigma} = \mathbf{0}$$

or in the coordinate form

$$n_x\sigma^{xx} + n_r\sigma^{rx} = 0$$

$$n_x\sigma^{xr} + n_r\sigma^{rr} = 0$$

where $\mathbf{n}(n_x, n_r)$—normal vector determined by the angle of tangent slope α (the tangent to the generatrix is meant here)—Fig. 20.1. Given σ^{xx} and α at the free surface, the values σ^{rx} and σ^{rr} are easily found (recall that σ^{xx} does not depend on r):

$$\sigma^{rx}(x, r_x) = -(n_x/n_r)\sigma^{xx}(x) = \sigma^{xx}(x)\mathrm{tg}\alpha \tag{20.4a}$$

$$\sigma^{rr}(x, r_x) = -(n_x/n_r)\sigma^{rx}(x, r_x) = \sigma^{xx}(x)\mathrm{tg}^2\alpha \tag{20.4b}$$

where r_x—radius of cross section s_x. The integration constant B_x is obtained from (20.3) and (20.4b):

$$B_x = \sigma^{xx}(x)\mathrm{tg}^2\alpha - \sigma^{xx}_{,xx}r_x^2/4$$

and this results in

$$\sigma^{rr} = \sigma^{xx}_{,xx}(r^2 - r_x^2)/4 + \sigma^{xx}(x)\mathrm{tg}^2\alpha \tag{20.5}$$

According to (20.2) and (20.5), the stress state of the body under all the hypotheses adopted is known if the axial component σ^{xx} is found. Consequently, the search of σ^{xx} is our next task. To address this task, some additional hypotheses have to be adopted.

According to (20.2), $\sigma^{rx}(x, r_x) = -\sigma^{xx}_{,x}r_x/2$ and this contradicts to (20.4a). Consequently, the difference $\delta_1 = |\sigma^{xx}_{,x}r_x/2 + \sigma^{xx}\mathrm{tg}\alpha|$ can be considered an error estimation of the analytical solution. There exists another variant of this estimation which will be considered further.

Turn to the friction law at the butt-ends. It has to agree the shear stress distribution (20.2) over a cross section, i.e., to be of the form

$$k = \sigma^{rx}/\sigma^{xx} = -\sigma^{xx}_{,x}r/2\sigma^{xx} \tag{20.6}$$

where k – coefficient of friction.

In the course of deformation, the stress tensor is not uniform over a cross section and if the yield criterion is of the form $I_2(\boldsymbol{\sigma}) = \sigma_y$ where I_2—second invariant and σ_y—yield stress then material points of any cross section appear to be in a plastic state at different instants. Therefore, one more hypothesis is adopted: components σ^{rx}, σ^{rr}, $\sigma^{\phi\phi}$ are excluded from the yield criterion such that it takes the form $|\sigma^{xx}| = \sigma_y$.

The next hypothesis is that a plane cross section S_X is mapped onto a plane cross section s_x. Finally, elastic strain is considered to be small.

Under all the hypotheses formulated, the mechanical balance equation holds exactly whereas the boundary conditions at the free surface and the state equation (hardening law)—only approximately.

20.3 Analytical Solution

Let us confine ourselves to materials with an exponential hardening law

$$\sigma_0 + H\left(1 - e^{-\alpha(\varepsilon - \varepsilon_e)}\right) = \sigma \tag{20.7}$$

where $\sigma = -\sigma^{xx} > 0$, σ_0—initial yield stress, H—hardening stress, $\alpha > 0$—numerical constant, ε—logarithmic strain of the form

$$\varepsilon = \ln\left(\frac{dx}{dX}\right)^{-1} = -\ln\frac{dx}{dX} \tag{20.8}$$

ε_e—small elastic strain entering the additive decomposition

$$\varepsilon = \varepsilon_p + \varepsilon_e$$

Let q be a lateral strain: $q = s_x/S_X$ where $x = x(X)$. If $\varepsilon_e = 0$ then the incompressibility condition $\varepsilon = \ln q$ holds. But, if $\varepsilon_e \neq 0$ then

$$\ln q = \varepsilon_p + 2\nu\varepsilon_e = \varepsilon - \varepsilon_e + 2\nu\varepsilon_e = \varepsilon - (1 - 2\nu)\varepsilon_e \tag{20.9}$$

where ν—Poisson coefficient. Recall that the lateral strain is here defined as the ratio of areas rather than radii. It follows that

$$q = e^{\varepsilon} e^{-(1-2\nu)\varepsilon_e}$$

and, taking into account smallness ε_e, this expression reads as

$$e^{\varepsilon} = q(1 + (1 - 2\nu)\varepsilon_e) \tag{20.10}$$

Let us rewrite hardening law (20.7) in the form (ε_e is small)

$$\sigma_0 + H\left(1 - (1 + \alpha\varepsilon_e)e^{-\alpha\varepsilon}\right) = \sigma$$

Substituting here $e^{-\alpha\varepsilon}$ expressed via q from (10), we obtain

$$\sigma_0 + H\left(1 - (1 + \alpha\varepsilon_e)(1 - \alpha(1 - 2\nu)\varepsilon_e)q^{-\alpha}\right) = \sigma$$

or, discarding the items of second order,

$$\sigma_0 + H\left(1 - (1 + 2\alpha\nu\varepsilon_e)q^{-\alpha}\right) = \sigma$$

Substituting here $\varepsilon_e = \sigma/E$ where E—Young modular and by denotation $\gamma = 2\alpha\nu/E$ we arrive to

$$\sigma_0 + H\left(1 - (1 + \gamma\sigma)q^{-\alpha}\right) = \sigma \tag{20.11}$$

Multiplying this equality by the cross section area, obtain the pressing force in the right hand side and this force is the same at all the cross sections:

$$\sigma_0 s_x + H\left(s_x - (s_x + \gamma F)q_x^{-\alpha}\right) = F$$

where the coordinate dependence is pointed by the lower index. Let us rewrite this relation in the Lagrange coordinates:

$$\sigma_0 S_X q_X + H\left(S_X q_X - (S_X q_X + \gamma F)q_X^{-\alpha}\right) = F$$

and multiply both the sides by q_X^{α} grouping the items:

$$S_X(\sigma_0 + H)q_X^{\alpha+1} - F q_X^{\alpha} - H S_X q_X - H\gamma F = 0 \tag{20.12}$$

It is an equation of power α + 1 having an analytical solution at α = 0, 1, 2, 3. The event α = 0 corresponds to ideal plasticity and is not of interest. Note that the term $\gamma\sigma = 2\alpha\nu\sigma/E$ in (20.11) is very small for metals and alloys and may be discarded that yields

$$S_X(\sigma_0 + H)q_X^{\alpha} - F q_X^{\alpha-1} - H S_X = 0$$

This equation has an analytical solution if α = 1, 2, 3, 4 but we confine ourselves to the event α = 2 which is simple and corresponds to some alloys [9]:

$$S_X(\sigma_0 + H)q_X^2 - F q_X - H S_X = 0$$

Discarding the negative root, obtain:

$$q_X = \frac{F + \sqrt{F^2 + 4H(\sigma_0 + H)S_X^2}}{2S_X(\sigma_0 + H)} = \frac{F}{2(\sigma_0 + H)}\left(\sqrt{S_X^{-2} + 4H(\sigma_0 + H)\big/F^2} + S_X^{-1}\right) \tag{20.13}$$

and it follows that

$$q_X^{-1} = \frac{2(\sigma_0 + H)S_X}{\sqrt{F^2 + 4H(\sigma_0 + H)S_X^2} + F} = \frac{F}{2H}\left(\sqrt{S_X^{-2} + 4H(\sigma_0 + H)/F^2} - S_X^{-1}\right) \tag{20.14}$$

The longitude coordinate x of any material point is determined as

$$x(X) = \int_0^x dx' = \int_0^X \frac{dx'}{dX'} dX' = \int_0^X e^{-\varepsilon} dX' \tag{20.15}$$

where ε is defined in (20.8).We are interested in large plastic strain therefore suppose that $F \geq \sigma_0 S_1$, i.e., the body is completely in plastic state. The shape of the body is not defined so far. If the generatrix is a straight line then we encounter the hypergeometric functions. But, if the inverse area S_X^{-1} is a linear function then we deal with elementary functions. Denote $u_X = S_X^{-1}$:

$$u_X = u_0 + (u_1 - u_0)X/L$$

where $u_0 = S_0^{-1}$ and $u_1 = S_1^{-1}$. This means that the shape of the body under consideration is specified by a hyperbolic generatrix. Taking into account $\varepsilon_e = \sigma/E = F/(Es_x)$, it follows from (20.10)

$$e^{-\varepsilon} = q_X^{-1} - (1-\nu)(F/E)q_X^{-2} S_X^{-1}$$

therefore integral (20.5) takes the form

$$x(X) = \int_0^X q_{X'}^{-1} dX' - (1-\nu)(F/E) \int_0^X q_{X'}^{-2} S_{X'}^{-1} dX'$$

In both the integrals replace X' by u. This leads to replacement dX' by $Ldu/(u_1 - u_0)$ and $X' \in [0, X]$ by $u \in [u_0, u_X]$. Taking into account $u_0 > u_X$, swap places the integration limits and rewrite the last equality in the form

$$x(X) = \frac{L}{u_0 - u_1} \left(\int_{u_X}^{u_0} q_u^{-1} du - (1-\nu)(F/E) \int_{u_X}^{u_0} q_u^{-2} u du \right)$$

The explicit expression for q_u^{-1} follows from (20.14) and both the integrals are calculated by means of standard ones №1786 and 1820 [10] such that this expression takes the form

$$x(X) = \frac{FL}{2H(u_0 - u_1)} \left(I_1 - \frac{(1-\nu)F^2}{2HE} I_2 \right) \tag{20.16}$$

where

$$I_1 = \int_{u_X}^{u_0} \left(\sqrt{u^2 + a} - u \right) du = \frac{a}{2} \left[u \left(u + \sqrt{u^2 + a} \right)^{-1} + \ln \left(u + \sqrt{u^2 + a} \right) \right] \Bigg|_{u_X}^{u_0}$$

$$I_2 = \int_{u_X}^{u_0} \left(2u^3 + au - 2u^2\sqrt{u^2+a}\right)du =$$

$$= \frac{1}{2}\left[u^2\left(u^2+a\right) - \frac{1}{2}\left(u\left(2u^2+a\right)\sqrt{u^2+a} - a^2\ln\left(u+\sqrt{u^2+a}\right)\right)\right]\Bigg|_{u_X}^{u_0}$$

and $a = 4H(\sigma_0 + H)/F^2$. The radial coordinate is simply

$$r = R\sqrt{q_X}$$

where q_X is given in (20.13). The stress tensor components are found from the axial component σ^{xx} and its axial derivatives:

$$-\sigma^{xx} = \sigma = F/s_x = F/(q_X S_X) = F q_X^{-1} S_X^{-1} = d\left(\sqrt{u^2+a} - u\right)u \quad (20.17)$$

where $d = F^2/2H$. By means of (20.10) one can find

$$\frac{du}{dx} = \frac{du}{dX}\left(\frac{dx}{dX}\right)^{-1} = \frac{u_1 - u_0}{L}e^{\varepsilon} = \frac{u_1 - u_0}{L}q_u\left(1 + \frac{1-2\nu}{E}\sigma_u\right) = bq_u(1 + c\sigma_u) \quad (20.18)$$

where $b = (u_1 - u_0)/L$, $c = (1 - 2\nu)/E$ and according to (20.13), q_u is of the form

$$q_u = e\left(\sqrt{u^2+a} - u\right) \quad (20.19)$$

where $e = F/2(\sigma_1 + H)$. By means of (20.17) and (20.9) one can find

$$\sigma_u' = d\left(\frac{2u^2+a}{\sqrt{u^2+a}} - 2u\right);$$

$$\sigma_u'' = d\left(\frac{2u^3+3au}{\left(u^2+a\right)^{3/2}} - 2\right)$$

$$q_u' = e\left(\frac{u}{\sqrt{u^2+a}} - 1\right)$$

So, the derivatives in x (see (20.18)) are as follows:

$$\sigma_x' = \sigma_u'\frac{du}{dx};$$

$$\sigma_x'' = \sigma_u''\left(\frac{du}{dx}\right)^2 + \sigma_u'(bq_u'(1+c\sigma) + bcq_u\sigma_u')\frac{du}{dx}$$

Substituting these expressions into (20.2) and (20.5), we obtain the stress tensor components σ^{rx} and σ^{rr}. This completes the analytical solution construction and the next task is estimation of its accuracy.

20.4 FEM Solution

In this Section, numerical solutions of the contact problem under consideration are adduced based on Finite Element Method. The geometrical parameters are as follows: $R_0 = 10$ mm, $R_1 = 15$ mm, $L = 100$ mm, 200 mm, and 500 mm, and the mechanical parameters are as follows: $E = 10^4$ [kgf/mm^2], $\nu = 0.33$, $\sigma_0 = 15$ [kgf/mm^2], $H = 76$ [kgf/mm^2], $\alpha = 2$ which correspond to annealed brass AISI C27400. Pressing force is equal to $F = 4\sigma_0 S_1 \approx 42{,}411$ [kgf]. The loading process is divided into 10 steps in the interval $0 \le F(t) \le \sigma_0 S_0$ and into 1000 steps in the interval $\sigma_0 S_0 \le F(t) \le 4\sigma_0 S_1$. The increase in number of steps up to 2000 does not appreciably affect the final result. To perform numerical analysis, an axisymmetric finite element [3] is adopted which is a counterpart of a nodal integrated stabilized tetrahedron [2]. The stabilization parameter p takes the two fixed values: $p = 0$ and $p = 0.05$. Note that the exponential hardening law (20.7) causes serious convergence difficulties therefore a linear item is augmented to alleviate this problem [4]. But, this item makes impossible an analytical solution and we are forced to reconcile ourselves to convergence violation.

The first numerical experiment is aimed at determining the necessary mesh refinement. For $L = 200$ mm, the finite element meshes 40×4, 80×8, 160×16, and 320×32 are adopted. As it was expected, convergence failed between 700 and 800 steps in the interval $\sigma_0 S_0 \le F(t) \le 4\sigma_0 S_1$ for any value of stabilization parameter p. Isoline patterns of the stress tensor components σ^{xx}, σ^{xr} and hydrostatic component σ^{h} are presented in Fig. 20.3 for $p = 0$ (σ^{rr} is omitted because of its smallness). The ten isolines are spaced with equal intervals between the minimum and maximum values of proper tensor components and the final length is denoted by $l = x(L)$. One can see considerable oscillations in σ^{h} which descend as the mesh is refined while oscillations in σ^{xr} are absent because the shear stress relates to the deviatory part of the stress tensor. In order to perform comparative analysis of the analytical and FEM solutions, the FEM solution has to be averaged because of its oscillatory character. The results of this analysis are presented in Fig. 20.4. The relative difference between the analytical and FEM solutions in displacement and axial stress at different mesh densities is shown in this figure and the mean-square $\overline{\delta}$ of this difference versus the mesh density is presented in Fig. 20.5. One can see that after the refinement 160×16, $\overline{\delta}$ ceases to descend. This means that the FEM solution appears to be sufficiently close to the exact one in order to obtain realistic error estimation for the analytical solution.

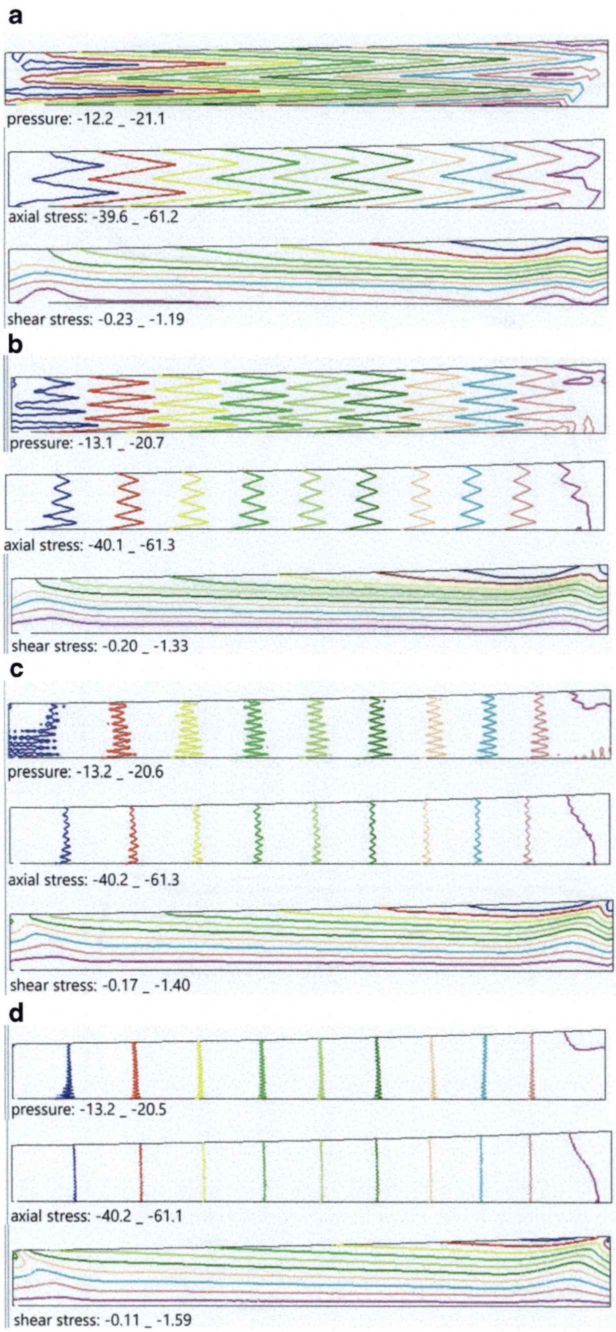

Fig. 20.3 Isoline patterns of σ^h, σ^{xx}, σ^{rx} for $L = 200$ and $p = 0$. **a** mesh 40 × 4 l = 141.4 mm 10 + 745 steps. **b** mesh 80 × 8 l = 141.3 mm 10 + 746 steps. **c** mesh 160 × 16 l = 141.2 mm 10 + 747 steps. **d** mesh 320 × 32 l = 141.25 mm 10 + 747 steps

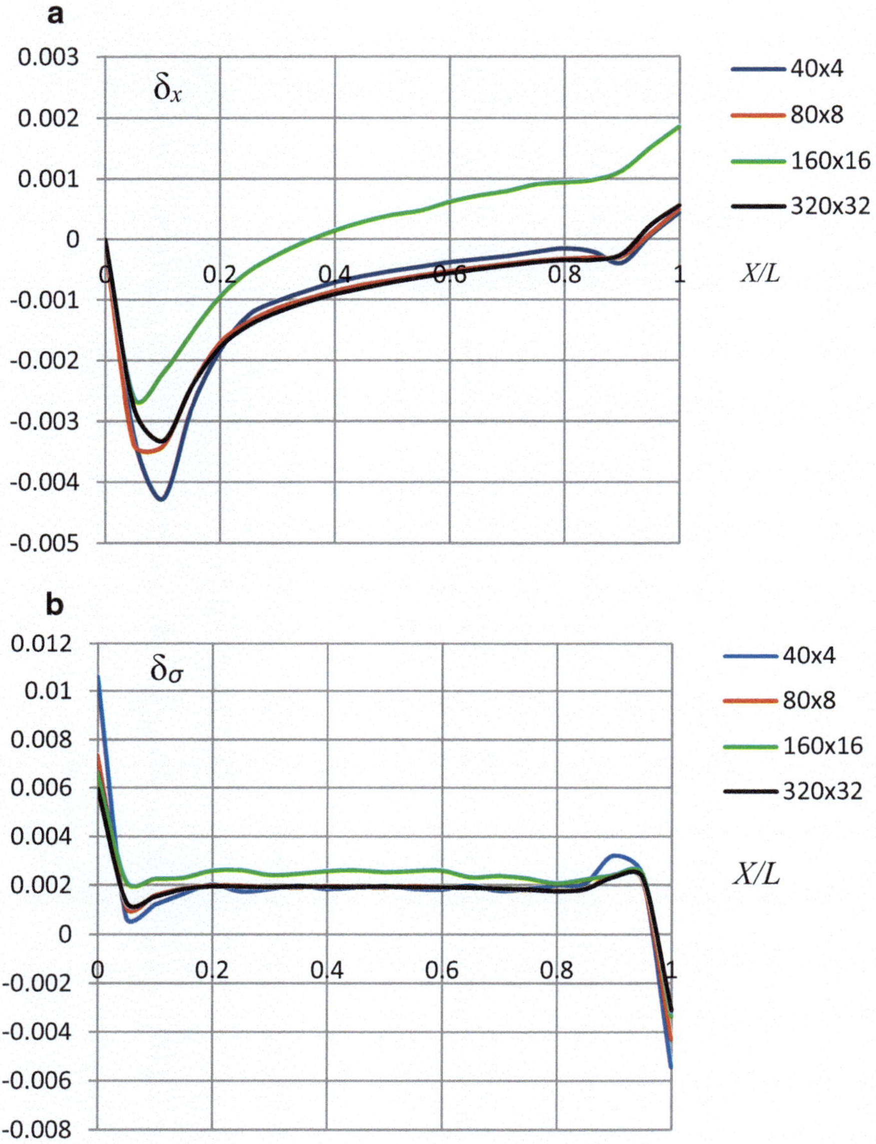

Fig. 20.4 **a** Relative difference in axial displacement between the analytical and FEM solutions vs the axial Lagrange coordinate. **b** Relative difference in axial stress between the analytical and FEM solutions vs the axial Lagrange coordinate

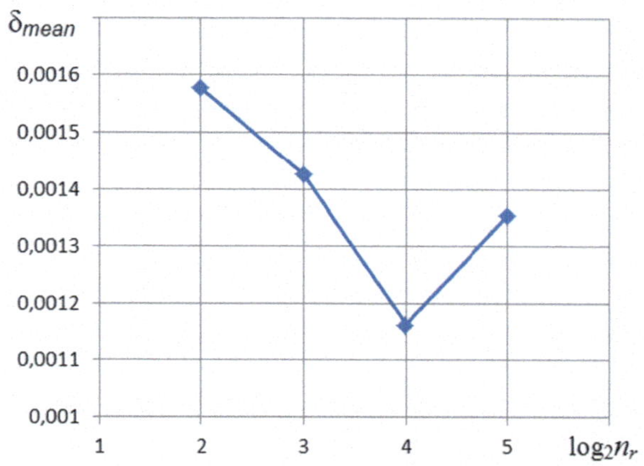

Fig. 20.5 Mean-square difference $\overline{\delta}_x$ of axial displacement between the analytical and FEM solutions versus the split number n_r of FE mesh in radial direction at $L = 200$ mm

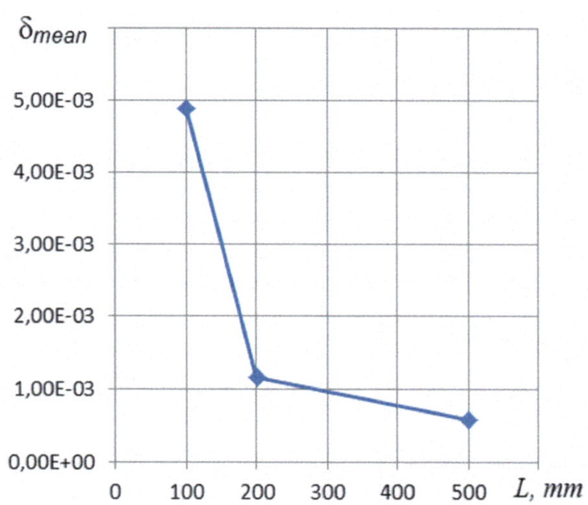

Fig. 20.6 Mean-square difference $\overline{\delta}_x$ of axial displacement between the analytical and FEM solutions versus L at fixed density of FE mesh for $n_r = 16$

This mesh density will be used further in numerical experiments with $L = 100$ mm and $L = 500$ mm and proper results can be considered an error estimation for the analytical solution.

In the second numerical experiment, the effect of stabilization parameter $p = 0.05$ was studied. The main result is that the difference between the analytical and FEM solutions appears to be much larger than for $p = 0$ and is equal to ≈5% for all the mesh densities. This means that stabilization of displacement-based finite elements is only valid for coarse meshes while for fine meshes this approach is inefficient.

The third numerical experiment is aimed at determination the analytical solution error depending on L. The mesh density corresponds to the mesh 160 × 16 from the first experiment. The results are presented in Fig. 20.6. One can see that at $L = 500$ mm the averaged error $\overline{\delta}_x$ reaches 5.7×10^{-4} and it is more than enough for testing program codes.

The final task to be solved is a-priory error estimation of the analytical solution without numerical experiments. Estimation δ_1 from Sect. 20.2 appears to be too small in comparison with error $\overline{\delta}_x$ obtained in the numerical experiments. More realistic estimation is of the form

$$\delta_2 = \max\left[0.5\left(\sigma^{rx}(x, r_x)/\sigma^{xx}\right)^2\right]$$

It follows from replacement the yield criterion $I_2(\sigma) = \sigma_0$ by $|\sigma^{xx}| = \sigma_0$ (see Sect. 20.2 where σ_y has to be replaced by σ_0 for monotonic processes) and also from the inequality $|\sigma^{rx}| >> |\sigma^{rr}|$. For $L = 100$, 200, and 500 mm, one obtains $\delta_2 = 4.1 \times 10^{-3}$, 1.0×10^{-3}, 1.6×10^{-4}. One can see that the first and second values are close to the numerical experiments data while the third one is understated by 3.5 times. More exact error estimation is a subject of further investigations.

20.5 Conclusion

1. The analytical solution obtained can be used as a benchmark for program codes in 3D contact mechanics when large elastic–plastic strain and friction are of importance.
2. Its accuracy is regulated by means of the geometric characteristics.
3. In the numerical experiments conducted, the mean-square error reached 5.7×10^{-4} in displacement.
4. Nodal integrated stabilized finite elements are only valid on coarse meshes whereas on fine meshes their accuracy remains at the same level ≈5% independently on mesh refinement.
5. A-priory error estimation of the analytical solution is proposed which may yield an understated by several times mean-square error value for very long bodies.

Acknowledgements The work has been carried out in Oryol State University named after I.S. Turgenev within the bounds of the Russian Federation government target No 075-01466-23-07 at 2023 year and at the planning period 2024–2025 years from 24.11.2023, project No FSGN-2023-002 (1023082800024-8-1.3.2).

References

1. Popov VL (2010) Contact mechanics and friction: physical principles and applications. Springer, Berlin, 362 p
2. Puso MA, Solberg J (2006) A formulation and analysis of a stabilized nodally integrated tetrahedral. Int J Numer Meth Eng 67:841–867
3. Morrev PG, Gordon VA (2018) An axisymmetric nodal averaged finite element. Lat Am J Solids Struct 15:e14

4. Areias P (2020) A finite strain mixed J2–u–p low-order tetrahedron. Finite Elem Anal Des 174:103401
5. Areias P, Rabczuk T, Carapau F, Carrilho Lopes J (2019) A continuous-stress tetrahedron for finite strain problems. Finite Elem Anal Des 165:52–64
6. Areias P, Tiago C, Carrilho Lopes J, Carapau F, Correia P (2020) A finite strain Raviart-Thomas tetrahedron. Eur J Mech/A Solids 80:103911
7. Weyler R, Oliver J, Sain T, Cante JC (2012) On the contact domain method: a comparison of penalty and Lagrange multiplier implementations. Comput Methods Appl Mech Eng 205–208:68–82
8. Storogevoi MV, Popov EA (1977) Theory of metal forming, 4th edn. Mechanical Engineering, Moscow, 424 p
9. Krokha VA (1980) Metal hardening by cold plastic deformation: handbook. Mechanical Engineering, Moscow, 160 p
10. Demidovich BP (2005) Collection of problems and exercises on mathematical analysis. Astrel, Moscow, 558 p

Chapter 21
Some Effects of the Dynamics of a Rigid Body Due to Its Viscoelastic Properties

Valery V. Novikov, Lyubov N. Fevralskikh, and Alexander K. Lyubimov

Abstract The work is devoted to the study of the influence of elastic and viscous properties of a body similar in shape to a ball on its free angular movements. In the presented problems, taking into account body deformations is fundamental, because leads to a qualitative restructuring of its dynamics. Using the example of a specific system, we consider the possibility of significant movement of the axis of stationary rotation to a new position in the body (global movement of the pole) with a small change in the angular velocity of rotation of the body. The change in the position of the axis is described geometrically. In the conservative case, using the integral of energy and the integral of the square of the kinetic momentum vector, the polodes (the trajectories of movement of the kinetic momentum vector relative to the body) are depicted. In the problem of nutation of a quasi-ball, it is shown that the periods of nutation of a body, with and without taking into account its elastic and viscous properties, can differ significantly. The idea of these effects of rigid body dynamics can be useful when considering the movement of the Earth's poles and its Chandler nutation.

Keywords Elastic quasi-ball · Free catch movements · Stability of stationary rotations · Pole displacement · Viscoelastic ball · Nutation · Chandler period

V. V. Novikov · L. N. Fevralskikh (✉) · A. K. Lyubimov
National Research Lobachevsky State University of Nizhny Novgorod, Nizhny Novgorod, Russia
e-mail: fevralskikh_ln@itmm.unn.ru

V. V. Novikov
e-mail: novikov@mm.unn.ru

A. K. Lyubimov
e-mail: ljubimov@mm.unn.ru

H. Altenbach and V. A. Eremeyev (eds.), *Dynamics of Discrete and Continuum Structures and Media*, Advanced Structured Materials 221,
https://doi.org/10.1007/978-3-031-75626-9_21

21.1 Introduction

The study of the angular motions of a rigid body and the identification of the dynamics determined by its elastic and viscous properties is important for solving some issues of the motion of celestial bodies, the orientation and stabilization of artificial satellites and spacecraft, and increasing the accuracy of precision instruments [1–8]. Works [9–11] are devoted to the consideration of free angular motions of a homogeneous solid body close in shape to a ball and to the identification of qualitative features of its dynamics due to the elastic and viscous properties of the body.

21.2 On the Possibility of Changing the Position of the Axis of Stationary Rotation in the Body, Due to Taking into Account the Elastic Properties of the Body

Let us consider the movement of a free homogeneous elastic body of volume V, bounded by a surface S, not much different from a sphere. The displacement of body points in a certain coordinate system $Ox_1x_2x_3$, rotating with angular velocity $\boldsymbol{\omega}(t)$, during its deformation is characterized by the vector $\mathbf{u}(\mathbf{r}, t)$. We require that the origin of the coordinate system $Ox_1x_2x_3$ coincide with the center of inertia of the body and that there are no small rigid displacements of the body relative to it. In this case, the following conditions must be met:

$$\int_V \mathbf{u}\,dV = 0, \quad \int_V [\mathbf{r}, \dot{\mathbf{u}}] = 0. \tag{21.1}$$

The equations of motion of a volume element of a body and the conditions for the absence of stress on the surface S of the body have the form:

$$\begin{gathered}(\kappa + 1)\mathbf{grad}\,div\,\mathbf{u} + \Delta\mathbf{u} = \\ = \varepsilon(\ddot{\mathbf{u}} + [\dot{\omega}, \mathbf{r} + \mathbf{u}] + [\boldsymbol{\omega}, [\boldsymbol{\omega}, \mathbf{r} + \mathbf{u}]] + 2[\boldsymbol{\omega}, \dot{\mathbf{u}}])\ in\ V,\end{gathered} \tag{21.2}$$

$$\left[\kappa u_{ll}\delta_{ij} + 2u_{ij}\right]n_j = 0\ on\ S,\ \ i = 1, 2, 3, \tag{21.3}$$

where $u_{ij} = \frac{1}{2}\left(\frac{\partial u_i}{\partial x_j} + \frac{\partial u_j}{\partial x_i}\right)$—small strain tensor, n_j—components of the normal vector to the body surface.

The characteristic size of the body R_0 is taken as the length scale, the time scale t_* is the characteristic time of motion of the body as a whole relative to the center of mass,

$$\varepsilon = \frac{\rho R_0^2}{\mu t_*^2}, \quad \kappa = \frac{\lambda}{\mu},$$

where ρ—body density, λ, μ—Lame elastic moduli.

We will assume that $\varepsilon << 1$, those the body is quite rigid, the period of elastic vibrations is much less than the time of their attenuation due to internal friction, which, in turn, is much less than the period of rotational movements t_*.

This allows you to represent the displacement vector as a series in a small parameter ε:

$$\mathbf{u}(\mathbf{r}, t) = \varepsilon \mathbf{u}'(\mathbf{r}, t) + \varepsilon^2 \mathbf{u}''(\mathbf{r}, t) + \cdots$$

and limit ourselves to considering $\mathbf{u}'(\mathbf{r}, t)$.

Problem (21.1), (21.2) in this formulation takes the following form for the displacement vector $\mathbf{u}'(\mathbf{r}, t)$:

$$\int_V \mathbf{u}' \, dV = 0, \quad \int_V [\mathbf{r}, \dot{\mathbf{u}}'] = 0. \tag{21.4}$$

$$\begin{gathered}(\kappa + 1)\mathbf{grad}\, div\, \mathbf{u}' + \Delta \mathbf{u}' = [\dot{\omega}, \mathbf{r}] + [\boldsymbol{\omega}, [\boldsymbol{\omega}, \mathbf{r}]] \; in \; V, \\ \left[\kappa u'_{ll} \delta_{ij} + 2u'_{ij}\right] n_j = 0 \; on \; S, \; i = 1, 2, 3.\end{gathered} \tag{21.5}$$

The components of the angular momentum vector $\mathbf{K}$ in the coordinate system $Ox_1x_2x_3$ are determined as follows:

$$K_i = I_{ij}\omega_j, \quad I_{ij} = \int_V [(x_l^2 + 2x_l u_l)\delta_{ij} - (x_i x_j + 2x_i u_j)] dV = I_{ij}^0 + \Delta_{ij},$$

where I_{ij}^0—inertia tensor of an undeformed body, $\Delta_{ij} \ll 1$—corrections due to the elastic properties of the body.

From here, in accordance with problem (21.4), (21.5), it follows that the components of the kinetic momentum are third-order polynomials of the projections of the angular velocity vector $\boldsymbol{\omega}$ and the energy E is a fourth-order polynomial:

$$K_i = I_{ij}^0 \omega_j + \varepsilon K_{ijkl} \omega_j \omega_k \omega_l, \quad E = \frac{1}{2} I_{ij}^0 \omega_i \omega_j + \frac{3}{4} \varepsilon K_{ijkl} \omega_i \omega_j \omega_k \omega_l.$$

In the conservative case, similar to the problem of the free rotation of an absolutely rigid body, one can find the polodes—the trajectories of the vector $\mathbf{K}$ relative to the body [12]—using two integrals $E = const$ and $\mathbf{K}^2 = const$.

The type of polodes in the case of regular polyhedra can be obtained from the following considerations. The direction of stationary rotations will correspond to the symmetry axes of the bodies. Consequently, the surface $E(\omega_1, \omega_2, \omega_3) = const$ has extrema along these directions and, as it were, repeats the surface of the body. In the approximation under consideration energy $E(\omega_1, \omega_2, \omega_3)$, being a polynomial of the fourth degree, cannot resolve the entire variety of forms. Bodies with a high degree of symmetry (for example, the dodecahedron and icosahedron) correspond to a spherical surface $E(\omega_1, \omega_2, \omega_3) = const$ just as in a rigid representation a body in the shape of a cube corresponds to a second-order spherical inertia tensor.

Forming a combination of integrals $E = const$ and $\mathbf{K}^2 = const$, in this case we obtain a spherical integral surface $\frac{4}{3}E - \frac{\mathbf{K}^2}{2I_0} = C$, the intersection of which with the surface of the body at different values of the constant C gives the appearance of polodes.

The influence of the elasticity of a body on its behavior in free motion is most clearly observed in cases where, in an undeformed state, the main moments of inertia of the body either coincide or are close. This is typical, in particular, for quasi-spherical homogeneous bodies, the surface of which can be specified in spherical coordinates r, θ, φ by the formula:

$$R(\theta, \varphi) = R_0 \left(1 + \delta \sum_{l=2,\ |m| \leq l}^{N} \alpha_{lm} Y_{lm}(\theta, \varphi) \right), \quad \delta \ll 1, \tag{21.6}$$

where $\alpha_{lm} = a_{lm} + ib_{lm}$, $\alpha_{lm} = (-1)^m \alpha^*_{l,-m}$, a_{lm}, b_{lm}—real numbers, $Y_{lm}(\theta, \varphi)$—spherical harmonics.

In the approximation under consideration (we limited ourselves to considering deformations of order ε) the dynamics of the body are determined only by harmonics with $m \leq 4$.

We will assume that the parameters ε and δ, which characterize the elastic properties and small deviation of the body surface from the sphere, are related by the inequality:

$$\varepsilon << \delta << 1.$$

This allows us to represent the displacement vector in problem (21.5) as:

$$\mathbf{u}'(\mathbf{r}) = \mathbf{u}_0(\mathbf{r}) + \delta \mathbf{u}_1(\mathbf{r}).$$

The displacement vector $\mathbf{u}_0(\mathbf{r})$ of the rotating ball with respect to the coordinate system $Ox_1x_2x_3$ with an arbitrary orientation of the angular velocity $\boldsymbol{\omega}$ is found from system (21.4), (21.5) under $\dot{\omega} = 0$ and boundary conditions specified on the sphere. We present corrections to the displacement vector due to the nonsphericity of the body (solutions to problems (21.4), (21.5)) in the following form:

$$\mathbf{u}'(\mathbf{r}, t) = \sum_{l,m,\nu} \left({}^{\nu}f^{0}_{lm}(r) + \delta^{\nu} f^{1}_{lm}(r) \right) \mathbf{Y}^{\nu}_{lm}(\theta, \varphi),$$

where are the spherical vectors $\mathbf{Y}^{\nu}_{lm}(\theta, \varphi)$ [13] are given by the expressions:

$$\mathbf{Y}^{+}_{lm}(\theta,\varphi) = (l+1)Y_{lm}(\theta,\varphi)\mathbf{e}_r - \mathbf{grad}\,Y_{lm}(\theta,\varphi),\ l = 0, 1, 2, \ldots,$$
$$\mathbf{Y}^{-}_{lm}(\theta,\varphi) = lY_{lm}(\theta,\varphi)\mathbf{e}_r + \mathbf{grad}\,Y_{lm}(\theta,\varphi),\ l = 1, 2, \ldots,$$
$$\mathbf{Y}^{0}_{lm}(\theta,\varphi) = \left[\mathbf{e}_r, \mathbf{grad}\,Y_{lm}(\theta,\varphi)\right],\ l = 1, 2, \ldots.$$

The solution to problems (21.4), (21.5) is given in [9]. The inertia tensor of an elastic quasi-ball in the considered approximation has the form:

$$I_{ij} = I^0_{ij} + \varepsilon\left(I'_{ij} + \delta I''_{ij}\right),$$

where I^0_{ij} corresponds to an absolutely rigid body, I'_{ij}, I''_{ij}—corrections due to elastic properties, and are, as follows from Eq. (21.5), quadratic functions of the angular velocity components.

The tensor I''_{ij} depends on the body shape parameters α_{lm} with indices $l \leq 4$. Higher harmonics affect the components of the inertia tensor when taking into account terms of higher order of smallness in ε.

Angular velocity $\boldsymbol{\omega}$ is determined by solving the equation for angular momentum:

$$\frac{d\mathbf{K}}{dt} + [\boldsymbol{\omega}, \mathbf{K}] = 0, \tag{21.7}$$

where in the considered approximation:

$$K_i = \left[I^0_{ij} + \varepsilon\left(I'_{ij} + \delta I''_{ij}\right)\right]\omega_j.$$

We will demonstrate the possibility of global movement of the axis of stable stationary rotation (poles) in a body using the example of a body, the deviation of the surface of which from a sphere is characterized by the initial ellipsoidality and fourth harmonics of the shape ($l = 4$). In this case, we assume the parameters α_{2i} are so smaller than α_{4i}, that when the body rotates, the main contribution to its deformation is associated with the fourth harmonic of the form.

Consider a body bounded by a surface, the equation of which has the form:

$$R(\theta,\varphi) = 1 + \delta\big(\alpha_{40}Y_{40}(\theta,\varphi) + \alpha_{44}\big(Y_{44}(\theta,\varphi) + Y_{4,-4}(\theta,\varphi)\big)\big).$$

We orient the ellipsoid of inertia of the undeformed body according to Fig. 21.1. Let us choose a system of main axes $Ox_1x_2x_3$ so that the axis Ox_3 passes through the main diagonal of the "cube". In the system $Ox_1x_2x_3$ the inertia tensor of the undeformed quasi-ball takes the form:

$$I^0_{ij} = (I_0 + \Delta_i)\delta_{ij}, \quad \Delta_i \ll I_0.$$

We will assume that Ox_3—the axis of the average moment of inertia: $\Delta_3 < \Delta_2 < \Delta_1$.

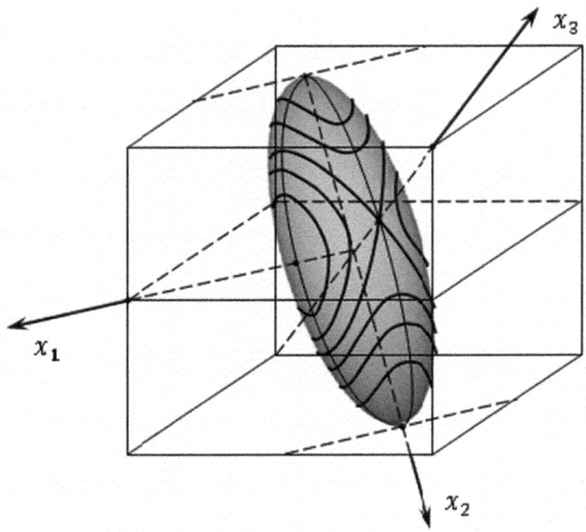

Fig. 21.1 Orientation of the ellipsoid of inertia of an undeformed body

The corrections to the inertia tensor corresponding to the deformation of the ball calculated from the solution of problem (21.5) have the form:

$$I'_{ij} = \frac{2}{35}\left[\left(-\frac{1}{3}(1-\xi_1) + 4\xi_2 + 28\xi_3\right)\omega^2\delta_{ij} + (1-\xi_1)\omega_i\omega_j\right],$$
$$\xi_1 = \frac{3\kappa+2}{2(19\kappa+14)}, \quad \xi_2 = \frac{1}{15(\kappa+2)}, \quad \xi_3 = \frac{1}{15(3\kappa+2)}.$$

The fourth harmonic of the shape contributes to the inertia tensor in the form of corrections:

$$I''_{11} = Q_2\left[\left(3\alpha_{40} + \sqrt{70}\alpha_{44}\right)\omega_1^2 + \left(\alpha_{40} - \sqrt{70}\alpha_{44}\right)\omega_2^2 - 4\alpha_{40}\omega_3^2\right],$$
$$I''_{22} = Q_2\left[\left(\alpha_{40} - \sqrt{70}\alpha_{44}\right)\omega_1^2 + \left(3\alpha_{40} + \sqrt{70}\alpha_{44}\right)\omega_2^2 - 4\alpha_{40}\omega_3^2\right],$$
$$I''_{33} = Q_2\left[-4\alpha_{40}\omega_1^2 - 4\alpha_{40}\omega_2^2 + 8\alpha_{40}\omega_3^2\right],$$
$$I''_{13} = -8Q_2\alpha_{40}\omega_1\omega_3, \quad I''_{23} = -8Q_2\alpha_{40}\omega_2\omega_3,$$
$$I''_{12} = 2Q_2\left(\alpha_{40} - \sqrt{70}\alpha_{44}\right)\omega_1\omega_2, \quad Q_2 = \frac{(\kappa+1)(35\kappa+26)}{35\sqrt{\pi}(19\kappa+14)^2}.$$

Taking into account the fact that I'_{ij}, I''_{ij} are quadratic functions of the angular velocity components, the components of the kinetic momentum vector **K** and energy E have the form:

$$K_i = I^0_{ij}\omega_j + \varepsilon K_{ijkl}\omega_j\omega_k\omega_l,$$

$$E = \frac{1}{2}I^0_{ij}\omega_i\omega_j + \frac{3}{4}\varepsilon K_{ijkl}\omega_i\omega_j\omega_k\omega_l,$$

$$K_{ijkl}\omega_i\omega_j\omega_k\omega_l = Q_1\omega^4 + \delta Q_2\Big[\Big(3\alpha_{40} + \sqrt{70}\alpha_{44}\Big)(\omega_1^4 + \omega_2^4) + 8\alpha_{40}\omega_3^4 +$$

$$+ 6\Big(\alpha_{40} - \sqrt{70}\alpha_{44}\Big)\omega_1^2\omega_2^2 - 24\alpha_{40}(\omega_1^2 + \omega_2^2)\omega_3^2\Big],$$

$$Q_1 = \frac{4}{35}\left[\frac{1}{3}(1 - \xi_1) + 2\xi_2 + 14\xi_3\right].$$

Let us bring the body into rapid rotation around an axis Ox_3 with angular velocity $\boldsymbol{\omega} = \omega_1\mathbf{e}_1 + \omega_2\mathbf{e}_2 + (\Omega + \omega_3)\mathbf{e}_3$ and consider small perturbations of this motion $\omega_i \ll \Omega,\ i = 1, 2, 3.$ $\omega_i \ll \Omega, i = 1{,}2, 3.$

We will describe the displacement of the axis of stable stationary rotation in the body geometrically, depicting the hodograph of the angular momentum vector $\mathbf{K}$ on the sphere $\frac{4}{3}E - \frac{\mathbf{K}^2}{2I_0} = const.$

At high rotation speeds, the quality of the body dynamics is determined by the fourth harmonic of the shape (Fig. 21.2). With a decrease in angular velocity Ω to a certain value Ω_1 the area of attraction of stable rotation decreases (Fig. 21.3). The adjacent cell, limiting the unstable focus, moves closer towards the position of stable rotation. The saddle point, which was initially located on the boundary of the attraction region, also behaves. The position of stable rotation, together with the saddle point, "turns" over time in the plane Ox_2x_3, thereby illustrating the redistribution of the influence of "cubicity" and the ellipsoidal distance of its shape on the dynamics of the body.

At a certain critical value of the angular velocity Ω_1 the cell adjacent to the attraction region disappears. Its boundary "collapses" into a segment that is part of the trajectory surrounding the axis of the smallest moment of inertia Ox_2. The place of the area of attraction is filled by trajectories covering the axis Ox_2 (Fig. 21.4). With further deceleration of rotation, the saddle will move towards the axle position Ox_3. A decrease in the angular velocity to a certain value Ω_2 leads to the disappearance of unstable foci contained in the loops of separatrices of the saddle point at the position of the axis of the greatest moment of inertia Ox_1, and the birth of a stable focus in the vicinity of this axis. From this moment, the ellipsoid of inertia of the body in an undeformed state completely determines the quality of its dynamics (Fig. 21.5).

Let us introduce small internal friction into consideration. Let the body perform angular motions near a certain direction of stable stationary rotation (in our case, around Ox_3). Over time, internal dissipation reduces the kinetic energy (angular velocity) of the body. However, its dynamics will not change qualitatively. Simultaneously with the decrease in angular velocity Ω the precession of the body fades, therefore, any movement in the body stops, and with it the internal dissipation of energy. The body returns to its original stationary rotation, only the area of its attraction is reduced.

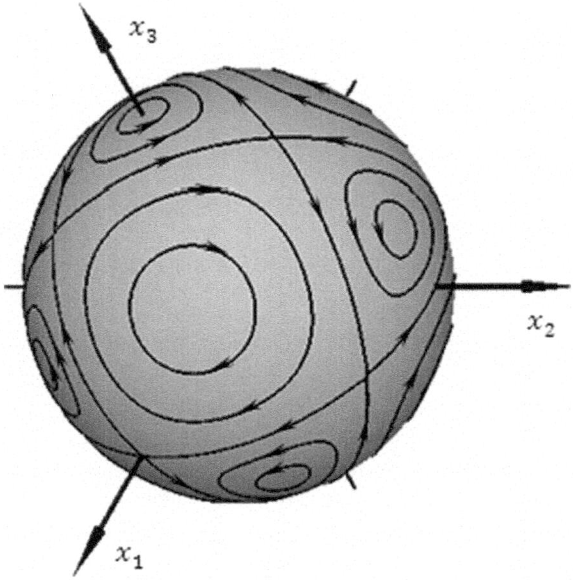

Fig. 21.2 Polodes at $\Omega >> \Omega_1$

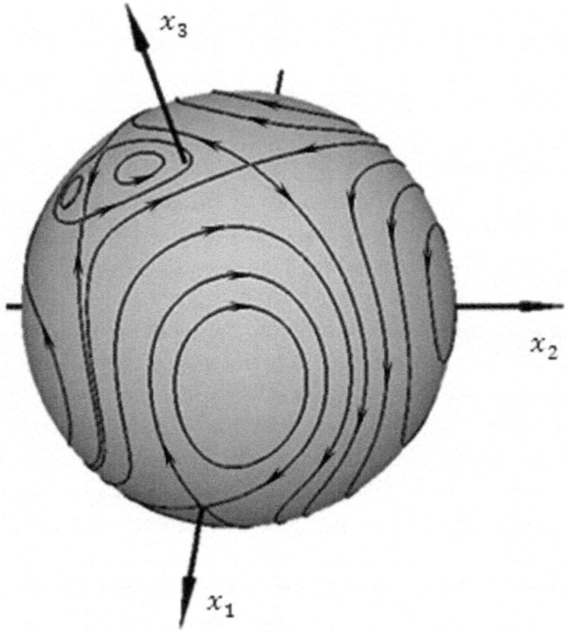

Fig. 21.3 Polodes at $\Omega > \Omega_1$

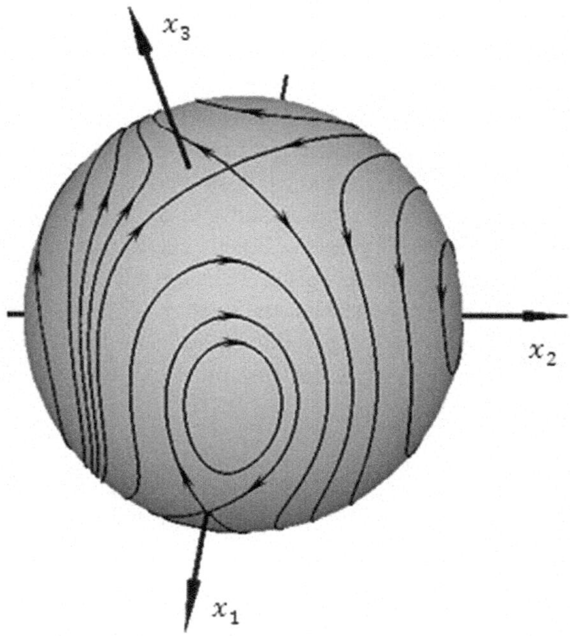

Fig. 21.4 Polodes at $\Omega_2 < \Omega < \Omega_1$

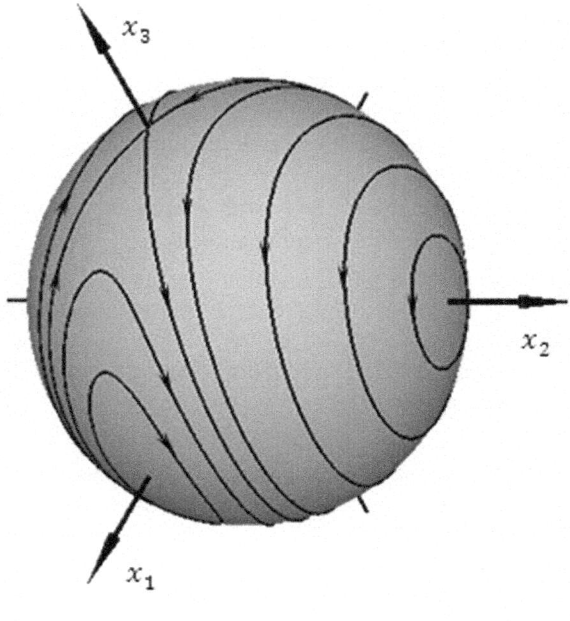

Fig. 21.5 Polodes at $\Omega < \Omega_2$

The situation is different when, along with the internal dissipation of energy, there is also a small external resistance to the movement of the body. The kinetic momentum is not conserved, but its change, due to its smallness, can be ignored. External resistance reduces the angular velocity of the body even in the absence of precession, and, therefore, over time changes the relationship between the factors that determine the dynamics of the body. As a result, under the $\Omega = \Omega_*$ the axis of stable stationary rotation shifts in the body to a new position. In the problems considered, the minimum of kinetic energy "moves" from Ox_3 to direction Ox_1.

The described mechanism of changing the position of the axis of stationary rotation in the body, otherwise, the rotation of the body in the inertial reference system relative to the constant position of the vector **K**, should be kept in mind to understand the phenomenon of global movement of the Earth's poles, supposedly observed in its history.

21.3 On Nutation of a Viscoelastic Homogeneous Body Close to a Sphere

We consider a homogeneous deformable body that obeys the equations of a linear elastic-viscous medium. The body is close in shape to a sphere and is rotated with angular velocity $\boldsymbol{\omega}_0$. A stationary stress state is established in it, characterized by a displacement vector $\mathbf{u}_0$. Having experienced a small deviation from this state, the body then undergoes nutation movement, which significantly depends on both its elastic and viscous properties. In this formulation, the problem may be of interest, in particular, in connection with the interpretation of the Chandler nutation of the Earth.

The movement of a body with a small nutation angle is considered in a system $Ox_1x_2x_3$, rotating with angular velocity:

$$\boldsymbol{\omega} = \omega_1\mathbf{e}_1 + \omega_2\mathbf{e}_2 + \omega_0\mathbf{e}_3, \quad \omega_1, \omega_2 \ll \omega_0.$$

The equation of motion of the volume element of the body and the conditions on its surface S have the form:

$$\begin{gathered}\gamma\big[(\kappa+1)\mathbf{grad}\,div\,\mathbf{u} + \Delta\mathbf{u}\big] + (\zeta+1)\frac{\partial}{\partial t}\mathbf{grad}\,div\,\mathbf{u} + \frac{\partial}{\partial t}\Delta\mathbf{u} = \\ = \delta(\ddot{\mathbf{u}} + [\dot{\omega}, \mathbf{r}+\mathbf{u}] + [\boldsymbol{\omega}, [\boldsymbol{\omega}, \mathbf{r}+\mathbf{u}]] + 2[\boldsymbol{\omega}, \dot{\mathbf{u}}]) \ in\ V, \\ \left[\left(\gamma\kappa + \zeta\frac{\partial}{\partial t}\right)u_{ll}\delta_{ij} + 2\left(\gamma + \frac{\partial}{\partial t}\right)u_{ij}\right]n_j = 0 \ on\ S,\ i = 1, 2, 3.\end{gathered} \tag{21.7}$$

For dimensionless parameters the following notation has been introduced:

$$\delta = \frac{\rho R_0^2}{\eta t_*^2}, \quad \gamma = \frac{\mu t_*}{\eta}, \quad \kappa = \frac{\lambda}{\mu}, \quad \zeta = \frac{\xi}{\eta}, \quad \varepsilon^2 = \frac{\delta}{\gamma} = \frac{\rho R_0^2}{\mu t_*^2},$$

where ξ, η characterize the viscous properties of the body.

The parameter ε is associated with the period of elastic vibrations of the body $T_K \sim \sqrt{\varepsilon} t_*$. In accordance with the assumption about the time scale of the processes occurring $\sqrt{\varepsilon} \ll 1$. The parameter δ that specifies the scale of processes caused by the viscosity of the body $T \sim \delta t_*$ is considered to be $\delta \ll 1$. In what follows, we will proceed from the fact that the relationship between the quantities ε and δ (therefore, and $\gamma = \mu t_*/\eta = \varepsilon^2/\delta$) can vary within wide limits.

Vector multiplication (21.7) by $\mathbf{r} + \mathbf{u}$ and subsequent integration of the resulting relations taking into account Eq. (21.1) leads to the equation for the angular momentum vector (21.6), where

$$K_i = I_{ij}\omega_j, \quad I_{ij} = \int\limits_V [(x_l^2 + 2x_l u_l)\delta_{ij} - (x_i x_j + 2x_i u_j)]dV, \quad i, j = 1, 2, 3.$$

Let us present the displacement vector of the body in the form $\mathbf{u} = \mathbf{u}_0 + \mathbf{u}'$, where $\mathbf{u}'$—is the deviation from the stationary state $\mathbf{u}_0$, due to rotation with angular velocity $\boldsymbol{\omega}_0$. Using the well-known solution to the problem of deformation of a rotating ball [9], we write down the nonzero components of the inertia tensor of the ellipsoid:

$$I_{11}^0 = I_{22}^0 = \frac{2}{5}(1+a) + \frac{2}{35}\varepsilon^2\left(-\frac{1}{3} + \frac{1}{3}\xi_1 + 4\xi_2 + 28\xi_3\right)\omega_0^2,$$
$$I_{33}^0 = \frac{2}{5}(1+c) + \frac{2}{35}\varepsilon^2\left(\frac{2}{3} - \frac{2}{3}\xi_1 + 4\xi_2 + 28\xi_3\right)\omega_0^2,$$

where the moment of inertia of a non-rotating ball in accepted units is $I_0 = 2/5$. Here:

$$\xi_1 = \frac{3\kappa + 2}{2(19\kappa + 14)}, \quad \xi_2 = \frac{1}{15(\kappa + 2)}, \quad \xi_3 = \frac{1}{15(3\kappa + 2)}$$

$\xi_1 = 1/2(3\kappa + 2)(19\kappa + 14)^{-1}$, $\xi_2 = 1/15(\kappa + 2)^{-1}$, $\xi_3 = 1/15(\xi + 2)^{-1}$ and it is assumed that in an undeformed state the body is a quasi-sphere with components of the inertia tensor $I_{11}^T = I_{22}^T = I_0(1 + a)$, $I_{33}^T = I_0(1 + c)$, and the values of a and c are comparable to the additions to the components of the inertia tensor associated with deformations $\mathbf{u}_0$.

Let us represent the deformation vector $\mathbf{u}'$ in the form $\mathbf{u}' = \delta\mathbf{u}^{(1)} + \delta^2\mathbf{u}^{(2)} + \cdots$ and restrict ourselves to the first term of this series. To determine $\mathbf{u}^{(1)}$ in accordance with (21.7), it is necessary to solve the following problem:

$$\gamma\left[(\kappa+1)\mathbf{grad}\,div\,\mathbf{u}^{(1)}+\Delta\mathbf{u}^{(1)}\right]+(\zeta+1)\mathbf{grad}\,div\,\dot{\mathbf{u}}^{(1)}+\Delta\dot{\mathbf{u}}^{(1)}=$$
$$=\delta([\dot{\omega}_1,\mathbf{r}]+[\boldsymbol{\omega}_0,[\boldsymbol{\omega}_1,\mathbf{r}]]+[\boldsymbol{\omega}_1,[\boldsymbol{\omega}_0,\mathbf{r}]])\ in\ V,$$
$$\left[\left(\gamma\kappa+\zeta\frac{\partial}{\partial t}\right)u_{ll}^{(1)}\delta_{ij}+2\left(\gamma+\frac{\partial}{\partial t}\right)u_{ij}^{(1)}\right]n_j=0\ on\ S,\ i=1,2,3. \tag{21.8}$$

where $\boldsymbol{\omega}_1=\omega_1\mathbf{e}_1+\omega_2\mathbf{e}_2$.

The solution to problem (21.8) is constructed in the form of polynomials in spatial variables containing linear and cubic terms. This is due to the fact that the left-hand sides of Eq. (21.8) are linear functions of spatial variables, and the equation and boundary conditions contain the second derivatives of the desired functions with respect to these variables. The solution to this problem takes the most compact form when the ratio of the parameters of elastic properties κ and the ratio of the characteristics of viscous properties ζ coincide. Therefore, in the future we will assume that $\kappa=\zeta$. This assumption greatly simplifies the consideration, but is not of fundamental importance for the phenomenon being studied.

Consideration of problem (21.8) leads to equations for the components of the vector $\mathbf{u}^{(1)}$: $\dot{u}_j^{(1)}+u_j^{(1)}=f_j,\ j=1,2,3$. Functions f_j are polynomials containing linear and cubic terms in spatial variables, the coefficients of which are linear functions ω_1 and ω_2. In [5], where the deformation of a rotating ellipsoid was studied without taking into account viscosity, the expressions f_j were obtained f_j. Using the results of this work, we find the corrections to the components of the inertia tensor associated with the vector $\mathbf{u}^{(1)}$:

$$I_{13,23}^{(1)}=I_0\delta d\omega_0e^{-\gamma t}\int_0^t\omega_{1,2}e^{\gamma\tau}d\tau,\quad d=\frac{1}{7}(1-\xi_1).$$

The remaining quantities $I_{ij}^{(1)}$ are absent in the considered approximation.

Let us write down the vector of angular momentum $\mathbf{K}=K_i\mathbf{e}_i$ in the system $Ox_1x_2x_3$, where:

$$K_1=I_1^0\omega_1+I_0\left(\frac{\varepsilon^2}{7}\left(-\frac{1}{3}+\frac{1}{3}\xi_1+4\xi_2+28\xi_3\right)\omega_1+\delta de^{-\gamma t}\int_0^t\omega_1e^{\gamma\tau}d\tau\right)\omega_0^2,$$

$$K_2=I_2^0\omega_2+I_0\left(\frac{\varepsilon^2}{7}\left(-\frac{1}{3}+\frac{1}{3}\xi_1+4\xi_2+28\xi_3\right)\omega_2+\delta de^{-\gamma t}\int_0^t\omega_2e^{\gamma\tau}d\tau\right)\omega_0^2,$$

$$K_3=I_3^0\omega_0+I_0\frac{\varepsilon^2}{7}\left(\frac{2}{3}-\frac{2}{3}\xi_1+4\xi_2+28\xi_3\right)\omega_0^3.$$

The nature of the movement of the axis of rotation can be judged by determining the disturbances ω_1,ω_2 in the angular velocity of the body. Let us write the Euler equation in projections on the axis of the coupled system:

$$\dot{\omega}_1 + (c-a)\omega_0\omega_2 + \varepsilon^2 d\omega_0^3\omega_2 - \delta d\omega_0^3 e^{-\gamma t}\int_0^t \omega_2 e^{\gamma\tau} d\tau = 0,$$

$$\dot{\omega}_2 - (c-a)\omega_0\omega_1 - \varepsilon^2 d\omega_0^3\omega_1 + \delta d\omega_0^3 e^{-\gamma t}\int_0^t \omega_1 e^{\gamma\tau} d\tau = 0.$$

n the limiting case of very high viscosity, when $\delta \to 0$, this system represents the equation of motion of an absolutely rigid ellipsoid and has a solution:

$$\omega = \omega_1 + i\omega_2 = z_0 e^{i\Omega_1 t}.$$

Here z_0 is a constant value, $i = \sqrt{-1}$, $\Omega_1 = \Omega_0 + \Omega$, where $\Omega_0 = (c-a)\omega_0$, $\Omega = \varepsilon^2 d\omega_0^3$.

Thus, in the case of very high viscosity, the nutation frequency of the body is equal to the nutation frequency of an absolutely rigid body with ellipsoidality, which is the sum of the initial ellipsoidality $\varepsilon^2 d\omega_0^2$ acquired during prolonged rotation.

When $\delta \neq 0$, looking for a solution to the system of equations in the form $\omega = \omega_1 + i\omega_2 = z_0 e^{pt}$ $\omega = \omega_1 + i\omega_2 = z_0 \exp pt$, we arrive at a characteristic equation, which can be conveniently written in the form:

$$p^2 + (\beta|\Omega_1| - i\Omega_1)p - i\alpha\beta\,\Omega_1|\Omega_1| = 0,$$

$$\alpha = \frac{\Omega_0}{\Omega_1} = \frac{c-a}{c-a+\varepsilon^2 d\omega_0^2}, \quad \beta = \frac{\gamma}{|\Omega_1|}.$$

In the absence of initial ellipsoidality $\alpha = 0$ the solutions of the characteristic equation have the form:

$$p_1 = -(\beta - i)\Omega_1, \quad p_2 = 0.$$

Regardless of $\beta \sim \mu/\eta$ the nutation frequency $\Omega_1 > 0$, the damping of this movement is determined by the value $\beta\,\Omega_1$. At sufficiently large values β nutation is not observed. The deformation of a ball with low intrinsic viscosity when the position of the body changes relative to the axis of rotation $\boldsymbol{\omega}_0$ "tracks" this change, therefore, there is no nutation.

The conducted study of the angular motions of a viscoelastic body can be applied to explain the nutation movements of the Earth. Assuming first that the Earth has a very high viscosity, we will have a period of the so-called unobservable Euler nutation $T_E = 2\pi/\Omega_1 = 306$ days. This period is calculated from the observed moments of inertia of the Earth, if we consider it absolutely solid. Knowing from observations the value of the Earth's nutation period $T_C = 428$ days—the Chandler period and the quality factor of these oscillations $Q = 25$ [14], it is possible to determine the viscosity coefficient from the dependence of the root of the characteristic equation p_1 p_1. This result is consistent with the estimates of the value of this coefficient in [14].

The value of the Chandler period is mainly determined by the ratio of the moments of inertia of the non-rotating Earth.

21.4 Conclusion

The work carried out a study of the influence of elastic and viscous properties of a body similar in shape to a ball on its free angular movements. In the considered problems it is shown that taking into account the deformations of the body leads to a qualitative change in its dynamics. The possibility of global displacement of the axis of stable stationary rotation to a new position in the body when the angular velocity of rotation of the body changes is demonstrated. In the problem of nutation of a quasi-ball, it is shown that the periods of nutation of a body with and without taking into account its elastic and viscous properties can differ significantly. An explanation of these rigid body dynamics effects can be useful in studying the motion of the Earth's poles and its Chandler nutation.

Acknowledgements The work was carried out with the financial support of the Ministry of Science and Higher Education of the Russian Federation (Project No. FSWR-2023-0034).

References

1. Chernousko FL (1979) The influence of self-elasticity and dissipation on the motion of a rigid body relative to the center of mass. In: Dynamics of continuum: collection of articles. Institute of Hydrodynamics SB AN USSR, Novosibirsk, vol 41. pp 118–122
2. Wilke VG (1983) Separation of motions and the averaging method in the mechanics of systems with an infinite number of degrees of freedom. In: Bulletin of Moscow state university, Mathematics-mechanics, no 5, pp 54–59
3. Markeev AP (1989) On the dynamics of an elastic body in a gravitational field. Space Res 27(2):163
4. Novikov VV (1987) Anisotropic elastic ball in free motion. Appl Math Mech 5(51):767–774
5. Denisov GG, Novikov VV (1987) On the free motions of an elastic ellipsoid. Izv. Acad Sci USSR. MTT 6:69–74
6. Novikov VV, Urman YM (2001) The influence of elasticity on the dynamics of a superconducting rotor rotating in a magnetic field. Appl Mech Tech Phys 3(42):162–167
7. Akulenko LD, Kumakshev SA, Markov YuG (2005) Perturbed rotation of the Earth. Izv RAS MTT 5:19–29
8. Akulenko LD, Kumakshev SA, Markov YG, Rykhlova LV (2005) Gravitational-tidal mechanism of oscillations of the Earth's pole. Izv RAS Astron J 10(82):950–960
9. Denisov GG, Novikov VV (1983) On the free movements of a deformable rigid body close to a ball. Proc USSR Acad Sci Mech Solids 3:43–50
10. Denisov GG, Novikov VV (1999) On the free motion of a viscoelastic quasi-sphere. Izv Acad Sci USSR Mech Solids 1:26–32
11. Novikov VV, Fevralskikh LN (2014) To the problem of free angular motions of a deformable rigid body close in shape to a ball. Uchen zap Kazan un-ta Ser Phys-Math Sci 156:109–116
12. Landau LD, Lifshits EM (1965) Theoretical physics, vol 1. Nauka, Mechanics. M., p 206

13. Petrashen GI (1949) Symmetry of rotation and spherical vectors. Scientific notes of Leningrad State University. Mathematical science mechanics, vol 17, pp 3–27
14. Monin AS (1972) Earth's rotation and climate. L.: Gidrometeoizdat, 112 p

Chapter 22
Differential Rotation of the Mantle and the Solid Core of the Earth as a Factor in the Seismic Wave Process

Valery V. Novikov, Lyubov N. Fevralskikh, and Alexander K. Lyubimov

Abstract Currently in seismology there is a concept of a wave seismic process (Vikulin in Dokl. Academy of Sciences of the USSR 4:821–824, 1990; Vikulin in Physics of wave seismic process, Publishing house of KSPU, Petropavlovsk-Kamchatsky, p. 151, 2003), according to which natural disasters such as earthquakes, volcanic eruptions, and the appearance of faults in rocks associated with stresses in the body of the Earth manifest themselves in the same region of the planet with a period of the order of several tens of years. This leads to the conclusion that the zones of greatest tension within the Earth periodically move. The purpose of the work is to draw attention to a new physical mechanism that explains this phenomenon. The fact is that, in recent decades, advanced rotation of the Earth's solid core relative to the mantle has been discovered (Song and Richards in Nature 382:221, 1996; Ovchinnikov et al. in Dokl. RAS 362:683–686, 1998). Based on this fact and assuming the ellipsoidality of the Earth's inner core, it has been shown (Denisov et al. in Dokl. RAS 400:625–629, 2005; Denisov et al. in Nature 3–10, 2013) that the gravitational interaction of the mantle and the inner core leads to observed long-period (about 100 years) fluctuations in the length of the day. This, in particular, serves as indirect evidence of the ellipsoidality of the inner core. In this work, a simple model is proposed that reflects some features of the internal structure of the Earth, which demonstrates the periodicity of movement of the zones of greatest stress. An elastic isotropic thick spherical shell rotating at a certain angular velocity is considered. In the cavity of the shell there is a body in the form of a dumbbell, which rotates ahead of the shell. The calculation of the distribution of stresses inside the shell was carried out depending on the influence of gravitational force from the dumbbell and the deformations of the shell associated with its angular movements. The zones of greatest stress are

---

V. V. Novikov (✉) · L. N. Fevralskikh · A. K. Lyubimov
National Research Lobachevsky State University of Nizhny Novgorod, Nizhny Novgorod, Russia
e-mail: novikov@mm.unn.ru

L. N. Fevralskikh
e-mail: fevralskikh_ln@itmm.unn.ru

A. K. Lyubimov
e-mail: ljubimov@mm.unn.ru

H. Altenbach and V. A. Eremeyev (eds.), *Dynamics of Discrete and Continuum Structures and Media*, Advanced Structured Materials 221,
https://doi.org/10.1007/978-3-031-75626-9_22

highlighted. It is shown that the relative rotation of the internal body leads to periodic movement of these zones. The results of the work may be useful for understanding the mechanism of the seismic wave process.

Keywords Elastic spherical shell · Ellipsoidal core · Long-period movement · Internal stress · Angular motion · Seismic activity

22.1 Introduction

The upper layers of the Earth consist of tectonic plates that move relative to each other and create tension at the joints. At a time when internal stresses in combination with external factors reach certain values, a stress discharge may occur at the junction of the plates, leading to seismic activity in this zone [2]. Analysis of data from global and regional seismicity catalogs indicates the presence of narrow seismic belts extending along the entire surface of the Earth for thousands and tens of thousands of kilometers. Displacements in time and space of the entire seismic belt, some of its parts and individual foci of the strongest earthquakes, and the repetition of the most powerful earthquakes in the same place at certain intervals have been established. These facts underlie the concept of the wave seismic process [1, 2], which makes it possible to analytically describe the occurrence of earthquakes and establish a connection between seismic activity and certain geophysical phenomena on a planetary scale. In particular, the relationship of the seismic process with the nutation of the pole and the Earth's natural oscillations was studied [2].

In connection with the concept of the wave seismic process, attention should be paid to the results on the dynamics of the Earth obtained in recent decades. Based on the analysis of seismic waves, advanced rotation of the Earth's solid core relative to the mantle was discovered [3, 4]. Based on this fact, it is shown [5, 6] that from the assumption of the ellipsoidality of the Earth's inner core it follows that under certain conditions the gravitational interaction of the mantle and the inner core leads to self-oscillations in this system. The mantle and the inner core periodically exchange kinetic momentum. This manifests itself as long-period (about 100 years) fluctuations in the length of the day. Such observed changes in the angular velocity of rotation of the mantle give rise to the conclusion that the Earth's inner core is ellipsoidal.

This work proposes a simple mechanical model for understanding the possible mechanism of the periodic movement of zones of the greatest stress in the body of the Earth, built on taking into account the peculiarities of the internal structure of the Earth and its dynamics, namely the advanced rotation of the inner core and the ellipsoidal shape of the core. The mantle is represented as a homogeneous elastic thick spherical shell, rotating at a certain constant angular velocity. There is a solid body in the cavity that performs advanced rotation relative to the shell. There may be a mismatch in the directions of rotation of these bodies, but this factor does not qualitatively change the result of the work. In gravitational interaction with the outer shell, the inner body is like a dumbbell—two point masses located in the equatorial plane and connected by a rigid massless rod. Relative to the elastic shell, the dumbbell slowly rotates in this plane. For a spherical external body, the distribution of radial stresses is constructed depending on the gravitational force from the dumbbell and the elastic deformations that arise during its rotation. It is shown that with increasing gravity, the position of zones of increased stress shifts from the poles to the equator. Taking into account the advanced rotation of the internal body leads to the migration of zones of increased stress with a period equal to half the period of its full rotation relative to the external body.

22.2 The Problem of the Deformation of an Elastic Rotating Ball Interacting with a Dumbbell Located in Its Cavity

Let us consider an elastic isotropic spherical shell bounded by surfaces $r = r_1$ and $r = r_2 > r_1$. We associate a coordinate system $Oxyz$ with the shell, and the point O is the center of the shell. Let us place a dumbbell in the cavity $r < r_1$, which represents two diametrically opposed point masses M, located in a plane Oxy at a distance $c \ll r_1$ from the center O (Fig. 22.1). The dumbbell rotates slowly at a constant angular velocity Ω around an axis Oz relative to the shell. A shell element of a unit volume is acted upon by a gravitational force from the side of a dumbbell, the projection of which in spherical coordinates, up to second-order terms in the parameter c/r, has the form:

$$F_r = -\frac{2GM\rho}{r^2}\left(1 - \frac{3}{2}\left(1 - 3\sin^2\theta\cos^2(\varphi - \Omega t)\right)\frac{c^2}{r^2}\right),$$

$$F_\theta = \frac{6GM\rho}{r^2}\sin\theta\cos\theta\cos^2(\varphi - \Omega t)\frac{c^2}{r^2},$$

$$F_\varphi = -\frac{6GM\rho}{r^2}\sin\theta\sin(\varphi - \Omega t)\cos(\varphi - \Omega t)\frac{c^2}{r^2}.$$

Let the coordinate system associated with the shell rotate with angular velocity $\boldsymbol{\omega}(t)$. The displacement of a body element during its deformation is characterized by the vector $\mathbf{u}(\mathbf{r}, t)$. The choice of a coupled system is carried out in accordance with the conditions of the absence in it of small translational movements and rotations of the body as a rigid whole:

$$\int_V \mathbf{u}\, dV = 0, \quad \int_V [\mathbf{r}, \dot{\mathbf{u}}] = 0.$$

The equation of motion of a shell element and the conditions for the absence of stress on its internal and external surfaces after the transition to dimensionless variables have the form:

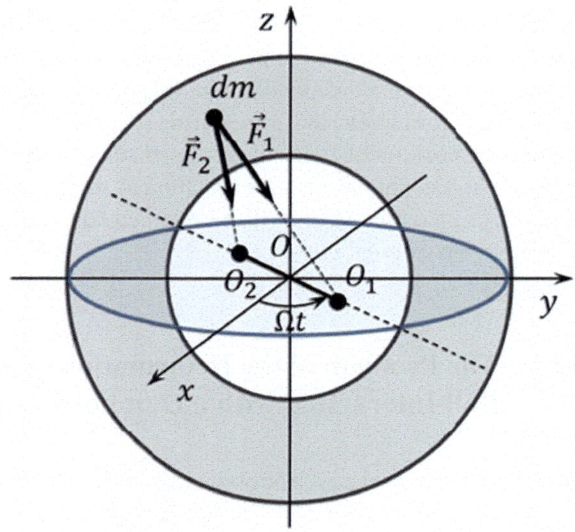

Fig. 22.1 A thick shell that interacts with the dumbbell located in its cavity

$$(\kappa + 1)\mathbf{grad}\, div\, \mathbf{u} + \Delta \mathbf{u} = [\boldsymbol{\omega}, [\boldsymbol{\omega}, \mathbf{r} + \mathbf{u}]] - \delta \mathbf{f}, \quad 1 < r < 1 + h, \tag{22.1}$$

$$\left[\kappa u_{ll} \delta_{ij} + 2u_{ij}\right] n_j = 0, \quad r = 1, \quad r = 1 + h, \quad i = 1, 2, 3, \tag{22.2}$$

where u_{ij}—small strain tensor, n_j—components of the surface normal vector, and **f**—dimensionless gravitational force acting on the shell element from the dumbbell side, $h = (r_2 - r)/R$.

The internal radius of the shell $R = r_1$ and the characteristic time of motion of the shell $t_* = 2\pi/|\omega|$ relative to the center of mass are taken as length and time scales, respectively.

For dimensionless parameters, the following notations are accepted:

$$\delta = \frac{GMt_*^2}{R^3}, \quad \kappa = \frac{\lambda}{\mu}, \quad a = \frac{c}{R},$$

where ρ—body density, and λ, μ—Lame constants.

Equation (22.1) corresponds to the case of a sufficiently rigid body, when the magnitude of the displacement vector is of the order of [7]

$\varepsilon = \frac{\rho R^2}{\mu t_*^2} \ll 1$.

Let us write the expression for the right side of Eq. (22.1) as a sum of the corresponding spherical vectors [8]:

$$
\begin{aligned}
[\boldsymbol{\omega}, [\boldsymbol{\omega}, \mathbf{r}]] - \delta\mathbf{f} = -4\sqrt{\pi}\left[\frac{1}{3}r\left(\omega_1^2 + \omega_2^2 + \omega_3^2\right) - \frac{\delta}{r^2}\right]\mathbf{Y}_{00}^{+} + \\
+ r\big[\beta_{20}\mathbf{Y}_{20}^{-}(\theta) + \beta_{21}\mathbf{Y}_{21}^{-}(\theta,\varphi) - \beta_{21}^{*}\mathbf{Y}_{2,-1}^{-}(\theta,\varphi) + \\
+ \beta_{22}\mathbf{Y}_{22}^{-}(\theta,\varphi) + \beta_{22}^{*}\mathbf{Y}_{2,-2}^{-}(\theta,\varphi)\big] - \\
- \delta\sqrt{\pi}\frac{a^2}{r^4}\Big(2\sqrt{3}\mathbf{Y}_{20}^{+}(\theta) - 3\sqrt{2}\big(\mathbf{Y}_{22}^{+}(\theta,\varphi - \Omega t) + \mathbf{Y}_{2,-2}^{+}(\theta,\varphi - \Omega t)\big)\Big),
\end{aligned}
$$

where

$$
\beta_{20} = -\frac{1}{3}\sqrt{2\pi}\left(\omega_1^2 + \omega_2^2 - 2\omega_3^2\right), \quad \beta_{21} = -\frac{2}{3}\sqrt{3\pi}\left(\omega_1 - i\omega_2\right)\omega_3,
$$
$$
\beta_{22} = \frac{1}{3}\sqrt{3\pi}\left(\omega_1^2 - \omega_2^2 - 2i\omega_1\omega_2\right),
$$

$\beta_{21}^{*}, \beta_{22}^{*}$—complex conjugate to β_{21}, β_{22} respectively.
We will look for a solution to problem (22.1), (22.2) in the form:

$$
\begin{aligned}
\mathbf{u} = \mathbf{u}_0 + \delta\mathbf{u}_1 = P_{00}(r)\mathbf{Y}_{00}^{+} + \\
+ \sum_{|m|\le 2}\left(P_{2m}(r)\mathbf{Y}_{2m}^{+}(\theta,\varphi) + M_{2m}(r)\mathbf{Y}_{2m}^{-}(\theta,\varphi)\right) + \\
+ \delta\big[p_{00}(r)\mathbf{Y}_{00}^{+} + p_{20}(r)\mathbf{Y}_{20}^{+}(\theta) + m_{20}(r)\mathbf{Y}_{20}^{-}(\theta) + \\
+ e^{-2i\Omega t}\left(p_{22}(r)\mathbf{Y}_{22}^{+}(\theta,\varphi) + m_{22}(r)\mathbf{Y}_{22}^{-}(\theta,\varphi)\right) + \\
+ e^{2i\Omega t}\left(p_{2,-2}(r)\mathbf{Y}_{2,-2}^{+}(\theta,\varphi) + m_{2,-2}(r)\mathbf{Y}_{2,-2}^{-}(\theta,\varphi)\right)\big].
\end{aligned}
\tag{22.3}
$$

Substituting (22.3) into Eq. (22.1) and grouping the expressions at the spherical vectors with the same indices, we obtain a system of differential equations for unknown radial functions $P_{lm}(r), M_{lm}(r), p_{lm}(r), m_{lm}(r)$ with zero boundary conditions. Then, using the found solution (22.3), the components of the strain tensor and stress tensor are calculated in spherical coordinates.

22.3 Distribution of Radial Stresses Near the Surface of the Ball

Let us discuss the nature of radial stresses in the shell at a certain fixed value of the radial coordinate. The distribution of radial stresses on a spherical surface depending on θ, φ is shown in Figs. 22.2, 22.3, and 22.4, where the areas of maximum stress are marked in red, and the areas of minimum stress are marked in blue.

In the absence of gravitational force, stresses are determined by the deformations of the shell that occur during its rotation. The shell is compressed along the axis of rotation, which leads to an increase in radial stresses in the area of the poles. Figure 22.2 shows the radial component of the stress vector in the case of a small

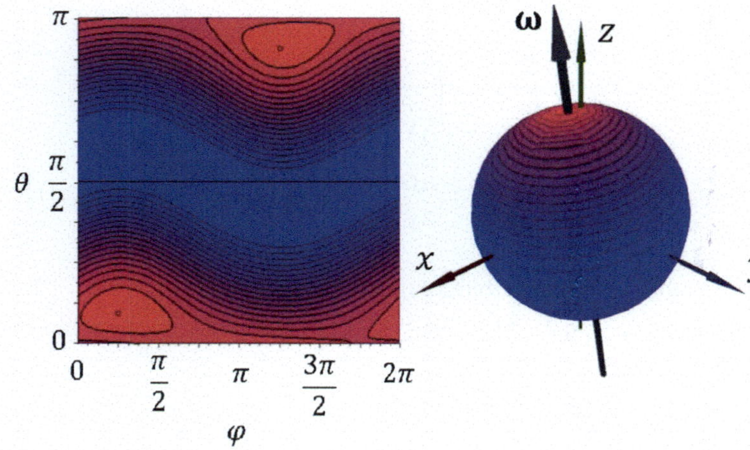

Fig. 22.2 Distribution of radial stresses in a rotating shell in the absence of gravitational interaction with a dumbbell

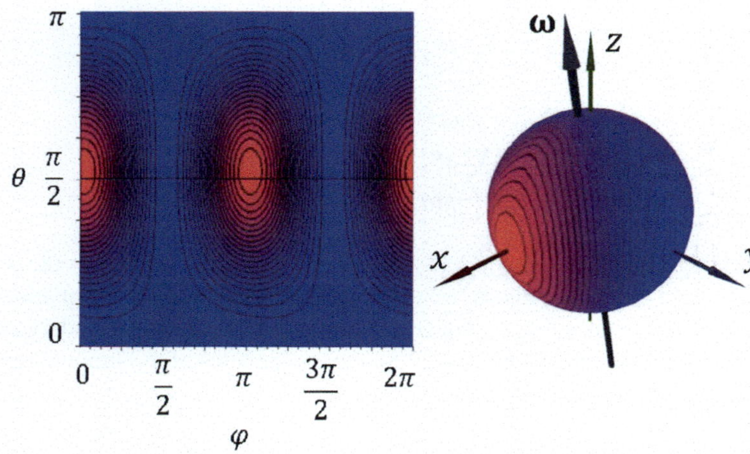

Fig. 22.3 Distribution of radial stresses in a non-rotating shell taking into account gravitational interaction with a dumbbell

deviation of the shell rotation axis from the direction Oz. Radial stresses are minimal in the plane perpendicular to the axis of rotation of the shell and increase as they approach the poles.

Let us consider separately the contribution of the gravitational force to the distribution of radial stresses. Let us fix the moment in time when the dumbbell was on the axis Ox. In this case, two diametrically opposite centers appear at the equator, in which the radial stress is maximum (Fig. 22.3). They are located on a line passing through the gravitating masses.

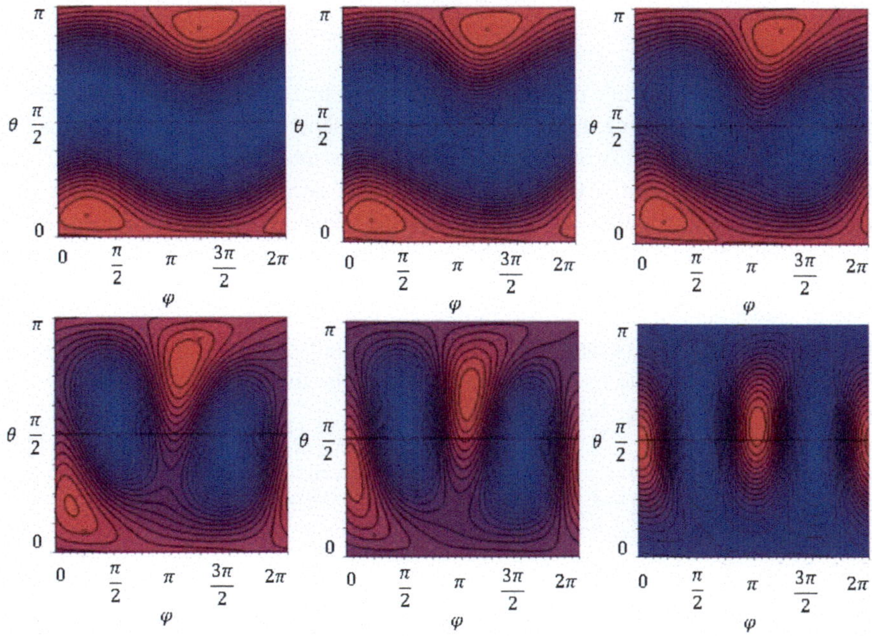

Fig. 22.4 Distribution of radial stresses in a rotating shell with increasing gravitational interaction with a dumbbell

As the influence of gravitational force on the deformation of the shell increases in comparison with the contribution of the centrifugal force, the stress distribution changes. Figure 22.4 shows the distribution of radial stresses in the order corresponding to the increase in the parameter δ characterizing the contribution of the gravitational force to the stressed state of the shell. It can be seen that when δ is low two zones of minimum stress appear in the equator region, located on opposite sides of the equator. The zones of maximum stress are still located at the poles. With a further increase in the parameter δ the areas of greatest stress gradually shift closer to the equator, tending to take positions in accordance with Fig. 22.3. Elastic deformations caused by the rotation of the body around a deflected axis lead to the fact that the zones of greatest stress are located in different hemispheres: one in the northern, the other in the southern.

Let us take into account the rotation of the dumbbell relative to the shell with a constant angular velocity Ω. Over time, the migration of zones of the highest radial stresses will occur, and while maintaining the orientation of the axis of rotation of the shell, one zone always remains above the equator, and the other below. The period of movement of these zones is equal to the time of full rotation of the dumbbell relative to the external body. In the case of a small influence of elastic deformations of the shell caused by angular movements, the areas of greatest stress move along the equator. The periodicity of this process is characterized by the half-cycle of the relative rotation of the dumbbell.

22.4 Conclusions

This work proposes a simple model that reflects some features of the internal structure of the Earth, demonstrating the periodicity of movement of zones of greatest stress in the body. The ellipsoidality of the body located in the internal cavity of the shell and its advanced rotation relative to the shell are taken into account. The stress distribution inside the shell was calculated depending on the influence of the attractive force from the ellipsoidal body and the deformations of the shell associated with its angular displacements. It is shown that the relative rotation of the internal body leads to periodic movement of the zones of highest stress near the surface of the shell, and the periodicity of this process is characterized by a half-cycle of the relative rotation of the internal body. The results of the work may be useful for understanding the mechanism of the seismic wave process.

Acknowledgements The work was carried out with the financial support of the Ministry of Science and Higher Education of the Russian Federation (Project No. FSWR-2023-0034).

References

1. Vikulin AV (1990) Phenomenological wave model of the seismic process. Dokl Acad Sci USSR 310(4):821–824
2. Vikulin AV (2003) Physics of wave seismic process. Publishing house of KSPU, Petropavlovsk-Kamchatsky, p 151
3. Song X, Richards RG (1996) Nature 382(6588):221
4. Ovchinnikov VM, Adushkin VV, An VA (1998) On the speed of relative rotation of the Earth's inner core. Dokl. RAS 362(5):683–686
5. Denisov GG, Novikov VV, Fedorov AE (2005) On long-period variations in the angular velocity of the Earth's rotation. Dokl. RAS 400(5):625–629
6. Denisov GG, Novikov VV, Fedorov AE (2013) How the solid core of the Earth changes the day. Nature 5:3–10
7. Denisov GG, Novikov VV (1983) On the free movements of a deformable rigid body close to a ball. Proc USSR Acad Sci. Mech Solids (3):43–50
8. Novikov VV, Fevralskikh LN (2020) Application of the apparatus of spherical vectors in problems of mechanics. Probl Strength Plasticity 82(3):317–327

Chapter 23
Analysis of Indentation Results Taking into Account the Plasticity of the Material

Ivan A. Panfilov, Sergei M. Aizikovich, and Alexander N. Litvinenko

Abstract Nanoindentation is a widely used method to control and determine the elastic properties of materials. For interpretation of the nanoindentation test results, one should solve the problem of constructing an adequate mathematical model. In most cases, analytical formulas obtained using a mathematical model of the foundation in the form of an elastic linearly deformable half-space are used. When formulating the problem, it is assumed that a non-deformable punch is indented in a homogeneous elastic half-space. The numerical formulation of the problem makes it possible to obtain and use a numerical solution obtained taking into account the complete plastic nonlinear behavior of the material. In the present work, we study contact problems on the indentation of a spherical punch into an elastoplastic homogeneous half-space. To verify the numerical solution, a comparison with known analytical solutions was carried out. Issues of convergence and tuning of numerical methods, the influence of plasticity and the applicability of analytical solutions are explored. Problems are solved numerically using the finite element method in the Ansys Mechanical software package.

Keywords Continuous contact · Contact mechanic · Contact problem · Indentation · Conical indenter · Spherical indenter · Finite element method

Nanoindentation is used for non-destructive testing of materials and obtaining their mechanical characteristics [1, 2]: hardness, elastic properties of bulk materials and coatings, etc. The method consists of pressing a rigid indenter (usually diamond or hard alloys) into the surface of the test sample and obtaining depth diagrams depending on the magnitude of the force on the nanometer scale, analyzing which the elastic properties are determined using the unloading segment of the diagram. The Field–Swain (spherical indenter) [3] and Oliver–Pharr (Berkovich indenter) [4]

I. A. Panfilov (✉) · S. M. Aizikovich · A. N. Litvinenko
Don State Technical University, 1 Gagarin Square, 344000 Rostov-On-Don, Russia
e-mail: panfilov.i@gs.donstu.ru

H. Altenbach and V. A. Eremeyev (eds.), *Dynamics of Discrete and Continuum Structures and Media*, Advanced Structured Materials 221,
https://doi.org/10.1007/978-3-031-75626-9_23

Table 23.1 Physical properties of materials

No.	Material	Young's modulus, GPa	Yield strength, GPa	Tangent modulus of elasticity for bilinear material, GPa
1	Diamond	1000	–	–
3	Aluminum	70	0.28	0.5

methods are based on solutions to contact problems of elasticity theory for spherical and parabolic indenters [5, 6]. The fundamental principles of contact interaction were laid in [1, 6, 7]; analytical dependencies for vertical force, displacement, contact radius, and pressure in the contact zone were obtained for a conical and spherical indenter. At the same time, the theory was based on hypotheses about small deformations and elastic behavior of the material.

Articles [8–10] are related to the assessment of the elastic properties of functionally graded coatings. In [11–13], the problems of indentation of a spherical indenter into an elastic half-space with a coating are considered, and analytical solutions are obtained using the asymptotic method. In [14–16], continuous and discontinuous contact problems for a functionally graded material are studied. Nanoindentation is widely used for the characterization of the mechanical properties of biological [17–21], biocompatible materials [22, 23] as well as organic films [24] for bioengineering applications.

In [25–31], approximate analytical solutions of axisymmetric contact problems of indentation of a spherical, conical, and cylindrical punch into an elastic half-space with a functionally graded coating were constructed. The possibility of using the developed models of indentation of bodies with coatings to interpret the nanoindentation of thin coatings was studied [32, 33].

The hypothesis of small indenter movements used in the presented works imposes serious restrictions on the use of analytical applied formulas, since even a small indenter displacement (less than 1%) causes the development of plastic deformations. The indentation force and contact area differ significantly from the analytical results [34].

This paper considers the problem of a spherical punch indentation into an elasto-plastic half-space in a static axisymmetric formulation. To implement plastic deformation, a bilinear material model with an elastic modulus (Young's modulus) and a tangential elastic modulus is used Table 23.1. The indenter is made of elastic diamond. The half-space is made of aluminum.

The problem is solved numerically using a finite element method in the Ansys software package. To verify the numerical calculations and compare them with analytical results, additional calculations were carried out on the rigid punch indentation into the elastic space of aluminum. The geometric parameters of the problem are shown in Table 23.2.

Contact problems are nonlinear problems due to the changing status of the contact and the stiffness matrix, and require special attention to the accuracy and convergence

Table 23.2 The geometric parameters of the problem

No.	Parameter	Values
1	Linear size of the half-space, mm	1
3	Sphere radius, mm	0.1
4	Cone opening angle, °	120
5	Indentation depth, mm	0.1

Table 23.3 Dimension of the finite element mesh

No.	Parameter	Values
1	Mesh size in the contact area, mm	0.01
2	Mesh size outside the contact zone, mm	0.02

of the solution. The settings of the numerical methods used in the Ansys finite element analysis package and applied in this calculation are listed below.

To implement the contact problem, the "Augmented Lagrange" contact algorithm is used. This is a modified contact algorithm of the common "Pure Penalty" method ("penalty function method"), characterized by the presence of an additional term λ in the expression of the contact force:

$$F_n = k_n \cdot x_p + \lambda, \tag{23.1}$$

where F_n—normal (contact) force, k_n—contact stiffness, and x_p—indentation depth. Contact stiffness, k_n, typically has a major impact on accuracy and convergence of the solution. A large value of stiffness provides high accuracy, but degrades convergence and vice versa. The use of the additional term λ allows one to reduce the sensitivity of the algorithm to contact stiffness k_n, and allows one to obtain acceptable results at a value $k_n = 1$, but also requires more iterations.

To "recognize" contacts, the "Gauss point detection" method is used, in which additional points are added on the edges of the elements. To improve convergence, it is possible to use "Normal from Contact" or "Normal to Target" recognition with an increase in the number of calculations. It is also important to mention about the shape and dimension of the elements of the finite element mesh. In this problem, to construct a finite element mesh, an 8-node element PLANE183 is used—a high-order element with intermediate nodes. In the area of contacts, the meshes are refined. The mesh dimensions for the half-space are presented in Table 23.3; for indenters, a larger mesh is used. In this case, the reference parameters, relative to which the mesh dimension and linear dimensions of the half-space should be set, are the contact area (implementation depth) and the coating thickness. Thus, in terms of the used partition, for example, in the elastic problem provides about 30 elements in the contact area for a spherical indenter.

For the problem of a spherical punch indentation, refinement of the mesh will obviously lead to more accurate results, and the problem will converge. For contact surfaces, elements such as CONTA172 and TARGE169 are used with automatic recognition and limitation of the contact area. Moreover, since our indenter stiffness is several times greater than the stiffness of the coating and substrate, the TARGE169 elements must be applied specifically to the indenter.

Ansys also allows to take into account geometric nonlinearity at large deformations by including a nonlinear strain tensor in the calculation $\boldsymbol{\varepsilon}(x, y, z)$ with respect to derived displacements $\boldsymbol{u}'(x, y, z)$ («Large deflection»):

$$\varepsilon_{ij} = \frac{1}{2}\left(\frac{\partial u_i}{\partial x_j} + \frac{\partial u_j}{\partial x_i} + \frac{\partial u_i}{\partial x_j} \cdot \frac{\partial u_j}{\partial x_i}\right). \tag{23.2}$$

Thus, in contrast to analytical theories, where only contact nonlinearity is taken into account under the hypothesis of small deformations, we can numerically take into account all three types of nonlinearity, namely, contact, geometric (large deformations), and physical (plasticity) ones.

23.1 Numerical Results

We assume that the punch, under the action of the applied load, moves vertically by 0.1 mm. Solving the problem, we find the force fields of deformation and stresses. The accuracy of the iterative algorithm was set according to the strength $1 \cdot 10^{-4}$ N, according to deformations—$1 \cdot 10^{-5}$ mm.

Analytical results for comparison were taken from the well-known formulas (23.1)–(23.7) for the implementation of rigid punches [1, 5–7]:

$$a_s = (Rd)^{\frac{1}{2}} \tag{23.3}$$

$$a_c = \frac{2d}{\pi \cdot \tan(\varphi)}, \tag{23.4}$$

$$F_s = \frac{4}{3} E^* R^{\frac{1}{2}} d^{\frac{3}{2}}, \tag{23.5}$$

$$F_c = \frac{2}{\pi} E \cdot \frac{d^2}{\tan(\varphi)}, \tag{23.6}$$

$$p_s = p_0\left(1 - \frac{r^2}{a^2}\right)^{\frac{1}{2}}, p_0 = \frac{2}{\pi} E^* \left(\frac{d}{R}\right)^{\frac{1}{2}}, \tag{23.7}$$

$$p_c = \frac{Ed}{(1 - \nu^2)\pi a} \ln\left(\frac{a}{r} - \left(\left(\frac{a}{r}\right)^2 - 1\right)^{\frac{1}{2}}\right), \tag{23.8}$$

$$E^* = \frac{E}{1 - \nu^2}. \tag{23.9}$$

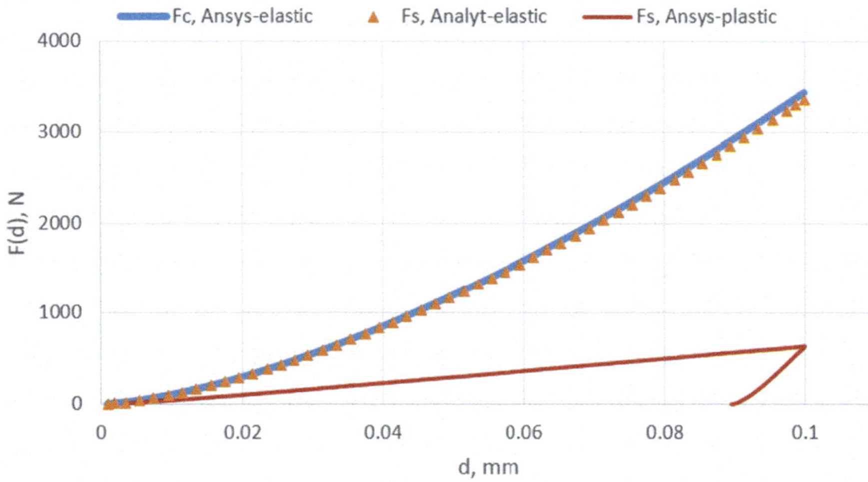

Fig. 23.1 Dependence of vertical force on displacement for a spherical indenter, mm

To take into account the stiffness of a diamond spherical indenter in expressions (23.3)–(23.9) for effective stiffness E^*, the following formula can be used:

$$\frac{1}{E^*} = \frac{1 - \nu_2^2}{E_1} + \frac{1 - \nu_2^2}{E_2}. \tag{23.10}$$

Here a—contact radius, d—punch penetration depth, φ—angle between the horizontal and lateral planes of the cone, r—vertical coordinate, E_1, E_2 and ν_1, ν_2—Young's modulus and Poisson's ratios of the indenter and half-space, respectively, F—vertical force, p—pressure in the contact area. The indices s and c here and below in the diagrams indicate a sphere and a cone, respectively.

Figures 23.1, 23.2, 23.3, 23.4, 23.5, and 23.6 demonstrate the results for indentation of a spherical punch into an elastic half-space made of aluminum.

23.2 Conclusions

For the elastic problem, the numerical method demonstrated almost complete agreement with the analytical results. The error increased with increasing deformation and for maximum values was about 2% for the vertical force. It is worth noting that these loads for the elastic problem significantly exceed the permissible ones in terms of the yield strength and were considered as model problems for comparing models. Within the limits of elastic deformations, the error in the numerical results was less than 0.1%.

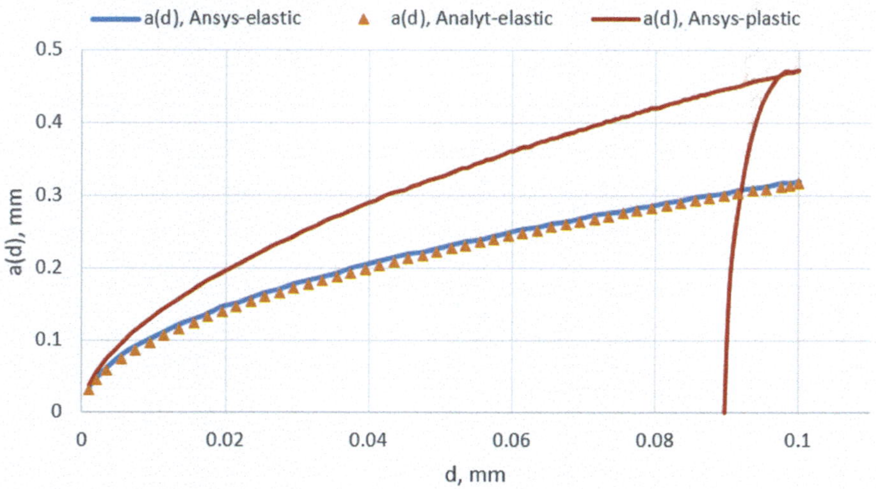

Fig. 23.2 Dependence of contact radius on displacement for a spherical indenter, mm

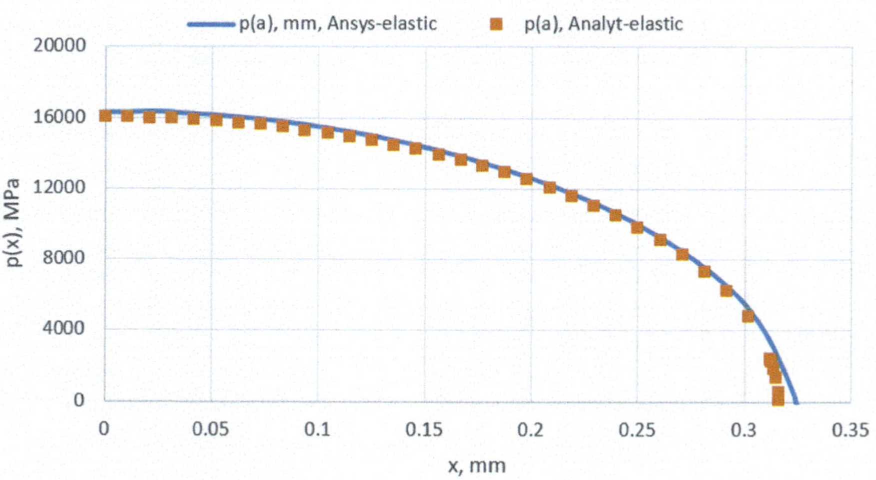

Fig. 23.3 Contact pressure in the contact area for a displacement of 0.1 mm for a spherical indenter, mm

When solving an elastoplastic problem numerically, we see a linear unloading segment, which is used to find Young's modulus using non-destructive testing methods (nanoindentation). The contact area for the elastic model and the elastoplastic model is significantly different. During plasticity, the material is "squeezed out" from under the indenters and significantly increases the contact zone (Figs. 23.5 and 23.6). For a spherical indenter for maximum displacement, the difference in contact radius was about 30%.

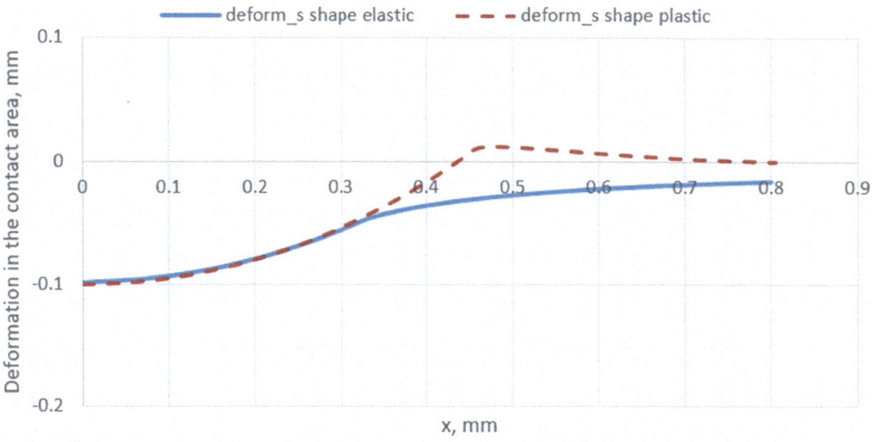

Fig. 23.4 Deformation in the contact area for a spherical indenter, mm

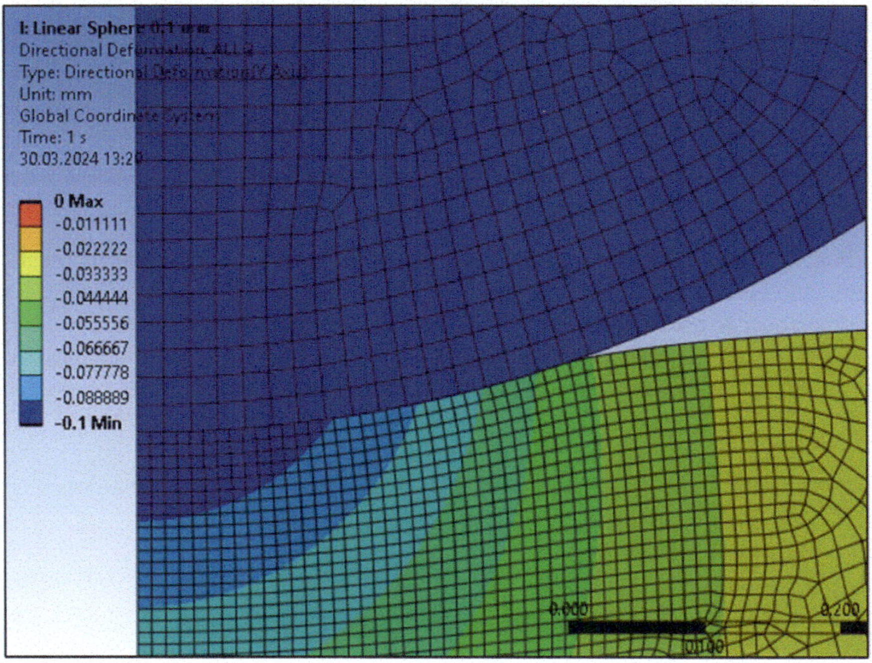

Fig. 23.5 Deformation field in the contact area for a spherical indenter for an elastic model

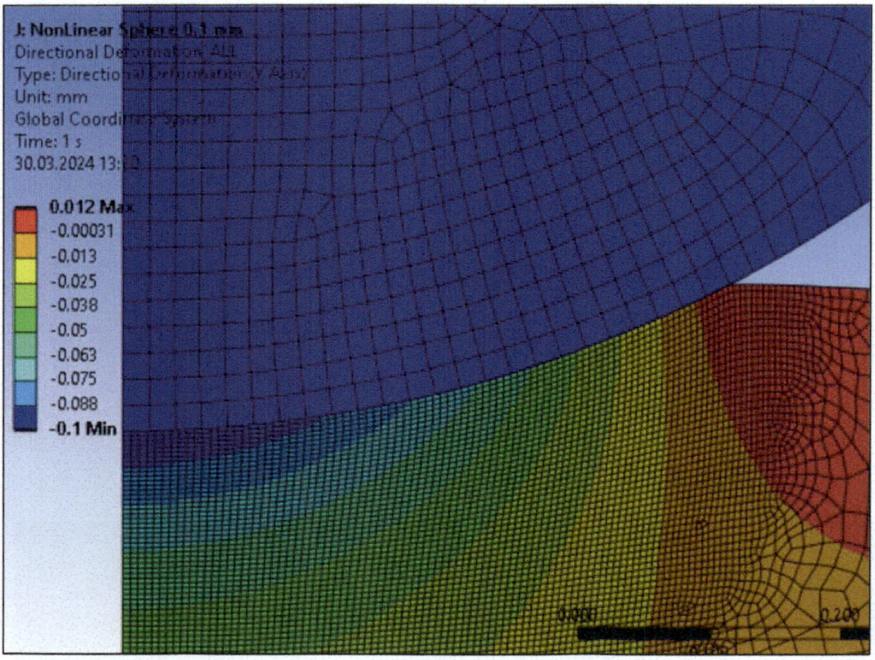

Fig. 23.6 Deformation field in the contact area for a spherical indenter for an elastic–plastic model

In general, the use of an elastic model can serve as a model problem for verification and calibration of numerical methods. However, for large deformations, it is recommended to use an elastoplastic deformation model, as well as a nonlinear strain tensor model (large deformation model). The study was carried out for the purpose of further use in assessing the accuracy of models used in identifying the properties of graded, multilayer, and composite materials with coatings.

Acknowledgements This research was supported by the Russian Science Foundation, grant number 22-49-08014, https://rscf.ru/en/project/22-49-08014/.

References

1. Bulychev SI, Alekhin VP (1990) Testing of materials by continuous indentation of an indenter. Moscow: Mech Eng 224 (in Russian)
2. Golovin YI (2009) Nanoindentation and its capabilities: a practical guide. Moscow: Mech Eng 312 (in Russian). ISBN 978-5-94275-476-1
3. Field JS, Swain MV (1993) A simple predictive model for spherical indentation. J Mater Res 8:297–306. https://doi.org/10.1557/JMR.1993.0297

4. Pharr GM, Oliver WC (1992) An improved technique for determining hardness and elastic modulus using load and displacement sensing indentation experiments //. J Mater Res 7(6):1564–1583
5. Hertz H (1881) Ueber die Beruhrung Fester Elastischer Korper. J Reine u Angew Math 92:156–171
6. Johnson KL (1987) Contact mechanics. Cambridge University Press
7. Galin LA (1953) Contact problems of the theory of elasticity. M.: GITTL 264
8. Aizikovich SM, Alexandrov VM, Kalker JJ, Krenev LI, Trubchik IS (2002) Analytical solution of the spherical indentation problem for a half-space with gradients with the depth elastic properties. Int J Solids Struct 39(10):2745–2772
9. Aizikovich SM, Vasiliev AS, Seleznev NM (2010) Inverse analysis for evaluation of the shear modulus of inhomogeneous media by torsion experiments. Int J Eng Sci 48:936–942. https://doi.org/10.1016/j.ijengsci.2010.05.013
10. Aizikovich SM, Krenev LI, Sevastianov IB, Trubchik IS, Evich LN (2011) Evaluation of the elastic properties of a functionally-graded coating from the indentation measurements. ZAMM Z Angew Math Mech 91(6):493–515. https://doi.org/10.1002/zamm.201000179
11. He LH, Swain MV (2007) Nanoindentation derived stress–strain properties of dental materials. Dental Mater 23(7):4–821. ISSN 0109-5641, https://doi.org/10.1016/j.dental.2006.06.017
12. Potelezhko VP, Filippov AP (1967) The contact problem for a slab lying on an elastic base. Soviet Appl Mech 3(1):49–51
13. Landau LD, Lifshts EM (1987) Theoretical physics. T. VII. Theory of elasticity: textbook, 4th edn. Spanish and additional - M.: Nauka, p 248 (in Russian)
14. Kral ER, Komvopoulos K, Bogy DB (1993) Elastic-plastic finite element analysis of repeated indentation of a half-space by a rigid sphere. ASME J Appl Mech 60:829–841
15. Vorovich II, Aleksandrov VM (ed) (2001) Mechanics of contact interactions M. FIZMATLIT, p 672 (in Russian)
16. Dub SN (1998) Curves of elasto-plastic deformation of thin coatings obtained in depth-sensing indentation experiments. MRS Symposium Proc 505:223–228
17. Zelentsov VB, Sadyrin EV, Mitrin BI, Swain MV (2023) Mathematical tools for recovery of the load on the fissure according to the micro-CT results. J Mech Behav Biomed Mater 138:105625. https://doi.org/10.1016/j.jmbbm.2022.105625
18. Sadyrin EV (2022) Correlating the mechanical properties to the mineral density of brown spot lesion in dentine using nanoindentation and X-ray micro-tomography. Adv Mater Modelling Mech, Med Biol Appl 389–398.https://doi.org/10.1007/978-3-030-81705-3_21
19. Sadyrin EV, Mitrin BI, Yogina DV, Swain MV (2021) Preliminary study of distribution of mechanical properties and mineral density by depth of liquid saturated carious dentine. IOP Conf Ser: Mater Sci Eng 1029(1):012056. https://doi.org/10.1088/1757-899X/1029/1/012056
20. Sadyrin EV, Yogina DV, Swain MV, Maksyukov SY, Vasiliev AS (2021) Efficacy of dental materials in terms of apparent mineral density restoration: composite resin, glass ionomer cement and infiltrant. Compos Part C: Open Access 6:100192. https://doi.org/10.1016/j.jcomc.2021.100192
21. Kislyakov EA, Karotkiyan RV, Sadyrin EV, Mitrin BI, Yogina DV, Kheygetyan AV, Maksyukov SY (2020) Nanoindentation derived mechanical properties of human enamel and dentine subjected to etching with different concentrations of citric acid. In: Modeling, synthesis and fracture of advanced materials for industrial and medical applications, pp 75–83. https://doi.org/10.1007/978-3-030-48161-2_5
22. Mitrin BI, Chapek SV, Sadyrin EV, Swain MV (2021) Mechanical properties and failure mechanisms of 3D-printed PLA scaffolds: a preliminary study. IOP Conf Ser: Mater Sci Eng 1029(1):012074. https://doi.org/10.1088/1757-899X/1029/1/012074
23. Krenev LI, Volkov SS, Sadyrin EV, Zubar' TI, Chizhik SA (2018) Mechanical material tests by the nanoindentation method at various indenter and specimen temperatures. J Eng Phys Thermophys 91:594–600. https://doi.org/10.1007/s10891-018-1780-5
24. Burlakova VE, Tyurin AI, Drogan EG, Sadyrin EV, Pirozhkova TS, Novikova AA, Belikova MA (2019) Mechanical properties and size effects of self-organized film. J Tribol 141(5):051601. https://doi.org/10.1115/1.4042678

25. El-Sherbiney MGD, Halling J (1996) The Hertzian contact of surfaces covered with metallic films. Wear 40(3):325–337
26. Pharr GM, Oliver WC (1992) Measurement of thin film mechanical properties using nanoindentation. MRS Bull 28–33
27. Aizikovich SM (1982) Asymptotic solutions of contact problems of elasticity theory for media inhomogeneous in depth. J Appl Math Mech 46(1):116–124
28. Vasiliev AS, Volkov SS, Aizikovich SM (2018) Approximated analytical solution of a problem on indentation of an electro-elastic half-space with inhomogeneous coating by a conductive punch. Doklady Phys—Pleiades Publishing 63:18–22
29. Volkov SS, Vasiliev AS, Aizikovich SM, Seleznev NM, Leontieva AV (2016) Stress-strain state of an elastic soft functionally-graded coating subjected to indentation by a spherical punch. PNRPU Mech Bull 4:20–34
30. Vasiliev AS, et al (2018) Indentation of an elastic half-space reinforced with a functionally graded interlayer by a conical punch. https://doi.org/10.18720/MPM.4022018_14
31. Vasiliev A, Volkov S, Aizikovich SM (2016) Indentation of an axisymmetric punch into an elastic transversely-isotropic half-space with functionally graded transversely-isotropic coating 28:11–15
32. Sadyrin E, Vasiliev A, Volkov S (2020) Mathematical modeling of experiment on berkovich nanoindentation of ZrN coating on steel substrate. Acta Polytechnica CTU Proc 27:18–21. https://doi.org/10.14311/APP.2020.27.0018
33. Volkov SS, Vasiliev AS, Sadyrin EV (2018) Contact problem on indentation of an elastic half-plane with functionally-graded coating in presence of tangential stresses on the surface. MATEC Web Conf 226:03018. https://doi.org/10.1051/matecconf/201822603018
34. Ovcharenko A, Halperin G, Verberne G et al (2007) In situ investigation of the contact area in elastic–plastic spherical contact during loading–unloading. Tribol Lett 25:153–160. https://doi.org/10.1007/s11249-006-9164-y

Chapter 24
On Solving Nonlinear Klein–Fock–Gordon Equations with First Derivatives

Yuri V. Pavlov

Abstract A method for solving a nonlinear wave equation with first derivatives and arbitrary nonlinearity on the right side is proposed. Examples of exact solutions for an autonomous equation with different nonlinearities are given.

Keywords Nonlinear wave equations · Functionally invariant solutions · Generalized sine-Gordon equation

24.1 Introduction

In this work we consider methods for solving the equations

$$\Box U + \sum_{i=0}^{n} \lambda_i \frac{\partial U}{\partial x_i} = F(U) \tag{24.1}$$

and

$$\Box U + \sum_{i=0}^{n} \lambda_i \frac{\partial U}{\partial x_i} = p(\mathbf{x}, t)\, F(U), \tag{24.2}$$

where

$$\Box U = \sum_{i=1}^{n} \frac{\partial^2 U}{\partial x_i^2} - \frac{1}{c^2} \frac{\partial^2 U}{\partial t^2}, \tag{24.3}$$

c is constant (the velocity of wave), $x_0 = ct$ and $\lambda_0, \lambda_1, \ldots, \lambda_n$ are constants, $F(U)$ is some given (in general nonlinear) function of U, and $p(\mathbf{x}, t)$ is some function of variables $\mathbf{x} = (x_1, \ldots, x_n)$ and t.

Y. V. Pavlov (✉)
Institute for Problems in Mechanical Engineering of Russian Academy of Sciences, 61 Bol'shoy, V.O., Saint Petersburg 199178, Russian Federation
e-mail: yuri.pavlov@mail.ru

H. Altenbach and V. A. Eremeyev (eds.), *Dynamics of Discrete and Continuum Structures and Media*, Advanced Structured Materials 221,
https://doi.org/10.1007/978-3-031-75626-9_24

Equation (24.1) with $F(U) = m^2U$ was introduced in the works [1–3]. It describes the scalar field of mass m and is widely known in theoretical physics as the Klein–Fock–Gordon equation. Therefore we will call Eqs. (24.1) and (24.2) autonomous and nonautonomous generalized nonlinear Klein–Fock–Gordon equations.

In the absence of the first derivatives Eq. (24.1) with $F(U) = \sin U$ is the well-known sine-Gordon equation. This equation and its generalizations often appear in physics and mechanics. The sine-Gordon equation describes dislocations in solids [4, 5], deformation of a crystal media with complex lattice [6–8], nonlinear wave processes [9]. It is used in field theory models and in elementary particle physics [10]. In the standard model of particle physics, Eq. (24.1) without first derivatives and with function $F(U) = p_1U^3 - p_2U$ is used to describe the Higgs mechanism of symmetry breaking [11]. Equations with first derivatives and variable amplitudes $p(\mathbf{x}, t)$ appear in the theory of nematic liquid crystals [12, 13]. Methods for solving nonlinear equations with variable amplitude $p(\mathbf{x}, t)$ are known only for a few special cases (see [14], for example).

Various efficient methods have been developed for solving the nonlinear equations: the Bäcklund transformation, the inverse scattering method [15–17], etc. The intensive use of nonlinear models in various fields of mechanics and physics requires the development of new methods for solving such equations.

Note that the first derivatives can be eliminated in Eqs. (24.1) and (24.2) using the substitution

$$U = \tilde{U} \exp \frac{1}{2} \left[\lambda_0 x_0 - \sum_{i=1}^{n} \lambda_i x_i \right]. \tag{24.4}$$

Equation (24.2) is transformed to the form

$$\Box \tilde{U} = \frac{1}{4} \left(\sum_{i=1}^{n} \lambda_i^2 - \lambda_0^2 \right) \tilde{U} + \\ p(\mathbf{x}, t) \exp \frac{1}{2} \left(\sum_{i=1}^{n} \lambda_i x_i - \lambda_0 x_0 \right) F \left[\tilde{U} \exp \frac{1}{2} \left(\lambda_0 x_0 - \sum_{i=1}^{n} \lambda_i x_i \right) \right]. \tag{24.5}$$

Thus, in the case of a nonlinear function $F(U)$, Eqs. (24.1) and (24.2) are not simplified. Next, we will consider a method for solving Eqs. (24.1) and (24.2), and therefore Eq. (24.5).

24.2 Solving the Nonlinear Equation

Let's look for a solution to Eq. (24.3) in the form $U = f[\Omega(\mathbf{x}, t)]$:

$$f''(\Omega)\left[\sum_{i=1}^{n}\left(\frac{\partial\Omega}{\partial x_i}\right)^2 - \left(\frac{\partial\Omega}{c\partial t}\right)^2\right] + f'(\Omega)\left[\Box\Omega + \sum_{i=0}^{n}\lambda_i\frac{\partial\Omega}{\partial x_i}\right] = p(\mathbf{x}, t)F[f(\Omega)]. \tag{24.6}$$

Equation (24.6) will be solved if

$$f''(\Omega)\,P(\Omega) + f'(\Omega)\,Q(\Omega) = F[f(\Omega)], \tag{24.7}$$

$$\sum_{i=1}^{n}\left(\frac{\partial\Omega}{\partial x_i}\right)^2 - \left(\frac{\partial\Omega}{c\partial t}\right)^2 = p(\mathbf{x}, t)P(\Omega), \tag{24.8}$$

$$\Box\Omega + \sum_{i=0}^{n}\lambda_i\frac{\partial\Omega}{\partial x_i} = p(\mathbf{x}, t)Q(\Omega), \tag{24.9}$$

where $P(\Omega)$, $Q(\Omega)$ are some arbitrary functions.

In the case $\sum_{i=1}^{n}\lambda_i^2 \neq 0$, let's choose $P(\Omega) = \Omega^2$, $Q(\Omega) = \Omega$ and enter the substitution

$$\Omega = e^{\phi}\omega, \quad \phi = \sum_{i=1}^{n} a_i x_i - a_0 x_0. \tag{24.10}$$

Equation (24.6) will be solved if

$$f''(\Omega)\,\Omega^2 + f'(\Omega)\,\Omega = F[f(\Omega)], \tag{24.11}$$

$$\sum_{i=1}^{n} a_i^2 = 1 + a_0^2, \tag{24.12}$$

$$\sum_{i=1}^{n} a_i\lambda_i = a_0\lambda_0, \tag{24.13}$$

$$\sum_{i=0}^{n}\lambda_i\frac{\partial\omega}{\partial x_i} = 0, \tag{24.14}$$

$$\sum_{i=0}^{n} a_i \frac{\partial \omega}{\partial x_i} = 0, \tag{24.15}$$

$$\Box\, \omega = (p(\mathbf{x}, t) - 1)\, \omega, \tag{24.16}$$

$$\sum_{i=1}^{n} \left(\frac{\partial \omega}{\partial x_i} \right)^2 - \left(\frac{\partial \omega}{c \partial t} \right)^2 = (p(\mathbf{x}, t) - 1)\, \omega^2. \tag{24.17}$$

In case of Eqs. (24.1), (24.16), and (24.17) transform to

$$\Box\, \omega = 0, \tag{24.18}$$

$$\sum_{i=1}^{n} \left(\frac{\partial \omega}{\partial x_i} \right)^2 - \left(\frac{\partial \omega}{c \partial t} \right)^2 = 0. \tag{24.19}$$

Equations (24.18) and (24.19) coincide with the system of equations defining functionally invariant solutions of the wave equation.

The solution $u(x_1, \dots x_n, t)$ is called functionally invariant if for arbitrary (twice) differentiable function f, the function $f[u(x_1, \dots x_n, t)]$ will also be the solution. D'Alembert's solution for one-dimensional wave equation is obviously functionally invariant. A number of functionally invariant solutions of the wave equation and electromagnetic field equations were obtained in [18]. V. I. Smirnov and S. L. Sobolev applied the functionally invariant solutions of wave equation for solving specific problems of mathematical physics [19]. In the three-dimensional case, all real and some part of complex functionally invariant solutions of the wave equation can be obtained from the expression [20]

$$\sum_{i=1}^{n} x_i\, l_i(\Omega) - t v^2 p(\Omega) + q(\Omega) = 0, \tag{24.20}$$

where $l_1(\Omega), \dots, l_n(\Omega), p(\Omega), q(\Omega)$ are the arbitrary twice differentiable functions satisfying the condition

$$\sum_{i=1}^{n} l_i^2(\Omega) = v^2 p^2(\Omega). \tag{24.21}$$

In the two-dimensional case ($n = 2$), such formulas give all functionally invariant solutions of the wave equation [21].

The functionally invariant solutions to Eq. (24.1) cannot exist for the case of nonlinear function $F(U)$. However, after the series of transformations the search for some solutions to the sine-Gordon equation and its generalizations has been reduced in [22, 23] to solving the auxiliary system of partial differential equations, which corresponds to the system of equations for functionally invariant solutions of the wave equation.

If $p(\mathbf{x}, t) = 1$ then the system of Eqs. (24.14), (24.15) (24.20), (24.21) for the function ω is such that $f(\omega)$ will also be a solution to these equations for any twice differentiable function f. So the solution of Eq. (24.1) can be written in the form

$$U(\mathbf{x}, t) = f\left[e^{\phi(\mathbf{x},t)}\psi(\omega(\mathbf{x}, t))\right] \tag{24.22}$$

with arbitrary twice differentiable function $\psi(\omega)$.

Denote that some analogue of the superposition principle takes place. The solution can be represented in the form

$$U(\mathbf{x}, t) = f\left[e^{\phi(\mathbf{x},t)}[\psi_1(\omega(\mathbf{x}, t)) + \ldots + \psi_k(\omega(\mathbf{x}, t))]\right], \tag{24.23}$$

where $\psi_1, \ldots, \psi_k$ are arbitrary double differentiable functions, and k is arbitrary integer number.

Equation (24.11) can be transformed, using substitution $z = \ln|\Omega|$, to the form

$$\frac{d^2 f}{dz^2} = F(f) \tag{24.24}$$

and after integration it takes the form

$$\int \frac{df}{\sqrt{V(f) + E}} = \sqrt{2}\ln\left(C_1|\Omega|^{\pm 1}\right), \tag{24.25}$$

where $V'(f) = F(f)$, and E and C_1 are constants. If the integral can be inverted, then the solution $U = f(\Omega)$ can be expressed in explicit form. Let's look at some examples.

$$\textbf{(1)}\ F(f) = \sin f,\quad V(f) = -\cos f,\quad E = 1 \ \Rightarrow\ U = 4\arctan C_1\Omega^{\pm 1}. \tag{24.26}$$

$$\textbf{(2)}\ F(f) = \sin f,\ V(f) = -\cos f,\ E > 1 \Rightarrow U = 2\,\mathrm{am}\left(\frac{1}{k}\ln C_1|\Omega|^{\pm 1}, k\right) + \pi, \tag{24.27}$$

where $\mathrm{am}\,(x, k)$ is Jacobi amplitude function, $k = \sqrt{2/(E+1)}$.

$$\textbf{(3)}\ \ F(f) = \sin f,\ \ V(f) = -\cos f,\ \ 1 > E > 0\ \Rightarrow$$

$$U = 2\arcsin\left[\sqrt{\frac{E+1}{2}}\,\mathrm{sn}\left(\ln C_1|\Omega|^{\pm 1}, \sqrt{\frac{E+1}{2}}\right)\right] + \pi, \tag{24.28}$$

where $\mathrm{sn}\,(x, k)$ is the Jacobi elliptic sine. Graphs of these solutions are presented in Fig. 24.1.

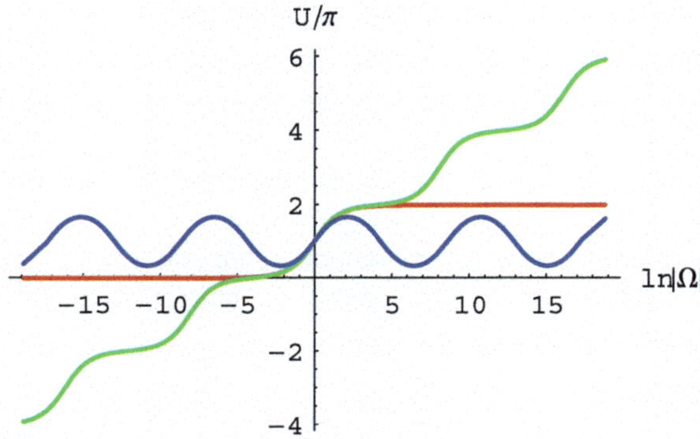

Fig. 24.1 Solutions U for the case (24.26) $E = 1$ (red line), the case (24.27) if $E = 1.01$ (green line), and the case (24.28) if $E = 0.5$ ($C_1 = 1$) (blue line)

$$(\mathbf{4})\ F(f) = \sinh f,\quad V(f) = \cosh f,\quad E = -1 \quad\Rightarrow\quad U = 4\,\mathrm{arc\,tanh}\, C_1\Omega^{\pm 1}. \tag{24.29}$$

$$(\mathbf{5})\ F(f) = \sinh f,\ V(f) = \cosh f,\ E = 0 \Rightarrow U = \mathrm{arc\,cosh\,cn}^{-2}\left(\ln C_1\Omega^{\pm 1}, \frac{1}{\sqrt{2}}\right), \tag{24.30}$$

where $\mathrm{cn}(x, a)$ is the Jacobi elliptic cosine.

$$(\mathbf{6})\ F(f) = \sinh f,\quad V(f) = \cosh f,\quad E = 1 \quad\Rightarrow\quad U = 2\log\tan\frac{\log(C_1\Omega^{\pm 1})}{2}. \tag{24.31}$$

$$(\mathbf{7})\ F(f) = \sinh f,\quad V(f) = \cosh f,\quad E > 1 \quad\Rightarrow$$

$$U = 2\,\mathrm{arc\,tanh\,sn}\left(\sqrt{\frac{E+1}{2}}\ln(C_1\Omega)^{\pm 1}, \sqrt{\frac{E-1}{E+1}}\right). \tag{24.32}$$

$$(\mathbf{8})\ \ F(f) = \frac{\sinh f}{\cosh^3 f},\quad V(f) = \frac{-1}{2\cosh^2 f},\quad E = \frac{1}{2} \quad\Rightarrow\quad U = \mathrm{arcsinh}\left(C_1\Omega^{\pm 1}\right). \tag{24.33}$$

$$(\mathbf{9})\ \ F(f) = \frac{\sinh f}{\cosh^3 f},\quad V(f) = \frac{-1}{2\cosh^2 f},\quad E > \frac{1}{2} \quad\Rightarrow$$

$$U = \mathrm{arc\,sinh}\left[C_1\Omega^{\pm\sqrt{2E}} + \frac{1-2E}{8EC_1}\Omega^{\mp\sqrt{2E}}\right]. \tag{24.34}$$

$$(\mathbf{10})\ \ F(f) = \frac{-\varphi''(f)}{(\varphi'(f))^3},\quad V = \frac{1}{2\,(\varphi'(f))^2},\quad E = 0 \ \Rightarrow\ U = \varphi^{-1}\left(\ln\left(C_1|\Omega|^{\pm 1}\right)\right), \tag{24.35}$$

where φ^{-1} is the inverse function for φ.

(11) $F(f) = \dfrac{-\sinh f}{\cosh^3 f}, \quad V(f) = \dfrac{1}{2\cosh^2 f}, \quad E > 0 \;\Rightarrow$

$$U = \operatorname{arcsinh}\left[\sqrt{\frac{1+2E}{2E}}\sinh\left(\sqrt{2E}\ln\left(C_1|\Omega|^{\pm 1}\right)\right)\right]. \tag{24.36}$$

The proposed replacement (24.10) does not allow to find real solutions depending on time in Eq. (24.1). Then from (24.14) and (24.17), we obtain $\omega = \text{const}$ and $a_0 = 0$. To obtain real solutions depending on time and space coordinates, in the case where there are only first partial time derivative, we can choose functions

$$P(\Omega) = 0, \quad Q(\Omega) = \Omega \tag{24.37}$$

instead of (24.10). Then from Eq. (24.7) one obtains

$$\int \frac{dF}{V'(F)} = \ln \Omega. \tag{24.38}$$

For the cases $V'(\Omega) = \sin \Omega$ $\;(V'(\Omega) = \sinh \Omega)$, we obtain in particular $F = 2\arctan \Omega$ $\;(F = 2\,\text{arc}\tanh \Omega)$.

The solution of (24.8), (24.9) (with $p = 1$) can be found in the form

$$\Omega = C_1 \exp\left(\sum_{i=0}^{n} a_i x_i\right), \tag{24.39}$$

where

$$\sum_{i=1}^{n} a_i^2 = a_0^2, \quad a_0 = \frac{1}{\lambda_0}. \tag{24.40}$$

24.3 Examples of Solutions

Consider the case of three spatial coordinates and further assume the speed $c = 1$. Let Eq. (24.1) have the form

$$\Box U + \frac{\partial U}{\partial t} + \frac{\partial U}{\partial x} = \sinh U. \tag{24.41}$$

The solutions (24.22) can be written as

$$U = f\left(e^{x+y-t}\psi(x-t)\right). \tag{24.42}$$

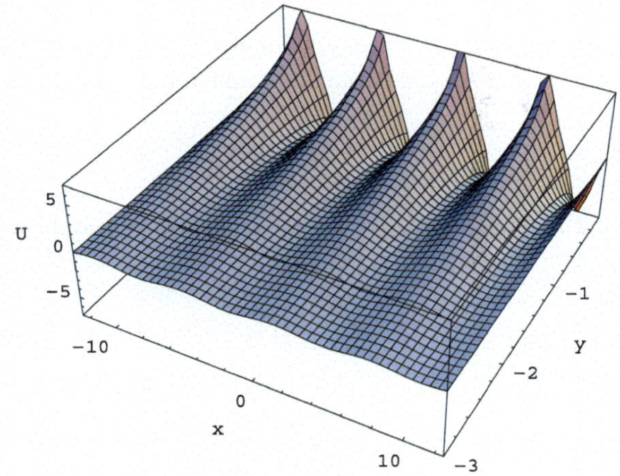

Fig. 24.2 Solutions U for the case (24.29), if $t = 0$

The dependence from z is absent in this case. This solution can be written in another form by redefining the function ψ

$$U = f\left(e^{y}\tilde{\psi}(x - t)\right). \tag{24.43}$$

Let's choose $\tilde{\psi}(x) = \sin x$. Then for the case (24.29) with $C_1 = 1$ and upper sign, we have

$$U = 4 \operatorname{arc\,tanh}\left[e^{y} \sin(x - t)\right]. \tag{24.44}$$

The plot of this solution is shown in Fig. 24.2:

$$\Box U + \sqrt{2}\frac{\partial U}{\partial t} + \frac{\partial U}{\partial x} + \frac{\partial U}{\partial y} = \sinh U. \tag{24.45}$$

The solutions (24.22) can be written as

$$U = f\left[e^{(x-y)/\sqrt{2}}\psi\left(x + y - \sqrt{2}t\right)\right]. \tag{24.46}$$

Choosing function ψ we can receive

$$U = f\left[e^{\sqrt{2}x - ct} + e^{ct - \sqrt{2}y}\right]. \tag{24.47}$$

Then the solution (24.32) with $C_1 = 1$ and upper sign has the form

$$U = 2 \operatorname{arc\,tanh} \operatorname{sn}\left[C_1 + \sqrt{\frac{E+1}{2}} \ln\left(e^{\sqrt{2}x - ct} + e^{ct - \sqrt{2}y}\right), \sqrt{\frac{E-1}{E+1}}\right], \tag{24.48}$$

where $E > 1$ and C_1 are arbitrary constants.

Let us give examples of solving the sine-Gordon equation with first derivatives

$$\Box U + \frac{\partial U}{\partial t} + \frac{\partial U}{\partial x} = \sin U. \tag{24.49}$$

From (24.26) we have

$$U = 4 \arctan\left[e^{y} \sin(x - t)\right]. \tag{24.50}$$

This solution represents waves moving along the (OX) axis with amplitude varying along (OY) axis as it shown in Fig. 24.3. The corresponding solution for the case (24.27) is

$$U = 2\,\mathrm{am}\left(\frac{1}{2k} \ln\left(e^{y} \sin(x - t)\right)^2), \; k\right) + \pi \tag{24.51}$$

(see Fig. 24.4).

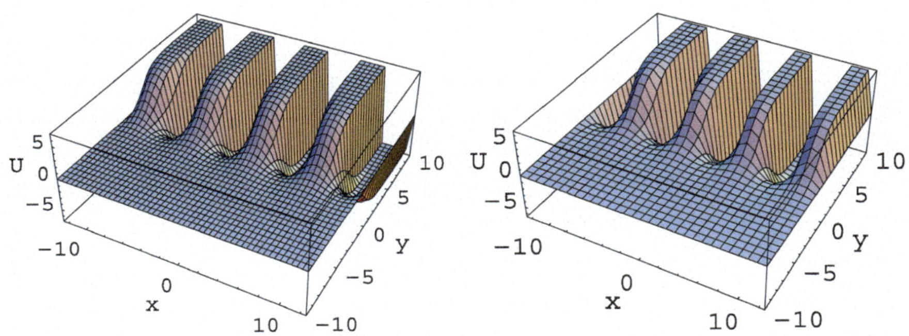

Fig. 24.3 Solutions U for the case (24.26), $t = 0$ (at the left), and $t = \pi$ (at the right side)

Fig. 24.4 Solutions U for the case (24.27), $t = 0$, and $E = 1.01$

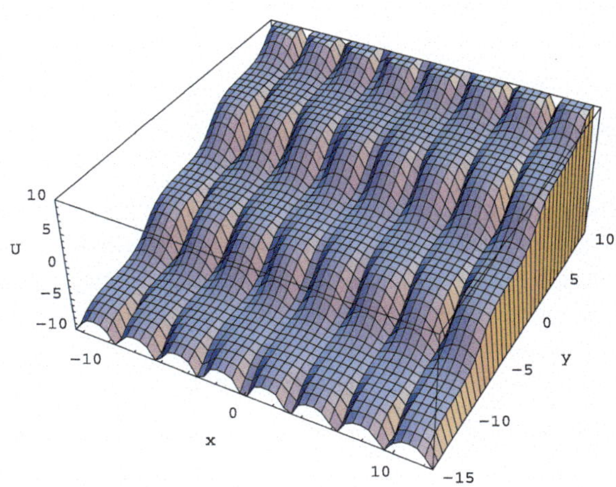

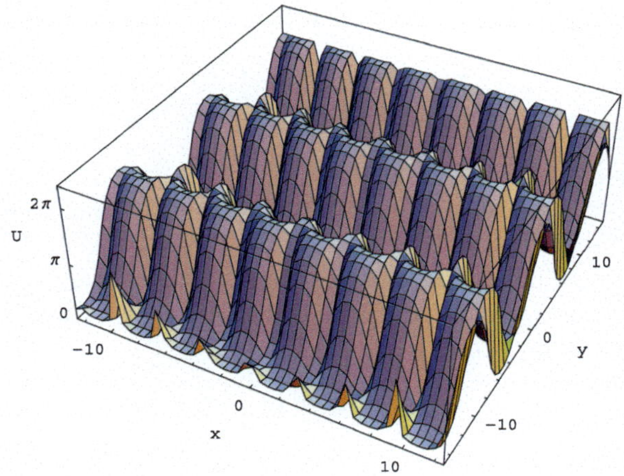

Fig. 24.5 Solutions U for the case (24.28), $t = 0$, and $E = 0.9$

The corresponding solution for the case (24.28) is

$$U = 2\arcsin\left[\sqrt{\frac{E+1}{2}}\,\mathrm{sn}\left(\frac{1}{2}\ln\left(e^{y}\sin(x-t)\right)^{2},\,\sqrt{\frac{E+1}{2}}\right)\right] + \pi \quad (24.52)$$

(see Fig. 24.5). The solutions (24.51) and (24.52) also represent waves moving along the (OX) axis with amplitude varying along (OY) axis.

The examples given refer to the case of an autonomous nonlinear Eq. (24.1) with first derivatives. Construction of specific solutions in the case of a non-autonomous Eq. (24.2) (in particular, the sine-Gordon equation with variable amplitude) requires solving the system of Eqs. (24.16), (24.17) with a given function $p(\mathbf{x}, t)$. This cannot be done for every function $p(\mathbf{x}, t)$. Methods for finding solutions (24.16), (24.17) and some examples of exact solutions are presented in the works [24, 25].

24.4 Conclusion

The paper presents a method for solving the nonlinear Klein–Fock–Gordon type equations with first derivatives. This method does not allow us to find a general solution to such equations. However, constructing a general solution explicitly in the case of a nonlinear partial differential equation is, as a rule, an unsolvable problem. The proposed method makes it possible to obtain exact solutions for different types of nonlinearities $F(U)$ and the presence of different first derivatives of the unknown solution. Therefore, the obtained exact solutions can be useful in the analysis of specific nonlinear models of mechanics and physics.

References

1. Klein O (1926) Quantentheorie und fünfdimensionale Relativitätstheorie. Z Physik 37:895–906. https://doi.org/10.1007/BF01397481
2. Fock V (1926) Über die invariante Form der Wellen- and der Bewegungsgleichungen für einen geladenen Massenpunkt. Z Physik 39:226–232. https://doi.org/10.1007/BF01321989
3. Gordon W (1926) Z Physik 40:117–133. Der Comptoneffekt nach der Schrödingerschen Theorie. https://doi.org/10.1007/BF01390840
4. Frenkel J, Kontorova T (1939) On the theory of plastic deformation and twinning. J Phys Acad Sci USSR 1:137–149
5. Braun OM, Kivshar YS (2004) The Frenkel-Kontorova model: concepts, methods, and applications. Springer, Berlin
6. Aero EL (2006) Micromechanics of a double continuum in a model of a medium with variable periodic structure. J Eng Math 55:81–95. https://doi.org/10.1007/s10665-005-9012-3
7. Aero EL, Bulygin AN (2007) Strongly nonlinear theory of nanostructure formation owing to elastic and nonelastic strains in crystalline solids. Mech Solids 42:807–822. https://doi.org/10.3103/S0025654407050147
8. Porubov AV, Aero EL, Maugin GA (2009) Two approaches to study essentially nonlinear and dispersive properties of the internal structure of materials. Phys Rev E 79:046608. https://doi.org/10.1103/PhysRevE.79.046608
9. Lonngren K, Scott A (eds) (1978) Solitons in action. Academic, New York
10. Takhtadzhyan LA, Faddeev LD (1974) Essentially nonlinear one-dimensional model of classical field theory. Theor Math Phys 21:1046–1057. https://doi.org/10.1007/BF01035551
11. Bogoliubov NN, Shirkov DV (1983) Quantum fields. Benjamin-Cammings, Massachusetts
12. Aero EL, Bulygin AN (1971) Equations of motion of nematic liquid-crystal media. J Appl Math Mech 35:831–843. https://doi.org/10.1016/0021-8928(71)90081-5
13. Aero EL (1995) Theory of Friedrichs transitions in nematic fluid crystals in curvilinear closed domains placed in an inhomogeneous magnetic field. Vysokomolek Soed Ser A 37:1286–1299
14. Aero EL (2002) Dynamic problems for the sine-Gordon equation with variable coefficients. Exact solutions. J Appl Math Mech 66:99–105. https://doi.org/10.1016/s0021-8928(02)00013-8
15. Miura RM (ed) (1976) Bäcklund transformations, the inverse scattering method, solitons, and their applications. Lecture Notes in Mathematics, vol 515. Springer, Berlin
16. Ablowitz MJ, Segur H (1981) Solitons and the inverse scattering transform. SIAM studies in applied mathematics, vol 4. SIAM, Philadelphia
17. Novikov S, Manakov SV, Pitaevskii LP, Zakharov VE (1984) Theory of solitons: the inverse scattering method. Springer, New York
18. Bateman H (1915) The mathematical analysis of electrical and optical wave-motion on the basis of Maxwell's equations. Cambridge University Press, Cambridge
19. Smirnoff V, Soboleff S (1932) Sur une méthode nouvelle dans le problème plan des vibrations élastiques. Trudy Seism Inst Leningrad Acad Sci USSR 20:1–37. [Engl. transl. in: Selected Works of Sobolev SL. Vol. I: Mathematical physics, computational mathematics, and cubature formulas. Demidenko GV, Vaskevich VL (eds) Springer, New York (2006), pp 45–80. https://doi.org/10.1007/978-0-387-34149-1_2]
20. Erugin NP (1948) On functionally-invariant solutions. Uch Zap Leningrad Univ ser Matem Vyp 15(96):101–134
21. Sobolev S (1934) Functionally-invariant solutions of wave equation. Tr Fiz-Mat Inst Steklova 5:259–264
22. Aero EL, Bulygin AN, Pavlov YuV (2009) Solutions of the three-dimensional sine-Gordon equation. Theor Math Phys 158:313–319. https://doi.org/10.1007/s11232-009-0025-3
23. Aero EL, Bulygin AN, Pavlov YuV (2011) New approach to the solution of the classical sine-Gordon equation and its generalizations. Diff Equat 47:1442–1452. https://doi.org/10.1134/S0012266111100077

24. Bulygin AN, Pavlov YuV (2019) Methods of finding of exact analytical solutions of nonautonomous nonlinear Klein-Fock-Gordon equation. In: Altenbach H et al (eds) Dynamical processes in generalized continua and structures. Springer, Cham, Switzerland, pp 147–161. https://doi.org/10.1007/978-3-030-11665-1_8
25. Bulygin AN, Pavlov YuV (2020) Some solutions of dynamic and static nonlinear nonautonomous Klein-Fock-Gordon equation. In: Altenbach H, Eremeyev VA, Pavlov IS, Porubov AV (eds) Nonlinear wave dynamics of materials and structures. Springer, Cham, Switzerland (2020), pp 107–120. https://doi.org/10.1007/978-3-030-38708-2_7

Chapter 25
Solution of Three-Dimensional Problems of the Dynamic Theory of Poroelasticity by the Boundary Element Method Using Parallel Computing

Andrey N. Petrov, Aleksandr A. Belov, Valeriy A. Bychenkov, and Elena I. Komarova

Abstract A direct approach of the boundary-element method to solving three-dimensional boundary-value problems of poroelastic dynamics is considered. To increase the efficiency of numerical modeling, elements of parallel computing are used. The results of computer experiments are presented.

Keywords Biot's theory · Boundary element method · Three-dimensional dynamics problems · Parallel computing

25.1 Introduction

The unique properties of porous materials have led to their widespread use in mechanical engineering, construction, etc. To study the propagation of waves in porous media, the model of a poroelastic medium proposed by Biot is most often used [1–4]. The boundary-element method (BEM) can be applied to such studies [5, 6]. The involvement of boundary-element technologies in the numerical modeling of the dynamics of poroelastic bodies is in an active development phase. The construction of modern numerical BEM schemes is impossible without the use of parallel computing.

A. N. Petrov · A. A. Belov (✉) · V. A. Bychenkov · E. I. Komarova
National Research Lobachevsky State University of Nizhny Novgorod, Nizhny Novgorod, Russia
e-mail: belov_a2@mech.unn.ru

A. N. Petrov
e-mail: andrey.petrov@mech.unn.ru

V. A. Bychenkov
e-mail: bychenkov@unn.ru

H. Altenbach and V. A. Eremeyev (eds.), *Dynamics of Discrete and Continuum Structures and Media*, Advanced Structured Materials 221,
https://doi.org/10.1007/978-3-031-75626-9_25

To create software that effectively uses computers built on multi-core processors, the theory of parallel computing is used. The idea behind this theory is to efficiently decompose a computational problem into a set of smaller tasks that can be performed independently of each other. In the process of creating a parallel program, the original sequential algorithm for solving a problem is subjected to a thorough analysis aimed at finding independent sections of the algorithm that allow parallel execution. It should be taken into account that too detailed decomposition of the original algorithm may not only improve the overall performance of the algorithm, but also make it lower. The reason for the decrease in performance is that the time required for data exchange between execution threads and their synchronization increases. Therefore, the greatest performance gain is observed when parallelization is introduced at a higher logical level, for example, external loops. An equally important factor in creating a parallel application is the choice of parallelization tool. To select the right tool, you should take into account not only the features of the algorithm, but also the features of the specific computer for which the application is being developed.

25.2 Biot's Poroelasticity

Equations of motion of Biot's model in Laplace domain for vanishing initial conditions are expressed as follows [6]:

$$\begin{aligned} G\overline{u}_{i,jj} + \left(K + \frac{G}{3}\right)\overline{u}_{j,ij} - (\alpha - \beta)\overline{p}_{,i} - s^2(\rho - \beta\rho_f)\overline{u}_i &= -\overline{F}_i, \\ \frac{\beta}{s\rho_f}\overline{p}_{,ii} - \frac{\varphi^2 s}{R}\overline{p} - (\alpha - \beta)s\overline{u}_{i,i} &= -\overline{a}, \end{aligned} \tag{25.1}$$

where $\overline{u}_i$ is the solid; $\overline{p}$ is the pore pressure; $G,\ K$ are elastic moduli; $\varphi = V^f/V$ is porosity; and $\overline{F}_i,\ \overline{a}$ are bulk body forces. The constants β, α, and R are introduced by

$$\beta = \frac{\kappa\rho_f\varphi^2 s}{\varphi^2 + s\kappa(\rho_a + \varphi\rho_f)},\ \alpha = 1 - \frac{K}{K_s}\ \textit{and}\ R = \frac{\varphi^2 K_f K_s^2}{K_f(K_s - K) + \varphi K_s(K_s - K_f)},$$

where κ is permeability; $\rho,\ \rho_a,\ \rho_f$ are, respectively, material density, apparent mass density, and filler density; and K_s, K_f are elastic bulk moduli of the skeleton and filler, respectively.

Equation (25.1) supplemented with boundary conditions:

$$\begin{aligned} \overline{u}'(x,s) &= \tilde{u}', \quad x \in \Gamma^u, \quad \overline{u}' = (\overline{u}_1, \overline{u}_2, \overline{u}_3, \overline{p}), \\ \overline{t}'_n(x,s) &= \tilde{t}'_n, \quad x \in \Gamma^\sigma, \quad \overline{t}' = (\overline{t}_1, \overline{t}_2, \overline{t}_3, \overline{q}), \end{aligned} \tag{25.2}$$

where Γ^u denotes Dirichlet boundary and Γ^σ denotes Neumann boundary, which fully describe the boundary problem in the representations of the 3D isotropic dynamic theory of poroelasticity.

25.3 The Boundary Element Method for Dynamic Poroelasticity

The integral representation for Eqs. (25.1) and (25.2) can be written for 3D domain in the absence of body force as

$$\int\limits_{\Gamma} \begin{bmatrix} \overline{U}_{ij}^{s} & -\overline{P}_{j}^{s} \\ \overline{U}_{i}^{f} & -\overline{P}^{f} \end{bmatrix} \begin{bmatrix} \bar{t}_i \\ \bar{q} \end{bmatrix} d\Gamma = \oint\limits_{\Gamma} \begin{bmatrix} \overline{T}_{ij}^{s} & -\overline{Q}_{j}^{s} \\ \overline{T}_{i}^{f} & -\overline{Q}^{f} \end{bmatrix} \begin{bmatrix} \overline{u}_i \\ \overline{p} \end{bmatrix} d\Gamma + \begin{bmatrix} {}_{ij} & 0 \\ 0 & c \end{bmatrix} \begin{bmatrix} \overline{u}_i \\ \overline{p} \end{bmatrix}, \tag{25.3}$$

where $\overline{U}_{ij}^{s}, \overline{U}_{i}^{f}, \overline{P}_{j}^{s}, \overline{P}^{f}, \overline{T}_{ij}^{s}, \overline{T}_{i}^{f}, \overline{Q}_{j}^{s}, \overline{Q}^{f}$—the components of the fundamental solutions and its double layer kernels and c_{ij} is the free term.

Approximation is required for obtaining numerical solutions of the previous presented boundary integral equations (BIE). Algebraic equation which can be solved computationally is derived via spatial discretization of the Laplace-domain BIEs shown in Eq. (25.3) by using the collocation method. Spatial discretization is applied by the following manner: each surface is discretized into bilinear (triangular or quadrilateral) boundary elements, over which generalized displacements and tractions are approximated with linear and constant shape functions, respectively. Thus, the following system of algebraic equations is obtained for a set of collocation points:

$$\frac{1-\alpha_\Omega}{2}\tilde{u}_i^m + \sum_{k=1}^{N}\sum_{l=1}^{4} A_{ij}^{m,k,l}\tilde{u}_j^{\chi(k,l)} = \sum_{k=1}^{N} B_{ij}^{m,k}\tilde{t}_j^k, \tag{25.4}$$

$$\frac{1-\alpha_\Omega}{8}\tilde{u}_i^m + \sum_{k=1}^{N}\sum_{l=1}^{4} A_{ij}^{m,k,l}\tilde{u}_j^{\chi(k,l)} = \sum_{k=1}^{N} B_{ij}^{m,k}\tilde{t}_j^k, \tag{25.5}$$

when the dense (full-populated) matrices A_{ij} and $B_{ij}^{m,k}$ arise from the integral of the kernels.

After the boundary conditions' imposition, the unknown coefficients of A_{ij} and $B_{ij}^{m,k}$ are collected into a vector and final system can be solved. The weak-singular integrals in (25.4) and (25.5) are integrated by subdividing the element into equal area subelements [7]. On the other hand, the subtraction of singularity method to evaluate strongly singular integrals is used [8]. The remaining regular integrals are performed with adaptive Gaussian quadrature and sub-elementation integration.

In the Laplace transform boundary integral equation method approach, the numerical inversion of the Laplace transform is very important. There are many powerful algorithms to compute the numerical inversion.

In order to get the results in time space, the modification of Durbin's algorithm is employed to transform the numerical results in Laplace space to real time space, which can be expressed by the following formulas:

$$f(0) \approx \sum_{k=1}^{n} \left[\frac{(F_{k+1} - F_k)\Delta_k}{2\pi} \right] \tag{25.6}$$

$$\begin{aligned} f(t) \approx \frac{e^{\alpha t}}{\pi t^2} \sum_{k=1}^{n} \Bigg[& \frac{F_{k+1} - F_k}{\Delta_k} (\cos(\omega_{k+1} t) - \cos(\omega_k t)) + \\ + & \frac{G_{k+1} - G_k}{\Delta_k} (\sin(\omega_{k+1} t) - \sin(\omega_k t)) \Bigg]. \end{aligned} \tag{25.7}$$

25.4 Parallel Computing

Analyzing the performance of the created parallel application is an important development stage. To determine performance, the concept of speed-up of a program is introduced [9]. Speed-up evaluates the performance of a parallel algorithm compared to a sequential algorithm:

$$S_p = \frac{T_1}{T_p},$$

where T is the program execution time using the fastest sequential implementation of the algorithm and T_p is the program execution time on p computing nodes.

Another value characterizing the performance of a program subjected to parallelization technology is parallel efficiency:

$$E_p = \frac{S_p}{p}.$$

Parallelization efficiency evaluates the efficiency of a program's use of resources. This assessment is useful from an economic point of view, as it allows one to predict the benefits of increasing computer computing power.

Many factors can affect the efficiency of parallelization: sections of sequential program code, synchronization costs, etc. There are estimates of the maximum performance of programs subjected to parallelization technology. For the maximum possible speed-up of the program $\overline{S}_p$ according to Amdahl's law [10], the following estimate is valid:

$$\overline{S}_p = \frac{p}{\alpha \cdot p + (1 - \alpha)},$$

where α expresses the fraction of calculations that can only be performed sequentially:

$$\alpha = \frac{T_{posl}}{T_1},$$

where T_{posl} is the time spent executing the sequential part of the application.

Thus, the speed-up obtained by parallelizing the program tends to the number of computing nodes p as the proportion of non-paralleled sections of the program decreases.

A similar estimate for maximum parallelization efficiency is as follows:

$$\overline{E}_p = \frac{\overline{S}_p}{p}.$$

The introduced characteristics make it possible to evaluate the performance and predict the scalability of the software implementation.

For each equation of system (25.4) and (25.5) written for the collocation node y^m, the corresponding coefficients $A_{ij}^{m,k,l}$ and $B_{ij}^{m,k}$ of the discrete analog are calculated independently of the coefficients of the equations written in other collocation nodes. Based on this remark, it is possible to construct a parallel algorithm for calculating the coefficients of discrete analogs.

Let us consider the construction of a parallel algorithm for calculating the coefficients for Eq. (25.4). The parallel algorithm is based on the sequential algorithm [11]. The block diagram of the sequential algorithm is shown in Fig. 25.1.

All boundary displacement approximation nodes are sequentially scanned in a loop and collocation nodes are found for which discrete analog coefficients are generated. Each iteration of the loop can be executed in parallel with other iterations. Based on this a parallel algorithm is constructed. The block diagram of the algorithm is shown in Fig. 25.1.

The execution thread E_k executes every k-th iteration of the sequential algorithm loop. The maximum number of threads E_k is equal to the number of iterations of the loop over nodes. Thus, the degree of parallelism of the algorithm is equal to the number of generalized displacement approximation nodes. A similar algorithm is used to calculate the coefficients of algebraic as given in Eq. (25.5). Its degree of parallelism is equal to the number of generalized tractions' approximation nodes.

The parallel algorithm is implemented in Fortran using the OpenMP parallel computing library [12] to implement execution threads. At the stage of program operation, the number of processor cores is determined and the same number of execution threads is launched to achieve the best performance on a given computer.

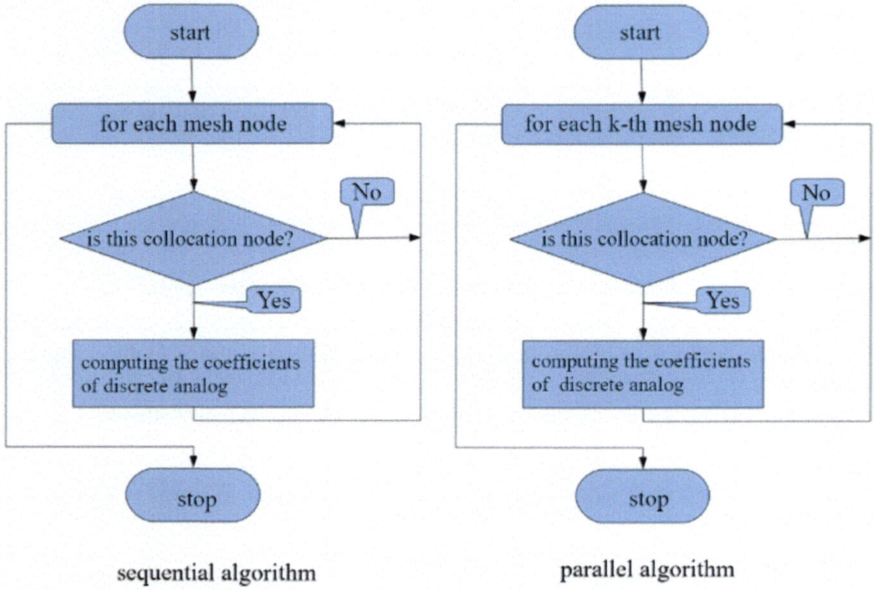

Fig. 25.1 Parallel solution flowchart

25.5 Numerical Example

The method was tested with a 3D transient poroelasto-dynamic problem. The problem of the effect of end force $F(t) = 1N/m^2$ on a 3 m-long 1 m-wide prismatic poroelastic body with a rigidly fixed end is considered. The boundary-value problem is presented in Fig. 25.2. The parameters of the partially saturated porous material correspond to those of sandstone: $K = 8 \cdot 10^9 N/m^2, G = 6 \cdot 10^9 N/m^2, \rho = 2458 kg/m^3, \varphi = 0.19, K_s = 3.6 \cdot 10^{10} N/m^2, \rho_f = 1000 kg/m^3, K_f = 3.3 \cdot 10^9 N/m^2, \kappa = 1.9 \cdot 10^{-10} m^4/(N \cdot \text{sec})$.

The problem was analyzed using three boundary-element meshes consisting of 224, 504, 896 square elements. Figure 25.3 shows these meshes marked as “a”, “b”,

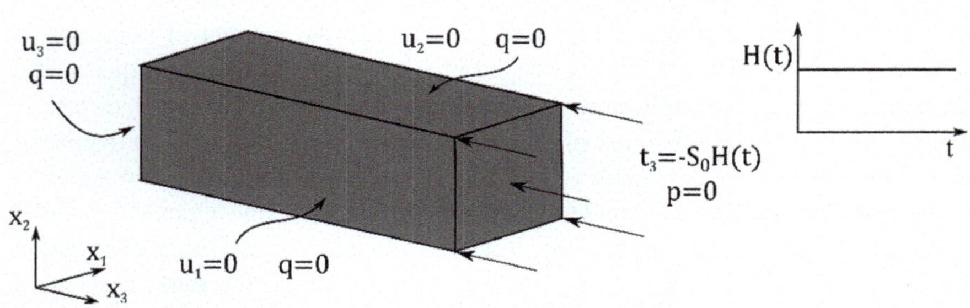

Fig. 25.2 Geometry and boundary conditions of a poroelastic column

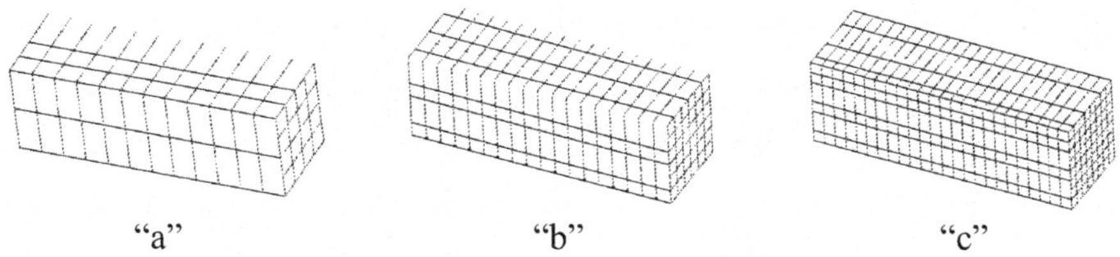

Fig. 25.3 Three types of mesh size used for boundary-element model

“c”, respectively. The program was run on three computers whose characteristics are presented in Table 25.1. The results are shown in Table 25.2.

The results obtained show that the use of a more productive computer in numerical experiments results in greater program acceleration. On all computers with the exception of computer No. 1, both acceleration and parallelization efficiency decrease with increasing number of elements. Computer No. 1 demonstrates lower productivity on grid “a” than on grid “b” despite the fact that grid “b” contains more elements and requires more time to calculate. This is due to the fact that on this computer the proportion of time spent on scheduling and synchronizing threads when calculating grid “a” is greater than when calculating on other grids. Therefore, the efficiency of solving the problem on grid “a” is somewhat lower than on other grids. It follows from this that for meshes with a small number of elements, it is more efficient in terms of resource consumption to use a smaller number of computational nodes. This is confirmed by the results of the experiment obtained on computers No. 2 and No. 3, where the efficiency of parallelization increases with a decrease in the number of grid

Table 25.1 Computer specifications

Number of computer	CPU name	Maximum frequency, GHz	Cores/threads	RAM, GB
1	Intel Core i7 2600	3.4	4/8	4
2	Intel Core i5 480 m	2.66	2/4	4
3	Intel Core 2 Duo E8400	3.00	2/2	2

Table 25.2 A comparison of execution times

	T_1			T_p			S_p			$\overline{S}_p$	E_p				$\overline{E}_p$
Mesh	a	b	c	a	b	c	a	b	c		a	b		c	
Computer															
No. 1	19	35	182	4	7	38	4.75	5.00	4.79	8	0.59	0.62		0.60	1
No. 2	30	106	268	12	44	118	2.50	2.41	2.27	4	0.62	0.60		0.57	1
No. 3	32	117	299	64	59	152	2.00	1.98	1.97	2	1.0	0.99		0.99	1

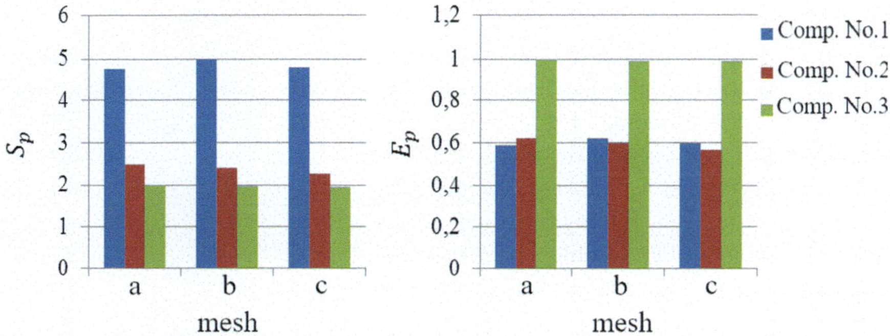

Fig. 25.4 Effect of parallelization on the resulting performance

elements. Figure 25.4 shows that computer No. 3 achieves the best results in terms of parallelization efficiency. This is due to the fact that computer No. 1 is built on a dual-core processor that does not implement Hyper Threading technology. In this case, full parallelism of execution threads is realized, when two execution threads work on different cores without competing for resources. On computers No. 1 and No. 2, processors support Hyper Threading technology and two execution threads are executed on each processor core. Threads begin to compete for the computing resources of their processor core as soon as their total computational load increases. Thus, full thread parallelism is not physically realized, which is reflected in the experimental results.

25.6 Conclusions

The work used a technique for parallelizing an algorithm based on the boundary-element method for the open standard OpenMP and created its software implementation. The parallelization of the computation of frequency steps in the Laplace transform domain was implemented. The method exploits the parallel capabilities of modern multi-core computers. A comparison of calculation time is presented using the example of a boundary-value problem of the dynamic theory of poroelasticity. Based on the results obtained, we can conclude that the method scales well and fully exploits a computer.

Acknowledgements The work was supported by the Russian Science Foundation grant № 22-19-00138.

References

1. Frenkel YI (1944) On the theory of seismic and seismically generated electric phenomena in wet soils. USSR Acad Sci (Izvestia), Ser Geopphys Geograph 8:134–149
2. Biot MA (1956) Theory of propagation of elastic waves in a fluid-saturated porous solid I. Low-frequency range. J Acoust Soc Am 28(2):168–78
3. Biot MA (1956) Theory of propagation of elastic waves in a fluid-saturated porous solid II. Higher frequency range. J Acoust Soc Am 28(2):179–91
4. Nikolaevskii VN (1996) Geomechanics and fluid dynamics with applications to the problems of gas and oil reservoirs. Moscow: Nedra, p 448 (in Russian)
5. Dunin SZ, Mikhailov DN, Nikolayevskii VN (2006) Longitudinal waves in partially saturated porous media: the effect of gas bubbles. J Appl Math Mech 70(2):251–263. ISSN 0021-8928, https://doi.org/10.1016/j.jappmathmech.2006.06.008
6. Schanz M (2001) Wave propagation in viscoelastic and poroelastic continua: a boundary element approach. Springer, Berlin
7. Lachat JC, Watson JO (1976) Effective numerical treatment of boundary integral equations: a formulation for three-dimensional elastostatics. Int J Numer Methods Eng 10:991–1005. https://doi.org/10.1002/nme.1620100503
8. Cruse TA, Aithal R (1993) Non-singular boundary integral equation implementation. Int J Nzmer Methods Eng 36:237–254
9. Ridgway Scott L, Clark T, Bagheri B (2005) Scientific parallel computing. Princeton University Press, p 374
10. Gustafson JL (1988) Reevaluating Amdahl's law. Commun ACM 31(5):532–533
11. Breshears C (2009) The art of concurrency. O'Reilly Media, Inc., p 302
12. Chapman B, Jost G, Van der Pas R (2008) Using OpenMP. Portable Shared Memory Parallel Programming. The MIT Press Cambridge, Massachusetts, London, England, p 353

Chapter 26
Numerical and Analytical Construction of Green's and Neumann's Functions of the Three-Dimensional Theory of Magnetoelectroelasticity

Andrey N. Petrov, Aleksandr A. Belov, Svetlana Yu. Litvinchuk, and Sergey A. Starikov

Abstract The system of equations of the anisotropic theory of magnetoelectroelasticity is considered. Green and Neumann functions were constructed. Two methods of constructing the Green and Neumann functions were used. A numerical example is given. Visualization of the Green and Neumann functions in the form of surfaces is presented.

Keywords Green and Neumann functions · Anisotropic magnetoelectroelasticity · Three-dimensional formulation · Numerical modeling

26.1 Introduction

Green's functions are used to find solutions to many applied problems in the study of cracks, defects, inclusions, etc. Much attention focused on the use of Green's functions in studies of anisotropic piezoelectric materials with electromechanical coupling [1–15].

A wide class of crystals [16] and a number of composite materials [17, 18], which simultaneously possess piezoelectric, piezomagnetic, and magnetoelectric effects, are classified as magnetoelectroelastic media. Based on the extended formalism,

A. N. Petrov · A. A. Belov (✉) · S. Yu. Litvinchuk · S. A. Starikov
National Research Lobachevsky State University of Nizhny Novgorod, Nizhny Novgorod, Russia
e-mail: belov_a2@mech.unn.ru

A. N. Petrov
e-mail: andrey.petrov@mech.unn.ru

S. Yu. Litvinchuk
e-mail: litvinchuk@mech.unn.ru

H. Altenbach and V. A. Eremeyev (eds.), *Dynamics of Discrete and Continuum Structures and Media*, Advanced Structured Materials 221,
https://doi.org/10.1007/978-3-031-75626-9_26

Pan [19] obtained three-dimensional Green's functions for anisotropic infinite, two-phase, and semi-infinite magnetoelectroelastic media. Among the works of authors who analyzed Green's functions for magnetoelectroelastic media, we note [20–23].

26.2 Problem Formulation

The basic equations of the three-dimensional anisotropic theory of elasticity have the form [24]:

$$\sigma_{ji,i} + F_i = 0, \sigma_{ij} = C_{ijlm}\varepsilon_{lm}, \varepsilon_{ij} = \frac{1}{2}\left(u_{i,j} + u_{j,i}\right), i, j, l, m = \overline{1,3}, \tag{26.1}$$

where σ_{ij} are the stress tensor components, F_i are the volume force density components, ε_{ij} are the strain tensor components, C_{ijlm} are the elastic moduli, u_i are the elastic displacement; repetition of indices (hereinafter) means summation, a comma means partial differentiation.

The basic equations of the three-dimensional linear theory of magnetoelectroelasticity are written in the form [25, 26]:

$$\sigma_{ji,j} + F_i = 0, D_{i,i} - Q = 0, B_{i,i} - M = 0, i, j = \overline{1,3} \tag{26.2}$$

$$\sigma_{ij} = C_{ijlm}\varepsilon_{lm} - e_{kij}E_k - q_{kij}H_k, D_i = e_{ijk}\varepsilon_{jk} + \tilde{\varepsilon}_{ij}E_j + \lambda_{ij}H_j \tag{26.3}$$

$$B_i = q_{ijk}\varepsilon_{jk} + \lambda_{ij}E_j + \mu_{ij}H_j, i, j, l, m = \overline{1,3} \tag{26.4}$$

$$\varepsilon_{ij} = \frac{1}{2}\left(u_{i,j} + u_{j,i}\right), E_i = -\varphi_{,i}, H_i = -\vartheta_{,i}, i, j = \overline{1,3}, \tag{26.5}$$

where σ_{ij}, D_i, and B_i are the components of the stress tensor, electric displacement vector, and magnetic induction, respectively; F, Q, and M are components of volume force densities, electric charge densities, and electric current densities; E_i are the components of the electric field vector; H_i are the components of the magnetic field vector; C_{ijlm}, e_{ijk}, $\tilde{\varepsilon}_{ij}$, μ_{ij}, q_{ijk}, λ_{ij}—elastic moduli, electroelastic coefficients, dielectric constants, magnetic conductivity moduli, piezomagnetic and magnetoelectric coefficients; u_i, φ, and ϑ—elastic displacement, electric, and magnetic potentials.

Let us denote the basic components for (26.2)–(26.5) as follows [27]:

$$\tilde{u}_i = \begin{cases} u_i, & i = 1, 2, 3, \\ \varphi, & i = 4 \\ \vartheta, & i = 5 \end{cases}, \hat{\varepsilon}_{ij} = \begin{cases} \varepsilon_{ij}, & i = 1, 2, 3, \\ -E_j, & i = 4, \\ -H_j, & i = 5, \end{cases}, \hat{\sigma}_{ij} = \begin{cases} \sigma_{ij}, & j = 1, 2, 3, \\ D_i, & j = 4, \\ B_i, & j = 5, \end{cases}$$

$$\hat{C}_{ijkl} = \begin{cases} C_{ijkl}, & j,k = 1,2,3, \\ e_{lij}, & j = 1,2,3,\ k = 4, \\ e_{ikl}, & j = 4,\ k = 1,2,3, \\ q_{lij}, & j = 1,2,3,\ k = 5, \\ q_{ikl}, & j = 5,\ k = 1,2,3, \\ -\lambda_{il}, & j = 4,\ k = 5;\ j = 5,\ k = 4, \\ -\varepsilon_{il}, & j,k = 4, \\ -\mu_{il}, & j,k = 5. \end{cases}$$

Then the physical equations and equilibrium equations of systems (26.2)—(26.5) will be written in the form (26.1):

$$\hat{\sigma}_{ij} = \hat{C}_{ijkl}\hat{\varepsilon}_{kl},\ \hat{\sigma}_{ij,i} + \hat{F}_j = 0$$

$$\hat{F}_j = \begin{cases} F_j, & j = 1,2,3, \\ -Q, & j = 4, \\ -M, & j = 5. \end{cases}$$

Consequently, for systems (26.2)–(26.5), it is possible to apply the approaches to constructing the Green and Neumann functions developed for (26.1) [28].

26.3 Obtaining Formulas for Representing Green and Neumann Functions

System of Eqs. (26.2)–(26.5) allows us to construct a differential equation:

$$\hat{C}_{ijkl}G_{km,li}(x) = -\delta_{jm}\delta(x),\quad i,l = \overline{1,3},\quad j,k,m = \overline{1,5}, \tag{26.6}$$

where δ_{jm} is the Kronecker delta, $\delta(x)$ is the Dirac delta function, and G_{km} are the components of Green's matrix.

Let's consider the integral method. The matrix G_{jk} using the inverse Radon transform can be written as follows [29]:

$$G_{jk} = \frac{1}{8\pi^2}\int\limits_{S^2}\left(M_{jk}^{zz}(z_i)\right)^{-1}\delta(z_i r_i)ds(z_i) = \frac{1}{8\pi^2 r}\int\limits_{S^2}\left(M_{jk}^{zz}(z_i)\right)^{-1}\delta\left(z_i r_i^0\right)ds(z_i),$$

$$r_i^0 = \frac{r_i}{r},\ \delta(z_i r_i) = \delta\left(r z_i r_i^0\right) = \frac{1}{r}\delta\left(z_i r_i^0,\right),\ M_{jk}^{zz} = \hat{C}_{ijkl}z_i z_l.$$

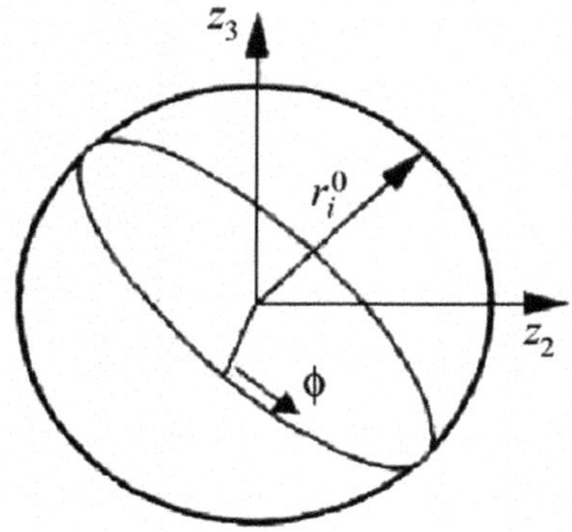

Fig. 26.1 Unit circle generated by the intersection of a sphere and a plane

The integral over the sphere S^2 :: $z_i z_i = 1$ is converted to the integral over the circle:

$$G_{jk} = \frac{1}{8\pi^2 r} \int_0^{2\pi} \left(M_{jk}^{zz}(z_i(\varphi))\right)^{-1} d\varphi \tag{26.7}$$

The unit circle is the intersection of the unit sphere with the plane $z_i r_i^0 = 0$ (Fig. 26.1).

The Neumann function—the first derivative of Green's matrix—has the form [29]:

$$G_{jk,s} = \frac{1}{8\pi^2 r^2} \int_0^{2\pi} (-r_s^0 (\mathcal{M}_{jk}^{kk})^{-1} - z_s F_{jk}) d\varphi, \tag{26.8}$$

$$D_{nq} = \mathcal{M}_{nq}^{r^0 z} + \mathcal{M}_{nq}^{zr^0}, \quad F_{jk} = (\mathcal{M}_{in}^{zz})^{-1} D_{nq} (\mathcal{M}_{qk}^{zz})^{-1},$$
$$\mathcal{M}_{ik}^{zr^0} = C_{ijkl} z_j r_l^0, \quad \mathcal{M}_{ik}^{r^0 z} = C_{ijkl} r_j^0 z_l.$$

Let's demonstrate the second method. Let's construct the inverse matrix $(M^{zz}(z))^{-1}$ using the matrix $A_{jk}(z)$ of algebraic additions $D(z)$ and the determinant, then Green's function will take the form [27]:

$$G_{jk} = \frac{1}{8\pi^2} \int_{S^2} \frac{A_{jk}(z_i)}{D(z_i)} \delta(z_i r_i) dS(z_i) = \frac{1}{4\pi^2 r} \int_{-\infty}^{+\infty} \frac{A_{jk}(p + \zeta q)}{D(p + \zeta q)} d\zeta. \tag{26.9}$$

The application of residue theory allows integral (26.8) to be written in the following form:

$$G_{jk}(x) = -\frac{\mathrm{Im}}{2\pi r}\sum_{m=1}^{5}\frac{A_{jk}(p+\zeta_m q)}{a_{11}(\zeta_m-\zeta_m^*)\prod\limits_{k=1,k\neq m}^{5}(\zeta_m-\zeta_k)(\zeta_m-\zeta_k^*)}, \tag{26.10}$$

where ζ_m are the roots of a polynomial of the tenth degree $D(p+\zeta q) = 0$ with respect to ζ; a_{11}—coefficient at ζ^{10}; $\mathrm{Im}\zeta_m > 0, m = \overline{1,5}$; and ζ_m^*—conjugate to the roots.

Thus, in contrast to the integral approach (26.7), the second method gives an explicit formula (26.10), which guarantees the efficiency of using $G(x)$ and the accuracy of the results obtained. Note that when constructing (26.10), all poles were assumed to be simple. In the case of multiple poles, small changes in the constants allow us to reduce the problem to simple poles with minor errors in the calculations of the extended Green's matrix [30]. When calculating, it is important that the matrix M_{jk}^{zz} is symmetric, as is its algebraic complement A_{jk}, and, therefore, the extended Green's matrix is symmetric and only 15 elements out of 25 need to be calculated.

Derivatives of Green's matrix can be calculated based on Lagrange interpolation [28]. The choice of interval h is critical. In this study, the values of this interval are taken $h = r \cdot 10^{-3}$, $h = r \cdot 10^{-2}$, where r is the distance between an arbitrary and starting points.

26.4 Numerical Experiments

For convenience, we write the elastic moduli tensor C in terms of the constants $c_{\alpha\beta}(\alpha = \overline{1,3}; \beta = \overline{1,6})$ [31]. The piezomagnetic coefficients q_{kij} can be written with two indices $q_{kp}(k = \overline{1,3}; p = \overline{1,6})$ as follows:

$$q_{k1} = q_{k11};\ q_{k2} = q_{k22};\ q_{k3} = q_{k33};\ q_{k4} = q_{k23} = q_{k32};$$
$$q_{k5} = q_{k13} = q_{k31};\ q_{k6} = q_{k12} = q_{k21}.$$

As an example, consider a magnetoelectroelastic material $BaTiO_3$ [25]. Non-zero elastic constants $c_{\alpha\beta}$ are given with an accuracy of a factor of $10^9 \mathrm{N/m^2}$ and are given in Table 26.1, piezoelectric coefficients e_{kp}—$/\mathrm{m}^2$ in Table 26.2, dielectric constants—$^2/\mathrm{Nm}^2$ in Table 26.3, piezomagnetic coefficients q_{kp}—N/Am in Table 26.4, and magnetic permeability modules—$\mathrm{Ns}^2/\mathrm{C}^2$ in Table 26.5. Magnetoelectric coefficients λ_{ij} are equal to zero for all $i, j = 1, 3$.

Table 26.1 Elastic constants

11	12	13	22	23	33	44	55	66
66,0	77,0	78,0	166,0	78,0	162,0	43,0	43,0	44,5

Table 26.2 Piezoelectric coefficients

e_{15}	e_{24}	e_{31}	e_{32}	e_{33}
11,6	11,6	−4,4	−4,4	18,6

Table 26.3 Dielectric constants

$\tilde{\varepsilon}_{11}$	$\tilde{\varepsilon}_{22}$	$\tilde{\varepsilon}_{33}$
11.2	11.2	12.6

Table 26.4 Piezomagnetic coefficients

q_{15}	q_{24}	q_{31}	q_{32}	q_{33}
550.0	550.0	580.3	580.3	699.7

Table 26.5 Magnetic permeability

μ_{11}	μ_{22}	μ_{33}
5.0	5.0	10.0

For point $x = (1, 1, 1)$, the components of the magnetoelectroelastic Green's matrix are presented in Table 26.6 for the integral method compared with the polynomial-based formula.

Table 26.6 Components $G_{pk,1}$

Component index values (p, k)	Integral method	Formula on polynomial bases
1,1	$7.714215538 \times 10^{-13}$	$7.714389635 \times 10^{-13}$
1,2; 2,1	$1.202690309 \times 10^{-13}$	$1.202687889 \times 10^{-13}$
1,3; 3,1	$1.294636801 \times 10^{-13}$	$1.294626181 \times 10^{-13}$
1,4; 4,1	$1.034752139 \times 10^{-3}$	$1.034750467 \times 10^{-3}$
1,5; 5,1	$-4.266314625 \times 10^{-6}$	$-4.266852604 \times 10^{-6}$
2,2	$7.714048991 \times 10^{-13}$	$7.714331183 \times 10^{-13}$
2,3; 3,2	$1.294614670 \times 10^{-13}$	$1.294676917 \times 10^{-13}$
2,4; 4,2	$1.034734450 \times 10^{-4}$	$1.034791046 \times 10^{-4}$
2,5; 5,2	$-4.266241639 \times 10^{-6}$	$-4.267020888 \times 10^{-6}$
3,3	$4.847872021 \times 10^{-13}$	$4.847808452 \times 10^{-13}$
3,4; 4,3	$5.007545247 \times 10^{-4}$	$5.007530930 \times 10^{-4}$
3,5; 5,3	$2.419569993 \times 10^{-5}$	$2.419607424 \times 10^{-5}$
4,4	$-3.404174605 \times 10^{6}$	$-3.404264448 \times 10^{6}$
4,5; 5,4	$2.974912499 \times 10^{4}$	$2.944956912 \times 10^{4}$
5,5	$-4.191966210 \times 10^{3}$	$-4.191996338 \times 10^{3}$

Table 26.7 Components $G_{pk,2}$

Component index values (p, k)	Integral method	Formula on polynomial bases, $h = r \cdot 10^{-3}$ (m)	Formula on polynomial bases, $h = r \cdot 10^{-2}$ (m)
1,1	$-9.8619624 \times 10^{-14}$	$-9.7185368 \times 10^{-14}$	$-1.0181226 \times 10^{-13}$
1,2; 2,1	$4.9196556 \times 10^{-15}$	$4.8202192 \times 10^{-15}$	$5.2111568 \times 10^{-15}$
1,3; 3,1	$3.8690715 \times 10^{-15}$	$4.0672418 \times 10^{-15}$	$3.4542521 \times 10^{-15}$
1,4; 4,1	4.7463037×10^{-6}	4.9050369×10^{-6}	4.4065895×10^{-6}
1,5; 5,1	$-1.4048214 \times 10^{-6}$	$-1.4116536 \times 10^{-6}$	$-1.3850792 \times 10^{-6}$
2,2	$-3.3913495 \times 10^{-13}$	$-3.3805723 \times 10^{-13}$	$-3.4160767 \times 10^{-13}$
2,3; 3,2	$-1.2558514 \times 10^{-13}$	$-1.2510472 \times 10^{-13}$	$-1.2665409 \times 10^{-13}$
2,4; 4,2	$-9.8722125 \times 10^{-5}$	$-9.8337499 \times 10^{-5}$	$-9.9584286 \times 10^{-5}$
2,5; 5,2	2.8617374×10^{-6}	2.8455247×10^{-6}	2.9027184×10^{-6}
3,3	$-1.9558380 \times 10^{-13}$	$-1.9485334 \times 10^{-13}$	$-1.9713804 \times 10^{-13}$
3,4; 4,3	$-1.8935050 \times 10^{-4}$	$-1.8858305 \times 10^{-4}$	$-1.9101680 \times 10^{-4}$
3,5; 5,3	$-1.1712346 \times 10^{-5}$	$-1.1692019 \times 10^{-5}$	$-1.1750239 \times 10^{-5}$
4,4	1.1785391×10^{6}	1.1737228×10^{6}	1.1893233×10^{6}
4,5; 5,4	-1.3080034×10^{4}	-1.3048614×10^{4}	-1.3143548×10^{4}
5,5	1.5679576×10^{3}	1.5620287×10^{3}	1.5805828×10^{3}

For point $x = (1, 1, 1)$, the components of the magnetoelectroelastic Neumann matrices $G_{pk,1}$, $G_{pk,2}$, $G_{pk,3}$ are presented in Tables 26.7, and 26.8 for the integral method in comparison with finite-difference analogs ($h = r \cdot 10^{-3}$,$h = r \cdot 10^{-2}$).

26.5 Visualization of Green's and Neumann's Functions

Let us describe the construction of Green's matrices in the form of two-dimensional surfaces based on an interpolation calculation scheme. Functions $G_{pk}(x)$ depend on three spatial variables x_1, x_2, and x_3. Let's move on to spherical coordinates (r, θ_1, θ_2). For a unit sphere, $G_{pk}(\theta_1, \theta_2)$ is a function of only two variables: the polar angle $0 \leq \theta_1$[unknown template] and the azimuth angle $0 \leq \theta_2$[unknown template]. The G_{pk} value for specific Tq and T1 is constructed using Lagrange interpolation [29].

Figures 26.2, 26.3, and 26.4 show the components of the magnetoelectroelastic Green's function. Figures 26.5, 26.6, and 26.7 show the components of the Neumann magnetoelectroelastic function for $BaTiO_3$.

Table 26.8 Components $G_{pk,3}$

Component index values (p, k)	Integral method (m)	Formula on polynomial bases,$h = r \cdot 10^{-3}$ (m)	Formula on polynomial bases, $h = r \cdot 10^{-2}$ (m)
1,1	$-3.3915701 \times 10^{-13}$	$-3.4364674 \times 10^{-13}$	$-3.3834666 \times 10^{-13}$
1,2; 2,1	$4.9476002 \times 10^{-15}$	$5.3059481 \times 10^{-15}$	$4.8881906 \times 10^{-15}$
1,3; 3,1	$-1.2558514 \times 10^{-13}$	$-1.2621220 \times 10^{-13}$	$-1.2547014 \times 10^{-13}$
1,4; 4,1	$-9.8722125 \times 10^{-5}$	$-9.9233223 \times 10^{-5}$	$-9.8630586 \times 10^{-5}$
1,5; 5,1	2.8617374×10^{-6}	2.8898501×10^{-6}	2.8582871×10^{-6}
2,2	$-9.8590072 \times 10^{-14}$	$-1.0208954 \times 10^{-13}$	$-9.7999566 \times 10^{-14}$
2,3; 3,2	$3.8864695 \times 10^{-15}$	$2.3695594 \times 10^{-15}$	$4.1571077 \times 10^{-15}$
2,4; 4,2	4.7599804×10^{-6}	3.5372712×10^{-6}	4.9753857×10^{-6}
2,5; 5,2	$-1.4052179 \times 10^{-6}$	$-1.3477601 \times 10^{-6}$	$-1.4134653 \times 10^{-6}$
3,3	$-1.9558045 \times 10^{-13}$	$-1.9779398 \times 10^{-13}$	$-1.9515677 \times 10^{-13}$
3,4; 4,3	$-1.8934726 \times 10^{-4}$	$-1.9171355 \times 10^{-4}$	$-1.8890744 \times 10^{-4}$
3,5; 5,3	$-1.1712146 \times 10^{-5}$	$-1.1766988 \times 10^{-5}$	$-1.1699631 \times 10^{-5}$
4,4	1.1785190×10^{6}	1.1937625×10^{6}	1.1758146×10^{6}
4,5; 5,4	-1.3079811×10^{4}	-1.3170731×10^{4}	-1.3061175×10^{4}
5,5	1.5679308×10^{3}	1.5858761×10^{3}	1.5645073×10^{3}

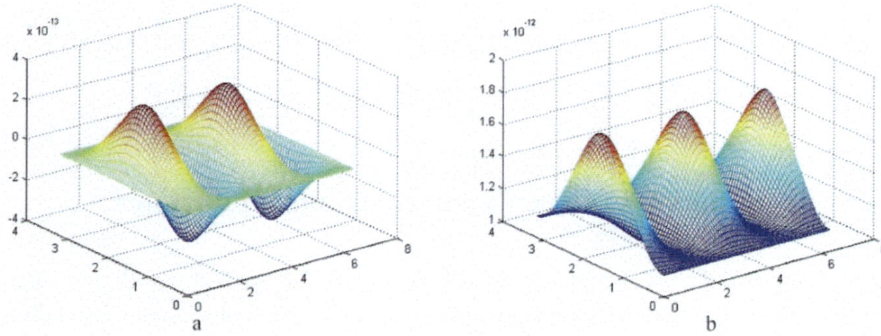

Fig. 26.2 Components G_{12} (**a**) and G_{11} (**b**)

26.6 Conclusions

The work used a technique for parallelizing an algorithm based on the boundary element method for the open standard OpenMP and created its software implementation. The parallelization of the computation of frequency steps in the Laplace transform domain was implemented. The method exploits the parallel capabilities of modern multi-core computers. A comparison of calculation time is presented using the example of a boundary value problem of the dynamic theory of poroelasticity.

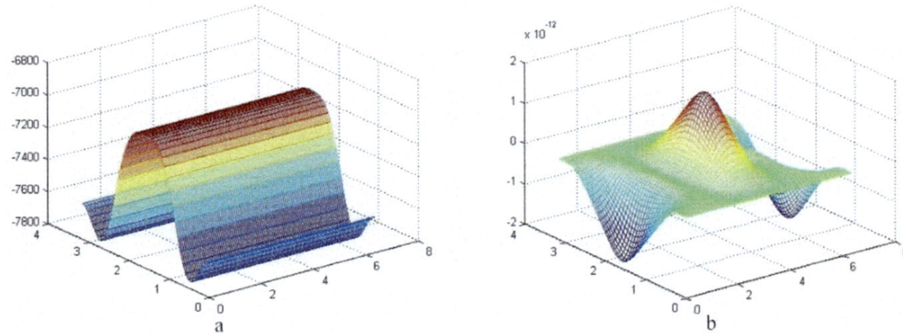

Fig. 26.3 Components G_{55} (**a**) and $G_{11,1}$ (**b**)

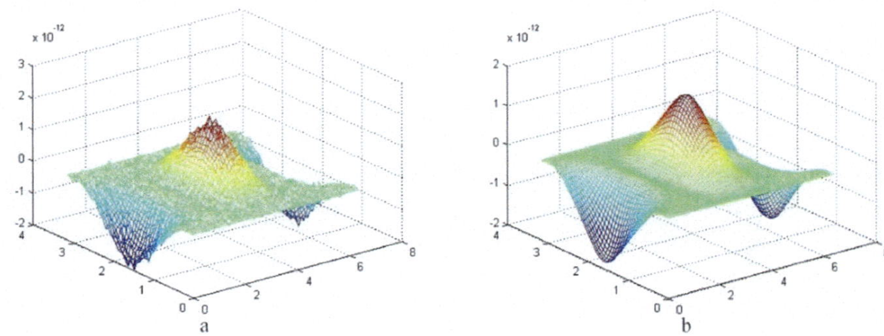

Fig. 26.4 Component $G_{11,1}$ for $h = r10^{-3}$ (**a**) and for $h = r10^{-2}$ (**b**)

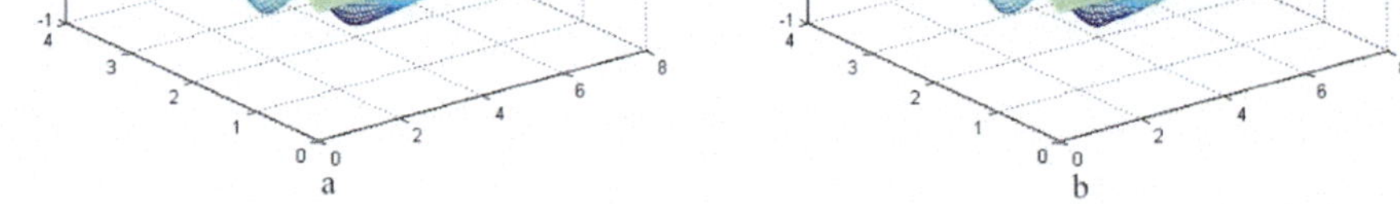

Fig. 26.5 Component $G_{12,2}$ via integral method (**a**) and via polynomial formula for $h = r10^{-3}$ (**b**)

Based on the results obtained, we can conclude that the method scales well and fully exploits a computer.

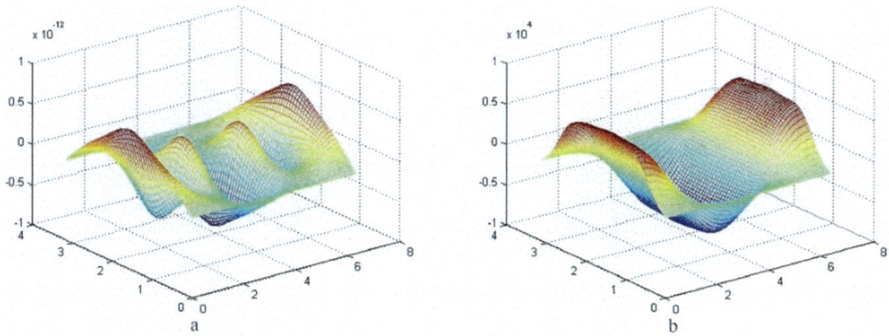

Fig. 26.6 Component $G_{12,2}$ via polynomial formula for $h = r10^{-2}$ (**a**) and $G_{55,1}$ (**b**) via integral method

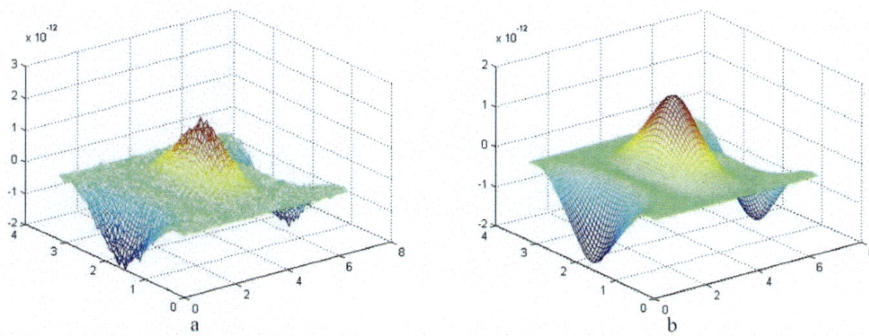

Fig. 26.7 Component $G_{55,1}$ for $h = r10^{-3}$ (**a**) and for $h = r10^{-2}$ (**b**)

Acknowledgements The work was carried out with the financial support of the Research and Education Mathematical Center «Mathematics for Future Technologies».

References

1. Deeg WF (1980) The analysis of dislocation, crack, and inclusion problem in piezoelectric solids: Ph.D. Dissertation. Stanford University
2. Wang B (1992) Three dimensional analysis of an ellipsoidal inclusion in a piezoelectric material. Int J Solids Struct 29:293–308
3. Benveniste Y (1992) The determination of the elastic and electric fields in a piezoelectric inhomogeneity. J Appl Phys (72):1086–1095
4. Chen TY (1993) Green's functions and the non-uniform transformation problem in a piezoelectric medium. Mech Res Commun 20:271–278
5. Chen TY, Lin FZ (1993) Numerical evaluation of derivatives of the anisotropic piezoelectric Green's functions. Mech Res Commun 20:501–506
6. Pan E, Tonon F (2000) Three-dimensional Green's functions in anisotropic piezoelectric solids. Int J Solids Struct 37:943–958

7. Pan E, Yuan FG (2000) Three-dimensional Green's functions in anisotropic piezoelectric biomaterials. Int J Eng Sci 38:1939–1960
8. Wang ZK, Chen GC (1994) A general solution and the application of space axisymmetric problem in piezoelectric materials. App Math Mech (15):615–626
9. Wang ZK, Zheng BL (1995) The general solution of three-dimensional problem in piezoelectric media. Int J Solids Struct 31: 105–115
10. Dunn ML, Electroelastic Green's functions for transversely isotropic piezoelectric media and their application to the solution of inclusion and inhomogeneity problems. Int
11. Dunn ML, Wienecke HA (1996) Green's functions for transversely isotropic piezoelectric solids. Int J Solids Struct 33:4571–4581
12. Dunn ML, Wienecke HA (1999) Half-space Green's functions for transversely isotropic piezoelectric solids. J App Mech (66):675–679
13. Ding HJ, Liang J, Chen B (1996) Fundamental solution for transversely isotropic piezoelectric media. Sci China (A) 39:766–775
14. Ding HJ, Chen B, Liang J (1997) On the Green's functions for two-phase transversely isotropic piezoelectric media. Int J Solids Struct 34:3041–3057
15. Dunn ML, Wienecke HA (1997) Inclusions and inhomogeneities in transversely isotropic piezoelectric solids. Int J Solids Struct 34:3571–3582
16. Sirotin YI, Shaskolskaya MP (1982) Fundamentals of crystal physics. Mir Publishers, Moscow
17. Huang JH, Kuo WS (1997) The analysis of piezoelectric/piezomagnetic composite materials containing an ellipsoidal inclusion. J App Phys (81):1378–1386
18. Li JY, Dunn ML (1998) Micromechanics of magnetoelectroelastic composite materials: average fields and effective behavior. J Intell Mater Syst Struct 7:404–416
19. Pan E (2002) Three-dimensional Green's functions in anisotropic magneto-electro-elastic biomaterials. Z Angew Math Phys (53):815–838
20. Wang X, Shen YP (2002) The general solution of three-dimensional problems in magnetoelectroelastic media. Int J Eng Sci 40:1069–1080
21. Alshits VI, Kirchner HOK, Ting TCT (1995) Angularly inhomogeneous piezoelectric piezomagnetic magnetoelectric anisotropic media. Philos Mag Lett 71:285–288
22. Kirchner HOK, Alshits VI (1996) Elastically anisotropic angularly inhomogeneous media II. The Green's function for piezoelectric piezomagnetic and magnetoelectric media. Philos Mag 74:861–885
23. Liu JX, Liu XL, Zhao YB (2001) Green's functions for anisotropic magneto-electro-elastic solids with an elliptical or a crack. Int J Eng Sci 39:1405–1418
24. Nowacki W (1970) Theory of elasticity. PWN, Warszawa [in Polish]
25. Igumnov LA, Litvinchuk SY, Pazin VP, Petrov AN (2010) Chislenno-analiticheskoe postroenie matrits Grina trekhmernykh teorii uprugosti i elektrouprugosti [The numerical-analytical construction of Green's matrices of 3-d elasticity and electro-elasticity theories]. Vestnik of Lobachevsky University of Nizhni Novgorod, no 3–1, pp 134–140
26. Bazhenov VG, Belov AA, Igumnov LA (2009) Boundary element modeling the dynamics of piecewise homogeneous media and constructions (Granichno-elementnoye modelirovaniye dinamiki kusochno-odnorodnyh sred i konstruktsiy). Lobachevsky State University of Nizhni Novgorod, Nizhny Novgorod (in Russian)
27. Tonon F, Pan E, Amadei B (2001) Green's functions and BEM formulation for 3D anisotropic media. Comput Struct (79):469–482
28. Ting TCT (1996) Anisotropic elasticity—theory and applications. Oxford University Press, NY
29. Bazhenov VG, Igumnov LA (2008) Methods of boundary integral equations and boundary elements in solving problems of three-dimensional dynamic theory of elasticity with conjugate fields. Fizmatlit, Moscow, 352p (in Russian)
30. Tonon F, Pan E, Amadei B (2001) Green's functions and BEM formulation for 3D anisotropic media. Comput Struct (79):469–482
31. Ting TCT (1996) Anisotropic elasticity—theory and applications. Oxford University Press, New York

Chapter 27
Analysis of Transient Response of Poroelastic Half-Space with Open Trench by BEM Using Runge–Kutta Convolution Quadrature

Andrey N. Petrov, Leonid A. Igumnov, Aleksandr A. Belov, and Vyacheslav V. Lukin

Abstract The development of a time-step boundary element scheme for computer simulation of solving boundary value problems of three-dimensional dynamics of poroelasticity is considered. The boundary element scheme is based on the time-step method of numerical inversion of the Laplace transform at the nodes of the Runge–Kutta scheme. A time-step scheme with a key is used, which arises from the integration of highly oscillating functions. The general approach is demonstrated by a numerical example. The problem of an elastic half-space with an open trench under the action of a suddenly applied force is considered примерах.

Keywords Poroelasticity · 3D · Dynamics · Boundary Element Method (BEM) · Laplace transform inversion · Convolution Quadrature Method (CQM) · Time-step methods · Runge–Kutta methods · Half-space · Open trench · Rayleigh wave

27.1 Introduction

In [1–3], an original approach to the construction of time-step schemes of the Boundary Element Method (BEM) is proposed. In [4–6], the BEM scheme was used, based on the stepwise method of numerical inversion of the Laplace transform. In [4, 7], a modification of the method with a variable integration step is formulated

A. N. Petrov · L. A. Igumnov (✉) · A. A. Belov · V. V. Lukin
National Research Lobachevsky State University of Nizhny Novgorod, Nizhny Novgorod, Russia
e-mail: igumnov@mech.unn.ru

A. N. Petrov
e-mail: andrey.petrov@mech.unn.ru

A. A. Belov
e-mail: belov_a2@mech.unn.ru

H. Altenbach and V. A. Eremeyev (eds.), *Dynamics of Discrete and Continuum Structures and Media*, Advanced Structured Materials 221,
https://doi.org/10.1007/978-3-031-75626-9_27

when calculating the weight multipliers of the quadrature sum. Another extension of the boundary element scheme is presented in articles [8, 9], suggesting the joint use of FEM with Runge–Kutta methods. Following [10], this article discusses the joint application of a step scheme at the nodes of the Runge–Kutta methods with BEM for solving three-dimensional elastic and poroelastic problems.

27.2 Problem Formulation

Biot's saturated poroelastic material model is used for description of porous materials [3]. Saturated poroelastic continuum assumes to be a two-phase continuum consisting of an elastic skeleton and compressible fluid or gas filler. Saturation means $V = V^f + V^s$, where V is the overall volume of the considered body, V^f is the summary pore (filler) volume, and V^s is the volume of the elastic skeleton. The filler can seep through the pores freely; closed pores are assumed to be a part of the skeleton.

There are several Biot's theory formulations distinguished by the number of unknowns: three skeleton displacements plus pore pressure or three filler displacements along with the skeleton ones [11, 12]. However, in 1987, G. Bonnet showed that four components are enough for saturated porous continuum description [13]. Boundary-value problem for full Biot's model of linear saturated poroelastic continuum in Laplace domain in terms of four unknowns—skeleton displacements $\overline{u}_i$ and pore pressure $\overline{p}$—takes the following form [3]:

$$
\begin{gathered}
G\overline{u}_{i,jj} + \left(K + \frac{G}{3}\right)\overline{u}_{j,ij} - (\alpha - \beta)\overline{p}_{,i} - s^2(\rho - \beta\rho_f)\overline{u}_i = -\overline{F}_i, \\
\frac{\beta}{s\rho_f}\overline{p}_{,ii} - \frac{\varphi^2 s}{R}\overline{p} - (\alpha - \beta)s\overline{u}_{i,i} = -\overline{a}, \quad x \in \Omega, \\
\overline{u}'(x, s) = \tilde{u}', \quad x \in \Gamma^u, \quad \overline{u}' = (\overline{u}_1, \overline{u}_2, \overline{u}_3, \overline{p}), \\
\overline{t}'_n(x, s) = \tilde{t}'_n, \quad x \in \Gamma^\sigma, \quad \overline{t}' = (\overline{t}_1, \overline{t}_2, \overline{t}_3, \overline{q}),
\end{gathered}
\tag{27.1}
$$

where Γ^u denotes the Dirichlet boundary and Γ^σ denotes the Neumann boundary, $G,\ K$ are the elastic moduli, $\varphi = V^f/V$ is the porosity, $\overline{F}_i,\ \overline{a}$ are the bulk body forces,

$$
\beta = \frac{\kappa\rho_f\varphi^2 s}{\varphi^2 + s\kappa(\rho_a + \varphi\rho_f)}, \alpha = 1 - \frac{K}{K_s} \text{ and } R = \frac{\varphi^2 K_f K_s^2}{K_f(K_s - K) + \varphi K_s(K_s - K_f)}
$$

are the constants reflecting interaction between skeleton and filler, κ is the permeability, $\rho,\ \rho_a,\ \rho_f$ are, respectively, the material density, apparent mass density, and filler density, and K_s, K_f are the elastic bulk moduli of skeleton and filler, respectively.

$\rho_a = C\varphi\rho_f$—apparent mass density—is an imaginary mass density introduced in [3] to describe dynamic interaction between the skeleton and filler phases. C

coefficient depends on the pores geometry and excitation frequency. G. Bonnet and J.-L. Auriault measured $C = 0.66$ for low-frequency excitation and a system of spherical pores distributed right up to each other [13].

BIE system in terms of elastic skeleton displacements and pore pressure can be represented in time domain only in conditions of $\kappa \to \infty$. This model corresponds to the filler passing through pores without friction. Sometimes incompressible Biot's model implying $\alpha = 1$ и $R \to \infty$ is applied; in this case, pore pressure is completely defined by elastic skeleton deformations and makes no degree of freedom anymore.

27.3 BEM Application

Boundary-value problem (27.1) can be reduced to the following BIE [3, 14, 15]:

$$\alpha_\Omega \bar{u}_k(x,s) + \int\limits_\Gamma \Big(\tilde{T}_{ik}(x,y,s)\bar{u}_i(y,s) - \tilde{T}^0_{ik}(x,y,s)\bar{u}_i(x,s) - \tilde{U}_{ik}(x,y,s)\bar{t}_i(y,s)\Big) d\Gamma = 0,$$

$$(x \in \Gamma), \quad \bar{t} = \left(\bar{t}_1, \bar{t}_2, \bar{t}_3, \bar{q}\right)^T, \quad \bar{u} = (\bar{u}_1, \bar{u}_2, \bar{u}_3, \bar{p})^T,$$

where $\tilde{U}(x,s)$, $\tilde{T}(x,s)$ are the fundamental and singular solutions, respectively, $\tilde{T}^0(x,s)$ contains the isolated singularities.

To approximate boundary of a piecewise homogeneous body, a set of quadrangular and triangular elements is used, where triangular elements are considered singular quadrangular. Quadrangular elements help to obtain a higher precision as compared to fully triangular approximation. Every boundary element is mapped to a reference one (canonical square $\xi = (\xi_1, \xi_2) \in [-1, 1]^2$ or triangle $0 \le \xi_1 + \xi_2 \le 1$, $\xi_1 \ge 0$, $\xi_2 \ge 0$) with the help of the formula:

$$y_i(\xi) = \sum_{l=1}^{8} N^l(\xi) y_i^{\beta(k,l)}, \qquad i = 1, 2, 3,$$

where $\beta(k,l)$ is the global node number, l—local node number in element k, and $N^l(\xi)$—shape functions. Boundary functions are interpolated by a subset of geometrical grid nodes. Local approximation of generalized boundary displacements and tractions follows Goldshteyn's displacement–stress matched model [4], meaning that generalized displacements are approximated by bilinear element and tractions—by constant.

After applying the collocation method on the set of boundary functions approximation nodes, the following system of linear equations is obtained per each homogeneous region of the whole compound body:

$$\frac{1-\alpha_\Omega}{2}u_i^m + \sum_{k=1}^{N}\sum_{l=1}^{4} A_{ij}^{m,k,l} u_j^{\chi(k,l)} = \sum_{k=1}^{N_1} B_{ij}^{m,k} t_j^k - \sum_{k=1}^{N_2}\sum_{l=1}^{4} D_{ij}^{m,k,l}(u_j^{\chi(k,l)} - u_j^{\overline{\chi}(\overline{k},l)}),$$

$$\frac{1-\alpha_\Omega}{8}u_i^m + \sum_{k=1}^{N}\sum_{l=1}^{4} A_{ij}^{m,k,l} u_j^{\chi(k,l)} = \sum_{k=1}^{N_1} B_{ij}^{m,k} t_j^k - \sum_{k=1}^{N_2}\sum_{l=1}^{4} D_{ij}^{m,k,l}(u_j^{\chi(k,l)} - u_j^{\overline{\chi}(\overline{k},l)}),$$

$$N = N_1 + N_2.$$

The first equation here is written in generalized displacements approximation nodes, the second one—in generalized tractions approximation nodes; the first right-side term in both equations is related to boundary elements with no contact, the second one satisfies the conditions of contact between elements k and $\overline{k}$ belonging to different regions; coefficients A, B, and D are computed with the help of the following formulae:

$$A_{ij}^{m,k,l} = \int_{-1}^{1}\int_{-1}^{1}\left[R^l(\xi)T_{ij}(x^m, y^k(\xi), s) - \delta_{\chi(k,l),}T_{ij}^0(x^m, y^k(\xi))\right]J^k(\xi)d\xi_1 d\xi_2,$$

$$B_{ij}^{m,k} = \int_{-1}^{1}\int_{-1}^{1} U_{ij}(x^m, y^k(\xi), s)J_k(\xi)d\xi_1 d\xi_2,$$

$$D_{ij}^{m,k,l} = \int_{-1}^{1}\int_{-1}^{1} R^l(\xi)U_{ij}(x^m, y^k(\xi), s)J_k(\xi)\alpha_j^{\chi\overline{\chi}}(y^k(\xi))d\xi_1 d\xi_2,$$

where s is the Laplace transform parameter. The integrals are calculated using Gaussian quadrature on the elements containing no singularities. Once an element contains singularity, singularity decreasing or eliminating algorithm is used [15]. An adaptive integration algorithm is used: the order of Gaussian quadrature in an element is chosen by the satisfaction of the prescribed precision if possible, otherwise the element is recursively subdivided to smaller elements for integration.

27.4 Laplace Transform Inversion

CQM is generally a method to calculate a convolution integral:

$$y(t) = f(t) * g(t) = \int_0^t f(t-\tau)g(\tau)d\tau$$

from one of the convoluted functions ($f(t)$) known in Laplace domain and some discrete values of another one ($g(t)$) [1–3]. The idea consists in rewriting the convolution integral in the following form with the help of the inverse Laplace transform formula:

$$\int_0^t f(t-\tau)g(\tau)d\tau = \frac{1}{2\pi i}\lim_{R\to\infty}\int_{c-iR}^{c+iR}\bar{f}(s)\underbrace{\int_0^t e^{s(t-\tau)}g(\tau)d\tau}_{x(t,s)}\,ds, \tag{27.2}$$

where $\bar{f}(s)$ denotes the function in Laplace domain.

Calculating $x(t, s)$ numerically is the only approximation to be used in scope of CQM. $x(t, s)$ is obviously the solution of the following Cauchy problem:

$$\frac{d}{dt}x(t,s) = sx(t,s) + g(t), x(0,s) = 0.$$

A scheme to solve this problem defines the particular case of CQM. Employing a linear multistep method to solve for $x(t, s)$, evaluating the outer integral in (27.2) with the help of residue theory and expanding $\bar{f}(\gamma(z)/\Delta t)$ in power series by z, we get the convolution integral in the following form [3]:

$$y(n\Delta t) = \sum_{k=0}^{n}\omega_{n-k}(\Delta t)g(k\Delta t), \quad n = 0, 1, ..., N. \tag{27.3}$$

Coefficients $\omega_n(\Delta t)$ can be computed by means of integral Cauchy formula:

$$\omega_n(\Delta t) = \frac{1}{2\pi i}\int_{|z|=R}\bar{f}\left(\frac{\gamma(z)}{\Delta t}\right)z^{-n-1}dz = \frac{R^{-n}}{2\pi}\int_0^{2\pi}\bar{f}\left(\frac{\gamma(R^{i\phi})}{\Delta t}\right)e^{-in\phi}d\phi. \tag{27.4}$$

R denotes radius of the analyticity region for $\bar{f}(\gamma(z)/\Delta t)$, and $\gamma(z)$ is characteristic function for the applied linear multistep method. Backward Differentiation Formula (BDF)-based methods of order ≤ 6 are applicable in scope of this solution scheme. For BDF-2 (traditional method, or second order Euler), we have $\gamma(z) = 3/2 - 2z + z^2/2$ [3].

Time-step method for Laplace transform numerical inversion is similar to CQM, but whereas CQM is based on the convolution theorem, it's based on the integration theorem, hence dedicated to the calculation of the original function integral. In order to get as a result, the original function $f(t)$ such as $f(0) = 0$ from $\bar{f}(s)$, we need to apply the theorem to its derivative:

$$f(t)=\int_0^t f'(\tau)d\tau.$$

In this case, (27.3), (27.4) will be rewritten as

$$f(0)=0,\quad f(n\Delta t)=\sum_{k=1}^{n}\omega_k(\Delta t),\quad n=1,...,N,$$

$$\omega_n(\Delta t)=\tfrac{R^{-n}}{L}\sum_{l=0}^{L-1}\overline{f}(s)se^{-in\phi},\quad s=\tfrac{\gamma(z)}{\Delta t},\quad z=R^{i\phi},\quad \phi=2\pi\tfrac{l}{L}.$$

Several modifications of the method can be derived from different ways to calculate $\omega_n(\Delta t)$. For varied step by ϕ and linear approximation of the integrand function, we obtain

$$\omega_n(\Delta t)=\frac{R^{-n}}{2\pi}\sum_{k=0}^{L-1}\left[\overline{f}(s_k)s_k e^{-in\phi_k}+\overline{f}(s_{k+1})s_{k+1}e^{-in\phi_{k+1}}\right]\frac{(\phi_{k+1}-\phi_k)}{2},\quad s_k=\frac{\gamma(R^{i\phi_k})}{\Delta t}$$

or, basing on the highly oscillatory quadrature [4]:

$$\omega_n(\Delta t)=\frac{R^{-n}}{2\pi}\sum_{k=0}^{L-1}\frac{\phi_{k+1}-\phi_k}{2}e^{-in\frac{\phi_k+\phi_{k+1}}{2}}\left[D_1(w)\overline{f}(s_k)s_k+D_2(w)\overline{f}(s_{k+1})s_{k+1}\right],$$

$$s_k=\frac{\gamma(R^{i\phi_k})}{\Delta t},$$

$$w=-n\frac{\phi_{k+1}-\phi_k}{2},\quad D_{1,2}(w)=\begin{cases}\dfrac{\sin w}{w}\pm\dfrac{w\cos w-\sin w}{w^2}i & \text{where } |w|>w_*,\\ e^{\mp wi} \text{ where } |w|\le w_*,\end{cases}$$

here w_* is specific key value, and the overall scheme using this formula is informally called "scheme with a key."

Substituting basic linear multistep method for Runge–Kutta method referred to as Butcher tableau $\frac{c|A^T}{b^T}$, $A\in R^{m\times m}$, $b, c\in R^m$, we obtain the following representation [8, 9]:

$$f(0)=0,\quad f(n\Delta t)=b^TA^{-1}\sum_{k=1}^{n}\omega_k(\Delta t),\quad n=1,...,N.$$

Formulae for coefficients $\omega_n(\Delta t)$ can still be used here taking into account that now they contain matrices instead of scalars. Characteristic function $\gamma(z)$ takes the following form [8, 9]:

$$\gamma(z)=A^{-1}-zA^{-1}[1]b^TA^{-1},\ [1]=(1,...,1)^T$$

Radau (Radau IIA) scheme [8] is chosen to further represent Runge–Kutta family.

27.5 Numerical Example

We consider the problem of the action of a vertical transient force $t_3 = t^0 f(t)$ of magnitude $t^0 = 1\text{N/m}^2$ on a surface element of a homogeneous poroelastic half-space with a width of 0.5 m (Fig. 27.1). The Heaviside function is taken as the law of change of the load on the area, i.e., $f(t) = H(t)$. At a distance $D = 1$ m from the loading region, there is a notch with dimensions $L = 1$ m, $W = 0.5$ m, $H = 1$ m. The surface of the half-space is free and permeable, i.e., the pore pressure is zero. The following material parameters (sandstone) are used for calculations: $K = 2.1 \cdot 10^8$ N/m^2, $G = 9.8 \cdot 10^7$ N/m^2, $\rho = 1884$ kg/m^3, $\varphi = 0.48$, $K_s = 1.1 \cdot 10^{10}$ N/m^2, $\rho_f =$ 1000 kg/m^3, $K_f = 3.3 \cdot 10^9$ N/m^2, $\kappa = 3.55 \cdot 10^{-9} m^4/(\text{N} \cdot \text{sec})$.

The problem was solved by the boundary-time element method using Euler and Radau schemes on boundary-element meshes of 588 (mesh 1), 720 (mesh 2), and 972 (mesh 3) elements, taking into account one plane of symmetry (Figs. 27.2, 27.3, and 27.4).

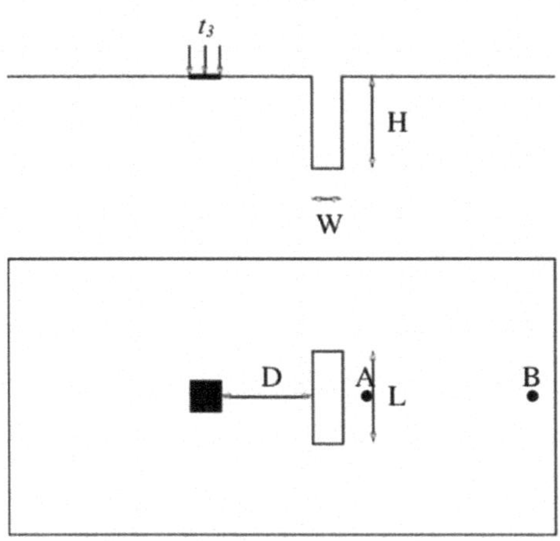

Fig. 27.1 Schematic of the problem of vibration isolation by an open trench

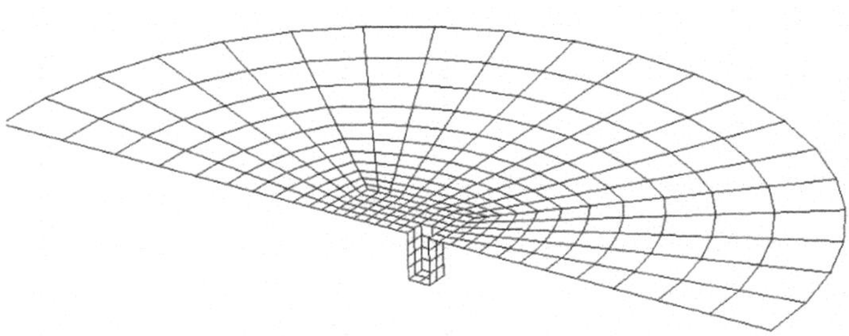

Fig. 27.2 Mesh consisting of 588 boundary elements

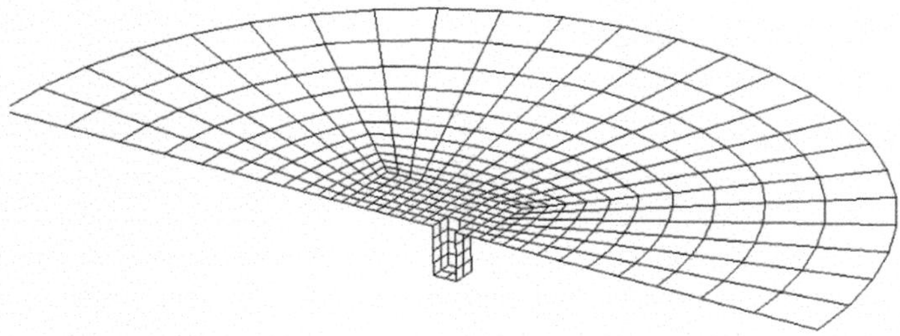

Fig. 27.3 Mesh consisting of 720 boundary elements

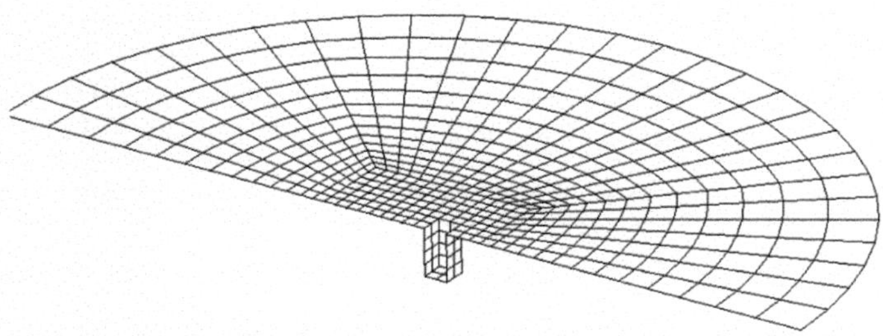

Fig. 27.4 Mesh consisting of 972 boundary elements

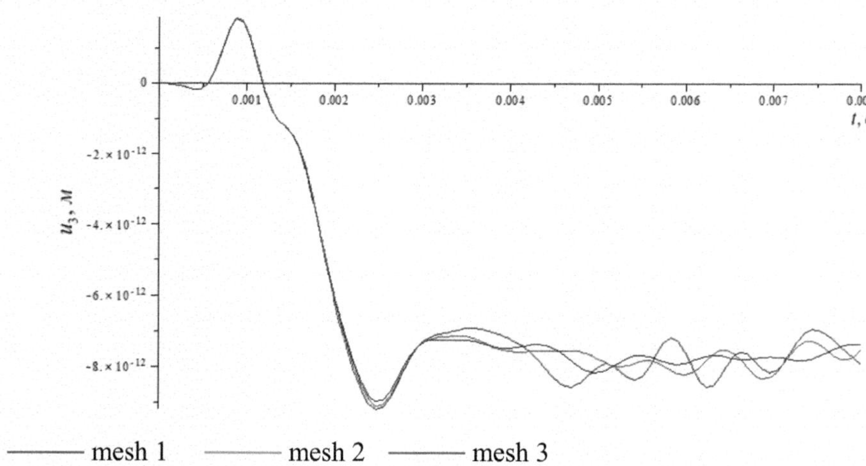

Fig. 27.5 Vertical displacement at point A

The dynamic responses of the vertical displacement component are considered at points at a distance of $l = 1.75$ m (point A), $l = 2.67$ m, $l = 4.94$ m, and $l = 6.83$ m from the load application zone, respectively. The following parameters are taken for calculations using the Euler scheme: $R = 0.997$; $L = 415$ nodes on the interval $[0,\ \pi]$, of which 400 are distributed uniformly on $[0,\ 2]$, 15—on $[2,\ \pi]$; number of time steps $N = 100$. The grid convergence of the solution using the scheme on the nodes of the Euler method in vertical displacements at point A is demonstrated in Fig. 27.5. Further, we will consider mesh 1 (588 elements).

Figures 27.6, 27.7, 27.8, and 27.9 show the results of the scheme study on the nodes of the Euler method to select the optimal step at the points $l = 1.75$ m, $l = 2.67$ m, $l = 4.94$ m, and $l = 6.83$ m, respectively.

Figures 27.10, 27.11, 27.12, and 27.13 show the solutions obtained using stepwise methods at the nodes of Euler and Radau schemes at points $l = 1.75$ m, $l = 2.67$ m, $l = 4.94$ m, and $l = 6.83$ m, respectively. The following parameters are taken for the calculations using the Radau scheme: $L = 121$ double nodes on the interval $[0, \pi]$,

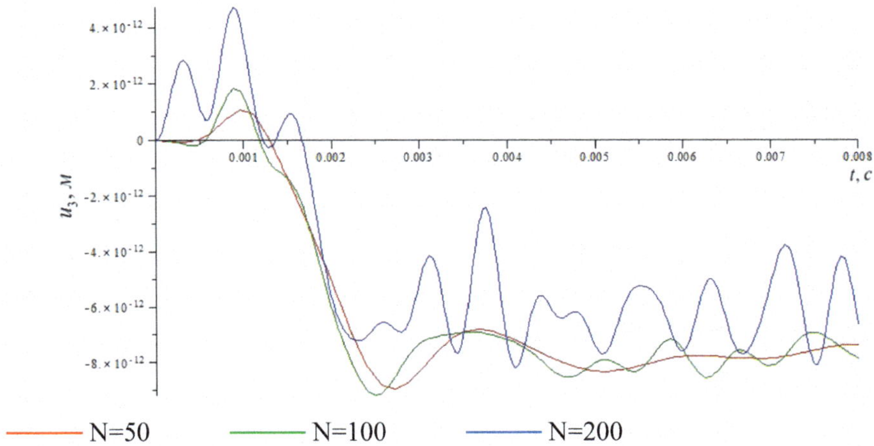

Fig. 27.6 Effect of N on the vertical displacement at point $l = 1.75$ m

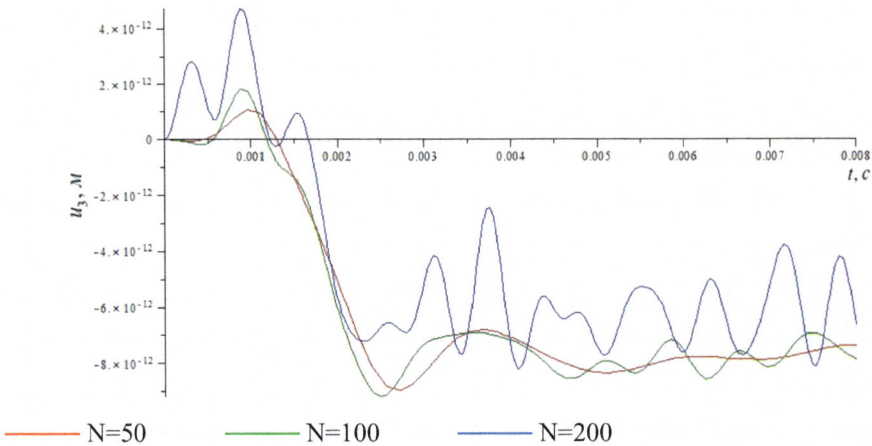

Fig. 27.7 Effect of N on the vertical displacement at point $l = 2.67$ m

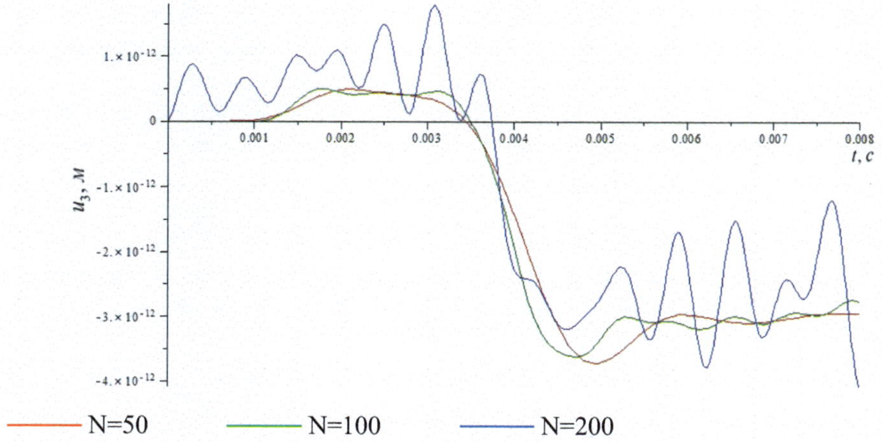

Fig. 27.8 Effect of N on the vertical displacement at point $l = 4.94$ m

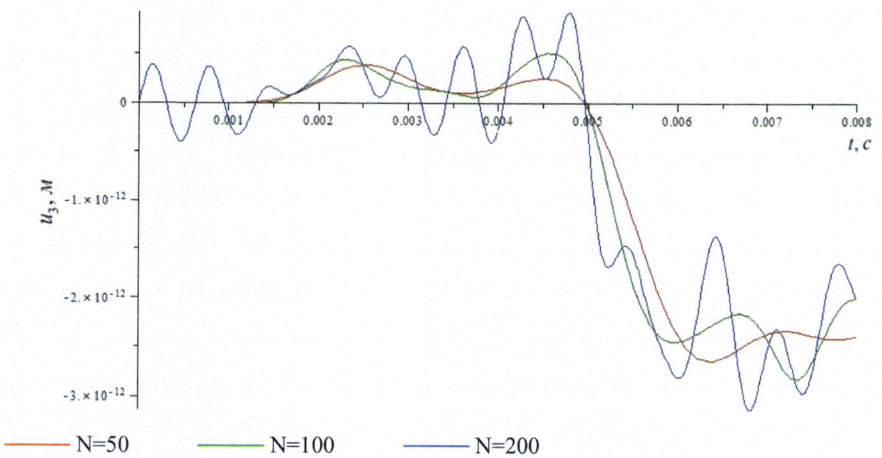

Fig. 27.9 Effect of N on the vertical displacement at point $l = 6.83$ m

of which 113 are distributed on [0, 2], 8 are distributed on [2, π]; $N = 30$ time steps.

A time-stepping BEM scheme on the Radau method nodes gives results close to solutions based on the traditional step scheme but requires much less cost: 30 time steps compared to 100 for the scheme on Euler method nodes, and 121 two-fold nodes by argument compared to 415 one-fold for the scheme on Euler method nodes.

Figures 27.14, 27.15, 27.16, 27.17, and 27.18 show the solutions for the space with a notch at points $l = 1.75$ m, $l = 2.67$ m, $l = 4.94$ m, and $l = 6.83$ m, respectively, compared to the results for the half-space without a notch.

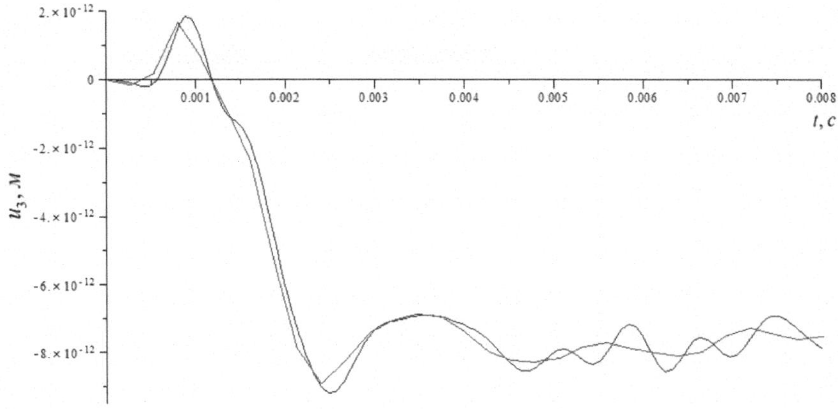

——— Euler, N=100, L=415 ——— Radau, N=30, Ł=242

Fig. 27.10 Vertical displacement at point $l = 1.75$ m

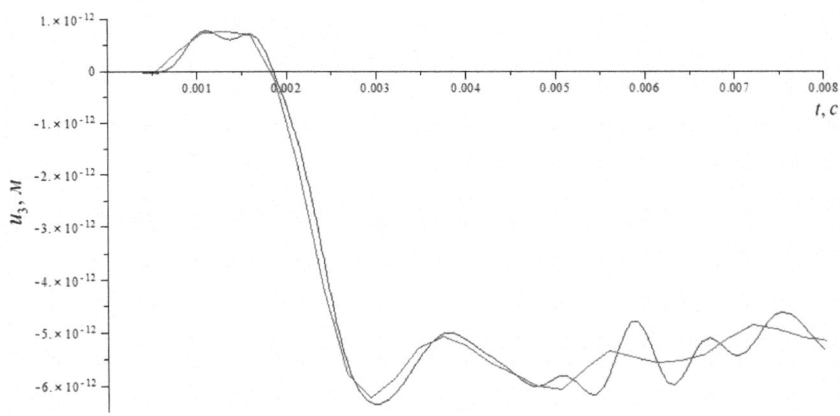

——— Euler, N=100, L=415 ——— Radau, N=30, Ł=242

Fig. 27.11 Vertical displacement at point $l = 2.67$ m

Figure 27.18 shows the dependence of the vertical displacement response on the distance of the observed point from the loading region.

27.6 Conclusions

The solution constructed using the step method based on the Euler scheme is noticeably inferior in approximating the displacement to the step method at the nodes of the Radau scheme.

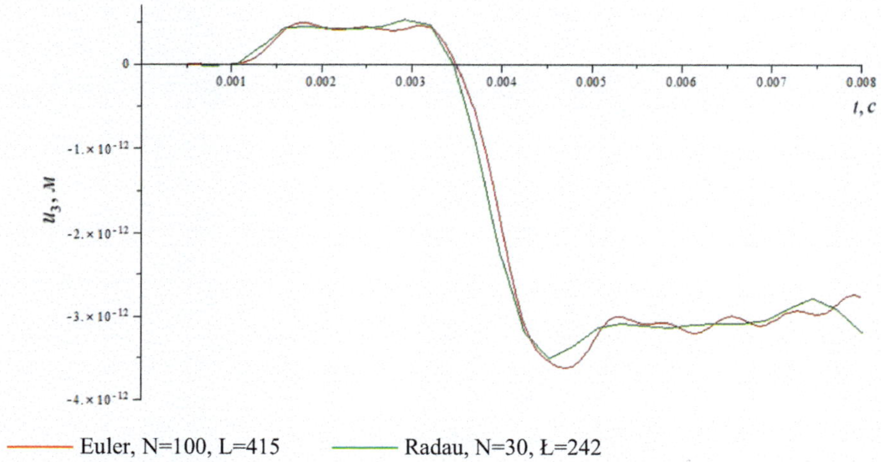

Fig. 27.12 Vertical displacement at point $l = 4.94$ m

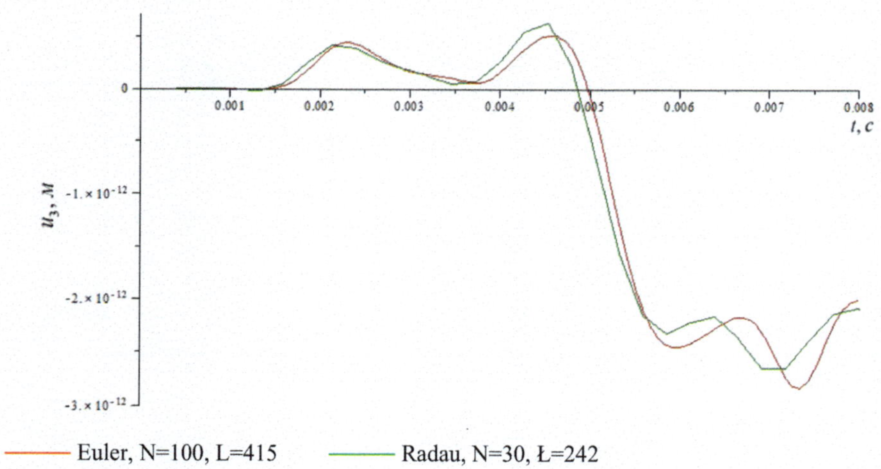

Fig. 27.13 Vertical displacement at point $l = 6.83$ m

In general, a time-step boundary element scheme with numerical reversal of the Laplace transform using the Radau scheme is more sensitive to computational errors than a scheme based on the Euler method, as a result of which great care must be taken in calculations. Nevertheless, with proper selection of parameters, it allows you to get more accurate results with a noticeable reduction in computational costs.

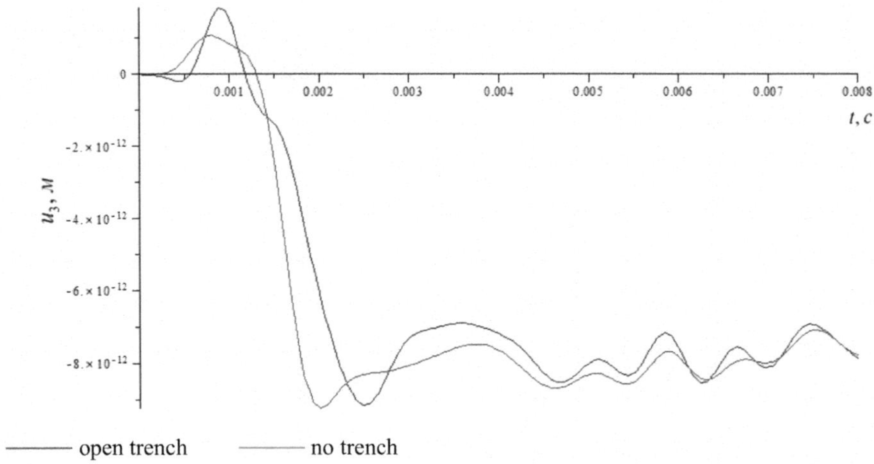

Fig. 27.14 Vertical displacement at point $l = 1.75$ m with open trench and without trench

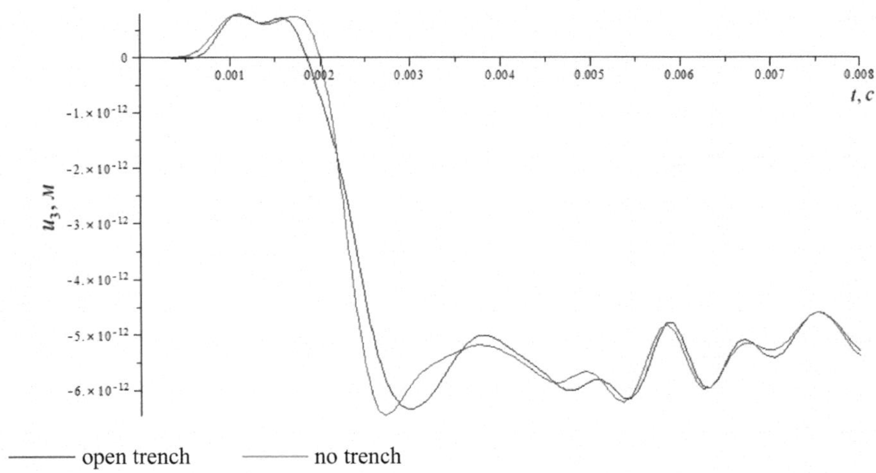

Fig. 27.15 Vertical displacement at point $l = 2.67$ m with open trench and without trench

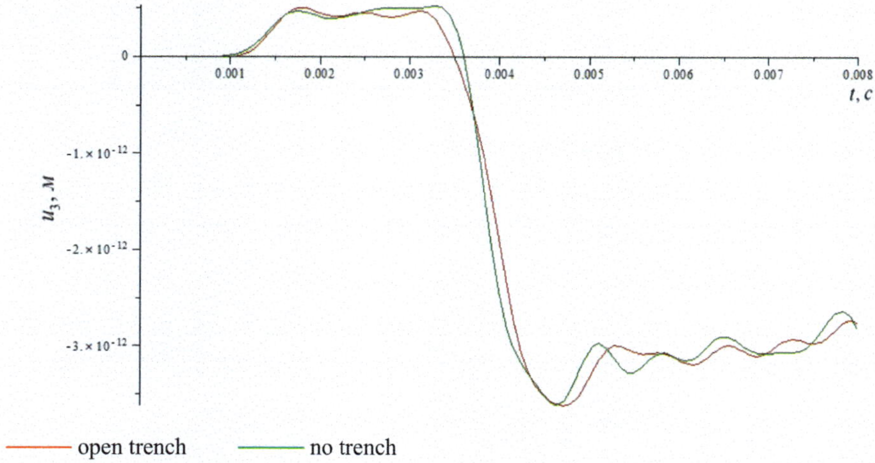

Fig. 27.16 Vertical displacement at point $l = 4.94$ m with open trench and without trench

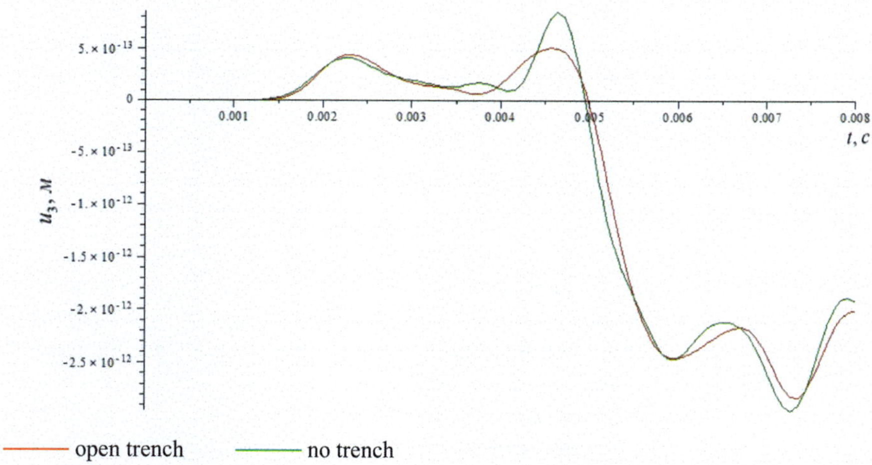

Fig. 27.17 Vertical displacement at point $l = 6.83$ m with open trench and without trench

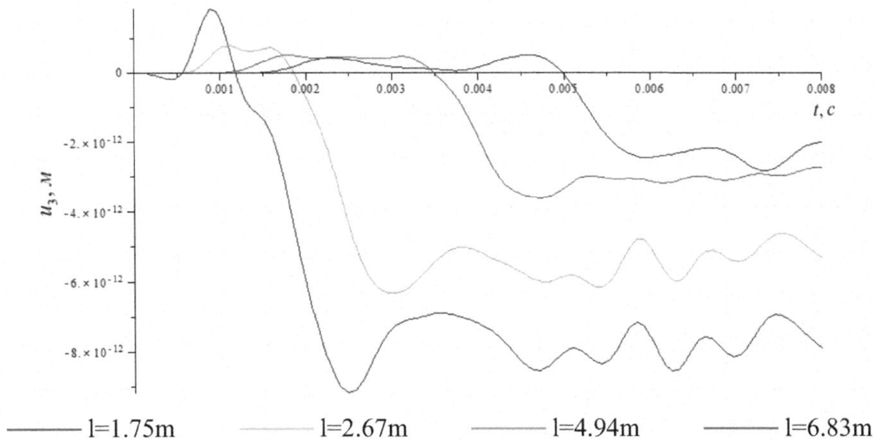

Fig. 27.18 Vertical displacement at the various points

Acknowledgements The work was supported by the Russian Science Foundation grant № 22-19-00138.

References

1. Lubich C (1988) Convolution quadrature and discretized operational calculus I. Numer Math 52:129–145. https://doi.org/10.1007/BF01398686
2. Lubich C (1988) Convolution quadrature and discretized operational calculus II. Numer Math 52:413–425. https://doi.org/10.1007/BF01462237
3. Schanz M (2001) Wave propagation in viscoelastic and poroelastic continua: a boundary element approach. Springer, Berlin
4. Bazhenov VG, Belov AA, Igumnov LA (2009) Boundary element modeling the dynamics of piecewise homogeneous media and constructions (Granichno-elementnoye modelirovaniye dinamiki kusochno-odnorodnyh sred i konstruktsiy). Lobachevsky State University of Nizhni Novgorod, Nizhny Novgorod (in Russian)
5. Amenitskiy AV, Belov AA, Igumnov LA, Litvinchuk SY (2008) Boundary-element analysis of the dynamic state of piecewise homogeneous elastic bodies using convolution quadratures. Comput Continuum Mech 1(3):5–14. https://doi.org/10.7242/1999-6691/2008.1.3.22
6. Petrov AN (2013) Modeling the dynamics of composite poroelastic bodies based on the boundary-time element method (Modelirovanie dinamiki sostavnyh porouprugih tel na osnove metoda granichno-vremennyh elementov). UNN
7. Belov AA, Igumnov LA, Litvinchuk SY (2008) Boundary element procedure based on modified convolution quadrature method in dynamic problems of elastic bodies. Probl Strength Plasticity 70:150–158 (in Russian)
8. Banjai L (2010) Multistep and multistage convolution quadrature for the wave equation: algorithms and experiments. SIAM J Sci Comput 32:2964–2994. https://doi.org/10.1137/090775981
9. Banjai L, Messner M, Schanz M (2012) Runge-Kutta convolution quadrature for the boundary element method. Comput Methods Appl Mech Eng 245–246:90–101. https://doi.org/10.1016/j.cma.2012.07.007

10. Igumnov LA, Rataushko YY (2013) Time-step method of laplace transformation numerical inversion based on the Runge-Kutta scheme nodes with a variable step of integration. Probl Strength Plasticity 75(4)
11. Biot MA (1956) Theory of propagation of elastic waves in a fluid-saturated porous solid I. Low-frequency range. J Acoust Soc Am 28(2):168–78
12. Bonnet G (1987) Basic singular solutions for a poroelastic medium in the dynamic range. J Acoust Soc Am 82:1758–1762
13. Bonnet G, Auriault JL (1985) Dynamics of saturated and deformable porous media: homogenization theory and determination of the solid-liquid coupling coefficients. In: Boccara N, Daoud M (eds) Physics of finely divided matter. Springer, Berlin, pp 306–316
14. Amenitskiy AA, Belov AA, Igumnov LA, Karelin IS (2009) Boundary integral equations for analyzing dynamic problems of 3-d porouselasticity. Probl Strength Plasticity 71:164–71
15. Goldshteyn RV (1978) Boundary integral equations method: numerical aspects and application in mechanics (Mechanics: New in Foreign Science) (Metod granichnyh integralnyh uravneniy: Vychislitelniye aspekty i prilozheniya v mehanike (Mehanika: novoye v zarubezhnoy nauke)): collected works, Mir, Moscow, p 183 (in Russian)
16. Bazhenov VG, Igumnov LA (2008) Boundary integral equations and boundary element methods in treating the problems of 3D elastodynamics with coupled fields (Metody granichnyh integralnyh uravneniy i granichnyh elementov v reshenii zadach tryohmernoy dinamicheskoy teorii uprugosti s sopryazhyonnymi polyami), PhysMathLit, Moscow (2008) (in Russian)

Chapter 28
The Response of Poroviscoelatic Layered Half-Space and Half-Space Weakened by Cavity to Transient Load by Boundary Element Approach

Andrey N. Petrov, Leonid A. Igumnov, Aleksandr A. Belov, and Valeriy V. Novikov

Abstract In this study, the dynamic response of a poroviscoelastic half-space to a transient surface load is investigated using Biot's dynamic theory of poroelasticity and direct boundary element method. The skeleton of a porous material is assumed to be viscoelastic material. A model of a standard viscoelastic solid is applied to describe the viscoelastic properties of a skeleton. Viscoelastic effects are taken into account by means of the elastic-viscoelastic correspondence principle. The initial boundary-value problem is reduced to a boundary-value problem by formal application of the Laplace transform. In a first stage, the solution is obtained in the frequency domain. The transient solutions are obtained, in a second stage, by applying numerical inversion of the Laplace transform which is applied based on the Durbin algorithm with a variable frequency step to the previously synthesized frequency domain solutions. The problem of the action of a vertical force in the form of the Heaviside function on the surface of a layered porous-elastic half-space and a half-space with a cavity is considered. A normal distributed uniform loading is applied over a rectangular region of the half-space surface, while the other parts are kept traction-free. The effect of the viscoelastic properties of the skeleton of a poroelastic material on vertical displacements on the half-space surface is investigated.

A. N. Petrov · L. A. Igumnov (✉) · A. A. Belov · V. V. Novikov
National Research Lobachevsky State University of Nizhny Novgorod, Nizhny Novgorod, Russia
e-mail: igumnov@mech.unn.ru

A. N. Petrov
e-mail: andrey.petrov@mech.unn.ru

A. A. Belov
e-mail: belov_a2@mech.unn.ru

H. Altenbach and V. A. Eremeyev (eds.), *Dynamics of Discrete and Continuum Structures and Media*, Advanced Structured Materials 221,
https://doi.org/10.1007/978-3-031-75626-9_28

28.1 Introduction

Currently, mathematical modeling is one of the main tools used for analysis and optimization of oil and gas field development processes in earthquake-resistant construction and bioengineering Hansen et al. [12], Phillips and Wheeler [17], Phillips and Wheeler [18], Sobhaniaragh et al. [20], Nazarova and Nazarov [14, 15]. In such problems, a poroelastic medium model is usually employed, which allows us to describe fluid filtration in pores together with a full-fledged mechanical model of the stress-strain state of the medium. In a modern form, such models were proposed by Biot [4–6]. The model describes the joint processes of deformation of an elastic medium and fluid flow in it. Within the framework of the model, it is assumed that the space enclosing the poroelastic medium is filled with a two-phase medium, with one phase corresponding to the elastic skeleton and the second to the fluid contained in the pores. Both phases are present at every point of physical space, and the distribution of phases in space is described by macroscopic quantities such as porosity. Application of the Biot theory allows solving a number of particular problems, among which the problems of wave propagation in homogeneous and layered porous half-spaces are of particular interest. However, the increasing complexity of the computational schemes of boundary-value problems requires the usage and development of new methods, such as the boundary element method (BEM). Possessing high accuracy and rigor of the approach, the BEM is the most suitable method when considering non-stationary processes in semi-infinite bodies, since it ensures the automatic fulfillment of the conditions for the behavior of solutions at infinity. Despite the noted advantages of the BEM, the possibility of modeling poroelastic dynamics with its help is mainly determined by the presence of the corresponding boundary-integral equations (BIE) and fundamental solutions. Fundamental solutions and BIE of the dynamic theory of poroelasticity were obtained in Bo and Hua [7], Banks et al. [2]; in Chapelle et al. [8], Theodorakopoulos and Beskos [21] various versions of boundary element schemes for solving the problems of particle dynamics were implemented, and the results of numerical experiments were presented. The results of boundary element modeling of the dynamics of poroelastic half-spaces are given in Gatmiri and Eslami [11], Ehlers and Markert [10] and Philippacopoulos [16]. The fundamental difference between the poroelastic formulation and the elastic and viscoelastic one is that it allows one to take into account the influence exerted by the fluid moving in the pores on the behavior of the medium as a whole. Historically, based on the Bio theory, the existence of two longitudinal waves in the porous medium was predicted—so-called fast and slow longitudinal waves. In addition, the frictional interaction of the viscous fluid and the skeleton leads to significant energy dissipation in the medium, which as a result exhibits viscoelastic behavior Theodorakopoulos and Beskos [21]. The viscoelastic behavior of a poroelastic medium can also be due to the viscoelastic properties of the skeleton Bo and Hua [7]. Some results of modeling wave processes in poroelastic solids using BIE and BEM and various models of viscoelastic behavior of the skeleton are presented in Schanz [19], Igumnov and Petrov [13]. In our work, the propagation of waves in homogeneous and non-homogeneous porous-viscoelastic semi-infinite solids is simulated. The Durbin method of numerical inversion of the Laplace integral transform is used. The problems of the action of an unsteady load on the surface of a half-space are solved, and the cases of layered half-space and half-space with a cavity are considered.

28.2 Mathematical Model

We consider a homogeneous solid Ω in three-dimensional space R^3, and $\Gamma = \delta\Omega$ is the boundary of Ω. Solid Ω is isotropic and poroviscoelastic.Basic poroelastic material is a two-phase material consisting of an elastic skeleton and compressible fluid or gas filler. Porous material of a volume V can be constructed as follows:

$$V = V^f + V^s \, , \tag{28.1}$$

where V is the total volume, V^f is the summary pore volume, and V^s is the volume of the skeleton. It is assumed that filler can openly seep through the pores, and all closed pores are assumed as a part of the skeleton. Then a correspondence principle is applied to the skeleton, so we extend poroelastic formulation to poroviscoelasticity. After formal applying of Laplace transform, we obtain a statement for a boundary-value problem for Biot's model of fully saturated poroelastic continuum in Laplace domain. The set of differential equations of poroelasticity for displacements $\bar{u}_i$ and pore pressure $\bar{p}$ for Biot's model of fully saturated poroelastic continuum in Laplace domain take the following form Schanz [19]:

$$G\bar{u}_{i,jj} + \left(K + \frac{1}{3}G\right)\bar{u}_{j,ij} - (\psi - \beta)\,\bar{p}_{,i} - s^2\,(\varrho - \beta\varrho_{\text{f}})\,\bar{u}_i = \bar{F}_i \, , \tag{28.2}$$

$$\frac{\beta}{s\varrho_{\text{f}}}\bar{p}_{,ii} - \frac{\phi^2 s}{R}\bar{p} - (\psi - \beta)\,s\bar{u}_{i,i} = \bar{a}_i \, , \tag{28.3}$$

where K and G are the elastic moduli of the porous material, p is the porous pressure of the filling material, and $\bar{F}_i$ is volume force density. Coefficient α is determined as

$$\psi = 1 - \frac{K}{K_{\text{s}}} \, , \tag{28.4}$$

where K_{s} is bulk modulus of the skeleton grains. In (28.3) R is parameter characterizing the relation between the solid body and the liquid:

$$R = \frac{\phi^2 K_{\text{f}} K_{\text{s}}^2}{K_{\text{f}}(K_{\text{s}} - K) + \phi K_{\text{s}}(K_{\text{s}} - K_{\text{f}})} \, , \tag{28.5}$$

$$\beta = \frac{\kappa \varrho_{\mathrm{f}} \phi^2 s^2}{\phi^2 s + s^2 \kappa (\varrho_{\mathrm{a}} + \phi \varrho_{\mathrm{f}})} ,$$

where $\varrho = \varrho_{\mathrm{s}}(1-\phi) + \phi\varrho_{\mathrm{f}}$, ϱ_s is density of the skeleton grains, κ is permeability, ϱ_{f} is density of the filling material, and K_{f} is the bulk modulus of the liquid.

A generalized displacement vector and a generalized force vector are additionally introduced as

$$\bar{u}(\mathbf{x}, s) = (\bar{u}_1, \bar{u}_2, \bar{u}_3, \bar{p}) \ \mathbf{x} \in \Omega,\ \Omega \subset R^3 , \tag{28.6}$$

$$\bar{t}(\mathbf{x}, s) = \left(\bar{t}_1, \bar{t}_2, \bar{t}_3, \bar{q}\right) \ \mathbf{x} \in \Omega,\ \Omega \subset R^3 . \tag{28.7}$$

Equations (28.2) and (28.3), supplemented with boundary conditions:

$$u(\mathbf{x}, s) = \widetilde{u},\ \mathbf{x} \in \Gamma^{\mathrm{u}} , \tag{28.8}$$

$$t(\mathbf{x}, s) = \widetilde{t},\ \mathbf{x} \in \Gamma^{\sigma} , \tag{28.9}$$

where Γ^{u} is Dirichlet's boundary and Γ^{σ} is Neumann's boundary, where corresponding generalized displacements and generalized tractions are prescribed.

Then the elastic-viscoelastic correspondence principle is applied to the skeleton, so we extend poroelastic formulation to poroviscoelasticity. Poroviscoelastic solution is obtained from poroelastic solution by replacing to skeleton's moduli K and G in Laplace domain with material creep functions $\bar{K}(s)$ and $\bar{G}(s)$.

In the present paper, we employ standard linear solid model:

$$\bar{K}(s) = K^{\infty} \cdot \left[(\theta - 1)\frac{s}{s+\gamma} + 1\right] , \tag{28.10}$$

$$\bar{G}(s) = G^{\infty} \cdot \left[(\theta - 1)\frac{s}{s+\gamma} + 1\right] , \tag{28.11}$$

$$\theta = \frac{K^0}{K^{\infty}} = \frac{G^0}{G^{\infty}} . \tag{28.12}$$

The equilibrium and instantaneous values of the relaxation function associated with material modules are connected as follows, and their values are denoted by '∞' and '0' respectively.

28.3 Solution Methods

Following regularized BIE system corresponding boundary-value problem (28.2) and (28.3) is considered:

$$\mathbf{C}(\mathbf{y})\mathbf{u}(\mathbf{y}, s) + \int_{\Gamma} \mathbf{T}(\mathbf{x}, \mathbf{y}, s)\mathbf{u}(\mathbf{x}, s)d\Gamma = \int_{\Gamma} \mathbf{U}(\mathbf{x}, \mathbf{y}, s)\mathbf{t}(\mathbf{x}, s)d\Gamma \ , \mathbf{x} \ , \mathbf{y} \in \Gamma \ , \tag{28.13}$$

where $\mathbf{U}(\mathbf{x}, \mathbf{y}, s)$ and $\mathbf{T}(\mathbf{x}, \mathbf{y}, \mathrm{s})$ are matrices of fundamental and singular solutions, respectively, $\mathbf{x}$ is an integration point, and $\mathbf{y}$ is an observation point. The values of the coefficients of matrix $\mathbf{C}$ are defined by the geometry of boundary Γ. A procedure for obtaining BIEs, based on the weighted residual method can be found in Schanz [19]. Some of the problems of the arising kernels of BIEs are discussed in Amenitsky et al. [1].

Equation (28.13) comprises singular integrals in the sense of Cauchy, which are quite difficult to compute. Use of the boundary properties of retarded potentials makes it possible, based on Bazhenov and Igumnov [3], to write down a regular representation of Eq. (28.13):

$$\int_{\Gamma} \left(\mathbf{T}(\mathbf{x}, \mathbf{y}, s)\mathbf{u}(\mathbf{x}, s) - \mathbf{T}^0(\mathbf{x}, \mathbf{y}, s)\mathbf{u}(\mathbf{y}, s) - \mathbf{U}(\mathbf{x}, \mathbf{y}, s)\mathbf{t}(\mathbf{x}, s)\right) d\Gamma = 0 \ , \mathbf{x} \ , \mathbf{y} \in \Gamma \ , \tag{28.14}$$

where $\mathbf{T}^0(\mathbf{x}, \mathbf{y}, s)$ is singularity matrix. Using Eq. (28.14), it is possible to construct a boundary element solution of the BIE.

In the result of spatial discretization, boundary Γ is represented with a set of K_E quadrangular eight-node boundary elements. The geometry of each element E_K is defined by biquadratic functions of form N_m and the global coordinates of nodes $\mathbf{x}_m^k$, related as Bazhenov and Igumnov [3]

$$\mathbf{x}(\zeta) = \sum_{m=1}^{8} N_m(\xi)\mathbf{x}_m^k \ , k = 1..K \ , \tag{28.15}$$

where $\xi = (\xi_1, \xi_2) \in (-1, 1)^2$ are local coordinates. According to correlated interpolation model, displacements are described using bilinear elements with the related bilinear functions of form $R_l(\xi)$, and surface generalized forces are described with constant boundary elements:

$$\mathbf{u}(\xi) = \sum_{l=1}^{4} R_l(\xi)\mathbf{u}_l^k \ , \tag{28.16}$$

$$\mathbf{t}(\xi) = \mathbf{t}^k \ , \tag{28.17}$$

where $\mathbf{u}_m^k$ and $\mathbf{t}^k$ are nodal values of displacements and tractions, respectively, over element E_k. To discretize the boundary surface eight-node biquadratic quadrilateral elements are used, generalized displacements and tractions—are approximated by linear and constant shape functions, respectively. Subsequent applying of collocation method leads to the system of linear equations. As the collocation nodes we take the approximation nodes of boundary functions. Gaussian quadrature is used to calculate integrals on regular elements. But if an element contains a singularity, algorithm of singularity avoiding or order reducing is applied. When singularity is excluded we use an adaptive integration algorithm. An appropriate order of Gaussian quadrature is chosen from primarily known necessary precision, and if it is impossible, the element is subdivided into smaller elements recursively.

28.4 Numerical Laplace Transform Inversion

The poroviscoelastic solution in time domain is obtained with the help of the Laplace transform inversion. Durbin's algorithm is used for numerical inversion of Laplace transform. The inverse Laplace transform is defined as the following contour integral:

$$L^{-1}\{\bar{f}(\mathbf{x}, s)\} = f(\mathbf{x}, t) = \frac{1}{2\pi i}\int_{\alpha-i\infty}^{\alpha+i\infty} \bar{f}(\mathbf{x}, s)e^{st}\, ds\ , \tag{28.18}$$

where $\alpha > 0$ is the arbitrary real constant greater than the real parts of all singularities in $\bar{f}(\mathbf{x}, s)$. When values of $\bar{f}(\mathbf{x}, s)$ are available only at the sample points, analytical evaluation of integral in Eq. (28.18) is impossible. Supposing $s = \alpha + i\omega$ we have the following expressions (for convenience the spatial variable $\mathbf{x}$ is omitted hereinafter):

$$f(0) = \frac{1}{\pi}\int_0^\infty \mathrm{Re}[\bar{f}(\alpha + i\omega)]\, d\omega\ , \tag{28.19}$$

$$f(0) = \frac{e^{\alpha t}}{\pi}\int_0^\infty \{\mathrm{Re}[\bar{f}(\alpha + i\omega)]\cos(\omega t) - \mathrm{Im}[\bar{f}(\alpha + i\omega)]\sin(\omega t)\}\, d\omega,\ t > 0. \tag{28.20}$$

Durbin [9] developed an approach, namely fast Laplace inverse transform (FLIT) for numerical evaluation of the integrals in Eqs. (28.19) and (28.20). In this section, we briefly review a modification recently proposed by Zhao [22] in order to overcome a drawback of constant integration step in FLIT. Let R be large real number so we can rewrite Eqs. (28.19) and (28.20) as follows:

$$f(0) = \frac{1}{\pi}\lim_{R\to\infty}\int_0^R \mathrm{Re}[\bar{f}(\alpha + i\omega)]\, d\omega\ , \tag{28.21}$$

$$f(0) = \frac{e^{\alpha t}}{\pi}\lim_{R\to\infty}\int_0^R \{\mathrm{Re}[\bar{f}(\alpha + i\omega)]\cos(\omega t) - \mathrm{Im}[\bar{f}(\alpha + i\omega)]\sin(\omega t)\}\, d\omega,\ t > 0. \tag{28.22}$$

Defining the nodes as $0 = \omega_1 < \omega_2 < \cdots < \omega_n < \omega_{n+1} = R$, we approximate Eqs. (28.21) and (28.22) as

$$f(0) = \frac{1}{\pi} \sum_{k=1}^{n} \int_{\omega_k}^{\omega_{k+1}} \mathrm{Re}[\bar{f}(\alpha + i\omega)]\, \mathrm{d}\omega \,, \tag{28.23}$$

$$f(0) = \frac{e^{\alpha t}}{\pi} \sum_{k=1}^{n} \int_{\omega_k}^{\omega_{k+1}} \{\mathrm{Re}[\bar{f}(\alpha + i\omega)] \cos(\omega t) - \mathrm{Im}[\bar{f}(\alpha + i\omega)] \sin(\omega t)\}\, \mathrm{d}\omega, t > 0. \tag{28.24}$$

In each segment $[\omega_k, \omega_{k+1}], k = \overline{1, n}$, the real and imaginary parts of $\bar{f}(s)$ are approximated with linear functions as follows:

$$\mathrm{Re}[\bar{f}(\alpha + i\omega)] \approx F_k + \frac{F_{k+1} - F_k}{\omega_{k+1} - \omega_k}(\omega - \omega_k) \,, \tag{28.25}$$

$$\mathrm{Im}[\bar{f}(\alpha + i\omega)] \approx G_k + \frac{G_{k+1} - G_k}{\omega_{k+1} - \omega_k}(\omega - \omega_k) \,, \tag{28.26}$$

where $F_k = \mathrm{Re}[\bar{f}(\alpha + i\omega_k)]$, $G_k = \mathrm{Im}[\bar{f}(\alpha + i\omega_k)]$. Substituting Eqs. (28.25) and (28.26) into Eqs. (28.23) and (28.24) and making direct integration, we obtain

$$f(0) \approx \sum_{k=1}^{n} \left[\frac{(F_{k+1} - F_k)\Delta_k}{2\pi} \right], \tag{28.27}$$

$$f(t) \approx \frac{e^{\alpha t}}{\pi t^2} \sum_{k=1}^{n} \left[\frac{(F_{k+1} - F_k)}{\Delta_k} (\cos(\omega_{k+1} t) - \cos(\omega_k t)) + \frac{(G_{k+1} - G_k)}{\Delta_k} (\sin(\omega_{k+1} t) - \sin(\omega_k t)) \right], \tag{28.28}$$

where $t > 0$, $\Delta_k = \omega_{k+1} - \omega_k$.

28.5 Numerical Examples

28.5.1 Layered Half-Space

The problem of the Heaviside-type load H(t) acting on the surface of a poroelastic layer located on a poroelastic half-space is considered (Fig. 28.1). Two variants of geometry problem are considered—with a layer thickness of 5 m and 10 m. A vertical load $t = -1000\mathrm{N/m^2} \cdot H(t)$ is specified on a surface area of 1 $\mathrm{m^2}$, the rest of the surface is traction-free and permeable. At the boundary between the layer and half-space, the flow, the force, displacements, and pore pressure are assumed to be unknown functions. The parameters of the poroelastic rock are $K = 8 \cdot 10^9\ \mathrm{N/m^2}$,

Fig. 28.1 Layered half-space under vertical load

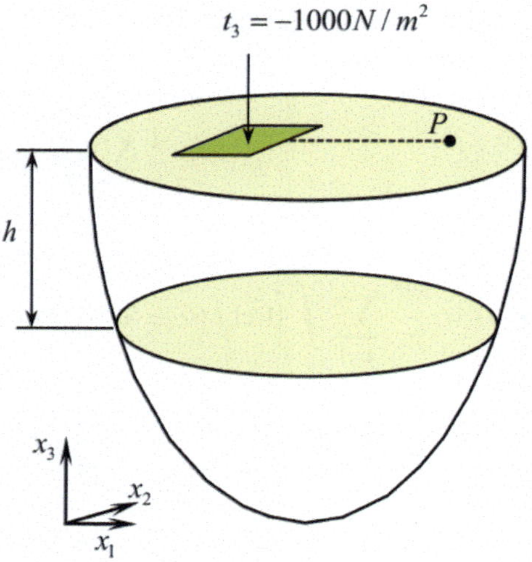

$G = 6 \cdot 10^9\,\mathrm{N/m^2}$, $K_s = 3.6 \cdot 10^{10}\,\mathrm{N/m^2}$, $K_f = 3.3 \cdot 10^9\,\mathrm{N/m^2}$, $\rho = 2458\,\mathrm{kg/m^3}$, $\rho_f = 1000\,\mathrm{kg/m^3}$, $\phi = 0.66$ and $\kappa = 1.9 \cdot 10^{-10}\,\mathrm{m^4/(N \cdot s)}$.

The parameters of the poroviscoelastic soil are $K = 2.1 \cdot 10^8\,\mathrm{N/m^2}$, $G = 9.8 \cdot 10^7\,\mathrm{N/m^2}$, $K_s = 1.1 \cdot 10^{10}\,\mathrm{N/m^2}$, $K_f = 3.3 \cdot 10^9\,\mathrm{N/m^2}$, $\rho = 1884\,\mathrm{kg/m^3}$, $\rho_f = 1000\,\mathrm{kg/m^3}$, $\phi = 0.48$ and $\kappa = 3.55 \cdot 10^{-9}\,\mathrm{m^4/(N \cdot s)}$.

Dynamic responses of vertical displacements u_3 at the point P, located at a distance of 10 m from the area of load application, are shown in Figs. 28.2, 28.3 and 28.4.

Figure 28.2 presents the comparison of vertical displacements, calculated for the layer with thicknesses of 5 m and 10 m and labeled u_3^{5m} and u_3^{10m}, respectively. Figure 28.2 also represents solutions for displacements (u_3^s), calculated for the same values of the half-space material parameters, which describe layer material. It can be seen that until the appearance of the fast longitudinal wave (t 0.01 s); all three curves are almost graphically indistinguishable; however, quantitative differences are observed further. In the moment of Rayleigh wave (t 0.047 s) appearance and up to the moment t 0.057 s, the displacement behavior is described by the relation $u_3^{5m} < u_3^s < u_3^{10m}$. The time moment t 0.057 s is marked by an increase in displacements u_3^{5m} and u_3^{10m}, and the displacements u_3^{5m} increase much more rapidly and exceed u_3^{10m} and u_3^s by amplitude. The observed effect is explained by the influence of longitudinal waves reflected from the half-space and is less noticeable in the case of a larger layer thickness due to the considerable wave dispersion in porous medium.

In order to obtain poroviscoelastic solutions standard linear solid model is employed. Parameter γ characterizes viscosity, and parameter $\theta = \frac{K^0}{K^\infty} = \frac{G^0}{G^\infty}$ characterizes dependence between equilibrium and instantaneous values of material modules. In further computations parameter $\theta = 4$.

Figures 28.3 and 28.4 show the effect of the viscoelastic properties of the skeleton on the dynamic response of vertical movements. Figures 28.3 and 28.4 shows

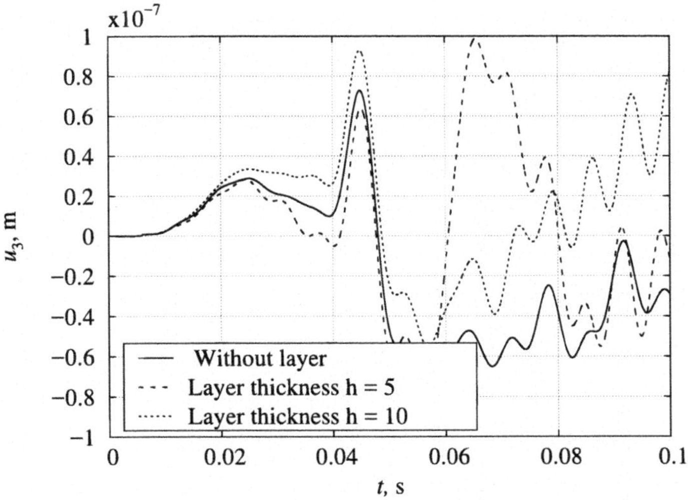

Fig. 28.2 BEM solutions for displacements $u_3(t)$ at point P versus time: Different depth of the layer

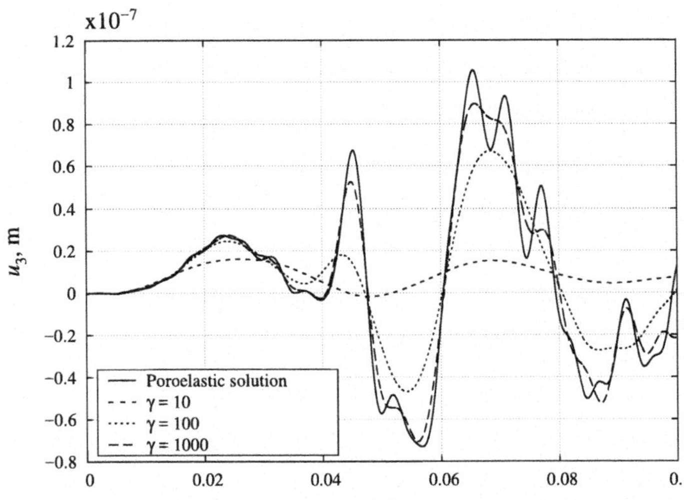

Fig. 28.3 Displacements $u_3(t)$ at point P versus time in case of poroviscoelastic layer thickness h = 5 m

the vertical displacements on the surface of a porous-viscoelastic layer located on a porous-viscoelastic base. It is seen that with increasing parameter γ, the viscoelastic solution approaches the poroelastic one, and with its decrease, a decrease in the displacements is observed. The influence of dissipative effects on the surface displacements of the five-meter layer is especially noticeable, when $\gamma = 100$ the

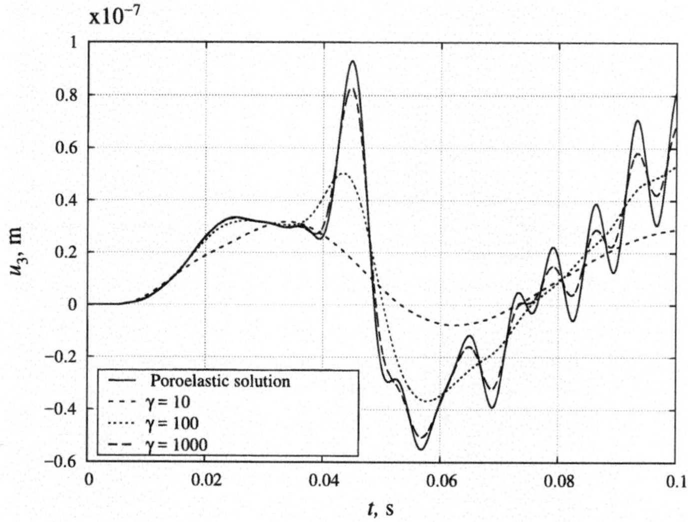

Fig. 28.4 Displacements $u_3(t)$ at point P versus time in case of poroviscoelastic layer thickness h = 10 m

amplitude of the displacements of the Rayleigh wave does not exceed the amplitude of displacements of the longitudinal wave

28.5.2 Half-Space with a Cavity

Problem of a vertical Heaviside-type load acting on a poroelastic half-space is considered. Two types of cavities are considered: cubic and spherical. Force $P(t)$ acting on a square $S = 1\ \text{m}^2$ on a surface of the half-space:

$$P(t) = P_0 f(t),\ f(t) = H(t),\ P_0 = 1000\ \text{N/m}^2 \tag{28.29}$$

Case of cubic cavity: center of the cavity is located at depth $h = 7.5$, and the edge of the cubic cavity is 10 m long (Fig. 28.5). Case of spherical cavity: center of the cavity is located at depth $h = 7.5$, and radius of the spherical cavity is $R = 5$ m (Fig. 28.6). We study the displacements on the boundary surface of the half-space at the distance of 15 m away from loading square.

Boundary element discretization is performed by taking into account two planes of symmetry. Boundary element mesh parameters: 864 elements for quarter of the half-space and 150 elements for quarter of the cavity. Basic poroelastic material parameters are $K = 8 \cdot 10^9\,\text{N/m}^2$, $G = 6 \cdot 10^9\,\text{N/m}^2$, $K_s = 3.6 \cdot 10^{10}\,\text{N/m}^2$, $K_f = 3.3 \cdot 10^9\,\text{N/m}^2$, $\rho = 2458\,\text{kg/m}^3$, $\rho_f = 1000\,\text{kg/m}^3$, $\phi = 0.66$ and $\kappa = 1.9 \cdot 10^{-10}\,\text{m}^4/(\text{N} \cdot \text{s})$.

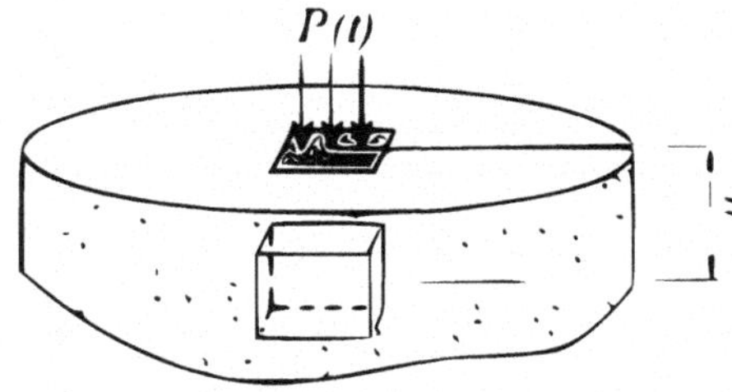

Fig. 28.5 Problem statement in case of cubic cavity

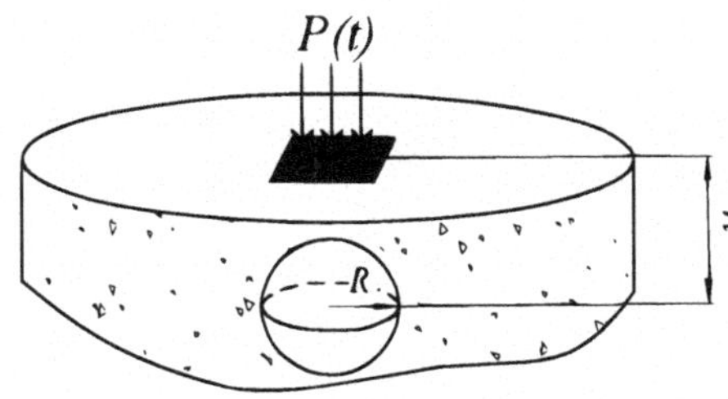

Fig. 28.6 Problem statement in case of spherical cavity

Poroviscoelastic solutions are obtained by using standard linear solid model:

$$\bar{K}(s) = K^{\infty} \cdot \left[(\theta - 1)\frac{s}{s+\gamma} + 1 \right], \tag{28.30}$$

$$\bar{G}(s) = G^{\infty} \cdot \left[(\theta - 1)\frac{s}{s+\gamma} + 1 \right], \tag{28.31}$$

$$\theta = \frac{K^0}{K^{\infty}} = \frac{G^0}{G^{\infty}}. \tag{28.32}$$

Equilibrium and instantaneous values are denoted by '∞' and '0' respectively. Parameters of standard model: $\theta = 4$, $\gamma = 1000$, 100 and 10.

On Figs. 28.7, 28.8, 28.9 and 28.10, we can observe the decrease of Rayleigh wave amplitude depending on viscosity.

28.6 Conclusions

In the present paper, a Biot model of poroelastic material is considered. The systems of equations of the theory of dynamical poroelasticity and the formulation of boundary-value problems in Laplace transforms are written. The simulation of a porous-viscoelastic material is based on the Bio theory of poroelasticity in combination with the principle of matching elastic and viscoelastic reactions. Methods of boundary integral equations and boundary elements are used to solve three-dimensional boundary-value problems. The viscous properties of the skeleton of a porous material are described by a model of a standard viscoelastic body. For

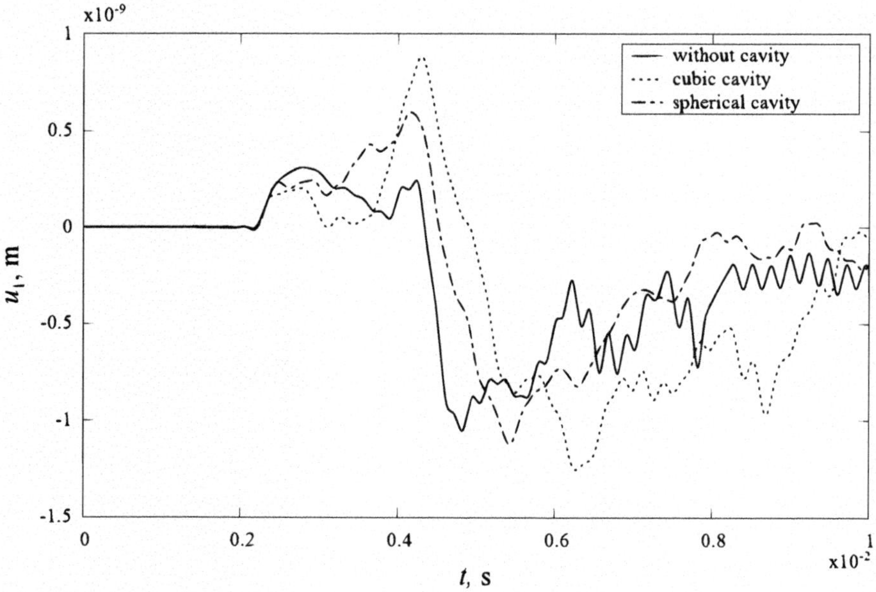

Fig. 28.7 Comparison of displacements u_1 in case of different cavity forms

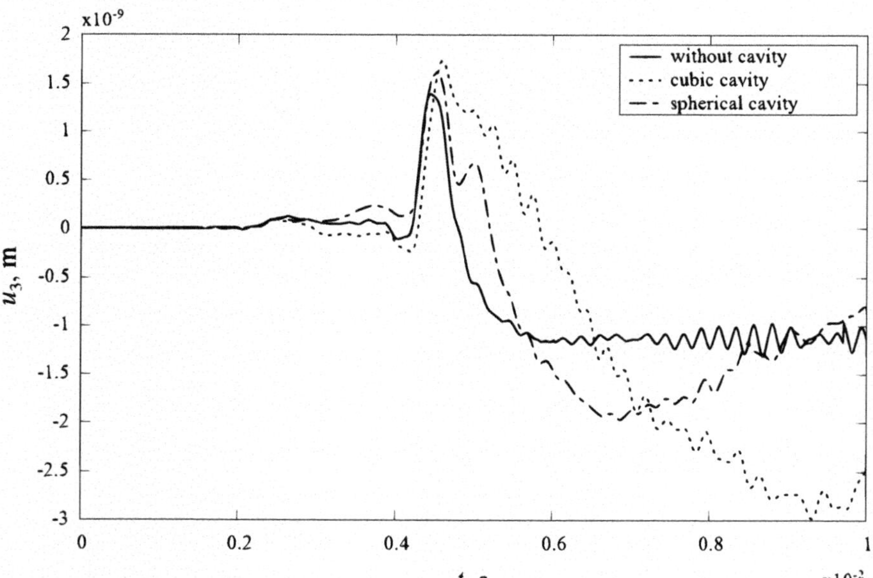

Fig. 28.8 Comparison of displacements u_3 in case of different cavity forms

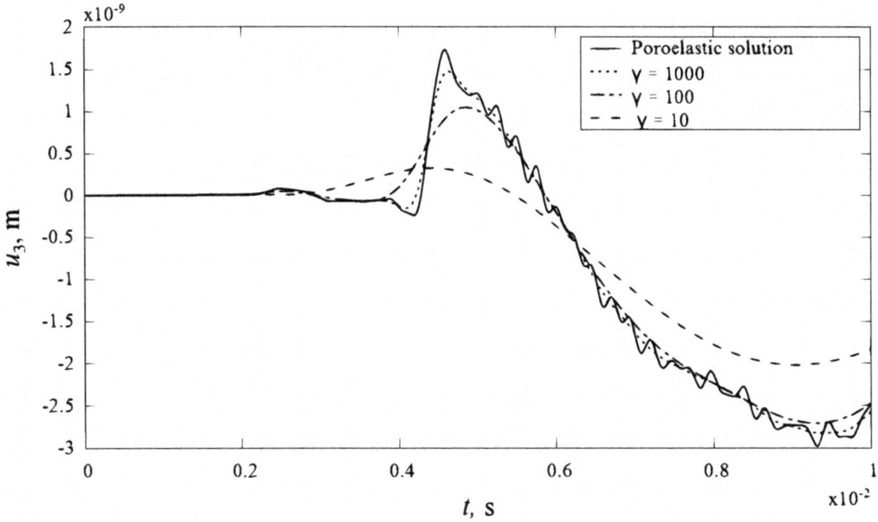

Fig. 28.9 Displacements u_1 in case of cubic cavity for various viscosity parameters

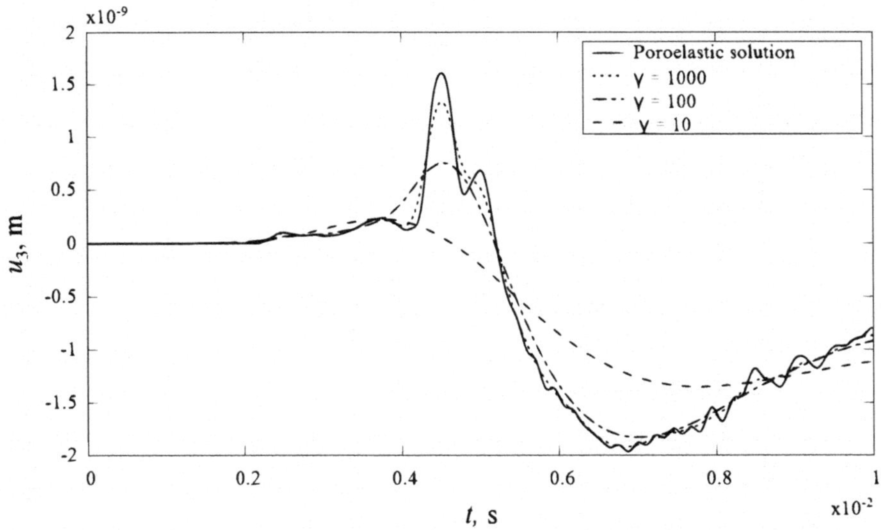

Fig. 28.10 Vertical displacements u_3 in case of cubic cavity for various viscosity parameters

the analysis of three-dimensional porous-viscoelastic problems, BEM formulation was presented in the Laplace transform space, based on integral representations of fundamental solutions. The modified Durbin method is used to reverse the Laplace transform and obtain a solution in time. The results of the boundary element modeling of the problem of the action of a vertical load on the surface of a poroelastic layer located on a poroelastic base are presented. The effect of the viscoelastic properties of the skeleton of a poroelastic material on vertical displacements on the layer surface is investigated. It is noted that the effect on the type of wave patterns is mainly exerted by dissipative effects due to the viscoelastic behavior of the layer.

Acknowledgements The work was carried out with the financial support of the Ministry of Science and Higher Education of the Russian Federation (Project No. FSWR-2023-0034).

References

1. Amenitsky AV, Belov AA, Igumnov LA, Karelin IS (2009) Granichnye integralnye uravneniya dlya resheniya dinamicheskikh zadach trekhmernoi teorii porouprugosti (Boundary integral equations for analyzing dynamic problems of 3-d porouselasticity (in Russian). Probl Strength Plast 71:164–171
2. Banks HT, Bekele-Maxwell K, Bociu L, Noorman M, Guidoboni G (2017) Sensitivity analysis in poro-elastic and poro-visco-elastic models with respect to boundary data. Q Appl Math 75:1. https://hdl.handle.net/1805/15774. Publisher: AMS
3. Bazhenov VG, Igumnov LA (2008) Metody granichnyh integral'nyh uravnenij i granichnyh elementov v reshenii zadach trekhmernoj dinamicheskoj teorii uprugosti s sopryazhennymi polyami (Boundary Integral Equations and Boundary Element Methods in Treating the Problems of 3D Elastodynamics with Coupled Fields (in Russian). PhysMathLit
4. Biot M (1956) Theory of propagation of elastic waves in a fluid-saturated porous solid I. Low-frequency range. J Acoust Soc Amer 28(2):168–178. https://doi.org/10.1121/1.1908239
5. Biot M (1956) Theory of propagation of elastic waves in a fluid-saturated porous solid II. Higher frequency range. J Acoust Soc Amer 28(2):179–191. https://doi.org/10.1121/1.1908241
6. Biot MA (1962) Mechanics of deformation and acoustic propagation in porous media. J Appl Phys 33(4):1482–1498. https://doi.org/10.1063/1.1728759. https://aip.scitation.org/doi/10.1063/1.1728759. Publisher: American Institute of Physics
7. Bo J, Hua L (1999) Vertical dynamic response of a disk on a saturated poroelastic half-space. Soil Dyn Earthquake Eng 18(6):437–443. https://doi.org/10.1016/S0267-7261(99)00013-5. https://www.sciencedirect.com/science/article/pii/S0267726199000135
8. Chapelle D, Gerbeau JF, Sainte-Marie J, Vignon-Clementel IE (2010) A poroelastic model valid in large strains with applications to perfusion in cardiac modeling. Comput Mech 46(1):91–101. https://doi.org/10.1007/s00466-009-0452-x
9. Durbin F (1974) Numerical inversion of Laplace transforms: an efficient improvement to Dubner and Abate's method. Comput J 17(4):371–376. https://doi.org/10.1093/comjnl/17.4.371. https://academic.oup.com/comjnl/article-pdf/17/4/371/1215519/17-4-371.pdf
10. Ehlers W, Markert B (2000) On the viscoelastic behaviour of fluid-saturated porous materials. GM 2(3):153–161. https://doi.org/10.1007/s100359900037
11. Gatmiri B, Eslami H (2007) Scattering of harmonic waves by a circular cavity in a porous medium: complex functions theory approach. Int J Geomech 7(5):371–381. https://doi.org/10.1061/(ASCE)1532-3641(2007)7:5(371). Publisher: American Society of Civil Engineers
12. Hansen A, Sinha S, Bedeaux D, Kjelstrup S, Gjennestad MA, Vassvik M (2018) Relations between seepage velocities in immiscible, incompressible two-phase flow in porous media. Transp Porous Med 125(3):565–587. https://doi.org/10.1007/s11242-018-1139-6
13. Igumnov L, Petrov A (2016) Dynamics of partially saturated poroelastic solids by boundary-element method. PNRPU Mech Bull 3:47–61. https://doi.org/10.15593/perm.mech/2016.3.03
14. Nazarova LA, Nazarov LA (2016) Evolution of stresses and permeability of fractured-and-porous rock mass around a production well. J Min Sci 52(3):424–431. https://doi.org/10.1134/S106273911603061X

15. Nazarova LA, Nazarov LA (2016) Evolution of stresses and permeability of fractured-and-porous rock mass around a production well. J Min Sci 52(3):424–431. https://doi.org/10.1134/S106273911603061X. http://link.springer.com/10.1134/S106273911603061X
16. Philippacopoulos AJ (1988) Waves in partially saturated medium due to surface loads. J Eng Mech 114(10):1740–1759. https://doi.org/10.1061/(ASCE)0733-9399(1988)114:10(1740). Publisher: American Society of Civil Engineers
17. Phillips PJ, Wheeler MF (2007) A coupling of mixed and continuous Galerkin finite element methods for poroelasticity I: the continuous in time case. Comput Geosci 11(2):131–144. https://doi.org/10.1007/s10596-007-9045-y
18. Phillips PJ, Wheeler MF (2007) A coupling of mixed and continuous Galerkin finite element methods for poroelasticity II: the discrete-in-time case. Comput Geosci 11(2):145–158. https://doi.org/10.1007/s10596-007-9044-z
19. Schanz M (2009) Poroelastodynamics: linear models, analytical solutions, and numerical methods. Appl Mech Rev 62(3):1–15
20. Sobhaniaragh B, Mansur WJ, Peters FC (2016) Three-dimensional investigation of multiple stage hydraulic fracturing in unconventional reservoirs. J Petrol Sci Eng 146:1063–1078. https://doi.org/10.1016/j.petrol.2016.07.019. https://www.sciencedirect.com/science/article/pii/S0920410516302777
21. Theodorakopoulos DD, Beskos DE (2006) Application of Biot's poroelasticity to some soil dynamics problems in civil engineering. Soil Dyn Earthquake Eng 26(6):666–679. https://doi.org/10.1016/j.soildyn.2006.01.016. https://www.sciencedirect.com/science/article/pii/S0267726106000170
22. Zhao X (2004) An efficient approach for the numerical inversion of laplace transform and its application in dynamic fracture analysis of a piezoelectric laminate. Int J Solids Struct 41(13):3653–3674. https://doi.org/10.1016/j.ijsolstr.2004.01.006. https://www.sciencedirect.com/science/article/pii/S0020768304000162

Chapter 29
Dynamic AFM Testing of the Roughness Parameters for Dental Materials Characterization

Evgeniy V. Sadyrin, Ivan O. Kharchevnikov, Ekaterina G. Drogan, Sergey S. Volkov, Tie-Jun Liu, Sergei S. Sherbakov, and Sergei M. Aizikovich

Abstract Dynamic atomic force microscopy testing is a powerful technique to study the microgeometry of the biomaterials. The data collected may be complementary to optical, scanning, and transmission microscopy as it provides the quantitative representation of the height profiles of the regions of interest of biological objects with the nanoscale resolution. In the present study, dynamic mode of atomic force microscopy was utilized to compare the roughness parameters of the tooth tissue surfaces with the three types of dental materials applied on the same teeth: composite resin, glass ionomer cement, and polymer infiltrant. A couple of repeating features of the dental materials as well as the artifacts of the internal structure were additionally analyzed.

Introduction

The dynamic mode of atomic force microscopy (AFM), also referred to as the semi-contact or tapping mode, is able to provide high-resolution scanning by combining the advantages of contact and non-contact modes by collecting information about the sample and vibration of the cantilever tip at or near the natural resonance frequency, which allows the cantilever tip to actuate per sample in the shortest time. Today both in the scientific institutions and industrial facilities AFM is used for characterization of a wide range of materials such as biomaterials, electrical circuits, semiconductors,

E. V. Sadyrin (✉) · I. O. Kharchevnikov · E. G. Drogan · S. S. Volkov · S. M. Aizikovich
Don State Technical University, 1 Gagarin Square, 344000 Rostov-on-Don, Russia
e-mail: evgeniy.sadyrin@gmail.com

T.-J. Liu
Inner Mongolia University of Technology, 49 Aimin Street, Hohhot, Xincheng District 010051, China
e-mail: liutiejun6204@163.com

S. S. Sherbakov
National Academy of Sciences of Belarus, 66 Independence Ave, 220072 Minsk, Republic of Belarus
e-mail: sherbakovss@presidium.bas-net.by

H. Altenbach and V. A. Eremeyev (eds.), *Dynamics of Discrete and Continuum Structures and Media*, Advanced Structured Materials 221,
https://doi.org/10.1007/978-3-031-75626-9_29

polymers, and coatings. [1–8]. Such a technique has become a standard method in the tribological researches [9, 10], including biotribology [11, 12].

Dynamic AFM testing of the tooth tissues and dental materials provides vital information on the microgeometrical parameters of the regions of interest complementary to optical, scanning, and transmission microscopy [13]. The potential of such a high-resolution imaging technique offers unique possibilities for tracing features characteristic for pathological alterations of the tissues or directly comparing the tissue roughness with the dental materials replacing them ex vivo [14–16]. Thus, Seredin et al. [17] demonstrated the process of formation of a biomimetic mineralizing layer on the surface of dental enamel using bioinspired nanocrystalline carbonate-substituted calcium hydroxyapatite. Solaymani et al. [18] utilized multifractal analysis to construct three-dimensional topography of enamel, inter enamel, inter dentin, and cementum of human canine teeth using dynamic AFM. The regional inhomogeneity of the periodontal ligament was explored by Connizo and Naveh [19] using AFM in the tapping mode. Leiva-Sabadini et al. [20] characterized the hydroxyapatite crystallites of the human enamel depending on the age of the patient.

A series of studies [21–27] was devoted to investigate the surface morphology and nanotopography of enamel specimens subjected to various pretreatment protocols involving acid etching. The influence of acidic drinks on enamel surface at the very beginning stage of etching process was shown by Li et al. [28], Finke et al. [29], and Barbour and Shellis [30] using AFM. The roughness parameters of the first clinically visible stage of caries were demonstrated in [31–34]. The roughness parameters of the human tooth cementum were studied in the works [35–38].

Dynamic AFM may be effectively used for characterizing teeth roughness parameters after dental treatments. Yilmaz and Kanik [39] performed investigation of surface roughness values of various restorative materials after brushing with blue covarine containing whitening toothpaste utilizing AFM. Saveanu et al. [40] analyzed the topographic structure of the enamel using AFM before and after the application of a varnish containing fluoride and xylitol. Hamze et al. [41] assessed the effect of an experimental sesame gel on the surface roughness of human enamel by AFM. Pradeep [42] investigated the surface roughness of two different bonding adhesives after debonding with AFM. Due to the ability of the AFM probe not to lose interaction for the height confidence interval between the sample parts within a scanning field in the dynamic mode, such devices may be used for examining interfaces in the biological materials. Thus, Yoshida et al. [43] proposed a method for scanning the soft-hard tissue interfaces in the biological objects. Sauro et al. [44] developed a method for studying the quality of the resin-dentin interfaces using confocal microscopy, AFM nano-indentation, and microtensile bond strength test. Using AFM, Lechner et al. [45] monitored the nanoscale surface morphology of human teeth at the dentine-enamel junction after performing successive demineralization steps with anacidic soft drink.

In the present study, we used dynamic AFM to compare the microgeometrical properties of the tooth tissue surfaces with the three types of dental materials applied on the same teeth: composite resin, glass ionomer cement, and polymer infiltrant. The closeness of the roughness parameters of the dental materials to the natural tissues may be important indication for establishing efficacy of the treatments under study. Some characteristic features of the dental materials as well as the artifacts of the internal structure were further analyzed.

29.1 Materials and Methods

Four teeth samples were extracted from patients for orthodontic reasons in the dental department of the Rostov State Medical University clinic (Rostov-on-Don, Russia). The local independent ethical committee of Rostov State Medical University approved the study, extract 14/21 from September 23, 2021, the patients provided informed consent.

The dental materials used were as follows: glass ionomer cement Vitremer (3M ESPE, St. Paul, USA)—1 sample, composite light-curing, radiopaque material Estelite Flow Quick (Tokuyama dental, Tokyo, Japan)—1 sample, and polymer infiltrant Icon (DMG Chemisch-Pharmazeutische, Berlin, Germany)—2 samples. Each material was applied by an experienced dental clinician according to protocol recommended by the manufacturer involving acid etching of the pathologically altered region. Longitudinal cross sections of the sample surface containing the areas under study were prepared using a precision saw. The grinding and polishing of the surface involved a series of procedures using SiC-based abrasive paper with the following grit: P800, P1200, P2000, P2500 with the subsequent use of oil suspensions containing 6 and 1 μm diameter diamond particles. Finish polishing was carried out using a sol-gel suspension based on 0.05 μm alumina particles. Each stage of preparation was followed by ultrasonic cleaning of the sample.

The AFM tests in microscale were conducted on Nanoeducator (NT-MDT, Zelenograd, Russia) device equipped with a tungsten probe. The scanning speed was 4–5 μm/s, and the resolution was 260 $\times$ 260 points for the scanning fields of 25 $\times$ 25 μm. Probe positioning was performed using the optical USB microscope mounted above the force interaction sensor of the device. Gwyddion software (Czech Metrological Institute, Brno, Czech Republic) was used for image processing. For each sample the following areas of interest were studied: filling/infiltrated enamel, and sound enamel on the medial opposite side of the tooth crown, dentine in the vicinity of filling/infiltrated enamel, sound dentine on the medial opposite side of the tooth crown.

Due to the large number of irregularities on the surface of the samples, the measurement of the roughness parameters (average surface roughness R_a, average maximum profile height R_z, maximum roughness height R_t, maximum roughness peak height R_p, and maximum roughness valley depth R_v) were measured in three directions: horizontal, vertical, and diagonal. Two profiles were constructed for each of the directions, and each profile was average one for the 20 adjacent profiles. After that, the mean value (totally from 120 profiles) with the standard deviation was calculated. The roughness height R_t was measured once more from the image as a whole (25 $\times$ 25 μm scanning field).

The AFM tests in nanoscale were conducted on Nano Compact (Phywe, Göttingen, Germany) device. equipped with monocrystalline Si probe with an Al coating, a resonance frequency of 190 ± 60 kHz, and cantilever stiffness of 48 N/m. Scanning velocity was 0.3 ms per line. The scanning resolution was 256 × 256 points. The quadratic mean noise level in the dynamic mode across the z axis (height) was 0.5 nm.

29.2 Results and Discussion

Figure 29.1 shows one image for each area of interest for the sample treated with the composite material obtained using AFM. The roughness parameters for this sample are given in Tables 29.1 and 29.2.

Analysis of the results of an AFM study of the areas of interest of the sample after installing a filling from a composite material made it possible to establish an increase in all the main microgeometrical characteristics for a composite filling compared to

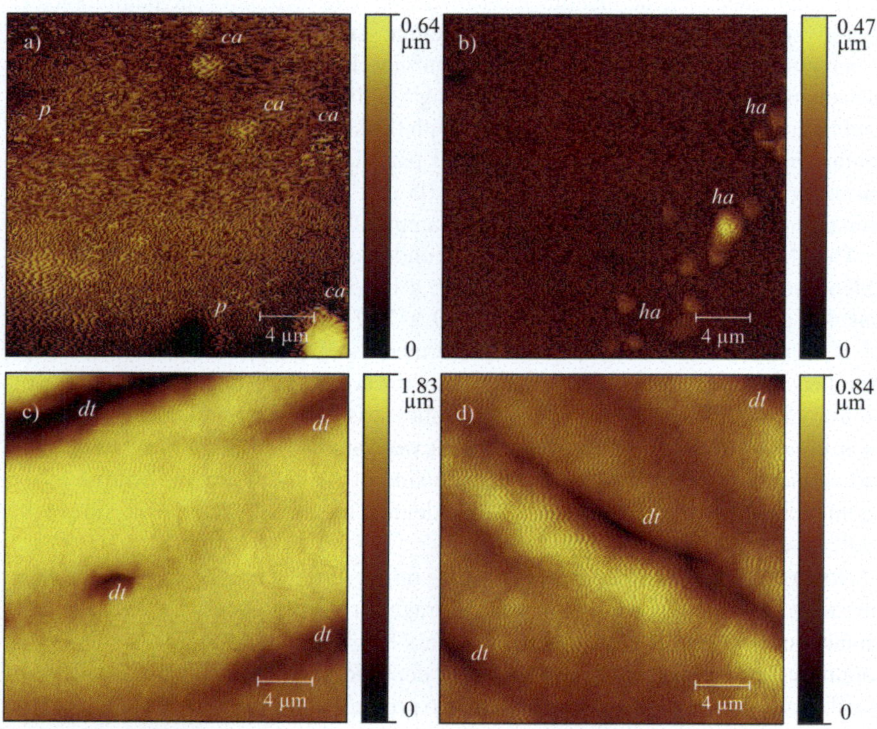

Fig. 29.1 Topography of the surface areas: **a–b** composite filling and sound enamel; **c–d** dentine in the vicinity of the filling and sound dentine; *ca*—composite agglomerates, *ha*—hydroxyapatite agglomerates, *dt*—dentinal tubule

Table 29.1 Roughness parameters R_a, R_z, R_p, R_v of the sample treated with the composite material

Area	R_a, nm	±	R_z, nm	±	R_p, nm	±	R_v, nm	±
Sound enamel	13.5	3.0	89.1	14.5	71.6	13.9	41.2	11.8
Composite filling	9.6	2.2	55.6	16.2	29.1	7.7	38.7	13.6
Sound dentine	69.3	18.9	259.5	106.3	108.4	20.7	272.9	105.4
Dentine in the vicinity of the filling	71.1	14.0	224.7	64.2	162.0	27.9	196.8	55.5

Table 29.2 Skewness, kurtosis, and R_t for the sample treated with the composite material

Area	R_{sk}	R_{ku}	R_t, nm	±	R_t (field), μm
Sound enamel	1.3	6.0	112.8	21.7	0.47
Composite filling	−0.4	3.7	67.8	13.7	0.64
Sound dentine	−1.0	4.0	381.3	111.1	1.83
Dentine in the vicinity of the filling	−0.3	2.7	358.8	61.8	0.84

sound enamel by 40.63%, 60.25%, 146.05%, 6.46%, 66.37%, respectively, for R_a, R_z , R_p, R_v, R_t. It should be noted that the above-surface elements of the images differ mainly, while the sub-surface (deepening) elements are generally similar. The peaks of microroughnesses for a filling are more symmetrical and flatter compared to the peaks of sound enamel. A slight increase in R_t (over the scanning field) for the filling is due to the presence of an artifact (presumably a conglomerate of composite material) in the lower right corner of the image. On the surface of the filling, pores and aggregates of the composite material were found, while on the surface of the enamel, small islands of agglomerate of hydroxyapatite crystals were found.

The roughness parameters of sound dentine and dentine in the vicinity of the filling are close, while the depths of the cavities are slightly higher in the sound dentine (R_v is higher by 38.67%), and the heights of the peaks, on the contrary, are higher in the dentine in the vicinity of the filling (R_p is higher by 49.4%), the peaks themselves are more symmetrical , but more gentle. Both investigated R_t values are lower for the case of dentine in the vicinity of the filling. A significant increase in R_t (over the scan field) for the sound enamel is due to the fact that a deep dentinal tube was captured in the upper left corner of the image, and it made a significant contribution to the growth of R_v on the sound tissue; otherwise, the heights in the images are similar. It can be concluded that the installation of the filling did not significantly affect the microgeometry of the dentine in its vicinity. In both images, depressions are clearly visible, corresponding to longitudinal sections of dentinal tubules.

Figure 29.2 demonstrates one image for each area of interest for the sample treated with the glass ionomer cement. The roughness parameters for this sample are given in Tables 29.3 and 29.4.

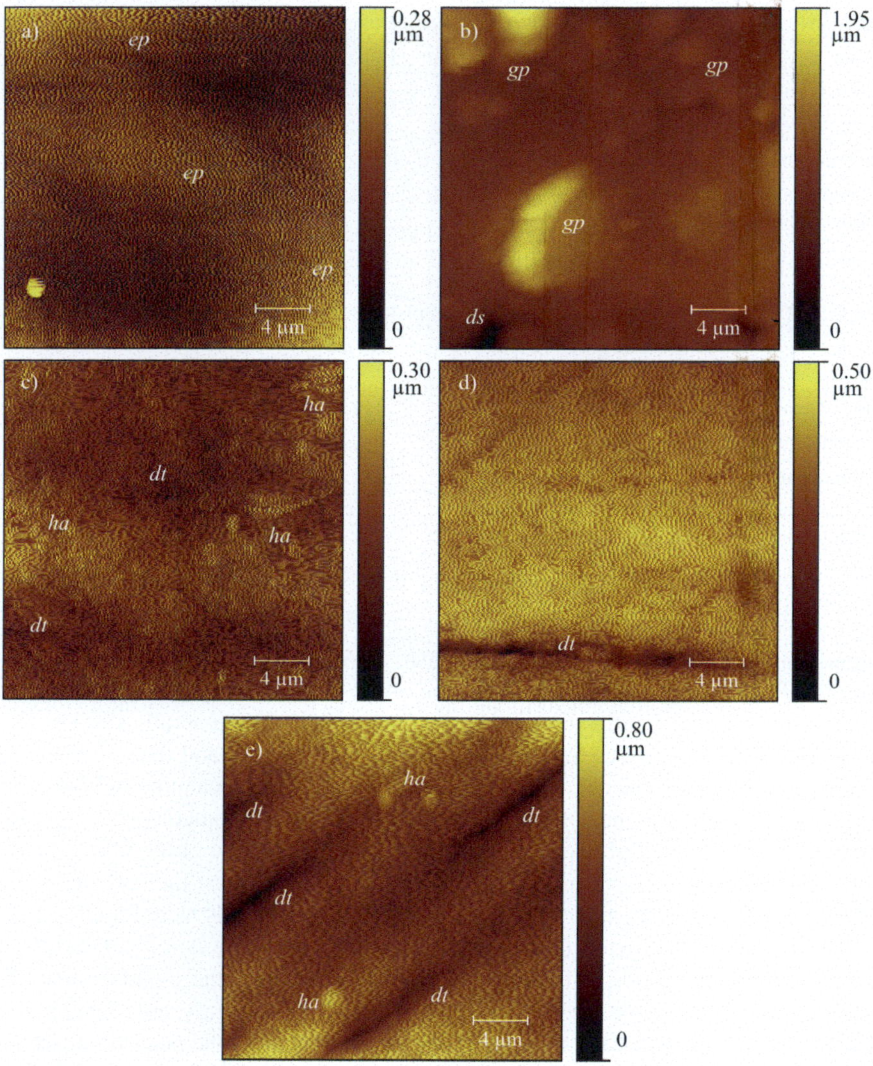

Fig. 29.2 Topography of the surface areas: **a–b** glass ionomer filling and sound enamel; **c–d** dentine in the vicinity of the filling and sound dentine; *ep*—enamel prism, *gp*—glass particle, *ha*—hydroxyapatite agglomerates, *dt*—dentinal tubule

Analysis of the AFM data for the sample after the installation of a GIC filling made it possible to establish an increase in all the main microgeometrical characteristics for a glass ionomer filling compared to enamel in 7.95, 6.55, 5.77, 7.26, 6.31, 6.96 times, respectively, for R_a, R_z , R_p, R_v, R_t and R_t across the scan field. The microroughness peaks of the glass ionomer cement are flatter, higher and less symmetrical compared to healthy enamel. The greatest contribution to the development of the relief of the

Table 29.3 Roughness parameters R_a, R_z, R_p, R_v of the sample treated with the glass ionomer cement

Area	R_a, nm	±	R_z, nm	±	R_p, nm	±	R_v, nm	±
Sound enamel	12.4	1.9	57.7	18.0	51.0	14.7	29.1	7.1
Glass ionomer cement	98.6	39.3	378.2	120.7	294.4	159.8	211.2	66.9
Sound dentine	13.9	4.0	55.0	7.3	49.0	21.0	38.5	10.1
Dentine in the filling vicinity—crown	73.8	15.8	300.0	62.1	171.8	66.4	214.1	66.6
Dentine in the filling vicinity—cervical	20.8	7.9	88.3	29.5	51.8	18.5	72.4	45.6

Table 29.4 Skewness, kurtosis, and R_t for the sample treated with the glass ionomer cement

Area	R_{sk}	R_{ku}	R_t, nm	±	R_t (field), μm
Sound enamel	0.5	3.4	80.1	19.2	0.28
Glass ionomer cement	0.6	2.9	505.6	197.6	1.95
Sound dentine	0.3	4.0	87.4	18.7	0.30
Dentine in the filling vicinity—crown	0	2.7	385.8	115.6	0.80
Dentine in the filling vicinity—cervical	−0.4	2.9	124.1	57.0	0.50

glass ionomer filling is made by glass particles, between which the separation surfaces are visualized. Also, an increase in all roughness parameters for dentine in the vicinity of the filling was found compared to the sound dentine: 5.31 and 1.50 times, 5.45 and 1.6 times, 3.51 times and 5.71%, 5.56 and 1.88, 2.67 and 1.67 times, respectively, for R_a, R_z, R_p, R_v, R_t and R_t across the scan field for the crown and cervical regions, respectively. The increase in values can be explained presumably in connection with shrinkage of the tissue caused by polymerization stresses inside the glass ionomer cement. Dentine near the filling in the crown part shows more symmetrical peaks than the dentine of the other two studied areas, while the peaks of sound dentine are sharper (with an average lower height). In the images, in addition to the cavities of the dentinal tubules, hydroxyapatite agglomerates, which appeared during the grinding and polishing of the sample, contribute to the growth of microroughness heights.

Figure 29.3 demonstrates one image for each area of interest for the first sample treated with the polymer infiltrant. The roughness parameters for this sample are given in Tables 29.5 and 29.6.

Dynamic AFM analysis for the areas of interest of the first sample after infiltration procedure revealed an increase in all the main microgeometrical characteristics for the infiltrated enamel compared to the sound one in 2.02, 1.86, 1.39, 3.33, 2.21, 3.16 times, respectively, for R_a, R_z, R_p, R_v, R_t and R_t across the scan field. Despite this, compared with filling materials, the growth turned out to be small, especially since

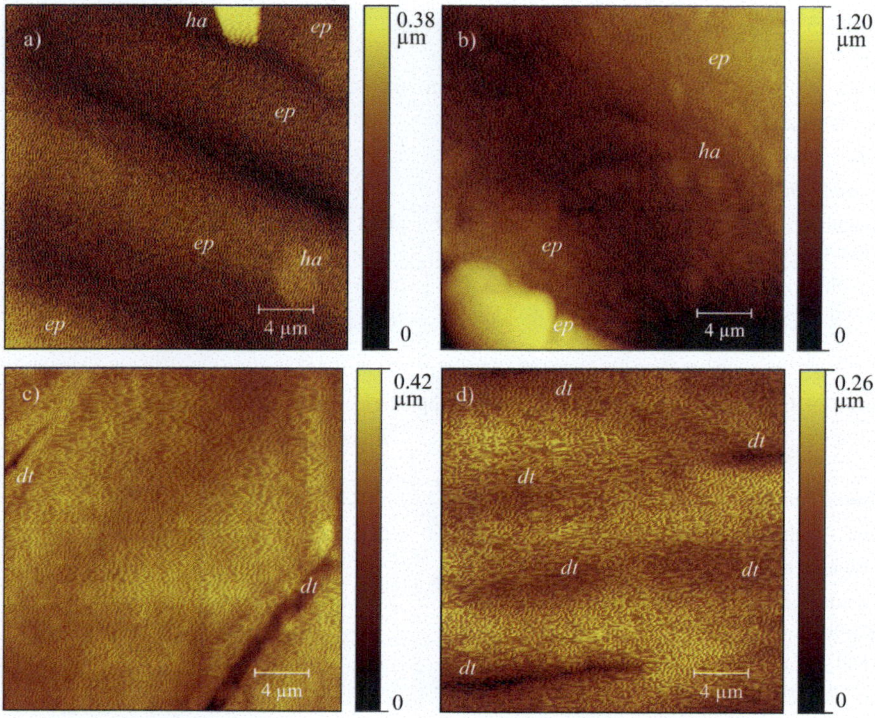

Fig. 29.3 Topography of the surface areas for the first infiltrated sample: **a**–**b** infiltrated enamel and sound enamel; **c**–**d** dentine in the vicinity of the infiltrated enamel and sound enamel; *ep*—enamel prism, *ha*—hydroxyapatite agglomerates, *dt*—dentinal tubule

Table 29.5 Roughness parameters R_a, R_z , R_p, R_v of the first sample treated with the infiltrant

Area	R_a, nm	±	R_z, nm	±	R_p, nm	±	R_v, nm	±
Sound enamel	16.7	1.2	62.2	13.8	53.9	5.7	39.5	7.1
Infiltrated enamel	33.8	19.6	115.4	79.4	74.7	47.5	131.7	85.2
Sound dentine	16.7	4.9	77.8	29.7	44.6	12.2	79.8	36.6
Dentine in the vicinity of infiltrated enamel	21.2	4.8	101.0	25.5	48.0	10.9	87.0	31.1

the peaks of microroughness in both cases are quite symmetrical and have a slight sharpening (more pronounced for the case of infiltrated enamel). Both enamel images show single agglomerates of hydroxyapatite crystals. Similarly, an increase in the main roughness parameters for dentine in the vicinity of infiltrated enamel compared to healthy dentin was found: 26.95%, 92.82%, 7.62%, 9.02%, 8.52%, respectively for R_a, R_z , R_p, R_v, R_t, except for R_t (across the scan field), where the value decreased by 61.54% due to the presence of a large depression in the dentinal tubule in the

Table 29.6 Skewness, kurtosis, and *R*t for the first sample treated with the infiltrant

Area	*R*sk	*R*ku	*R*t, nm	±	*R*t (field), μm
Sound enamel	0.4	2.7	93.4	10.7	0.38
Infiltrated enamel	−0.7	3.9	206.4	126.1	1.20
Sound dentine	−0.6	4.9	124.4	35.9	0.42
Dentine in the vicinity of infiltrated enamel	−0.7	3.7	135.0	24.1	0.26

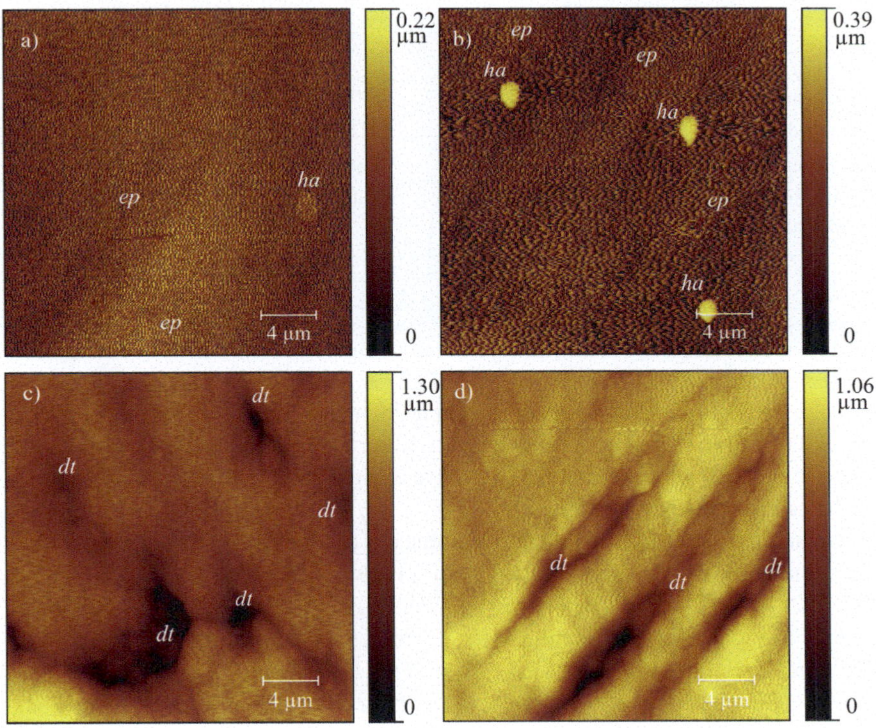

Fig. 29.4 Topography of the surface areas for the second infiltrated sample: **a**–**b** infiltrated enamel and sound enamel; **c**–**d** dentine in the vicinity of the infiltrated enamel and sound enamel; *ep*—enamel prism, *ha*—hydroxyapatite agglomerates, *dt*—dentinal tubule

lower right part of the image of sound dentine. The microroughness peaks in both cases have a slight skewness and a fairly close degree of sharpening of the peaks.

Figure 29.4 demonstrates one image for each area of interest for the second sample treated with the polymer infiltrant. The roughness parameters for this sample are given in Tables 29.7 and 29.8.

Analysis of the AFM study of the areas of interest of the second sample after enamel infiltration made it possible to establish an increase in all the main microgeo-

Table 29.7 Roughness parameters *R*a, *R*z , *R*p, *R*v of the second sample treated with the infiltrant

Area	*R*a, nm	±	*R*z, nm	±	*R*p, nm	±	*R*v, nm	±
Sound enamel	7.4	1.5	39.0	10.8	22.6	3.9	27.7	6.8
Infiltrated enamel	7.9	1.8	41.9	6.7	26.2	10.3	31.6	13.1
Sound dentine	88.5	21.6	381.7	144.1	183.3	41.7	351.2	115.2
Dentine in the vicinity of infiltrated enamel	117.5	26.8	508.8	185.5	186.5	12.7	438.5	104.3

Table 29.8 Skewness, kurtosis, and *R*t for the second sample treated with the infiltrant

Area	*R*sk	*R*ku	*R*t, nm	±	*R*t (field), μm
Sound enamel	−0.1	3.0	50.3	10.2	0.22
Infiltrated enamel	−0.2	4.3	57.8	14.1	0.39
Sound dentine	−1.0	4.2	539.4	132.3	1.30
Dentine in the vicinity of infiltrated enamel	−1.1	3.8	624.9	102.3	1.06

metrical characteristics for infiltrated enamel compared to the sound counterpart by 6.76%, 7.44%, 15.93%, 14.08%, 14.91%, 77.27%, respectively, for *R*a, *R*z , *R*p, *R*v, *R*t and *R*t (over the scan field). Compared with filling materials, the growth turned out to be small (even smaller than in the first infiltration case), the peaks of microroughness in both cases are quite symmetrical and have a slight sharpening (more pronounced for the case of infiltrated enamel).

An increase in the main roughness parameters for dentine in the vicinity of infiltrated enamel was found in comparison with the sound dentine by 32.77%, 33.3%, 1.75%, 24.86%, 15.85%, respectively for *R*a, *R*z , *R*p, *R*v, *R*t, with the exception of *R*t (over the scan field), where the value decreased by 22.64% due to the presence of a large depression in the dentinal tubule in the lower central part of the image of sound dentine. The microroughness peaks in both cases have a slight skewness and a very close average sharpening of the peaks.

Visualization of individual sound hydroxyapatite crystallites for first sample treated with the polymer infiltrant is given on Fig. 29.5.

The typical width and length for the hydroxyapatite crystallites were about 35–40 and 70–80 nm respectively. The profiles for one such typical crystallite are shown in Fig. 29.6. Visualization of enamel in nanoscale for the first sample treated with the polymer infiltrant is given on Fig. 29.7 in the phase contrast mode. Analysis of the dentine surface before (Fig. 29.8a, b) and after the infiltration procedure (Fig. 29.8c, d) in phase-contrast mode reveals similar features of tissue structure at the nanoscale: dense packed hydroxyapatite crystallites with a slightly increased spaces between them in case of the infiltrated enamel. Such spaces are typical for the enamel [46] that undergo demineralization stage as weakened crystallites (either at their cores or on the periphery [47, 48]) are easier subjected to break in course of grinding and polishing

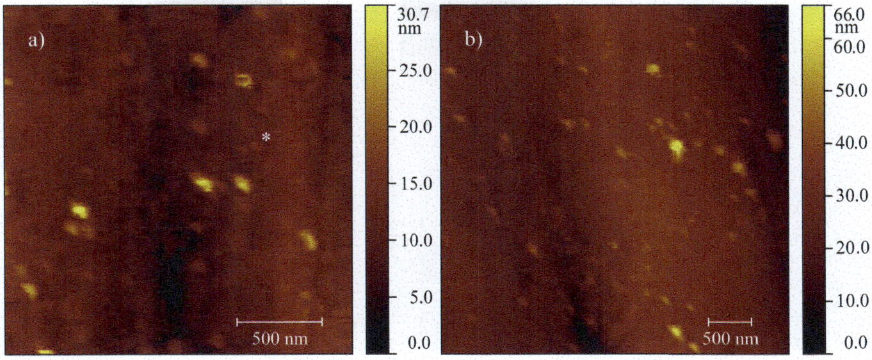

Fig. 29.5 Individual sound hydroxyapatite crystallites for first sample treated with the polymer infiltrant on the scanning fields: **a** 2 μm, the typical crystallite for profile construction was marked with *; **b** 4 μm

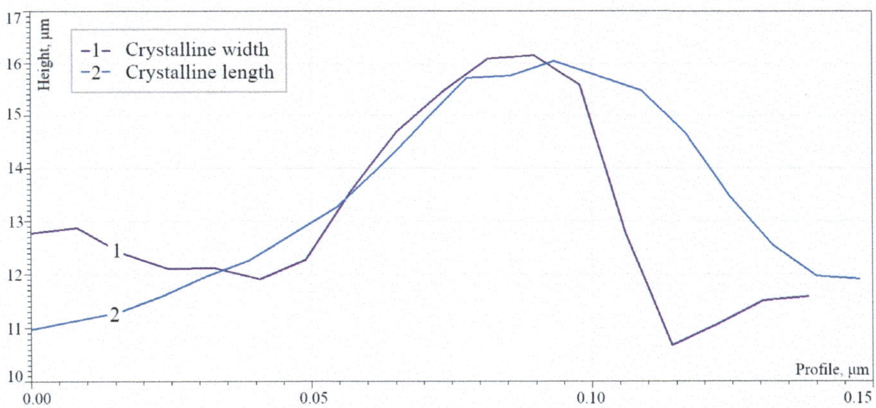

Fig. 29.6 Profiles of the length and width for the crystallite from Fig. 29.5

procedures. Note that the dentine crystallites both before and after treatment are undistinctive in terms of entirety of crystallites—no one visualized crystallite after infiltration on dentine demonstrated loss of its components.

Despite the fact that both surfaces of the fillings were different by their roughness parameters than the sound enamel, the composite materials demonstrated lesser divergence than the glass ionomer cement. Moreover, composite filling contained smaller artifacts of internal structure—their diameters were 2.79 ± 1.15 μm—whereas the glass ionomer contained the structures of 3.29 ± 1.73 μm diameter. Moreover, the properties for the dentine in the vicinity of the composite filling was pretty close to those for the sound dentine; on the other hand the properties for the dentine near the glass ionomer filling differed dramatically, presumably due to the shrinkage of the dentine tissue caused by polymerization stresses inside the glass ionomer cement [49].

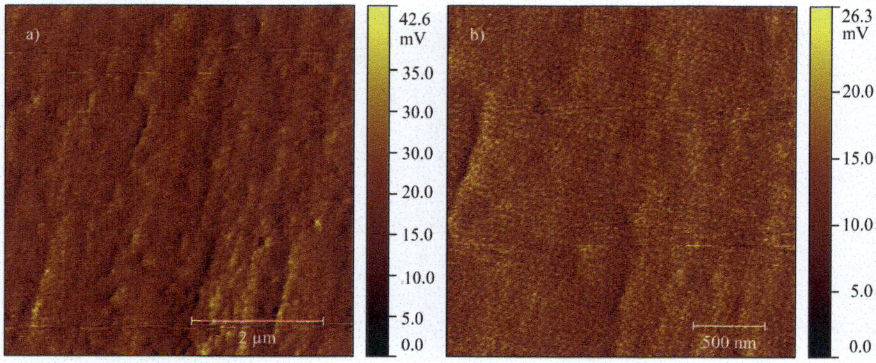

Fig. 29.7 Enamel surface in nanoscale (phase contrast mode) for first sample treated with the polymer infiltrant on the scanning fields: **a** 5 μm; **b** 2.5 μm

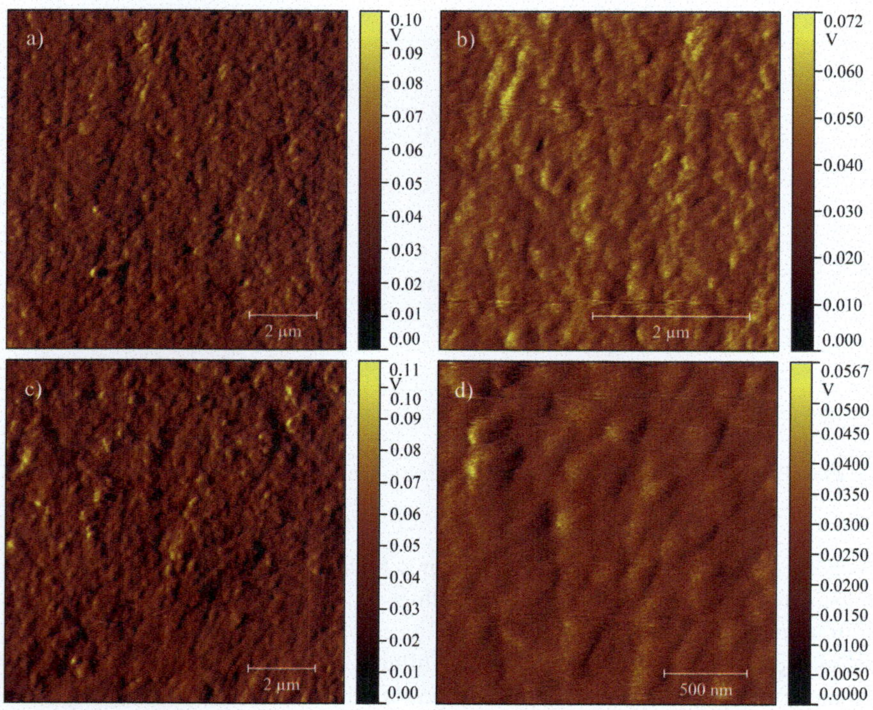

Fig. 29.8 Dentine surface for first sample treated with the polymer infiltrant in phase contrast mode before (**a–b**) and after the infiltration (**c–d**)

The roughness parameters for the tissues treated with the polymer infiltrant were close to those for the sound tissues: especially dentine in the first case and enamel in the second one. This observation indicates that by modifying the enamel, the infiltrant as a whole preserves the structure of the enamel more close to the natural tissue compared to the filling materials under study. However, more statistical data is needed complemented by the measurements of the physical and mechanical properties of dental materials [50–55] as well as mathematical modeling [56–59] for the reliable prediction of the strength of dental materials, especially on the interfaces with the living tissues and occlusal surfaces [60–63].

29.3 Conclusion

In the present work, dynamic mode of AFM was used for comparison of the roughness parameters of the tooth tissue surfaces with the three types of dental materials applied on the same teeth on the 25 × 25 μm scanning fields. The size of the hydroxyapatite crystallites was measured for the tooth enamel in nanoscale. The closeness of the parameters to the natural tissues indirectly indicate the higher efficacy of the composite resin compared to the glass ionomer cement among the two fillings. The number of the artifacts of the internal structure was higher for the glass ionomer cement; moreover, the artifacts themselves were bigger in diameter. The roughness parameters for the tissues treated with the polymer infiltrant were close to those for the sound tissues resembling the relief features and preserving the enamel tissue.

Acknowledgements This research was supported by the Russian Science Foundation, grant number 22-19-00732 (https://rscf.ru/en/project/22-19-00732/). Sample preparation and AFM study in microscale were conducted in the Nanocenter of Don State Technical University (http://nano.donstu.ru). AFM imaging at nanoscale was performed in the Chemistry Department of Don State Technical University. The authors acknowledge the help of Maksyukov S. Yu. and Yogina D. V. (Rostov State Medical University, Rostov-on-Don, Russia) for the help with *ex vivo* treatment of the teeth samples.

References

1. Jalili N, Laxminarayana K (2004) A review of atomic force microscopy imaging systems: application to molecular metrology and biological sciences. Mechatronics 14(8):907–945. https://doi.org/10.1016/j.mechatronics.2004.04.005
2. Ilyasov VV, Meskhi BC, Ryzhkin AA, Yershov IV (2012) Ultrathin carbon films on sapphire grown by laser ablation: synthesis and AFM-study. Adv Eng Res 12(1):31–35
3. Zhou ZR, Jin ZM (2015) Biotribology: recent progresses and future perspectives. Biosurf Biotribology 1(1):3–24. https://doi.org/10.1016/j.bsbt.2015.03.001
4. Xu K, Liu Y (2022) Studies of probe tip materials by atomic force microscopy: a review. Beilstein J of Nanotech 13(1):1256–1267. https://doi.org/10.3762/bjnano.13.104

5. Melnikova G, Kuznetsova T, Lapitskaya V, Petrovskaya A, Chizhik S, Zykova A, Safonov V, Aizikovich S, Sadyrin E, Sun W, Yakovin S (2021) Nanomechanical and nanotribological properties of nanostructured coatings of tantalum and Its compounds on steel substrates. Nanomater 11(9):2407. https://doi.org/10.3390/nano11092407
6. Rajabifar B, Meyers GF, Wagner R, Raman A (2022) Machine learning approach to characterize the adhesive and mechanical properties of soft polymers using peakforce tapping AFM. Macromolecules 55(19):8731–8740. https://doi.org/10.1021/acs.macromol.2c00147
7. Lapitskaya VA, Kuznetsova TA, Khudoley AL, Khabarava AV, Chizhik SA, Aizikovich SM, Sadyrin EV (2021) Influence of polishing technique on crack resistance of quartz plates. Int J Fract 231:61–77. https://doi.org/10.1007/s10704-021-00564-5
8. Nikolaev AL, Mitrin BI, Sadyrin EV, Zelentsov VB, Aguiar AR, Aizikovich SM (2020) Mechanical properties of microposit S1813 thin layers. In: Aizikovich S, Altenbach H, Eremeyev V, Swain M, Galybin A (eds) Modeling, synthesis and fracture of advanced materials for industrial and medical applications. Adv Struct Mater, 136, Cham: Springer, pp 137–146. https://doi.org/10.1007/978-3-030-48161-2_9
9. Kuprin AS, Gilewicz A, Kuznetsova TA, Lapitskaya VA, Tolmachova GN, Warcholinski B, Aizikovich SM, Sadyrin EV (2021) Structure and properties of ZrON coatings synthesized by cathodic arc evaporation. Materials 14(6):1483. https://doi.org/10.3390/ma14061483
10. Burlakova VE, Tyurin AI, Drogan EG, Sadyrin EV, Pirozhkova TS, Novikova AA, Belikova MA (2019) Mechanical properties and size effects of self-organized film. J Tribol 141(5):051601. https://doi.org/10.1115/1.4042678
11. Ilyasov VV, Meskhi BC, Ryzhkin AA, Yershov IV (2012) Adhesion of amorphous ultrafine carbon films on sapphire: scratch testing. Adv Eng Res 7(68):27–32
12. Swain MV, Michael S, Thai LA, Lucas PW (2017) Influence of indenter geometry on the frictional sliding resistance of tooth enamel. Biosurf and Biotribology 3(4):224–237. https://doi.org/10.1016/j.bsbt.2017.11.003
13. Nezafat NB, Ghoranneviss, M, Elahi, SM, Shafiekhani A, Ghorannevis Z, Solaymani S. Microstructure, micromorphology, and fractal geometry of hard dental tissues: Evaluation of atomic force microscopy images. Micro Res Tech 82(11):1884–1890. https://doi.org/10.1002/jemt.23356
14. Machoy M, Wilczyński S, Szyszka-Sommerfeld L, Woźniak K, Deda A, Kulesza S (2021) Mapping of nanomechanical properties of enamel surfaces due to orthodontic treatment by AFM method. Appl Sci 11(9):3918. https://doi.org/10.3390/app11093918
15. Solomon SM, Timpu D, Forna DA, Stefanache MAM, Martu S, Stoleriu S (2016) AFM comparative study of root surface morphology after three methods of scaling. Mater. Plast 53:546–549
16. Sadyrin EV, Yogina DV, Volkov SS, Aizikovich SM (2022) Evaluation of density and microgeometrical characteristics of glass-ionomer cement and composite resin fillings: a biomechanical ex vivo study. Russ J Biomech 26(2):67–73. https://doi.org/10.15593/RZhBiomeh/2022.2.06
17. Seredin P, Goloshchapov D, Kashkarov V, Emelyanova A, Buylov N, Barkov K, Ippolitov Y, Khmelevskaia T, Mahdy IA, Mahdy MA, Prutskij T (2022) Biomimetic mineralization of tooth enamel using nanocrystalline hydroxyapatite under various dental surface pretreatment conditions. Biomimetics 7(3):111. https://doi.org/10.3390/biomimetics7030111
18. Solaymani S, ălu, Ghoranneviss M, Elahi SM, Shafiekhani A, Hantehzadeh M, Nezafat NB, (2020) Multifractal analysis of human canine teeth at nano scale: atomic force microscopy studies. Int Nano Let 10:15–22. https://doi.org/10.1007/s40089-019-00293-7
19. Connizzo BK, Naveh GR (2020) In situ AFM-based nanoscale rheology reveals regional non-uniformity in viscoporoelastic mechanical behavior of the murine periodontal ligament. J Biomech 111:109996. https://doi.org/10.1016/j.jbiomech.2020.109996
20. Leiva-Sabadini C, Schuh CM, Barrera NP, Aguayo S (2022) Ultrastructural characterisation of young and aged dental enamel by atomic force microscopy. J Microscopy 288(3):185–192. https://doi.org/10.1111/jmi.13126
21. Lee YL, Li KC, Yiu CKY, Boyd DH, Ekambaram M (2021) Evaluation of developmentally hypomineralised enamel after surface pretreatment with Papacarie Duo gel and different etching modes: an in vitro SEM and AFM study. Eur Arch Paediatric Dent 1–15. https://doi.org/10.1007/s40368-021-00671-w

22. Sadyrin EV, Kislyakov EA, Karotkiyan RV, Yogina DV, Drogan EG, Swain MV, Maksyukov SYu, Nikolaev AL, Aizikovich SM (2020) Influence of citric acid concentration and etching time on enamel surface roughness of prepared human tooth: in vitro study. In: Altenbach H, Brünig M, Kowalewski Z (eds) Plasticity, Damage and Fracture in Advanced Materials. Adv Struct Mater, 121. Springer, Cham, pp 135–150. https://doi.org/10.1007/978-3-030-34851-9_8
23. Kislyakov EA, Karotkiyan RV, Sadyrin EV, Mitrin BI, Yogina DV, Kheygetyan AV, Maksyukov SY (2020) Nanoindentation derived mechanical properties of human enamel and dentine subjected to etching with different concentrations of citric acid. In: Aizikovich S, Altenbach H, Eremeyev V, Swain M, Galybin A (eds) Modeling, synthesis and fracture of advanced materials for industrial and medical applications. Adv Struct Mater, 121. Springer, Cham, pp 75–83. https://doi.org/10.1007/978-3-030-48161-2_5
24. Wong J, Tsujimoto A, Fischer NG, Baruth AG, Barkmeier WW, Johnson EA, Samuel SM, Takamizawa T, Latta MA, Miyazaki M (2020) Enamel etching for universal adhesives: examination of enamel etching protocols for optimization of bonding effectiveness. Oper dent 45(1):80–91. https://doi.org/10.2341/18-275-L
25. Sorozini M, dos Reis Perez C, Rocha GM (2018) Enamel sample preparation for AFM: influence on roughness and morphology. Microscop Res Tech 81(9):1071–1076. https://doi.org/10.1002/jemt.23073
26. Zhao S, Zhang Y, Chen Y, Xing X, Wang Y, Wu G (2023) Evaluation of Chitosan-Oleuropein nanoparticles on the durability of Dentin bonding. Drug Design Develop Therapy 17:167–180
27. Tsujimoto A, Fischer N, Barkmeier W, Baruth A, Takamizawa T, Latta M, Miyazaki M (2017) Effect of reduced phosphoric acid pre-etching times on enamel surface characteristics and shear fatigue strength using universal adhesives. J Adhes Dent 19(3):267–275. https://doi.org/10.3290/j.jad.a38437
28. Li P, Oh C, Kim H, Chen-Glasser M, Park G, Jetybayeva A, Yeom J, Kim H, Ryu J, Hong S (2020) Nanoscale effects of beverages on enamel surface of human teeth: an atomic force microscopy study. J Mech Behavior Biomed Mater 110:103930. https://doi.org/10.1016/j.jmbbm.2020.103930
29. Finke M, Jandt KD, Parker DM (2000) The early stages of native enamel dissolution studied with atomic force microscopy. J Colloid Interface Sci 232(1):156–164. https://doi.org/10.1006/jcis.2000.7200
30. Barbour ME, Shellis RP (2007) An investigation using atomic force microscopy nanoindentation of dental enamel demineralization as a function of undissociated acid concentration and differential buffer capacity. Phys Med Biol 52(4):899. https://doi.org/10.1088/0031-9155/52/4/003
31. Cuy JL, Mann AB, Livi KJ, Teaford MF, Weihs TP (2002) Nanoindentation mapping of the mechanical properties of human molar tooth enamel. Arch Oral Biol 47(4):281–291. https://doi.org/10.1016/S0003-9969(02)00006-7
32. Ma X, Lin X, Zhong T, Xie F (2019) Evaluation of the efficacy of casein phosphopeptide-amorphous calcium phosphate on remineralization of white spot lesions in vitro and clinical research: a systematic review and meta-analysis. BMC Oral Health 19(1):1–11. https://doi.org/10.1186/s12903-019-0977-0
33. Sadyrin EV, Yogina DV, Vasiliev AS, Aizikovich SM (2022) Evaluation of the influence of white spot lesion on the mechanical properties of human tooth enamel and dentine. Izv Saratov Univ Math Mech Inform 22(3):346–359. https://doi.org/10.18500/1816-9791-2022-22-3-346-359
34. Hidaka K, Nishimura K, Miyazawa K, Miwa H, Goto S (2011) Effects of acid etchants used for bonding orthodontic brackets on the remineralization of enamel white-spot lesions. Orthodon Waves 70(4):126–135. https://doi.org/10.1016/j.odw.2011.08.002
35. Sadyrin E, Lapitskaya V, Kuznetsova T, Yogina D, Maksyukov S, Aizikovich S (2022) Nanoindentation and Atomic Force Microscopy Derived Mechanical and Microgeometrical Properties of Tooth Root Cementum. Micro 2(4):575–588. https://doi.org/10.3390/micro2040038
36. Ho SP, Goodis H, Balooch M, Nonomura G, Marshall SJ, Marshall G (2004) The effect of sample preparation technique on determination of structure and nanomechanical proper-

ties of human cementum hard tissue. Biomater 25(19):4847–4857. https://doi.org/10.1016/j.biomaterials.2003.11.047
37. Jang AT, Lin JD, Choi RM, Choi EM, Seto ML, Ryder MI, Gansky SA, Curtis DA, Ho SP (2014) Adaptive properties of human cementum and cementum dentin junction with age. J Mech Behavior Biomed Mater 39:184–196. https://doi.org/10.1016/j.jmbbm.2014.07.015
38. Hurng JM, Kurylo MP, Marshall GW, Webb SM, Ryder MI, Ho SP (2011) Discontinuities in the human bone-PDL-cementum complex. Biomater 32(29):7106–7117. https://doi.org/10.1016/j.biomaterials.2011.06.021
39. Yılmaz C, Kanık Ö (2022) Investigation of surface roughness values of various restorative materials after brushing with blue covarine containing whitening toothpaste by two different methods: AFM and profilometer. Microscopy Res Tech 85(2):521–532. https://doi.org/10.1002/jemt.23925
40. Saveanu CI, Dragos O, Anistoroaei D, Bobu LI, Saveanu AE, Armencia A, Solomon SM, Tanculescu O (2022) Xylitol fluoride varnish: In Vitro effect analysis on enamel by atomic force microscopy. Biomedicines 10(8):1900. https://doi.org/10.3390/biomedicines10081900
41. Hamze F, Ghasemi L, Kamalinejad M (2022) Evaluating the effect of an experimental Sesame Gel on human enamel: atomic force microscopy study. J Dent 23(1). https://doi.org/10.30476/DENTJODS.2021.87878.1293
42. Pradeep MR (2023) A comparative evaluation of enamel surface roughness of two different bonding adhesives after debonding with atomic force microscopy. J Adv Sci 2(1):11–16. https://doi.org/10.58935/joas.v2i1.22
43. Yoshida Y, Meerbeek BV, Snauwaert J, Hellemans L, Lambrechts P, Vanherle G, Pashley DH (1999) A novel approach to AFM characterization of adhesive tooth-biomaterial interfaces. J Biomed Mater Res 47(1):85–90. https://doi.org/10.1002/(SICI)1097-4636(199910)47
44. Sauro S, Osorio R, Watson TF, Toledano M (2012) Assessment of the quality of resin-dentin bonded interfaces: An AFM nano-indentation, μTBS and confocal ultramorphology study. Dent Mater 28(6):622–631. https://doi.org/10.1016/j.dental.2012.02.005
45. Lechner BD, Röper S, Messerschmidt J, Blume A, Magerle R (2015) Monitoring demineralization and subsequent remineralization of human teeth at the dentin-enamel junction with atomic force microscopy. ACS Appl Mater Interf 7(34):18937–18943. https://doi.org/10.1021/acsami.5b04790
46. Sadyrin E, Swain M, Mitrin B, Rzhepakovsky I, Nikolaev A, Irkha V, Yogina D, Lyanguzov N, Maksyukov S, Aizikovich S (2020) Characterization of enamel and dentine about a white spot lesion: mechanical properties, mineral density, microstructure and molecular composition. Nanomaterials 10(9):1889. https://doi.org/10.3390/nano10091889
47. Yanagisawa T, Miake Y (2003) High-resolution electron microscopy of enamel-crystal demineralization and remineralization in carious lesions. Microscopy 52(6):605–613. https://doi.org/10.1093/jmicro/52.6.605
48. Ogiwara M, Miake Y, Yanagisawa T (2008) Changes in dental enamel crystals by bleaching. J Hard Tissue Biol 17(1):11–16. https://doi.org/10.2485/jhtb.17.11
49. Ausiello P, Ciaramella S, Garcia-Godoy F, Gloria A, Lanzotti A, Maietta S, Martorelli M (2017) The effects of cavity-margin-angles and bolus stiffness on the mechanical behavior of indirect resin composite class II restorations. Dent Mater 33(1):e39–e47. https://doi.org/10.1016/j.dental.2016.11.002
50. Sadyrin EV (2022) Correlating the mechanical properties to the mineral density of brown spot lesion in dentine using nanoindentation and X-ray micro-tomography. In: Altenbach H, Eremeyev VA, Galybin A, Vasiliev A (eds) Advanced Materials Modelling for Mechanical, Medical and Biological Applications. Adv Struct Mater, 155. Springer, Cham, pp 389–398. https://doi.org/10.1007/978-3-030-81705-3_21
51. Shahmoradi M, Wan B, Zhang Z, Swain M, Li Q (2022) Mechanical failure of posterior teeth due to caries and occlusal wear-A modelling study. J Mech Behavior Biomed Mater 125:104942. https://doi.org/10.1016/j.jmbbm.2021.104942
52. Sadyrin EV (2023) Influence of a polymeric infiltrant on the density of enamel white spot lesions. Izv Saratov Univ Math Mech Inform 23(1):83–94. https://doi.org/10.18500/1816-9791-2023-23-1-83-94

53. Koldehoff J, Swain MV, Schneider GA (2020) The geometrical structure of interfaces in dental enamel: a FIB-STEM investigation. Acta Biomater 104:17–27. https://doi.org/10.1016/j.actbio.2019.12.040
54. Zaytsev D, Panfilov P (2015) Anisotropy of the mechanical properties of human dental enamel. Mater Lett 159:428–431. https://doi.org/10.1016/j.matlet.2015.07.057
55. Sadyrin EV, Mitrin BI, Yogina DV, Swain MV (2021) Preliminary study of distribution of mechanical properties and mineral density by depth of liquid saturated carious dentine. IOP Conf Ser: Mater Sci Eng 1029(1). https://doi.org/012056.10.1088/1757-899X/1029/1/012056
56. Lin M, Xu F, Lu TJ, Bai BF (2010) A review of heat transfer in human tooth-experimental characterization and mathematical modeling. Dent Mater 26(6):501–513. https://doi.org/10.1016/j.dental.2010.02.009
57. Izyumov RI, Rusakov SV, Zuev AL, Gileva OS, Muraveva MA (2014) Mathematical modeling of caries initiation and progression occurring in dental enamel. Key Eng Mater 592:362–365. https://doi.org/10.4028/www.scientific.net/KEM.592-593.362
58. Kabanova AV, Zaitsev DV, Grigor'ev SS, Panfilov PE (2020) Effect of a fluid on the deformation behavior of human dentin during diametral compression. Russ Metall 2020:1177–1181. https://doi.org/10.1134/S0036029520100109
59. Ilie O, Van Turnhout AG, Van Loosdrecht MCM, Picioreanu C (2014) Numerical modelling of tooth enamel subsurface lesion formation induced by dental plaque. Caries Res 48(1):73–89. https://doi.org/10.1159/000354123
60. Zelentsov VB, Sadyrin EV, Mitrin BI, Swain MV (2023) Mathematical tools for recovery of the load on the fissure according to the micro-CT results. J Mech Behavior Biomed Mater 138:105625. https://doi.org/10.1016/j.jmbbm.2022.105625
61. Sadyrin EV (2023) Modeling the mechanism of reducing enamel mineral density in the vicinity of the fisure tip. Russ J Biomech 1:24–31. https://doi.org/10.15593/RZhBiomech/2023.1.03
62. Wu C, Fang J, Zhang Z, Entezari A, Sun G, Swain MV, Li Q (2020) Fracture modeling of brittle biomaterials by the phase-field method. Eng Fract Mech 224:106752. https://doi.org/10.1016/j.engfracmech.2019.106752
63. Wan B, Shahmoradi M, Zhang Z, Shibata Y, Sarrafpour B, Swain M, Li Q (2019) Modelling of stress distribution and fracture in dental occlusal fissures. Sci Rep 9(1):4682. https://doi.org/10.1038/s41598-019-41304-z

Chapter 30
In Situ Mechanical Testing and Microtomography of 3D-Printed Hard Tissue Scaffolds for Bioengineering Applications

Evgeniy V. Sadyrin, Roman V. Karotkiyan, Andrey L. Nikolaev, Sergei V. Chapek, Polina A. Lapina, and Yun-Che Wang

Abstract Understanding the evolution of the network of cracks in scaffolds meant to replace the hard tissues during loading is of crucial importance in the medical and bioengineering practice. To investigate the changes in the microstructure of the scaffolds associated with the dynamical compression process, we conducted uniaxial compression experiments on scaffold samples that were fabricated using digital light processing. Additionally, we used X-Ray microtomography to obtain a 3D visualization of the fracture pattern. The compression process was divided into three stages: compaction, elastic limit, and tensile strength. An increase in stress led to an increase in the concentration and unevenness of cracks, which demonstrated the propagation of initial microcracks in a certain direction and the subsequent process of their merging. These results provide important information about the internal structure and behavior of the scaffold under stress, facilitating the development of more effective scaffold structural design strategies.

E. V. Sadyrin (✉) · R. V. Karotkiyan · A. L. Nikolaev · P. A. Lapina
Laboratory for Mechanics of Biomaterials, Don State Technical University, 1 Gagarin Square, 344000 Rostov-on-Don, Russian Federation
e-mail: evgeniy.sadyrin@gmail.com

R. V. Karotkiyan
e-mail: andreynicolaev@eurosites.ru

P. A. Lapina
e-mail: polina_azarova86@mail.ru

S. V. Chapek
The Smart Materials Research Institute, Southern Federal University, 178/24 Sladkova Street, 344090 Rostov-on-Don, Russian Federation
e-mail: chapek@sfedu.ru

Y.-C. Wang
National Cheng Kung University, 1 University Road, Tainan 70101, Taiwan
e-mail: yunche@ncku.edu.tw

H. Altenbach and V. A. Eremeyev (eds.), *Dynamics of Discrete and Continuum Structures and Media*, Advanced Structured Materials 221,
https://doi.org/10.1007/978-3-031-75626-9_30

30.1 Introduction

The management of critical size defects of the hard tissues represents a serious challenge in clinical practice. In order to overcome the drawbacks of the conventional treatment modalities utilizing autografts or allografts such as restricted availability, complications at the donor site, and potential transmission of diseases, and rejection of foreign bodies, and various synthetic materials and methodologies involving 3D printing are being intensively developed [1, 2].

Scaffolds are 3D porous or fibrous structures designed to serve as a mechanical support for cellular growth and tissue regeneration [3]. Ideally, scaffolds should possess specific properties conducive to the development of functional bone tissue. These properties include an adhesive surface that facilitates cell adhesion, proliferation, and differentiation; biocompatibility to prevent immunological rejection; non-toxicity; controlled biodegradability synchronized with tissue growth; and an optimal pore size to enable proper cell distribution, vascularization, nutrient diffusion, and waste removal [4]. Multiple literature reviews provide an overview of diverse printing methodologies utilized in the fabrication of bone scaffolds composed of metallic [5], ceramic [6], and polymeric [7] materials, emphasizing a range of synthesis and processing pathways.

The widespread adoption and utilization of the polymeric materials is facilitated by the rapid advancements and standardization of 3D printing methods, such as selective laser sintering and fused filament fabrication [8]. Among these methods, stereolithography (SLA) stands out for its ability to produce intricate details with high precision, surface quality, and cost-effectiveness. The SLA method, which relies on light sources and imaging technology, encompasses two main techniques: digital light processing (DLP) and laser-based stereolithography. In DLP, a light ray passes through a digital micro-mirror device to project an image onto the resin vat, while laser-based stereolithography utilizes a laser beam directed by a swinging mirror to cure photopolymer resin based on the image of each layer [9, 10].

Dental bone pathologies are prevalent conditions that can arise from various factors such as trauma, infections, neoplasms, congenital anomalies, and mechanical wear. An illustrative example includes the resorption of alveolar bone following tooth extraction, leading to alterations in the original volume of the alveolar ridge and the development of bone defects. Post-extraction, an average loss of approximately 30% (vertically) and 40–50% (horizontally) of alveolar bone occurs within a six-month period [11]. Without intervention, this bone loss progresses, resulting in reductions of 40–60% in bone crest volume within three years [12]. Subsequent bone remodeling creates a bone defect that can impede or prevent the placement of dental implants depending on their size and location [13]. To mitigate the aesthetic and functional consequences of intraoral bone defects, clinical utilization of scaffolds as preventive and regenerative measures has become widespread. This facilitates the successful installation of dental implants and subsequent implant-supported prosthesis rehabilitation [14]. For effective bone regeneration, an ideal scaffold should possess biodegradable and biocompatible properties, along with osteoconductive, osteoinductive, and osteogenic capabilities [15, 16].

Polymeric materials are comprised of elongated chains of monomers connected by covalent bonds. Polylactic acid (PLA) is a polyester produced through the condensation of hydroxyl and carboxyl groups within the lactic acid monomer or by the opening of the lactide ring [17]. Apart from its biocompatibility and biodegradability, PLA exhibits low rigidity, excellent processability, and thermal stability. Polyglycolic acid (PGA) is a biodegradable thermoplastic polymer, serving as the simplest linear aliphatic polyester [18]. Its synthesis can commence from glycolic acid through polycondensation or ring-opening polymerization processes. However, the production of high-molecular-weight PGA faces challenges due to its instability and susceptibility to degradation compared to other synthetic polymers. Polylactic-co-glycolic acid (PLGA) emerges as a biocompatible biomaterial that is easily synthesized and degrades into non-toxic by-products [19]. This copolymer finds application in various therapeutic devices, with the specific form of PLGA depending on the ratio of PLA to PGA utilized during polymerization. Polycaprolactone (PCL) comprises hexanoate units and serves as a significant biodegradable aliphatic polymer. Its synthesis involves the polycondensation of 6-hydroxyhexanoic acid and ring-opening polymerization of ϵ-caprolactone [20]. Polyvinyl alcohol (PVA) is a synthetic biodegradable polymer obtained through the hydrolysis of poly(vinyl acetate). The unique characteristics of PVA, including solubility, flexibility, biocompatibility, biodegradability, and mechanical strength, position it as a valuable choice for polymeric scaffolds in bone tissue engineering [21]. Polypropylene fumarate (PPF) fulfills several medical requirements such as biocompatibility, mechanical properties, osteoconductivity, sterilizability, and degradation through hydrolysis of its ester bonds [22].

In the field of bioengineering, nanofiber scaffolds have garnered significant interest owing to their exceptional biomimetic properties. The processing of polymers into nanofibers predominantly employs three key techniques: electrospinning, phase separation, and self-assembly [23]. A variety of nanostructured materials may be used for the scaffold design [24, 25].

A variety of techniques are utilized for the mechanical characterization of scaffolds including compression/tension tests [26, 27], nanoindentation [28], scanning electron [29], and atomic force microscopy (including scratch testing [30]) as well as X-ray computed microtomography (micro-CT) [31]. Multidisciplinary methods [32–35] including mathematical modeling for contact problems of complex media [36–38] for assessment of the characteristic of bone and dental tissues are crucial as well for helping a practicing clinician to predict the development of the pathological alterations.

In the present study, the uniaxial compression experiments on scaffold samples that were fabricated using DLP technique were performed with the subsequent micro-CT observations of the 3D fracture patterns. The results obtained provide important information about the internal structure and behavior of the scaffold under stress, facilitating the development of more effective scaffold structural design strategies as well as a sturdy basis for the mathematical modeling of the process of compression of the scaffolds.

30.2 Materials and Methods

To carry out an experimental program a CT5000TEC compression/tension machine (Deben UK Ltd, UK), which allows loading the material with a maximum force of up to 5 kN at a constant loading rate in the range from 0.03 to 3.00 mm/min, was prepared to conduct experimental studies of compression in a micro-CT device by developing and manufacturing special equipment from radiolucent material for the compression plates of the device. The compression plates printed from acryl made it possible to minimize the number of X-ray image artifacts that occur near the compression plates and interfere with thorough data analysis.

In situ mechanical tests of the scaffold samples for uniaxial compression were carried out using an Xradia Versa 520 micro-CT (Carl Zeiss X-ray Microscopy, Inc., Pleasanton, USA) and a prepared experimental setup with a choice of loading profiles in the form of loading force-time diagrams. This method requires capturing of multiple (typically more than 1000) projection images of the sample as it is rotated. Those projections are later reconstructed into 3D volumes composed of elements (voxels). The entire volume or its part can be further processed in the digital environment [39–41].

To perform the work, scaffolds with dimensions of 12.5 × 12.5 × 7.5 mm and wall thicknesses from 0.5 to 1.5 mm, with a pitch of 0.2 mm, were used, which were combined into groups. The scaffolds were manufactured using digital light processing. The material used for creation was transparent photopolymer resin with ultra-high resolution from Nano Clear (FunToDo, Kotka, Finland).

At the first stage, experiments were carried out on uniaxial compression of scaffold samples at a loading rate of 0.05 mm/min. The yield strength was calculated as $\sigma_T = F_T/A_0$, and it is the stress at which the material "flows", i.e. the deformation of the sample increases almost without increasing the load. Tensile strength $\sigma_P = F_{max}/A_0$ is the stress corresponding to the highest value of force that the sample can withstand, where A_0 is the cross-sectional area (mm^2).

The second stage of the experiment was as follows:

1. To ensure contact of the compression plates with the edges of the scaffold, uniaxial compression was carried out at a loading rate of 0.05 mm/min. The loading was stopped when the uniaxial stress reached a minimum value, which was characterized by the contact of the jaws with the sample without applying an actual load. The first micro-CT observation was carried out at this step. Microtomograms under load are subsequently compared with it, and an experimental study of the porosity of the material was carried out according to histograms of gray values using VGSTUDIO Max 3.5 software (Volume Graphics, Heidelberg, Germany).
2. The loading rate of 0.05 mm/min was kept constant, and a second micro-CT scan was performed when the axial stress reached the yield point.
3. Damage to the scaffold was observed in real time, and a third micro-CT scan was performed when the uniaxial stress reached its breaking point. Note that after the stage, the yield stress usually occurred earlier than for samples without preload. The full yield strength at this stage was recorded only for samples with

wall thicknesses of 1.3 and 1.5 mm. For the remaining samples, the limit occurred when the relative movement of the installation plate was 0.85, 0.2, 0.75, 0.25 mm, respectively, for samples with wall thicknesses of 0.5, 0.7, 0.9, 1.1 mm.

It should also be noted that scans were performed after the specified limits were reached. In all microtomography observations, the following scanning parameters were used: X-ray tube voltage 40 kV, current 2.5 W, voxel size 28.0 μm, exposure time 5 s, and X-ray tube filter LE1; 1301 projections were collected for each scan. An additional set of 10 projections was also collected with identical parameters, but with a filter on the LE3 X-ray tube for subsequent correction of ring artifacts.

For each group of scaffolds at all stages of loading, the obtained data was reconstructed into sets of virtual sections in the XRMReconstructor 12.0.8086.19558 software with center displacement values determined manually, and a Gaussian blur filter (0.5) was used. During reconstruction, correction of sample drift during scanning, as well as ring artifacts, was performed. Next, Phong rendering [42] was performed, obtaining virtual sections of the internal structure of the materials under study, and constructing volumetric reconstructions of architectural samples.

30.3 Results and Discussion

The relative displacement versus load plot for each scaffold at the first stage of the experiments (uniaxial compression) is shown in Fig. 30.1. The compression pressure was maintained using the servo control function. The results indicate an increase in

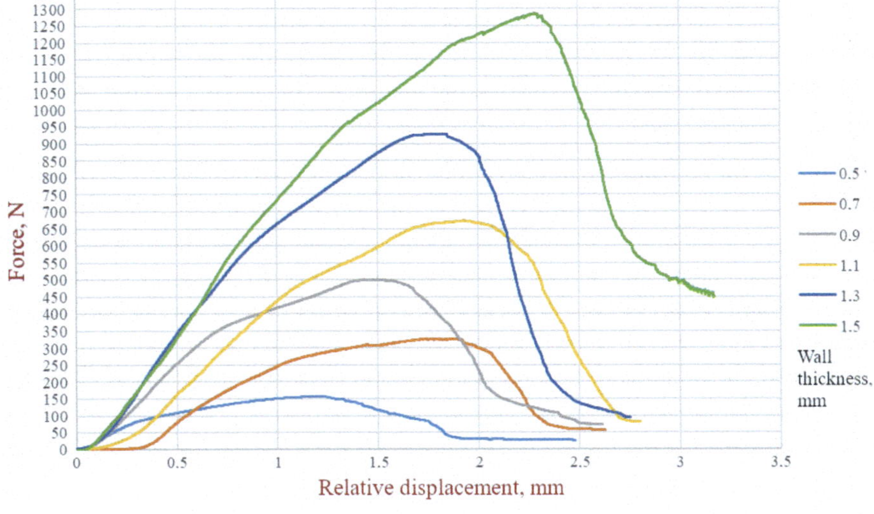

Fig. 30.1 Deformation diagrams of the samples under study

Table 30.1 Values of yield and tensile strength for each group of scaffolds

Groups by thickness, mm	Yield strength, σ_T MPa	Tensile strength, σ_P MPa
0.5	0.64	1.02
0.7	1.28	2.08
0.9	2.24	3.2
1.1	3.2	4.03
1.3	4.48	5.89
1.5	5.12	7.81

the load resistance of the scaffolds depending on the wall thickness. A scaffold with a wall thickness of 1.5 mm had the greatest resistance, and the least was observed for the thickness of 0.5 mm. The yield and strength limits were also recorded for each of the groups of scaffolds that were used in the further course of the work, the results are listed in Table 30.1.

Figure 30.2a shows a microtomogram of a scaffold with a wall thickness of 0.5 mm before loading. For this sample, a number of printing artifacts are noted in the form of small spherical voids with diameters of 0.087–0.185 mm. There are also small microcracks and defects such as chips along the periphery of the sample. Figure 30.2b shows the same scaffold when loaded to the yield strength point. In this case, a number of cracks and single ruptures of stiffening ribs up to 0.352 mm in length are removed, mainly along the periphery of the sample, while most of the print artifacts remain unchanged. Figure 30.2c shows the same scaffold loaded to its tensile strength point. Multiple fracture sites were visualized both in the ribs and in the nodes of the architectural material; a small part of the nodes remained intact. At the same time, fracture developed outside the print artifacts.

Figure 30.3a shows a microtomogram of a scaffold with a wall thickness of 0.7 mm before loading. For this sample, a number of isolated printing artifacts are noted in the form of small spherical growths up to 0.25 mm in diameter, as well as small step artifacts along the edges of the scaffold surface. Figure 30.3b shows the same scaffold when loaded to the yield strength point. In this case, a number of cracks up to 0.625 mm in depth (almost the entire width of the rib) and single ruptures of the stiffeners up to 0.447 mm in length are observed, mainly along the periphery of the sample. Locations with spherical nodules remained unchanged compared to the case before the load. Figure 30.3c shows the same scaffold when loaded to the tensile strength point. Multiple fracture sites were visualized both in the ribs and in the nodes of the architectural material; the width of the tears was comparable to the width of the ribs. At the same time, fracture sites developed outside the print artifacts.

Figure 30.4a shows a microtomogram of a scaffold with a wall thickness of 0.9 mm before loading. For this sample, a number of single printing artifacts are noted in the form of rectangular growths on the stiffening ribs with edge lengths up to 0.479 mm, as well as microcracks located at an angle of 25° with a depth of up to 0.356 mm. Figure 30.4b shows a microtomogram of the same scaffold when loaded to the yield

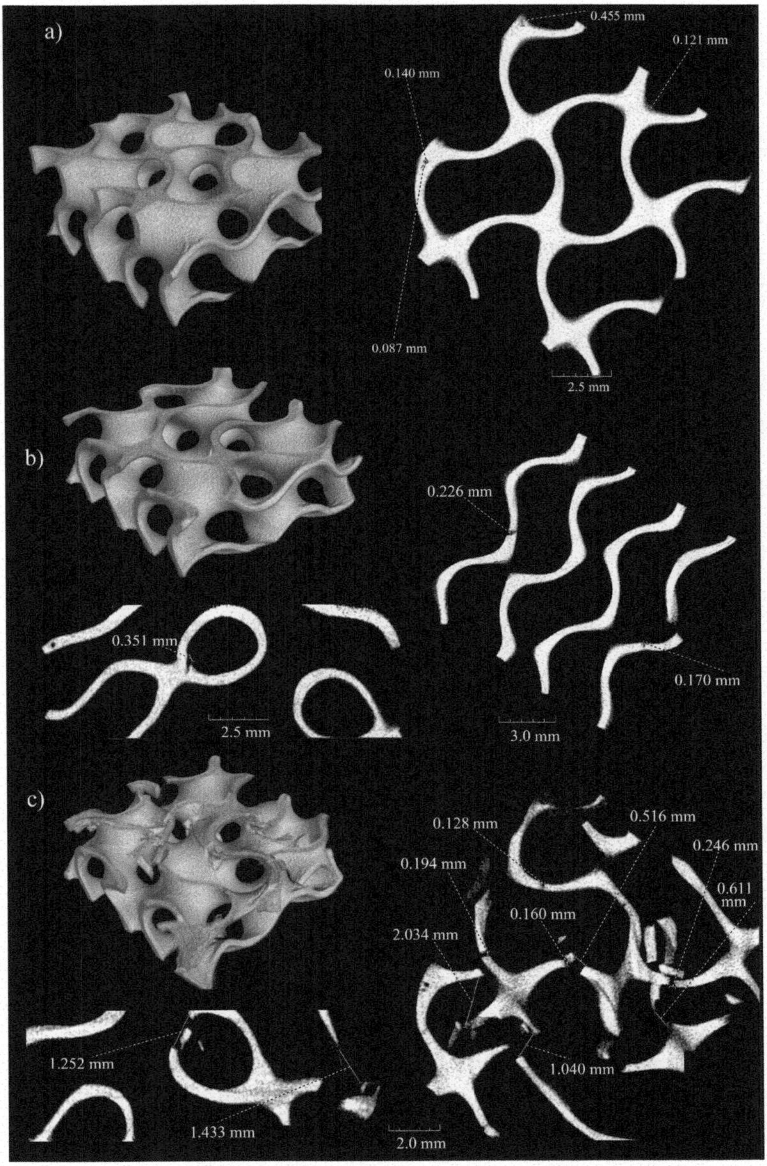

Fig. 30.2 Microtomogram of a scaffold with a wall thickness of 0.5 mm: **a** before loading; **b** loaded to the yield strength; **c** loaded to the tensile strength

strength point. In this case, a number of cracks up to 0.561 mm in depth (almost the entire width of the rib) and single ruptures of the stiffeners up to 0.472 mm in length are observed, mainly along the periphery of the sample. Figure 30.4c shows a microtomogram of the same scaffold when loaded to the tensile strength point.

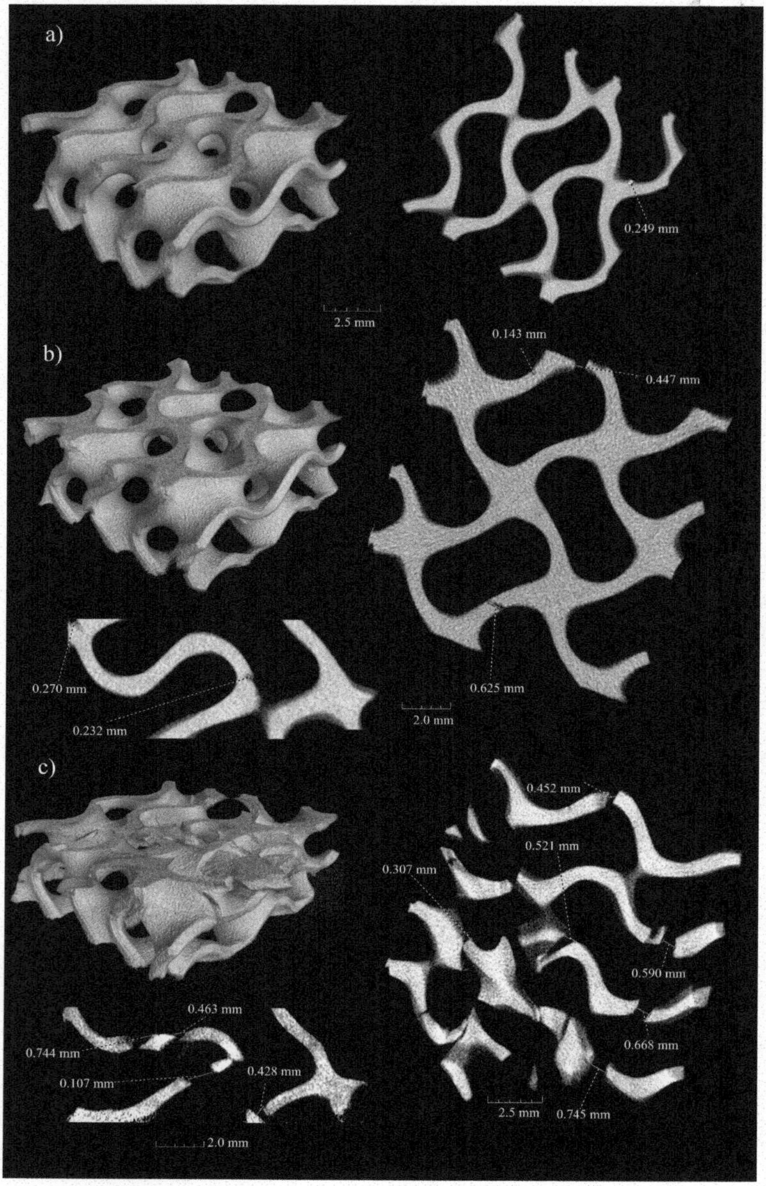

Fig. 30.3 Microtomogram of a scaffold with a wall thickness of 0.7 mm: **a** before loading; **b** loaded to the yield strength; **c** loaded to the tensile strength

Multiple damages were visualized both in the ribs and in the nodes of the architectural material; the width of the tears was comparable to the width of the ribs and more, up to 1.7 mm. The integrity of the structure is impaired; part of the structure is

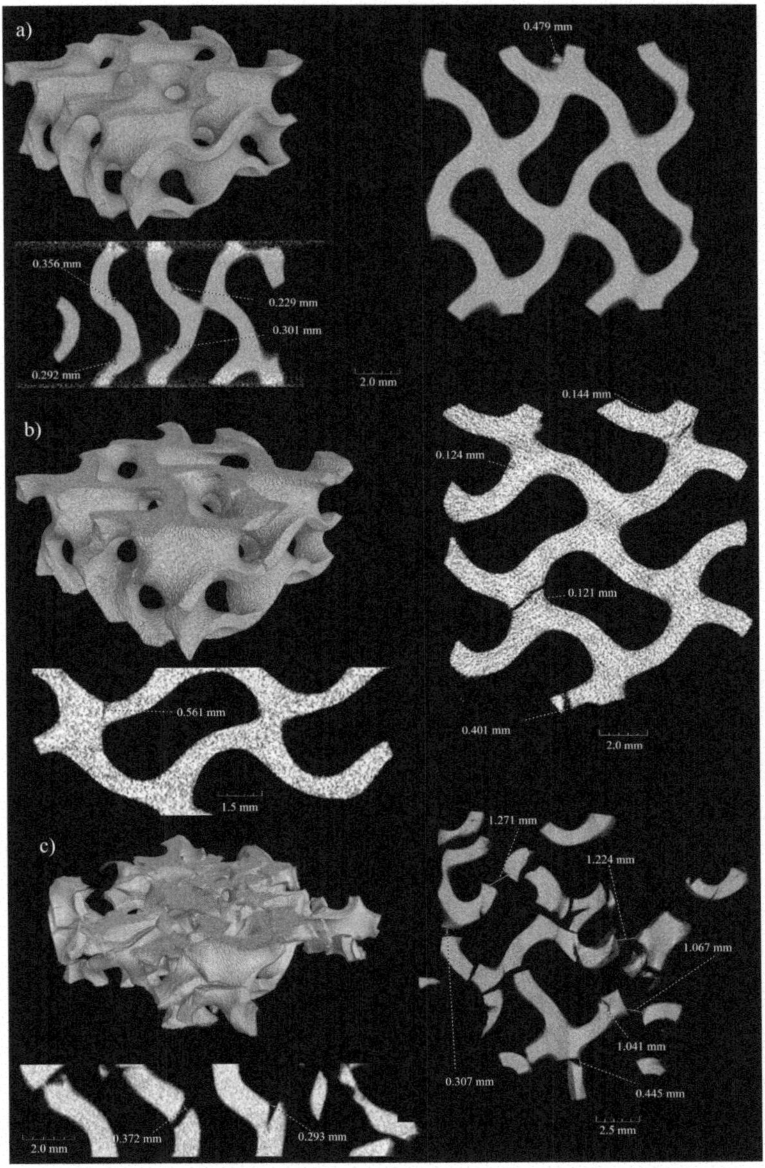

Fig. 30.4 Microtomogram of a scaffold with a wall thickness of 0.9 mm: **a** before loading; **b** loaded to the yield strength; **c** loaded to the tensile strength

completely separated from the main scaffold. The scaffold filling pattern is only partially preserved in the central and lower parts of the scaffold.

Figure 30.5a shows a microtomogram of a scaffold with a wall thickness of 1.1 mm before loading. For this sample, a number of single printing artifacts are noted in

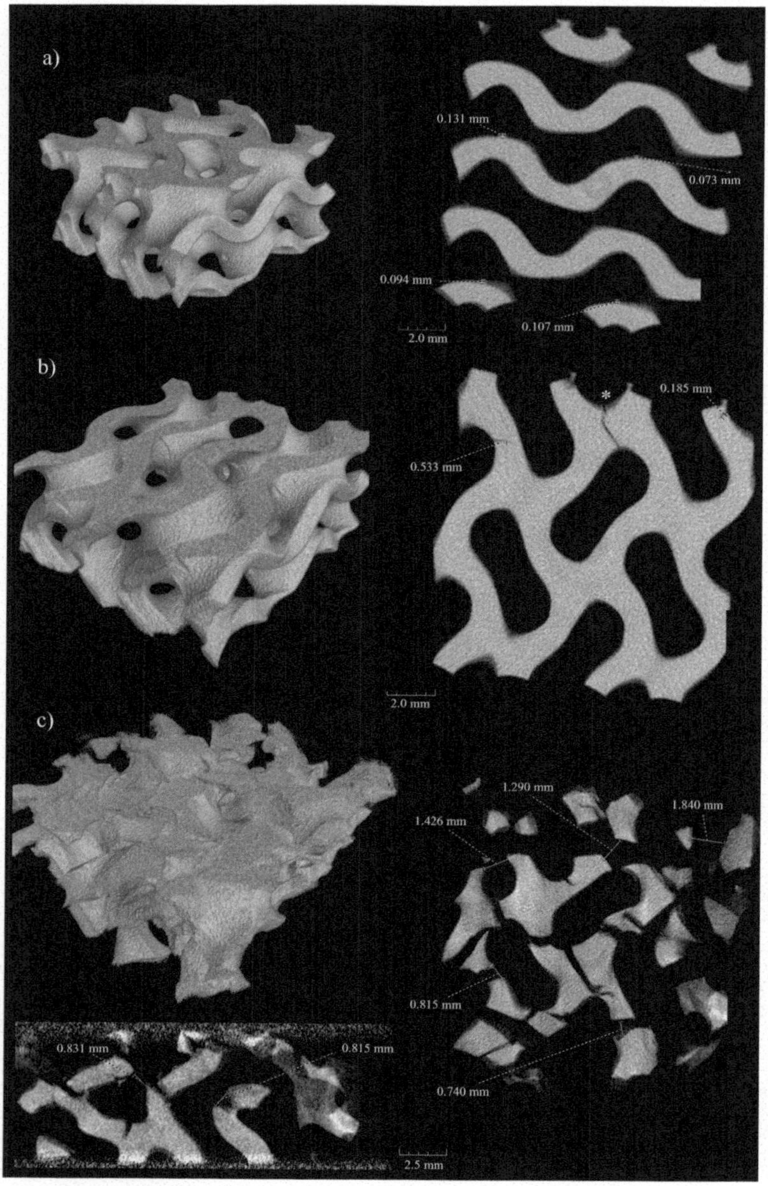

Fig. 30.5 Microtomogram of a scaffold with a wall thickness of 1.1 mm: **a** before loading; **b** loaded to the yield strength; **c** loaded to the tensile strength

the form of ellipsoidal growths on the scaffold stiffeners with a height of 0.073–0.131 mm. Figure 30.5b shows a microtomogram of the same scaffold when loaded to the yield strength point. In general, we note an insignificant number of damages to the sample. In particular, there are a number of cracks along its periphery in depth from 0.085 mm to the thickness of the ribs. Figure 30.5c shows a microtomogram of the same scaffold when loaded to the tensile strength point. Multiple fracture sites were visualized both in the ribs and in the nodes of the architectural material; the width of the tears was comparable to the width of the ribs and more, up to 1.7 mm. The integrity of the structure is compromised, part of the structure is completely separated from the main scaffold, and the structure has acquired a noticeable slope. The filling pattern of the scaffold was preserved only in the very bottom part of the scaffold at a height of up to 0.8 mm.

Figure 30.6a shows a microtomogram of a scaffold with a wall thickness of 1.3 mm before loading. For this sample, a number of single printing artifacts are noted in the form of spherical growths with a diameter of up to 0.415 mm and internal voids in the form of ellipsoidal cylinders up to 0.701 mm in length and 0.209 mm in diameter. Figure 30.6b shows a microtomogram of the same scaffold when loaded to the yield strength point. With an insignificant number of damages to the sample, there is a number of cracks along its periphery with a depth predominantly not exceeding 0.374 mm, but in two cases completely intersecting the stiffeners of the structure. Figure 30.6c shows a microtomogram of the same scaffold when loaded to the tensile strength point. Multiple damages were visualized both in the ribs and in the nodes of the architectural material. The scaffold filling pattern was preserved in the lower part of the scaffold at a height of up to 2.47 mm. The structures in the upper part of the scaffold are massively damaged, and some of the structural elements are completely separated from the main part of the material.

Figure 30.7a shows a microtomogram of a scaffold with a wall thickness of 1.5 mm before loading. This sample has no internal structure artifacts. However, at the end of the ribs of one of its faces, growths in the form of parallelipipeds up to 0.7 mm high, up to 0.37 mm wide, and up to 0.51 mm long were discovered, probably formed due to imperfections in the technological process of product production. Figure 30.7b shows a microtomogram of the same scaffold when loaded to the yield strength point. With an insignificant number of damages to the sample, there is a number of cracks along its periphery (mainly in the upper part of the sample) with a depth predominantly not exceeding 0.670 mm, but in four cases completely intersecting the stiffeners of the structure. Figure 30.7c shows a microtomogram of the same scaffold when loaded to the tensile strength point. Multiple fracture sites were visualized both in the ribs and in the nodes of the scaffold; however, some of the central nodes, especially starting from the middle of the scaffold thickness and downwards, were intact. The scaffold filling pattern was preserved in the lower part of the scaffold to a greater extent than in the upper part, and in general to a greater extent than for the cases of wall thicknesses of 1.1 and 1.3 mm. However, small fragments were still separated from the main part of the scaffold.

In general, the compression process can be divided into compaction stage, linear elastic stage, and peak failure stage. Judging by the results of computed tomography of the points, as the uniaxial stress increases, a network of cracks gradually begins to nucleate, develop, and penetrate into the sample. Large cracks can be observed in the structure of the sample, but due to the redistribution of stresses, it retains a high

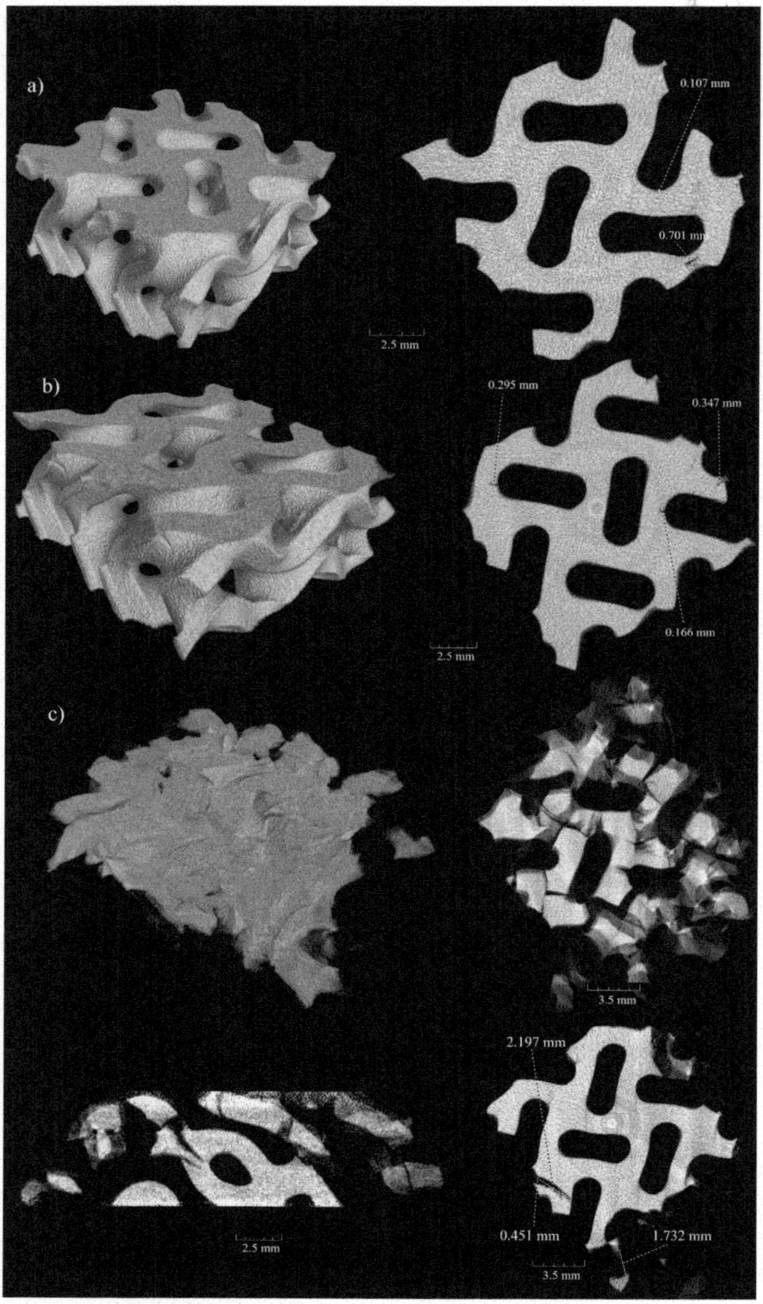

Fig. 30.6 Microtomogram of a scaffold with a wall thickness of 1.3 mm: **a** before loading; **b** loaded to the yield strength; **c** loaded to the tensile strength

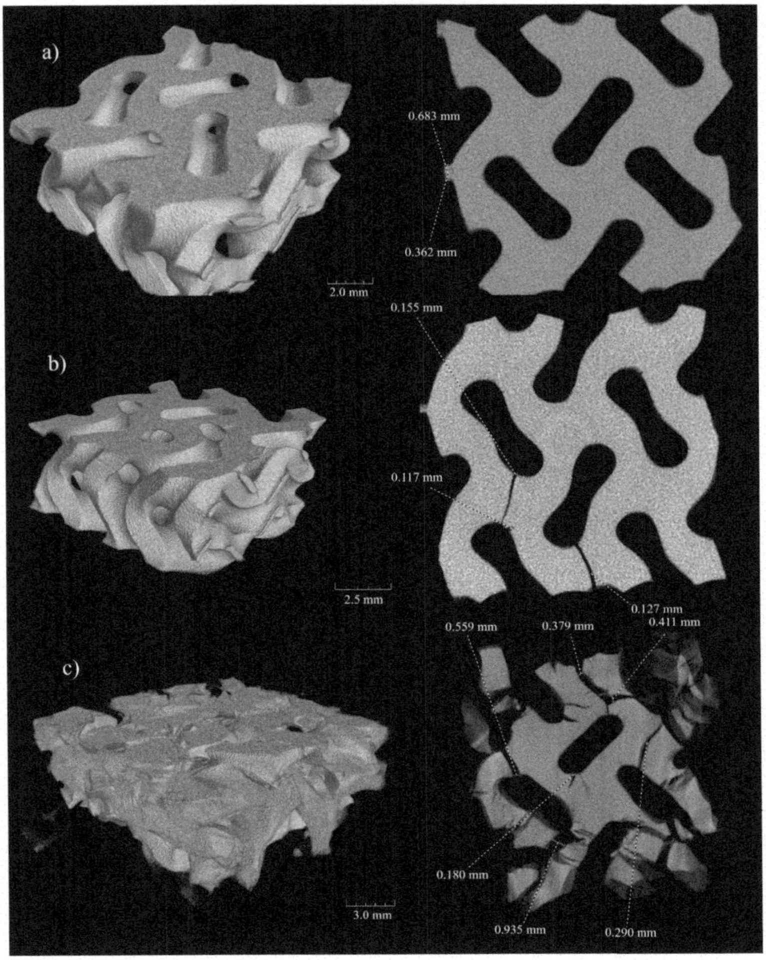

Fig. 30.7 Microtomogram of a scaffold with a wall thickness of 1.5 mm: **a** before loading; **b** loaded to the yield strength; **c** loaded to the tensile strength

load-bearing capacity. Subsequently, the line grew approximately linearly, but due to an increase in stress, cracks continued to develop and the weakest sections of the structure broke off. The main cracks continued to propagate until they penetrated the sample, causing its failure and complete loss of load-bearing capacity.

Comparison of the reconstruction of scaffolds during loading with volumetric reconstructions before loading, highlighting displacements of material surfaces using VGSTUDIO Max 3.5 software made it possible to identify the most weakened areas of the scaffold structure, as well as to estimate the relative elongation l of each type of scaffold on different stages of loading (along the frontal projection, Table 30.2).

Table 30.2 Relative elongation of scaffolds under load (along the frontal projection)

Groups by thickness, mm	l at σ_T, mm	l at σ_P, mm
0.5	0.608	2.665
0.7	0.439	3.165
0.9	0.201	2.760*
1.1	0.377	3.901
1.3	0.112	4.513
1.5	0.282	2.375

* The separated area was not taken into account

Table 30.3 Scaffold porosity

Groups by thickness, mm	P, %
0.5	87.88
0.7	79.15
0.9	69.98
1.1	62.25
1.3	54.30
1.5	50.13

Due to the exceptional significance of the pore size on the process of osseointegration [43, 44] the porosity values P (in %) were determined as the ratio of the pore volume to the volume of the scaffold without applying a load, and the results are listed in Table 30.3. To calculate the pore volume, the volume of the parallelepiped into which the scaffold is inscribed was calculated, then using the Region growing tool in VGSTUDIO Max 3.5 software, the scaffold was segmented, excluding the plates' experimental setup, and finally the volume of the scaffold was subtracted from the volume of the parallelepiped.

30.4 Conclusions

To study the deformation behavior of the polymeric DLP printed scaffolds with varying wall thicknesses and the dynamic evolution of internal fracture patterns under different loading conditions, the samples were subjected to uniaxial compression with in situ microtomography. An experimental study of the porosity of architectural materials was carried out according to histograms of gray values of the corresponding microtomograms. Reconstruction of the obtained data, rendering, obtaining virtual sections of the internal structure of the materials under study, constructing volumetric reconstructions of samples at different stages of the compression test, comparing them with volumetric reconstructions before loading, and identifying displacements of material surfaces made it possible to study the deformation behavior of the bioengineering structures with varying wall thickness.

For each type of scaffold, the yield and tensile strength points were experimentally determined. The study showed that scaffolds with wall thicknesses from 0.5 to 0.9 mm better withstand peak loads and do not lose load-bearing capacity even after the material reaches its breaking point. Scaffolds with wall thicknesses of 1.1–1.5 mm generally accumulate less damage at the yield strength point stage, but are more severely damaged when the tensile strength is reached.

With increasing wall thickness in scaffolds, a decrease in artifacts of the internal structure (pores, voids, cracks, growths, etc.) formed during the printing process is observed. Moreover, in most cases, failure under uniaxial compression is initiated not in the artifacts, but in stiffeners along the periphery of the sample. In this case, the stiffeners are damaged first, and only then the nodes in which they are connected.

Competing Interests The authors have no conflicts of interest to declare that are relevant to the content of this chapter.

Acknowledgements P. A. Lapina acknowledges the support of the grant of the Russian Science Foundation, grant number 22-49-08014. Y. C. Wang acknowledges the support of the Russia—Taiwan Bilateral Project, contract No. MOST 111-2923-E-006-003-MY3. All micro-CT experiments were conducted in the Resource Center of Collective Usage, Don State Technical University (http://nano.donstu.ru).

References

1. Koons GL, Diba M, Mikos AG (2020) Materials design for bone-tissue engineering. Nat Rev Mater. https://doi.org/10.1038/s41578-020-0204-2
2. Gupta K, Meena K (2023) Artificial bone scaffolds and bone joints by additive manufacturing: a review. Bioprint. https://doi.org/10.1016/j.bprint.2023.e00268
3. Mirkhalaf M, Men Y, Wang R, No Y, Zreiqat H (2023) Personalized 3D printed bone scaffolds: a review. Acta Biomater. https://doi.org/10.1016/j.actbio.2022.04.014
4. Kuznetsova DS, Timashev PS, Bagratashvili VN, Zagainova EV (2014) Scaffold- and cell system-based bone grafts in tissue engineering (review). Mod Tech Med 6:201–212 (in Russian)
5. Velásquez-García LF, Kornbluth Y (2021) Biomedical applications of metal 3D printing. Ann Rev Biomed Eng. https://doi.org/10.1146/annurev-bioeng-082020-032402
6. Ma H, Feng C, Chang J, Wu C (2018) 3D-printed bioceramic scaffolds: from bone tissue engineering to tumor therapy. Acta Biomater. https://doi.org/10.1016/j.actbio.2018.08.026
7. Szymczyk-Ziółkowska P, Łabowska MB, Detyna J, Michalak I, Gruber P (2020) A review of fabrication polymer scaffolds for biomedical applications using additive manufacturing techniques. Biocybern Biomed Eng. https://doi.org/10.1016/j.bbe.2020.01.015
8. Gardan J (2017) Additive manufacturing technologies: state of the art and trends. Int J Product Res. https://doi.org/10.1080/00207543.2015.1115909
9. Saed AB, Behravesh AH, Hasannia S, Ardebili SAA, Akhoundi B, Pourghayoumi M (2020) Functionalized poly l-lactic acid synthesis and optimization of process parameters for 3D printing of porous scaffolds via digital light processing (DLP) method. J Manufact Proc. https://doi.org/10.1016/j.jmapro.2020.04.076

10. Naghieh S, Ravari MK, Badrossamay M, Foroozmehr E, Kadkhodaei M (2016) Numerical investigation of the mechanical properties of the additive manufactured bone scaffolds fabricated by FDM: The effect of layer penetration and post-heating. J Mech Behavior Biomed Mater. https://doi.org/10.1016/j.jmbbm.2016.01.031
11. Van der Weijden F, Dell'Acqua F, Slot DE (2009) Alveolar bone dimensional changes of post-extraction sockets in humans: a systematic review. J Clin Periodontol. https://doi.org/10.1111/j.1600-051X.2009.01482.x
12. Tallgren A (2003) The continuing reduction of the residual alveolar ridges in complete denture wearers: a mixed-longitudinal study covering 25 years. J Prosthet Dent. https://doi.org/10.1016/S0022-3913(03)00158-6
13. Araújo MG, Lindhe J (2005) Dimensional ridge alterations following tooth extraction. An experimental study in the dog. J Clinic Periodontol. https://doi.org/10.1111/j.1600-051X.2005.00642.x
14. Özcan M, Hotza D, Fredel MC, Cruz A, Volpato CAM (2021) Materials and manufacturing techniques for polymeric and ceramic scaffolds used in implant dentistry. J Composit Sci. https://doi.org/10.3390/jcs5030078
15. Johnson AJW, Herschler BA (2011) A review of the mechanical behavior of CaP and CaP/polymer composites for applications in bone replacement and repair. Acta Biomater. https://doi.org/10.1016/j.actbio.2010.07.012
16. Girón J, Kerstner E, Medeiros T, Oliveira L, Machado GM, Malfatti CF, Pranke P (2021) Biomaterials for bone regeneration: an orthopedic and dentistry overview. Brazil J Med Biol Res. https://doi.org/10.1590/1414-431X2021e11055
17. Carrasco F, Pages P, Gamez-Perez J, Santana OO, Maspoch ML (2010) Processing of poly (lactic acid): characterization of chemical structure, thermal stability and mechanical properties. Polymer Degrad Stab. https://doi.org/10.1016/j.polymdegradstab.2009.11.045
18. Carothers WH, Dorough GL, Natta FV (1932) Studies of polymerization and ring formation. X. The reversible polymerization of six-membered cyclic esters. J Amer Chem Soc. https://doi.org/10.1021/ja01341a046
19. Silva ATCR, Cardoso BCO, e Silva MESR, Freitas RFS, Sousa RG, (2015) Synthesis, characterization, and study of PLGA copolymer in vitro degradation. J Biomater Nanobiotech. https://doi.org/10.4236/jbnb.2015.61002
20. Amiryaghoubi N, Fathi M, Pesyan NN, Samiei M, Barar J, Omidi Y (2020) Bioactive polymeric scaffolds for osteogenic repair and bone regenerative medicine. Med Res Rev. https://doi.org/10.1002/med.21672
21. Chandra RUSTGI, Rustgi R (1998) Biodegradable polymers. Prog Polymer Sci 23:1273–1335
22. Yaszemski MJ, Payne RG, Hayes WC, Langer R, Mikos AG (1996) In vitro degradation of a poly (propylene fumarate)-based composite material. Biomater. https://doi.org/10.1016/0142-9612(96)00008-7
23. Abdelaziz AG, Nageh H, Abdo SM, Abdalla MS, Amer AA, Abdal-Hay A, Barhoum A (2023) A review of 3D polymeric Scaffolds for bone tissue engineering: principles, fabrication techniques, immunomodulatory roles, and challenges. Bioeng. https://doi.org/10.3390/bioengineering10020204
24. Pryjmaková J, Kaimlová M, Hubáek T, Švorík V, Siegel J, (2020) Nanostructured materials for artificial tissue replacements. Int J Molecular Sci. https://doi.org/10.3390/ijms21072521
25. Melnikova G, Kuznetsova T, Lapitskaya V, Petrovskaya A, Chizhik S, Zykova A, Safonov V, Aizikovich S, Sadyrin E, Sun W, Yakovin S (2021) Nanomechanical and nanotribological properties of nanostructured coatings of tantalum and its compounds on steel substrates. Nanomater. https://doi.org/10.3390/nano11092407
26. Mitrin BI, Chapek SV, Sadyrin EV, Swain MV (2021) Mechanical properties and failure mechanisms of 3D-printed PLA scafolds: a preliminary study. Materials Science and Engineering, IOP Conference Series. https://doi.org/10.1088/1757-899X/1029/1/012074
27. Zhang XY, Fang G, Zhou J (2017) Additively manufactured scaffolds for bone tissue engineering and the prediction of their mechanical behavior: a review. Materials. https://doi.org/10.3390/ma10010050

28. Shahgholi M, Oliviero S, Baino F, Vitale-Brovarone C, Gastaldi D, Vena P (2016) Mechanical characterization of glass-ceramic scaffolds at multiple characteristic lengths through nanoindentation. J Eur Ceram Soc. https://doi.org/10.1016/j.jeurceramsoc.2016.01.042
29. Hakimi F, Jafari H, Hashemikia S, Shabani S, Ramazani A (2023) Chitosan-polyethylene oxide/clay-alginate nanofiber hydrogel scaffold for bone tissue engineering: preparation, physical characterization, and biomimetic mineralization. Int J Biol Macromolecules. https://doi.org/10.1016/j.ijbiomac.2023.123453
30. Marrese M, Guarino V, Ambrosio L (2017) Atomic force microscopy: a powerful tool to address scaffold design in tissue engineering. J Funct Biomater. https://doi.org/10.3390/jfb8010007
31. Sadyrin EV, Nikolaev AL, Chapek SV, Nazarenko DV, Aizikovich SM, Wang YC (2023) Manufacturing quality evaluation of photopolymer resin 3D-printed scaffolds using microtomography. Sixty shades of generalized continua: dedicated to the 60th Birthday of Prof. Victor A. Eremeyev. https://doi.org/10.1007/978-3-031-26186-2_38
32. Sadyrin EV, Mitrin BI, Yogina DV, Swain MV (2021) Preliminary study of distribution of mechanical properties and mineral density by depth of liquid saturated carious dentine. Materials Science and Engineering, IOP Conference Series. https://doi.org/10.1088/1757-899X/1029/1/012056
33. Zelentsov VB, Sadyrin EV, Mitrin BI, Swain MV (2023) Mathematical tools for recovery of the load on the fissure according to the micro-CT results. J Mech Behavior Biomed Mater. https://doi.org/10.1016/j.jmbbm.2022.105625
34. Sadyrin EV, Kislyakov EA, Karotkiyan RV, Yogina DV, Drogan EG, Swain MV, Maksyukov SYu, Nikolaev AL, Aizikovich SM (2020) Influence of citric acid concentration and etching time on enamel surface roughness of prepared human tooth: in vitro study. Plastic Damage Fract Adv Mater. https://doi.org/10.1007/978-3-030-34851-9_8
35. Sadyrin EV (2022) Correlating the mechanical properties to the mineral density of brown spot lesion in dentine using nanoindentation and X-ray microtomography. Adv Mater Model Mech Med Biol App. https://doi.org/10.1007/978-3-030-81705-3_21
36. Kennedy SM, Amudhan K, Robert RJ, Vasanthanathan A, Pandian AVM (2023) Experimental and finite element analysis on the effect of pores on bio-printed polycaprolactone bone scaffolds. Bioprint. https://doi.org/10.1016/j.bprint.2023.e00301
37. Zelentsov VB, Sadyrin EV, Sukiyazov AG, Shubchinskaya NY (2018) On a method for determination of Poisson's ratio and Young modulus of a material. MATEC Web Conf. https://doi.org/10.1051/matecconf/201822603027
38. Volkov SS, Vasiliev AS, Aizikovich SM, Sadyrin EV (2018) Contact problem on indentation of an elastic half-plane with an inhomogeneous coating by a flat punch in the presence of tangential stresses on a surface. AIP Conference Proceedings. https://doi.org/10.1063/1.5034712
39. Tkachev SY, Mitrin BI, Karnaukhov NS, Sadyrin EV, Voloshin MV, Maksimov AY, Goncharova AS, Lukbanova EA, Zaikina EV, Volkova AV, Khodakova DV, Mindar MV, Yengibarian MA, Protasova TP, Kit SO, Ermakov AM, Chapek SV, Tkacheva MS (2021) Visualization of different anatomical parts of the enucleated human eye using X-ray micro-CT imaging. Exp Eye Res. https://doi.org/10.1016/j.exer.2020.108394
40. Huang TTY, Jones AS, He LH, Darendeliler MA, Swain MV (2007) Characterisation of enamel white spot lesions using X-ray micro-tomography. J Dent. https://doi.org/10.1016/j.jdent.2007.06.001
41. Sadyrin EV, Yogina DV, Swain MV, Maksyukov SYu, Vasiliev AS (2021) Efficacy of dental materials in terms of apparent mineral density restoration: composite resin, glass ionomer cement and infiltrant. Compos Part C: Open Access. https://doi.org/10.1016/j.jcomc.2021.100192
42. Jeong T, Shin HJ (2014) An approximation technique for real-time rendering of Phong reflection model with image-based lighting. J Korea Comp Graphics Soc 20:13–19 (in Korean)
43. Wang Z, Wang C, Li C, Qin Y, Zhong L, Chen B, Li Z, Liu H, Chang F, Wang J (2017) Analysis of factors influencing bone ingrowth into three-dimensional printed porous metal scaffolds: a review. J Alloys Compd 717:271–285. https://doi.org/10.1016/j.jallcom.2017.05.079

44. Kilina P, Kuchumov AG, Sirotenko L, Vassilouk V, Golovin S, Drozdov A, Sadyrin EV (2024) Influence of porous titanium-based jaw implant structure on osseointegration mechanisms. J Mech Behav Biomed Mater 160:106724. https://doi.org/10.1016/j.jmbbm.2024.106724

Chapter 31
Surface Layer as a Source of Electrical Adhesive Forces

Vladimir S. Shorkin, Gennady V. Barsukov, Sergey N. Romashin, and Olga G. Kozhus

Abstract A variant of description of the role of the surface layer in the appearance of adhesion forces is proposed. The reason for the appearance of these forces is the piezo effect—polarization of the dielectric material caused by the stress–strain state that arises during the formation of a free surface. The forces of adhesion are the Coulomb forces. They are generated by the inhomogeneous electric field generated near the dielectric surface by the material polarized due to deformation. They can affect both individual particles and elementary particles of solid or liquid material polarizable under their influence. Since the polarization vector is considered to be proportional to the strain tensor, the main attention in this work is given to the construction of a model of the surface layer. The model of its deformed state is built on the basis of ideas about nonlocal pair and triple interactions of particles of continuous materials, which the authors used earlier to describe adhesive interactions of electrically neutral media. A qualitative confirmation of the fairness of the results obtained in the work is considered to be the experimentally established effect of increasing the viscosity of polar liquid—water near the surface of the quartz channel. Theoretical constructions of the work are proposed to be used for estimation of the influence of coatings of channel walls with surface-active substances on the viscosity of liquid flows contacting with them.

Keywords Piezoelectric effect · Surface layer · Nonlocal interaction · Gradient model · Viscosity anomaly near a solid wall

V. S. Shorkin (✉) · G. V. Barsukov · S. N. Romashin · O. G. Kozhus
Orel State University named after I.S. Turgenev, 29, Naugorskoe Shosse, 302020 Orel, Russia
e-mail: v.s.shorkin@yandex.ru

G. V. Barsukov
e-mail: awj@list.ru

S. N. Romashin
e-mail: sromashin@yandex.ru

O. G. Kozhus
e-mail: tila_olya@mail.ru

H. Altenbach and V. A. Eremeyev (eds.), *Dynamics of Discrete and Continuum Structures and Media*, Advanced Structured Materials 221,
https://doi.org/10.1007/978-3-031-75626-9_31

31.1 Introduction

Adhesive interactions between solid and liquid media play a major role (both positive and negative) in a wide variety of life activities. Therefore, the theoretical study of this phenomenon, creation of mathematical models of adhesion of the most diverse solid and liquid materials adequate to real processes is an urgent task. The complexity of chemical composition and atomic-molecular structure of interacting media makes the solution of this problem by methods of solid-state physics difficult. To a certain extent, these difficulties can be avoided by methods of continuum mechanics. In this way, a number of models have been created, within the framework of which a number of practically important problems have been solved with sufficient accuracy (e.g., [1–3]).

The adhesive effect of the material of one body on the particles of another body is realized by means of long-range forces. It is known from physics [4, 5] that their cause can be electric fields, which affect electrically charged particles, and van der Waals force fields, which affect electrically neutral particles. It was noted in [5] that the nature of the interparticle interactions, whether pair or triple, affects their outcome.

A model of adhesion of solid bodies, which is based on the consideration of nonlocal potential interaction of their electrically neutral particles, is presented in [6]. Here, it is shown that for certain combinations of materials of interacting bodies B and B^* for pair or triple interactions of their macroparticles dB and dB^* (these are representative particles of continua B and B^*, consisting of real microparticles q_j and q_j^*—atoms or molecules, real materials) the result of interaction can be different. The approach used in [6] to construct a model of the adhesive interaction between bodies B and B^* (as shown by the comparison of theoretical and experimental data) allowed us to construct results adequate to reality.

The necessity of using a unified approach to modeling adhesive interactions of continua B and B^* (solid or liquid) both by means of electric fields and long-range interaction forces of their electrically neutral particles is obvious.

In this paper, we propose a mathematical model of the emergence of an external electric field near the boundary surface A of a solid body B due to the deformation of the dielectric in the resulting surface layer during its formation (piezo effect [7]). The dielectric deformation causes polarization of the material. The polarized material forms an electric field. The possibility of this process exists for a number of dielectrics.

According to the physical concept of the piezo effect, the vector of polarization induced by the material deformation is proportional to the tensor describing it or the stresses that lead to its occurrence. The coefficient of proportionality is a well-known material tensor. Therefore, the main difficulty in describing the effect on the surface layer is to describe the stress–strain state of the surface layer. In this paper, the gradient model, which is a local (weakly local) approximation of the nonlocal model based on the description of pair and triple potential interactions of particles of a continuous medium [3, 6], is used for this purpose.

The model is relevant for practical applications. In particular, in combination with the model of adhesive interaction of two continuums, based on the ideas of nonlocal potential interactions of their electrically neutral particles, it is proposed to apply it to construct a quantitative model of the process of the working medium movement along the metal channel of waterjet metal cutting devices [8]. The working medium is a mixture of quartz sand water. The use of the proposed set of models can help to improve the efficiency of the process of waterjet cutting of metals.

31.2 Mathematical Model of the Material

31.2.1 Discrete Material Model

The atomic-molecular structure of the investigated body B is modeled by a discrete system $\{q_j\}_N$ $(j = 1, ..., N)$ of a large number N of microparticles—models of individual atoms or molecules. In this paper, we consider a dielectric for which microparticles q_j can be dipoles in the reference configuration or become them in the process of deformation.

When a free surface A of a body B is formed due to the breaking of interatomic bonds in the material layer adjacent to it, the atomic-molecular structure is rearranged due to the recombination of interatomic bonds. The configuration of the system $\{q_j\}_N$ after formation A (actual configuration) is characterized by vectors $\mathbf{x}_1, ..., \mathbf{x}_N$; $\delta_1, ..., \delta_N$. Before this event, the configuration was characterized by a set of vectors $\mathbf{X}_1, ..., \mathbf{X}_N; \mathbf{D}_1, ..., \mathbf{D}_N$ (reference configuration). The rearrangement of the structure is characterized by the transition

$$(\mathbf{X}_1, ..., \mathbf{X}_N;\ \mathbf{D}_1, ..., \mathbf{D}_N) \to (\mathbf{x}_1, ..., \mathbf{x}_N;\ \delta_1, ..., \delta_N). \tag{31.1}$$

It is accompanied by the emergence of inequalities

$$\mathbf{X}_j - \mathbf{x}_j \neq 0,\ \mathbf{D}_j - \delta_j \neq 0\ (j = 1, ..., N). \tag{31.2}$$

These inequalities indicate translational and orientational deformations of the discrete structure $\{q_j\}_N$. In the process of these deformations, relative charge displacement is possible (e.g., in ionic crystals). It is believed that this leads to polarization of the material. It is also assumed that the values $|\mathbf{X}_j - \mathbf{x}_j|$ and $|\mathbf{D}_j - \delta_j|$ are small values. So, the work of changing each of the set values $\mathbf{X}_1, ..., \mathbf{X}_N;\ \mathbf{D}_1, ..., \mathbf{D}_N$ at transition (31.1) is independent of each other.

As one moves away from the free surface A into the depth of the body B, the influence of surface recombination decreases. In this case

$$(\mathbf{x}_1, ..., \mathbf{x}_N;\ \delta_1, ..., \delta_N) \to (\mathbf{X}_1, ..., \mathbf{X}_N;\ \mathbf{D}_1, ..., \mathbf{D}_N), \tag{31.3}$$

$$\mathbf{X}_j - \mathbf{x}_j \to 0,\ \mathbf{D}_j - \delta_j \to 0.$$

The thermodynamic properties of the system under study are determined using a statistical sum. This sum is defined through the Hamiltonian of the system, which (for both solid and liquid medium) at any moment of time t is expressed through the kinetic and configurational parts [9, 10]. The configurational part is the potential energy of the system. The potential energy W_B of the whole body B is determined by the interaction of energy $\Phi_N = \Phi_N(\mathbf{x}_1, ..., \mathbf{x}_N;\ \delta_1, ..., \delta_N)$ of all microparticles q_j of the system $\{q_j\}_N$ among themselves. It is considered to be the sum of two summands—translational energy $W_{B(tr)}$ and polarization energy $W_{B(P)}$. The first of them is equal to the sum of all pair and triple interactions of microparticles q_j in the assumption that they are material points.

$$W_{B(tr)} = \sum_{j}^{N}\sum_{>k}^{N} \Phi_{ij} + \sum_{i}^{N}\sum_{>j}^{N}\sum_{>k}^{N} \Phi_{ijk}. \tag{31.4}$$

The second kind of energy is proposed to equate the polarization energy of the considered dielectric caused by the appearance of the surface layer.

In the transition to the continuum approximation, the system $\{q_j\}_N$ is divided into a set K of subsystems $\{q_j\}_k$ $(k = 1, ..., K)$ containing N_k microparticles each. Each subsystem $\{q_j\}_k$ corresponds to a representative macroparticle $\Delta B_k \to dB$ of the continuum B. So, the continuum B is a union of them.

31.2.2 Continuum Representations of Materials

Kinematic Properties

A body B is said to be a continuum filling a region Λ of Euclidean space Ω, in which there is an orthonormalized basis $\mathbf{e}_1$, $\mathbf{e}_2$, $\mathbf{e}_3$ of Cartesian coordinate system $y_1 \equiv y$, y_2, y_3. It is a union of non-intersecting macroparticles dB occupying regions $d\Lambda$ with volumes dV. Vectors $\mathbf{R}$ and $\mathbf{r}$ define the positions of their centers of inertia in the reference and current configurations, respectively. They differ from each other by a displacement vector dB equal to the difference $\mathbf{u} = \mathbf{r} - \mathbf{R}$.

One of the differences between the current configuration and the reference configuration is the change of the position of an arbitrary macroparticle dB_2 with respect to another macroparticle $dB \equiv dB_1$ from vector $\mathbf{L} \equiv \mathbf{L}_{(12)} = \mathbf{R}_2 - \mathbf{R}$ to vector $\mathbf{l}_{(12)}(\mathbf{L})$ by vector

$$\Delta\mathbf{u} \equiv \Delta\mathbf{u}_{12} = \mathbf{u}_2(\mathbf{L}) - \mathbf{u} = \mathbf{l}_{12}(\mathbf{L}) - \mathbf{L}. \tag{31.5}$$

It is assumed that function (31.5) can be represented as the following sum (gradient model [3, 6])

$$\Delta \mathbf{u} = (\nabla \mathbf{u}) \cdot \mathbf{L} + \frac{1}{2!}(\nabla^2 \mathbf{u}) \cdot \cdot \mathbf{L}^2 + ... + \frac{1}{m!}(\nabla^m \mathbf{u}) \overbrace{\cdot ... \cdot}^{(m-1)\ times} \mathbf{L}^n \tag{31.6}$$

This sum contains the first $\nabla \mathbf{u}$, second $\nabla^2 \mathbf{u}$, etc. gradients of the displacement vector defined at $\mathbf{L} = 0$ (at the point(at the point $\mathbf{R} = \mathbf{R}_1$). A convolution operation is performed over them respectively with vector $\mathbf{L}$, dyad $\mathbf{L}^2 = \mathbf{LL} \equiv \mathbf{L} \otimes \mathbf{L}$, etc. up to n—multiple of tensor product ($\otimes$—its sign) $\mathbf{L}^n = \overbrace{\mathbf{L}...\mathbf{L}}^{n\ times} \equiv \overbrace{\mathbf{L} \otimes ... \otimes \mathbf{L}}^{n\ times}$. The convolution is performed according to the rule:

$$(\nabla^j \mathbf{u}) \overbrace{\cdot ... \cdot}^{(j-1)\ times} \mathbf{L}^j =$$

$$= u_{i,k_1...k_j}(\mathbf{e}_i \otimes \mathbf{e}_{k_1} \otimes ... \otimes \mathbf{e}_{k_j}) \overbrace{\cdot ... \cdot}^{(j-1)\ times} (\mathbf{e}_{kj} \otimes ... \otimes \mathbf{e}_{k_1}) L_{k_1} ... L_{k_j} = \tag{31.7}$$

$$= (u_{i,k_1...k_j} L_{k_1} ... L_{k_j}) \mathbf{e}_i (\mathbf{e}_{k_j} \cdot \mathbf{e}_{k_j}) ... (\mathbf{e}_{k_1} \cdot \mathbf{e}_{k_1}).$$

The index after the decimal point means differentiation on the coordinate with this number, (.) is the sign of the scalar product—convolution.

During the deformation process, the distance between $dB \equiv dB_1$ and dB_2 changes from $L = L_{(12)}$ to $|\mathbf{l}| = l \equiv l_{(12)} = |\mathbf{r}_2 - \mathbf{r}|$ $||\nabla \mathbf{u}|| = \sqrt{(\nabla \mathbf{u}) \cdot \cdot (\nabla \mathbf{u})^T} << 1$, this change is characterized by means of a tensor $(\nabla \mathbf{u})^S = (1/2)((\nabla \mathbf{u}) + (\nabla \mathbf{u})^T)$ (the upper index (^{T}) means transpose).

Energy of Nonlocal Interactions of Continuum Particles

For a continuum B, it is assumed that its potential energy is equal to the energy of a discrete system modeling a real body. This energy consists of two summands $W_{B(tr)}$ and $W_{B(P)}$. Each of them is equal to its analog of the discrete system.

It is assumed that the potential energy $W_{B(tr)}$ of the system of pair and triple interactions of microparticles, defined by equality (31.4), is equal to the potential energy of the continuum B, which is defined by the following expressions:

$$W_{B(tr)} = \int_{\Lambda} w_{(tr)} dV, \tag{31.8}$$

$$w_{(tr)} = \Phi_0^{(1)}\varphi^{(1)} +$$

$$+ \Phi_0^{(2)} \int\limits_{\Lambda} \left[\varphi^{(2)}(\beta l_{(12)}) + \Phi_0^{(3)} \int\limits_{\Lambda} \varphi^{(3)}(\beta l_{(12)}, \beta l_{(13)}) dV_3 \right] dV_2 \qquad (31.9)$$

Here, $w_{(tr)}$ is the volume density of the internal potential energy of the continuum B. The quantities $\Phi_0^{(1)}$, $\Phi_0^{(2)}$, $\Phi_0^{(3)}$, β are material parameters. The first of them is the volume density of the internal potential energy of a single macroparticle, equal to the potential energy of interaction of the microparticles composing it in the absence of deformation. The value $\varphi^{(1)}$ characterizes the dependence of this energy under deformation. In the paper, $\varphi^{(1)} = 1$ is assumed. The parameters $\Phi_0^{(2)}$ and $\Phi_0^{(3)}$ characterize the "reaction" of the continuum to pair and triple interactions between its macroparticles $dB \equiv dB_1$ and dB_2 in the first case, and $dB \equiv dB_1, dB_2$, and dB_3 in the second case, respectively. The parameter β characterizes the "rate" of change of the interaction potential of two macroparticles when the distance $l_{(12)} = |\mathbf{l}_{(12)}|$ or $l_{(13)} = |\mathbf{l}_{(13)}|$ between them changes. The functions $\varphi^{(2)}(\beta l_{(12)})$ and $\varphi^{(3)}(\beta l_{(12)}, \beta l_{(13)})$ characterize the dependence of pair and triple potentials on the interparticle distances indicated in brackets. Their form is set empirically. It is similar to the form of the corresponding functions for the interaction potentials of microparticles. A general requirement is the stability condition of the system of macroparticles dB, forming a body B (discussed in [11, 12]). The values of $\Phi_0^{(1)}$, $\Phi_0^{(2)}$, $\Phi_0^{(3)}$, and β are determined on the basis of experimental data.

The integrand functions in expressions (31.8) and (31.9) describe the nonlocal stress–strain state of a continuous nonpolarized body B, characterizes the state of the medium at a constant temperature.

Local Description of the Stress–Strain State

The stress–strain state of the continuum B is described in a local (weakly nonlocal) approximation in the form of an m-order gradient model of materials [13]. This is necessary to take into account the influence of neighboring infinitesimal particles on each other in the surface layer (cohesive zone). The method is proposed and justified in [14].

The kinematic characteristics in the proposed model are a set of gradients of different orders $\{\nabla^n \mathbf{u}\}$. All of them are defined at the center of inertia $\mathbf{R} \equiv \mathbf{R}_1$ of the macroparticle $dB \equiv dB_1$.

Expression (31.9) of the bulk density $w_{(tr)}$ of potential energy $w_{(tr)}dV$ of the macroparticle $dB \equiv dB_1$ is represented by a second degree polynomial with respect to the set of gradients $\{\nabla^n \mathbf{u}\}$.

$$w_{(tr)} = \sum_{n=1}^{m} P_0^{(n)} \cdot\cdot\cdot\cdot\cdot (\nabla^n \mathbf{u})^T + \sum_{n=1}^{m}\sum_{k=1}^{m} (\nabla^n \mathbf{u}) \cdot \ldots \cdot C^{(n,k)} \cdot\cdot\cdot\cdot\cdot (\nabla^k \mathbf{u})^T. \tag{31.10}$$

The coefficients of this polynomial $P_0^{(n)}$, $C^{(n,k)}$, are expressed through the functions $\varphi^{(2)}(\beta l_{(12)})$ and $\varphi^{(3)}(\beta l_{(12)}, \beta l_{(13)})$ using the expressions [3]:

$$P_0^{(n)} = \Phi_0^{(2)} \int_{\Lambda} \left(\frac{d\varphi^{(2)}}{d\mathbf{L}_{(12)}}\right) \mathbf{L}_{(12)} dV_2 +$$

$$+ \Phi_0^{(3)} \sum_{k=2}^{3} \int_V \left[\int_V \left(\frac{d\Phi^{(3)}}{d\mathbf{L}_{(1k)}}\right) \mathbf{L}_{(1k)} dV_2 \right] dV_3 \tag{31.11}$$

$$C^{(n,k)} = \Phi_0^{(2)} \int_{\Lambda} \mathbf{L}^n_{(12)} \left(\frac{d^2\varphi^{(2)}}{d\mathbf{L}^2_{(12)}}\right) \mathbf{L}^k_{(12)} dV_2 +$$

$$+ \Phi_0^{(3)} \sum_{p,q=2}^{3} \int_V \left[\int_V \mathbf{L}^n_{(1k)} \left(\frac{d^2\varphi^{(3)}}{d\mathbf{L}_{(1p)}\mathbf{L}_{(1q)}}\right) \mathbf{L}^k_{(1q)} dV_2 \right] dV_3 \tag{31.12}$$

In an infinite medium $\Lambda = \Omega$, due to its isotropy (dependence of functions only on interparticle distances), the tensors are $P_0^{(n)}$—at even value of n, and $C^{(n,k)}$—at odd sum of $n + k$. In addition, the equality $C^{(n,k)} = C^{(k,n)}$ is satisfied.

From definitions (31.11) and (31.12), we see that the tensors $P_0^{(n)}$ and $C^{(n,k)}$ depend on the position of the point of their definition—the center of inertia $\mathbf{R} \equiv \mathbf{R}_1$ of the macroparticle $dB \equiv dB_1$, relative to the boundary $\partial\Lambda$ of the region Λ occupied by the body B. If the body B is infinite ($\Lambda = \Omega$), the coefficients $P_0^{(n)}$ and $C^{(n,k)}$ are constant.

The characteristics of the stress state of the continuum B are introduced as follows.

It is considered that there are no external volumetrically distributed surface (along the boundary A) distributed influences on the body B. The stress–strain state in each macroparticle dB develops under the action on it of contact forces $\mathbf{f}_A$ internal along the boundary surface ΔA. Their action on the macroparticle dB is replaced by their equal force $d\mathbf{\Psi} = \phi dV$. The vector ϕ is defined by the equality

$$\phi = \lim_{\Delta V \to 0} \int_{\partial(\Delta\Lambda)} \mathbf{f}_A d(\Delta A). \tag{31.13}$$

Under the action of this force, the macroparticle dB, considered as a material point, can move with velocity $\mathbf{v} = \partial\mathbf{u}/\partial t$ and acceleration $\mathbf{a} = \partial^2\mathbf{u}/\partial t^2$.

For the vector ϕ, there exists a tensor Π, for which the equality [15] holds:

$$\phi = \nabla \cdot \Pi. \tag{31.14}$$

The tensor Π is an internal stress tensor. The representation for it is taken to be

$$\Pi = P^{(1)} - \nabla \cdot (P^{(2)} - \nabla \cdot (P^{(3)} - \cdots \nabla \cdot P^{(s)}) \cdots). \tag{31.15}$$

In view of the above, the equation of motion of the macroparticle dB takes the following form: $\boldsymbol{\Psi} \equiv \rho \dfrac{\partial^2 \mathbf{u}}{\partial t^2} = \phi$ or

$$\rho \frac{\partial^2 \mathbf{u}}{\partial t^2} = \nabla \cdot \Pi = \nabla \cdot (P^{(1)} - \nabla \cdot (P^{(2)} - \nabla \cdot (P^{(3)} - \cdots \nabla \cdot P^{(s)}) \cdots)). \tag{31.16}$$

Here, $\boldsymbol{\Psi} \equiv \rho \dfrac{\partial^2 \mathbf{u}}{\partial t^2}$ is the vector of volume density of inertia forces.

The work of the vector $(\boldsymbol{\Psi} - \phi)$ on the possible displacements $\partial \mathbf{u}$ of all macroparticles of the body B in the absence of external influences on it can be represented as

$$\delta Q = \int_{\Lambda} (\boldsymbol{\Psi} - \phi) \cdot \delta \mathbf{u} dV =$$

$$= \int_{\Lambda} \left[P^{(1)} \cdot\cdot \delta(\nabla \mathbf{u})^T + \cdots + P^{(s)} \cdot\cdot\cdot\cdot\cdot \delta(\nabla^s \mathbf{u})^T \right] dV. \tag{31.17}$$

At constant temperature, the tensors $P^{(1)}$,..., $P^{(s)}$, are considered as internal potential generalized forces, which do work on variations of generalized displacements $\delta(\nabla \mathbf{u})$,..., $\delta(\nabla^s \mathbf{u})$.

The condition of the minimum of the potential energy of the body B is the equality

$$\delta W_{B(tr)} = \int_{\Lambda} \delta w_{(tr)} dV = 0. \tag{31.18}$$

The conditions that this minimum is reached at work of generalized forces $P^{(1)}$,..., $P^{(s)}$ on generalized displacements $(\nabla \mathbf{u})$,..., $(\nabla^s \mathbf{u})$ are the equations

$$\delta Q = \delta W_{B(tr)}, \; P^{(k)} = \frac{\partial w_{(tr)}}{\partial(\nabla^k \mathbf{u})}, \ldots, \; P^{(s)} = \frac{\partial w_{(tr)}}{\partial(\nabla^s \mathbf{u})} \; k = 1, 2, \ldots, s. \tag{31.19}$$

It can be seen that the presented reasoning makes the assumption that only $s \leq m$ of the set of m gradients of $\{\nabla^n \mathbf{u}\}$ give a minimum to $W_{B(tr)}$. It is considered that variations $(m - s)$ of the gradients $(\nabla^{s+1} \mathbf{u})$,..., $(\nabla^m \mathbf{u})$ do not affect the minimum of the value $W_{B(tr)}$.

Taking into account the representation (31.10), the tensors $P^{(1)}$,..., $P^{(s)}$ are defined by the following equations:

$$P^{(n)} = \frac{\partial w_{(tr)}}{\partial(\nabla^n \mathbf{u})} = P_0^{(n)} + \sum_{k=1}^{m} C^{(n,k)} \cdot\cdot\cdot\cdot\cdot (\nabla^k \mathbf{u})^T;\ k = 1, 2, ..., s;\ s \leq m. \tag{31.20}$$

After substituting these expressions into (31.16), the equation of motion of the continuum B in displacements is obtained.

In [6, 11, 15], Eq. (31.20) was used to describe the stress–strain state in an infinitely extended medium Ω in two cases. In the first case, the homogeneous stress state of isotropic material was considered for equality $m = s = 1$. In the second case for equality $s = 2$, $m = 3$, the dispersion law for plane acoustic waves of high frequency was considered. The experimental data on the characteristics of the stress–strain state available in the literature for these cases allowed us to make quantitative estimates of the parameters $\Phi_0^{(1)}$, $\Phi_0^{(2)}$, $\Phi_0^{(3)}$, β for a number of materials at empirically chosen types of functions $\varphi^{(2)}(\beta l_{(12)})$ and $\varphi^{(3)}(\beta l_{(12)}, \beta l_{(13)})$.

31.3 Surface Layer and Piezoelectric Effect

31.3.1 Surface Layer

To simplify the reasoning and make it more clear, it is assumed that the body B under study is semi-infinite and its boundary surface A is a plane. They occupy, respectively, the areas

$$\Lambda :\ 0 < y_1 \equiv y < \infty,\ -\infty < y_j < +\infty,\ i = 2, 3, \tag{31.21}$$

$$\partial\Lambda :\ y_1 \equiv y = 0.$$

The vector of unit normal to the plane A is $\mathbf{n} = -\mathbf{e}_1$.

The deformation process is assumed to be one-dimensional, occurring along $\mathbf{n} = -\mathbf{e}_1$, depending on only one coordinate $y_1 \equiv y$. So, $\mathbf{u} = (u(y), 0, 0)$. It is also assumed that the tensors $P_0^{(n)}$ $C^{(n,k)}$, are computed.

To simplify the notation, the following notations are used: σ, $p^{(n)}$, $p_0^{(n)}$, $c^{(n,k)}$—for the coordinate, respectively, of the tensors: the tensor Π, $P^{(n)}$, $P_0^{(n)}$, $C^{(n,k)}$, all numbers of which are equal to one, and also ε—for the derivative of $\frac{du}{dx}$. Taking this into account, the expressions for the coordinates $P_{1111}^{(1)}$, $P_{111111}^{(2)}$, ... of the stress tensors $P^{(1)}$, $P^{(2)}$, ... take the following form:

$$P_{1111}^{(1)} \equiv p^{(1)} = p_0^{(1)} + c^{(1,1)}\frac{du}{dy} + 2c^{(1,2)}\frac{d^2u}{dy^2} + 2c^{(1,3)}\frac{d^3u}{dy^3} + \dots + 2c^{(1,m)}\frac{d^mu}{dy^m}, \tag{31.22}$$

$$P_{111111}^{(2)} \equiv p^{(2)} = p_0^{(2)} + 2c^{(2,1)}\frac{du}{dy} + c^{(2,2)}\frac{d^2u}{dy^2} +$$

$$+ 2c^{(2,3)}\frac{d^3u}{dy^3} + 2c^{(2,4)}\frac{d^4u}{dy^4} + \dots + 2c^{(1,m)}\frac{d^mu}{dy^m}$$

In this problem, it is assumed that $m = 4$ $s = 2$. The choice of these values is conditioned by the necessity to use the mentioned method of determining the values of material parameters on the one hand, and on the other hand, to obtain such a solution, which satisfies the condition of decreasing in the process of removal from the boundary to the depth of the body and asymptotic aspiration to a constant value of the perturbation caused by the occurrence of the free surface A.

In the stated formulation, deformation of the material along the normal to the boundary plane does not cause deformations along this plane. The coordinates of the tensors $P^{(1)}$ and $P^{(2)}$, associated with the directions along the boundary plane are not considered.

In expressions (31.22), in accordance with the stability conditions of continuum B [11, 12] and definitions (31.10)–(31.12), the following statements are satisfied:

$$(n = 1, 2) \Rightarrow p^{0(n)} > 0, \tag{31.23}$$

$$(m + n = 2k,\ k = 1, 2, \ldots) \Rightarrow$$

$$\Rightarrow \left(\frac{dc^{(n,m)}}{dy} > 0,\ c^{(n,m)}(0) = \frac{1}{2}c_{(\Omega)}^{(n,m)} > 0,\ \lim_{x\to\infty} c^{(n,m)}(y) = c_{(\Omega)}^{(n,m)}\right),$$

$$2c^{(1,3)} - c^{(2,2)} > 0,$$

$$(m + n = 2k + 1) \Rightarrow \left(\frac{dc^{(n,m)}}{dy} < 0,\ c^{(n,m)}(0) = c_0^{(n,m)} > 0,\ \lim_{x\to\infty} c^{(n,m)}(y) = 0\right).$$

Hereinafter, an index $_{(\Omega)}$ means the definition of a quantity with that index at $y \to \infty$.

In the one-dimensional case proposed for consideration, the problem of determining displacements and deformations of surface layer particles together with expressions (31.22) and comments (31.23) takes the form:

$$\frac{d}{dy}\left(p^{(1)} - \frac{d}{dy}p^{(2)}\right) = 0. \tag{31.24}$$

$$y = 0:\ u = 0,\ p^{(1)} = 0,\ p^{(2)} = p_0^{(2)}. \tag{31.25}$$

$$y \to \infty:\ \frac{dp^{(1)}}{dy} \to 0,\ p^{(1)} \to p_0^{(1)}. \tag{31.26}$$

The coefficients $p_0^{(n)}$ and $c^{(n,k)}$ used in the problem (31.24)–(31.25) are variables. This significantly complicates obtaining its analytical solution. This, in turn, complicates the qualitative analysis of the solution results. Therefore, we further construct a function convenient for qualitative analysis, which we propose to consider as an approximate solution of the problem (31.22), (31.24)–(31.25). Its construction is based on the following reasoning.

From expression (31.23) follows the constancy of the value σ for the whole body B.

$$\sigma = p^{(1)} - \frac{d}{dy}p^{(2)} = D_1 = const. \tag{31.27}$$

Here, D_1 is an as-yet unknown constant of integration.

To take into account the changes in the atomic structure of the material caused by recombination of interatomic bonds during the formation of a free surface [5], the following assumptions are made.

It is considered that material parameters are variable in a thin near-surface layer with an effective thickness h $(0 < y < h)$ much less than the effective thickness of the surface layer, in which a stress–strain state arises by a change in potential energy described by the function (31.10).

It is considered that, beyond the surface layer, the material parameters are constant. In [15], it is shown that the conditions are fulfilled here:

$$P_0^{(n)} = 0\ (n \geq 2),\ (m + n = 2k + 1) \Rightarrow (c^{(n,m)}(y) = 0). \tag{31.28}$$

It is considered that, as an initial approximation, the variability of material characteristics can be described (instead of dependences of parameters on distance $0 < y < h$, constructed on the basis of definitions (31.11) and (31.12)) by means of the Heaviside function. Let, for example, the function be

$$H(y) = (1/2)(1 + th(\gamma y))\ (\gamma \to \infty),\ \gamma \approx 1/h. \tag{31.29}$$

As a result of this step, it turns out that the following approximate equations are satisfied inside the boundary layer.

$$p^{0(1)}(y) = (p^{0(1)}_{(\Omega)}/2)(1 + th(\gamma y)) = p^{0(1)}_{(\Omega)} H(y), \tag{31.30}$$

$$p^{0(2)}(y) = (p^{0(2)}(0))[1 - th(\gamma y)] = (2p^{0(2)}(0))[1 - H(y)],$$

$$c^{(1,1)}(y) = c^{(1,1)}_{(\Omega)} H(y),\ c^{(2,2)}(y) = c^{(2,2)}_{(\Omega)} H(y),$$

$$c^{(1,3)}(x) = c^{(1,3)}_{(\Omega)} H(y),\ c^{(2,4)}(x) = c^{(2,4)}_{(\Omega)} H(y).$$

The lower index $_{(\Omega)}$ means the definition of the object to which it refers, for an infinite unperturbed and unaffected medium.

Such a step made it possible to take into account the requirements to the solution describing the main property of the surface layer—the surge of internal stresses and deformations near the boundary with its subsequent decrease as it moves away from the source of the surge. It turns out that at $y = 0$ and $y \to \infty$ the values of material parameters turn out to be defined exactly.

Approximations for the dependence $c^{(1,2)}(y)$ are not presented, as they are not used further.

As a result, it turns out that outside the layer $0 < y < h$ the stresses are determined by the equations having constant coefficients with values $p^{0(1)} = p^{0(1)}_{(\Omega)}$ and $c^{(m,n)} = c^{(m,n)}_{(\Omega)}$ which are characteristic of an infinite medium.

$$p^{(1)} = p^{0(1)} + c^{(1,1)}\frac{du}{dy} + 2c^{(1,3)}\frac{d^3u}{dy^3}, \tag{31.31}$$

$$p^{(2)} = c^{(2,2)}\frac{d^2u}{dy^2} + 2c^{(2,4)}\frac{d^4u}{dy^4}. \tag{31.32}$$

The use of (31.31) and (31.32) leads the equilibrium equation to the form:

$$\varepsilon + a_2\frac{d^2\varepsilon}{dy^2} - a_4\frac{d^4\varepsilon}{dy^4} = f,\ \varepsilon = \frac{du(y)}{dy}. \tag{31.33}$$

Here denoted by

$$f = \frac{D_1 - p^{0(1)}}{c^{(1,1)}},\ a_2 = \frac{2c^{(1,3)} - c(2,2)}{c(1,1)} > 0,\ a_4 = \frac{2c^{(2,4)}}{c^{(1,1)}} > 0; \tag{31.34}$$

Solution of Eq. (31.33):

$$\varepsilon(x) = f + D_2 e^{-kx},\ u(x) = f\,x - \frac{D_2}{k_{(1)}} e^{-ky} + D_3, \tag{31.35}$$

$$k = \sqrt{\frac{a_2 + \sqrt{a_2^2 + 4a_4}}{2a_4}}. \tag{31.36}$$

The justification for the choice of the values $m = 4, s = 2$ follows from the analysis of the solution of the characteristic equation for the first of the Eq. (31.33). Among its three roots, there is only one real root $k = k_{(1)} < 0$, which corresponds to the boundary layer description. When displacement gradients of less than fourth order are not used, this effect cannot be described.

Using the first of the expressions (31.35) outside the layer $0 < y < h$, the equations are obtained.

$$p^{(1)} = D_1 + (c^{(1,1)} + 2k^2 c^{(1,3)}) D_2 e^{-ky}, \tag{31.37}$$

$$p^{(2)} = -(c^{(2,2)} + 2c^{(2,4)} k^2) k D_2 e^{-ky}.$$

The integration constants D_1 and D_2 are determined as follows. Taking into account the variability of the function $p^{(2)}(y)$ on the basis of the conditions at $y = 0$, the following equality must be fulfilled: $p^{(2)}\Big|_{y=0} = p_0^{(2)}$. In this case, the values of the coefficients in the second Eq. (31.37) are known: $c^{(2,2)}(0) = \frac{1}{2} c_{(\Omega)}^{(2,2)}$, $c^{(2,4)}(0) = \frac{1}{2} c_{(\Omega)}^{(2,4)}$. On the basis of these data, we can obtain

$$D_2 = -\frac{2p_0^{(2)}}{k(c_{(\Omega)}^{(2,2)} + 2k^2 c_{(\Omega)}^{(2,4)})}. \tag{31.38}$$

After that, taking into account the dependencies (31.30) for the material parameters used and the first of the expressions (31.37), the formula for the constant D_1 is

$$D_1 = \frac{p_0^{(1)}}{2} + \frac{(c_{(\Omega)}^{(1,1)} + 2k^2 c_{(\Omega)}^{(1,3)}) p_0^{(2)}}{k(c_{(\Omega)}^{(2,2)} + 2k^2 c_{(\Omega)}^{(2,4)})}. \tag{31.39}$$

After that, the parameter is determined

$$f = \frac{D_1 - p^{0(1)}}{c^{(1,1)}} = \frac{1}{c_{(\Omega)}^{(1,1)}} \left[\frac{(c_{(\Omega)}^{(1,1)} + 2k^2 c_{(\Omega)}^{(1,3)}) p_0^{(2)}}{k(c_{(\Omega)}^{(2,2)} + 2k^2 c_{(\Omega)}^{(2,4)})} - \frac{p_0^{(1)}}{2} \right] \neq 0. \tag{31.40}$$

For the constant D_3 on the basis of the condition of immobility of the boundary plane: $u = 0$ is taken: $D_3 = 0$.

It follows from conditions (31.35) and (31.40) that the appearance of the surface layer is accompanied in the depth of the material by a homogeneous deformation $\varepsilon = \varepsilon_{(\Omega)}$ along the normal to the boundary plane. In further considerations, this circumstance is not considered, since the reason for the inequality $\varepsilon = \varepsilon_{(\Omega)} \neq 0$ may be the approximation of the function taken as a solution.

31.3.2 Piezo Effect and External Electric Field

Description of the Effect

Solid-state physics proposes [7] to associate the polarized state to the characteristics of the stress–strain state described by the traditional continuum theory of elasticity by linear dependences of two types:

$$\mathbf{T} = \hat{R}_\sigma \cdot\cdot \boldsymbol{\Pi}_{(cl)} \ or \ \mathbf{T} = \hat{R}_\xi \cdot\cdot \hat{\xi} \tag{31.41}$$

Here $\mathbf{T}$ is polarization vector; $\hat{R}_\sigma$ and $\hat{R}_\xi$ are piezoelectric coefficients in connection of $\mathbf{T}$ with stresses and strains, respectively—tensors of the third rank, which is a material characteristic of the investigated material; $\Pi_{(cl)}$ is classical stress tensor, in the considered one-dimensional case its component σ is tensile or compressive stress; $\hat{\xi} = (\nabla \mathbf{u})^S$ is symmetric part of the tensor—displacement gradient.

In this paper, the second of definitions (31.41) is used to characterize the piezo effect caused by the appearance of a free surface. The following reasoning serves as a basis.

Macroparticles dB are representative particles of the dielectric. They are formed by a set $\{q_j\}$ of microparticles q_j, which contain both positive and negative charges. The average positions of the positive and negative charges of the set $\{q_j\}$ coincide. Therefore, the dipole moment of a macroparticle is zero. Affected by external influences, the positions of its positive and negative charges in the macroparticle dB cease to coincide. It becomes polarized and acquires a dipole moment.

The value $\sigma = const$, defined by expression (31.27), is the tension or compression stress. It can cause non-uniform distribution of positive and negative charges inside the set $\{q_j\}$ and polarization of dB. The form of expressions (31.37) defining the value of σ, indicates that the deformed states of different macroparticles dB and dB^*, characterized by the same value of σ, can be very different. In this case, the unambiguity of the relation between the polarized and deformed states will be violated. This will not happen in the case when the component ε of the tensor $\hat{\xi} = (\nabla \mathbf{u})^S$ is taken as the cause of the mismatch of the mean positions of positive and negative charges in this paper. Thus, further for the polarization vector, the second of expressions (31.41) is taken as the defining expression. In our one-dimensional case:

$$\mathbf{T} = \tau \mathbf{e}_1, \ \ \hat{R}_\xi = \Re(\mathbf{e}_1 \otimes \mathbf{e}_1), \ \ \tau = \Re\, \varepsilon. \tag{31.42}$$

Consideration of a semi-infinite body means only that in a bounded body all boundaries, except the one involved in the consideration, are distant from the part of the body, which is assumed to be semi-infinite, by large distances. The above reasoning applies equally to any section of infinitely distant boundaries.

The formation of each such section leads (if the statement $\varepsilon = \varepsilon_{(\Omega)} \neq 0$ is correct) to the same in any direction relative strain $\varepsilon = \varepsilon_{(\Omega)} \neq 0$. That is, the deformations arising far from the boundaries in the depth of a finite-sized body possess spherical symmetry. Therefore (due to the initial isotropy of the properties of the investigated material and the properties of the piezo effect), they are not considered in the analysis of the causes of material polarization.

Electric Field Outside the Body

In this paper, the cause of the electric field outside the body B near its free boundary surface A is considered to be the polarization of the dielectric surface layer during its formation.

Multiplying the first of expressions (31.35), whose coefficients f, D_2, k are expressed through the material parameters of the material under study, by the parameter $\mathfrak{R}$ we can obtain the distribution of the magnitude of the polarization vector $\tau = |\mathbf{T}|$.

$$\tau(y) = \chi e^{-ky}, \ \chi = \mathfrak{R} D_2. \tag{31.43}$$

Due to polarization, each macroparticle dB acquires a dipole moment

$$d\delta = \mathbf{T} dV, \ \mathbf{T} = \tau(y)\mathbf{e}_1. \tag{31.44}$$

Here $\mathbf{T} = \tau(y)\mathbf{e}_1$ is the polarization vector.

Each such dipole creates an electric field around itself, including outside the body B. By summing up their potentials and voltages outside the area occupied by the body B, one can obtain the following potentials and voltages of the external electric field depending on the distance h to the boundary plane A:

$$\varphi = \varphi(h) = \frac{\chi}{2\varepsilon_0 k} e^{-kh}, \ \mathbf{E} = \mathbf{E}(h) = -\nabla\varphi(h) = \frac{\chi}{2\varepsilon_0 k} e^{-kh}. \tag{31.45}$$

This electric field is inhomogeneous, the values of its potential and intensity rapidly decrease with distance from the body surface B.

Thus, it is shown that near the dielectric surface, in which the piezo effect is possible, there is an inhomogeneous electric field. The maximum values of its potential and intensity are maximal at the boundary surface and decrease with distance from it.

It is known [16] that an inhomogeneous electric field (obtained near a solid surface) is capable of "pulling in" microparticles—dipoles, which are part of the dielectric B^*

caught in this field. The Coulomb forces acting on these particles are adhesion forces. Formulas (31.45) are the basis for constructing expressions that determine the adhesion forces of bodies B and B^*. As a technique for constructing these expressions, we can use the reasoning presented in [11].

31.3.3 Correspondence of the Obtained Results with Their Known Analogs

A qualitative comparison of the obtained results with their known analogs is proposed.

Quartz is considered as an example of a dielectric in which the piezo effect develops during deformation. It is known that the piezoelectric coefficient for it is $\Re > 0$ [17]. In this case, in accordance with expression (31.38), the value $D_2 < 0$. So the coefficient $\chi = \Re D_2 < 0$, and the potential $\varphi(h)$, defined by formula (31.45) is negative. This conclusion coincides with the results of experiments reported in [18].

The relationship between the properties of the quartz surface and polar liquid–water—is characteristic. In [19] it is shown that in nanopores of amorphous silicon dioxide, orientational ordering of dipole moments of individual water molecules is possible. The field, which is described by formulas (31.45) (follows from [12]) also orders the orientation of the dipole molecules.

Of interest is the connection between the experimentally observed increase in the viscosity of water near the quartz surface and the resulting prediction of the ability of the quartz surface to attract water molecules, increasing their activation energy [20] in the surface layer.

Here it is necessary to take into account the possibility of considering nonlocal interactions of macroparticles not only in solid but also in liquid media. This possibility follows from the following data.

The thermodynamic properties of the system of microparticles q_j are defined by means of the Hamiltonian [9], which has the following form

$$H = \sum_{j=1}^{N} \left(\frac{\overline{\pi}_j^2}{2m_j} \right) + \Phi_N(\mathbf{X}_1, ..., \mathbf{X}_N). \tag{31.46}$$

Here $\overline{\pi}_j$ is momentum of a microparticle q_j with mass m_j; $\Phi_N(\mathbf{X}_1, ..., \mathbf{X}_N)$ is potential energy (potential) of interaction of all microparticles q_j of the system located at the given moment of time t in points with radius vectors $\mathbf{X}_j \in \Lambda$.

For the function $\Phi_N(\mathbf{X}_1, ..., \mathbf{X}_N)$, as well as for solid media, it is possible to introduce the representation (31.4) in the form of the sum of energies of pair and triple interactions and further to introduce into consideration similar representations for liquid—continuum.

What has been said about the potential energy of a liquid medium makes it possible to speak at a fixed moment of time about a group of microparticles as a representative macroparticle of the continuum modeling the medium. This macroparticle or a group of such particles can enter into adhesive nonlocal interaction with particles of another body and among themselves.

Due to the presence of bonding potential energy, liquid molecules cannot move freely [20]. This accounts for its viscosity. The coefficient of dynamic viscosity of a liquid is determined by the formula

$$\eta = \eta_0 \exp\left(\frac{G_0}{RT}\right). \tag{31.47}$$

Here T is absolute temperature, R is universal gas constant, G_0 is activation energy of one mole of liquid—energy required for transition from the initial state to the fluid state at given pressure and temperature away from the solid surface.

Let ΔG be a positive or negative additive to the energy G_0. It can be calculated as the energy of adhesion of the macroparticle to the quartz surface, which is per mole. For this purpose, the results of this paper on the interaction between polarized bodies and macroparticles, and the results of [6, 9] for electrically neutral macroparticles of solid and liquid media are used. The expression determining the local viscosity value in this case takes the form:

$$\eta = \eta_0 \exp\left(\frac{G_0 + \Delta G}{RT}\right). \tag{31.48}$$

Then in the case when $\Delta G > 0$ (the macroparticle dB^* is attracted to the body B) the viscosity of the liquid near the wall should be greater than the bulk value. Otherwise, the opposite picture should be observed. According to the results obtained in this paper, water molecules are attracted to the surface of quartz. So, for them $\Delta G > 0$, it means that the result obtained in the paper predicts an increase in the viscosity of water when flowing along the quartz surface. This conclusion coincides with the experimental data discussed, for example, in [21].

A quantitative assessment of the electric field near the surface of quartz has been made. Firstly, it should be noted that the form of the found potential dependence $\varphi(x)$, described by the first Eq. (31.45), exactly coincides with the electric potential distribution derived in Ref. [23] near a solid surface using the Gouy–Chapman model. The values of the parameters k and $\varphi_0 = \chi/(2\varepsilon_0 k)$ in the potential dependence $\varphi(x)$ are estimated. The assumptions about interparticle potential interactions introduced in Ref. [12] are used for this purpose. According to these assumptions, in the reference and current configurations, the function $\varphi^{(3)}(\beta l_{(12)}, \beta l_{(13)}) \equiv \varphi^{(3)}(\xi_{(12)}, \xi_{(13)})$ is represented as a product of the following:

$$\varphi^{(3)}(\xi_{(12)}, \xi_{(13)}) = \varphi^{(2)}(\xi_{(12)})\varphi^{(2)}(\xi_{(13)}). \tag{31.49}$$

Following [12], as one of the options, we used the dependence

$$\varphi^{(2)}(\xi) = \begin{cases} \varphi_1(\xi) = \sum\limits_{s=0}^{S} a_s \xi^s, \ 0 \le \xi \le \xi_0 \\ \varphi_2(\xi) = e^{-2(\xi-\xi_0)} - 2e^{-(\xi-\xi_0)}, \ \xi_0 < \xi < \infty \end{cases}, \ \zeta_0 \to 0 \quad (31.50)$$

Here a_S are the coefficients, the values and number S of which are determined by the stability conditions of a deformed continuous infinitely extended material within the framework of the local approximation of the nonlocal model being developed and the conditions for pairing the branches $\varphi_1(\xi)$ and $\varphi_2(\xi)$ set forth in [12].

The parameters of pair and triple potentials β, $\Phi_0^{(2)}$ and $\Phi_0^{(3)}$ were determined in accordance with the methodology outlined in [6]. For this purpose, the material characteristics of quartz–Lame parameters—were used; $\lambda = 74 \cdot 10^{-9}$ N/m^2, $\mu = 63 \cdot 10^{-9}$ N/m, average intermolecular distance $l_{mot} = 9 \cdot 10^{-10}$ m and piezoelectric coefficient $\Re_\sigma = 2.31 \cdot 10^{-12}$ K/N [17]. After this, estimates of the parameter values φ_0 and k are made.

For the surface potential we obtained: $\varphi_0 = -0.2\,\mathrm{V/m^2}$. The data from [24], which were obtained experimentally for hydrophilic molecularly smooth surfaces, indicate that the surface potential of a quartz capillary is negative and approximately equal to $(-0.2)\,\mathrm{V/m^2}$. This result coincides in sign and is comparable in magnitude with the calculated value.

The value obtained for the rate of decrease of the electrostatic field potential is $k \approx 1.1 \cdot 10^9\ 1/m$. The value $1/k \approx 0.9 \cdot 10^{-9}$ m can be considered as the effective width of the region of the near-surface region of space where the forces of the electric field created by quartz act. In accordance with [24], this quantity is of the order of $1 \cdot 10^{-9}$ m.

31.4 Practical Application of the Results Obtained

The ideas about different types of adhesion forces are in agreement with the explanation of the decrease in the apparent viscosity of the water flow in the metal channel due to the use of surface-active substances (surfactants) and can be used, in particular, in the construction of a quantitative model of the process of movement of the working medium through the metal channel of devices for waterjet cutting of metals. The working medium is a mixture of quartz sand water [8]. It is experimentally found that the use of surface-active agents (surfactants) reduces the apparent viscosity of the mixture flow.

Here, water is a polar dielectric, the metal wall of the channel and quartz sand grains are materials whose surfaces form an electric field. In the first—due to the presence of a double electric layer, in the second—due to the piezo effect. A surfactant molecule has both a hydrophilic component (polar functional group) and a hydrophobic part (non-polar hydrocarbon radical) in its composition.

Surfactant molecules adsorb on the metal walls of the channel and on the surfaces of quartz sand grains. As a result, the existence of a solid surfactant layer on the walls can be assumed. This layer on the surface in contact with the metal is polarized by its electric field. It, and with it the whole layer, is attracted to the metal surface and held near it by electrical adhesion forces. The outer part of the surfactant layer has a structure characteristic of auxetic materials [22]. They are characterized by an equal to zero or negative value of the Poisson's ratio. In [6], based on the model of adhesive forces generated by pair and triple nonlocal interactions of electrically neutral macroparticles, it was shown that the repulsion forces generated by their triple interactions prevail in the interaction of such materials with macroparticles of other materials. In this case, the activation energy of water macroparticles G_0 receives a negative addition: $\Delta G < 0$. Therefore, according to formula (57), the viscosity of water near the surfactant layer covering the metal wall will decrease in $\exp\left(\frac{-\mathrm{G}}{\mathrm{RT}}\right)$. As a result, there will be an apparent "slip" effect along the surfactant-covered channel walls.

31.5 Conclusion

A description of the role of the surface layer in the appearance of adhesion forces, the reason for the appearance of which are electrical forces, has been proposed. Adhesion forces are generated by the inhomogeneous electric field created near the dielectric surface by the polarized material. They can act on the material polarized under their influence. The cause of polarization is the stress–strain-induced piezo effect of the surface layer. It is assumed that the model presented in this paper complements the existing model of solid-body adhesion, which relies on the consideration of nonlocal interactions between electrically neutral particles of elastic materials. To describe the mechanical properties of the material in describing the stress–strain state of the surface layer, the approach based on the consideration of nonlocal multiparticle interactions, which is unified with the previously constructed models of the elastic medium, is used.

Acknowledgements This research was supported by the Russian Science Foundation, grant No 23-79-10245.

References

1. Deryagin BV (1987) Modern state of the investigation of long-range surface forces. Langmuir 3(5):601–606. https://doi.org/10.1021/la00077a001
2. Johnson KL (1997) Adhesion and friction between a smooth elastic spherical asperity and plane surface. Proc Roy Soc Lond Ser A 453:163–179
3. Shorkin VS, Frolenkova LY (2017) Surface energy and adhesion energy of elastic bodies. Mech Solids 52(1):62–74. https://doi.org/10.3103/S0025654417010083
4. Partensky M (1979) Self-consistent electron theory of a metallic surface. Uspekhi Fizicheskih Nauk 128(5):69. https://doi.org/10.3367/UFNr.0128.197905c.0069

5. Vekilov YK, Verner VD, Samsonova MB (1987) Electron structure of surfaces of nontransition metals. Sov Phys Usp 30:172–192. https://doi.org/10.1070/PU1987v030n02ABEH002814, https://ufn.ru/en/articles/1987/2/g/
6. Barsukov G, Zhuravleva T, Kozhus O (2017) Quality of hydroabrasive waterjet cutting machinability. Proc Eng 206:1034–1038. https://doi.org/10.1016/j.proeng.2017.10.590
7. Romashin SN, Shorkin VS (2021) Variant of the relationship between the mechanical and adhesive properties of solid materials. Mech Solids 55:1392–1405. https://doi.org/10.3103/S0025654420080233
8. Gautschi G (2002) Piezoelectric sensorics: force, strain, pressure, acceleration and acoustic emission sensors. Materials and amplifiers. Springer, Berlin. 978-3-662-04732-3. https://doi.org/10.1007/978-3-662-047323
9. Croxton CA (1947) Liquid state physics - A statistical mechanical introduction. Cambridge University Press, Cambridge
10. Girifalco LA (1973) Statistical physics of materials. Wiley, New York
11. Presnetsova VY, Romashin SN, Frolenkova LY, Shorkin VS, Yakushina SI (2018) A variant of the description of the adhesion interaction in the probe-sample system of an atomic force microscope. Nanosci Technol: Int J 9(4):299–323. https://doi.org/10.1615/NanoSciTechnolIntJ.2018026714
12. Shorkin VS, Vilchevskaya EN, Altenbach H (2023) Linear theory of micropolar media with internal nonlocal potential interactions. Zeitschrift fur angewandte Mathematik und Mechanik 103(11):e202300099. https://doi.org/10.1002/zamm.202300099
13. Toupin RA (1964) Theories of elasticity with couple-stress. Arch Ration Mech Anal 17(5):85–112
14. Eringen AC (2002) Nonlocal continuum field theories. Springer, New York. https://doi.org/10.1007/b97697
15. Shorkin VS (2011) Nonlinear dispersion properties of high-frequency waves in the gradient theory of elasticity. Mech Solids 6:898–912. https://doi.org/10.3103/S0025654411060094
16. Sivukhin DV (2002) General course of physics, vol III. Electricity. M.: FIZMATLIT; MIPT Publishing house. Moscow
17. Smagin AG, Yaroslavskij MI (1970) Quartz piezoelectricity and quartz resonators (in Russ.). Energiya. Moscow
18. Tatarintsev AA, Zykova EY, Ieshkin AE, Orlikovsaya NG, Rau EI (2023) Electrization of the quartz glass surface by electron beams. Phys Solid State 65(8):1236–1243. https://doi.org/10.21883/PSS.2023.08.56567.79
19. Milovidova SD, Rogazinskaya OV, Darinskii BM, Nichugovskaya ES (2019) Dielectric characteristics of a silica-water nanocomposite. Bull Russ Acad Sci Phys 83(9):1091–1093. https://doi.org/10.3103/S1062873819090168
20. Frenkel J (1946) Kinetic theory of liquids. The Clarendon press, Oxford
21. Churayev NV, Sobolev VD, Zorin ZM (1970) Measurement of viscosity of liquids in quartz capillaries. Spec Discuss Faraday Soc 1:213–220. https://doi.org/10.1039/SD9700100213
22. Goldstein RV, Gorodtsov VA, Lisovenko DS (2009) About negativity of the Poisson's ratio for anisotropic materials. Dokl Phys 54(12):546–548. https://doi.org/10.1134/S1028335809120064
23. Derjagin BV (1989) Theory of stability of colloids and thin films. Springer, New York
24. Churayev NV (2004) Surface forces and physicochemistry of surface phenomena. Russ Chem Rev 73(1):25–36. https://doi.org/10.1070/RC2004v073n01ABEH000867

Chapter 32
Displacements of a FGM-Coated Half-Space Surface Caused by Spherical Indentation

Sergei S. Volkov, Sergei M. Aizikovich, Andrey L. Nikolaev, Evgeniy V. Sadyrin, and Vasilina A. Lapitskaya

Abstract An analytical expression for radial and normal displacements of the coating surface during an axisymmetric indentation by a spherical punch is obtained the first time using the bilateral asymptotic method. These expressions allow one to determine the surface displacements of homogeneous, layered, and functionally graded coatings of any thickness. As an example, radial and normal displacements are constructed for a homogeneous coating (titanium nitride on a silicon substrate) and two FGM coatings with harmonic variation on Young's modulus over the coating thickness. Effect of a relative coating thickness is analyzed.

Keywords Axisymmetric contact problem · FGM-coating · Indentation · Surface displacements · Analytical expressions

Introduction

To improve the performance of structural elements and machine parts, various types of coatings are used. For construction of the protective coatings, a variety of materials may be used, namely homogeneous, layered, and functionally graded ones. The latter are gaining popularity for various engineering applications. Functionally graded

S. S. Volkov (✉) · S. M. Aizikovich · A. L. Nikolaev · E. V. Sadyrin
Research and Education Center "Materials", Don State Technical University, 1 Gagrina sq., 344000 Rostov-on-Don, Russia
e-mail: fenix_rsu@mail.ru

A. L. Nikolaev
e-mail: andreynicolaev@eurosites.ru

E. V. Sadyrin
e-mail: e.sadyrin@sci.donstu.ru

V. A. Lapitskaya
A.V. Luikov Heat and Mass Transfer Institute of the National Academy of Sciences of Belarus, Minsk, Belarus
e-mail: vasilinka.92@mail.ru

H. Altenbach and V. A. Eremeyev (eds.), *Dynamics of Discrete and Continuum Structures and Media*, Advanced Structured Materials 221,
https://doi.org/10.1007/978-3-031-75626-9_32

materials are multiphase composites characterized by a continuous change in the percentage of fractions of their constituents by volume of the material. These changes lead to a inhomogeneous microstructure with continuously changing mechanical and physical properties of the material. In particular, elastic modules in such materials are continuous functions of spatial coordinates [1]. A large number of modern researches have been devoted to the study of deformation of samples consisting of functionally graded materials and bodies with functionally graded or composite coatings [2–11, 13, 14, 16–19]. Contact interaction was investigated [2–10, 16–18], taking into account temperature [4, 20], electro and electromagnetic effects [6, 13–15], acoustic effects [11], wear [20], lubrication [21–23] and other types of influences. The problems of bending of functionally graded plates [13] were studied as well. However, due to the serious mathematical difficulties most contact problems are usually solved with some restrictions on the values of geometric and physical parameters. For example, in a number of works devoted to the contact of bodies with functionally graded coatings, special cases of the laws of variation of elastic properties by depth of the coating are considered [6–8, 10]. In most cases, the methods used to solve integral equations are effective only in a narrow range of changes in the characteristic geometric parameter of the problem, equal to the relative thickness of the coating.

In the present paper, analytical formulas were obtained for the normal and radial displacements of the surface appearing in the axisymmetric contact problem on indentation of a functionally graded coating by a spherical punch. For that purpose, we use approximated analytical solution of problem obtained earlier [2, 21, 24] and the idea of the bilateral asymptotic method [25]. Similar results were obtained earlier for different types of loading: circular and strip electrode on the surface of piezoelectric FGM coating [26, 27] and thermal heating of the coating surface in a circular area [28].

32.1 Contact Problem Statement

Let us consider a homogeneous elastic half-space with a functionally graded coating. Thickness of the coating is H. We introduce a cylindrical coordinate system (r, ϕ, z) so that the $z-$axis is normal to the coating surface and passes through the center of the punch. The equilibrium equations are as follows:

$$\begin{aligned} &\frac{\partial \sigma_r}{\partial r} + \frac{\partial \tau_{zr}}{\partial z} + \frac{\sigma_r - \sigma_\varphi}{r} = 0, \\ &\frac{\partial \tau_{zr}}{\partial r} + \frac{\partial \sigma_z}{\partial z} + \frac{\tau_{zr}}{r} = 0. \end{aligned} \tag{32.1}$$

Constitutive equations of linear elasticity:

$$\begin{gathered}\sigma_r = 2\mathrm{M}(z)\frac{\partial u}{\partial r} + \Lambda(z)\theta,\ \sigma_\varphi = 2\mathrm{M}(z)\frac{u}{r} + \Lambda(z)\theta, \\ \sigma_z = 2\mathrm{M}(z)\frac{\partial w}{\partial z} + \Lambda(z)\theta,\ \tau_{rz} = \mathrm{M}(z)\left(\frac{\partial u}{\partial z} + \frac{\partial w}{\partial r}\right), \\ \theta = \frac{\partial u}{\partial r} + \frac{u}{r} + \frac{\partial w}{\partial z}.\end{gathered} \tag{32.2}$$

Here u and w are radial and normal displacements, respectively, σ_r, σ_φ, σ_z, τ_{zr}—components of the stress tensor; M and Λ are Lame coefficients which are related to Young's modulus E and Poisson's ratio ν by the following expressions:

$$\Lambda = \frac{E\nu}{(1+\nu)(1-2\nu)},\quad M = \frac{E}{2(1+\nu)}. \tag{32.3}$$

We assumed that Young's modulus and Poisson's ratio of the half-space vary in depth according to the following rule:

$$\{E,\nu\} = \begin{cases}\{E_c(z)\ ,\nu_c(z)\}\ , & -H \le z \le 0, \\ \{E_s,\nu_s\} = \text{const}, & -\infty < z < -H.\end{cases} \tag{32.4}$$

where $E_c(z)\ ,\nu_c(z)$ are continuously differentiable or piecewise constant functions describing variation of properties in depth of the coating; E_s, ν_s are values of elastic properties of the homogeneous substrate. Here and further, the indices (c) and (s) indicate the characteristics corresponding to the coating and substrate.

By force P applied to the center of a spherical indenter of radius R, the punch is pressed in the surface $z = 0$ of the elastic half-space to a depth of δ. Radius of the contact area is denoted as a. The spherical punch is approximated by a paraboloid of revolution. Thus the following mixed boundary conditions are satisfied:

$$z = 0:\quad \tau_{rz} = 0,\quad \begin{cases}\sigma_z = 0, & r > a \\ w = -\left(\delta - \frac{r^2}{2R}\right) & r \le a\end{cases}. \tag{32.5}$$

We assume that the displacements vanish at infinity, i.e., $u, w \to 0$ at $z \to -\infty$.

Continuity of mechanical characteristics is assumed at the coating/substrate interface:

$$z = -H:\quad w^{(c)} = w^{(s)}, u^{(c)} = u^{(s)}, \sigma_z^{(c)} = \sigma_z^{(s)}, \tau_{zr}^{(c)} = \tau_{zr}^{(s)}. \tag{32.6}$$

It is required to determine the distribution of contact normal stresses under the punch

$$\sigma_z\,|_{z=0} = -p_a(r),\quad r \le a. \tag{32.7}$$

and distribution of normal and radial displacements of the coating surface.

32.2 Approximate Analytical Solution of the Problem

Let us introduce the displacements as an integral Hankel's transform:

$$\begin{aligned} w(z,r) &= \int_0^\infty \bar{w}(z,\alpha) J_0(\alpha r)\alpha d\alpha, \\ u(z,r) &= -\int_0^\infty \bar{u}(z,\alpha) J_1(\alpha r)\alpha d\alpha. \end{aligned} \tag{32.8}$$

Following representation of Hankel's transforms of displacements on the coating surface is used:

$$\{\bar{u}, \bar{w}\}(\alpha, 0) = -\left\{\frac{L_{11}(\alpha)}{E_{11}^{(c)}}, \frac{L_{13}(\alpha)}{E_{13}^{(c)}}\right\} \frac{\bar{p}_a(\alpha)}{\alpha}. \tag{32.9}$$

$E_{ij}^{(c)}$ are effective elastic moduli of the coating surface ($z = 0$), $\bar{p}_a(\alpha)$ are the Hankel transform of contact stresses on the surface of the coating, $L_{ij}(\alpha)$ are the compliance functions [29, 30] describing the elastic behavior of FGM-coated half-space under normal loading. Applying (32.8) to the boundary conditions of (32.5) taking into account (32.9), we obtain a dual integral equation of the problem:

$$\begin{cases} \int_0^\infty \bar{p}(\gamma) L_{13}(\lambda\gamma) J_0(r\gamma) d\gamma = E_{13}^{(c)} \left(\frac{\delta}{a} - \frac{ar}{2R}\right), & 0 \le r \le 1 \\ \int_0^\infty \bar{p}(\gamma) J_0(r\gamma)\gamma d\gamma = 0, & r > 1 \end{cases} \tag{32.10}$$

Following dimensionless variables are used above (the primes are omitted):

$$r' = \frac{r}{a}, \lambda = \frac{H}{a}, \ p(r') = p_a(r'a), L'_{kj}(\alpha) = L_{kj}\left(\frac{\alpha}{H}\right). \tag{32.11}$$

We use the bilateral asymptotic method [25] to solve (32.10). For that purpose, we approximate the compliance functions by the following expressions:

$$L_{ij}(\lambda\alpha) \approx \Pi_{ij}^{+}(\lambda\alpha), \quad \Pi_{ij}^{\pm}(\lambda\alpha) = \prod_{n=1}^{N_{ij}} \frac{\left(A_{ijn}^2 \pm \lambda^2\alpha^2\right)}{\left(B_{ijn}^2 \pm \lambda^2\alpha^2\right)}. \tag{32.12}$$

Then the solution of the approximated dual integral equation was obtained [2] in a closed analytical form:

$$\bar{p}(\gamma) = \frac{4aE_{13}^{(s)}}{\pi R}\bar{p}_0(\gamma),$$
$$\bar{p}_0(\gamma) = \left\{-\frac{\gamma\cos\gamma - \sin\gamma}{\gamma^3} - \frac{1}{2}\left(1 - \delta\frac{R}{a^2} + 2\lambda^2\sum_{i=1}^{N_{13}}(A_{13i}^{-2} - B_{13i}^{-2})\right)\frac{\sin\gamma}{\gamma} + \right.$$
$$\left. + \sum_{i=1}^{N_{13}} C_{13i}\frac{\gamma\sin\gamma\cosh(A_{13i}\lambda^{-1}) + A_{13i}\lambda^{-1}\cos\gamma\sinh(A_{13i}\lambda^{-1})}{\gamma^2 + A_{13i}^2\lambda^{-2}}\right\}. \tag{32.13}$$

$E_{ij}^{(s)}$ are effective elastic moduli of the substrate.

After converting the Hankel's transform in (32.13) taking into account the condition $p(1) = 0$, we obtain an approximate expression for contact pressures under a spherical punch

$$p(r) = \frac{4aE_{13}^{(s)}}{\pi R}\left[\sqrt{1 - r^2} - \sum_{i=1}^{N_{13}}\frac{C_{13i}A_{13i}}{\lambda}\int_r^1\frac{\sinh\left(A_{13i}\lambda^{-1}t\right)}{\sqrt{t^2 - r^2}}dt\right], 0 \le r \le 1. \tag{32.14}$$

The relation between displacement of the punch, radius of the contact area, and radius of the punch has the form

$$\delta = \frac{2a^2}{R}\left(-\sum_{i=1}^{N_{13}} C_{13i}\cosh(A_{13i}\lambda^{-1}) + \frac{1}{2} + \sum_{i=1}^{N_{13}}\left(A_{13i}^{-2}\lambda^2 - B_{13i}^{-2}\lambda^2\right)\right), \tag{32.15}$$

The resulting solution (32.14) is asymptotically accurate for $\lambda \to 0$ and $\lambda \to \infty$ [25]. Constants C_{13i} $(i = 1,..,N)$ are determined from the system of linear algebraic equations below

$$\sum_{i=1}^{N_{13}} C_{13i}\left[\frac{A_{13i}\sinh\left(A_{13i}\lambda^{-1}\right) + B_{13n}\cosh\left(A_{13i}\lambda^{-1}\right)}{\left(A_{13i}^2 - B_{13n}^2\right)} - \frac{\cosh\left(A_{13i}\lambda^{-1}\right)}{B_{13n}}\right]$$
$$= \frac{(B_{13n}\lambda^{-1} + 1)\lambda^2}{B_{13n}^3}, \quad n = 1, 2, \ldots, N. \tag{32.16}$$

32.3 Analytical Expressions for Radial and Normal Displacements of the Coating Surface

Substituting (32.9) into (32.8), we obtain expressions for displacements in the form of quadratures:

$$
\begin{aligned}
w(z,r) &= -\frac{4a^2 E_{13}^{(s)}}{\pi R E_{13}^{(c)}} \int\limits_0^\infty L_{13}(\alpha\lambda)\, \bar{p}_0(\alpha)\, J_0(\alpha r) d\alpha, \\
u(z,r) &= \frac{4a^2 E_{13}^{(s)}}{\pi R E_{11}^{(c)}} \int\limits_0^\infty L_{11}(\alpha\lambda)\, \bar{p}_0(\alpha)\, J_1(\alpha r) d\alpha.
\end{aligned} \tag{32.17}
$$

Let us introduce the following notation

$$
\{u_s, w_s\}(r) = \frac{R\left\{u(r,0)E_{11}^{(s)},\, w(r,0)E_{13}^{(s)}\right\}}{2a^2 E_{13}^{(s)}}, \quad \beta_{ij} = \frac{E_{ij}^{(s)}}{E_{ij}^{(c)}}. \tag{32.18}
$$

Using approximations (32.12) for compliance functions, calculating quadratures in (32.18), we obtain expressions for dimensionless displacements in the form:

$$
\begin{aligned}
w_s(r) &= -\left(w_{\text{hom}}(r) + S_0 I_1(r) + \beta_{13} w_{\text{inh}}(r)\right), \\
u_s(r) &= \tfrac{2}{\pi}\left(u_{\text{hom}}(r) + S_0 I_2(r) + \beta_{11} u_{\text{inh}}(r)\right),
\end{aligned} \tag{32.19}
$$

Terms with the index "hom" describe displacements in the case of a similar problem for a homogeneous half-space without a coating:

$$
\begin{aligned}
u_{\text{hom}}(r) &= \begin{cases} \frac{1}{3r}, & r > 1 \\ \frac{1}{3r}\left(r^2\sqrt{1-r^2} - \sqrt{1-r^2} + 1\right), & 0 < r \le 1, \end{cases} \\
w_{\text{hom}}(r) &= \begin{cases} \frac{1}{4}\left(2 - r^2\right), & 0 < r \le 1, \\ \frac{1}{2\pi}\left(\sqrt{1-r^2} - \left(r^2 - 2\right)\arcsin\left(\frac{1}{r}\right)\right), & r > 1 \end{cases}
\end{aligned} \tag{32.20}
$$

where $S_0 = \sum\limits_{i=1}^{N} C_i \cosh(A_i \lambda^{-1})$, expressions $u_{\text{inh}}(r)$ and $w_{\text{inh}}(r)$ have the form

$$
\begin{aligned}
w_{\text{inh}}(r) &= \sum_{n=1}^{N_{13}} P_{13n} I_{12}\left(B_{13n}\lambda^{-1}, r\right) + \sum_{n=1}^{N_{13}} F_{13n} I_{11}\left(B_{13n}\lambda^{-1}, r\right) \\
&\quad + \sum_{n=1}^{N_{13}} \frac{\Pi_{13n}(i B_{13n})}{B_{13n}^4 \lambda^{-4}} I_1(r),
\end{aligned} \tag{32.21}
$$

$$u_{\mathrm{inh}}(r)=\sum_{n=1}^{N_{11}}P_{11n}I_{22}\left(B_{11n}\lambda^{-1},r\right)+\sum_{n=1}^{N_{11}}F_{11n}I_{21}\left(B_{11n}\lambda^{-1},r\right)$$
$$+\sum_{i=1}^{N_{13}}C_{13i}\Pi_{11}^{-}(A_{13i})\left(\cosh\left(\frac{A_{13i}}{\lambda}\right)I_{21}\left(\frac{A_{13i}}{\lambda},r\right)+\frac{A_{13i}}{\lambda}\sinh\left(\frac{A_{13i}}{\lambda}\right)I_{22}\left(\frac{A_{13i}}{\lambda},r\right)\right)$$
$$+\sum_{n=1}^{N_{11}}\frac{\Pi_{11n}(iB_{11n})}{B_{11n}^{4}\lambda^{-4}}I_2(r). \tag{32.22}$$

It can be easily verified that the boundary condition (32.5) is satisfied at $r\le 1$. The following designations were used above:

$$I_1(r)=\begin{cases}-1, & 0<r\le 1,\\ -\frac{2}{\pi}\arcsin\left(\frac{1}{r}\right), & r>1\end{cases},\quad I_2(r)=\begin{cases}-\frac{1}{r}, & r>1\\ -\frac{r}{1+\sqrt{1-r^2}}, & 0<r\le 1\end{cases},$$

$$I_{11}(A,r)=\begin{cases}e^{-A}I_0(Ar), & 0<r\le 1,\\ e^{-A}I_0(Ar)-\frac{2}{\pi}\int\limits_1^r\frac{\cosh(A(t-1))dt}{\sqrt{r^2-t^2}}, & r>1\end{cases},$$

$$I_{12}(A,r)=\begin{cases}\frac{1}{A}e^{-A}I_0(Ar), & 0<r\le 1,\\ \frac{1}{A}\left(e^{-A}I_0(Ar)-\frac{2}{\pi}\int\limits_1^r\frac{\sinh(A(t-1))}{\sqrt{r^2-t^2}}\right), & r>1,\end{cases},$$

$$I_{21}(r,B)=\begin{cases}\sinh(B)K_1(Br), & r>1\\ \sinh(B)K_1(Br)-\frac{1}{r}\int\limits_r^1\frac{t\cosh(B(1-t))}{\sqrt{t^2-r^2}}dt, & 0<r\le 1\end{cases},$$

$$I_{22}(r,B)=\begin{cases}\frac{1}{B^2}\left(\frac{1}{r}-B\cosh(B)K_1(Br)\right)+\frac{1}{Br}\int\limits_r^1\frac{t\sinh(B(1-t))}{\sqrt{t^2-r^2}}, & 0<r\le 1\\ \frac{1}{B^2}\left(\frac{1}{r}-B\cosh(B)K_1(Br)\right), & r>1\end{cases},$$

$$P_{ijn}=\sum_{k=1}^{N_{13}}C_{13k}\frac{\lambda A_{13k}\Pi_{ijn}(iB_{ijn})\sinh\left(A_{33k}\lambda^{-1}\right)}{A_{13k}^2-B_{ijn}^2}+\frac{\Pi_{ijn}(iB_{ijn})}{B_{ijn}^2\lambda^{-2}},$$
$$F_{ijn}=\sum_{k=1}^{N_{13}}C_{13k}\frac{\lambda^2\Pi_{ijn}(iB_{ijn})\cosh\left(A_{33k}\lambda^{-1}\right)}{A_{13k}^2-B_{ijn}^2}+\frac{\Pi_{ijn}(iB_{ijn})}{B_{ijn}^2\lambda^{-2}}\left(S_0+\frac{1}{B_{ijn}^2\lambda^{-2}}\right),$$

$$\Pi_{ijn}(\alpha)=\left(A_{ijn}^2-B_{ijn}^2\right)\prod_{m=1,m\ne n}^{N_{ij}}\frac{A_{ijm}^2-\alpha^2}{B_{ijm}^2-\alpha^2}.$$

32.4 Numerical Results and Discussion

As an example, let us consider three coatings: coating 1 is a homogeneous coating of titanium nitride, Young's modulus of which is obtained from a nanoindentation experiment and is equal to $E_c^{(1)}=376$ GPa, and the Poisson's ratio $\nu_c = 0.27$; for coating 2 Young's modulus varies according to the following law $E_c^{(2)}(z) = \frac{1}{2}\left(E_c^{(1)} + 300 + (E_c^{(1)} - 300)\cos(2\pi z)\right)$, $\nu_c = 0.27$; coating 3—$E_c^{(3)}(z) = \frac{1}{2}\left(E_c^{(1)} + 452 + (E_c^{(1)} - 452)\cos(2\pi z)\right)$, $\nu_c = 0.27$. All three coatings are rigidly connected to a silicon substrate 170 GPa, $\nu_s = 0.25$. It is convenient to analyze the effect of the coating type on the change in the displacements of the coating surface in relative terms:

$$w_{\text{rel}} = \frac{w_s}{w_{\text{hom}}}, \quad u_{\text{rel}} = \frac{u_s}{u_{\text{hom}}}.$$

Figure 32.1a, b shows diagrams of relative normal and radial displacements for coating 1, respectively.

It can be seen from the diagrams in Fig. 32.1a that for large and small values of the λ parameter, the value w_{rel} approaches 1, which indicates that the normal displacements reach the values of displacements characteristic of a homogeneous half-space with elastic modules of the coating and substrate, respectively. The largest difference between normal displacements and displacements of a homogeneous half-space is observed for λ in the range from 0.1 to 5. In general, a similar behavior is observed for diagrams of relative radial displacements on Fig. 32.1b, however u_{rel} approaches 1 for small values of the λ parameter, and for large values it approaches $\beta_{11}/\beta_{13} \approx 0.945$. At the same time, it can be seen from the diagram $\lambda = 0.01$ that under the indenter the

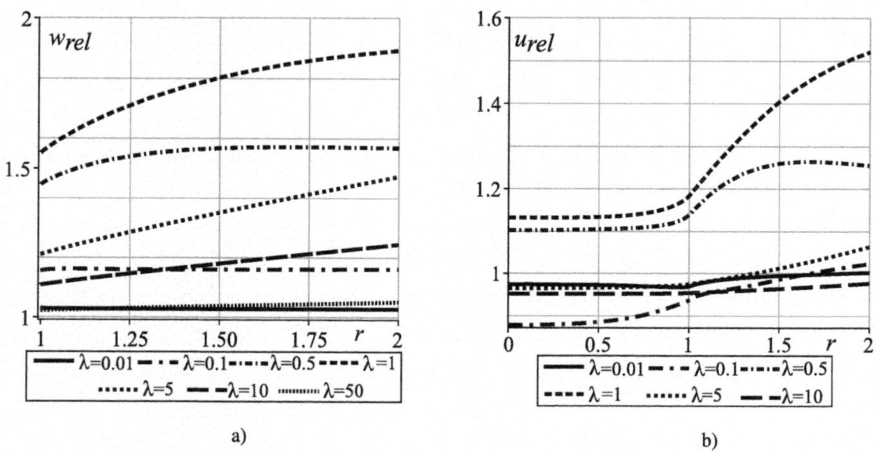

Fig. 32.1 Relative normal (**a**) and radial (**b**) displacements of the surface for coating 1 and a set of λ parameters

difference from the displacements for a homogeneous half-space is more significant than outside it.

Figures 32.2 and 32.3 show diagrams of the relative normal and radial displacements of the coating surface for coatings 1–3 and the values of λ parameter 0.5 and 5. From the diagrams, one can see how the law of inhomogeneity and the value of the λ parameter affects the redefinition of the displacement. Generally, the diagrams for all three coatings behave in a similar way, but differ significantly quantitatively. A significant qualitative difference in the behavior of the diagrams can be seen in Fig. 32.3b: for coating 2 after $r = 1$, the relative radial displacements grow faster at a sharper angle, while for coating 3 there is a smoother increase in them.

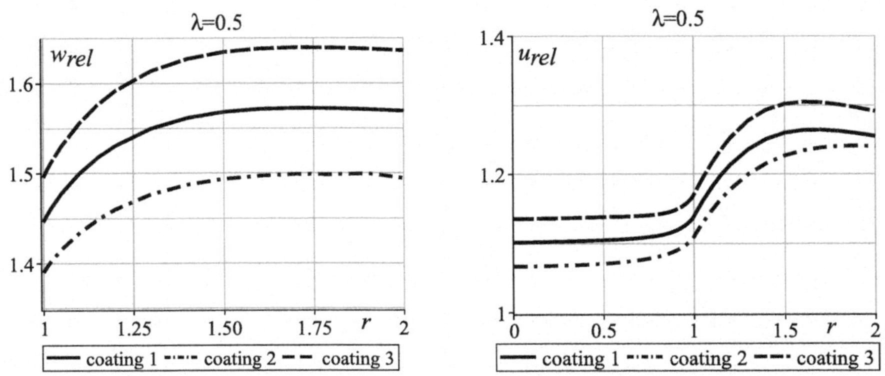

a) b)

Fig. 32.2 Relative normal (**a**) and radial (**b**) displacements of the surface of coatings 1–3 for $\lambda = 0.5$

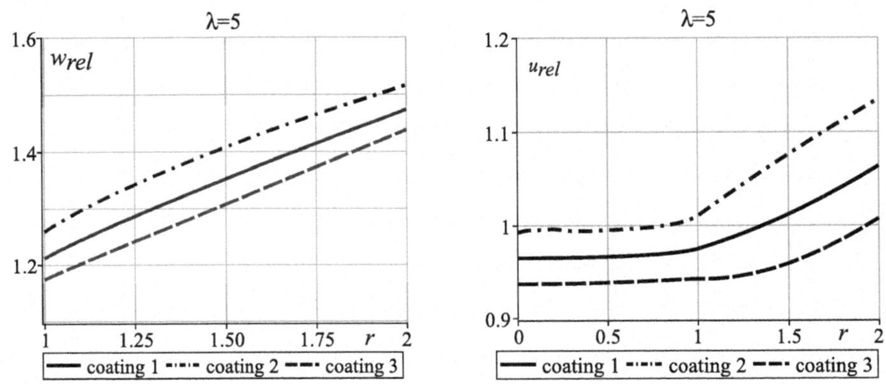

a) b)

Fig. 32.3 Relative normal (**a**) and radial (**b**) displacements of the surface of coatings 1–3 for $\lambda = 5$

32.5 Conclusion

Approximated analytical expressions are obtained for the normal and radial displacements of the coating surface caused by a spherical punch. The formulas obtained are effective for a wide range of geometrical parameter of the problem and arbitrary continuous variation of properties in depth of FGM coating as well as piecewise constant variation. Numerical examples illustrating features of relative radial and normal displacements are provided for a homogeneous titanium nitride coating on a silicon substrate and for the case of FGM coating with nonmonotonic harmonic variation of Young's modulus. Effect of relative coating thickness and type of variation of properties in depth of the coating are discussed.

Acknowledgements The study was supported by the Russian Science Foundation grant No. 23-49-10062, https://rscf.ru/project/23-49-10062/. Lapitskaya V. A. acknowledges the support of the Belarus Republican Foundation for Fundamental Research grant No. T23RNF-132.

References

1. Niino M, Maeda S (1990) Recent development status of functionally gradient materials. ISIJ Int 30:699–703. https://doi.org/10.2355/isijinternational.30.699
2. Aizikovich S, Alexandrov VM, Kalker JJ, Krenev L, Trubchik I (2002) Analytical solution of the spherical indentation problem for a half-space with gradients with the depth elastic properties. Int J Solids Struct 39:2745–2772. https://doi.org/10.1016/S0020-7683(02)00124-5
3. Volkov S, Litvinenko A, Aizikovich S, Wang Y-C, Vasiliev A (2016) Axisymmetric bending of a circular plate with stiff edge on a soft FGM layer. Struct. Eng. Mech. 59:227–241. https://doi.org/10.12989/sem.2016.59.2.227
4. Liu J, Ke L-L, Wang Y-S, Yang J, Alam F (2012) Thermoelastic frictional contact of functionally graded materials with arbitrarily varying properties. Int J Mech Sci 63:86–98. https://doi.org/10.1016/j.ijmecsci.2012.06.016
5. Liu T-J, Wang Y-S, Zhang C (2008) Axisymmetric frictionless contact of functionally graded materials. Arch Appl Mech 78:267–282. https://doi.org/10.1007/s00419-007-0160-y
6. Ma J, Ke L-L, Wang Y-S (2014) Frictionless contact of a functionally graded magneto-electro-elastic layered half-plane under a conducting punch. Int J Solids Struct 51:2791–2806. https://doi.org/10.1016/j.ijsolstr.2014.03.028
7. Kucuksucu A, Guler M, Avci A (2015) Mechanics of sliding frictional contact for a graded orthotropic half-plane. Acta Mech 226:3333. https://doi.org/10.1007/s00707-015-1374-7
8. Alinia Y, Beheshti A, Guler M, El-Borgi S, Polycarpou A (2016) Sliding contact analysis of functionally graded coating/substrate system. Mech Mater 94:142–155. https://doi.org/10.1016/j.mechmat.2015.11.017
9. Suresh S, Giannakopoulos AE, Alcala J (1997) Spherical indentation of compositionally graded materials: theory and experiments. Acta Mater 45:1307–1321. https://doi.org/10.1016/s1359-6454(96)00291-1
10. Awojobi AO (1976) On the hyperbolic variation of elastic modulus in a non-homogeneous stratum. Int J Solids Struct 12:739–748. https://doi.org/10.1016/0020-7683(76)90039-1
11. Glushkov E, Glushkova N, Eremin A (2011) Forced wave propagation and energy distribution in anisotropic laminate composites. J Acoust Soc Am 129:2923–34. https://doi.org/10.1121/1.3559699

12. Altenbach H, Eremeyev V (2008) Direct approach-based analysis of plates composed of functionally graded materials. Arch Appl Mech 78:775–794. https://doi.org/10.1007/s00419-007-0192-3
13. Su J, Ke L-L, Wang Y-S (2016) Axisymmetric frictionless contact of a functionally graded piezoelectric layered half-space under a conducting punch. Int J Solids Struct 90. https://doi.org/10.1016/j.ijsolstr.2016.04.011
14. Su J, Ke L-L, Wang Y-S (2017) Axisymmetric partial slip contact of a functionally graded piezoelectric coating under a conducting punch. J Intell Mater Syst Struct 28:1925–1940. https://doi.org/10.1177/1045389X16682849
15. Vasiliev AS (2019) Penetration of a spherical conductive punch into a piezoelectric half-space with a functionally graded coating. Int J Eng Sci 142:230–241. https://doi.org/10.1016/j.ijengsci.2019.06.006
16. Su J, Ke L-L, Wang Y-S, Xiang Y (2017) The axisymmetric torsional contact problem of a functionally graded piezoelectric coated half-space. Acta Mech Sin 33. https://doi.org/10.1007/s10409-016-0627-6
17. Su J, Ke L-L, Wang Y-S, Xiang Y (2017) Axisymmetric torsional fretting contact between a spherical punch and an FGPM coating. Appl Math Model 52:576–589. https://doi.org/10.1016/j.apm.2017.08.010
18. Su J, Ke L-L, El-Borgi S, Xiang Y, Wang Y-S (2018) An effective method for the sliding frictional contact of a conducting cylindrical punch on FGPMs. Int J Solids Struct 141–142. https://doi.org/10.1016/j.ijsolstr.2018.02.017
19. Liu T-J, Ke L-L, Wang Y-S, Xing Y-M (2014) Stress analysis for an elastic semispace with surface and graded layer coatings under induced torsion. Mech Based Des Struct Mach 43:74–94. https://doi.org/10.1080/15397734.2014.928222
20. Zelentsov VB, Lapina PA, Nikolaev AL (2023) Grinding of functionally graded coating taking into account wear and friction heating. In: Altenbach H, Eremeyev V (eds) Advances in linear and nonlinear continuum and structural mechanics. Advanced structured materials, vol 198. Springer, Cham. https://doi.org/10.1007/978-3-031-43210-1_32
21. Kudish I, Volkov S, Vasiliev A, Aizikovich S (2017) Lubricated point heavily loaded contacts of functionally graded materials. Part 1. Dry contacts. Math Mech Solids 23. https://doi.org/10.1177/1081286517704689
22. Kudish II, Volkov SS, Vasiliev AS, Aizikovich SM (2018) Lubricated point heavily loaded contacts of functionally graded materials. Part 2. Lubricated contacts. Math Mech Solids 23:1081–1103. https://doi.org/10.1177/1081286517704690
23. Kudish II, Pashkovski E, Volkov SS, Vasiliev AS, Aizikovich SM (2020) Heavily loaded line EHL contacts with thin adsorbed soft layers. Math Mech Solids 25:1011–1037. https://doi.org/10.1177/1081286519898878
24. Volkov S, Vasiliev A, Aizikovich SM, Seleznev NM, Leontieva AV (2016) Stress-strain state of an elastic soft functionally-graded coating subjected to indentation by a spherical punch. PNRPU Mech Bull 20–34. https://doi.org/10.15593/perm.mech/2016.4.02
25. Aizikovich SM (1990) An asymptotic solution of a class of coupled equations. J Appl Math Mech 54:719–24
26. Vasiliev A, Volkov S, Kislyakov E, Irkha V (2021) Analytical expressions for the displacements of a surface of piezoelectric FGM-coated half-plane with a strip electrode. Contin Mech Thermodyn 33:1–12. https://doi.org/10.1007/s00161-021-00991-8
27. Volkov S, Vasiliev AS, Aizikovich SM (2022) Displacements of the surface of a piezoelectric half-space with functionally-graded coating and circle electrode on the surface. Comput Contin Mech 15:45–55. https://doi.org/10.7242/1999-6691/2022.15.1.4
28. Volkov SS (2021) Displacements of a surface of FGM-coated half-space heated in a circular area taking into account imperfect coating-substrate interface. IOP Conf Ser Mater Sci Eng 1029:012062. https://doi.org/10.1088/1757-899X/1029/1/012062

29. Il'man VM, Privarnikov AK (1971) The effect of a system of punches on an elastic multilayer base. Sov Appl Mech 7:602–6
30. Vasiliev AS (2019) Compliance functions of electromagnetoelastic piezoelectric and piezomagnetic half-plane and half-space with functionally graded or layered coatings. J Samara State Tech Univ Ser Phys Math Sci 23:475–96

Chapter 33
Modeling Processes of Complex Plastic Deformation of Structural Alloys on Flat Trajectories of Arbitrary Curvature

Ivan A. Volkov, Leonid A. Igumnov, Andrey I. Volkov, and Anna I. Yudintseva

Abstract The processes of complex plastic deformation of polycrystalline structural alloys along flat trajectories of arbitrary curvature are considered. The version of the equations of state of thermoplasticity used in this article is based on the concept of a yield surface and the principle of gradient of the plastic strain rate vector to the yield surface at the loading point. This version of the equations of state reflects the main effects of elastoplastic deformation of the material for arbitrary complex deformation trajectories. Particular attention is paid to the issues of modeling the processes of elastoplastic deformation of polycrystalline structural alloys for complex deformation processes accompanied by rotation of the main areas of stress tensors, total and plastic deformations. Material parameters and scalar functions included in the constitutive relations of the mathematical model of thermoplasticity for steel 45 used in this work were obtained. To assess the degree of reliability and determine the limits of applicability of the constitutive relations of thermoplasticity, numerical studies of complex plastic deformation of steel 45 with a yield plateau along flat trajectories of arbitrary curvature were carried out: flat piecewise broken trajectory of the "hourglass" type; two flat trajectories, symmetrical with respect to the bisector of a right angle, representing a quarter of a circle and a straight section with a trajectory bend of 90°. It is shown that the thermoplasticity model used in the work qualitatively and with the accuracy necessary for practical calculations quantitatively describes the main effects of complex plastic deformation of steel 45 with a yield plateau under loading along flat trajectories of arbitrary curvature.

Keywords Plasticity · Complex loading · Modeling · Numerical experiment · Natural experiment · Plane and spatial deformation trajectories

---

I. A. Volkov · L. A. Igumnov (✉) · A. I. Volkov · A. I. Yudintseva
National Research Lobachevsky State University of Nizhny Novgorod, Nizhny Novgorod, Russia
e-mail: igumnov@mech.unn.ru

H. Altenbach and V. A. Eremeyev (eds.), *Dynamics of Discrete and Continuum Structures and Media*, Advanced Structured Materials 221,
https://doi.org/10.1007/978-3-031-75626-9_33

33.1 Introduction

In [1–3], an original approach to the construction of time-step schemes of the boundary element method (BEM) is proposed. In [4–6], the BEM scheme was used, based on the stepwise method of numerical inversion of the Laplace transform. In [4, 7], a modification of the method with a variable integration step is formulated when calculating the weight multipliers of the quadrature sum. Another extension of the boundary element scheme is presented in articles [8, 9], suggesting the joint use of FEM with Runge–Kutta methods. Following [10], this article discusses the joint application of a step scheme at the nodes of the Runge–Kutta methods with BEM for solving three-dimensional elastic and poroelastic problems.

33.2 Problem Formulation

Calculation of the durability of structural elements operating under conditions of non-stationary thermomechanical loading requires an analysis of the processes of complex plastic deformation of structural materials since the history of deformation determines the rate of damage accumulation as a result of non-stationary plastic deformation of the material, and the accuracy of the calculated estimates of their strength depends on the reliability of the modeling of these processes and durability [1–3]. For this reason, to date, numerous experiments have been carried out aimed at elucidating the basic laws of the processes of elastoplastic deformation of polycrystalline structural alloys under various thermomechanical loading conditions (mainly under conditions of uniaxial tension–compression and alternating torsion of cylindrical samples).

The formulation of models of elastoplastic media and their verification is carried out on the basis of an analysis of the results of experimental studies of the processes of deformation of laboratory samples under loads corresponding to the operating conditions of the structural material as part of a structural element [2, 4].

Studies of the laws of elastoplastic deformation of structural materials are also of fundamental importance for the development of the fundamental principles of the theory of plasticity [4, 5]. To verify the physical reliability of the constitutive relations of elastoplastic media, it is necessary to carry out calculations and compare the obtained numerical results with the experimental data available in the literature on a wide class of deformation trajectories [6–16]. The results of such studies can also be useful in determining (specifying) the material parameters of scalar functions and functionals included in the constitutive relations of plasticity models [2, 4, 5].

In this article, to describe the processes of complex plastic deformation of steel 45 with a yield plateau along flat and spatial trajectories of arbitrary curvature, a flow theory model with kinematic and isotropic strengthening is used [4]. The obtained numerical results are compared with the data of full-scale experiments obtained on automated testing machines of a high accuracy class [17–19]. The qualitative and quantitative agreement between numerical and experimental data is shown in describing the effects of complex loading of steel 45 with a yield plateau during deformation along flat trajectories of arbitrary curvature.

33.3 Constitutive Relations of Plasticity

The thermoviscoplasticity equations formulated below are aimed at describing the main effects that are essential for the development of damage processes in structural elements under thermal-force unsteady loading:

- monotonic and cyclic hardening under proportional and disproportionate loading, including transient cyclic processes and stabilized cyclic behavior of the material;
- local anisotropy of plastic deformation during a break in the deformation trajectory;
- non-isothermal viscoplastic deformation of the material under various laws of temperature change and mechanical deformation;

The constitutive relations of plasticity are based on the following basic principles [2, 4]:

- components of strain tensors e_{ij} and strain rates $\dot{e}_{ij}$ include elastic e^e_{ij}, $\dot{e}^e_{ij}$ and plastic e^p_{ij}, $\dot{e}^p_{ij}$ components, that is, reversible and non-reversible components;
- the initial yield surface for different temperatures is described by the surface in the Mises form. The evolution of a change in the yield surface is described by a change in its radius C_p and a displacement of its center ρ^p_{ij};
- change in body volume elastically;
- initially, isotropic media are considered, only anisotropy caused by plasticity processes is taken into account;
- deformation processes characterized by small deformations are considered.

In the elastic region, the connection between the spherical and deviatoric components of the stress and strain tensors and their velocities is established using Hooke's law:

$$\sigma = 3Ke, \quad \sigma'_{ij} = 2Ge'^{e}_{ij}, \quad \dot{\sigma} = 3K\dot{e} \quad \dot{\sigma}'_{ij} = 2Ge'^{e}_{ij} \tag{33.1}$$

where K is the bulk modulus of elasticity of P. Bridgman, and G is the shear modulus.

To describe the effects of monotonic and cyclic deformation, a yield surface is introduced in the Mises form [4]:

$$\mathrm{F}_s = \mathrm{S}_{\mathrm{ij}},\mathrm{S}_{\mathrm{ij}} - \mathrm{C}^2_{\mathrm{p}} = 0, \tag{33.2}$$

To describe complex deformation modes in the deformation space, a cyclic "memory" surface is introduced [4]:

$$F_\rho = \rho_{ij}\rho_{ij} - \rho_{max}^2 = 0 \qquad (33.3)$$

where ρ_{max}—maximum module in the loading history ρ_{ij}.

Let us assume that the structure of the evolution equation for the radius of the flow surface has the form [4]:

$$\dot{C}_p = \left[q_\chi H(F_\rho) + a(Q_s - C_p)(F_\rho)\right]\dot{\chi} + q_3\dot{T}, \qquad (33.4)$$

$$C_p = C_p^0 + \int_0^t \dot{C}_p dt, \quad \dot{\chi} = \left(\frac{2}{3}\dot{e}_{ij}^p\dot{e}_{ij}^p\right)^{1/2},$$
$$\chi_m = \int_0^t \dot{\chi}H(F_p)dt, \quad \chi = \int_0^t \dot{\chi}dt, \qquad (33.5)$$

$$q_\chi = \frac{q_2A\psi_1 + (1-A)q_1}{A\psi_1 + (1-A)}, \quad Q_s = \frac{Q_2A\psi_2 + (1-A)Q_1}{A\psi_2 + (1-A)}, \quad 0 \le \psi_i \le 1, \quad i = 1, 2$$

$$A = 1 - \cos^2\Theta, \quad \cos\Theta = n_{ij}^e n_{ij}^s, \quad n_{ij}^e = \frac{\dot{e}'_{ij}}{(\dot{e}'_{ij}\dot{e}'_{ij})^{1/2}}, \quad n_{ij}^s = \frac{S_{ij}}{(S_{ij}S_{ij})^{1/2}}$$

$$H(F_p) = \begin{cases} 1, & F_p = 0 \wedge \rho_{ij}\dot{\rho}_{ij} > 0, \\ 0, & F_p < 0 \wedge \rho_{ij}\dot{\rho}_{ij} \le 0, \end{cases} \quad (F_p) = 1 - H(F_p) \qquad (33.6)$$

Here, $q_1, \ q_2, \ q_3$—isotropic hardening modules;$Q_1, \ Q_2$—modules of cyclic isotropic hardening; a *is* a constant that determines the rate of stationary process of the hysteresis loop of cyclic deformation of the material; Q_s *is* the stationary value of the radius of the yield surface at given ρ_{max} and temperature re T;C_p^0 is the initial value of the radius of the yield surface. The operators $H(F_\rho)$ and (F_ρ) make it possible to automatically separate monotonic and cyclic deformation processes.

The evolution of the internal variable ρ_{ij} *is* taken as

$$\dot{\rho}_{ij} = f(\chi_m)\left[g_1\dot{e}_{ij}^p - g_2\rho_{ij}\dot{\chi}\right] + g_T\rho_{ij}\langle\dot{T}\rangle + \dot{\rho}_{ij}^*, \quad \rho_{ij} = \int_0^t \dot{\rho}_{ij}dt \qquad (33.7)$$

$$f(\chi_m) = 1 + k_1(1 - \exp(-k_2\chi_m)), \quad \dot{\rho}^* = g_3\dot{e}_{ij}^p H(F_\rho) - g_4\rho_{ij}^*\dot{\chi}(F_\rho)\langle\cos\gamma\rangle, \qquad (33.8)$$

$$\cos\gamma = \frac{\dot{\rho}_{ij}\rho_{ij}^*}{\left(\dot{\rho}_{ij}\dot{\rho}_{ij}\right)^{1/2}\left(\rho_{ij}^*\rho_{ij}^*\right)^{1/2}}, \qquad (33.9)$$

where $g_1,\ g_2,\ g_3,\ g_4,\ k_1,\ k_2$ are experimentally determined material parameters; for any value B enclosed in angle brackets, the following conditions are met:

$$< \mathrm{B} >= \begin{cases} \mathrm{B},\ \mathrm{B} > 0, \\ 0,\ \mathrm{B} \le 0. \end{cases}$$

When $g_3 = g_4 = k_1 = 0$ from Eq. (33.7), we obtain a special case—the Armstrong—Frederick—Kadashevich equation:

$$\dot{\rho}_{\mathrm{ij}} = g_1 \dot{e}_{ij}^p - g_2 \rho_{ij} \dot{\chi}. \tag{33.10}$$

To describe the evolution of the "memory" surface, it is necessary to formulate an equation for $\rho_{\max}\rho_{\max}$

$$\dot{\rho}_{\max} = \frac{\left(\rho_{ij}\dot{\rho}_{ij}\right)H\left(F_\rho\right)}{\left(\rho_{mn}\rho_{mn}\right)^{1/2}} - g_2\rho_{\max}\dot{\chi} - g_2\rho_{\max}\dot{T}. \tag{33.11}$$

The components of the plastic deformation rate tensor obey the law of gradient of the plastic deformation rate vector to the yield surface at the loading point:

$$\dot{e}_{ij}^p = \lambda S_{ij}, \tag{33.12}$$

where λ is the proportionality coefficient determined from the condition of the passage of the new yield surface through the end of the stress deviator vector at the end of the loading stage.

When the corresponding material parameters are set to zero, from the general case of a model of thermoplastic deformation of a material, particular variants of the model (a system of "nested" models) are obtained: models with linear and nonlinear isotropic hardening, models with linear and nonlinear kinematic hardening, a model of an ideal plastic medium, etc.

33.4 Comparison of Numerical Results with Experimental Data

In [13, 14, 17–19], the results of experimental studies of proportional and disproportionate cyclic elastoplastic deformation of structural materials (metals and their alloys) are presented, which were obtained in the mechanical testing laboratory of the Department of Strength of Materials, Theory of Elasticity and Plasticity of Tver State Technical University on an automated computational and experimental complex of high accuracy class under the general supervision of Professor V.G. Zubchaninova. The experiments were carried out at a normal temperature $T = 20°C$ on thin-walled cylindrical samples with a working part length of $l = 110$ mm, a wall thickness of $h = 1$ mm, and a radius of the middle surface $R = 15,5$ of mm, made of steel 45 with a yield plateau as delivered.

The samples were loaded according to a given test program with axial force P, internal pressure, intensity q and torque M. The components of the stress and strain tensors in the working part of the sample were calculated using the formulas:

$$\sigma_{11} = \frac{P}{2\pi Rh}, \quad \sigma_{22} = \frac{qR}{h}, \quad \sigma_{12} = \frac{M}{2\pi R^2 h}, \quad \sigma_{33} = \sigma_{13} = \sigma_{23} = 0,$$
$$e_{11} = \frac{\Delta l}{l}, \quad e_{22} = \frac{\Delta R}{R}, \quad e_{12} = \frac{\varphi R}{2l}, \tag{33.13}$$

where φ—twist angle.

The experiments measured deformations e_{11}, e_{22}, e_{12} and voltage σ_{11}, σ_{22}, σ_{12}. Stress vectors $\overline{\sigma}'$ and deformations $\overline{e}'$ in the five-dimensional vector space of deviators A.A. Ilyushin with a common fixed reference $\{\overline{e}_k\}$ have the form:

$$\overline{\sigma}' = S_k \overline{e}_k, \quad \overline{e}' =_k \overline{e}_k \quad (k = 1, 2, 3, 4, 5) \tag{33.14}$$

Components S_k, $_k$ vectors of stress $\overline{\sigma}'$, strain $\overline{e}'$ and their modules σ', e' were calculated using the formulas:

$$S_1 = \sqrt{\frac{3}{2}}(\sigma_{11} - \sigma) = \sqrt{\frac{3}{2}}\left[\sigma_{11} - \frac{1}{2}(\sigma_{22} - \sigma_{33})\right], \quad S_2 = \frac{1}{\sqrt{2}}(\sigma_{11} - \sigma_{22}), \quad S_3 = \sqrt{2}\sigma_{12},$$

$$S_4 = \sqrt{2}\sigma_{23} = 0, \quad S_5 = \sqrt{2}\sigma_{13} = 0,$$

$$_1 = \sqrt{\frac{3}{2}}(e_{11} - e), \quad _2 = \sqrt{2}\left(e_{22} + \frac{1}{2}e_{11} - \frac{3}{2}e\right), \quad _3 = \sqrt{2}e_{12}, \tag{33.15}$$

$$_4 = \sqrt{2}e_{23} = 0, \quad _5 = \sqrt{2}e_{13} = 0,$$

$$\sigma' = \sqrt{S_1^2 + S_2^2 + S_3^2}, \quad e' = \sqrt{_1^2 + _2^2 + _3^2},$$

where

$$e = \frac{1}{3}(e_{11} + e_{22} + e_{33}), \quad \sigma = \frac{1}{3}(\sigma_{11} + \sigma_{22} + \sigma_{33}).$$

When processing experimental data, the incompressibility condition was assumed $e_{11} + e_{22} + e_{33} = 0$ and it was believed that $\sigma_{33} = 0$ [13, 14, 17–19].

In publications [13, 14, 17–19], plane and spatial trajectories of constant and variable curvature in the five-dimensional deviatoric space of deformations by A.A.

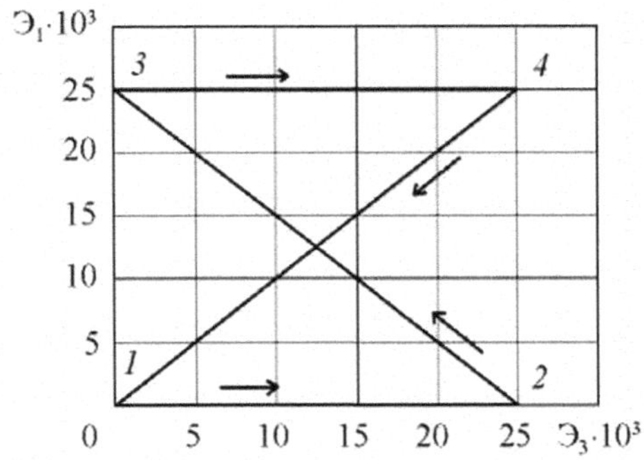

Fig. 33.1 Hourglass-type deformation trajectories in deformation space

Table 33.1 Basic physical and mechanical characteristics and parameters of the MPS steel 45 model

K, MPA	G, MPa	C_p^0, MPa	g_1, MPa	g_2	g_3, MPa	g_4	k_1	k_2	a
175,000	80,770	305	19,000	224	1500	0	0	0	5

Ilyushin were studied as complex loading trajectories. The experiments were carried out according to the "hard loading" scheme.

In the numerical simulation of the indicated experimental processes using the developed constitutive relations of elastoplasticity [2, 4], the experimental trajectories of deformations were given (patterns of change in the components of the strain tensor $e_{ij}(t)$, and the stress trajectories were obtained as a result of integrating the constitutive relations of thermoplasticity over a given history of change e_{ij}. The results of the numerical simulation were compared with experimental data.

An analytical "hourglass" deformation trajectory consisting of four straight piecewise broken sections was studied as complex loading trajectories (Fig. 33.1):

- in the first straight section, proportional torsion was realized along the coordinate $_3$ at $_1 = 0$ to value $^*_3 = 0, 025$;
- in the second section, when the deformation trajectory was bent at an angle, $\theta = 135°$ combined tension with torsion was realized up to values $^*_3 = 0$ And $^*_1 = 0, 025$;
- in the third section when the deformation trajectory breaks at an angle $\theta = 135°$ when $^*_1 = 0, 025$ torsion was carried out up to $^*_3 = 0, 025$;
- in the fourth section, when the deformation trajectory broke again at $\theta = 135°$, combined compression with torsion was realized with the arrival at the origin of coordinates.

The main physical and mechanical characteristics of steel 45 and the material parameters of the modified plasticity model with translational isotropic hardening are given in Tables 33.1 and 33.2.

Figure 33.2 shows the response to the realized deformation trajectory in the form of a stress trajectory in the plane S_3S_1 of the combined deviatoric subspace E_5.

Table 33.2 Modulus of monotonic hardening q_1 steel 45

q_1, MPa	−23,505	−17,527	−8244	−982	5800	7892	3000	0	120	720
χ	0	0.00175	0.006	0.009	0.013	0.0165	0.02	0.03	0.05	0.08

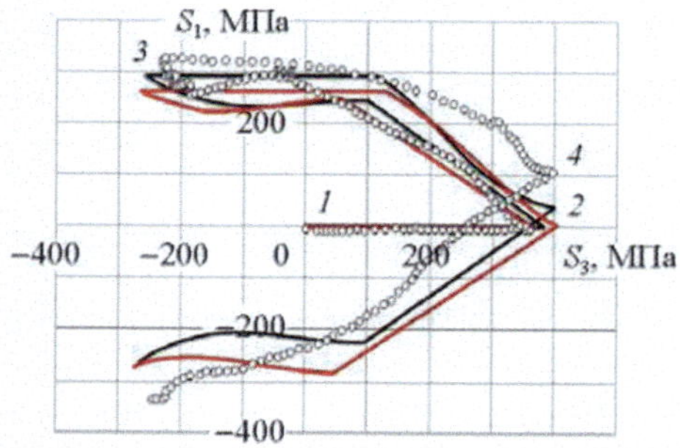

Fig. 33.2 Response to the realized deformation trajectory in stress space (markers—experiment, black solid lines—calculation using the model used, red solid lines—calculation using the Armstrong—Frederick—Kadashevich equations)

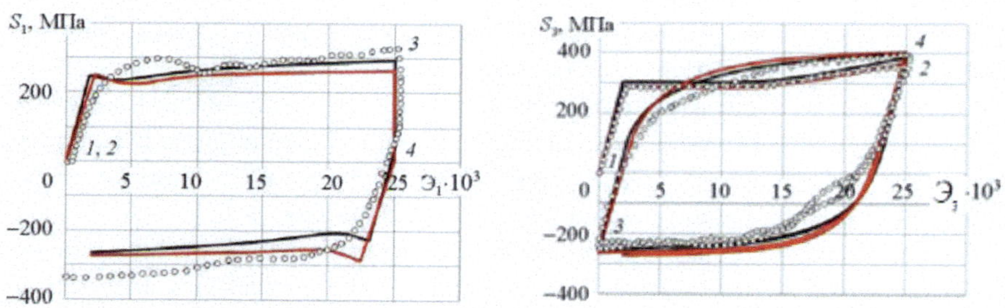

Fig. 33.3 Local deformation diagrams $S_1 \sim_1$ and $S_3 \sim_3$ (markers—experiment, black solid lines—calculation using the model used, red solid lines—calculation using the Armstrong—Frederick—Kadashevich equations)

Figure 33.3 shows local deformation diagrams.

Figure 33.4 shows the results of a comparison of calculated experimental data for a diagram of tracing the process of complex loading $\sigma' \sim \chi$, characterizing the scalar properties of materials depending on the length of the arc *5* of complex deformation.

Numbers *1–4* in Figs. 33.1, 33.2, 33.3, and 33.4 indicate the starting points of the corresponding sections of the implemented piecewise broken deformation trajectory of the "hourglass" type. Here and below, experimental data are marked with dots, and model-calculated data are marked with black solid lines. Red solid lines indicate calculated data obtained using the Armstrong–Frederick–Kadashevich Eq. (33.10).

One can see a qualitative and quantitative agreement acceptable for engineering calculations between experimental and model representations for plane "hourglass"

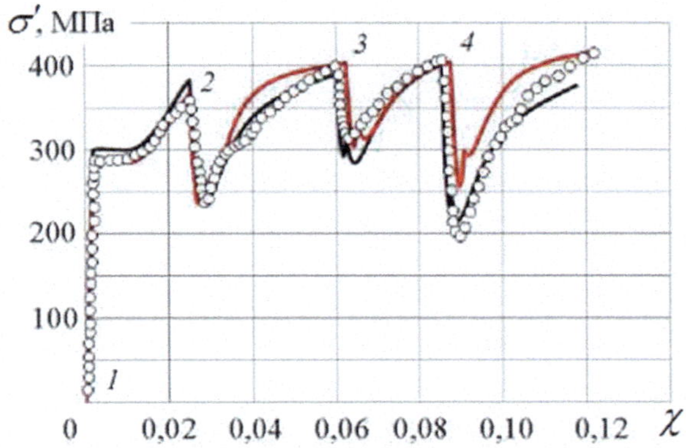

Fig. 33.4 Diagram of the complex loading process $\sigma' \sim \chi$(markers—experiment, black solid lines—calculation using the model used, red solid lines—calculation using the Armstrong—Frederick—Kadashevich equations)

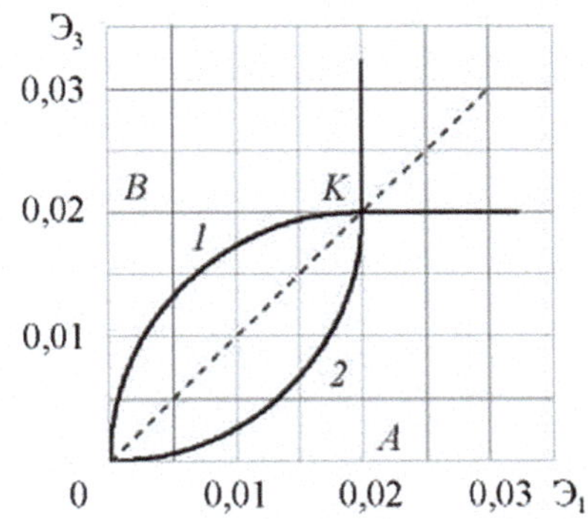

Fig. 33.5 Trajectory of deformation in the space of deformation (1—the first trajectory, consisting of two piecewise broken lines; 2—mirrored trajectory of deformation)

deformation trajectories according to the theory of plastic flow with translational isotropic hardening, which indicates the correctness of modeling the process of complex deformation of a material and finding material parameters and scalar functions constitutive relations of plasticity. It should also be noted that the modified plasticity model more accurately describes the kinetics of the stress–strain state in the vicinity of the breakpoints of the deformation trajectory than the constitutive relations of plasticity Yu.G. Korotkikh using the Armstrong–Frederick–Kadshevich equation.

To assess the reliability of the plasticity model and verify the isotropy postulate, A.A. Ilyushin, two broken nontrivial deformation trajectories were considered, which mirror each other relative to the bisector of a right angle on the plane $Э_1$ $Э_3$[18].

The first of two deformation trajectories (trajectory 1 in Fig. 33.5) consists of two piecewise broken sections:

- the first of the sections is a quarter circle with a radius of 0.02 with the center of curvature at point *A* (0.02; 0), on which a combined tensile loading with torsion was implemented up to point *K* (0.02; 0.02);
- in the second section, when the deformation trajectory breaks, $\theta = 90°$ torsional loading is carried out;

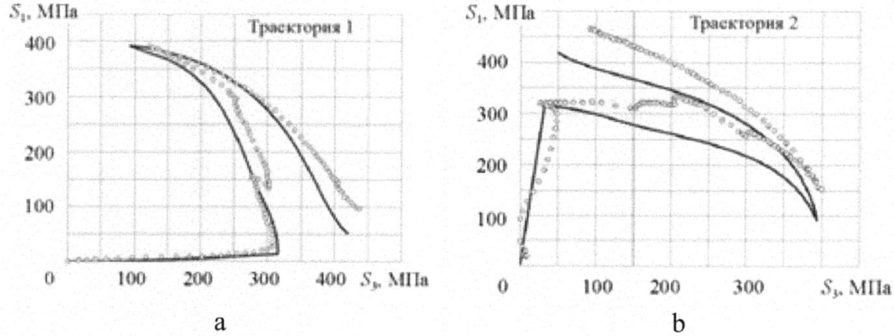

Fig. 33.6 Response to the realized deformation trajectories in stress space: **a**—trajectory 1, **b**—trajectory 2 (markers—experiment, black solid lines—calculation using the model used)

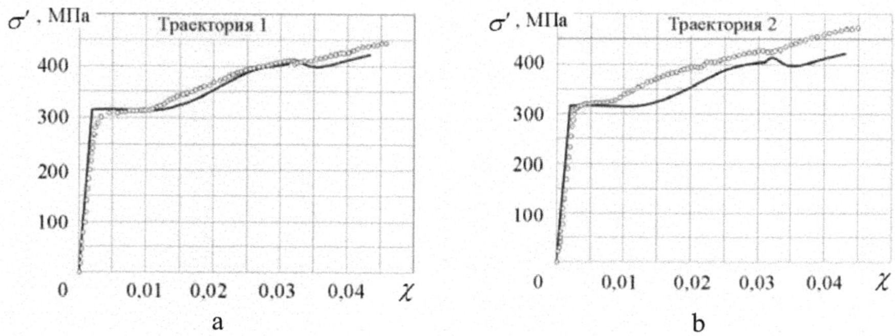

Fig. 33.7 Diagram of complex plastic deformation $\sigma' \sim \chi$: **a**—trajectory 1, **b**—trajectory 2 (markers—experiment, black solid lines—calculation using the model used)

- mirror image of the deformation trajectory (trajectory 2 in Fig. 33.5):
- the first of the sections is a quarter of a circle with a radius of 0.02 with the center of curvature at point *B* (0; 0.02), on which combined stretching with torsion was realized to point *K* (0.02; 0.02);
- in the second section, at a break in the deformation trajectory $\theta = 90°$, subsequent monotonic tensile loading is carried out.

The results of numerical calculations and their comparison with experimental data are presented in Figs. 33.6a and 33.7a for the first trajectory, and in Figs. 33.6b and 33.7b for the second, respectively.

Figure 33.6 shows the response to the realized deformation trajectories (see Fig. 33.6) in the form of stress trajectories in the plane of the combined deviatoric stress space A.A. Ilyushin.

Figure 33.7 shows the results of a comparison of calculated and experimental data for a diagram of tracing the process of complex loading along trajectories *1* and *2*. It can be seen that for the two types of deformation trajectories under consideration, the isotropy postulate of A.A. Ilyushin is confirmed experimentally and by calculation.

33.5 Conclusions

The modified constitutive relations of the theory of flow with isotropic and kinematic hardening [2, 4] were verified by comparing the results of numerical experiments with experimental data on the elastoplastic deformation of steel 45 with a yield plateau on flat deformation trajectories of arbitrary curvature. The verification results indicate the correctness of modeling proportional and disproportionate plastic deformation of structural steels along arbitrary flat deformation trajectories.

Acknowledgements The work was supported by the Russian Science Foundation grant № 22-19-00138.

References

1. Mitenkov FM, Kaidalov VB, Korotkikh YG (2007) Methods for justifying the service life of nuclear power plants. M.: Mech Eng 448
2. Volkov IA, Korotkikh YG (2008) Equations of state of viscoelastoplastic media with damage. M.: Fizmatlit, p 424
3. Collins J (1984) Damage to materials in structures. Analysis. Prediction. Prevention. M:. Mir, p 624
4. Mitenkov FM, Volkov IA, Igumnov LA, Korotkikh YG and others (2015) Applied theory of plasticity. M.: Fizmatlit, p 284
5. Zubchaninov VG (2010) Mechanics of plastic processes. M.: Fizmatlit, p 352
6. Degtyarev VP (1967) Plasticity and creep of mechanical engineering structures. M.: Mech Eng 130
7. Zhukov AM (1961) Some features of the behavior of materials under elastoplastic deformation. Questions of the theory of plasticity. M.: USSR Acad Sci 30–57
8. Lamba GS, Sidebottom OM (1978) Plasticity under cyclic loading along disproportionate trajectories. In: Theoretical foundations of engineering calculations, vol 100, no 1, pp 108–117
9. McDowell DL (1985) Experimental study of the structure of governing equations for non-proportional cyclic plasticity. In: Theoretical foundations of engineering calculations, vol 107, no 4, pp 98–111
10. Ohashi I, Kawai M, Kaito T (1985) Inelastic behavior of 316 steel under multiaxial non-proportional cyclic loading at elevated temperature. In: Theoretical foundations of engineering calculations, vol 107, no 2, pp 6–15
11. Tanaka E, Murakami S, Ooka M (1985) Effects of plastic strain amplitudes on non-proportional cyclic plasticity. Acta Mechanica 57:167–182
12. Tanaka E, Murakami S, Ooka M (1985) Effects of strain path shapes on non-proportional cyclic plasticity. J Mech Phys Solids 33(6):559–575
13. Zubchaninov VG, Okhlopkov NL, Garannikov VV (2003) Experimental plasticity. Book 1. Processes of complex deformation. Tver: TSTU, p 172
14. Zubchaninov VG, Okhlopkov NL, Garannikov VV (2004) Experimental plasticity. Book 2. Processes of complex loading. Tver: TSTU, p 184

15. Ohashi I, Tanaka E, Ooka M (1985) Plastic deformation of type 316 stainless steel under the influence of out-of-phase deformation cycles. In: Theoretical foundations of engineering calculations, vol 107, no 4, pp 61–73
16. Zakhovaiko AA, Mozharovsky NS (1982) On the characteristics of strength and ductility under complex loading paths. Strength Probl 1:90–93
17. Zubchaiiiov VG, Alekseev AA, Guyotyaev VI (2017) Modeling the processes of elastoplastic deformation of materials along multi-link piecewise broken rectilinear trajectories. Bull Perm Natl Res Polytechnic Univ. Mech (3):203–215
18. Zubchaninov VG, Alekseev AA, Alekseeva EG (2017) Gultiaev VI experimental verification of postulate of isotropy and mathematical modeling of elastoplastic deformation processes following the complex angled nonanalytic trajectories. Mater Phys Mech 32(3):298–304
19. Zubchaninov VG, Gultiaev VI, Alekseev AA, Garanikov VV (2017) Subbotin SL testing of steel 45 under complex loading along the cylindrical screw trajectories of deformation. Mater Phys Mech 32(3):305–311

Chapter 34
Series Solutions for Clamped Peridynamic Beams Using Fourth-Order Eigenfunctions

Ziyu Wang and Ivan C. Christov

Abstract We propose an analytical approach to solving nonlocal generalizations of the Euler–Bernoulli beam. Specifically, we consider a version of the governing equation recently derived under the theory of peridynamics. We focus on the clamped–clamped case, employing the natural eigenfunctions of the fourth derivative subject to these boundary conditions. Static solutions under different loading conditions are obtained as series in these eigenfunctions. To demonstrate the utility of our proposed approach, we contrast the series solution in terms of fourth-order eigenfunctions to the previously obtained Fourier sine series solution. Our findings reveal that the series in fourth-order eigenfunctions achieve a given error tolerance (with respect to a reference solution) with ten times fewer terms than the sine series. The high level of accuracy of the fourth-order eigenfunction expansion is due to the fact that its expansion coefficients decay rapidly with the number of terms in the series, one order faster than the Fourier series in our examples.

34.1 Introduction

Peridynamics is a recent reformulation of continuum mechanics introduced by Silling [21, 23] to overcome the limited applicability of classical continuum mechanics to problems with discontinuities. The idea is that any material point in the continuum can interact with other points within a finite *horizon* of itself, giving peridynamic theory a "nonlocal" character. Different approaches to nonlocality are also discussed in the context of generalized continua, see, e.g., [2]. Due to peridynamic theory's ability to model the initiation and growth of cracks in materials without the need for additional techniques to handle such discontinuities, peridynamics is often used to simulate the deformation of elastic materials until failure. For example, Kim et al. [9] simulated

Z. Wang · I. C. Christov (✉)
School of Mechanical Engineering, Purdue University, West Lafayette, Indiana 47907, USA
e-mail: christov@purdue.edu

Z. Wang
e-mail: wang5600@purdue.edu

H. Altenbach and V. A. Eremeyev (eds.), *Dynamics of Discrete and Continuum Structures and Media*, Advanced Structured Materials 221,
https://doi.org/10.1007/978-3-031-75626-9_34

the deformation, damage, and failure of an elastic band under fluid flow loading in a fluid–structure interaction framework. They demonstrated that a peridynamics-enabled computational model could capture how the flow forces not only deform but also tear apart the elastic band, a simulation that would not be possible within a fluid–structure interaction framework based on classical continuum mechanics. Although peridynamics theory is typically implemented as a computational framework, there is nevertheless fundamental interest in finding analytical solutions of the governing equations of peridynamics and peristatics [8, 11, 12, 14].

Deriving reduced models, such as for beams, plates, and shells, is a time-honored research area within classical continuum mechanics [1, 10, 26]. More recently, there has been an interest in doing the same for peridynamic beams [15, 27, 28, 30, 31], membranes [22], plates [6, 13, 16, 25, 29], and shells [16]. Once such models are developed, it is of interest to find analytical solutions to basic problems (if possible) [3, 27–31]. In this work, we continue the quest, providing new analytical peristatic solutions for the bending of clamped–clamped beams using fourth-order eigenfunctions.

Perhaps for obvious reasons, the eigenfunctions of the fourth-derivative operator are often termed *beam* functions (e.g., [17, 19]). Historically, Lord Rayleigh [24, §§170-175] derived these functions in the context of the lateral vibrations of elastic bars, including clamped–clamped, clamped–free, and loaded boundary conditions among other combinations. Chandrasekhar [4, §133] also derived these orthogonal eigenfunctions and employed them to study the linear stability of thermal convection.

Motivated by the recent work of Yang et al. [28–30], we continue the search for closed-form analytical solutions to peridynamic problems. Yang et al. [30] used Castigliano's theorem to replace the clamped–clamped boundary conditions with moment conditions on a supported beam, a case that can be solved by a Fourier sine series. On the other hand, in this work, we demonstrate that the fourth-order (beam) eigenfunctions, which naturally satisfy the clamped boundary conditions, can also be used to obtain analytical series solutions without resorting to Castigliano's theorem. Moreover, we show that the series in terms of fourth-order eigenfunctions converges significantly faster than a Fourier sine series, which means the series solutions derived by our approach provide a more accurate representation of peridynamic beam bending (for a fixed number of terms in the series).

34.2 The Fourth-Order "Beam" Eigenfunctions

First, we review the problem of the dynamic bending of an Euler–Bernoulli beam, originating from "classical continuum mechanics," as depicted in the schematic in Fig. 34.1. The beam has mass per area m_s, bending rigidity B, and applied pressure load $p(x, t)$. When it is clamped at both ends, its vertical displacement $u(x, t)$ obeys the fourth-order initial-boundary-value problem (IBVP):

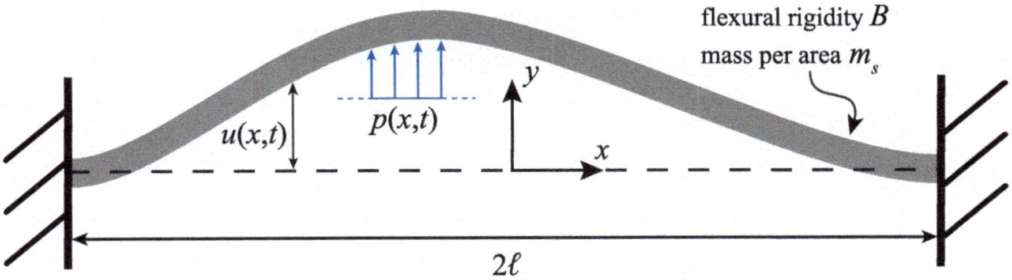

Fig. 34.1 Schematic of a clamped–clamped beam bending due to a pressure load

$$\begin{cases} m_s \dfrac{\partial^2 u}{\partial t^2} + B \dfrac{\partial^4 u}{\partial x^4} = p(x,t), & -\ell < x < \ell, \quad t > 0, \quad (34.1a) \\ u|_{x=\pm\ell} = \left.\dfrac{\partial u}{\partial x}\right|_{x=\pm\ell} = 0, & t > 0, \quad (34.1b) \\ u(x,t) = u^0(x), & -\ell \le x \le \ell. \quad (34.1c) \end{cases}$$

We make the IBVP (34.1) dimensionless by introducing the following variables based on the typical load scale p_0:

$$T = \frac{t}{\sqrt{m_s \ell^4 / B}}, \quad X = \frac{x}{\ell}, \quad U(X,T) = \frac{u(x,t)}{p_0 \ell^4 / B}, \quad P(X,T) = \frac{p(x,t)}{p_0}, \tag{34.2}$$

to obtain

$$\begin{cases} \dfrac{\partial^2 U}{\partial T^2} + \dfrac{\partial^4 U}{\partial X^4} = P(X,T), & -1 < X < 1, \quad t > 0, \quad (34.3a) \\ U|_{X=\pm 1} = \left.\dfrac{\partial U}{\partial X}\right|_{X=\pm 1} = 0, & T > 0, \quad (34.3b) \\ U(X,T) = U^0(X), & -1 \le X \le 1. \quad (34.3c) \end{cases}$$

IBVP (34.3) is referred to as "classical beam theory" (CBT) in previous works.

Following the notation of Papanicolaou et al. [17, 19], the fourth-order Sturm–Liouville eigenvalue problem (EVP) on $X \in [-1, +1]$ associated with the IBVP (34.3) is

$$\begin{cases} +\dfrac{d^4 \psi}{dX^4} = \lambda^4 \psi, & -1 < X < 1, \quad (34.4a) \\ \psi|_{X=\pm 1} = \left.\dfrac{d\psi}{dX}\right|_{X=\pm 1} = 0. & (34.4b) \end{cases}$$

The EVP (34.4) is self-adjoint and there exist two linearly independent sets of eigenfunctions $\{\psi_m^s, \psi_m^c\}$ with associated eigenvalues $\{\lambda_m^s, \lambda_m^c\}$ [17, 19]:

$$\begin{cases} \psi_m^s(X) = \dfrac{1}{\sqrt{2}}\left[\dfrac{\sinh(\lambda_m^s X)}{\sinh(\lambda_m^s)} - \dfrac{\sin(\lambda_m^s X)}{\sin(\lambda_m^s)}\right], & \coth\lambda_m^s - \cot\lambda_m^s = 0, \quad (34.5a)\\ \psi_m^c(X) = \dfrac{1}{\sqrt{2}}\left[\dfrac{\cosh(\lambda_m^c X)}{\cosh(\lambda_m^c)} - \dfrac{\cos(\lambda_m^c X)}{\cos(\lambda_m^c)}\right], & \tanh\lambda_m^c + \tan\lambda_m^c = 0. \quad (34.5b) \end{cases}$$

Here, $\psi_m^s(-X) = -\psi_m^s(X)$ are the odd, "sine" ("s"), eigenfunctions, while $\psi_m^c(-X) = \psi_m^c(X)$ are the even, "cosine" ("c"), eigenfunctions. The respective sets of eigenvalues λ_m^s and λ_m^c are a countable ($m = 1, 2, \ldots$), real set of numbers, and satisfy the transcendental relations given in Eqs. (34.5).

As shown in [17, 19] (and the references therein, in particular [4, §133]), the eigenfunctions ψ_m^s and ψ_m^c *each* are orthonormal (for any $m, n = 1, 2, \ldots$):

$$\int_{-1}^{+1} \psi_m^s(X)\psi_n^s(X)\,dX = \delta_{m,n}, \tag{34.6a}$$

$$\int_{-1}^{+1} \psi_m^c(X)\psi_n^c(X)\,dX = \delta_{m,n}, \tag{34.6b}$$

where $\delta_{m,n}$ is the Kronecker delta symbol, and *mutually* orthogonal (for any $m = 1, 2, \ldots$):

$$\int_{-1}^{+1} \psi_m^s(X)\psi_m^c(X)\,dX = 0. \tag{34.6c}$$

Then, by standard theorems of ordinary differential equations [5], it follows that the eigenfunctions (34.5) form a complete set in the space of square-integrable functions, $L^2[-1, +1]$, under the standard inner product. Thus, any function $U(X) \in L^2[-1, +1]$ has the following expansion:

$$U(X) = \sum_{m=1}^{\infty} a_m^s \psi_m^s(X) + a_m^c \psi_m^c(X), \tag{34.7a}$$

$$a_m^s \overset{\text{def}}{=} \int_{-1}^{+1} U(X)\psi_m^s(X)\,dX, \tag{34.7b}$$

$$a_m^c \overset{\text{def}}{=} \int_{-1}^{+1} U(X)\psi_m^c(X)\,dX. \tag{34.7c}$$

For the unsteady beam problem (34.3), $U = U(X, T)$, and the coefficients may depend on time as $a_m^s = a_m^s(t)$ and $a_m^c = a_m^c(t)$.

34.3 Peridynamic Beam Problem Formulation

Under the theory of peridynamics [21, 23], Yang et al. [28] showed that the classical beam equation (34.1a) is generalized as the integrodifferential equation:

$$m_s \frac{\partial^2 u}{\partial t^2} - \frac{B}{\delta^2} \int_{-\delta}^{+\delta} \frac{1}{\xi^2} \left[\int_{-\delta}^{+\delta} \frac{u(x+\eta) - u(x)}{\eta^2} d\eta \right. \\ \left. - \int_{-\delta}^{+\delta} \frac{u(x+\xi+\eta) - u(x+\xi)}{\eta^2} d\eta \right] d\xi = p(x,t). \tag{34.8}$$

The new feature of this equation is the *horizon* size $\delta \ll \ell$. The horizon refers to the finite domain of influence for any material point x, which is the nonlocal feature of the theory. A larger δ value defines a bigger neighborhood around the material point x that influences the behavior there, introducing interactions with more distant material points. The horizon size should be chosen based on the nature of the specific problem being addressed [21, 23], for example, a nonzero δ should be used to study a nonlocal beam.

Again, introducing dimensionless variables based on the typical load scale p_0,

$$T = \frac{t}{\sqrt{m_s \ell^4 / B}}, \quad X = \frac{x}{\ell}, \quad \Xi = \frac{\xi}{\ell}, \quad H = \frac{\eta}{\ell}, \quad \Delta = \frac{\delta}{\ell}, \\ U(X,T) = \frac{u(x,t)}{p_0 \ell^4 / B}, \quad P(X,T) = \frac{p(x,t)}{p_0}, \tag{34.9}$$

Equation (34.8) becomes

$$\frac{\partial^2 U}{\partial T^2} - \frac{1}{\Delta^2} \int_{-\Delta}^{+\Delta} \frac{1}{\Xi^2} \left[\int_{-\Delta}^{+\Delta} \frac{U(X+H) - U(X)}{H^2} dH \right. \\ \left. - \int_{-\Delta}^{+\Delta} \frac{U(X+\Xi+H) - U(X+\Xi)}{H^2} dH \right] d\Xi = P(X,T). \tag{34.10}$$

The IBVP comprised of Eqs. (34.10), (34.3b), and (34.3c) is termed "peridynamic beam theory" (PBT) in this work.

Henceforth, we consider steady problems, so $U = U(X)$ and $P = P(X)$ only, and the governing equation is

$$\frac{1}{\Delta^2}\int_{-\Delta}^{+\Delta}\frac{1}{\Xi^2}\left[\int_{-\Delta}^{+\Delta}\frac{U(X+H)-U(X)}{H^2}\,dH - \int_{-\Delta}^{+\Delta}\frac{U(X+\Xi+H)-U(X+\Xi)}{H^2}\,dH\right]d\Xi = -P(X). \tag{34.11}$$

For sufficiently smooth $U(X)$, it may be verified by Taylor series expansions that the left-hand side of Eq. (34.11) reduces $-\partial^4 U/\partial X^4$ as $\Delta \to 0^+$, consistent with CTB (34.3a).

34.4 New Solutions for Clamped–Clamped Peridynamic Beams

In this section, we show how Eq. (34.11) can be solved via a series of the form given in Eq. (34.7).

34.4.1 Odd Solutions

Consider an odd load such that $P(-X) = -P(X)$. Then, from the structure of the boundary-value problem, it is expected that $U(-X) = -U(X)$ is also an odd function. Thus, $a_m^c = 0\ \forall m$, and from Eq. (34.7a) the solution can be expressed as

$$U(X) = \sum_{m=1}^{\infty} a_m^s \psi_m^s(X). \tag{34.12}$$

Next, substituting Eq. (34.12) into Eq. (34.11), we obtain

$$\frac{1}{\sqrt{2}}\sum_{m=1}^{\infty} a_m^s \left\{\frac{\sinh(\lambda_m^s X)}{\sinh(\lambda_m^s)}\left[\mathcal{I}_{1,m}^s(\Delta)\right]^2 - \frac{\sin(\lambda_m^s X)}{\sin(\lambda_m^s)}\left[\mathcal{I}_{2,m}^s(\Delta)\right]^2\right\} = P(X). \tag{34.13}$$

The full details of the calculation, including the definitions of $\mathcal{I}_{1,m}^s$ and $\mathcal{I}_{2,m}^s$, are given in the Appendix. Then, multiplying both sides of Eq. (34.13) by $\psi_n^s(X)$ and integrating over the domain yields

$$\sum_{m=1}^{\infty} A_{n,m}^s a_m^s = P_n^s, \qquad n = 1, 2, \ldots, \tag{34.14}$$

where

$$A^s_{n,m} \stackrel{\text{def}}{=} \frac{1}{\sqrt{2}} \int_{-1}^{+1} \left\{ \frac{\sinh(\lambda^s_m X)}{\sinh(\lambda^s_m)} \left[\mathcal{I}^s_{1,m}(\Delta)\right]^2 - \frac{\sin(\lambda^s_m X)}{\sin(\lambda^s_m)} \left[\mathcal{I}^s_{2,m}(\Delta)\right]^2 \right\} \times \left[\frac{\sinh(\lambda^s_n X)}{\sinh(\lambda^s_n)} - \frac{\sin(\lambda^s_n X)}{\sin(\lambda^s_n)} \right] dX, \tag{34.15}$$

and

$$P^s_n \stackrel{\text{def}}{=} \int_{-1}^{+1} P(X) \left[\frac{\sinh(\lambda^s_n X)}{\sinh(\lambda^s_n)} - \frac{\sin(\lambda^s_n X)}{\sin(\lambda^s_n)} \right] dX. \tag{34.16}$$

The integral in Eq. (34.15) is evaluated analytically and given in the Appendix, Eq. (34.36).

The series in Eq. (34.12) and the summation in Eq. (34.14) can be truncated at M terms, leading to a square system of linear equations for the odd series coefficients $\{a^s_m\}_{m=1}^M$:

$$\begin{pmatrix} A^s_{1,1} & A^s_{1,2} & \cdots & A^s_{1,M} \\ A^s_{2,1} & A^s_{2,2} & \cdots & A^s_{2,M} \\ \vdots & \vdots & \ddots & \vdots \\ A^s_{M,1} & A^s_{M,2} & \cdots & A^s_{M,M} \end{pmatrix} \begin{pmatrix} a^s_1 \\ a^s_2 \\ \vdots \\ a^s_M \end{pmatrix} = \begin{pmatrix} P^s_1 \\ P^s_2 \\ \vdots \\ P^s_M \end{pmatrix}, \tag{34.17}$$

where the square matrix, denoted by $\mathbf{A^s}$, is non-symmetric but diagonally dominant as shown in Fig. 34.2a for $\Delta = 0.1$.

Consider the example load, $P(X) = -X$, in which case

$$P^s_n = -\frac{2}{(\lambda^s_n)^2} \left[\lambda^s_n \cot(\lambda^s_n) + \lambda^s_n \coth(\lambda^s_n) - 2 \right]. \tag{34.18}$$

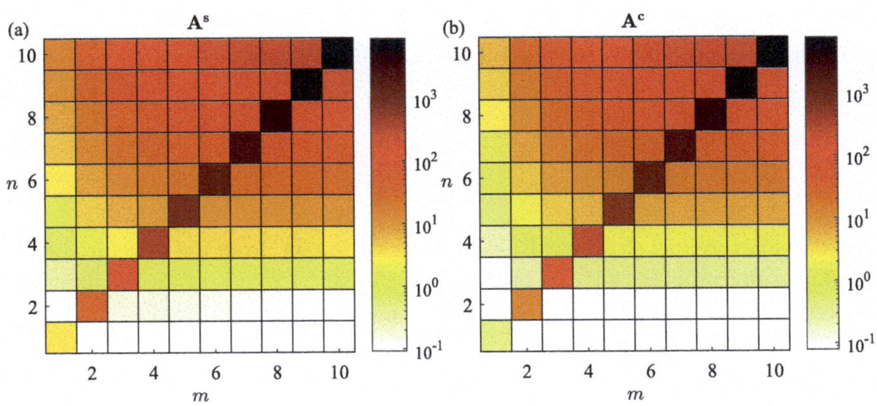

Fig. 34.2 Visualization of matrices **a**, $\mathbf{A^s}$, and **b**, $\mathbf{A^c}$, showing the magnitude of their entries, for a truncation at $M = 10$ terms with $\Delta = 0.1$

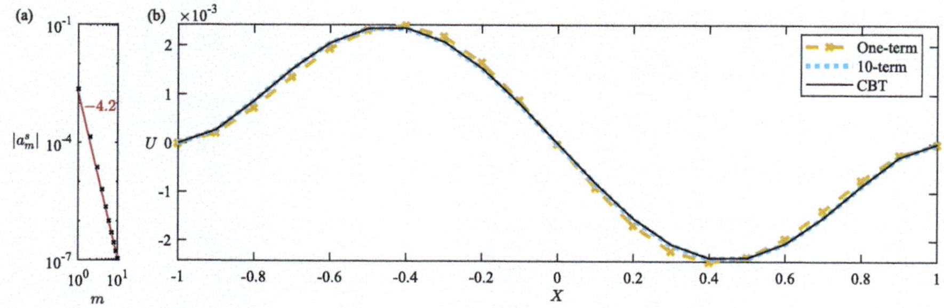

Fig. 34.3 Beam bending under the load $P(X) = -X$, which allows using only odd eigenfunctions. **a** The absolute values of odd series coefficients, $|a^s_m|$. **b** Peridynamic beam theory (PBT) deflections calculated using a one-term truncation of the series (34.12), a ten-term truncation of the series (34.12), for $\Delta = 0.1$, and the classical beam theory (CBT) solution (34.19)

The beam deflection can be represented as an odd eigenfunction series. The absolute values of the first ten expansion coefficients are plotted in Fig. 34.3a, alongside a reference line of slope -4.2. The coefficients' rapid decay at a rate $\mathcal{O}(m^{-4})$ highlights the efficiency of using fourth-order eigenfunction series to solve for the deflection under PBT, as the series (34.12) converges quite rapidly.

In Fig. 34.3b, we compare the beam deflection computed using only the first term with that calculated using the first ten terms of the series (34.12) and the deflection from the attendant CBT solution, i.e., the solution of the steady BVP (34.3) with the same load $P(X) = -X$, which is easily found to be

$$U_{\mathrm{CBT}}(X) = -\frac{1}{120}\left(X^5 - 2X^3 + X\right). \tag{34.19}$$

From Fig. 34.3b, it can be seen that, surprisingly, a single-term approximation can provide a satisfactory solution. For the value $\Delta = 0.1$ of the horizon chosen in this figure, the difference between the CBT and the PBT is small.

34.4.2 Even Solutions

Consider an even load such that $P(-X) = P(X)$. Again, from the structure of the boundary-value problem, it is expected that $U(-X) = U(X)$ is also an even function. Thus, $a^s_m = 0 \; \forall m$, and from Eq. (34.7a) the solution can be expressed as

$$U(X) = \sum_{m=1}^{\infty} a^c_m \psi^c_m(X). \tag{34.20}$$

Next, substituting Eq. (34.20) into Eq. (34.11), we obtain

$$\frac{1}{\sqrt{2}}\sum_{m=1}^{\infty} a_m^c \left\{ \frac{\cosh(\lambda_m^c X)}{\cosh(\lambda_m^c)} \left[\mathcal{I}_{1,m}^c(\Delta)\right]^2 - \frac{\cos(\lambda_m^c X)}{\cos(\lambda_m^c)} \left[\mathcal{I}_{2,m}^c(\Delta)\right]^2 \right\} = P(X). \quad (34.21)$$

The full details of the calculation, including the definitions of $\mathcal{I}_{1,m}^c$ and $\mathcal{I}_{2,m}^c$, are given in the Appendix. Then, multiplying both sides of Eq. (34.21) by $\psi_n^c(X)$ and integrating over the domain yields

$$\sum_{m=1}^{\infty} A_{n,m}^c a_m^c = P_n^c, \qquad n = 1, 2, \ldots, \quad (34.22)$$

where

$$A_{n,m}^c \stackrel{\text{def}}{=} \frac{1}{\sqrt{2}} \int_{-1}^{+1} \left\{ \frac{\cosh(\lambda_m^c X)}{\cosh(\lambda_m^c)} \left[\mathcal{I}_{1,m}^c(\Delta)\right]^2 - \frac{\cos(\lambda_m^c X)}{\cos(\lambda_m^c)} \left[\mathcal{I}_{2,m}^c(\Delta)\right]^2 \right\} \times \left[\frac{\cosh(\lambda_n^c X)}{\cosh(\lambda_n^c)} - \frac{\cos(\lambda_n^c X)}{\cos(\lambda_n^c)} \right] dX, \quad (34.23)$$

and

$$P_n^c \stackrel{\text{def}}{=} \int_{-1}^{+1} P(X) \left[\frac{\cosh(\lambda_n^c X)}{\cosh(\lambda_n^c)} - \frac{\cos(\lambda_n^c X)}{\cos(\lambda_n^c)} \right] dX. \quad (34.24)$$

The integral in Eq. (34.23) is evaluated analytically and given in the Appendix, Eq. (34.39).

As before, in Eq. (34.20) and the summation in Eq. (34.22) can be truncated at M terms, leading to a square system of linear equations for the even series coefficients $\{a_m^c\}_{m=1}^M$:

$$\begin{pmatrix} A_{1,1}^c & A_{1,2}^c & \cdots & A_{1,M}^c \\ A_{2,1}^c & A_{2,2}^c & \cdots & A_{2,M}^c \\ \vdots & \vdots & \ddots & \vdots \\ A_{M,1}^c & A_{M,2}^c & \cdots & A_{M,M}^c \end{pmatrix} \begin{pmatrix} a_1^c \\ a_2^c \\ \vdots \\ a_M^c \end{pmatrix} = \begin{pmatrix} P_1^c \\ P_2^c \\ \vdots \\ P_M^c \end{pmatrix}, \quad (34.25)$$

where the square matrix, denoted by $\mathbf{A^c}$, is non-symmetric but diagonally dominant as shown in Fig. 34.2b for $\Delta = 0.1$.

Consider the example load, $P(X) \equiv 1$, in which case

$$P_n^c = -\frac{2}{\lambda_n^c}\left[\tan(\lambda_n^c) - \tanh(\lambda_n^c)\right]. \quad (34.26)$$

The beam deflection can be represented as an even eigenfunction series (34.20). We calculated and plotted the absolute values of the first ten expansion coefficients in Fig. 34.4a, alongside a reference line with slope of −5.6. The coefficients' rapid decay at a rate $\mathcal{O}(m^{-5})$, which is significantly faster than fourth order, again highlights the

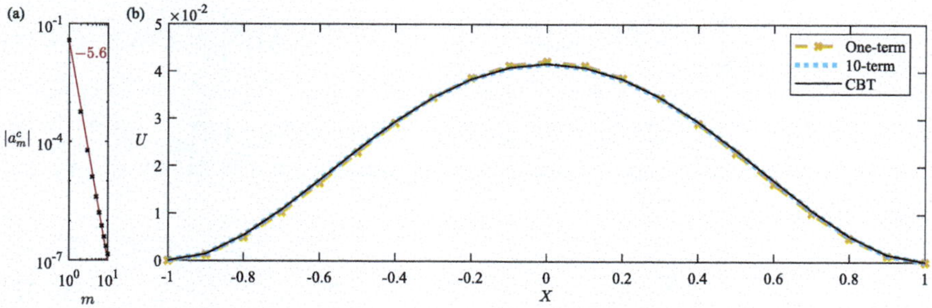

Fig. 34.4 Beam bending under the uniform load $P(X) \equiv 1$, which allows using only even eigenfunctions. **a** The absolute values of even series coefficients, $|a_m^c|$. **b** Peridynamic beam theory (PBT) deflections calculated using a one-term truncation of the series (34.20), a ten-term truncation of the series (34.20), for $\Delta = 0.1$, and the classical beam theory (CBT) solution (34.27)

efficiency of using fourth-order eigenfunction series to solve for the deflection under PBT.

In Fig. 34.4b, we compare the PBT deflection computed using only the first term with that calculated using the first ten terms of the series (34.20) and the deflection from the attendant CBT solution, i.e., the solution of the steady BVP (34.3) with the same load $P(X) \equiv 1$, which is easily found to be:

$$U_{\mathrm{CBT}}(X) = \frac{1}{24}(X-1)^2(X+1)^2. \tag{34.27}$$

Once again, it can be seen from Fig. 34.4b that a single-term approximation based on the fourth-order eigenfunctions can provide a satisfactory solution.

34.4.3 General Case

As the third example, consider the deflection of a beam caused by a step load, represented as the Heaviside unit step function: $P(X) = H(X)$. In this case,

$$P_n^s = \frac{1}{\lambda_n^s}\big[\cot(\lambda_n^s) - \csc(\lambda_n^s) + \tanh(\lambda_n^s/2)\big], \tag{34.28a}$$

$$P_n^c = -\frac{1}{\lambda_n^c}\big[\tan(\lambda_n^c) - \tanh(\lambda_n^c)\big]. \tag{34.28b}$$

Now, the solution $U(X)$ is neither odd nor even, and *both* the odd and even series coefficients have to be computed. Hence, $U(X)$ is expressed as in Eq. (34.7a). Substituting Eq. (34.7a) into Eq. (34.11), we obtain

$$\frac{1}{\sqrt{2}}\sum_{m=1}^{\infty} a_m^s \left\{ \frac{\sinh(\lambda_m^s X)}{\sinh(\lambda_m^s)} \left[\mathcal{I}_{1,m}^s(\Delta)\right]^2 - \frac{\sin(\lambda_m^s X)}{\sin(\lambda_m^s)} \left[\mathcal{I}_{2,m}^s(\Delta)\right]^2 \right\}$$
$$+ \frac{1}{\sqrt{2}}\sum_{m=1}^{\infty} a_m^c \left\{ \frac{\cosh(\lambda_m^c X)}{\cosh(\lambda_m^c)} \left[\mathcal{I}_{1,m}^c(\Delta)\right]^2 - \frac{\cos(\lambda_m^c X)}{\cos(\lambda_m^c)} \left[\mathcal{I}_{2,m}^c(\Delta)\right]^2 \right\} = P(X). \tag{34.29}$$

Multiplying both sides of Eq. (34.29) by $\psi_n^s(X)$ and integrating over the domain yields

$$\sum_{m=1}^{\infty} A_{n,m}^s a_m^s = P_n^s, \qquad n = 1, 2, \ldots, \tag{34.30}$$

where $A_{n,m}^s$ is given in Eq. (34.36) in the Appendix. Likewise, multiplying both sides of Eq. (34.29) by $\psi_n^c(X)$ and integrating over the domain yields

$$\sum_{m=1}^{\infty} A_{n,m}^c a_m^c = P_n^c, \qquad n = 1, 2, \ldots, \tag{34.31}$$

where $A_{n,m}^c$ is given in Eq. (34.39) in the Appendix. Note that, upon truncating the series in this case, we have to solve *two* linear matrix equations, one for the odd coefficients and another for the even coefficients.

The absolute values of the odd and even coefficients are plotted in Fig. 34.5a and in Fig. 34.5b, respectively. As before, the odd coefficients exhibit a decay rate of $\mathcal{O}(m^{-4})$, whereas the even coefficients diminish at a rate of $\mathcal{O}(m^{-5})$, as shown by the reference lines of slope −4.8 and −5.6, respectively. The closed-form analytical solution for the beam deflection based on CBT, i.e., the solution of the steady

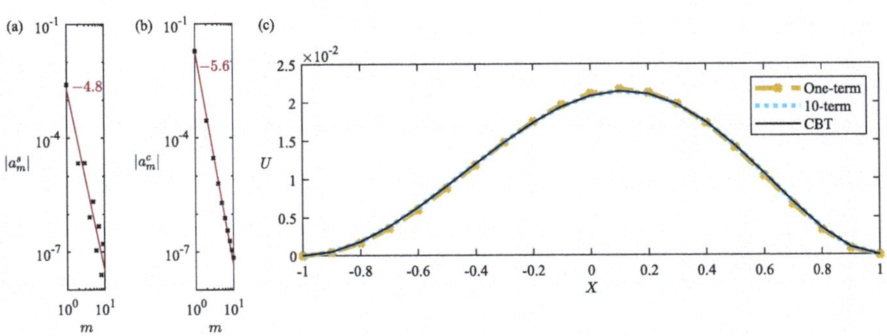

Fig. 34.5 Beam bending under the step load $P(X) = H(X)$, using both even and odd fourth-order eigenfunctions. **a** The absolute values of odd coefficients, $|a_m^s|$. **b** The absolute values of even coefficients, $|a_m^c|$. **c** Peridynamic beam theory (PBT) deflections calculated using a one-term truncation of the series (34.7a), a ten-term truncation of the series (34.7a), for $\Delta = 0.002$, and the classical beam theory (CBT) solution (34.32)

BVP (34.3) with the same load $P(X) = H(X)$, is easily found as

$$U_{\mathrm{CBT}}(X) = \frac{1}{96}\big[4H(X)X^3 - 3X^2 - 4X + 3\big]. \tag{34.32}$$

In Fig. 34.5c, we compare the beam deflection computed using only the first term (of both the even and odd functions, so two terms total) with that calculated using the first ten terms of the series (of both the even and odd functions, so twenty terms total) and the deflection from the attendant CBT solution. Once again, it can be seen that a single-term approximation can yield a satisfactory solution.

34.4.4 Comparison to the Fourier Sine Series Solution

Yang et al. [30] employed a Fourier sine series to determine the deflection of a peridynamic beam (34.10) subjected to the point load $P(X) = (3/2)\delta(3X + 1)$, where here $\delta(\cdot)$ is the Dirac delta, *not* the dimensional horizon. In this case,

$$P_n^s = \frac{3}{2}\left[\frac{\sin(\lambda_n^s/3)}{\sin(\lambda_n^s)} - \frac{\sinh(\lambda_n^s/3)}{\sinh(\lambda_n^s)}\right], \tag{34.33a}$$

$$P_n^c = \frac{3}{2}\left[\frac{\cosh(\lambda_n^c/3)}{\cosh(\lambda_n^c)} - \frac{\cos(\lambda_n^c/3)}{\cos(\lambda_n^c)}\right]. \tag{34.33b}$$

We made the solution from [30] dimensionless using our variables from Eq. (34.9) and transformed it to the domain $X \in [-1, +1]$. To align with the setup of Yang et al. [30] (Sect. 5.1 therein), we take $\Delta = 0.002$. In Fig. 34.6a, Yang et al.'s solution, truncated at 1000 terms, is plotted alongside our solution based on the fourth-order eigenfunctions series (34.7), truncated at 10 terms. Clearly, the two series solutions agree well.

The decay rate of the absolute values of the Fourier sine series' coefficients is $\mathcal{O}(m^{-3})$, as illustrated by a reference line in Fig. 34.6b. In contrast, $|a_m^s|$ and $|a_m^c|$ from our study, also plotted in Fig. 34.6b, have a convergence rate of $\mathcal{O}(m^{-4})$. Notably, while the Fourier series' coefficients decrease to 10^{-7} after 100 terms, the fourth-order eigenfunction series' coefficients achieve this magnitude after just 10 terms.

Furthermore, we calculated the root mean square error (RMSE) between the m-term sum truncation and 1000-term sum truncation for both Yang et al.'s and our series solutions. These results are shown in Fig. 34.6c. The reference lines indicate that the sine series' partial sums converge at a rate $\sim m^{-2.5}$, whereas the fourth-order eigenfunction series' partial sums converge one order faster at a rate $\sim m^{-3.5}$. This observation means the proposed fourth-order eigenfunction series solution (34.7) can achieve a comparable precision with far fewer terms (roughly 10 times fewer for large m).

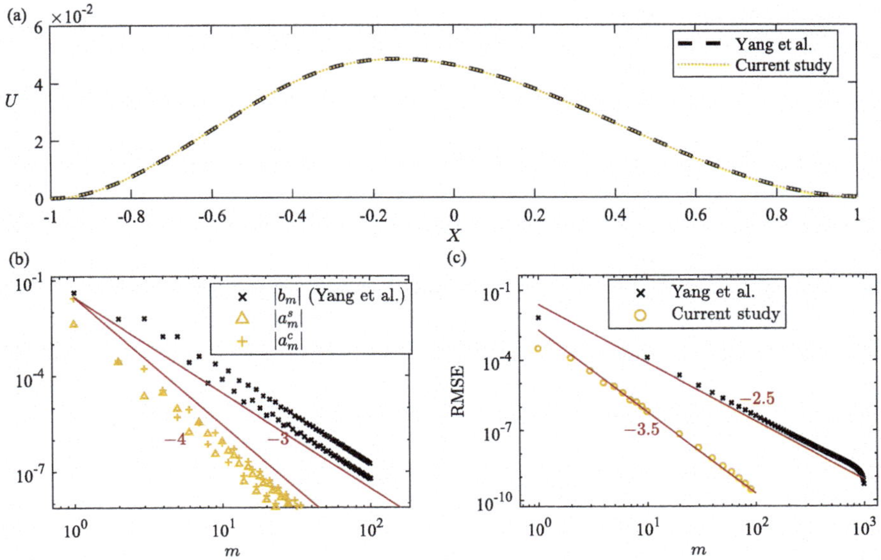

Fig. 34.6 **a** Comparison of different solutions for the peridynamic beam theory (PBT) deflection under the point-load $P(X) = (3/2)\delta(3X + 1)$: the sine series of Yang et al. [30] (dashed) and beam function series of this work (dotted). **b** The absolute values of the series' coefficients from each approach. **c** The root-mean-squared error (RMSE) as a function of the number of terms m in the truncated sum for each series

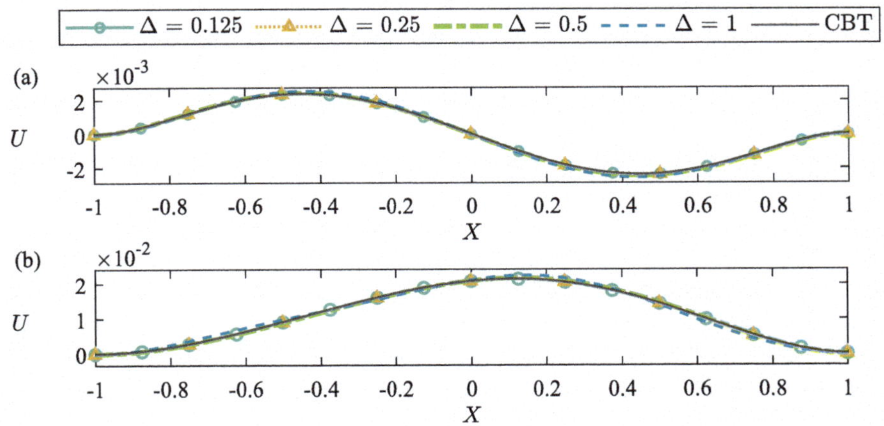

Fig. 34.7 The effect of varying the horizon on the deflection under the peridynamic beam theory (PBT), computed as a beam function series and truncated at 5 terms, and as compared to the corresponding solution under the classical beam theory (CBT), under the loads **a** $P(X) = -X$ and **b** $P(X) = H(X)$

34.4.5 Effect of the Peridynamic Horizon

Finally, we investigate the influence of the horizon size Δ on the solutions involving both a continuous and a discontinuous load, specifically $P(X) = -X$ and $P(X) = H(X)$, from Sects. 34.4.2 and 34.4.3, respectively. The results are shown in Fig. 34.7a, b, respectively. To provide a benchmark, the analytical solutions obtained from CBT, namely Eqs. (34.19) and (34.32), are plotted alongside for both cases. It is expected that as $\Delta \to 0^+$, the PBT should reduce to the CBT. Our analysis indeed reveals that diminishing Δ leads to convergence of the peridynamic solution to the classical one. Remarkably, even a relatively "large" value of $\Delta = 1$ provides a solution that looks similar to the CBT one, though some differences are apparent on zooming in. While it may be that such a large Δ value is unrealistic for some material being described by peridynamic theory, here we are just performing a mathematical exercise and exploring the outcomes.

34.5 Conclusion

In this work, we proposed a new analytical approach for solving for the displacement of clamped–clamped nonlocal generalizations of the Euler–Bernoulli beam under a peridynamic theory. Specifically, in Sect. 34.2, we introduced the "natural" fourth-order eigenfunctions that satisfy the clamped–clamped boundary conditions arising from the classical beam theory. Using these eigenfunctions, we derived solutions for static bending of the peridynamic beam introduced in Sect. 34.3 subject to several representative loads—including ones that engage only the "odd" or "even" eigenfunctions, as well as both.

In Sect. 34.4, we compared our series solutions based on the fourth-order eigenfunctions to those previously obtained as Fourier sine series. The series solutions based on fourth-order eigenfunctions converge much faster than the Fourier sine series. Specifically, for an example of an offset point load on the beam, we showed that the expansion coefficients of the series based on fourth-order eigenfunctions decay as the number of terms to the fourth power, as opposed to the third power for the sine series. Further, the partial sums of the series based on the fourth-order eigenfunction were found to decay one order faster than the sine series. Thus, despite the fact that the fourth-order eigenfunctions do not diagonalize the peridynamic operator (like sines do), the matrix that has to be inverted to obtain the coefficients of an accurate series solution is small.

The peridynamic model discussed in this work is essentially linear but nonlocal. An interesting generalization would be to consider *nonlinear* nonlocal interactions with a horizon, perhaps along the lines of the idea introduced by Porubov et al. [20] within the context of lattice dynamics. In future work, it would also be of interest to couple the peridynamic beam equation (34.8) with the flow of a thin film underneath it (a fluid–structure interaction problem), which may also lead to an IBVP that can be solved analytically using eigenfunction expansions [18].

Competing Interests The authors have no conflicts of interest to declare that are relevant to the content of this chapter.

Acknowledgements This research was supported by the U.S. National Science Foundation under grant No. CMMI-2245343. We dedicate this work to the 60th birthday of Prof. Alexey V. Porubov. We also thank Prof. Victor A. Eremeyev for his kind invitation to contribute to this volume and for his efforts in co-organizing it. I.C.C. further acknowledges many insightful discussions with Prof. Nectarios C. Papanicolaou on beam eigenfunctions.

Appendix

In this Appendix, we provide the formulas for evaluating the integrals in the peridynamic beam equation (34.11), when the displacement U is expanded in a series (34.7) in terms of the fourth-order eigenfunctions (34.5).

First, for the case of the odd function expansion in Eq. (34.12), we have

$$\frac{1}{\Delta}\int_{-\Delta}^{+\Delta}\frac{U(X+H)-U(X)}{H^2}\,dH = \frac{1}{\sqrt{2}}\sum_{m=1}^{\infty} a_m^s\left[\frac{\sinh(\lambda_m^s X)}{\sinh(\lambda_m^s)}\mathcal{I}_{1,m}^s(\Delta) - \frac{\sin(\lambda_m^s X)}{\sin(\lambda_m^s)}\mathcal{I}_{2,m}^s(\Delta)\right], \tag{34.34a}$$

and

$$\frac{1}{\Delta}\int_{-\Delta}^{+\Delta}\frac{U(X+\Xi+H)-U(X+\Xi)}{H^2}\,dH = \frac{1}{\sqrt{2}}\sum_{m=1}^{\infty} a_m^s\left[\frac{\sinh\big(\lambda_m^s(X+\Xi)\big)}{\sinh(\lambda_m^s)}\mathcal{I}_{1,n}^s(\Delta) - \frac{\sin\big(\lambda_m^s(X+\Xi)\big)}{\sin(\lambda_m^s)}\mathcal{I}_{2,m}^s(\Delta)\right], \tag{34.34b}$$

where

$$\mathcal{I}_{1,m}^s(\Delta) \stackrel{\text{def}}{=} \frac{1}{\Delta}\int_{-\Delta}^{+\Delta}\frac{1}{H^2}\big[\cosh(\lambda_m^s H)-1\big]\,dH \tag{34.35a}$$

$$= \frac{2}{\Delta^2}\big[\Delta\lambda_m^s\,\mathrm{Shi}(\Delta\lambda_m^s) - \cosh(\Delta\lambda_m^s) + 1\big] \to (\lambda_m^s)^2 \quad \text{as} \quad \Delta \to 0^+,$$

$$\mathcal{I}_{2,m}^s(\Delta) \stackrel{\text{def}}{=} \frac{1}{\Delta}\int_{-\Delta}^{+\Delta}\frac{1}{H^2}\big[\cos(\lambda_m^s H)-1\big]\,dH \tag{34.35b}$$

$$= -\frac{2}{\Delta^2}\big[\Delta\lambda_m^s \operatorname{Si}(\Delta\lambda_m^s) + \cos(\Delta\lambda_m^s) - 1\big] \to -(\lambda_m^s)^2 \quad \text{as} \quad \Delta \to 0^+,$$

and $\operatorname{Si}(z) \stackrel{\text{def}}{=} \int_0^z \zeta^{-1} \sin\zeta \, d\zeta$ and $\operatorname{Shi}(z) \stackrel{\text{def}}{=} \int_0^z \zeta^{-1} \sinh\zeta \, d\zeta$ are the sine and hyperbolic sine integrals, respectively [7, §6.2].

The integral defining $A_{n,m}^s$ in Eq. (34.15) can be evaluated to yield

$$A_{n,m}^s = \frac{1}{\sqrt{2}} \begin{cases} \big[\mathcal{I}_{1,m}^s(\Delta)\big]^2\big[\coth(\lambda_m^s)/\lambda_m^s + 1 - \coth^2(\lambda_m^s)\big] \\ \quad +\big[\mathcal{I}_{2,m}^s(\Delta)\big]^2\big[-\cot(\lambda_m^s)/\lambda_m^s + 1 + \cot^2(\lambda_m^s)\big], & \text{if } n = m, \\ 2\big[\mathcal{I}_{1,m}^s(\Delta)\big]^2\big\{\big[\lambda_n^s \coth(\lambda_n^s) - \lambda_m^s \coth(\lambda_m^s)\big]/\big[(\lambda_n^s)^2 - (\lambda_m^s)^2\big] \\ \quad -\big[\lambda_m^s \coth(\lambda_m^s) - \lambda_n^s \cot(\lambda_n^s)\big]/(\lambda_n^{s\,2} + \lambda_m^{s\,2})\big\} \\ +2\big[\mathcal{I}_{2,m}^s(\Delta)\big]^2\big\{\big[\lambda_m^s \cot(\lambda_m^s) - \lambda_n^s \cot(\lambda_n^s)\big]/\big[(\lambda_n^s)^2 - (\lambda_m^s)^2\big] \\ \quad -\big[\lambda_n^s \coth(\lambda_n^s) - \lambda_m^s \cot(\lambda_m^s)\big]/\big[(\lambda_n^s)^2 + (\lambda_m^s)^2\big]\big\}, & \text{if } n \neq m. \end{cases} \tag{34.36}$$

From the definitions in Eq. (34.35) and the eigenvalue relation in Eq. (34.5a), it is easy to see that $A_{n,m}^s = \sqrt{2}(\lambda_m^s)^4\delta_{n,m}$ as $\Delta \to 0^+$, as expected, noting that we multiplied through by $\sqrt{2}$ in Eq. (34.14).

Second, for the case of the even function expansion in Eq. (34.20), we have

$$\frac{1}{\Delta}\int_{-\Delta}^{+\Delta} \frac{U(X+H) - U(X)}{H^2}\,dH = \frac{1}{\sqrt{2}}\sum_{m=1}^{\infty} a_m^c \left[\frac{\cosh(\lambda_m^c X)}{\cosh(\lambda_m^c)}\mathcal{I}_{1,m}^c(\Delta) - \frac{\cos(\lambda_m^c X)}{\cos(\lambda_m^c)}\mathcal{I}_{2,m}^c(\Delta)\right], \tag{34.37a}$$

and

$$\frac{1}{\Delta}\int_{-\Delta}^{+\Delta} \frac{U(X+\Xi+H) - U(X+\Xi)}{H^2}\,dH = \frac{1}{\sqrt{2}}\sum_{m=1}^{\infty} a_m^c \left[\frac{\cosh\big(\lambda_m^c(X+\Xi)\big)}{\cosh(\lambda_m^c)}\mathcal{I}_{1,m}^c(\Delta) - \frac{\cos\big(\lambda_m^c(X+\Xi)\big)}{\cos(\lambda_m^c)}\mathcal{I}_{2,m}^c(\Delta)\right], \tag{34.37b}$$

where

$$\mathcal{I}_{1,m}^c(\Delta) \stackrel{\text{def}}{=} \frac{1}{\Delta}\int_{-\Delta}^{+\Delta} \frac{1}{H^2}\big[\cosh(\lambda_m^c H) - 1\big]\,dH, \tag{34.38a}$$

$$= \frac{2}{\Delta^2}\big[\Delta\lambda_m^c \operatorname{Shi}(\Delta\lambda_m^c) - \cosh(\Delta\lambda_m^c) + 1\big] \to (\lambda_m^c)^2 \quad \text{as} \quad \Delta \to 0^+,$$

$$\mathcal{I}^c_{2,m}(\Delta) \stackrel{\text{def}}{=} \frac{1}{\Delta}\int_{-\Delta}^{+\Delta} \frac{1}{H^2}\big[\cos(\lambda^c_m H) - 1\big]\,dH \tag{34.38b}$$
$$= -\frac{2}{\Delta^2}\big[\Delta\lambda^c_m \,\mathrm{Si}(\Delta\lambda^c_m) + \cos(\Delta\lambda^c_m) - 1\big] \to -(\lambda^c_m)^2 \quad \text{as} \quad \Delta \to 0^+.$$

The integral definining $A^c_{n,m}$ in Eq. (34.23) can be evaluated to yield

$$A^c_{n,m} = \frac{1}{\sqrt{2}}\begin{cases} \big[\mathcal{I}^c_{1,m}(\Delta)\big]^2\big[\tanh(\lambda^c_m)/\lambda^c_m + 1 - \tanh^2(\lambda^c_m)\big] \\ \quad +\big[\mathcal{I}^c_{2,m}(\Delta)\big]^2\big[\tan(\lambda^c_m)/\lambda^c_m + 1 + \tan^2(\lambda^c_m)\big], & \text{if } n = m, \\ 2\big[\mathcal{I}^c_{1,m}(\Delta)\big]^2\big\{\big[\lambda^c_n \tanh(\lambda^c_n) - \lambda^c_m \tanh(\lambda^c_m)\big]/\big[(\lambda^c_n)^2 - (\lambda^c_m)^2\big] \\ \quad -\big[\lambda^c_m \tanh(\lambda^c_m) + \lambda^c_n \tan(\lambda^c_n)\big]/\big[(\lambda^c_n)^2 + (\lambda^c_m)^2\big]\big\} \\ +2\big[\mathcal{I}^c_{2,m}(\Delta)\big]^2\big\{\big[-\lambda^c_m \tan(\lambda^c_m) + \lambda^c_n \tan(\lambda^c_n)\big]/\big[(\lambda^c_n)^2 - (\lambda^c_m)^2\big] \\ \quad -\big[\lambda^c_n \tanh(\lambda^c_n) + \lambda^c_m \tan(\lambda^c_m)\big]/\big[(\lambda^c_n)^2 + (\lambda^c_m)^2\big]\big\}, & \text{if } n \neq m. \end{cases} \tag{34.39}$$

From the definitions in Eq. (34.38) and the eigenvalue relation in Eq. (34.5b), it is easy to see that $A^c_{n,m} = \sqrt{2}(\lambda^c_m)^4\delta_{n,m}$ as $\Delta \to 0^+$, as expected, noting that we multiplied through by $\sqrt{2}$ in Eq. (34.22).

References

1. Altenbach H, Zhilin PA (2004) The theory of simple elastic shells. In: Kienzler R, Ott I, Altenbach H (eds) Theories of plates and shells: critical review and new applications. Springer, Berlin/Heidelberg, pp 1–12. https://doi.org/10.1007/978-3-540-39905-6_1
2. Altenbach H, Maugin GA, Erofeev V (2011) Mechanics of generalized continua, advanced structured materials, vol 7. Springer, Berlin/Heidelberg. https://doi.org/10.1007/978-3-642-19219-7
3. Challamel N (2018) Static and dynamic behaviour of nonlocal elastic bar using integral strain-based and peridynamic models. Comptes Rendus - Mecanique 346:320–335. https://doi.org/10.1016/j.crme.2017.12.014
4. Chandrasekhar S (1981) Hydrodynamic and hydromagnetic stability. Dover Publications, New York, NY
5. Coddington EA, Levinson N (1955) Theory of ordinary differential equations. McGraw-Hill, New York, NY
6. Diyaroglu C, Oterkus E, Oterkus S, Madenci E (2015) Peridynamics for bending of beams and plates with transverse shear deformation. Int J Solids Struct 69–70:152–168. https://doi.org/10.1016/j.ijsolstr.2015.04.040
7. DLMF (2024) In: Olver FWJ, Olde Daalhuis AB, Lozier DW, Schneider BI, Boisvert RF, Clark CW, Miller BR, Saunders BV, Cohl HS, McClain MA (eds) NIST digital library of mathematical functions. Release 1.2.0 of 2024-03-15. https://dlmf.nist.gov
8. Huang Z (2018) The singularity in the state-based peridynamic solution of uniaxial tension. Theor Appl Mech Lett 8:351–354. https://doi.org/10.1016/j.taml.2018.05.008
9. Kim KH, Bhalla APS, Griffith BE (2023) An immersed peridynamics model of fluid-structure interaction accounting for material damage and failure. J Comput Phys 493:112466. https://doi.org/10.1016/j.jcp.2023.112466
10. Kraus H (1967) Thin elastic shells. Wiley, New York, NY

11. Mikata Y (2012) Analytical solutions of peristatic and peridynamic problems for a 1D infinite rod. Int J Solids Struct 49:2887–2897. https://doi.org/10.1016/j.ijsolstr.2012.02.012
12. Mikata Y (2023) Analytical solutions of peristatics and peridynamics for 3D isotropic materials. Eur J Mech A/Solids 101:104978. https://doi.org/10.1016/j.euromechsol.2023.104978
13. Naumenko K, Eremeyev VA (2022) A non-linear direct peridynamics plate theory. Compos Struct 279:114728. https://doi.org/10.1016/j.compstruct.2021.114728
14. Nishawala VV, Ostoja-Starzewski M (2017) Peristatic solutions for finite one- and two-dimensional systems. Math Mech Solids 22:1639–1653. https://doi.org/10.1177/1081286516641180
15. O'Grady J, Foster J (2014) Peridynamic beams: a non-ordinary, state-based model. Int J Solids Struct 51:3177–3183. https://doi.org/10.1016/j.ijsolstr.2014.05.014
16. O'Grady J, Foster J (2014) Peridynamic plates and flat shells: a non-ordinary, state-based model. Int J Solids Struct 51:4572–4579. https://doi.org/10.1016/j.ijsolstr.2014.09.003
17. Papanicolaou NC (2003) A Galerkin spectral method for fourth-order boundary value problems. PhD thesis, University of Louisiana at Lafayette. https://www.proquest.com/dissertations-theses/galerkin-spectral-method-fourth-order-boundary/docview/305212891/se-2
18. Papanicolaou NC, Christov IC (2023) Orthonormal eigenfunction expansions for sixth-order boundary value problems. J Phys: Conf Ser 2675:012016. https://doi.org/10.1088/1742-6596/2675/1/012016. 2308.00673
19. Papanicolaou NC, Christov CI, Homsy GM (2009) Galerkin technique based on beam functions in application to the parametric instability of thermal convection in a vertical slot. Int J Num Meth Fluids 59:945–965. https://doi.org/10.1002/fld.1845
20. Porubov AV, Osokina AE, Michelitsch TM (2018) Nonlocal approach to square lattice dynamics. In: Altenbach H, Pouget J, Rousseau M, Collet B, Michelitsch T (eds) Generalized models and non-classical approaches in complex materials 1. Springer International Publishing, Cham, Switzerland, pp 641–654. https://doi.org/10.1007/978-3-319-72440-9_34
21. Silling SA (2000) Reformulation of elasticity theory for discontinuities and long-range forces. J Mech Phys Solids 48:175–209. https://doi.org/10.1016/S0022-5096(99)00029-0
22. Silling SA, Bobaru F (2005) Peridynamic modeling of membranes and fibers. Int J Non-Linear Mech 40:395–409. https://doi.org/10.1016/j.ijnonlinmec.2004.08.004
23. Silling SA, Lehoucq RB (2010) Peridynamic theory of solid mechanics. Adv Appl Mech 44:73–168. https://doi.org/10.1016/S0065-2156(10)44002-8
24. Strutt JW (1877) The Theory of Sound, vol 1. Macmillan and Co., London. http://books.google.com/books?id=GyI5AAAAMAAJ&oe=UTF-8
25. Taylor M, Steigmann DJ (2015) A two-dimensional peridynamic model for thin plates. Math Mech Solids 20:998–1010. https://doi.org/10.1177/1081286513512925
26. Timoshenko S, Woinowsky-Krieger S (1959) Theory of plates and shells, 2nd edn. McGraw-Hill, New York, NY
27. Yang Z, Oterkus E, Oterkus S (2020) A state-based peridynamic formulation for functionally graded Euler-Bernoulli beams. Comput Model Eng Sci 124:527–544. https://doi.org/10.32604/cmes.2020.010804
28. Yang Z, Naumenko K, Altenbach H, Ma CC, Oterkus E, Oterkus S (2022) Some analytical solutions to peridynamic beam equations. Z Angew Math Mech (ZAMM) 102:e202200132. https://doi.org/10.1002/zamm.202200132
29. Yang Z, Naumenko K, Ma CC, Altenbach H, Oterkus E, Oterkus S (2022b) Some closed form series solutions to peridynamic plate equations. Mech Res Commun 126:104000. https://dx.doi.org/j.mechrescom.2022.104000
30. Yang Z, Naumenko K, Ma CC, Chen Y (2023) Closed-form analytical solutions for the deflection of elastic beams in a peridynamic framework. Appl Sci 13:10025. https://doi.org/10.3390/app131810025
31. Yang Z, Naumenko K, Ma CC, Oterkus E, Oterkus S (2023) Peridynamic analysis of curved elastic beams. Eur J Mech A/Solids 101:105075. https://doi.org/10.1016/j.euromechsol.2023.105075

Chapter 35
Interfacial Wave on a Composite Membrane Resting on a Winkler Foundation

Peter Wootton, Danila Prikazchikov, and Julius Kaplunov

Abstract Time-harmonic interfacial waves on a composite elastic membrane supported by a weak Winkler foundation are studied. The membrane is comprised of two identical semi-infinite parts connected by a narrow strip having distinct material properties. The associated dispersion relation is derived, turning out to be a relatively simple one compared to those known for interfacial waves. The existence of the sought for localised solution over the long-wave low-frequency domain is addressed, revealing a dependence solely on the ratios of material parameters. The effect of elastic foundation on the lowest cut-off frequency is investigated.

35.1 Introduction

Analysis of interfacial waves in elastic media usually assumes tedious algebraic manipulations and non-trivial existence conditions, see the original contribution by Stoneley [1] for two joint isotropic half-spaces, with existence analysed by Scholte [2]. The effects of more sophisticated material properties lead to even more technical treatment, see, e.g. [3, 4]. An important step forward was provided by implementation of Stroh formalism, see [5, 6]. Another notable contribution to the general theory of surface and interfacial waves was proposed by Chadwick [7], who suggested a representation of the eigensolution in terms of a single plane harmonic function, see also a more recent generalisation of this approach to 3D, see [8]. We also mention specialised asymptotic hyperbolic-elliptic models for surface and interfacial waves, see [9] and references therein.

P. Wootton (✉) · D. Prikazchikov · J. Kaplunov
School of Computer Science and Mathematics, Keele, Staffordshire, ST5 5BG, UK
e-mail: p.t.wootton@keele.ac.uk

D. Prikazchikov
e-mail: j.kaplunov@keele.ac.uk

J. Kaplunov
e-mail: d.prikazchikov@keele.ac.uk

H. Altenbach and V. A. Eremeyev (eds.), *Dynamics of Discrete and Continuum Structures and Media*, Advanced Structured Materials 221,
https://doi.org/10.1007/978-3-031-75626-9_35

Analysis of bending interfacial waves in Kirchhoff plate theory, see [10], is not much easier than the conventional Stoneley wave, due to rather sophisticated dispersive behaviour, and equally non-trivial existence conditions. Interfacial waves in thin elastic shells provide a more advanced example, combining the features of in-plane and bending waves in plates. Remarkably, there are special low-frequency interfacial waves, due to the effect of the curvature, see [11, 12], and also [13, 14].

In this paper, we are attempting to present a simpler example of an interfacial structural wave, enabling explicit treatment. We consider a composite membrane resting on the Winkler foundation modelling, for example, a thin film on a substrate. A string on a Winkler foundation has been earlier studied for illustrating an important class of trapped modes, see [15, 16] and references therein. The membrane analysed below is comprised of two semi-infinite parts, connected through by a membrane of finite small width, with different material parameters. First, we mention that, as known from [17, 18], a membrane is associated with a leading order low-frequency approximation for a thin pre-stressed elastic layer. Another point which is worth noting is that 2D plate models in case of elastic contact with the Winkler foundation are valid only if the ratio of foundation and plate stiffness is small, see [19–21]. As a result, in the present paper we also assume that the properly normalised ratio of the tension in the membrane and the Winkler stiffness constant is a small parameter. In this case, we may expect that the interfacial wave of interest may have a cut-off frequency similarly to the bending edge wave on a Kirchhoff plate supported by elastic foundations, see also [22, 23].

In what follows, we derive a relatively simple dispersion relation for the sought for localised wave. In contrast to the current state of art, the associated existence conditions take a straightforward explicit form. Two-term asymptotic expansion of the dispersion relation is obtained, valid only when the above-mentioned existence conditions are satisfied. Numerical examples, illustrating the effect of problem parameters on the analysed wave, are presented.

35.2 Formulation of the Problem and Dispersion Relation

Consider an infinite membrane with an infinite rectangular section of small width $2h$ replaced by a similar membrane but with distinct mass density and internal tension. If this membrane rests on the surface of a Winkler foundation then we obtain governing equations of the out-of-plane displacement inside the strip u_1 and outside the strip u_2 respectively, as (Fig. 35.1)

$$T_1(u_{1,11} + u_{1,22}) - \rho_1 u_{1,tt} = \beta u_1 \tag{35.1}$$

$$T_2(u_{2,11} + u_{2,22}) - \rho_2 u_{2,tt} = \beta u_2 \tag{35.2}$$

where at $|x_2| = h$ there are junction conditions

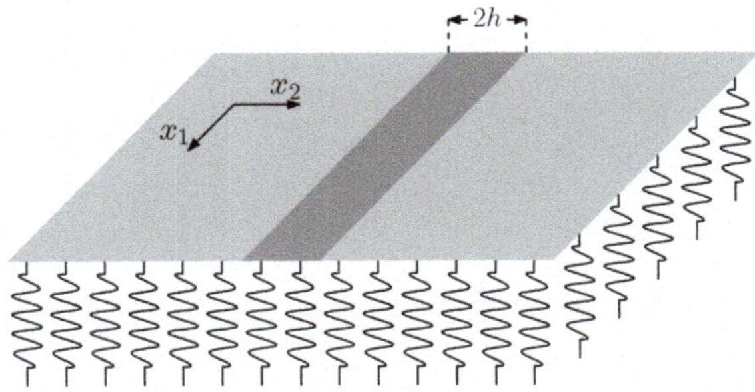

Fig. 35.1 Schematic of a Winkler foundation supporting an infinite elastic membrane with vertical displacement u_2 and an inserted infinite rectangular strip of small width $2h$ filled with a similar membrane with distinct material parameters and displacement u_1

$$u_1 = u_2, \qquad T_1 u_{1,2} = T_2 u_{2,2}. \tag{35.3}$$

Define non-dimensional angular frequency and wavenumber for the travelling wave

$$\frac{\omega h}{c_2} = \Omega, \qquad kh = K \tag{35.4}$$

along with non-dimensional material parameters

$$B = \frac{\beta h^2}{T_2}, \qquad T = \frac{T_1}{T_2}, \qquad P = \frac{\rho_1}{\rho_2}, \qquad C = \frac{c_1}{c_2} = \sqrt{\frac{T}{P}}. \tag{35.5}$$

Let us consider symmetric interfacial waves with the profile along the strip membrane ($-\infty < x_1 < \infty$, $-h \le x_2 \le h$), given by, see schematic in Fig. 35.2,

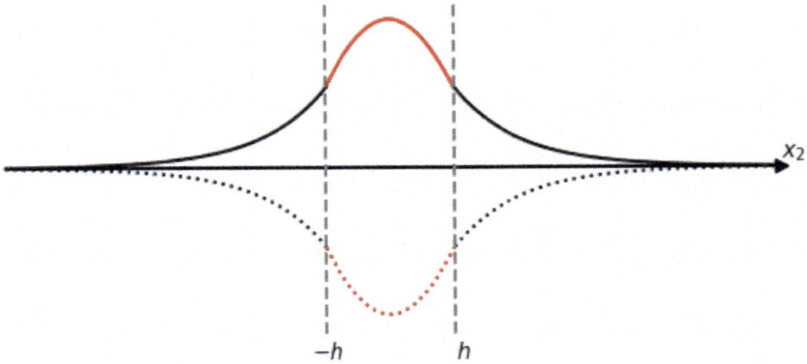

Fig. 35.2 Wave profile of vertical displacement along the x_2 direction for symmetric wave modes for the region of the strip $|x_2| < h$ and exponential decay outside of the strip $|x_2| > h$

$$u_1^{(s)} = A_1 \exp(i(kx_1 - \omega t)) \cos \mu x_2, \tag{35.6}$$

where

$$\mu^2 = \frac{\omega^2}{c_1^2} - \frac{\beta}{T_1} - k^2, \tag{35.7}$$

or, in dimensionless form,

$$(\mu h)^2 = \frac{\Omega^2}{C^2} - \frac{B}{T} - K^2. \tag{35.8}$$

Since we are seeking a wave which is localised in the vicinity of the strip and decays outside, then for the exterior of the strip we can assume the solutions

$$u_2 = A_2 \exp(i(kx_1 - \omega t)) \exp(-\lambda |x_2|) \tag{35.9}$$

where

$$\lambda^2 = k^2 + \frac{\beta}{T_2} - \frac{\omega^2}{c_2^2} \tag{35.10}$$

or

$$(\lambda h)^2 = K^2 + B - \Omega^2. \tag{35.11}$$

Hence, since $\lambda > 0$, we deduce $\Omega^2 - K^2 < B$ assuming that wavenumber K is real. Then, on applying boundary conditions at $|x_2| = h$, we obtain

$$A_1 \cos \mu h = A_2 \exp(-\lambda h) \quad ; \quad A_1 \mu \sin \mu h = A_2 \lambda \exp(-\lambda h) \tag{35.12}$$

which results in the dispersion relation

$$T \tan(\mu h) = \frac{\lambda}{\mu} \tag{35.13}$$

admitting an infinite succession of modes. Prior to analysis of this dispersion relation, we discuss the problem parameters. First of all, we assume that the dimensionless quantity B associated with the relative stiffness of the foundation is small ($B \ll 1$). Indeed, this assumption underlies the validity of the model above, see also asymptotic considerations in [19–21], concerned with justification of existing structural models in presence of a Winkler foundation. We also suppose that $\Omega \ll 1$ and $K \ll 1$, operating in the long-wave low-frequency regime. This is in line with the main interest to interfacial wave generated by the effect of a narrow membrane strip. In addition, we have two independent dimensionless parameters T and P, expressing the ratios of tensions and density, respectively. Both of them are assumed to be of order unity. Below we show that the relation between them influences drastically the sought for wave.

35.3 Parametric Analysis

Let us first obtain a shortened form of the dispersion relation (35.13) under the assumptions on the problem parameters made in the previous section. Keeping the leading order term at the Taylor expansion for tan, we have

$$\sqrt{K^2 - \Omega^2 + B} = T\left(\frac{\Omega^2}{C^2} - K^2 - \frac{B}{T}\right), \tag{35.14}$$

or, equivalently,

$$\sqrt{K^2 - \Omega^2 + B} = P\Omega^2 - TK^2 - B. \tag{35.15}$$

At $K = 0$ the associated cut-off frequency Ω_0 appears to exist only for $P > 1$ and is given by the asymptotic expansion in the small parameter B

$$\Omega_0^2 = B - (P-1)^2 B^2 + \dots . \tag{35.16}$$

Next, we investigate (35.15) at $\Omega \sim \Omega_0$, assuming that $\Omega^2 = \Omega_*^2 B$, where Ω_* is of order unity. In this case we readily obtain from (35.15)

$$K^2 = (\Omega_*^2 - 1)B + \left(\Omega_*^2(P-T) + T - 1\right)^2 B^2 + \dots . \tag{35.17}$$

which corresponds to the interfacial wave of interest, provided that

$$\Omega_*^2 - 1 > 0, \qquad \Omega_*^2(P-T) + T - 1 > 0. \tag{35.18}$$

These conditions follow from the aposteriori inspection of the formulae in this section. Finally, we arrive at the classification of all the possible low-frequency scenarios, see Fig. 35.3.

(i) $1 < P,\ T < P$
Formula (35.17) is valid for $\Omega > \Omega_0,\ K > 0$.

(ii) $T < P < 1$
Formula (35.17) is valid for $\Omega > \Omega_1,\ K > K_1$, with

$$\Omega_1^2 = \frac{1-T}{P-T} B > \Omega_0^2, \quad K_1^2 = \frac{1-P}{P-T} B > 0. \tag{35.19}$$

(iii) $P < T < 1$
Formula (35.17) is not valid, i.e. the interfacial wave does not exist.

(iv) $1 < P < T$
Formula (35.17) is valid for $\Omega_0^2 < \Omega^2 < \Omega_1^2$ and $0 < K^2 < K_1^2$.

At $\Omega_0^2 \ll \Omega^2 \ll 1$ the effect of the foundation does not appear at the leading order. As a result, we have from (35.15) the estimate $K^2 \sim \Omega^2$ valid at $P > T$.

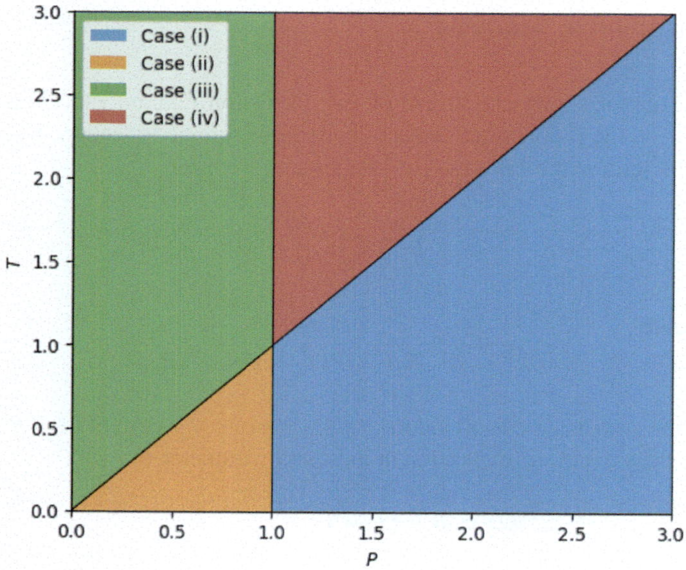

Fig. 35.3 Visualisation of the required values of T and P for the interfacial wave cases (i)–(iv)

35.4 Dispersion Curves

Below are shown a selection of dispersion curves for different choices of material parameters for the two membranes and the foundation, illustrating Cases (i)–(iv) detailed above. For these curves we consider only the long-wave and low-frequency regime (Figs. 35.4, 35.5, 35.6 and 35.7).

From each can be seen the permissible regimes described above; these regimes can be qualitatively described based on the behaviour of the curves $\lambda = 0$ and $\mu = 0$. It is clear from these plots that the permissible region is bounded by $\mu = 0$ above and $\lambda = 0$ below and as there is at most one intersection between these curves, there can be only one permissible region for the wave modes for a given set of material parameters.

For the selection of parameters for $B = 0.1$ each of these four regimes can be seen. For $T = 1.6$, $P = 0.8$ there is no possible wave mode as there is no intersection and the $\lambda = 0$ curve is above the $\mu = 0$ curve. Conversely for $T = 0.4$, $P = 1.4$ there is a wave mode which has a cut-off frequency for $K = 0$ and no upper frequency bound. The two-term approximation of the dispersion curve is close to numerical solution, particularly at the intersection with the dispersion curve with the Ω axis at $K = 0$. For $T = 0.4$, $P = 0.8$ there is an intersection of the curves so there is a cut-off frequency and wavenumber and an increasing dispersion curve with no upper bound. Finally for $T = 1.6$, $P = 1.4$ there is an intersection so the permissible region has an upper frequency and wavenumber band, so this solution represents a frequency pass-band for interfacial waves.

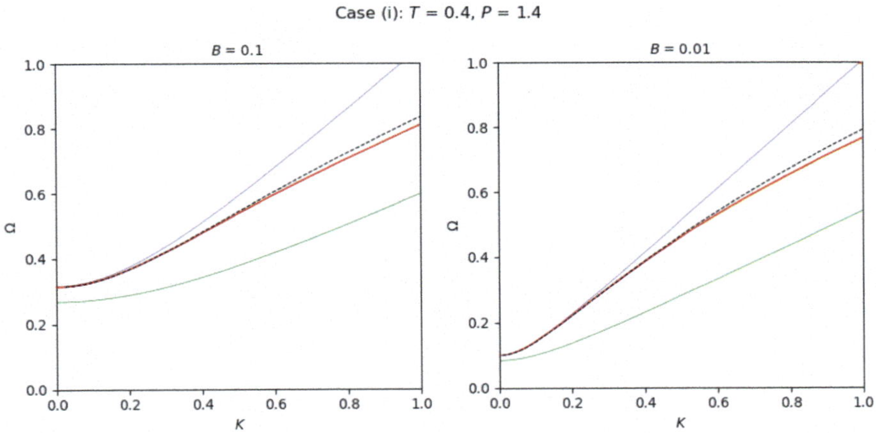

Fig. 35.4 A set of dispersion curves with material parameter contrast $T = 0.4$ and $P = 1.4$ for foundation relative stiffness of $B = 0.1$ (left) and $B = 0.01$ (right). The red line represents the symmetric wave mode solution and the purple and green lines respectively represent the curves for $\lambda = 0$ and $\mu = 0$. The black dashed line is the two-term approximation (35.17)

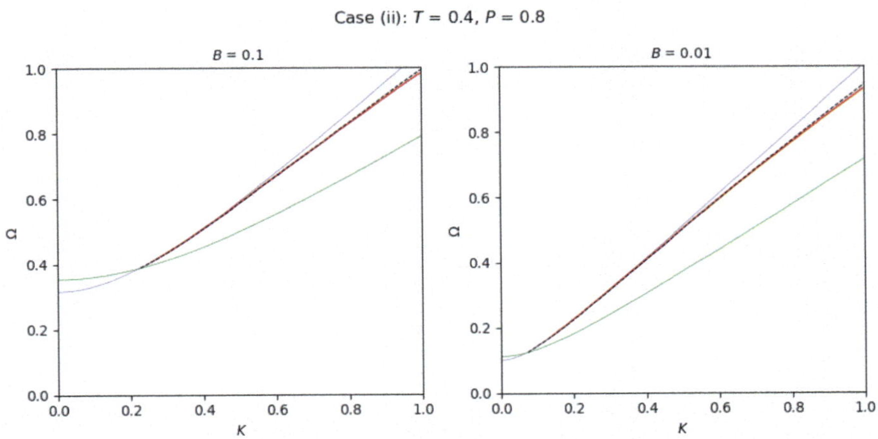

Fig. 35.5 A set of dispersion curves with material parameter contrast $T = 0.4$ and $P = 0.8$ for foundation relative stiffness of $B = 0.1$ (left) and $B = 0.01$ (right). The red line represents the symmetric wave mode solution and the purple and green lines respectively represent the curves for $\lambda = 0$ and $\mu = 0$. The black dashed line is the two-term approximation (35.17)

The dispersion curves for $B = 0.01$ are shown for the same parameters as above. Since B represents the contribution of the foundation, as B tends to 0 the waves will behave as if there is no foundation, and the dispersion curves will pass through the origin. Hence, as B decreases to 0 so will the cut-off frequency. For parameter choices where there is an intersection between the curves $\lambda = 0$ and $\mu = 0$, this intersection is directly proportional to B so similarly the cut-off wavenumber will tend to 0.

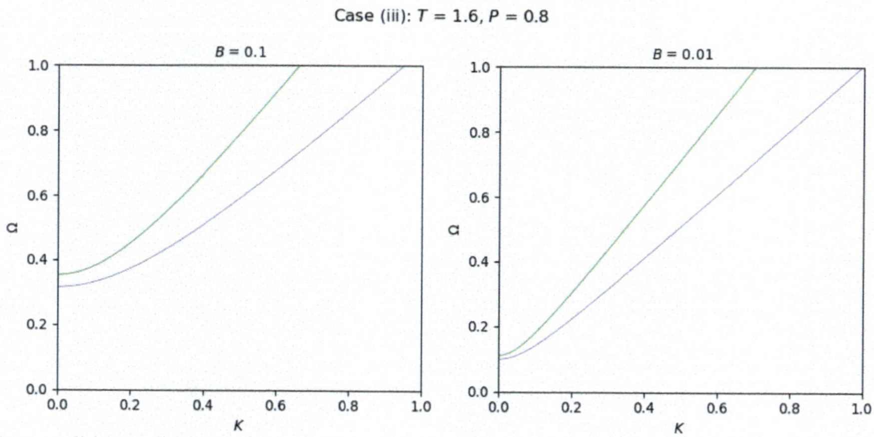

Fig. 35.6 A set of dispersion curves with material parameter contrast $T = 1.6$ and $P = 0.8$ for foundation relative stiffness of $B = 0.1$ (left) and $B = 0.01$ (right). The red line represents the symmetric wave mode solution and the purple and green lines respectively represent the curves for $\lambda = 0$ and $\mu = 0$. The black dashed line is the two-term approximation (35.17)

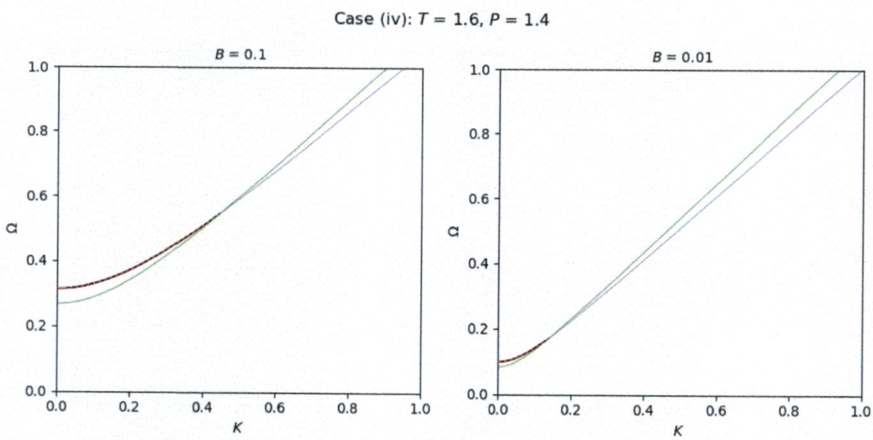

Fig. 35.7 A set of dispersion curves with material parameter contrast $T = 1.6$ and $P = 1.4$ for foundation relative stiffness of $B = 0.1$ (left) and $B = 0.01$ (right). The red line represents the symmetric wave mode solution and the purple and green lines respectively represent the curves for $\lambda = 0$ and $\mu = 0$. The black dashed line is the two-term approximation (35.17)

35.5 Conclusions

Low-frequency interfacial wave on a composite membrane resting on a relatively soft Winkler foundation is considered. Explicit existence conditions are given in terms of the problem parameters. Two-term asymptotic formula (35.16) for the cut-off frequency caused by the effect of foundation is obtained. All scenarios for the sought for wave manifestation are elucidated in Sect. 35.3. Asymptotic approximation for the dispersion curve (35.17) is compared with the numerical solution of the full dispersion relation (35.15).

The developed approach can be extended to the anti-symmetric interfacial wave, as well as to more sophisticated models of the elastic foundation, e.g. see [23].

References

1. Stoneley R (1924) Elastic waves at the surface of separation of two solids. Proc Roy Soc Lond. Ser A, Contain Papers Math Phys Character 106(738):416–428
2. Scholte JG (1947) The range of existence of Rayleigh and Stoneley waves. Geophys Suppl Mon Notices Roy Astronom Soc 5(5):120–126
3. Dowaikh MA, Ogden RW (1991) Interfacial waves and deformations in pre-stressed elastic media. Proc Roy Soc Lond. Ser A: Math Phys Sci 433(1888):313–328
4. Destrade M (2007) Interface waves in pre-stressed incompressible solids. Waves in nonlinear pre-stressed materials. Springer Vienna, Vienna, pp 63–102
5. Barnett DM, Lothe J, Gavazza SD, Musgrave MJP (1985) Considerations of the existence of interfacial (Stoneley) waves in bonded anisotropic elastic half-spaces. Proc Roy Soc Lond. A. Math Phys Sci 402(1822):153–166
6. Barnett DM (2000) Bulk, surface, and interfacial waves in anisotropic linear elastic solids. Int J Solids Struct 37:45–54
7. Chadwick P (1976) Surface and interfacial waves of arbitrary form in isotropic elastic media. J Elast 6(1):73–80
8. Kiselev AP, Parker DF (2010) Omni-directional Rayleigh, Stoneley and Schölte waves with general time dependence. Proc Roy Soc A: Math Phys Eng Sci 466(2120):2241–2258
9. Kaplunov J, Prikazchikov DA (2017) Asymptotic theory for Rayleigh and Rayleigh-type waves. Adv Appl Mech 50:1–106
10. Zilbergleit AS, Suslova IB (1983) Contact flexural waves in thin plates. Sov Phys Acoust 29(2):108–111
11. Kaplunov JD, Wilde MV (2000) Edge and interfacial vibrations in elastic shells of revolution. Zeitschrift für angewandte Mathematik und Physik ZAMP 51:530–549
12. Kaplunov JD, Wilde MV (2002) Free interfacial vibrations in cylindrical shells. J Acoust Soc Am 111(6):2692–2704
13. Ghulghazaryan GR, Ghulghazaryan LG, Mikhasev GI (2018) Free interfacial and boundary vibrations of thin elastic circular cylindrical shells with free ends. Izvestia NAS Armenia, Mech 71(1):61–78
14. Ghulghazaryan G, Ghulghazaryan L (2023) Free Localized Vibrations of a Thin Elastic Composite Panel. Mechanics of High-Contrast Elastic Solids: Contributions from Euromech Colloquium 626. Springer International Publishing, Cham, pp 91–118
15. Kaplunov JD, Sorokin SV (1995) A simple example of a trapped mode in an unbounded waveguide. J Acoust Soc Am 97(6):3898–3899
16. Shishkina EV, Gavrilov SN, Mochalova YA (2019) Non-stationary localized oscillations of an infinite Bernoulli-Euler beam lying on the Winkler foundation with a point elastic inhomogeneity of time-varying stiffness. J Sound Vib 440:174–185
17. Kaplunov JD, Nolde EV, Rogerson GA (2000) A low-frequency model for dynamic motion in pre-stressed incompressible elastic structures. Proc Roy Soc Lond. Ser A: Math Phys Eng Sci 456(2003):2589–2610
18. Pichugin AV, Rogerson GA (2002) An asymptotic membrane-like theory for long-wave motion in a pre-stressed elastic plate. Proc Roy Soc Lond. Ser A: Math Phys Eng Sci 458(2022):1447–1468

19. Erbaş B, Kaplunov J, Nobili A, Kiliç G (2018) Dispersion of elastic waves in a layer interacting with a Winkler foundation. J Acoust Soc Am 144(5):2918–2925
20. Erbaş B, Kaplunov J, Elishakoff I (2022) Asymptotic derivation of a refined equation for an elastic beam resting on a Winkler foundation. Math Mech Solids 27(9):1638–1648
21. Erbaş B, Kaplunov J, Kiliç G (2022) Asymptotic analysis of 3D dynamic equations in linear elasticity for a thin layer resting on a Winkler foundation. IMA J Appl Math 87(5):707–721
22. Kaplunov J, Prikazchikov DA, Rogerson GA, Lashab MI (2014) The edge wave on an elastically supported Kirchhoff plate. J Acoust Soc Am 136(4):1487–1490
23. Kaplunov J, Nobili A (2017) The edge waves on a Kirchhoff plate bilaterally supported by a two-parameter elastic foundation. J Vib Control 23(12):2014–2022

Chapter 36
Torsion, Inflation, and Tension of Hollow Elastic Cylinders with Distributed Dislocations Under Large Deformations

Anastasia A. Zelenina and Leonid M. Zubov

Abstract We discuss the influence of distributed dislocations on the nonlinear deformation of a hollow circular elastic cylinder. Hydrostatic pressures are set on the inner and outer surfaces of the cylinder. Longitudinal force and torque act on the ends of the tube. The system of equations of the nonlinear continuum theory of dislocations is solved by the semi-inverse method. For the axisymmetric distribution of edge and screw dislocations in the case of an isotropic elastic material, the initial problem is reduced to a nonlinear boundary value problem for a system of ordinary differential equations. In the framework of the isotropic incompressible material model, several exact quadrature solutions of the problem of cylinder equilibrium with dislocations under large deformations are founded. These solutions are universal in the class of isotropic elastic bodies. In a number of practically important cases, we have obtained simple explicit formulas. These formulas describe the effect of dislocations on the nonlinear resistance of the cylinder to external loads.

36.1 Introduction

Dislocations are a common and important element of the solids structure. The effect of dislocations on the nonlinear deformation of structures can be significant for many modern materials [1, 2]. Dislocations are associated with such phenomena as plastic flow, internal friction, fatigue and fracture, creep, etc. Using the dislocation distribution, we can model various other defects in solids. Intensive reproduction of defects is observed, for example, during radiation exposure of nuclear power structures. In this work, we apply the nonlinear continuum dislocations theory, constructed in [3–6]. The equations of this theory are quite complicated. So, only a small number of exact

A. A. Zelenina (✉) · L. M. Zubov
Southern Federal University, Milchakova Ave. 8a, 344092 Rostov-on-Don, Russia
e-mail: a.zelenina@gmail.com

L. M. Zubov
e-mail: zubovl@yandex.ru

H. Altenbach and V. A. Eremeyev (eds.), *Dynamics of Discrete and Continuum Structures and Media*, Advanced Structured Materials 221,
https://doi.org/10.1007/978-3-031-75626-9_36

solutions to problems of nonlinear deformation of elastic bodies with distributed dislocations are currently known [7–13].

Particularly, in [7, 9, 12, 14, 15] a number of problems on large deformations of a solid circular cylinder with dislocations have been solved. In this paper, we find new exact solutions to the nonlinear dislocation theory for a hollow circular cylinder with different distributions of screw and edge dislocations. With the help of these solutions, quantitative and qualitative effects due to distributed dislocations in solids are discovered.

36.2 Basic Relations

The system of equations describing large static deformations of a nonlinearly elastic body consists [16] of equilibrium equations for the stresses

$$\mathrm{div}\mathbf{P} = 0 \tag{36.1}$$

constitutive relations

$$\mathbf{P} = \frac{dW}{d\mathbf{F}}, \quad W = W\left(\mathbf{F}\cdot\mathbf{F}^T\right) \tag{36.2}$$

and the geometric relations

$$\mathbf{F} = \mathrm{grad}\mathbf{R} \tag{36.3}$$

Here **P** is the non-symmetric Piola stress tensor called also the first Piola-Kirchhoff tensor, **R** is the radius-vector of an elastic body point in a deformed state, **F** is the deformation gradient, W is the deformation specific energy. In (36.1) it is assumed that it is possible to neglect the action of mass forces. The differential gradient, rotor, and divergence operators in the reference configuration are introduced as follows [16],

$$\mathrm{grad}\mathbf{K} = \mathbf{r}^s \otimes \frac{\partial \mathbf{K}}{\partial q^s}, \quad \mathrm{rot}\mathbf{K} = \mathbf{r}^s \times \frac{\partial \mathbf{K}}{\partial q^s}$$
$$\mathrm{div}\mathbf{K} = \mathbf{r}^s \cdot \frac{\partial \mathbf{K}}{\partial q^s}, \quad \mathbf{r}^s = \mathbf{i}_k \frac{\partial q^s}{\partial x_k}, \quad s, k = 1, 2, 3 \tag{36.4}$$

where $\mathbf{i}_k$ is the constant orts of Cartesian coordinates x_k of the body reference configuration, q^s is the some curvilinear Lagrangian coordinates, **K** is the arbitrary differentiable tensor field.

If there is a continuous distribution of translational dislocations in an elastic body, then the vector field $\mathbf{R}(q^s)$ does not exist. Just as there is no displacement field $\mathbf{u} = \mathbf{R} - x_k\mathbf{i}_k$. In this case, the relation (36.3) is replaced by the incompatibility equation

$$\mathrm{rot}\mathbf{F} = \boldsymbol{\alpha} \tag{36.5}$$

where $\boldsymbol{\alpha}$ is the dislocation density tensor, which must satisfy the solenoidality equation

$$\mathrm{div}\boldsymbol{\alpha} = 0 \tag{36.6}$$

The derivation of the incompatibility Eq. (36.5) and the introduction of the dislocation density tensor are described in [5, 6, 14, 17–19]. The concept of the dislocation density tensor arises as a result of the limit transition from a discrete set of dislocations to their continuous distribution. From a physical point of view, the dislocation density can be interpreted as a second rank tensor field $\boldsymbol{\alpha}$, the flow of which through any surface gives the total Burgers vector of dislocations crossing this surface [18–20].

In the presence of distributed dislocations in the body, the tensor $\mathbf{F}$ is called the distortion tensor.

Note that the first-order incompatibility Eq. (36.5) can be transformed [21] to nonlinear second-order incompatibility equations with respect to the components of the metric tensor $\mathbf{G} = \mathbf{F} \cdot \mathbf{F}^T$. These equations form the basis of the classical continuum theory of dislocations [3–5] and, from a differential-geometric point of view, establish the relationship of the dislocation density tensor with the Cartan torsion of the metric connectivity space. The second-order incompatibility equations are much more complicated than equations (36.5) and will not be used in this work.

The elastic material will be considered isotropic in the future. The constitutive relations of such a material have the form [16]

$$\mathbf{P} = 2\left(\frac{\partial W}{\partial I_1} + I_1\frac{\partial W}{\partial I_2}\right)\mathbf{F} - 2\frac{\partial W}{\partial I_2}\mathbf{G} \cdot \mathbf{F} + 2I_3\frac{\partial W}{\partial I_3}\mathbf{F}^{-T} \tag{36.7}$$

$$\begin{gathered} \mathbf{G} = \mathbf{F} \cdot \mathbf{F}^T, \quad \mathbf{F}^{-T} = \left(\mathbf{F}^{-1}\right)^T = \left(\mathbf{F}^T\right)^{-1} \\ I_1 = \mathrm{tr}\mathbf{G}, \quad I_2 = \frac{1}{2}\left(\mathrm{tr}^2\mathbf{G} - \mathrm{tr}\mathbf{G}^2\right), \quad I_3 = \det\mathbf{G} = \det{}^2\mathbf{F} \end{gathered} \tag{36.8}$$

Here $\mathbf{G}$ is the metric tensor, also called the Cauchy strain measure, $W(I_1, I_2, I_3)$ is the specific elastic energy of an isotropic body, I_1, I_2, I_3 is the deformation invariants.

36.3 Dislocations Distribution in a Hollow Circular Cylinder

Consider an elastic body in the form of a hollow circular cylinder, the axis of which has the direction of the ort $\mathbf{i}_3$. Let's introduce cylindrical coordinates in the reference configuration (i.e., undeformed state) $q^1 = r$, $q^2 = \varphi$, $q^3 = z$ according to the formulas $x_1 = r\cos\varphi$, $x_2 = r\sin\varphi$, $x_3 = z$. The area occupied by the body is given by the inequality $r_1 \leq r \leq r_0, 0 >\leq \varphi \leq 2\pi, 0 \leq z \leq l$. The unit vectors tangent to the coordinate lines are denoted by $\mathbf{e}_r$, $\mathbf{e}_\varphi$, $\mathbf{i}_3$. The relations are right

$$\mathbf{e}_r = \mathbf{i}_1 \cos\varphi + \mathbf{i}_2 \sin\varphi, \quad \mathbf{e}_\varphi = -\mathbf{i}_1 \sin\varphi + \mathbf{i}_2 \cos\varphi \tag{36.9}$$

As is known [14, 16], in the absence of dislocations, the final deformations of inflating, torsion, and tension of the cylinder are described by the following transformation

$$R = R(r), \quad \Phi = \varphi + \psi z, \quad Z = \lambda z \tag{36.10}$$

Here R, Φ, z are cylindrical coordinates of cylinder points after deformation (i.e., Eulerian coordinates), ψ is the twist angle, $\lambda > 0$ is the multiplicity of axial elongation. The distortion tensor corresponding to the coordinate transformation (36.10) has the form

$$\mathbf{F} = \frac{dR}{dr}\mathbf{e}_r \otimes \mathbf{e}_R + \frac{R}{r}\mathbf{e}_\varphi \otimes \mathbf{e}_\Phi + \psi R \mathbf{i}_3 \otimes \mathbf{e}_\Phi + \lambda \mathbf{i}_3 \otimes \mathbf{i}_3 \tag{36.11}$$

$$\mathbf{e}_R = \mathbf{e}_r \cos\psi z + \mathbf{e}_\varphi \sin\psi z, \quad \mathbf{e}_\Phi = -\mathbf{e}_r \sin\psi z + \mathbf{e}_\varphi \cos\psi z, \tag{36.12}$$

Since formulas (36.10) specify the displacement field of an elastic body, the tensor (36.11), as it is easy to check, satisfies the compatibility condition $\mathrm{rot}\mathbf{F} = 0$.

Now suppose that continuously distributed dislocations are present in the cylinder, and the dislocation density tensor has the form

$$\boldsymbol{\alpha} = \alpha_{11}(r)\mathbf{e}_r \otimes \mathbf{e}_R + \alpha_{22}(r)\mathbf{e}_\varphi \otimes \mathbf{e}_\Phi + \alpha_{23}(r)\mathbf{e}_\varphi \otimes \mathbf{i}_3 + \alpha_{33}(r)\mathbf{i}_3 \otimes \mathbf{i}_3 \tag{36.13}$$

where $\alpha_{11}(r)$, $\alpha_{22}(r)$, $\alpha_{23}(r)$, $\alpha_{33}(r)$ are the given functions. The requirement of solenoidality (36.6) leads to the following limitation on these functions

$$\frac{d}{dr}(r\alpha_{11}) = \alpha_{22} \tag{36.14}$$

Scalar dislocation densities α_{23} and α_{33} can be arbitrary functions of the radial coordinate r.

In the presence of dislocations, the distortion tensor must satisfy the tensor equation of incompatibility (36.5). We will look for the tensor field of distortion in the form generalizing the expression (36.11)

$$\mathbf{F} = \frac{dR(r)}{dr}\mathbf{e}_r \otimes \mathbf{e}_R + \frac{R(r)}{r}\mathbf{e}_\varphi \otimes \mathbf{e}_\Phi + F_{23}(r)\mathbf{e}_\varphi \otimes \mathbf{i}_3 + F_{32}(r)\mathbf{i}_3 \otimes \mathbf{e}_\Phi + F_{33}(r)\mathbf{i}_3 \otimes \mathbf{i}_3 \tag{36.15}$$

where R, F_{23}, F_{32}, F_{33} are unknown functions of one variable. With the help of (36.4), (36.9), (36.12), (36.15) it can be verified that the incompatibility equation (36.5) will be fulfilled if and only if the following equalities are true

$$\begin{aligned} &F_{23}(r) = \tfrac{1}{r}\textstyle\int_{r_1}^{r} \alpha_{33}(\rho)\,\rho d\rho, \quad F_{32}(r) = \psi R(r) - r\alpha_{11}(r) \\ &F_{33}(r) = \lambda - \int_{r_1}^{r} \alpha_{23}(\rho)\,d\rho, \quad \lambda = \mathrm{const} \end{aligned} \tag{36.16}$$

Note that although it is impossible to determine the displacement field in general in a body with distributed dislocations, in the considered special case of the dislocations distribution (36.13) there is a radial component of displacement. Therefore the function $R(r)$ in expression (36.15) retains the meaning of the radial coordinate of the deformed cylinder points.

The function α_{11} in (36.13) represents the density of screw dislocations with a radial axis, α_{22} is the density of screw dislocations with a circular axis, and α_{33} is the density of screw dislocations whose axes are parallel to the cylinder axis. The α_{23} component describes the distribution of edge dislocations with a circular axis.

Further, we will assume that the following inequalities are satisfied when deforming a cylindrical tube

$$\frac{dR}{dr} > 0, \quad \frac{R}{r} F_{33} > F_{23} F_{32} \tag{36.17}$$

Since according to (36.15)

$$\det \mathbf{F} = \frac{dR}{dr}\left(\frac{R}{r} F_{33} - F_{23} F_{32}\right) \tag{36.18}$$

inequalities (36.17) provide the requirement $\det\mathbf{F} > 0$, which is necessary for the physical feasibility of an elastic medium deformation.

Based on (36.8), (36.16) we find

$$\begin{aligned} \mathbf{G} &= \left(\tfrac{dR}{dr}\right)^2 \mathbf{e}_r \otimes \mathbf{e}_r + \left(\tfrac{R^2}{r^2} + F_{23}^2\right) \mathbf{e}_\varphi \otimes \mathbf{e}_\varphi \\ &+ \left(\frac{R}{r} F_{32} + F_{23} F_{33}\right) (\mathbf{e}_\varphi \otimes \mathbf{i}_3 + \mathbf{i}_3 \otimes \mathbf{e}_\varphi) + \left(F_{32}^2 + F_{33}^2\right) \mathbf{i}_3 \otimes \mathbf{i}_3 \end{aligned} \tag{36.19}$$

$$I_1 = \left(\frac{dR}{dr}\right)^2 + \frac{R^2}{r^2} + F_{23}^2 + F_{32}^2 + F_{33}^2$$

$$I_2 = \left(\frac{dR}{dr}\right)^2 \left(\frac{R^2}{r^2} + F_{23}^2 + F_{32}^2 + F_{33}^2\right) + \left(\frac{R^2}{r^2} + F_{23}^2\right) F_{33}^2 - \frac{2RF_{23}F_{32}^2}{r}$$

$$I_3 = \left(\frac{dR}{dr}\right)^2 \left(\frac{R}{r} F_{33} - F_{23} F_{32}\right)^2$$

$$\begin{aligned} \mathbf{F}^{-T} &= \left(\tfrac{dR}{dr}\right)^{-1} \mathbf{e}_r \otimes \mathbf{e}_R + \left(\tfrac{R}{r} F_{33} - F_{23} F_{32}\right)^{-1} (F_{33} \mathbf{e}_\varphi \otimes \mathbf{e}_\Phi \\ &- F_{32} \mathbf{e}_\varphi \otimes \mathbf{i}_3 - F_{23} \mathbf{i}_3 \otimes \mathbf{e}_\Phi + \frac{R}{r} \mathbf{i}_3 \otimes \mathbf{i}_3) \end{aligned} \tag{36.20}$$

It follows from (36.19), (36.20) and the isotropic material constitutive relation that the Piola stress tensor will take the form

$$\mathbf{P} = P_{11}(r)\mathbf{e}_r \otimes \mathbf{e}_R + P_{22}(r)\mathbf{e}_\varphi \otimes \mathbf{e}_\Phi + P_{23}(r)\mathbf{e}_\varphi \otimes \mathbf{i}_3 + P_{32}(r)\mathbf{i}_3 \otimes \mathbf{e}_\Phi + P_{33}\mathbf{i}_3 \otimes \mathbf{i}_3 \tag{36.21}$$

With the help of (36.4), (36.9), (36.12), (36.21) we will get

$$\mathrm{div}\mathbf{P} = \left(\frac{dP_{11}}{dr} + \frac{P_{11} - P_{22}}{r} - \psi P_{32} \right) \mathbf{e}_R \tag{36.22}$$

If a hydrostatic pressure q is applied on the surface of an elastic body, the boundary conditions for the Piola stress tensor are written as follows [16]

$$\mathbf{n} \cdot \mathbf{P} = -q(\mathrm{det}\mathbf{F})\mathbf{n} \cdot \mathbf{F}^{-T} \tag{36.23}$$

where n is the unit normal to the body surface in the reference configuration. Using (36.18), (36.20), (36.21), (36.23) the boundary conditions on the outer and inner surfaces of the hollow cylinder, which are loaded with pressures q_0 and q_1, respectively, will take the form

$$\begin{aligned} P_{11} &= -q_0 \left(\tfrac{R}{r} F_{33} - F_{23} F_{32} \right) \quad r = r_0 \\ P_{11} &= -q_1 \left(\frac{R}{r} F_{33} - F_{23} F_{32} \right) \quad r = r_1 \end{aligned} \tag{36.24}$$

According to (36.16), the distortion components $F_{23}(r)$ and $F_{33}(r)$ are completely determined by a given dislocation distribution. The function $R(r)$ remains unknown, through which the remaining components of the distortion tensor in (36.15) are expressed, as well as the components of the tensor $\mathbf{P}$ in (36.21). The function $R(r)$ is determined from the vector equilibrium equation (36.1), which by virtue of (36.22) is reduced to a single scalar equation

$$\frac{dP_{11}}{dr} + \frac{P_{11} - P_{22}}{r} - \psi P_{32} = 0 \tag{36.25}$$

Since the Piola stresses are expressed using (36.7), (36.15), (36.19), (36.20) through the function $R(r)$ and its derivative, then (36.25) is an ordinary nonlinear second-order differential equation with respect to $R(r)$ with boundary conditions (36.24). Thus, due to the formulation (36.15), the problem of large deformations of a cylindrical tube with distributed dislocations (36.13) is reduced for any isotropic material to a nonlinear boundary value problem for an ordinary differential equation, i.e., to a one-dimensional boundary value problem.

If the twisting angle ψ and the multiplicity of elongation λ are considered given parameters, then the longitudinal force Q and the torque M acting at the ends of the cylinder $z = 0$ and $z = l$ are expressed through the solution of the indicated one-dimensional problem according to the formulas

$$Q = 2\pi \int_{r_1}^{r_0} P_{33}(r) r dr, \quad M = 2\pi \int_{r_1}^{r_0} P_{32}(r) R(r) r dr \tag{36.26}$$

Note that the reason for the twisting of the tube is not only the torque M, but also the distribution of screw dislocations. If at least one of the functions $\alpha_{11}, \alpha_{22}, \alpha_{33}$ is not identically zero, then at $M = 0$ the twist angle ψ will be different from zero.

36.4 Incompressible Elastic Material

If the isotropic cylinder material is incompressible, then the function $R(r)$ is determined from the incompressibility condition $\det \mathbf{F} = 1$, and the constitutive equation (36.7) is modified as follows [16]

$$\begin{aligned} \mathbf{P} &= 2\left(\tfrac{\partial W}{\partial I_1} + I_1 \tfrac{\partial W}{\partial I_2}\right)\mathbf{F} - 2\tfrac{\partial W}{\partial I_2}\mathbf{G}\cdot\mathbf{F} - p\mathbf{F}^{-T} \\ W &= W(I_1, I_2) \end{aligned} \tag{36.27}$$

Here p is the pressure in an incompressible body that is not expressed in terms of deformation. In our cylinder problem, $p = p(r)$ is an unknown function determined from the equilibrium Eq. (36.25).

Next, we consider two special cases in which it is possible to find an exact explicit solution to the problem of one-dimensional deformations of a cylinder with dislocations. In the first case, there are no screw dislocations in the direction of the cylinder axis, i.e., $\alpha_{33} = 0$. Then by virtue of (36.15), (36.16) we have

$$\begin{aligned} \mathbf{F} &= \tfrac{dR}{dr}\mathbf{e}_r \otimes \mathbf{e}_R + \tfrac{R}{r}\mathbf{e}_\varphi \otimes \mathbf{e}_\Phi + F_{32}\mathbf{i}_3 \otimes \mathbf{e}_\Phi + F_{33}\mathbf{i}_3 \otimes \mathbf{i}_3 \\ F_{32} &= \psi R - r\alpha_{11}, \quad F_{33} = \lambda - \int_{r_1}^{r} \alpha_{23}(\rho) d\rho \end{aligned} \tag{36.28}$$

The incompressibility condition according to (36.18), (36.28) leads to the equation

$$\frac{dR}{dr} = \frac{r}{RF_{33}}$$

which has such a solution

$$R^2(r) = 2\int_{r_1}^{r} \frac{\rho d\rho}{F_{33}(\rho)} + R_1^2, \quad R_1 = R(r_1) \tag{36.29}$$

where R_1 is an unknown constant.

Further, on the basis of (36.28) we have

$$\mathbf{F}^{-T} = \left(\tfrac{dR}{dr}\right)^{-1} \mathbf{e}_r \otimes \mathbf{e}_R + \tfrac{r}{R}\mathbf{e}_\varphi \otimes \mathbf{e}_\Phi - \tfrac{rF_{32}}{RF_{33}}\mathbf{e}_\varphi \otimes \mathbf{i}_3 + \tfrac{1}{F_{33}}\mathbf{i}_3 \otimes \mathbf{i}_3$$
$$\operatorname{div}\left(\mathbf{F}^{-T}\right) = \frac{-R\alpha_{23}}{r}\mathbf{e}_R \tag{36.30}$$

By entering the designation

$$\mathbf{P}^* = 2\left(\tfrac{\partial W}{\partial I_1} + I_1 \tfrac{\partial W}{\partial I_2}\right)\mathbf{F} - 2\tfrac{\partial W}{\partial I_2}\mathbf{G} \cdot \mathbf{F} =$$
$$= P^*_{11}\mathbf{e}_r \otimes \mathbf{e}_R + P^*_{22}\mathbf{e}_\varphi \otimes \mathbf{e}_\Phi + P^*_{23}\mathbf{e}_\varphi \otimes \mathbf{i}_3 + P^*_{32}\mathbf{i}_3 \otimes \mathbf{e}_\Phi + P^*_{33}\mathbf{i}_3 \otimes \mathbf{i}_3 \tag{36.31}$$

we see that by virtue of (29) $\mathbf{P}^*$ is a known function of r. From (36.25), (36.27), (36.30) a differential equation is derived for the function $p(r)$

$$F_{33}\frac{dp}{dr} - \alpha_{23}p = \frac{r}{R}\left(\frac{dP^*_{11}}{dr} + \frac{P^*_{11} - P^*_{22}}{r} - \psi P^*_{32}\right) \tag{36.32}$$

Equation (36.32) has the following solution

$$p(r) = exp\left(\textstyle\int_r^{r_0} L(\rho)d\rho\right)\left[p(r_0) - \textstyle\int_r^{r_0} N(r')exp(-\textstyle\int_{r'}^{r_0} L(\rho)d\rho)dr')\right]$$
$$L(r) = -\frac{\alpha_{23}}{F_{33}}, \quad N(r) = \frac{r}{RF_{33}}\left(\frac{dP^*_{11}}{dr} + \frac{P^*_{11} - P^*_{22}}{r} - \psi P^*_{32}\right) \tag{36.33}$$

Thus, for $\boldsymbol{\alpha} = \alpha_{11}\mathbf{e}_r \otimes \mathbf{e}_R + \alpha_{22}\mathbf{e}_\varphi \otimes \mathbf{e}_\Phi + \alpha_{23}\mathbf{e}_\varphi \otimes \mathbf{i}_3$ an exact quadrature solution has been found for the problem of large strains of tension-compression, torsion and inflation of a hollow cylinder with dislocations. This solution is suitable for any isotropic incompressible elastic materials.

As a simple example, consider a solid cylinder ($r_1 = 0$) with an unloaded surface $r = r_0$, assuming that $\boldsymbol{\alpha} = \alpha_0(\mathbf{e}_r \otimes \mathbf{e}_R + \mathbf{e}_\varphi \otimes \mathbf{e}_\Phi)$, $\alpha_0 = \text{const}$. This dislocations distribution satisfies the condition of solenoidality (36.14). We will use the model of a neo-Hookean incompressible material, for which the relations are valid [16]

$$W = \frac{1}{2}\mu\left(I_1 - 3\right), \quad \mathbf{P} = \mu\mathbf{F} - p\mathbf{F}^{-T}, \quad \mu = \text{const} \tag{36.34}$$

In this example, using formulas (36.27)–(36.33), we obtain

$$R = \tfrac{r}{\sqrt{\lambda}}, \quad \mathbf{F} = \tfrac{1}{\sqrt{\lambda}}\left(\mathbf{e}_r \otimes \mathbf{e}_R + \mathbf{e}_\varphi \otimes \mathbf{e}_\Phi\right) + \psi_0 r\mathbf{i}_3 \otimes \mathbf{e}_\Phi + \lambda\mathbf{i}_3 \otimes \mathbf{i}_3$$
$$\mathbf{F}^{-T} = \sqrt{\lambda}\left(\mathbf{e}_r \otimes \mathbf{e}_R + \mathbf{e}_\varphi \otimes \mathbf{e}_\Phi\right) - \frac{\psi_0 r}{\sqrt{\lambda}}\mathbf{e}_\varphi \otimes \mathbf{i}_3 + \frac{1}{\lambda}\mathbf{i}_3 \otimes \mathbf{i}_3 \tag{36.35}$$

$$p = -\frac{\mu\psi\psi_0\left(r^2 - r_0^2\right)}{2\sqrt{\lambda}} + \frac{\mu}{\lambda}, \quad \psi_0 = \frac{\psi}{\sqrt{\lambda}} - \alpha_0 \tag{36.36}$$

$$P_{11} = P_{22} = \tfrac{1}{2}\mu\psi\psi_0\left(r^2 - r_0^2\right), \quad P_{23} = \tfrac{\psi_0 r}{\sqrt{\lambda}} p$$
$$P_{32} = \mu\psi_0 r, \quad P_{33} = \frac{\mu\psi\psi_0\left(r^2 - r_0^2\right)}{2\lambda\sqrt{\lambda}} + \mu\left(\lambda - \frac{1}{\lambda^2}\right) \tag{36.37}$$

In addition to the Piola stress tensor, the symmetric Cauchy stress tensor **T** [16] is of interest

$$\mathbf{T} = \mathbf{F}^T \cdot \mathbf{P} = \sigma_R \mathbf{e}_R \otimes \mathbf{e}_R + \sigma_\Phi \mathbf{e}_\Phi \otimes \mathbf{e}_\Phi + \tau_{\Phi Z}\left(\mathbf{e}_\Phi \otimes \mathbf{i}_3 + \mathbf{i}_3 \otimes \mathbf{e}_\Phi\right) + \sigma_Z \mathbf{i}_3 \otimes \mathbf{i}_3$$

Using (36.35)–(36.37), we obtain expressions of normal and tangential Cauchy stresses

$$\sigma_R = \tfrac{\mu\psi\psi_0\left(r^2 - r_0^2\right)}{2\sqrt{\lambda}}, \quad \tau_{\Phi Z} = \mu\lambda\psi_0 r$$
$$\sigma_\Phi = \mu{\psi_0}^2 r^2 + \frac{\mu\psi\psi_0\left(r^2 - r_0^2\right)}{2\sqrt{\lambda}} \tag{36.38}$$
$$\sigma_Z = \mu\left(\lambda^2 - \frac{1}{\lambda}\right) + \frac{\mu\psi\psi_0\left(r^2 - r_0^2\right)}{2\sqrt{\lambda}}$$

The longitudinal force and torque according to (36.26) are given by the formulas

$$Q(\lambda, \psi, \alpha_0) = \pi\mu r_0^2\left(\lambda - \tfrac{1}{\lambda^2} - \tfrac{\psi\psi_0 r_0^2}{4\lambda\sqrt{\lambda}}\right)$$
$$M(\lambda, \psi, \alpha_0) = \frac{\pi\mu\psi_0 r_0^4}{2\sqrt{\lambda}} \tag{36.39}$$

Using (36.39), we analyze the effect of dislocations on the dependence of the longitudinal force on the twist angle in the absence of longitudinal deformation, i.e., when $\lambda = 1$. This dependence is shown in Fig. 36.1.

The appearance of a longitudinal force during torsion of the cylinder under conditions that prevent axial displacement of the ends is a manifestation of the Poynting effect [16, 22]. At $\alpha_0 = 0$, the longitudinal force is negative, i.e., compressive. The occurrence of a compressive longitudinal force at $\lambda = 1$ obviously means that the length of the same cylinder will increase with free torsion, i.e., at $Q = 0$. If $Q > 0$ with $\lambda = 1$, then the cylinder will shorten with free torsion. As can be seen from Fig. 36.1, at $\alpha_0 \neq 0$ in a certain range of the parameter ψ, the force Q turns out to be tensile. Thus, the presence of dislocations can change the sign of the Poynting effect.

Analyzing formulas (36.38), it can be seen that at $\psi = \sqrt{\lambda}\alpha_0$. i.e., at $\psi_0 = 0$, the stresses $\sigma_R, \sigma_\Phi, \tau_{\Phi Z}$ turn to zero, and the stress σ_Z is constant. There is a homogeneous uniaxial stress state. The cylinder is twisted without the appearance of tangential stresses and without the application of external torque. It is not difficult to show that such a phenomenon occurs for any isotropic material and a rod of arbitrary cross-section. Indeed, for $\psi_0 = 0$ according to (36.35) we have

$$\mathbf{F} = \frac{1}{\sqrt{\lambda}}\left(\mathbf{e}_r \otimes \mathbf{e}_R + \mathbf{e}_\varphi \otimes \mathbf{e}_\Phi\right) + \lambda \mathbf{i}_3 \otimes \mathbf{i}_3 \tag{36.40}$$

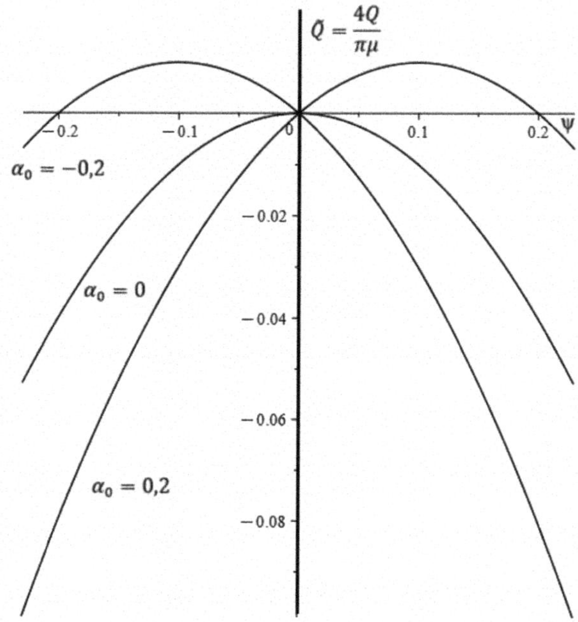

Fig. 36.1 The figure assumes $r_0 = 1$

Using (36.9), (36.12), we proceed to (36.40) to the constructed basis of Cartesian coordinates $\mathbf{i}_k$ and represent the distortion tensor (36.40) as a product

$$\mathbf{F} = \left[\frac{1}{\sqrt{\lambda}}(\mathbf{i}_1 \otimes \mathbf{i}_1 + \mathbf{i}_2 \otimes \mathbf{i}_2) + \lambda \mathbf{i}_3 \otimes \mathbf{i}_3\right] \cdot [(\mathbf{i}_1 \otimes \mathbf{i}_1 + \mathbf{i}_2 \otimes \mathbf{i}_2)\cos\psi z + (\mathbf{i}_1 \otimes \mathbf{i}_2 - \mathbf{i}_2 \otimes \mathbf{i}_1)\sin\psi z + \mathbf{i}_3 \otimes \mathbf{i}_3] \quad (36.41)$$

So we have

$$\mathbf{G} = \frac{1}{\lambda}(\mathbf{i}_1 \otimes \mathbf{i}_1 + \mathbf{i}_2 \otimes \mathbf{i}_2) + \lambda^2 \mathbf{i}_3 \otimes \mathbf{i}_3 = \text{const} \quad (36.42)$$

The dislocation density tensor (36.13) for $\alpha_{23} = \alpha_{33} = 0$, $\alpha_{11} = \alpha_{22} = \alpha_0$ and $\psi_0 = 0$ in the Cartesian basis takes the form

$$\boldsymbol{\alpha} = \frac{\psi}{\sqrt{\lambda}}[(\mathbf{i}_1 \otimes \mathbf{i}_1 + \mathbf{i}_2 \otimes \mathbf{i}_2)\cos\psi z + + (\mathbf{i}_1 \otimes \mathbf{i}_2 - \mathbf{i}_2 \otimes \mathbf{i}_1)\sin\psi z] \quad (36.43)$$

From formulas (36.41)–(36.43) it is deduced that for $\psi_0 = 0$ for any isotropic incompressible material and a rod of arbitrary cross-section, there is a uniaxial stress state $\mathbf{T} = \sigma_Z \mathbf{i}_3 \otimes \mathbf{i}_3$, where $\sigma_Z = \text{const}$. An inhomogeneous rotation field at the ψz angle caused by dislocations is superimposed on this state of homogeneous deformation. This phenomenon, which can exist only in the presence of distributed dislocations, can be called a quasi-homogeneous deformed state, since in this case the metric tensor $\mathbf{G}$ is constant, i.e., it does not depend on coordinates, and the distortion tensor $\mathbf{F}$ and the rotation tensor $\left(\mathbf{F}\cdot\mathbf{F}^T\right)^{-1/2}\cdot\mathbf{F}$ are variables.

36.5 Nonlinear Lame Problem for a Cylinder with Distributed Dislocations

To analyze the effect of dislocations on the resistance of the tube to hydrostatic loads on cylindrical surfaces, we consider the following distribution of dislocations

$$\boldsymbol{\alpha} = \frac{a}{r}\mathbf{e}_r \otimes \mathbf{e}_R, \quad a = \text{const} \tag{36.44}$$

The tensor field (36.44) satisfies the condition of solenoidality (36.14). Based on (36.28)–(36.30) in this case we have

$$\lambda R\frac{dR}{dr} = r, \quad R(r) = \sqrt{R_1^2 + \lambda^{-1}\left(r^2 - r_1^2\right)} \tag{36.45}$$

$$\begin{aligned}
\mathbf{F} &= \tfrac{r}{\lambda R}\mathbf{e}_r \otimes \mathbf{e}_R + \tfrac{R}{r}\mathbf{e}_\varphi \otimes \mathbf{e}_\Phi + (\psi R - a)\mathbf{i}_3 \otimes \mathbf{e}_\Phi + \lambda\mathbf{i}_3 \otimes \mathbf{i}_3 \\
\mathbf{F}^{-T} &= \frac{\lambda R}{r}\mathbf{e}_r \otimes \mathbf{e}_R + \frac{r}{R}\mathbf{e}_\varphi \otimes \mathbf{e}_\Phi + \frac{(a-\psi R)}{\lambda R}\mathbf{e}_\varphi \otimes \mathbf{i}_3 + \frac{1}{\lambda}\mathbf{i}_3 \otimes \mathbf{i}_3
\end{aligned} \tag{36.46}$$

Next, using (36.8), (36.45), (36.46) we will get

$$\begin{aligned}
&\mathbf{G} = \tfrac{r^2}{\lambda^2 R^2}\mathbf{e}_r \otimes \mathbf{e}_r + \tfrac{R^2}{r^2}\mathbf{e}_\varphi \otimes \mathbf{e}_\varphi + \tfrac{(\psi R - a)R}{r}(\mathbf{e}_\varphi \otimes \mathbf{i}_3 + \mathbf{i}_3 \otimes \mathbf{e}_\varphi) \\
&+\left[\lambda^2 + (a-\psi R)^2\right]\mathbf{i}_3 \otimes \mathbf{i}_3 \\
&I_1 = \frac{r^2}{\lambda^2 R^2} + \frac{R^2}{r^2} + \lambda^2 + (a-\psi R)^2 \\
&I_2 = \frac{\lambda^2 R^2}{r^2} + \frac{r^2}{R^2} + \frac{1}{\lambda^2} + \frac{(a-\psi R)^2 r^2}{\lambda^2 R^2}
\end{aligned} \tag{36.47}$$

$$\begin{aligned}
&\mathbf{G}\cdot\mathbf{F} = \tfrac{r^3}{\lambda^3 R^3}\mathbf{e}_r \otimes \mathbf{e}_R + \tfrac{R}{r}\left[(a-\psi R)^2 + \tfrac{R^2}{r^2}\right]\mathbf{e}_\varphi \otimes \mathbf{e}_\Phi - \tfrac{(a-\psi R)\lambda R}{r}\mathbf{e}_\varphi \otimes \mathbf{i}_3 \\
&-(a-\psi R)\left[\frac{R^2}{r^2} + \lambda^2 + (a-\psi R)^2\right]\mathbf{i}_3 \otimes \mathbf{e}_\Phi + \lambda\left[\lambda^2 + (a-\psi R)^2\right]\mathbf{i}_3 \otimes \mathbf{i}_3
\end{aligned} \tag{36.48}$$

With (36.31), (36.44)–(36.48) Eq. (36.32) will take the form

$$\frac{dp}{dr} = \frac{r}{\lambda R}\left(\frac{dP_{11}^*}{dr} + \frac{P_{11}^* - P_{22}^*}{r} - \psi P_{32}^*\right) \tag{36.49}$$

$$P_{11}^* = 2\left(\frac{\partial W}{\partial I_1} + I_1\frac{\partial W}{\partial I_2}\right)\frac{r}{\lambda R} - 2\frac{\partial W}{\partial I_2}\frac{r^3}{\lambda^3 R^3}$$
$$P_{22}^* = 2\left(\frac{\partial W}{\partial I_1} + I_1\frac{\partial W}{\partial I_2}\right)\frac{R}{r} - 2\frac{\partial W}{\partial I_2}\left[\frac{(a-\psi R)^2 R}{r} + \frac{R^3}{r^3}\right]$$
$$P_{32}^* = 2\left(\frac{\partial W}{\partial I_1} + I_1\frac{\partial W}{\partial I_2}\right)(\psi R - a) - 2\frac{\partial W}{\partial I_2}(\psi R - a)\left[\frac{R^2}{r^2} + \lambda^2 + (\psi R - a)^2\right] \tag{36.50}$$

Let's enter instead of $p(r)$ a new unknown function

$$\sigma(r) = \frac{r}{\lambda R}P_{11} = \frac{r}{\lambda R}P_{11}^* - p \tag{36.51}$$

Then the boundary conditions (36.24) will take the form

$$\sigma(r_0) = -q_0, \quad \sigma(r_1) = -q_1 \tag{36.52}$$

Equality follows from (36.49), (36.51), and (36.52)

$$q_1 - q_0 = \int_{r_1}^{r_0} \frac{1}{\lambda R}\left(P_{22}^* - \frac{r^2}{\lambda R^2}P_{11}^* + \psi r P_{32}^*\right) dr \tag{36.53}$$

By virtue of (36.45), (36.47), (36.50) given the values of the parameters λ, ψ, a, the right part of (36.53) has a known function of R_1. Therefore, formula (36.53) gives an explicit expression for the tube loading diagram, i.e., the dependence of the pressure difference $q_1 - q_0$ on the radius of the cavity of the deformed cylinder R_1.

36.6 Cylinder with Distributed Edge Dislocations and Screw Dislocations of the Axial Direction

In this section, we consider the second case of dislocation distribution, which allows an exact solution to the problem of an elastic cylinder equilibrium. The dislocation density tensor will be considered as given in the form

$$\boldsymbol{\alpha} = \alpha_{23}(r)\mathbf{e}_\varphi \otimes \mathbf{i}_3 + \alpha_{33}(r)\mathbf{i}_3 \otimes \mathbf{i}_3 \tag{36.54}$$

Referring to (36.15), (36.16) we get

$$\mathbf{F} = \frac{dR(r)}{dr}\mathbf{e}_r \otimes \mathbf{e}_R + \frac{R(r)}{r}\mathbf{e}_\varphi \otimes \mathbf{e}_\Phi + F_{23}(r)\mathbf{e}_\varphi \otimes \mathbf{i}_3 + \\ +\psi R\mathbf{i}_3 \otimes \mathbf{e}_\Phi + F_{33}(r)\mathbf{i}_3 \otimes \mathbf{i}_3 \tag{36.55}$$

where the functions F_{23} and F_{33} are expressed by formulas (16).

The incompressibility equation is written as

$$R\frac{dR}{dr}\left(\frac{F_{33}}{r} - \psi F_{23}\right) = 1$$

and has the following solution

$$R^2 = 2\int_{r_1}^{r} \frac{\rho d\rho}{F_{33}(\rho) - \psi\rho F_{23}(\rho)} + R_1^2 \tag{36.56}$$

For the pressure function $p(r)$, a linear differential equation of the first order is obtained, the solution of which is written in the same way as formula (36.33).

Next, we will limit ourselves to the case of a solid cylinder ($r_1 = R_1 = 0$) and such a distribution of dislocations: $\alpha_{33} = \beta r^{-1}, \alpha_{23} = \gamma$, where β, γ—are constants. Based on (36.16) we obtain $F_{23} = \beta$, $F_{33} = \lambda - \gamma r$, and the expression (36.56) takes the form

$$R^2 = -\frac{2\lambda}{(\gamma + \beta\psi)^2}\ln\left[1 - \frac{(\gamma + \beta\psi)r}{\lambda}\right] - \frac{2r}{\gamma + \beta\psi} \tag{36.57}$$

The pressure function $p(r)$ now satisfies the equation

$$\frac{dp}{dr} - \frac{\gamma + \beta\psi}{\lambda - r(\gamma + \beta\psi)}p = \frac{dR}{dr}\left(\frac{dP_{11}^*}{dr} + \frac{P_{11}^* - P_{22}^*}{r} - \psi P_{32}^*\right) \tag{36.58}$$

Next, let's focus on analyzing the effects caused only by edge dislocations, i.e., let's put $\beta = 0$. The cylindrical rod will be considered loaded only by the longitudinal force Q. Then there will be no torsion, which means $\psi = 0$. The density of edge dislocations is subject to the constraint $\gamma < \lambda r_0^{-1}$. Now according to (36.55), (36.57) we have

$$\mathbf{F} = \frac{r}{(\lambda - \gamma r)R}\mathbf{e}_r \otimes \mathbf{e}_r + \frac{R}{r}\mathbf{e}_\varphi \otimes \mathbf{e}_\varphi + (\lambda - \gamma r)\mathbf{i}_3 \otimes \mathbf{i}_3 \tag{36.59}$$

$$R^2(r) = -\frac{2\lambda}{\gamma^2}\ln\left(1 - \frac{\gamma r}{\lambda}\right) - \frac{2r}{\gamma} \tag{36.60}$$

To determine the stress field, we use the Bartenev-Khazanovich incompressible material model (Varga) [16], for which the relations are valid

$$W = 2\mu(\mathrm{tr}\mathbf{U} - 3), \quad \mathbf{P} = 2\mu\mathbf{A} - p\mathbf{F}^{-T}, \quad \mu = \mathrm{const} \tag{36.61}$$

Here $\mathbf{U}$ is a symmetric positive definite stretching tensor, $\mathbf{A}$ is the properly-orthogonal rotation tensor. Tensors $\mathbf{U}$ and $\mathbf{A}$ form a polar decomposition of the distortion tensor: $\mathbf{F} = \mathbf{U} \cdot \mathbf{A}$. For the distortion tensor (36.59), we obviously have $\mathbf{U} = \mathbf{F}, \mathbf{A} = \mathbf{I}$, where $\mathbf{I}$ is the unit tensor. So $P_{11}^* = P_{22}^* = 2\mu$. The solution of equation (36.58) for this material at $\psi = 0$, $\beta = 0$ has the form

$$p = \frac{2\mu r_0}{(\lambda - \gamma r) R_0}, \quad R_0 = R(r_0) \tag{36.62}$$

This solution satisfies the boundary condition $P_{11}(r_0) = 0$. From (36.59), (36.61), (36.62) we obtain the expression of the Piola stress tensor

$$\begin{aligned} &\mathbf{P} = P_{11}\mathbf{e}_r \otimes \mathbf{e}_r + P_{22}\mathbf{e}_\varphi \otimes \mathbf{e}_\varphi + P_{33}\mathbf{i}_3 \otimes \mathbf{i}_3 \\ &P_{11} = 2\mu\left(1 - \tfrac{r_0 R}{r R_0}\right), \quad P_{22} = 2\mu\left[1 - \tfrac{r_0 r}{(\lambda-\gamma r) R_0 R}\right] \\ &P_{33} = 2\mu\left[1 - \tfrac{r_0}{(\lambda-\gamma r)^2 R_0}\right] \end{aligned} \tag{36.63}$$

The Cauchy stresses are expressed as follows

$$\sigma_R = \frac{r P_{11}}{(\lambda - \gamma r) R}, \quad \sigma_\Phi = \frac{r}{R} P_{22}, \quad \sigma_Z = (\lambda - \gamma r) P_{33} \tag{36.64}$$

Thus, the uniform distribution of edge dislocations with a circular axis creates an inhomogeneous stress field in the cylinder.

The longitudinal force is calculated using (36.26) and (36.63) and has the form

$$Q = 2\mu\pi\left[r_0^2 + \frac{r_0}{\lambda R_0}\left(R_0^2 - \frac{2r_0^2}{\lambda - \gamma r_0}\right)\right] \tag{36.65}$$

$$R_0 = \sqrt{-\frac{2\lambda}{\gamma^2}\ln\left(1 - \frac{\gamma r_0}{\lambda}\right) - \frac{2r_0}{\gamma}}$$

As can be seen from Fig. 36.2, in the absence of a longitudinal force ($Q = 0$), the cylinder lengthens ($\lambda > 1$) in the case of a positive dislocation density γ and shortens ($\lambda < 1$) in the case of a negative γ. There are such negative values γ for which the cylindrical equilibrium shape of a Bartenev-Khazanovich rod is not possible without the application of a tensile longitudinal force.

36.7 Conclusion

In this paper, the problem of large deformations of torsion, inflation, and stretching-compression of a hollow circular elastic cylinder containing an axisymmetric distribution of edge and screw dislocations is solved. By means of a special substitution in the system of equations of the nonlinear theory of dislocations, the initial three-dimensional problem in the case of an isotropic material is reduced to a nonlinear one-dimensional boundary value problem, i.e., to a problem for ordinary differential equations. It has been established that the twisting of the cylinder is caused not only by external torque, but also by the presence of distributed edge dislocations.

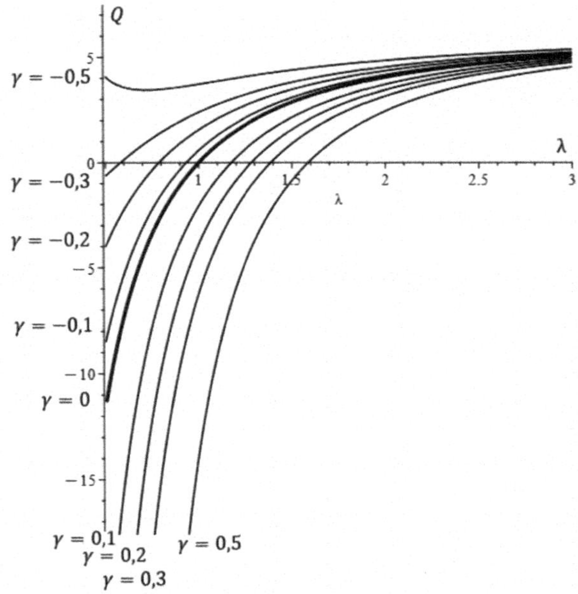

Fig. 36.2 The figure assumes $r_0 = 1$, $\mu = 1$

For a model of an isotropic incompressible material of a general type, a number of exact solutions have been found in quadratures about the equilibrium of a cylindrical tube with dislocations. For an arbitrary isotropic incompressible material, an explicit formula is derived describing the diagram of tube loading by external and internal pressures, taking into account distributed screw dislocations of the radial direction. The phenomenon of quasi-homogeneous deformations of elastic bodies has been discovered. In a quasi-homogeneous state of equilibrium, which can exist only in the presence of dislocations, the strain tensor is constant, i.e., it does not depend on coordinates, and the distortion and rotation tensors are variable. In a number of practically important cases, simple explicit formulas have been obtained describing the effect of dislocations on the nonlinear resistance of the cylinder to external loads. These ratios are written in terms of the quantities available for measurement in the experiment.

Acknowledgements The reported study was funded by the Russian Science Foundation, project number 23-21-00123, https://rscf.ru/en/project/23-21-00123/.

References

1. Clayton JD (2011) Nonlinear mechanics of crystals. Springer, Dordrecht
2. Gutkin MY, Ovid'ko IA (2004) Plastic deformation in nanocrystalline materials. Springer, Berlin

3. Kondo K (1952) On the geometrical and physical foundations in the theory of yielding. In: Proceedings of 2nd Japanese national congress of applied mechanics, Tokyo, pp 41–47
4. Bilby BA, Bullough R, Smith E (1955) Continuous distribution of dislocations: a new application of the methods of non-Riemannian geometry. Proc Roy Soc Lond A 231:263–273
5. Kröner E (1960) Allgemeine Kontinuumstheorie der Versetzungen und Eigenspannungen. Arch Ration Mech Anal 4:273–334
6. Eshelby JD (1956) The continuum theory of lattice defects. In: Seitz F, Turnbull D (eds) Solid state physics, vol 3. Academic, New York, pp 79–144
7. Zelenina AA, Zubov LM (2009) Bending and twisting of nonlinear elastic bodies with continuously distributed dislocations. Vestn Yuzhn Nauchn Tsentr RAN 5(4):15–22
8. Yavary A, Goriely A (2012) Riemann-Cartan geometry of nonlinear dislocations mechanics. Arch Ration Mech Anal 205:59–118
9. Zelenina AA, Zubov LM (2013) Nonlinear effects during the tension, bend and torsion of elastic bodies with distributed dislocations. Dokl Phys 58(8):354–357
10. Zubov LM (2014) Spherically symmetric solutions in the nonlinear theory of dislocations. Dokl Phys 59(9):419–422
11. Goloveshkina EV, Zubov LM (2019) Universal spherically symmetric solution of nonlinear dislocation theory for incompressible isotropic elastic medium. Arch Appl Mech 89(3):409–424
12. Zubov LM (2020) Universal solutions of nonlinear dislocation theory for elastic cylinder. Mech Solids 55:701–709. https://doi.org/10.3103/S0025654420050167
13. Goloveshkina EV, Zubov LM (2021) Spherically symmetric tensor fields and their application in nonlinear theory of dislocations. Symmetry. 13:830. https://doi.org/10.3390/sym13050830
14. Zubov LM (1997) Nonlinear theory of dislocations and disclinations in elastic bodies. Springer, Berlin
15. Galasko AA, Zubov LM (2015) Nonlinear theory of torsion of elastic cylinder with a screw dislocation. Izv Vyssh Uchebn Zaved Sev-Kavk Reg Estestv Nauki 4:35–43
16. Lurie AI (1990) Nonlinear theory of elasticity. North-Holland, Amsterdam
17. Eremeyev VA, Cloud MJ, Lebedev LP (2018) Applications of tensor analysis in continuum mechanics. World Scientific, New Jersey
18. Nye JF (1953) Some geometrical relations in dislocated crystals. Acta Metall 1(2):153–162
19. Landau LD, Lifshitz EM (1975) Theory of elasticity. Theoretical physics, vol 7. Pergamon, Oxford
20. Vakulenko AA (1991) Relationship between micro- and macroproperties in elastic media (in Russian). Adv Sci Technol. Mech Solid. 22 (VINITI, Moskow):3–54
21. Derezin SV, Zubov LM (2011) Disclinations in nonlinear elasticity. ZAMM. 91:433–442
22. Zubov LM (2001) Direct and inverse Poynting effects in elastic cylinders. Dokl Phys 46:675–677

The manufacturer's authorised representative in the EU is Springer Nature Customer Service Centre GmbH, Europaplatz 3, 69115 Heidelberg, Germany. If you have any concerns regarding our products, please contact ProductSafety@springernature.com

Printed and bound by CPI Group (UK) Ltd, Croydon, CR0 4YY
05/10/2026
02228127-0019